D1796351

Integrated Science

Volume 6

Editor-in-Chief

Nima Rezaei, Tehran University of Medical Sciences, Tehran, Iran

The **Integrated Science** Series aims to publish the most relevant and novel research in all areas of Formal Sciences, Physical and Chemical Sciences, Biological Sciences, Medical Sciences, and Social Sciences. We are especially focused on the research involving the integration of two of more academic fields offering an innovative view, which is one of the main focuses of Universal Scientific Education and Research Network (USERN), science without borders.

Integrated Science is committed to upholding the integrity of the scientific record and will follow the Committee on Publication Ethics (COPE) guidelines on how to deal with potential acts of misconduct and correcting the literature.

Nima Rezaei

Editor

Multidisciplinarity and Interdisciplinarity in Health

 Springer

Editor
Nima Rezaei (iD)
Universal Scientific Education
and Research Network (USERN)
Stockholm, Sweden

ISSN 2662-9461   ISSN 2662-947X (electronic)
Integrated Science
ISBN 978-3-030-96816-8   ISBN 978-3-030-96814-4   (eBook)
https://doi.org/10.1007/978-3-030-96814-4

This Springer imprint is published by the registered company Springer Nature Switzerland AG
The registered company address is: Gewerbestrasse 11, 6330 Cham, Switzerland

This book series would not have been possible without the continuous encouragement of my family. I dedicate this book series to my daughters, Ariana and Arnika, hoping that integrated science could solve complex problems and make a brighter future for the next generation.

Preface

The Integrated Science book series was set up in 2020. The goals of this series are health, science, education, and integration—while encouraging scientists toward a higher level of collaboration—universal engagement, and offering students a higher level of education—real-world thinking.

The Universal Scientific Education and Research Network (USERN) is an independent, non-governmental, non-profit organization and network where scientists are aware of the demarcation problem and act beyond geographical and political boundaries between countries. Networking and teamwork at the USERN are the principles to pair science with research, education, and then performance and support knowledge-producing activities that guarantee to make humans healthier and happier universally. The Integrated Science book series is the initiative of USERN to integrate human knowledge.

USERN interest groups invest in research and education. They provide a unique opportunity that allows junior students to communicate with senior professionals doing *borderless research* in many fields, including immunology and cancer immunology, neuroimaging, pharmacology, astrobiology, biomaterials, health and art, epidemiology, nanomedicine, nutrition, genetics and immunogenetics, transforming education, biology, chemistry, biochemistry, photomedicine, nuclear medicine, toxicology, tissue engineering, clinical reasoning, humanities, medical history, artificial intelligence, aesthetics, neuroscience, metacognition, psychotherapy, to name some of them.

The three volumes of the book series are resources that represent the world of complex problems. Their corresponding solutions show the functions of fields of knowledge listed above when integrated. The unique themes of these books are:

  i. *Integrated Science: Science Without Borders*
 ii. *Integrated Science: Transdisciplinary*
iii. *Integrated Science: Multidisciplinarity and Interdisciplinarity in Health*

Science Without Borders[1] first appeared. It introduces disciplinary and transdisciplinarity methodologies and models as a problem-solving necessity and the

---

[1]https://link.springer.com/book/10.1007/978-3-030-65273-9.

need to integrate sciences and humanities, sciences and integral human knowledge, and arts and sciences. *Integrated Science: Transdisciplinary* holds the potential to pose problems and move across disciplines to identify the most exciting definitions and solutions relevant to the problem. Its discussions embrace a range of concepts, including multi-agent world, chaos theory, augmented reality, film theory, land management, resilient communities, policymaking, systems view, automotive industry, cultural heritage, viral marketing, ecodemocracy, landscape, biosimilars, justice, meeting in art, geoethics, cosmopolitan localism, twin earth, protein structure, molecular dynamics, smart agriculture, and green cities. Finally, you are a reader of *Multidisciplinarity and Interdisciplinarity in Health*. This volume is for those who are working with neuroscience, psychology, sociology, engineering, medical sciences, public health, arts, artificial intelligence, and education but are willing to gather other knowledge and experiences to use them to make a smarter movement above their counterparts who are specialized to absorb knowledge from the same ground.

For a pragmatic approach to what is outlined in the book:

i. Cognitive neuroscience and social neuroscience, the second to tenth chapters begin with an interdisciplinary approach to cognitive science and continue with brain physics. Then, the mechanisms that underpin empathy, lifestyle behaviors (irrational choices and medical inadherence), and neuropsychiatry symptoms are discussed;

ii. Medical sciences and public health, the eleventh to seventeenth chapters mainly include intricacies of pandemics, addressing how super-spreading, environmental factors and social factors shape human behavior, making some individuals susceptible more than others. Also, interdisciplinary approaches to drug discovery, nuclear medicine, and dentistry are explored;

iii. Health and art, the eighteenth to twenty-second chapters give evidence of the function of different forms of arts as aids and therapies, especially in difficult-to-treat conditions, like neuropsychiatric disorders, and also during critical periods, like a pandemic;

iv. Health and artificial intelligence, the twenty-third to twenty-fifth chapters review integrated artificial intelligence in health. This review also addresses how much it can be helpful depending on support from healthcare settings and also society;

v. Health informatics and education, the twenty-sixth and twenty-seventh chapters include integrated information technology and education approaches in medical settings;

vi. Social science, ethics, and health, the twenty-eighth and twenty-ninth chaters introduce the concept of social capital and health, followed by a discussion of discrimination in medical research.

And we will make the convention in the last chapter. It is to share all the book authors' opinions for thirty years later from now.

Welcome to *Integrated Science: Multidisciplinarity and Interdisciplinarity in Health.*

Tehran, Iran   Nima Rezaei, M.D., Ph.D.
November 2021

Acknowledgments

I would like to express my gratitude to the Editorial Assistant of this book series, Dr. Amene Saghazadeh. Without doubt, the book would not have been completed without her contribution.

Contents

Introduction on Integrated Science: Multidisciplinarity and Interdisciplinarity in Health

Nima Rezaei and Amene Saghazadeh

"Specialized science, whether physical or social, inevitably passes into a stage of uncorrelated scientific piece-work. In this stage of dismemberment, science is as inconclusive through its lack of coherence as it was in an earlier period from its superficiality. That is, it then had breadth without depth, it now has depth without breadth."

Albion Small

Summary

Health is a complex problem for scientists; we have made a 30-year effort to expect a seven-year increase in life length; on the other hand, a crisis suddenly comes and causes a two-to-nine-year decrease in life length, simply overturning our expectation. We argue that it is not the causing pathogen that is perfect; it is the health science that is imperfect. This is an introduction to the multidisciplinary (MD) and interdisciplinary (ID) in health, scrolling through their probable timeline. The need to implement MD/ID teams in health care and health education is introduced. We elaborate on some solutions to meet challenges posed by their implementation and then find the merits of the ID approach to health research. Why integrated ID in health is necessary is in our

N. Rezaei (✉) · A. Saghazadeh
Integrated Science Association (ISA), Universal Scientific Education and Research Network (USERN), Tehran, Iran

N. Rezaei · A. Saghazadeh
Research Center for Immunodeficiencies, Children's Medical Center, Tehran University of Medical Sciences, Tehran, Iran

N. Rezaei
Department of Immunolochagy, School of Medicine, Tehran University of Medical Sciences, Tehran, Iran

© The Author(s), under exclusive license to Springer Nature Switzerland AG 2022
N. Rezaei (ed.), *Multidisciplinarity and Interdisciplinarity in Health*,
Integrated Science 6, https://doi.org/10.1007/978-3-030-96814-4_1

subsequent discussion, followed by using artificial intelligence and arts to explain how integrated ID in health can be facilitated. Then, it is reasonable to integrate ID in education. It is a rapid discussion, pointing out that the ID approach can be integrated with both higher education and next-generation education, and cognitive-behavioral therapy, design thinking, and interfaces allow for this possibility. Finally, we list further research fields, disciplines, and concepts. It just gives representative reasoning for each of them to imply that they are ID.

Cercis siliquastrum, also known as Judas-tree, is a tree grown in some European and Asian regions. In Persian, it is called "Arghavan," written as ارغوان.In Iran, this tree is mainly found in two provinces: Ilam and Mazandaran. The picture shows the Strait of Arghavan in Ilam, Iran. Unfortunately, it has declined as a result of climate change impact during the recent years. The poem, says Molana Jalal-e-Din Mohammad Molavi Rumi (1207–1273) and translates Nevit Oguz Ergin, *O gardener, O gardener, Autumn has come, Autumn has come. Who dress the green of the grasses and meadows? Where is Arghavan? Where is Arghavan?* Arghavan has heart-like flowers with a beautiful color so named rose-purple color (ارغوانی)). Indeed, the poet prospectively and correctly thought that humankind would not be a good gardener for Arghavan.

[Photo adapted from https://www.baeghtesad.com; *Nasta'līq prepared by* http://nastaliqonline.ir/*].*

Keywords

Health · Imperfection · Interdisciplinary

1   Introduction

1.1   That the Larger, the Less Imperfect

The larger the knowledge base is, the more distance we can dance out, allowing us to wonder that there remains at least a simple solution linked with every complex problem in the world—the solutions we are apt to overlook when walking within disciplines. The linkage between different disciplines is more supported in inter-disciplinary (ID) than multidisciplinary (MD); however, both deal with two or more disciplines simultaneously that would ensure blurring the boundaries between the disciplines involved, resulting in the development of space of knowledge within the framework of disciplinary research (Table 1). Transdisciplinarity can accomplish the goal of expanding and building upon the knowledge between, across, and beyond different disciplines. Thus, disciplinary and transdisciplinary methodologies use different knowledge sources, and both would be necessary to assemble the whole knowledge. Indeed, *when they escape from their own imperfection,* Mawlana Jalal Al-Din Muhammad Rumi writes, *they make a dance.* The Integrated Science is a book series dedicated to integration, a remedy to a highly specialized science in isolation, identifying a science applicable to complex, undecidable problems with interactions that are beyond local [1]. Transdisciplinary views across different disciplines have been maintained in the second volume of the Integrated Science book series. In this volume, MD and ID works of health are discussed.

Table 1 Methodologies of MD thinking and ID thinking in science and their concerns and goals [141]

Methodology	Concern	Goal
Multidisciplinarity	– Studying a research topic in several disciplines simultaneously – Bringing a plus to the discipline in question – Overflowing the disciplinary boundaries	– Limited to the framework of disciplinary research
Interdisciplinarity (interprofessional)	– Transferring methods from one discipline to another – Overflowing the disciplinary boundaries – Generating new disciplines	– Limited to the framework of disciplinary research

1.2   That the Longer, the More Imperfect

The people expect to live longer than previously thought. The global life expectancy at birth increased by 10.6%, corresponding to about seven years from 1990 to 2017, as shown in systematic estimations [2]. Not only is this the case for those possessing a healthy body and a happy mind, but also for those suffering from different conditions. Suppose Duchenne muscular dystrophy (DMD), a neuromuscular genetic disease associated with a high burden. For one with DMD, there was no chance of survival to age 25 in the 1960s, whereas it was estimated to be about 10% and greater than 50% in the 1980s and 1990s—the decade that nocturnal ventilation was offered [3]. And this is not at all surprising that a boy just born with DMD will experience his 40th birthday [4]. It might be, however, difficult to enjoy living with an aging-related disability or chronic condition. Moreover, life itself remains vulnerable to being influenced by different, complex challenges. The pandemic of coronavirus disease (COVID) [5] is a live example of such vulnerability; it is estimated to cause life expectancy reduction between two and nine years varying by the interaction of disease prevalence with country development status [6].

Whether consequently, shall we judge or not, that the health arena is far from that can sustain the health, happiness, and human development, we shall accept that the man's effort to improve life expectancy in the last three decades was worth as long as an unprecedented global crisis happened. To investigate the process humankind has made to observe a gain in life expectancy, we begin to see the application of MD and ID in health, obtained by searching PubMed, a major medical database. A curve can measure the number of publications versus time (Fig. 1). It is steady—from inception (i.e., the 1940s for ID and 1950s for MD) through 1990s—and becomes exponential from 1990s for both ID and MD. The curve has one peak in the year 2020 when the COVID emerged.

2   Healthcare Teams

2.1   Old World and Wars

For effectively dynamic health care, a team approach that lets professionals have communication and cooperation is versatile. The team-based health care will ensure "the provision of health services to individuals, families, and/or their communities by at least two health providers who work collaboratively with patients and their caregivers—to the extent preferred by each patient—to accomplish shared goals within and across settings to achieve coordinated, high-quality care" [7]. No document explicitly declares who originated the idea of the team approach in health care. However, it is traceable to India before 1900, when mission hospitals, Royer says, led teams consisting of auxiliaries, nurses, and doctors focused on health care in remote communities [8]. During the 1910s, the idea of providing health services

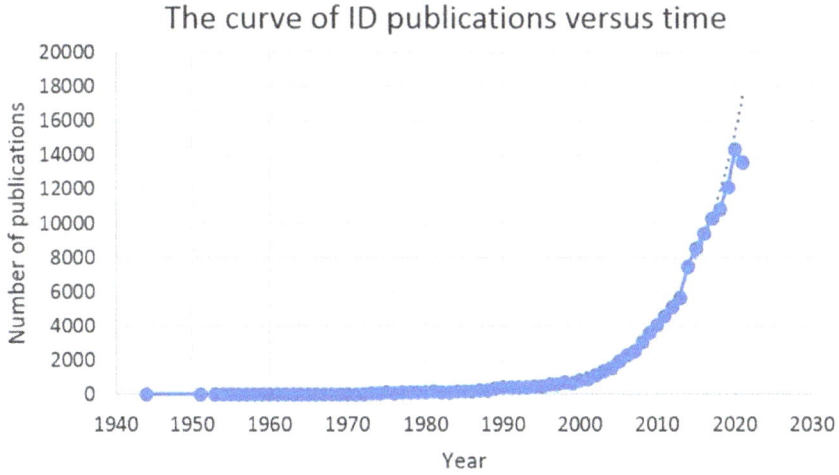

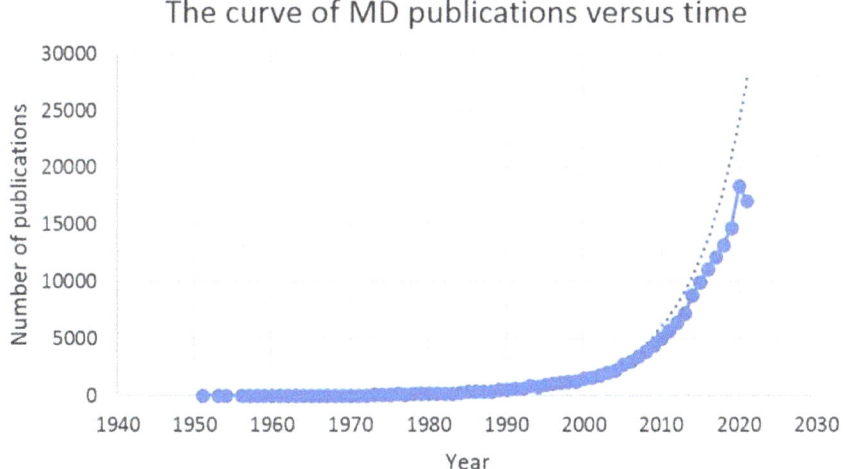

Fig. 1 Growth of interdisciplinary (ID) and multidisciplinary (MD) research in health over time

by teams of doctors, educators, and social workers was made in the USA and extended to include clinical "efficiency" and "social teamwork." Team-based care was then employed to provide primary care and home care services in military, community, inpatient, and outpatient settings in other geographic locations (Table 2). It is important to notice that these pioneering works were mainly focused on the local scale. World War II (1939–1945) was the starting point of the public importance of MD teams that became much more applicable for medical–surgical

Table 2 Timeline of the origin of team-based health care [8]

Team	Leader	Leader's responsibilities	Members' roles	Communication	Coordination	Outcomes
Multidisciplinarity	Gatekeeper	Figures out which disciplines need to work in an independent, discipline-specific team	– Services are assessed, planned, and provided separately – Decisions are made independently	Shared	Little to none	Discipline-specific parameters
Interdisciplinarity	Chief	– Develops and applies interdisciplinary curriculum and teaching strategies that meet students' needs – Develops and applies intervention and management strategies according to students' learning and behavioral styles – Coordinates communication on teaching and learning	– Problem-solving is not confined by disciplines – Decisions are made in an interdependent manner	Collaborative	Great	Beyond discipline parameters

conditions and established for mental health care as well as for education [8]. A few years later, an early form of modern ID teams came from the Montefiore Hospital in the USA with the goals to provide family-centered health for disease prevention, known as the Family Health Maintenance Demonstration Project, while at the University of Washington's Child Health Center, teams of faculty and students, including different health disciplines, were assembled to support the experience-based ID education. Next, many more efforts were made to develop MD and ID programs, associations, and offices. Some of the known examples, like "Great Society," "War on Poverty," the American Medical Student Association (AMSA), and the Office of Economic Opportunity (OEO), are reviewed in [8].

2.2 New World and Non-communicable Diseases

Recognizing the sustainable development goal (SDG) 3.4, that is "by 2020, reduce by one-third premature mortality from non-communicable diseases through prevention and treatment, and promote mental health and well-being," programs have been engineered to prevent premature mortality from non-communicable diseases (NCDs) by adjusting modifiable risk factors and promoting healthier lifestyle and behaviors. This process engages the individuals who adopt healthy choices, their families, and the range of professionals such as doctors, dieticians, nurses, occupational therapists, pharmacists, physiotherapists, physical activity specialists, psychologists, and social workers that establish and encourage healthy behavior changes [9]. All the people getting involved are related to each other in navigating the target population, identifying the complex patients, thinking about what they do need and want, and attempting to deliver the healthcare services that effectively satisfy their needs while being sensitive to their desires. Nine targets dominated by the World Health Organization (WHO) Global Monitoring Framework will be important to reach SDG 3.4 [10]. They direct a threefold focus on:

i. the most common NCDs, mainly including cancers, cardiovascular diseases (CVD), chronic respiratory diseases, and diabetes mellitus;
ii. shared risk factors, e.g., tobacco smoking, hypertension, high blood cholesterol, alcohol use, psychological problems, unhealthy diets, low levels of physical activity, environmental factors, and obesity; and
iii. the availability and affordability of treatments and technologies.

Moreover, social factors play a deterministic role in the premature mortality of NCDs that socially disadvantaged people are highly vulnerable.

3   MD/ID Approach

3.1   MD/ID Teams and Health Care

When faced with SDG 3.4, we will inevitably be in a position to understand and work on several aspects listed above. Moreover, we must give importance not only to the prevention but also to the early diagnosis and treatment of and recovery from diseases. It emphasizes the key role of MD and ID teams in managing health care. Table 3 mentions the major roles of one who leads the team called "gatekeeper" for an MD team and "chief" for an ID team and differences in members' roles, the kind of communication, and the amount of coordination between an MD and ID team. An ideal ID healthcare team for a chronic disease project includes the four main dimensions of well-being, e.g., social function, physiological function, physical function, and emotional function [11]. It also needs to be compliant with the healthcare transition (HCT) process domains and related outcomes at the individual,

Table 3   Team approach in the health care: multidisciplinary teams versus interdisciplinary teams [142, 143]

Leader	Date, location of the establishment	Team goal	Setting
Quoted by Royer	Before 1900, India	The provision of health services	Mission hospitals
Richard Cabot	1915, USA	The provision of health services	The outpatient department of Massachusetts General Hospital
Michael Davis and Andrew Warner	1918, USA	The advocacy of clinical 'efficiency' and 'social teamwork'	Boston Dispensary
Quoted by Royer	1920, Great Britain	The provision of 'front-line' primary care and a triage system	Military
Peckham Experiment	1920s, London	The development of collaborative health care teams and the idea of a 'positive health model'	London's Pioneer Health Centre
Martin Cherkasky	1948, New York	The provision of home care services	Montefiore Hospital in NewYorkCity
Sidney Kark	1951, South Africa (then Israel)	The implementation of primary health teams	Community
Robert Deisher and De Witt C. Baldwin	1953, USA	The development of teams of faculty and students focused on the linkage between experience and education	The University of Washington's Child Health Center

Table 4 Interdisciplinary healthcare transition (HCT) domains and respective outcomes [13]

HCT domain	Outcome
Individual	Life quality
	Understanding conditions, characteristics, and complications
	Medication knowledge
	Self-management
	Medication adherence
	Clarity of health insurance
Health care system	Attending medical appointments
	Having a medical home
	Avoiding unnecessary hospitalization
Social support	Belonging to a social network

the family/social support, the healthcare system, and the environment level [12, 13]; see Table 4. Social workers are important for both the ID healthcare teams and patients to achieve the common team goal (CTG) of effective healthcare delivery. They offer better communications and discussions in terms of directive decision-making, directive completion, and end-of-life family planning issues, especially in hospitalized older patients [14]. Typically, they go through a multi-phase advance directive process consisting of "initiation of advance directives, disclosure of information, identification of a surrogate, discussion of treatment options, elicitation of patient values, interaction with family members and significant others, and collaboration with other healthcare professionals." In geriatrics, the ID approach has been shown to reduce the length of hospital stay, admission to a nursing home, and mortality in the year following discharge [15]. For the efficacy of the MD/ID approach in other examples of the medical context, e.g., chronic diseases, CVD, diabetes, high-risk pregnancy, oncology, primary health care, and stroke rehabilitation, see [16, 17]. Also, for the importance of such approach in oral health, a neglected health issue, see Chap. 12.

MD and ID teams need to work in both frontstage and backstage truly. The doctor–patient interaction is on the frontstage [15], while a "place, relative to a given performance, where the impression fostered by the performance is knowingly contradicted" gives way to the backstage of health care [18]. It is backstage in which the "capacity of a performance to express something beyond itself may be painstakingly fabricated ... illusion and impressions are openly constructed. Here stage prons and items of personal from can be stored in a kind of compact collapsing of whole repertoires." In the frontstage, the physician and patient make communications corresponding to information gathering and giving, relationship building, social talk, and positive and negative talks [19]. Communication processes in backstage include "informal impression and information sharing, checking clinic progress, relationship building, space management, training students, handling interruptions, formal reporting" [15]. When these two stages do work in an ID manner, our healthcare system will be compatible with the human values planned, particularly commitment to integrity and commitment to excellence [20]. We, then, expect to take the idealization of the healthcare delivery as the ID work supplied by communications is *perfectly* synchronized between the frontstage and backstage.

3.2 MD/ID Teams and Health Education

To work together means to learn together; in practice, the successful implementation of the MD and ID approach is a reasonable expectation if we can hold relevant materials in the education setting. ID education is referred to "a process of teaching health professional students the knowledge, attitudes, and skills needed for the interdisciplinary practice of health care" [21]. It encompasses either of the following criteria [22];

    i. Faculty level, "students in one health profession taught by faculty from more than one profession";
    ii. Student level, "students from more than one health profession taught by faculty from one health profession";
    iii. Faculty–student level, "students from more than one health profession taught by faculty from more than one profession."

    ID health education is not new: The University of Washington's Child Health Center and the Saint Louis University Health Sciences Center carried out the ID programs during the 1950s and 1960s, respectively. Since the 1960s, it was possible to make models, courses, and diverse international perspectives on ID education to observe it being learned/taught in public schools. This [21] is for further reading, in history, the progress of ID education along with looking at some of the issues in communication and group practices. Also, this [23] is a systematic review that highlights heterogeneity across studies investigating the effect of ID education on clinical decision-making and collaborative skills in the healthcare setting, so calling for further research.

3.3 Challenges to ID Healthcare and Education Teams

While contributing as part of an ID team is one of the major competencies to gain in practice [24], team-based ID health care may be economically, historically, politically, and socially not feasible, especially when dealing with fragmented health services and settings under such an existing changing world [25]. As on *it's all about relationships* [26], challenges facing coworkers in ID health teams mostly arise from cultural differences, perceived inequity (unfairness) in performance and expectations, and lack of mutual understanding [27]. The members subsequently get an intuitive, unsatisfying feeling for what they are asking: should we trust this relationship? Worse, it is maintaining a balance between competition and complementation [24] and between friendship and collegiality [27] when opposing forces are applied—one, as a friend, feels affection, favoritism, and openness, while the other, as a colleague, is committed to instrumentality, impartiality, and closeness. As against *team vs. no team*, the teamwork health project can succeed; its success depends on the team's kind, goals, size, and the conditions under which the team works [8]. More precisely, it is easy to get confused when working in team sizes with larger organizations and more heterogenous team compositions. Worse of all is one occupied with role ambiguity, role conflict, role overload, and stress. Suppose, for example, the ID team of physicians and nonphysician clinicians, e.g., nurses. The two most reported contradictions are [27];

i. Autonomy vs. interdependence, because of a lack of professional role understanding and flexible role enactment;
ii. Uniqueness vs. predictability, because of proactive problem-solving and action learning.

These contradictions will have consequences for the healthcare consumer, the profession, medical education, professional associations, research, and administration by negatively affecting participation rates of team members and their collegiality [11, 25]. Behavioral strategies that help tackle such oppositions fall into two broad categories [27]; see Table 5.

And, a kind of inertia and ultimately silence might emerge whenever people become part of an ID team, and find different sorts of professions, personalities, work styles, and cultures, hindering them from practicing collaboration. Particularly, the ID education of healthcare professional teams is faced with two broad categories of challenges: system issues and content issues [28]; see Table 6. Team members require ID education materials, such as curriculums and schedules, to teach them how to control this inertia fully and achieve and sustain sufficient

Table 5 Strategies used to overcome oppositions in working relationships in ID healthcare teams [27]

Category	Strategy	Mechanism	Example
Dysfunctional	Denial	Deny a contradiction is present	A coworker that does not communicate at all with a friend in the workplace
	Disorientation	View contradictions as negative events	A coworker that discourages the other coworkers from being initiative at work
Functional	Recalibration	Use a strategy that transcends the contradiction	A coworker that provides other coworkers with opportunities to work independently
	Reaffirmation	View contradictions as positive events	A coworker that believes in differences as inevitable result of working relationships and pave the way to explore and effectively discuss contradictions with other coworkers

Table 6 Challenges to ID education of healthcare teams [28]

System issues	Availability of interdisciplinary education
	Timing of the interdisciplinary education intervention
	Non-traditional teaching methods
	Need for faculty development
	Institutional support
	Participants' characteristics
Content issues	Professional role demarcation versus role blurring
	Group skills
	Communication skills
	Conflict resolution skills
	Leadership skills

collaboration. Thus, the development of a sustainable ID health team is a long process of dealing with complex visible or invisible phenomena. There should be time enough for regular team meetings and training [24]. Table 1 in [24] lists examples of topics of the team training sessions that provide members with the opportunities, making them able to share observations and information, communicate confidently, enhance their teamwork knowledge, and make sense of human actions in terms of how our personality and personal traits interact. Baseline–endline (2 years) comparison surveys reveal that the unity with the teamwork is developed after the first year of teamwork is complete [24]. At the same time, team members have a certain kind of loyalty— "dedication and faithfulness to the organization"—whose appropriateness determines whether the individual is effective in teamwork [24]. In contrast, the overrepresentation of self-related beliefs, e.g., self-protection, self-interest first, and self-sufficiency, results in a team functioning-limiting effect [24]. The top ten principles of a good ID health team and associated themes that emerged from the available literature and workshops are listed in Table 4 of [29], among which are leadership and management. Managing meetings, either formal or informal, would effectively let the managers measure the function spectrum of the team. The analysis described in [30] shows ID team functioning as a function of CTGs, the ability of team members to work together to accomplish CTGs, decision-making, relationships, and communication (Table 7). Accordingly, the two ends of the spectrum are good functioning teams (*positive* side) and dysfunctional teams (*negative* side), so that the role of the manager is in identifying dysfunctional dimensions to make timely interventions that boost team functioning for the relevant aspects, narrowing the team's dysfunction to the goal, good functioning; the team hadn't died at all, that it was dysfunctional for a while!

Table 7 Five dimensions of ID healthcare team functioning and relevant aspects [30]. A good functioning team is *positive*, whereas a dysfunctional group is *negative* for allthesepoints

Dysfunctional team functioning		Good team functioning
—	**Common team goals**	+
	Clarity of team goals	
	Consistently focused on common team goals	
	Congruency with organization and program's mission, vision, and strategic directions	
	Clear roles and responsibilities and coordination of activities contribute positively to goal attainment	
—	**Ability to work together to achieve team goals**	+
	Timely completion of tasks of high-quality outcomes	
	Satisfaction by team members and capacity for continued teamwork	
	Appropriate skills and knowledge	
	Positive attitude that contributes positively to achievement of team goals and function	
	Assist others to achieve team goals	
	Commitment to group task	
—	**Decision-making**	+
	Decisions focused on common team goals	
	Consistent with organization goals and policies	
	Timely decision making	
	Ability to reach consensus	
	Resolution of conflict in a professional and respectful manner	
	Positive contribution to decision making by all team members	
—	**Relationships**	+
	Professional and collegial relationships	
	Mutual respect	
	Trust that team members will work together for the common good of the team	
	Constructive, sensitive feedback done privately	
	Individual dysfunction which impacts negatively on team function	
—	**Communication**	+
	Effective formal communication	
	Effective informal communication	
	Comfort with discussion of different opinions and styles, and able to resolve conflict in a constructive way	
	Participation in communication by all team members	
	Positive feedback	
	Appropriate communication	

4 ID Approach to Health Research

ID health research scholars can carry out the study(ies) involving more than one specific scientific discipline. They do research on a conceptual model formed by a linkage between or integration of theoretical frameworks from the respective disciplines. Their research is an opportunity to make use of a design and methodology that is not discipline-specific and identify an array of skills and perspectives as they move through different research phases [31]. Once ID research scholars are competent [32], they will reveal some outstanding abilities at;

i. conducting research, ID research scholars can develop integrated frameworks and concepts and apply them in designing ID research protocols, carry out ID-based hypothesis investigation, develop joint proposals for funding ID research projects, and disseminate their ID research results within and outside their respective discipline;
ii. communicating, ID research scholars are initiatives within a field of study through the advocacy of ID research, treat other disciplines' perspectives with respect, read journals outside their respective discipline, regularly communicate with other disciplines' scholars, make a language meaningful in an ID team to share their respective discipline research, modify their research plan to account for interactions with other disciplines' colleagues, and present their ID research effectively in terms of two or more disciplines;
iii. interacting with others, ID research scholars invite colleagues from other disciplines to give their views on research problems, make interactive training exercised with scholars from other disciplines, attend presentations with members from other disciplines, use respectful and equitable collaboration with scholars from other disciplines in designing ID research frameworks, and publish with scholars from other disciplines.

So, doing an ID research program, we are most likely to have access to the best available resources, approaches, organizations, and technologies, turning out to be one who comes up with solutions that are not blinded or biased but are multifaceted to fit the complexity of health issues. It lets us produce synergistic results of high quality and generalize them to the target population faster than a similar project without the aid of the ID approach. Despite its high strength, there is a long way for the ID research collaboration to surmount the absence of a clear communicative language meaningful to all disciplines and reduce difficulty in being agreed on research goals/objectives and methodologies—that arise due to variability in terminology, frameworks, approaches, and methodologies. More precisely, the study [33] denotes the boundaries of ID collaboration, divided by cognitive and political boundaries. Cognitive boundaries are mainly referred to the incommensurability of knowledge between disciplines, regarding lack of a common language, methods, and knowledge for using new instrumentation. In the political context where institutional resources and rewards are crucial concerns, we see that motivation and certainty of gaining ID knowledge and doing ID work are insufficient; the engine does not work! Symbiont practices seem to satisfy these boundaries [33], enabling doctoral students to acquire different capabilities. Moreover, an ID research scholar is required to enthusiastically gain/enhance her/his skills in collaboration, management, planning, and project administration, and has a great dedication to integrating these skills into her/his research so as to vividly compensate his/her lack of understanding of other disciplines' communication and research agenda [34].

5 Why Integrated ID

It is an old debate on the inclusion of integration in interdisciplinary studies (IDS): the integrationists group sees integration as an integral part of IDS, whereas, from the generalists' view, the integration is faced with obscure and minimum capacities in IDS, rejecting the integrationist ones. But why integration?

Let the mind answer. Cognitive ontologies directly search for human cognition, at the center of which is the brain–body concept. Research is conducted at different ontologies, mainly mental functioning, cognitive paradigm, neural electromagnetic, neuroscience information, cognitive atlas, neurological diseases, neuro behavioral, and neuropsychological [35]. They seek mental processes to yield a satisfactory account of cellular and molecular properties of these processes to sensory input and motor function. These essential aspects are connected in a powerful way, yet invisible to the eye, and thus would be difficult to investigate. Neuroimaging methods could facilitate such connections when they are normal or abnormal, yet difficulties continue to override. Primarily, we deal with the mind; it is not a planet for physicists that can observe its structure. It is not a matter for chemists that can determine the elements it contains. It is not a disease for biologists that can diagnose and treat it [36]. Then, it might be due to a lack of standard criteria to define a mental process and standard methods and tools to harmonize multimodal data and reach formal agreements, and the methodologies by which all these definitions and data can be tested at large-scale to become generalizable and reproducible [35]. Having such a deep history of heterogeneity lies in different approaches to the brain, which, in turn, depends on what discipline is studying it. Table 8 lists a few of these approaches [36]. It is worthwhile to compare how these approaches define the mind with that of ID approaches. SiMA is, for example, an interdisciplinary, functional; approach to the problem of reading the mind [37]. It suggests three layers:

i. the first layer (lowest), it consists of neural activities;
ii. the second layer (middle), it executes a process symbolizing or desymbolizing neural inputs; and
iii. the third layer (topmost), it is a machine acting on what is symbolized, making the *psyche*.

Table 8 Approaches to the problem of mind [36]

Approach	Age	Method or resource	Debate or union	Questions addressed
The philosophical approach	Old (Ancient Greeks)	Deductive reasoning and inductive reasoning	Free will–determinism debate, are our actions fully predictable/known?	What is mind? How do we come to know things? How is mental knowledge organized?
The psychological approach	Nineteenth century	Valid knowledge about the world	Structuralist-functionalistic-generalist, what mind is?	How does internal space (mind) work? How do external behaviors form?
The cognitive approach	1960s	Modularity	A shift from a discussion of external behaviorists' emphasis to cognitivism' emphasis (from around external behaviors to internal mental processes)	How does the cognitive processes (attention, memory, imagery, problem-solving, and pattern recognition) take place?
The neuroscientific approach	Relatively young	Machines measuring brain function	Brain as the focus described in terms of individual neurons, synaptic transmission, neural activity patterns at local/large scale	What are neural mechanisms underlying cognitive processes?
The network approach	Relatively young	Artificial neural networks	Mind as the focus described in terms of a collection of individual computing units and the connectivity of these units	How are neuronal units connected?
The evolutionary approach	1859 (by Charles Darwin)	Naturalselection	Evolutionary psychology, evolutionary commuting, and Darwinism	Which the selection forces do we inherit from our ancestors that cause cognitive structures?
The linguistic approach	Old	Elective	The development of language elements, properties, and systems and their relationship with thinking processes	What is language? How do we acquire language? What parts of the brain underlie language use?
The AI approach	Relatively young	Computer algorithms, programs, and devices	The application of AI-based techniques to carry out human intelligence-requiring functions	How human mental processes are operated?
The robotic approach	Relatively young	Autonomous or semi-autonomous devices	The design of a top-down approach for a robot to have human-like actions	How are the kinds of mind and behavior associated?

This comparison reveals to us that non-ID approaches often consider one idea as the core of the scope of their approaches, whereas an ID approach to the mind would consider some of the associated ideas, as well. Accordingly, it calls into question the cognitive sciences that nothing like boundaries in literature, cultures, and theories should happen at all [38]. Cognitive psychologists can see IDS integrated with theories of common ground, called cognitive interdisciplinarity [39], running through interests of both scientific and humanistic disciplines, from the philosophy of mind and anthropology to artificial intelligence (AI), computer science, neuroscience, psychology, and linguistics [40], for understanding the human mind of the twentieth century. Chapters 2 and 4 provide ID insights of the cognitive sciences.

Now let us refer again to the present-day problem. In the middle of COVID-19 lay an exceptional opportunity for collaboration as an ID researcher to be in motion in the winding road of understanding a new global challenge [41–43]. Different disciplines collaborated on COVID-19 research, from epidemiology, virology, omics, imaging, and genetics to biomedicine, molecular sciences, clinical medicine, and pharmacy [44]. ID integrated research in COVID-19 could make it possible to manage the pandemic by defining the causing pathogen, a coronavirus called SARS-CoV-2 and developing screening methods, diagnostics, therapeutics, and vaccines.

Suppose health issues focusing on mental health that is a part of SDG, target 3.4, as quoted earlier. It is a multidimensional concept involving biomedical sciences and sociomedical sciences [45]. It is a complex issue since its etiology is multifactorial and not precisely understood. It is a global challenge, and men and women suffering from mental illnesses are mapped to the world.

Say climate change. The scientific consensus on global climate change has emerged that the greenhouse effects we ourselves produce are responsible for most of the observed warming of the Earth's surface. It is now 30 years since this awareness occurred [46]. However, climate change is still affecting the globe, and we, as scientists, are predicting the extinction risk [47]. It calls to take effective actions with regards to the assessment of climate change impact on infrastructure as well as engineering approaches to make resilient infrastructure [48]; see Chap. 15.

As another example, say poverty. The problem is the development of an indicator to rate poverty. The single-item measures include the head-count ratio and the income gap ratio [49]. The head-count ratio (H) is the proportion of a population whose income or consumption falls below the poverty line. The income gap ratio (I) is the average deviation from the poverty line for those living below the poverty line. None of these ratios are able to reflect changes in income redistribution that takes from a poor person and gives to another poor person. Given this and its relations with aspects of psychological health and well-being, poverty is now argued to account as an unidimensional concept, so it can only be approximated using an unidimensional measurement [50]. New approaches emerged to take poverty in its actual place as a multidimensional concept. Particularly since 2010, the global multidimensional poverty index (MPI)—which is an adjusted H ratio—was formulated [51], based on three fundamental dimensions of poverty and related indicators, including health, education, and living standards. With this multidimensional definition, it is more effective to recognize the interconnections that govern the problem complexity, and so the likelihood of the problem to be decidable would be increased.

It would be reasonable for all these situations that an ID approach for health care is more likely to be impactful than a non-ID approach [52]. Also, innovative integrated research can lead to science breakthroughs by providing international mentorship and interdisciplinary collaborations and communications; for example, see one recent paper on the "Integrated Innovation™" approach [53]. Implementing this approach requires a paradigm shift in health education, and research settings so become universally connected [54]. Collaborative interdisciplinary global health tracks (CIGHTs) are, for example, a novel ID health program planned to address different challenges to achieve global health, with the inclusion of ID faculty, curriculum, and international exchange programs [55].

6 How Integrated ID

6.1 Art

The origins of arts are diverse, so are their forms and applications. The ID research wants to see all arts as integrated into the sciences. However, to integrate art and sciences, finding common ground is necessary to share knowledge and questions and achieve shared goals [56]. To exemplify, arts are favorable to evolutionary biology; to survive means to make high-quality artworks. It implies that talents, skilled art workers, and people possessing aesthetic characteristics are more likely to fit the environment. From a cognitive perspective, arts are the result of thinking processes; to make an artwork means to think in different modes, e.g., symbolic, abstract, and referential. From a psychological perspective, the effects of art can be traced to the brain and mind. Exposure to artwork has been shown to be courageous, so it makes the mind appreciate (art appreciation), and the brain activates several brain regions (aesthetic experience) [56]. For a detailed discussion, see Chap. 3. And many other perspectives have already been developed, concluding that arts are a means of interaction between neuroscience, biology, evolution, etc., as reviewed in [57]. That would extend interdisciplinary cooperation and communication and sociocultural groups [58], which are vital for scientists to manage the complex global changes and challenges and practice science without borders [59]. Computational aesthetics, for example, results from applying computational methods to the expression of aesthetic-related qualities. Enormous capacity lies in this field, ranging from photography and fine arts to industrial design, a process that all service systems need to undertake to design, interestingly, medical products and devices and healthcare research programs [60].

Science-inspired arts convey a language meaningful to the public, making the knowledge acquisitor unpressurized [61]. They can be included in research papers and collaborative projects and raise awareness. At the public level, it is especially helpful to critical and complex discussions: for example, the illustration of hygiene practices during the critical period of the COVID-19 pandemic [62], which can be generalized to other crises, and the illustration of signaling pathways and molecular mechanisms underpinning a complex medical condition [63]. Focused on health, arts help improve healthcare quality by affecting both parties, patients, caregivers, and healthcare providers, and helping them find an individual meaning of health and health care. Visualization, in particular, improves observation skills. Visual arts can be used as a part of ID science education in different healthcare settings, e.g., from anatomy and microbiology laboratories to nursing schools [64–66]. Exploring the meaning of artworks is a task that is the core of visual thinking strategies (VTS). Medical students who participated in VTS sessions reported increased empathy and tolerance for ambiguity [67]. Art workshops that provide a space for storytelling about the patient narratives and professional health care have shown a similar effect [68]. Also, storytelling and expressive writing enhance collaborative and creative skills, making dialogue, and practice-related quality of life, and can be incorporated into ID programs of professions, like nurses [69]. Different types of art therapies have, therefore, been implemented in healthcare environments, including art, dance, drama, music, and expressive therapies, with the purpose of providing *aesthetic healing environments*. Art directives are also inspiring. A randomized study has shown the effect of art directives, e.g., viewing an artwork, in mood enhancement for patients and empathy enhancement for hospital staff [70]. However, future research is required to establish common ID logics between health and art to facilitate the ID science-art education and practice. Chapters 18–22 represent the integration of art in health sciences.

6.2 AI

AI is a science of engineering, both a means of integration and a means of inter-disciplinarity. For this purpose, computers (machines) are required to perform analysis of human intelligence, identify behavior patterns, and then build models. AI-based approaches can reduce the human-made interventions, and instead of or complementary to them, undertake analytical thinking and learning to problem-solving, decision-making, and acting. Medical AI applications are of two broad categories: virtual and physical [71]. Virtual AI mainly involves health informatics. AI and related approaches, such as deep learning, machine learning, and neural network algorithms, undertake data analysis and make predictions about the diagnosis, prognosis, and treatment. The AI-based bioinformatic analysis offers great help to personalized medicine, personalized care, or precision medicine; terms interchangeably used to refer to medical models that incorporate individual profiles to determine individual care. What data are analyzed vary between different specialties. For neuropsychiatry, the neuroimaging data and pharmacogenomics data are important [72], while for oncology, histopathological image analysis and genetic data are required. Additionally, there is an AI-based physical assistant. This assistant can be a robot that interacts with the patient and/or physician or a nanorobot that helps with drug delivery. In this manner, medical AI applications can reduce the healthcare workflow while providing precise predictions and assistance. Medical AI application is, however, faced with social, ethical, and economic challenges. In summary, the issues are of psychological and methodological category: the attitudes of both healthcare recipients and providers with regard to AI application are not psychologically prepared, and resources for AI algorithms that need to be methodologically situation-specific are yet being developed [73]. Of particular importance is that AI is lacking in the current medical curricula, and AI is needed to be integrated into ID medical education [74]. For details, see Chapters 23–25.

7 To Do so

The development of an ID education program is an engineering design problem [75]. From this perspective, we need to address;

   i. what the problem is, as defined by an expert or a team of experts;
   ii. how much variables are integrated (multidisciplinary, interdisciplinary, and transdisciplinary); and
   iii. where the learning environment is.

Above are the keys for addressing the ID education problem; however, these keys have subkeys we might encounter and pass them. A systematic review summarizes four keys and respective ten subkeys as follows [76];

   i. ID thinking, it requires disciplinary and ID knowledge. Higher-order cognitive and communicative skills play a role in this context, as well;
   ii. Student, personal characteristics and prior social and educational experiences are important;
   iii. The learning environment, which involves:

   - disciplinary and ID curriculum;
   - an intellectually ID-focused teacher interested in team teaching and development;
   - pedagogy with the purpose of ID, active learning, and collaboration; and
   - assessment of intellectual maturity and of ID performance in students; and

   iv. Learning process, the learning pattern can be phased, linear, iterative, and milestones. Learning activities need to improve ID performance and reflective thinking.

Above the above, we see a manager who is an expert in ID thinking and making sense [77]. Considering the above points, ID should see no obstacle to integration in education so achievable, even in monodisciplinary structures [78]. The vast majority of instructional materials are, often, discipline-specific in higher education, and general ID education is a less considered issue. Yet, it is with the ID education that students will know the global knowledge economy [79], the economic benefits gained by "knowledge management technologies," and be taught how it is integrated along with their own knowledge, having them prepared for solving real-world problems and global challenges beyond their respective discipline. It, therefore, defines a shift from merely discipline-specific traditional education to the integration of ID curriculums into higher education. The following are a part of the preparations for such an ID educative journey [79];

i. *induction and preparation of students for entry into new disciplines;*
ii. *language checklists;*
iii. *cognitive maps;*
iv. *benchmarking disciplinary knowledge;*
v. *fostering interdisciplinary exchanges;*
vi. *decentering programs;*
vii. *evaluating interdisciplinarity.*

Summarizing, a general ID education is based on a set of instructions in terms of skills-oriented teaching and using social science-based and humanistic science-based methodologies [80]. It can engage learners' reflective thinking of and solving problems of various topics. Becoming generally ID educated, one can, based on learning, acquire habits, such as problem-posing and problem-solving, confidence, mastering, reasoning, abstract thinking, and conceptualization that give the mind the quality of "dispassionate analyzer." ID topics of learning are different in a health setting, including teamwork, primary health care, problem-solving, chronic diseases, clinical skills, communicative skills, health behavior, improvement, therapeutics, and labor and delivery [81]; see Chap. 27. Importantly, ID education for pre-professionals, including medical and social work students, has been shown to improve social skills useful for palliative care [82]. Also, ID approaches can be a part of educational psychology for next-generation education to promote the development of executive functions and social skills. Kindergarten children who underwent an eight-month ID intervention were better in academic abilities as a function of autonomy, language, mathematics, and peer communication in first grade than children using traditional education [83]. ID has been embedded into STEM (science, technology, engineering and mathematics) programs for high-school-age children. Particularly, a real-world project-based or problem-based learning approach could promote the positive effect of ID integration on learning outcomes. However, a recently published systematic review has raised concerns about insufficient data and the heterogeneity of available data [84], so future high-quality investigations are necessary before large-scale ID-STEM integration can be used.

Below are a few strategies to enhance ID learning.

7.1 Interface

Table 9 lists features to contrast behaviorism with cognitivism regarding their viewpoints of learning [85]. These two foundations have many differences but also a common desire to apply multimedia technologies universally. The method of application of procedural ID models again occurs differently, as a function of the model, learning process, planning, the need to work under experts' supervision, the cause-effect relationship between program objectives and design/development, and the method of evaluation. Yet, both models again arrive similarly at being improved by implementing interface design guidelines and integrating them into ID models. With such a framework, we say with certainty which kind of external learning instruction would cause an internal cognitive strategy; from "gain attention" to "reception," from "tell learners the objective" to "expectancy," from "stimulate recall of prior learning" to "retrieval to working memory," from "present stimulus with distinctive features, that is, tell or show the students what they are to do" to "selective perception," from "provide learning guidance" to "semantic encoding," from "elicit performance" to "learner response," from "provide feedback" to "reinforcement," from "assess performance" to "retrieval and reinforcement," and finally, from "enhance retention and transfer of learning" to "retrieval and spaced review."

7.2 Cognitive-Behavioral Therapy

Integrating ID cognitive approaches in tools and interventions can help manage challenging conditions associated with behavioral risk factors. For example, chronic pain is an intricate property that can be hidden from the sight of evaluators in older people with limited ability to communicate. ID teams might help with diagnosis and, therefore, treatment of pain in this vulnerable population [86]. Moreover, chronic pain is a difficult-to-treat condition that tends to persist and recur, correlating with considerable distress and disability, especially in those suffering from mental health issues, like depression and anxiety [87]. ID cognitive-behavioral programs offer to manage pain, improve physical function and somatic awareness, and alleviate anxiety [88]. Obesity is another example. Diet and physical activity are among lifestyle habits obese people lose control over them, and so maintenance of long-term weight loss is not an easy project to do. Undergoing ID cognitive-behavioral programs, they can learn to set goals, outline actions, identify and manage barriers, and regulate and monitor self-related behaviors. M.O.B.I.L.I.S. is such an intervention proved to increase weight loss maintenance, regular exercise, and self-efficacy [89]. At the end of two years, people were more able to make a good food choice and set and achieve their goals [89]. Hypertension is also a behavior-related disease. With a short, four-week ID education program, patients could achieve a more decrease in systolic blood pressure compared to those undergoing standard care [90]. Chapters 8 and 10 discuss social cognition in relation to irrational choices about their health.

Table 9 Behavioristic versus cognitivist' points of view of learning [85]

		Behaviorism	Cognitivism
The mind–body	Mental activity	Body, action	Mind, consciousness, schema, knowledge structure, and duplex memory
Structuring	Method	Explicit objectives, tasks, and subtasks	Advance organizers
	Complexity	Relatively lower	Relatively higher
	Reflective thinking	Relatively lower	Relatively higher
	Learning level	Relatively lower	Relatively higher
Tutoring	Method	Examination, performance analysis, and remedial or extended instruction	Coaching and scaffolding
Assessment	Testing	Multiple choice tests and problem-solving questions with right answer	Authentic
	Integrated into activities	−	+
Motivation	Factor	Success	Learning
	Reward	Extrinsic	Intrinsic
Control	Method	Unassisted learner control, program control, or adaptive instruction	Assisted leaner control
	Support	Relatively lower	Relatively higher
Multimedia application	Analysis, decomposition, and simplification of tasks	+	+
	Using devices	+	+
	Engaging learner	+	+
	Interactive decision-making	+	+
	Meaningful learning	+	+
	Realistic learning	+	+
Procedural ID models	Model	Objective-rational models to constructivist-interpretivist models	Constructivist models
	Learning process	Sequential and linear	Recursive, non-linear, or chaotic
	Planning	Top down and systematic	Organic developmental, reflective, and collaborative

(continued)

Table 9 (continued)

	Behaviorism	Cognitivism
Experts	–	+
Objectives-development relationship	Objectives  development and design	Development and design  objectives
Goal	Delivery of preselected knowledge	Personal understanding
Summative evaluation	+	–

7.3 Design Thinking

ID research itself is seen as a creativity-based design process. From inspiration to ideation to implementation, creativity is the whole story of design thinking. Therefore, design thinking is very discussed as an intermediate between creativity and ID working—applications of design thinking range from in school classrooms to higher education and medical education. To give an example, in the medical education setting, design thinking courses were administered to teams of students from medicine, neuroscience, psychiatry, art, psychology, social sciences, with the goals to design/redesign one or more intervention(s), strategy(ies), ward(s), or environment aimed at solving different problems, e.g., obesity prevention, management of attentional disorders, etc.; see details in [91]. The course effectively enhanced holistic strategies and collaborative skills, creative thinking, and academic achievement.

7.4 Simulation-Based Training

Crisis resource management (CRM) indicates the healthcare providers' awareness of human factors in crisis management [92]. It is especially essential for emergency medicine and anesthesia professionals who work in acute care settings [93, 94]. Despite its significance that necessitates CRM being integrated into medical education, CRM training has not been conducted effectively and systematically in many countries [92]. CRM training consists of ID healthcare teams. Its outcomes include, for trainees, attitude modification, knowledge acquisition, and behavioral changes; for patients, health or well-being improvement; and for trainers, return on costs. A systematic review of CRM team training has shown that a simulation-based training approach could be higher in improving team behaviors and acquisition of skills than a didactic case-based approach and simulation without training [95].

8 Further Works for Health

ID integration is of importance to biomedical sciences and of interest to this volume, too.

8.1 Biomedical Engineering (Bioengineering)

Biomedical engineering is an ID discipline that really happens upon life sciences through engineering. Biomedical engineering has accomplished a lot to what we see today in the modern healthcare settings [96], from the electron microscope and iron lung to hearing aids (Chap. 21) and computers for laboratories; for a review of its golden accomplishments in history, see [97]. By definition (not always the same), it applies engineering principles to biological systems to explain their behavior, so they can be changed and controlled [98]. Biomedical engineering contributes to activities, of which some are listed in Table 10. In summary, the development of:

i. diagnostics that include imaging modalities (below will be discussed), monitoring biosensors and transducers, and instruments and devices for clinical laboratory;
ii. therapeutics that apply to sensory/motor problems (nerve stimulation, intraocular lens, and cochlear implants), cardiovascular diseases (stents, balloon pumps, catheter-based ablation, defibrillators, and pacemakers), renal diseases (dialysis), respiratory diseases (artificial lungs and devices for blood-gas exchange), and tissue or organ failure (materials, organ transplantation, prostheses, and implants); and
iii. devices and procedures that aid in surgery and rehabilitation [99].

occurs under the umbrella of biomedical engineering. Moreover, tissue engineering, biotechnology, and biomaterials, that involve the development of biodegradables, monoclonal antibodies, vaccines, and gene therapy are of biomedical engineering careers [99]. Besides those who work as engineer-scientists

Table 10 Biomedical engineering activities [98]

Area	Career
Engineering system analysis	Physiologic systems modeling, simulation, and control
Monitoring physiological signals	Biosensors and biomedical instrumentation
Diagnostics	Bioelectrical signal processing
Therapeutics and rehabilitation	Rehabilitation engineering
Repairing and replacing body parts	Artificial organs
Clinical data analysis and decision making	Medical informatics and artificial intelligence
Evaluating physiological functions	Biomedical imaging

and technological entrepreneurs, biomedical engineers can become members of ID healthcare teams, known as clinical engineers, help develop cost-effective approaches, make decisions regarding medical equipment selection, apply standardized organizational procedures, and train the healthcare staff. To this end, clinical engineers need to interact effectively with regulatory agencies, patients, doctors, allied health professionals, nurses, vendors, etc.; see Fig. 1.10 in [98]. Biomedical engineering courses have to incorporate inclusive education programs and integrated ID teaching approaches while maintaining the balance between comprehensiveness versus depth [99–101].

8.1.1 Biomedical Imaging

Biomedical imaging techniques are used to maintain an image of the target organ that is:

i. functional (fMRI, functional magnetic resonance imaging; SPECT, single-photon emission computed tomography), these methods are useful to measure the function of the target organ;
ii. metabolic (PET; SPECT), these methods are useful to monitor molecular processes implicated in the pathogenesis of diseases, corresponding to disease activity;
iii. molecular (MRS, magnetic resonance spectroscopy; PMRS, phosphor magnetic resonance spectroscopy; scintigraphy; PET; SPECT), these techniques are useful to determine the profile cellular and molecular markers associated with pathologies;
iv. functional and metabolic (dual modalities, e.g., SPECT/CT, PET/CT, and PET/MRI), these methods are the most efficient imaging techniques used in oncology [102].

Other biomedical imaging techniques include biomedical optical imaging, computer-aided detection (CAD), and image-guided intervention and therapy. Therefore, the roots of biomedical imaging techniques are the measurement and analysis of physiological, molecular, cellular, and biochemical processes. To emphasize, they directly integrate rules of physics, chemistry, and biology into imaging, and on the other side, specialists and researchers from many fields and disciplines apply these imaging techniques to their health-related practice and research. Chapter 11 provides an ID approach to nuclear medicine imaging methods. Chapters 3, , 6, , 7, and 9 include fMRI studies for the understanding of empathy, aesthetic experience, sensorimotor integration, and psychiatric symptoms.

8.2 Data Mining

Its job is to examine massive data and find patterns and information useful to practical applications (or simply as digitalization). A modern variant of data mining (DM), distributed data mining (DDM), is powerful enough to resolve limitations and challenges that traditional forms were faced with, e.g., information processing on centrally collected or centralized data that were resident in-memory and static. DDM can do information processing on large-scale data obtained from distributed places in a system [103]. So, many areas benefit from the DM method. For the healthcare setting, there are various DM-based technologies and software tools that the data is digitally available, in the form of the internet of things (IoT), fabric and flexible sensors, mobile smartphones, social media, and electronic medical records (EHRs). Smart devices (smartwatch, smart glass, Fitbit), wristband sensors, and health monitors (ECG sensor, EEG sensor, PPG sensor) are IoT devices that can be worn and provide remote healthcare monitoring. Examples of flexible sensors are skin-on patches used for monitoring blood chemistry, blood pressure, ultraviolet (UV) exposure, and cardiac rhythm. Nonwearable IoT cannot be worn but is equipped with different sensors, e.g., door, motion, object contact, pressure, sound, and video sensors, to monitor patients' daily behavior and activities. These sensors can help the health care of older adults and those with memory impairment (known as assisted living). Mobile smartphone applications (mHealth) do work useful for both professionals and patients, in many areas, from time management, communication, information curation, medical education and training, medical decision-making, and diagnosis and treatment management. Examples of mHealth apps include search portals, textbooks, clinical practice guidelines, medication adherence, and references. Easy engagement is an important advantage of these apps to patients that underlies successful behavioral change [104]. Chapter 9 discusses medication adherence from a multidisciplinary perspective. EHRs digitally organize the data and information about imaging scans, vital signs, pathology reports, clinical notes and history, radiographs. EHRs are accessible, so healthcare providers can search for and extract any data required. Chapter 23 is about big data and AI in the healthcare setting. The study [105] reviews how DM technologies use digital applications and services, collectively referred to as digital data foundations, to facilitate health care when and where required for patients. Also, digital procurement techniques, such as electronic data interchange (EDI), e-procurement systems, and blockchain, that increase accountability, anti-corruption, and transparency are important ID solutions to the global issue of pharmaceutical corruption [106]. DM has contributed to the ID research in several health conditions, including, but not limited to, cardiac diseases, HIV/AIDS, blood bank sector, cancer, tuberculosis, diabetes mellitus, kidney dialysis, dengue, dementia, and in vitro fertilization (IVF), with classification and clustering techniques useful for diagnosis and risk predication [107–109]. For a review of DM techniques in medical data classification, see [110].

It is not within the scope of this introductory chapter. But you may recognize it as related to integrated ID research; what other important health-related implications do DM methodologies have? Here are a few examples. Geospatial DM studies have established significant associations between certain chemical mixtures of air pollutants and adverse birth outcomes [111]. It is a warning for pregnant women living in regions exposed to particulate matter (PM), carbon monoxide (CO), xylene, toluene, and methyl ethyl ketone, 2-butoxyethanol and n-butyl alcohol

[111]. During the COVID-19 pandemic, DM was integrated into ID research in COVID-19, and the major themes of DM studies were spatiotemporal, health and social geography, and web-based mappings. These studies, for example, found an ID association between the number of confirmed COVID-19 with social behavior and the country's ruling party in terms of a higher number of Republican voters in countries with less rigorous social distancing measures. It revealed how much influence have the political leaders along with the social distancing behaviors [112]. Also, studies showed how closely the spread of COVID-19 and public transportation-related behaviors could be (see Chap. 16). DM using a Web mapping service could provide recommendations to policymakers with regards to transportation mode and means, venue types visited, time to check-in venues, preferences over origin–destination (OD) distance, and patterns of OD; see Chap. 17 [113]. Working with such a broad array of social, behavioral, and phenomenological events, DM is, thus, a technique truly applying to our ID health ecosystem [114].

8.3 Nanotechnology

It is an ID science of particles of the nanoscale ($10^{-9} - 10^{-7}$). These particles, having favorable features and functions, especially due to their small size, are what many need for their investigations. Therefore, nanotechnology has been simply fused with different disciplines, e.g., biology, dentistry, engineering, chemistry, medicine, nanotechnology, nutritional sciences, and physics, to name a few. And this fusion is expected to be increasing as the new generations and themes of nanotechnology are developed. We have seen four generations up to 2015, with the themes ranging from passive nanotechnology and active nanotechnology to nanosystem technology and molecular nanosystem. Each of these themes corresponds to an important event in the application of nanotechnology; see Table 1 in [115]. Health-related nanotechnology applications are in medical imaging (contrast agents), healthcare diagnostics and therapeutics, drug delivery, and tissue engineering. Particularly, nanoformulations apply to cancer therapy, vaginal therapy, wound infections, and skin diseases, nanoparticles to the treatment of prostate cancer and parasites, and nanotechnology-based drug delivery to Alzheimer's disease and eye diseases [116]. ID nanotechnology courses have been pronounced since the 2010s; for example, [117–120].

8.4 Health Informatics

As mentioned earlier, achieving precision medicine is not hopeless, with information processing methods by AI, data mining, health informatics, and big data. For the purpose of this subsection, we need to return to the state that the data is readily available, but before any of these processors run, *how to deal with the different types of health data for making meaningful information.*

A systems approach to medicine means integrating different types of data regarding a health challenge. This integration should fit its context and related data. For example, if we have a collection of "omics" data, it is based on concepts, statistics, or models. For the different types of data to become integratable require:

i. cross-referenced data, different types of data are linked, for example, in a patient-specific manner;
ii. a single platform, through which we have access to the cross-referenced data; and
iii. optimal platform and data format, which enable us to generate systematic queries.

Accordingly, a systems medicine approach might deviate from its goal when a, internal data formats are not standardized; b, access to external data resources is not good; c, different types of data from different ontologies are not cross-referenced; and d, the records of intermediate results or data are not stored. Therefore, health informatics is an ID discipline providing tools and platforms to meet the data efficiently; see Chap. 26 [121].

8.5 Diagnosis and Treatment

Why we need an ID approach to health was earlier discussed for some cases, e.g., global challenges, NCDs, and mental health issues. Also, we previously explained that the main difficulty with complex medical conditions is that we cannot define their *borders*, and so we cannot diagnose and manage them in a specialty-specific manner [122]. An ID approach is, therefore, required for their diagnosis and treatment. Consider interdisciplinary and diagnosis as keywords (previously, we know this search is not comprehensive but is enough effort to prove the relevance). At the top of search results are different types of cancer. We add the other keyword, cancer, find that it is within the scope of more than 20% of publications on ID diagnosis. Then, we take melanoma as the condition to go deeper. Cutaneous melanoma is the deadliest skin cancer. The diagnostic approach to melanoma includes clinical, dermoscopic, and histopathologic findings, along with information that may come from additional staging examinations [123]. The therapeutic approach varies by disease stage and includes various options, including surgical therapy, radiation therapy, adjuvant therapy, chemotherapy, and chemoimmunotherapy. As such, *in practice,* melanoma management must have an ID-based group of specialists, dermatologists, medical oncologists, surgical oncologists, radiotherapists, and pathologists. And to develop these diagnostics and therapeutics would need a larger body of basic scientists, clinical scientists, and translational scientists *in research*. Another example is autism spectrum disorders (ASD), neurodevelopmental disorders associated with a high economic burden. We previously know the genetics and epigenetics of ASD [124, 125]. Systematically reviewing the literature has proposed a collective of environmental factors, e.g., parental age, maternal obesity and diabetes, birth complications, nutrient deficiencies, and toxic elements [126–128]. Each of these factors for ASD prevention/treatment maintains a value that needs to be evaluated by discipline-specific scientists. However, the disease results from the interaction between these values, that is, where ID scientists are crucial to making diagnostic tools and therapeutic interventions (Chaps. 5 and 6).

8.6 Drug Delivery

For therapeutics destined for delivery to the desired target, drug delivery systems (DDS) have been developed. The "target" ranges from subcellular organs and cells to tissues and body organs. The last decade's progress has made DDS optimal to increase the performance of drugs while decreasing their side effects; for review, see [129]. DDSs apply to diseases such as cancer, respiratory diseases, inflammatory diseases, eye diseases, skin diseases, and neurodegenerative diseases [130–135]. Different sciences are important to make a good ground for developing strategies and techniques useful for DDS, particularly biomedical sciences, material sciences, pharmaceutical sciences, pharmaceutical companies, and industrial scientists; see Chap. 13.

8.7 Pharmacokinetics

It is a science that dynamically deals with drugs throughout the processes of absorbing, distributing, and eliminating. The knowledge of drug dynamics is useful for drug research and development, toxicity estimation, and drug prescription. Particularly, it guides the therapy by determining the administration route, dose regimen, and dose individualization [136]. So, ID teams of pharmacokinetic scientists and medical doctors would be helpful for medication management, especially in critically ill patients [137].

8.8 Social Capital and Health

It is not easy to define social capital directly, so instead, we refer to two major forms of social capital:

  i. Bonding, it is on the horizontal direction held in the bonds of community and voluntary groups;
 ii. Bridging, it is said to have the two directions of horizontal (above) and vertical. The vertical direction acts as a bridge between the community and voluntary groups and organizations (statutory and non-statutory).

Bonding, therefore, occurs at the micro-level of individual, family, or household, whereas bridging involves the macro-level of organizations on the local and national to global scale. Both consist of two major components: structural and cognitive. The structural component has social networks for bonding form and is embedded in accessible goods and services for the bridging form. The cognitive component gives the social control, social efficacy, common values, social trust, and reciprocity norms for bonding form. It gives participation, a sense of belonging, and decision-making for the bridging form.

Social conditions negatively charge health equity (Chaps. 28 and 29), and therefore, social capital and health emerged as a concept to weave together these two complex concepts [138]. This concept is, in particular, of interest to the ID field of medical geography, environmental health (urban), social psychology/psychiatry, sociology, anthropology, health policy, and political sciences [139]. The relationship between social capital and health, in general, seems rather positive, but it varies as a function of the level, approach, and setting of social capital and health outcomes [140]. The gap of this knowledge lies in a lack of integrated qualitative and quantitative data [139].

9 Conclusion

Since the 1990s, the MD/ID approach has been increasingly integrated with health care, health education, and health research. Such an approach is compatible with SDG 3.4 to improve physical and mental health; we emphasize it is a multidimensional approach. Additionally, an MD/ID approach can be optimized to support the healthcare transition process to involve all the individual, the family/social support, the healthcare system, and the environment levels; we call it a multipole approach. MD/ID teams are helpful to both frontstage and backstage and become even more ideal when linking these stages in the healthcare setting. There are numerous challenges associated with MD/ID teams within health care and education; however, they can be managed effectively so to see a dead team becomes alive. ID approach to research allows scholars to become competent in conducting research, communicating, and interacting with others. It is not simple, of course. There are cognitive and political barriers and no way to escape. An ID research scholar is required to enthusiastically gain/enhance her/his skills in collaboration, management, planning, and project administration, and has a great dedication to integrating these skills into her/his research so as to vividly compensate his/her lack of understanding of other disciplines' communication and research agenda.

ID, the mind responds, wants integration—considering that reading the mind is another complex problem, the mind is a fair representative one to ask. A simple comparison reveals that non-ID approaches often consider one idea as the core of the scope of their approaches, whereas an ID approach to the mind would consider some of the associated ideas, as well. Another representative is the present-day problem. It was the integration of ID research in COVID-19 that made it possible to manage (not fully yet, but we get vaccinated) the pandemic by defining the causing pathogen, a coronavirus called SARS-CoV-2, and developing screening methods, diagnostics, therapeutics, and vaccines.

AI and arts can help its integration. AI-based approaches can reduce the human-made interventions, and instead of or complementary to them, undertake analytical thinking and learning to problem-solving, decision-making, and acting. Also, science-inspired arts convey a language meaningful to the public, making the knowledge acquisitor unpressurized. They can be included in research papers and collaborative projects and raise awareness; on the other hand, the arts help improve healthcare quality by affecting both parties, patients and healthcare providers, and helping them find an individual meaning of health and health care.

It is not easy to integrate ID into education programs. An engineering design approach might help address the problem effectively. Studies report the integration of ID approaches into both general education and next-generation education. Design thinking, interfaces, and cognitive-behavioral therapy can be useful in this context. Yet, it is a problem again!

Core Messages

- MD/ID approach is a multidimensional approach compatible with the SDG 3.4 to improve physical and mental health.
- MD/ID is a multipole approach so can be optimized to support the healthcare transition process.
- Challenges associated with MD/ID teams within health care, education, and research can be managed effectively.
- AI and arts can help integrate ID in research.
- It is not easy to integrate ID into education programs; it is a problem again.

It is not easy to integrate ID into education programs. An engineering design approach might help address the problem effectively. Studies report the integration of ID approaches into both general education and next-generation education. Design thinking, interfaces, and cognitive-behavioral therapy can be useful in this context. Yet, it is a problem again!

References

1. Rezaei N (2021) Integrated Science. Integrated Science, vol 1, 1 edn. Springer, Cham. doi: https://doi.org/10.1007/978-3-030-65273-9
2. Kyu HH, Abate D, Abate KH, Abay SM, Abbafati C, Abbasi N et al (2018) Global, regional, and national disability-adjusted life-years (DALYs) for 359 diseases and injuries and healthy life expectancy (HALE) for 195 countries and territories, 1990–2017: a systematic analysis for the Global Burden of Disease Study 2017. The Lancet 392(10159):1859–1922. https://doi.org/10.1016/S0140-6736(18)32335-3

3. Eagle M, Baudouin SV, Chandler C, Giddings DR, Bullock R, Bushby K (2002) Survival in Duchenne muscular dystrophy: improvements in life expectancy since 1967 and the impact of home nocturnal ventilation. Neuromuscul Disord 12(10):926–929. https://doi.org/10.1016/S0960-8966(02)00140-2

4. Landfeldt E, Thompson R, Sejersen T, McMillan HJ, Kirschner J, Lochmüller H (2020) Life expectancy at birth in Duchenne muscular dystrophy: a systematic review and meta-analysis. Eur J Epidemiol 35(7):643–653. https://doi.org/10.1007/s10654-020-00613-8

5. Rezaei N (2021) Coronavirus Disease-COVID-19. Springer

6. Marois G, Muttarak R, Scherbov S (2020) Assessing the potential impact of COVID-19 on life expectancy. PLoS One 15 (9):e0238678

7. Naylor MD, Coburn KD, Kurtzman ET, Prvu Bettger JA, Buck H, Van Cleave J et al. Inter-professional team-based primary care for chronically ill adults: State of the science. In, 2010. pp 24–25

8. Baldwin DWC (1996) Some historical notes on interdisciplinary and interprofessional education and practice in health care in the USA. J Interprof Care 10(2):173–187

9. Jennings C, Astin F (2017) A multidisciplinary approach to prevention. European journal of preventive cardiology 24 (3_suppl):77–87

10. NCDs WEOf (2017) The WHO global monitoring framework on noncommunicable diseases: progress towards achieving the targets for the WHO European Region. WHO Regional Office for Europe, Copenhagen

11. Feiger SM, Schmitt MH (1979) Collegiality in interdisciplinary health teams: Its measurement and its effects. Social Science & Medicine Part A: Medical Psychology & Medical Sociology 13:217–229

12. Betz CL, Smith KA, Van Speybroeck A, Hernandez FV, Jacobs RA (2016) Movin'on up: an innovative nurse-led interdisciplinary health care transition program. J Pediatr Health Care 30(4):323–338

13. Fair C, Cuttance J, Sharma N, Maslow G, Wiener L, Betz C et al (2016) International and interdisciplinary identification of health care transition outcomes. JAMA Pediatr 170(3):205–211

14. Black K (2005) Advance directive communication practices: Social workers' contributions to the interdisciplinary health care team. Soc Work Health Care 40(3):39–55

15. Ellingson LL (2003) Interdisciplinary health care teamwork in the clinic backstage. J Appl Commun Res 31(2):93–117

16. Benagiano G, Brosens I (2014) The multidisciplinary approach. Best Pract Res Clin Obstet Gynaecol 28(8):1114–1122

17. Clarke DJ, Forster A (2015) Improving post-stroke recovery: the role of the multidisciplinary health care team. J Multidiscip Healthc 8:433

18. Goffman E (1978) The presentation of self in everyday life, vol 21. Harmondsworth London,

19. Roter DL, Hall JA, Katz NR (1988) Patient-physician communication: A descriptive summary of the literature. Patient Educ Couns 12(2):99–119. https://doi.org/10.1016/0738-3991(88)90057-2

20. Rider EA, Kurtz S, Slade D, Longmaid HE, Ho M-J, Pun JK-H et al (2014) The International Charter for Human Values in Healthcare: An interprofessional global collaboration to enhance values and communication in healthcare. Patient Educ Couns 96(3):273–280. https://doi.org/10.1016/j.pec.2014.06.017

21. Lavin MA, Ruebling I, Banks R, Block L, Counte M, Furman G et al (2001) Interdisciplinary health professional education: a historical review. Adv Health Sci Educ 6(1):25–47

22. Pellegrino ED (1972) Interdisciplinary education in the health professions: Assumptions, definitions, and some notes on teams.

23. Lapkin S, Levett-Jones T, Gilligan C (2013) A systematic review of the effectiveness of interprofessional education in health professional programs. Nurse Educ Today 33(2):90–102. https://doi.org/10.1016/j.nedt.2011.11.006

24. Cashman SB, Reidy P, Cody K, Lemay CA (2004) Developing and measuring progress toward collaborative, integrated, interdisciplinary health care teams. J Interprof Care 18(2): 183–196
25. Jansen L (2008) Collaborative and interdisciplinary health care teams: ready or not? J Prof Nurs 24(4):218–227
26. Nair KM, Dolovich L, Brazil K, Raina P (2008) It's all about relationships: A qualitative study of health researchers' perspectives of conducting interdisciplinary health research. BMC Health Serv Res 8(1):1–10
27. Martin DR, O'Brien JL, Heyworth JA, Meyer NR (2008) Point counterpoint: the function of contradictions on an interdisciplinary health care team. Qual Health Res 18(3):369–379
28. Hall P, Weaver L (2001) Interdisciplinary education and teamwork: a long and winding road. Med Educ 35(9):867–875
29. Nancarrow SA, Booth A, Ariss S, Smith T, Enderby P, Roots A (2013) Ten principles of good interdisciplinary team work. Hum Resour Health 11(1):19. https://doi.org/10.1186/1478-4491-11-19
30. Blackmore G, Persaud DD (2012) Diagnosing and improving functioning in interdisciplinary health care teams. Health News 31(3):195–207
31. Aboelela SW, Larson E, Bakken S, Carrasquillo O, Formicola A, Glied SA et al. (2007) Defining interdisciplinary research: Conclusions from a critical review of the literature. Health services research 42 (1p1):329–346
32. Gebbie KM, Mason Meier B, Bakken S, Carrasquillo O, Formicola A, Aboelela SW et al (2008) Training for interdisciplinary health research defining the required competencies. J Allied Health 37(2):65–70
33. Kaplan S, Milde J, Cowan RS (2017) Symbiont practices in boundary spanning: Bridging the cognitive and political divides in interdisciplinary research. Acad Manag J 60(4):1387–1414
34. Bindler RC, Richardson B, Daratha K, Wordell D (2012) Interdisciplinary health science research collaboration: strengths, challenges, and case example. Appl Nurs Res 25(2): 95–100
35. Hastings J, Frishkoff GA, Smith B, Jensen M, Poldrack RA, Lomax J et al (2014) Interdisciplinary perspectives on the development, integration, and application of cognitive ontologies. Front Neuroinform 8:62
36. Friedenberg J, Silverman G, Spivey MJ (2021) Cognitive science: an introduction to the study of mind. Sage Publications
37. Schaat S, Wendt A, Kollmann S, Gelbard F, Jakubec M Interdisciplinary Development and Evaluation of Cognitive Architectures Exemplified with the SiMA Approach. In, 2015. Citeseer,
38. Crane MT, Richardson A (1999) Literary studies and cognitive science: Toward a new interdisciplinary. Mosaic: a journal for the interdisciplinary study of literature:123–140
39. Repko A, Navakas F, Fiscella J (2007) Integrating interdisciplinarity: How the theories of common ground and cognitive interdisciplinarity are informing the debate on interdisciplinary integration. Issues in Interdisciplinary Studies
40. Thagard P (2005) Being interdisciplinary: Trading zones in cognitive science. Interdisciplinary collaboration: An emerging cognitive science:317–339
41. Moradian N, Ochs HD, Sedikies C, Hamblin MR, Camargo CA, Martinez JA et al (2020) The urgent need for integrated science to fight COVID-19 pandemic and beyond. J Transl Med 18(1):1–7
42. Moradian N, Moallemian M, Delavari F, Sedikides C, Camargo CA Jr, Torres PJ et al (2021) Interdisciplinary Approaches to COVID-19. Adv Exp Med Biol 1318:923–936
43. Momtazmanesh S, Samieefar N, Uddin LQ, Ulrichs T, Kelishadi R, Roudenok V et al (2021) Socialization During the COVID-19 Pandemic: The Role of Social and Scientific Networks During Social Distancing. Adv Exp Med Biol 1318:911–921

44. Er Saw P, Jiang S (2020) The significance of interdisciplinary integration in academic research and application. BIO Integration 1(1):2–5
45. Bemme D, Kirmayer LJ (2020) Global Mental Health: Interdisciplinary challenges for a field in motion. Transcult Psychiatry 57(1):3–18. https://doi.org/10.1177/1363461519898035
46. Oreskes N (2004) The scientific consensus on climate change. Science 306(5702):1686–1686
47. Urban MC (2015) Accelerating extinction risk from climate change. Science 348(6234):571. https://doi.org/10.1126/science.aaa4984
48. Jaroszweski D, Chapman L, Petts J (2010) Assessing the potential impact of climate change on transportation: the need for an interdisciplinary approach. Journal of Transport Geography 18 (2)
49. Morse S (2013) Indices and indicators in development: An unhealthy obsession with numbers. Routledge,
50. Chakravarty SR (2019) An axiomatic approach to multidimensional poverty measurement via fuzzy sets. In: Poverty, Social Exclusion and Stochastic Dominance. Springer, pp 123–141
51. Alkire S, Foster J (2011) Counting and multidimensional poverty measurement. J Public Econ 95(7):476–487. https://doi.org/10.1016/j.jpubeco.2010.11.006
52. Agamuthu P, Ragossnig AM, Velis C (2019) Publishing impactful interdisciplinary waste-related research on global challenges: Circular economy, climate change and plastics pollution. Waste management & research : the journal of the International Solid Wastes and Public Cleansing Association, ISWA 37(4):313–314. https://doi.org/10.1177/0734242x19837785
53. Logie C, Dimaras H, Fortin A, Ramón-García S (2014) Challenges faced by multidisplinary new investigators on addressing grand challenges in global health. Glob Health 10:27. https://doi.org/10.1186/1744-8603-10-27
54. Herzig Van Wees SL, Målqvist M, Irwin R (2019) Achieving the SDGs through interdisciplinary research in global health. Scandinavian journal of public health 47(8):793–795. https://doi.org/10.1177/1403494818812637
55. McHenry MS, Baenziger JTH, Zbar LG, Mendoza J, den Hartog JR, Litzelman DK et al (2020) Leveraging Economies of Scale via Collaborative Interdisciplinary Global Health Tracks (CIGHTs): Lessons From Three Programs. Academic medicine : journal of the Association of American Medical Colleges 95(1):37–43. https://doi.org/10.1097/acm.0000000000002961
56. Pepperell R (2018) Art, energy, and the brain. Prog Brain Res 237:417–435. https://doi.org/10.1016/bs.pbr.2018.03.022
57. Zaidel DW (2013) Cognition and art: the current interdisciplinary approach. Wiley interdisciplinary reviews Cognitive science 4(4):431–439. https://doi.org/10.1002/wcs.1236
58. Zaidel DW (2013) Art and brain: the relationship of biology and evolution to art. Prog Brain Res 204:217–233. https://doi.org/10.1016/b978-0-444-63287-6.00011-7
59. Dahm R, Byrne J, Wride MA (2019) Interdisciplinary Communication Needs to Become a Core Scientific Skill. BioEssays : news and reviews in molecular, cellular and developmental biology 41(9):e1900101. https://doi.org/10.1002/bies.201900101
60. Bo Y, Yu J, Zhang K (2018) Computational aesthetics and applications. Visual computing for industry, biomedicine, and art 1(1):6. https://doi.org/10.1186/s42492-018-0006-1
61. Harrower J, Parker J, Merson M (2018) Species Loss: Exploring Opportunities with Art-Science. Integr Comp Biol 58(1):103–112. https://doi.org/10.1093/icb/icy016
62. Rezaei N, Vahed A, Ziaei H, Bashari N, Afkham SA, Bahrami F et al (2021) Health and Art (HEART): Integrating Science and Art to Fight COVID-19. Adv Exp Med Biol 1318:937–964
63. Engholm DH, Kilian M, Goodsell DS, Andersen ES, Kjærgaard RS (2017) A visual review of the human pathogen Streptococcus pneumoniae. FEMS Microbiol Rev 41(6):854–879. https://doi.org/10.1093/femsre/fux037

64. Adkins SJ, Rock RK, Morris JJ (2018) Interdisciplinary STEM education reform: dishing out art in a microbiology laboratory. FEMS microbiology letters 365 (1). https://doi.org/10.1093/femsle/fnx245

65. Klugman CM, Beckmann-Mendez D (2015) One thousand words: evaluating an interdisciplinary art education program. J Nurs Educ 54(4):220–223. https://doi.org/10.3928/01484834-20150318-06

66. Grogan K, Ferguson L (2018) Cutting Deep: The Transformative Power of Art in the Anatomy Lab. J Med Humanit 39(4):417–430. https://doi.org/10.1007/s10912-018-9532-2

67. Bentwich ME, Gilbey P (2017) More than visual literacy: art and the enhancement of tolerance for ambiguity and empathy. BMC Med Educ 17(1):200. https://doi.org/10.1186/s12909-017-1028-7

68. Kelly M, Rivas C, Foell J, Llewellyn-Dunn J, England D, Cocciadiferro A et al (2015) Unmasking quality: exploring meanings of health by doing art. BMC Fam Pract 16:28. https://doi.org/10.1186/s12875-015-0233-x

69. Singer R, Kruse K (2019) Art and health care: A dialog about interdisciplinary collaboration. Nurs Forum 54(3):403–409. https://doi.org/10.1111/nuf.12347

70. Ho RT, Potash JS, Fang F, Rollins J (2015) Art viewing directives in hospital settings effect on mood. HERD 8(3):30–43. https://doi.org/10.1177/1937586715575903

71. Hamet P, Tremblay J (2017) Artificial intelligence in medicine. Metabolism: clinical and experimental 69s:S36-s40. https://doi.org/10.1016/j.metabol.2017.01.011

72. Lin E, Lin CH, Lane HY (2020) Precision Psychiatry Applications with Pharmacogenomics: Artificial Intelligence and Machine Learning Approaches. International journal of molecular sciences 21 (3). https://doi.org/10.3390/ijms21030969

73. Jiang Y, Yang M, Wang S, Li X, Sun Y (2020) Emerging role of deep learning-based artificial intelligence in tumor pathology. Cancer communications (London, England) 40 (4):154–166. https://doi.org/10.1002/cac2.12012

74. Lang J, Repp H (2020) Artificial intelligence in medical education and the meaning of interaction with natural intelligence - an interdisciplinary approach. GMS journal for medical education 37 (6):Doc59. https://doi.org/10.3205/zma001352

75. Klaassen RG (2018) Interdisciplinary education: a case study. Eur J Eng Educ 43(6):842–859

76. Spelt EJH, Biemans HJA, Tobi H, Luning PA, Mulder M (2009) Teaching and learning in interdisciplinary higher education: A systematic review. Educ Psychol Rev 21(4):365–378

77. Saghazadeh A, Khaksar R, Rezaei N (2019) The Manager's Sixth Sense: An Art in Organizational, Educational, Moral, and Expert Thinking. In: Biophysics and Neurophysiology of the Sixth Sense. Springer, pp 345–350

78. Lindvig K, Lyall C, Meagher LR (2019) Creating interdisciplinary education within monodisciplinary structures: the art of managing interstitiality. Stud High Educ 44(2):347–360

79. Davies M, Devlin M (2010) Interdisciplinary higher education. Emerald Group Publishing Limited,

80. Hursh B, Haas P, Moore M (1983) An interdisciplinary model to implement general education. The journal of higher education 54(1):42–59

81. Cooper H, Carlisle C, Gibbs T, Watkins C (2001) Developing an evidence base for interdisciplinary learning: a systematic review. J Adv Nurs 35(2):228–237

82. Fineberg IC, Wenger NS, Forrow L (2004) Interdisciplinary education: evaluation of a palliative care training intervention for pre-professionals. Acad Med 79(8):769–776

83. Hermida MJ, Segretin MS, Prats LM, Fracchia CS, Colombo JA, Lipina SJ (2015) Cognitive neuroscience, developmental psychology, and education: Interdisciplinary development of an intervention for low socioeconomic status kindergarten children. Trends in Neuroscience and Education 4(1–2):15–25

84. White D, Delaney S (2021) Full STEAM Ahead, but Who Has the Map for Integration?–A PRISMA Systematic Review on the Incorporation of Interdisciplinary Learning into Schools. LUMAT: International Journal on Math, Science and Technology Education 9 (2):9–32

85. Deubel P (2003) An investigation of behaviorist and cognitive approaches to instructional multimedia design. Journal of educational multimedia and hypermedia 12(1):63–90

86. Apinis C, Tousignant M, Arcand M, Tousignant-Laflamme Y (2014) Can adding a standardized observational tool to interdisciplinary evaluation enhance the detection of pain in older adults with cognitive impairments? Pain Med 15(1):32–41

87. Ólason M, Andrason RH, Jónsdóttir IH, Kristbergsdóttir H, Jensen MP (2018) Cognitive behavioral therapy for depression and anxiety in an interdisciplinary rehabilitation program for chronic pain: a randomized controlled trial with a 3-year follow-up. Int J Behav Med 25 (1):55–66

88. Eccleston C, Malleson PN, Clinch J, Connell H, Sourbut C (2003) Chronic pain in adolescents: evaluation of a programme of interdisciplinary cognitive behaviour therapy. Arch Dis Child 88(10):881–885

89. Göhner W, Schlatterer M, Seelig H, Frey I, Berg A, Fuchs R (2012) Two-year follow-up of an interdisciplinary cognitive-behavioral intervention program for obese adults. J Psychol 146(4):371–391

90. Lauzière TA, Chevarie N, Poirier M, Utzschneider A, Bélanger M (2013) Effects of an interdisciplinary education program on hypertension: A pilot study. Canadian Journal of Cardiovascular Nursing 23 (2)

91. van de Grift TC, Kroeze R (2016) Design thinking as a tool for interdisciplinary education in health care. Acad Med 91(9):1234–1238

92. Rall M, Lackner CK (2010) Crisis Resource Management (CRM). Notfall + Rettungsmedizin 13 (5):349–356. https://doi.org/10.1007/s10049-009-1271-5

93. Carne B, Kennedy M, Gray T (2012) Crisis resource management in emergency medicine. Emerg Med Australas 24(1):7–13

94. Gaba DM (2010) Crisis resource management and teamwork training in anaesthesia. Oxford University Press

95. Fung L, Boet S, Bould MD, Qosa H, Perrier L, Tricco A et al (2015) Impact of crisis resource management simulation-based training for interprofessional and interdisciplinary teams: a systematic review. J Interprof Care 29(5):433–444

96. Enderle J (2012) Introduction to biomedical engineering. Academic press,

97. Nebeker F (2002) Golden accomplishments in biomedical engineering. IEEE Eng Med Biol Mag 21(3):17–47

98. Bronzino J (2005) Biomedical engineering: a historical perspective. In: Introduction to Biomedical Engineering. Elsevier, pp 1–29

99. Herz L, Russo MJ, Ou-Yang HD, El-Aasser M, Jagota A, Tatic-Lucic S et al. (2011) Development of an interdisciplinary undergraduate bioengineering program at Lehigh University. Advances in Engineering Education 2 (4):n4

100. Hashimoto S, Ohsuga M, Yoshiura M, Tsutsui H, Akazawa K, Mochizuki S et al. Parallel Curriculum of Biomedical Engineering Subjects with Rotational Experimental Project for Interdisciplinary Study Field. In, 2007. pp 39–44

101. Coger RN, De Silva HV (1999) An integrated approach to teaching biotechnology and bioengineering to an interdisciplinary audience. Int J Eng Educ 15(4):256–264

102. Sun Z, Ng KH, Ramli N (2011) Biomedical imaging research: a fast-emerging area for interdisciplinary collaboration. Biomedical imaging and intervention journal 7 (3)

103. Gan W, Lin JCW, Chao HC, Zhan J (2017) Data mining in distributed environment: a survey. Wiley Interdisciplinary Reviews: Data Mining and Knowledge Discovery 7 (6): e1216

104. Serrano KJ, Coa KI, Yu M, Wolff-Hughes DL, Atienza AA (2017) Characterizing user engagement with health app data: a data mining approach. Translational behavioral medicine 7(2):277–285
105. Jayaraman PP, Forkan ARM, Morshed A, Haghighi PD, Kang YB (2020) Healthcare 4.0: A review of frontiers in digital health. Wiley Interdisciplinary Reviews: Data Mining and Knowledge Discovery 10 (2):e1350
106. Mackey TK, Cuomo RE (2020) An interdisciplinary review of digital technologies to facilitate anti-corruption, transparency and accountability in medicines procurement. Glob Health Action 13(sup1):1695241
107. Mia MR, Hossain SA, Chhoton AC, Chakraborty NR A Comprehensive Study of Data Mining Techniques in Health-care, Medical, and Bioinformatics. In, 2018. IEEE, pp 1–4
108. Hancock JT, Khoshgoftaar TM (2020) CatBoost for big data: an interdisciplinary review. Journal of big data 7(1):1–45
109. Ayatollahi H, Gholamhosseini L, Salehi M (2019) Predicting coronary artery disease: a comparison between two data mining algorithms. BMC Public Health 19(1):1–9
110. Lashari SA, Ibrahim R, Senan N, Taujuddin N Application of data mining techniques for medical data classification: a review. In, 2018. EDP Sciences, p 06003
111. Serrano-Lomelin J, Nielsen CC, Jabbar MSM, Wine O, Bellinger C, Villeneuve PJ et al. (2019) Interdisciplinary-driven hypotheses on spatial associations of mixtures of industrial air pollutants with adverse birth outcomes. Environment international 131:104972
112. Franch-Pardo I, Napoletano BM, Rosete-Verges F, Billa L (2020) Spatial analysis and GIS in the study of COVID-19. A review. Science of The Total Environment 739:140033
113. Huang J, Wang H, Fan M, Zhuo A, Sun Y, Li Y Understanding the impact of the COVID-19 pandemic on transportation-related behaviors with human mobility data. In, 2020. pp 3443–3450
114. Mens T, Adams B, Marsan J (2017) Towards an interdisciplinary, socio-technical analysis of software ecosystem health. arXiv preprint arXiv:171104532
115. Tarafdar JC, Sharma S, Raliya R (2013) Nanotechnology: Interdisciplinary science of applications. African Journal of Biotechnology 12 (3)
116. Rai M, dos Santos CA (2017) Nanotechnology applied to pharmaceutical technology. Springer
117. Hoover E, Brown P, Averick M, Kane A, Hurt R (2009) Teaching small and thinking large: Effects of including social and ethical implications in an interdisciplinary nanotechnology course. Journal of Nano Education 1(1):86–95
118. Porter LA (2007) Chemical Nanotechnology: A Liberal Arts Approach to a Basic Course in Emerging Interdisciplinary Science and Technology. J Chem Educ 84(2):259. https://doi.org/10.1021/ed084p259
119. Jiao L, Barakat N (2011) Balanced depth and breadth in a new interdisciplinary nanotechnology course. J Educ Technol Syst 40(1):75–87
120. Roco MC (2011) The long view of nanotechnology development: the National Nanotechnology Initiative at 10 years. In: Nanotechnology research directions for societal needs in 2020. Springer, pp 1–28
121. Bauer CR, Knecht C, Fretter C, Baum B, Jendrossek S, Rühlemann M et al (2017) Interdisciplinary approach towards a systems medicine toolbox using the example of inflammatory diseases. Brief Bioinform 18(3):479–487
122. Rezaei N, Saghazadeh A (2021) Introduction on Integrated Science: Science Without Borders. In: Integrated Science. Springer, pp 1–37
123. Garbe C, Peris K, Hauschild A, Saiag P, Middleton M, Spatz A et al (2010) Diagnosis and treatment of melanoma: European consensus-based interdisciplinary guideline. Eur J Cancer 46(2):270–283
124. Vorstman JAS, Parr JR, Moreno-De-Luca D, Anney RJL, Nurnberger JI Jr, Hallmayer JF (2017) Autism genetics: opportunities and challenges for clinical translation. Nat Rev Genet 18(6):362–376

125. Waye MMY, Cheng HY (2018) Genetics and epigenetics of autism: A Review. Psychiatry Clin Neurosci 72(4):228–244
126. Modabbernia A, Velthorst E, Reichenberg A (2017) Environmental risk factors for autism: an evidence-based review of systematic reviews and meta-analyses. Molecular autism 8:13. https://doi.org/10.1186/s13229-017-0121-4
127. Saghazadeh A, Rezaei N (2017) Systematic review and meta-analysis links autism and toxic metals and highlights the impact of country development status: Higher blood and erythrocyte levels for mercury and lead, and higher hair antimony, cadmium, lead, and mercury. Prog Neuropsychopharmacol Biol Psychiatry 79:340–368
128. Saghazadeh A, Ahangari N, Hendi K, Saleh F, Rezaei N (2017) Status of essential elements in autism spectrum disorder: systematic review and meta-analysis. Rev Neurosci 28(7):783–809
129. Li C, Wang J, Wang Y, Gao H, Wei G, Huang Y et al (2019) Recent progress in drug delivery. Acta pharmaceutica sinica B 9(6):1145–1162
130. Mehta M, Tewari D, Gupta G, Awasthi R, Singh H, Pandey P et al (2019) Oligonucleotide therapy: An emerging focus area for drug delivery in chronic inflammatory respiratory diseases. Chem Biol Interact 308:206–215
131. Akhtar A, Andleeb A, Waris TS, Bazzar M, Moradi A-R, Awan NR et al. (2020) Neurodegenerative diseases and effective drug delivery: A review of challenges and novel therapeutics. Journal of Controlled Release
132. He W, Kapate N, Shields Iv CW, Mitragotri S (2020) Drug delivery to macrophages: A review of targeting drugs and drug carriers to macrophages for inflammatory diseases. Adv Drug Deliv Rev 165:15–40
133. Thakur AK, Chellappan DK, Dua K, Mehta M, Satija S, Singh I (2020) Patented therapeutic drug delivery strategies for targeting pulmonary diseases. Expert Opin Ther Pat 30(5):375–387
134. Carter P, Narasimhan B, Wang Q (2019) Biocompatible nanoparticles and vesicular systems in transdermal drug delivery for various skin diseases. Int J Pharm 555:49–62
135. Kaji H, Nagai N, Nishizawa M, Abe T (2018) Drug delivery devices for retinal diseases. Adv Drug Deliv Rev 128:148–157
136. Urso R, Blardi P, Giorgi G (2002) A short introduction to pharmacokinetics. Eur Rev Med Pharmacol Sci 6:33–44
137. Phe K, Heil EL, Tam VH (2020) Optimizing pharmacokinetics-pharmacodynamics of antimicrobial management in patients with sepsis: a review. The Journal of Infectious Diseases 222 (Supplement_2):S132-S141
138. Bird CE, Conrad P, Fremont AM, Timmermans S (2010) Handbook of medical sociology. Vanderbilt University Press,
139. Almedom AM (2005) Social capital and mental health: An interdisciplinary review of primary evidence. Soc Sci Med 61(5):943–964
140. Ehsan A, Klaas HS, Bastianen A, Spini D (2019) Social capital and health: a systematic review of systematic reviews. SSM-population health 8:100425
141. Nicolescu B (2014) Methodology of transdisciplinarity. World Futures 70(3–4):186–199
142. Dyer JA (2003) Multidisciplinary, interdisciplinary, and transdisciplinaryeducational models and nursing education. Nurs Educ Perspect 24(4):186–188
143. Garner HG (1995) Teamwork models and experience in education. Allyn & Bacon

Nima Rezaei gained his medical degree (M.D.) from Tehran University of Medical Sciences (TUMS) in 2002 and subsequently obtained an M.Sc. in Molecular and Genetic Medicine in 2006 and a Ph.D. in Clinical Immunology and Human Genetics in 2009 from the University of Sheffield, UK. He also spent a short-term fellowship in Pediatric Clinical Immunology and Bone Marrow Transplantation in the Newcastle General Hospital. Since 2010, Prof. Rezaei has worked at the Department of Immunology and Biology, School of Medicine, TUMS; he is now the Full Professor and Vice Dean of International Affairs, School of Medicine, TUMS, and the co-founder and Head of the Research Center for Immunodeficiencies. He is also the founding President of Universal Scientific Education and Research Network (USERN). He has edited more than 40 international books, has presented more than 500 lectures/posters in congresses/meetings, and has published more than 1000 articles in international scientific journals.

Amene Saghazadeh gained her M.D. from the Tehran University of Medical Sciences in 2019. She does research on clinical immunology, genetics, epigenetics, and nutrition at the Research Center for Immunodeficiencies, Children's Medical Center, TUMS. She is the manager of Integrated Science Association (ISA) in the Universal Scientific Education and Research Network (USERN).

Cognitive Sciences as a Naturalistic Model of Interdisciplinary Approaches

2

Antonino Pennisi and Donata Chiricò

> *"Etenim, quid Corpus possit, nemo hucusque determinavit, hoc est, neminem hucusque experientia docuit, quid Corpus ex solis legibus naturæ, quatenus corporea tantum consideratur, possit agere, & quid non possit, nisi a Mente determinetur. Nam nemo hucusque Corporis fabricam tam accurate novit, ut omnes ejus functiones potuerit explicare [...]. Quod satis ostendit, ipsum Corpus ex solis suæ naturæ legibus multa posse, quæ ipsius Mens admiratur."*
>
> B. Spinoza, Ethica Ordine Geometrico Demonstrata, 1677

Summary

Cognitive sciences have a history of almost a century. The reflection on the nature of intelligence born in the cybernetic field has won the interest of all human, social, and life sciences. This process has certainly represented an enrichment and an important evolution. However, it was not painless. Initially the cognitive sciences have proved convinced that the activity of the mind can be entirely simulated by algorithmic procedures. After all, there is no doubt that this idea survives in neuroscience, neuropsychology, and much

A. Pennisi (✉)
Department of Cognitive Sciences, University of Messina, Messina, Italy
e-mail: apennisi@unime.it

Catania, Sicily, Italy

D. ChiricòDepartment of Cultures, education and society, University of Calabria, Rende, Italy
e-mail: donata.chirico@unical.it

Soveria Simeri (CZ), Calabria, Italy

© The Author(s), under exclusive license to Springer Nature Switzerland AG 2022   41
N. Rezaei (ed.), *Multidisciplinarity and Interdisciplinarity in Health*,
Integrated Science 6, https://doi.org/10.1007/978-3-030-96814-4_2

philosophy of the mind. In recent decades, cognitive neurosciences, on the one hand, have brought the naturalism of the brain–body to the center of the debate, but on the other, they have enhanced the cerebrocentrism that studies the brain in computational terms and completely neglects the role of non-brain–body. Yet, the natural sciences have taught us that the body is the protagonist of all the abilities of human beings, animals, and machines. On the contrary, the complex series of philosophies recognized today under the name of embodied cognition believes that understanding the brain means taking into account the fact that it is instanced throughout the living organism and that living organisms are instanced in communities of conspecifics. The great lesson of contemporary evolutionism and ethology is that natural selection transforms morphological and, subsequently, functional structures, gradually and filtered by the genetic pool. Knowing the brain, therefore, means studying its relationships with the whole set of structures and functions that characterize the fixation of a species.

Education: a rural school in Central Sicily
[Photography by Antonino Pennisi]
The code of this chapter is *01110100 01101111 01101100 01000011 01100101 01100011 01101100 01100101 01101001 01110110.*

Keywords

Body · Brain · Chronological causalism · Cognitive science ·
Embodied cognition · Evolutionism · Naturalistic models ·
Populational thinking

1 Introduction

The culture of the new millennium has opened up under the banner of cognitive science. There is no branch of scientific and even humanistic disciplines (as far as this distinction is still valid) has not been affected, positively or negatively, by the invasiveness of the cognitive paradigm. Today, neuroscience is the best example of team research: modular, interactive, effective because it is cooperative, the true science without borders. If we want to use a metaphor, we could compare a team of cognitive neuroscience to an integrated crime research department where organized information from specialized state bodies flows into a single database. As if by magic, by linking separate data from different intelligences, an increasingly accurate, likely, identikit of the culprit is reconstructed, probably resembling the real criminal.

The study of aphasia, for example, has been able to connect the specific knowledge of neurologists, linguists, neuropsychologists, language philosophers, rehabilitators, and speech therapists. The results have been surprising both theoretically and therapeutically: aphasia is no longer an irreversible catastrophic event but a starting point for reconstructing not only the physiological machine of language but the individual's linguistic life, the one that Heidegger called *das Haus des Seins* (*the house of being*).

Similar phenomena occur in robotics, where the application of artificial models of empirical knowledge makes it possible to create devices that are increasingly useful for work and interaction with humans, allowing them to engage in creative activities. Economics, marketing, and stock exchange trading also increasingly use the cooperation between artificial calculation models and cognitive psychology models and the reasoning developed by game theories. The potential of this method in deriving new homogeneous images from continuous comparisons between apparently heterogeneous data is therefore unimaginable.

Cognitive sciences, after all, are born with one of the most ambitious projects that have ever been advanced in the history of western culture. They investigate the nature and functioning of the mind in any *thinking system*, natural or artificial. It is a research program that aims to explain mental processes in such a transparent way that even a machine can then reproduce them, simulating the procedures of our activities: infer, deduce, argue, but also be aware, believe, imagine, and desire.

Different disciplines have occupied themselves with similar issues: philosophy, psychology, neurology, linguistics, and artificial intelligence (AI). The history of cognitive science, however, cannot be identified as the history of any of these disciplines. Rather than a general program of explanation, cognitive science is an interdisciplinary method that has the merit of having solved problems that had become insoluble, closed as they were, within isolated disciplines.

From not long ago, cognitive sciences began to have a respectable history of almost a century. The reflection on the nature of cognition, born in the cybernetic field, has now spread like wildfire on all human and natural sciences. From Turing, Simon, Newell, we have come today to Chomsky, Fodor, Dennett, Pinker,

Gazzaniga, Kandell, and Damasio, but also, among others, to the anthropology of Dan Sperber, the philosophy of the mind of Ned Block, the psychology of neuromarketing by Daniel Kahneman and Vernon Smith, the neuroesthetics of David Freedberg and the studies on the performing arts, not to mention the decisive contribution that embodied cognition (EC) and the modern synthesis of evolutionary biology are making. It was an enrichment and a transformation, but not a painless process. The idea that all cognitive skills can be simulated by algorithmic procedures belongs, in fact, to the prehistory of cognitive sciences. However, this idea survives in other forms, both in neuroscience and in the philosophy of mind.

On the other hand, neurosciences have developed a tendency to become the fulcrum of all causal explanations: with a prevalent word in recent decades, they have become "cerebrocentric." The cerebrocentric theories focus only on the computational function of the brain–body and overlook the significant contribution of the non-brain–body to all the capacities of humans, animals, and machines, downgrading performances to a mere executive function.

Hurley [1] resorts to "the sandwich model" metaphor to outline this idea. The traditional description for the mind, the "highest" function, emphasizes the role of the mind in elaborating information, while perception is limited to carrying inputs, and action transforms these inputs into output. It is exacerbated when, in the *philosophy of language*, Chomsky depicts the morphology and semantics of language as "externalization devices" [2] independently of the cognitive nature of language itself. In a recent book, Chomsky claimed: "the externalization of narrow syntax like the printer attached to a computer, rather than the computer's CPU" [3, 4] has named "separability thesis" the main idea permeating the field of cognitive neuroscience, according to which "from knowledge of mental properties it is impossible to predict properties of the body. Therefore, a human like mind could very well exist in a nonhumanlike body" (cfr. p. 167): this is the opposite of what happens in evolution, and it is a position alarmingly close to Putnam's hypothesis of "the brain in a vat" [5].

With these artificial and anthropocentric residues, both the set of new philosophies of the mind that go by the name of EC and the evolutionary biology that has promoted the position of the brain are being measured in the last twenty years—the ascending *metaphor of the current cognitive sciences*—within *the living organism*. The greatest in the topics of contemporary evolutionism is, in fact, "organisms and individuals fit, and not their structures in isolation, as if they were self-sufficient" [6]. Therefore, "to study the asymmetrical brain of a species means to study also its relations with the feet and hands, with the muscle-skeletal system, with the structure of the circulatory, respiratory, digestive, nervous systems: in short with all the patterns that have set during the evolutionary history of the physiological type of the species. The same is true in terms of functions. An animal that is able to speak not only communicates differently but also perceives in differently, thinks and remembers in a different way, wants, gets excited and acts in a different way, is differently related with its conspecifics: he came to this condition through the inexorable interplay between chance and natural selection" (ib.).

2 What Does "Naturalism" Mean Today

The ever-wider expansion of the field of cognitive sciences, which we have just seen in its young twentieth-century history, has led to serious identity problems within it. When not only the traditionally "hard" sciences but also a good part of the human sciences began to explore the virtues of interdisciplinarity, it seemed to many that the nature of the cognitive method could reveal cracks and contradictions. If we wanted to identify three fundamental identifying principles on which one cannot but agree by practicing any "content" meaning of the cognitive sciences, we could identify them in: i, heuristic principle instead of the descriptive principle; ii, monistic principle instead of the dualistic principle; and iii, experimental principle instead of the speculative principle.

The first point comes back to the founding act of cognitive sciences. In fact, they were born as a reaction to the behaviorist hegemony of the mid-twentieth century. In Skinner's famous review of Verbal Behavior (1959), Chomsky affirmed for the first time in linguistics the idea that describing language behaviors and ways of functioning does not mean explaining them. More precisely, he affirmed the idea that a linguistic theory cannot explain its object of study through the recognition and recording of speakers' stimuli and responses, but that it is necessary: *"To know in depth the internal structure of the organism and the ways in which it processes information and organizes its behaviors (...): a complicated product of an innate structure, a genetically determined maturation process and past experience"* [7].

The cognitive change impressed, not only on linguistics but on the whole culture of the twentieth century, since the end of behaviorism, is one of those points of no return that characterize the history of science. In this birth certificate of cognitive sciences, there is already expressed; however, the fundamental contradiction between an internalist philosophy (which presupposes—therefore—only the existence of mental rules) and computationalism of the mind (the black box that processes information algorithmically) and the biological and evolutionary nature codified through the phylogenetic structure and realized in individuals through ontogenetic development.

The second point has even older reasons. Already in the seventeenth century, Spinoza had caught in the dualism between res cogitans and res extensa the most evident contradiction of the Cartesian idea of science: "he had conceived mind as so distinct from body that he could assign no one cause either of this union or of the mind itself; and found it necessary to have recourse to the cause of the entire universe, that is, God" [8]. The Spinozian heritage cannot be denied not only by the cognitive sciences but by all the contemporary epistemology of science. Anyone who believes in doing science cannot suppose the existence of two different substances in any way they are identified. The extension should not be confused with the "visibility" of a substance. Visibility can always emerge with technology that changes our idea of the infinitely large and infinitely small (Lovejoy). Especially in the field of cognitive sciences, it is the brain's neuroelectric activity that generates the organization of information that we call the "mind." The human brain contains more than one hundred billion neurons, emerging into *a combination of mental states* that exceeds *the number of particles* elementaries of *the known universe* [9]. Ultimately, the complexity of the mind must be attributed to the infinity combinatorics of neural connections, that is, to the unlimited creativity that only one substance can be capable of.

Finally, the third point concerns the method, which in cognitive sciences are always experimental and never speculative (in the sense of not being based on the evaluation and comparison of the data collected). Also, in this case, the relationship between the principle of the experimental method and the potential of the analysis enters into a direct relationship with the technologies. Technologies constitute not only the eye with which we can look at data but also the procedures with which we can produce it. Is respect for these three principles enough to make cognitive sciences a chapter of natural sciences? Or to justify an entirely naturalistic approach to the scientific knowledge of cognition? The answer to this question lies in the meaning we give to the term "naturalism."

2.1 Physicalist Naturalism

Throughout the twentieth century, the term "naturalism" almost always coincided with a physicalist orientation. The main coordinates of this orientation must be inscribed in a complex cultural matrix that intertwines the great season of the epistemology of science—which culminated in the sixties with the works of Popper and Khun—the primacy of analytical philosophy, in particular with the Quine's approach, the first computational season of cognitive sciences and generative-transformational grammar. Each of these scientific battleships has helped to dictate a piece of that great philosophical paradigm that has imposed its hegemony for the whole last century. The philosophy of science has traced the perimeter within which it had to move: a forced synchronic, formal, analytical-deductive paradigm. The "linguistic turn," began with Wittgenstein, has given him the chrism of logical-formal self-reference. The "linguistic turn," which began with Wittgenstein, has provided him with the chrism of logical-formal self-reference, Turing with the cognitive efficiency of recursive and decidable systems and Chomsky with the mentalistic and biologic nature. All the pieces of this mosaic have often interacted with each other and, together, have been inspired, as far as possible, by philosophical traditions of the past: Platonic essentialism, the mathematical philosophy of the Pythagoreans, the Galilean-Newtonian method, Boolean logical laws, Cartesian dualism and iatromechanics, and, in some cases, the epistemological circularity of idealistic rationalism.

In what sense can this coherent philosophical paradigm consolidated throughout the twentieth century be defined as "naturalistic"?

Firstly, because the knowledge produced within it is modeled as "natural laws," which tend to be analogous to those of physics. Therefore, general laws, controlled on an experimental and repetitive basis, formalized in a falsifiable theory, and expressed in a mathematizable language. In the linguistic field, for example,

Fregean semantics, formal grammars, generative rules of rewriting, and very abstract ones of universal grammar are some examples, differently graded, of theories that express "natural laws" in a physicalist sense.

Secondly, which is the one covered by the so-called Chomskyan "biology," physicalism can be interpreted naturalistically because its laws would conform to human nature in the very specific sense of innate devices, genetically predisposed and which manifest themselves without exception (pathologies apart) in human ontogenesis. Every child, ad., e.g., would manifest a constant pattern of language acquisition, of reaching specific stages such as motor learning, the appearance of syntax, etc.

Finally, this approach can be defined as naturalistic because it is based on a method traditionally attributed to the Galilean idea of "nature," assimilated to a language of knowledge in which all terms are previously linked to deductive definitions, "in the language of mathematics" whose characters are "triangles, circles and other geometric figures, without which means it is impossible to understand humanly words, without these it is a wandering around for a dark labyrinth" (Galileo Galilei 1564–1642). From this idea derives the exclusion of any other scientific form that is not a specific mold of mechanics: the only known model of science in which it is possible to reduce knowledge to calculations without residues.

Taking on such a strong epistemological statute led to very high prices to be paid in terms of the empirical adequacy of the theories. For example, to always remain in the field of linguistics, expose yourself to the continuous reformulation of an entire theory because it cannot explain a given form that manifests itself in a certain language and that comes to conflict with its axiomatic principles bring to a devaluation of linguistic variability.

A second price paid to physicalist naturalism is purely philosophical. From Descartes onwards, in fact, the clear separation between totally mechanizable and mathematizable sciences and holistic and non-deterministic sciences has been able to establish itself through the specific philosophical stratagem of dualism. The res extensa would concern the components reducible to mechanistic analysis, and the res cogitans would remain the ontologically distinct domain of psychic reality. In computational hypotheses, this solution has proved to be perfectly suited to the distinction between hardware and software. The first models of AI, however, proved unsuitable to explain even the simplest semantic uses, and the state of the linguistic simulations of the artificialist models has never exceeded the stochastic level of the syntactic parsing of the sentences.

In the philosophy of the mind, the physicalist approach has even prevented us from tackling central problems of linguistic cognition, starting from the relationship between consciousness and language. It seemed completely intuitive to those who had taken on the onerous task of proposing a hybrid hypothesis on conscience to take refuge in some form of "naturalistic" [10] or "attributive" dualism [11]. In Chomsky, the dualistic assumption—very strong since Cartesian linguistics—ended up taking the forms of the epistemological distinction between problems and mysteries of language [12].

2.2 The humanistic Naturalism of Evolutionary Biology

There is a second form of naturalism: the "humanistic" form of evolutionary biology. To one of Darwin's most trusted heirs, Ernst Mayr, this possibility had not only not escaped but, at the end of his career, in a testament book written more than a hundred years old, it had seemed like a founding awareness: "Considering how similar evolutionary biology is to historical science and how different it is from physics in conceptualization and methodology, it is not surprising that drawing a definite line between the natural sciences and the humanities is so difficult, indeed nearly impossible" [13].

What is this similarity of biology-based more on *Geistenswissenschaften* than on exact sciences (id.: 34) from an epistemological point of view? What kind of naturalism is Darwinian, and why would it be so pertinent to cognitive studies so that a "part of the philosophy of humans can therefore by merged with biophilosophy" (id.: 55)?

The first point we should address to answer these questions is the Darwinian concept named by Mayr "populational thinking:" a notion that has been mainly ignored, so far, by those who work in the field of cognitive sciences, and instead deeply explored "by biologists who have applied the Darwinian thought to population genetics to reconstruct the bottom layer of knowledge about the speakers of historical-natural languages" [14, 15]. Actually, this topic is mostly unfamiliar and too philosophical for those interested in a naturalistic perspective rather than in a physicalist one. How populational thinking is in essence and defined by the biological and philosophical terms is summarized in Table 1.

3   The Neo-naturalistic Model of Cognitive Sciences

Undoubtedly, the cognitive sciences can be considered naturalistic sciences. However, the history of their naturalistic status must be considered as a sway with a progressive shift from physicalism to biology. The beginnings are all skewed toward the computational paradigm, the cornerstone of mid-century physicalism.

3.1 The Computational Paradigm

In 1936, Alan Mathison Turing, in the famous article—*On Computable Numbers, with an Application to the Entscheidungsproblem*—proposed a virtual machine whose operating rules appear simple but capable of simulating any calculable function. Its reliability is very high because it operates through deterministic procedures. Its operating principles are recursion (i.e., the possibility of applying rules that recall themselves for an indeterminate number of times) and the finiteness of the number of logical states in which it can be located and the tape on which it writes and reads the results of the elaborations. To function, however, it is essential

Table 1 How is populational thinking in essence and defined by the biological and philosophical terms

	Populational thinking
In essence	Populational thinking indicates that evolution always occurs within population: it is the effect of reception accorded by the environment to individual mutations, also helped by chance, which are advantageous for the individual in which they manifest themselves. Thus, it allows selection to be directed not toward the idealization of the best "essences" (as in the platonic-cartesian paradigm) but toward processes specifically related to reproductive success
Current biology	Populational thinking coincides with the concept of a "gene pool" that has taken the place of the classification of species based on morphological characteristics, similarities, or affinities of any kind. It is the set of all alleles of the genes belonging to all individuals of a population p at a given time t. A genetic pool, therefore, always contains an inherent possibility of very high variation that—associated to events caused by mutations, drift, gene flow, and by innumerable randomness factors related to the concrete realization of chromosomal transformations (for example, crossing over in meiosis)—basically makes regular but non-deterministic the reproductive process of individuals and entirely connects the affirmation of progressively adapted populations to natural selection
Philosophy	Populational thinking opens a new epistemological model to naturalism. It is no longer a typological universe of eidos, essences, formally defined classes treatable through reductionist formal processes that tend to be predictable, as in several physicalist paradigms, but it consists in events strongly constrained by structural trends and even more conditioned by concrete performance

that the problems submitted to it are solvable: that is, that the calculations necessary to find the solutions are made up of a number, perhaps very high, but nevertheless finished with steps.

The problem of the halting problem of a Turing machine is, however, Turing undecidable; that is, it cannot be predicted a priori. That the halting problem of a Turing machine is undecidable appears equivalent to limiting theorems of the formal systems of Kurt Gödel. Both tell us a very simple story at the end: not all human knowledge can be subjected to formalization processes. But we have to trust that part that we can formalize.

A few years later, Turing tries again. In *Computing Machinery and Intelligence,* published on "Mind" in 1950, he challenges the world to demonstrate his ability to distinguish only from external manifestations (the answers given to certain questions by an agent A to an agent B places in separate rooms) if reasoning comes from a machine or a human. In that important article, he demonstrates the impossibility of solving the problem with logical methods. The Turing test thus becomes the symbol of the first phase of cognitive sciences that in which the idea triumphs that everything we can preach about human intelligence can also simulate it through computers. Modern AI is born.

Today, AI is a mainly computer science discipline that seeks to develop algorithms that allow machines (computers or robots) to perform practical tasks performed up to now by human agents but no longer attempting to imitate the logic of human thought (a striking example is the use of *Deep Learning* [16]). By the mid-twentieth century, many thinkers had, however, cultivated the hope that by studying these procedures, cognitive scientists could identify the formal principles that govern the whole human reasoning and behavior (the so-called AI-strong principle). Turing was agnostic on this point. He showed neither skepticism nor optimism. Alongside the awareness that problems—still debated—such as self-awareness, but also transcendence ("theological objection"), common sense, naïve psychology, etc. (id.: 129–147) play in the formation of human thought without being able to be addressed in the context of an explicit mechanistic theory of the mind. Turing also has a clear understanding of the problematic nature of the real discriminating question for AI: the biological dimension of human cognitiveness.

Neurology, physiology, and morphogenesis have been some of his major interests since the late 1940s. In 1952, he published *The Chemical Basis of Morphogenesis* in which he tried to apply his formal models to the development of embryos. These are problems of "formidable mathematical complexity" [17], in which the guiding idea of the role of simplification, idealization, and, therefore, of the potential application of reductionist methods to biological reality is strengthened. The development of embryos, as well as the evolution of individuals, seems to Turing an event that can be simulated by machines that learn: "there is an obvious connection between this process and evolution" [18] as schematically presented in Fig. 1.

The limit of this assimilation between the machine–child and the biological child becomes, however, the insurmountable limit, in the first phase of the cognitive sciences, of the specificity of the cognitive embodiment: "to understand the Turing model of the brain, it was crucial to realize that it considered the physics and chemistry, including all the arguments of quantum mechanics (…), as essentially irrelevant. (…) The claim was that whatever a brain did, it did it by virtue of its structure as a logical system and not because it was in the head of a person, or is a spongy tissue made up of a particular type of biological cell formation" [19].

The theoretical efforts of the newborn cognitive sciences have thus been oriented toward making explicit systems, relations, and operative processes of the mind, relegating their bodily implementations to any kind of hardware, artificial or biological, in the background.

3.2 The Hegemony of Neuroscience and the Resilience of the Philosophy of the Mind

The basic idea that accompanied the conversion from the computational origins of cognitive sciences toward the neuroscientific approach is the naturalization of the mind. The computational mind is, in fact, an artificial mind. In a sense, however, no scholar, even among the most extremist supporters of strong AI, has ever truly believed that the computer metaphor could be anything but a method, a philosophy. On the contrary, in the neuroscientific paradigm, mind and brain really coincide.

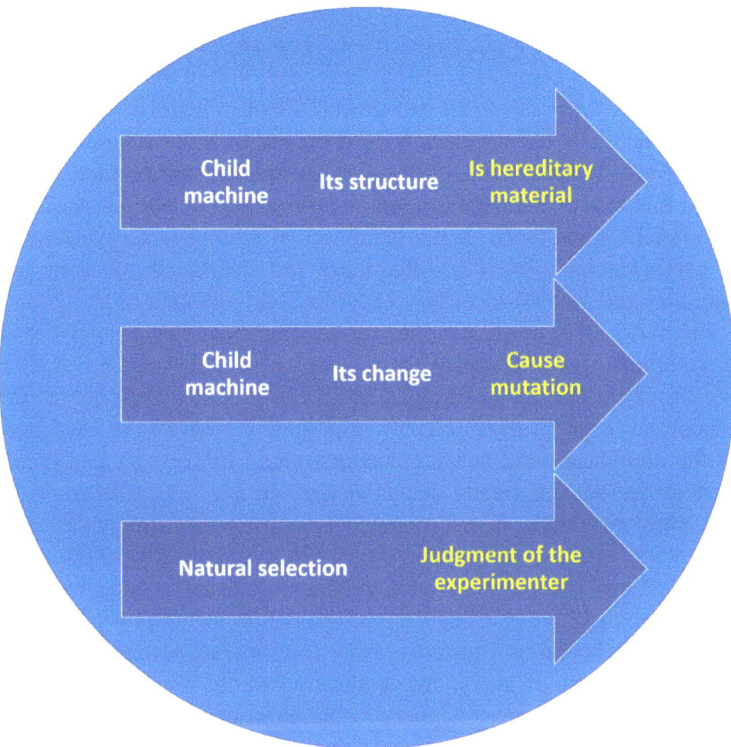

Fig. 1 Evolution and learning [18]

The brain is no longer a metaphor for the mind. In a sense, "it is" the mind itself. Although full of technical problems, the neuroscientific hypothesis is finally radically monistic: the mental process is resolved entirely in the brain process.

It is now a matter of *"simply"* mapping the correspondences between brain processes and mental events, associating neuronal sites with behaviors. The tendency to naturalize all the knowledge related to cognitive sciences is rooted in this hypothesis. In general, it can be said that the relationship with the two main naturalistic approaches that we have previously described varies with the type of cognitive investigation carried out, with the researcher's "job."

For those who deal exclusively with studying and measuring body phenomena, this statement is quite obvious: a neuroscientist studies the brain through histological, neurobiological, biochemical, neurophysiological, and instrumental analysis (brain imaging). His method is linked to meticulous and procedurally very rigorous experimental activity. In very different ways, even for the psychologist, the naturalization of the field is nothing more than an implementation of the experimental method applied to the attempt to identify the relationship between behaviors and neuropsychological hypotheses. The specificity of psychology, however, already complicates the transparency of cognitive analysis operations. In fact, how can we identify behavior if not as sets of observations on what subjects perform in terms of more or less complex functions that need, first of all, to be pre-defined in order to be considered falsifiable constructs? Finally, linguists also operate in a domain in which it is not excessively problematic to adopt a knowledge naturalization program. Linguists are also called to monitor the relationship between observable material phenomena (phonemes, words, sentences, speeches) and categorical classifications (such as the definition of grammatical, syntactic, and semantic competencies). However, the latter are connected to brain activities; therefore, they are, at least in theory, experimentally demonstrable.

The integration of neuroscience, neuropsychology, and neurolinguistics is today, in fact, completely satisfactory in the research practice of cognitive sciences. It operates through a true methodological synergy: the circumscription of the "sub- and neo-cortical areas in neuroscientific topography would have been impossible without the identification of actual behaviors, linguistic (or not), and their "intangible" interpretation—i.e. purely deductive—in terms of an explanatory theory of the interconnected functioning of systems of competencies (inference, mind reading, perception, syntax, semantics, etc.)" [20, 21]. Finally, it is particularly important that this type of integration produces relevant application results: the whole sector of cognitive and linguistic rehabilitation therapies is a clear example.

But if the rule is that as you move away from a craft centered on the material nature of your object of study, the application of the term naturalism appears increasingly problematic, what happens in the philosophical context in which you have to do directly not with the material substrates of ideas and concepts, but with the ideas and concepts themselves? It is the biggest problem of the philosophy of the mind and of all the collateral disciplines that revolve around cognitive neuroscience. What does it mean, in these cases, to consider mental behaviors natural behaviors?

For Quine [22], Goldman [23], and Dretske [24]—to whom we owe the first complete formulation of philosophical naturalism—it meant practicing philosophical euthanasia: philosophy must only dissolve in scientific knowledge by adopting the methods of the same natural sciences, especially physics. A more moderate formulation supports a gap between the philosophical problems that can be treated through the methods of the natural sciences and those that do not appear reducible to them. Supporters of this "liberal naturalism," including McDowell, Millikan, and Sellars, tend to place themselves at the ideal center of the dispute but risk using a blanket that leaves both the feet and the head of the debate uncovered.

Opening up to a dimension of thought inaccessible to naturalistic methods can mean, in fact, returning to dualistic solutions. However, such a hypothesis cannot coexist with the cognitive science program, at least from its second phase onwards.

There are two possible loopholes that the philosophy of the mind could take to avoid or mask the dualisms implicit in untreatable problems. The first would consist in the reduction of non-treatable problems to treatable problems. For example, consider the problem of the soul, subjectivity, consciousness, etc., as linguistic-conceptual problems. The second is to replace much more simplified versions of these problems with the problems themselves; for example, to replace the study of neurophysiological automatisms with that of the problem of extended consciousness, that of states of epileptiform alteration in mystical crises with that of the problem of religious sense (see Newberg and d'Aquili [25]), etc. For the purposes that we propose here, it is sufficient to say that only a few scholars have taken the first road, while most have poured on the second.

Complicating the tasks of the philosophy of the mind is the thorny question of the nature of language and its role in human cognition, which emerged above all with the "linguistic turn" in the philosophy of the first half of the twentieth century. The expression is of the American philosopher Richard Rorty, who in the 1960s advanced the idea according to which the analysis of language constitutes the method for solving all philosophical problems. This hypothesis, in addition to highlighting a lack of confidence in the autonomy of the physical and biological sciences, has also led to a sort of anthropocentric resilience which, by denying language to non-human animals, has contributed to amplifying its cognitive distance with humans and disadvantaging thus attempts to naturalize the history of evolution.

In many respects, therefore, the naturalistic program of neuroscience encounters resistance and generates problems, especially in the philosophical component of cognitive sciences, that is, in the philosophy of mind and language. At the same time, however, these philosophies pose a challenge to the development of cognitive naturalism, relaunching those problems which today seem to go beyond the threshold of scientific knowledge but which Turing considered, as we have seen, potentially solvable.

Could there, however, exist a naturalistic philosophy of cognitive science that transforms these challenges into a significant new step forward? In other words, is it possible to go beyond the neuroscientific phase that is currently dominating the scene in a naturalistic direction? According to Nobel Prize winner Gerald Edelman, this is possible only if cognitive sciences become definitively hinged in the theoretical and methodological contexts of contemporary biology. It is his program for the third phase of the new paradigm: "we must incorporate biology into our theories of knowledge and language. (…) We must develop what I have called a biologically based epistemology—an account of how we know and how we are aware in light of the facts of evolution and developmental biology" [26].

3.3 The Embodied Cognition and the Neo-naturalism of Contemporary Cognitive sciences

Edelman was one of the first to glimpse the crisis of the cognitivist paradigm at the very moment when it seemed to have reached its most successful point: "the blend of psychology, computer science, linguistics, and philosophy known as cognitive science." As with all vigorous efforts, ill-founded or not, much has emerged that is of great interest to scientists and non-scientists. Not the least of the positive results has been the routing of simple-minded behaviorism, but at the same time, "an extraordinary misconception of the nature of thought, reasoning, meaning, and of their relationship to perception has developed that threatens to undermine the whole enterprise" [26]. He reproaches at early cognitive sciences an exaggerated hyper-formalism that reduced representations of meaning to a logical-syntactic combinatorics. To tell the truth, animal cognitive systems are biological in nature. A genuine return to Darwin, therefore, means, for Edelman, the adoption of a populational and empirical biological conception. Therefore, not only "the mind is in the body," but "certain dictates of the body must be followed by the mind" (ibid. p 239).

EC is a set of theories recently developed as a reaction to the "cerebrocentrism" crisis denounced by Edelman. Its general principles can be summarized in the following points:

1. It is impossible for cognitive sciences to neglect the involvement of body structures in cognitive processes [27, 28].
2. Different body structures correspond to different cognitive systems [4].
3. Cognitive processes are not confined to the brain but involve wide-body structures and interaction with the environment [29–32].

Beyond these extremely important merits, however, the embodied perspective has soon split into different positions, to the point that Shaun Gallagher and Mark Rowlands coined the term "4E cognitions," that is: embedded, embodied, enacted, and extended (Fig. 2).

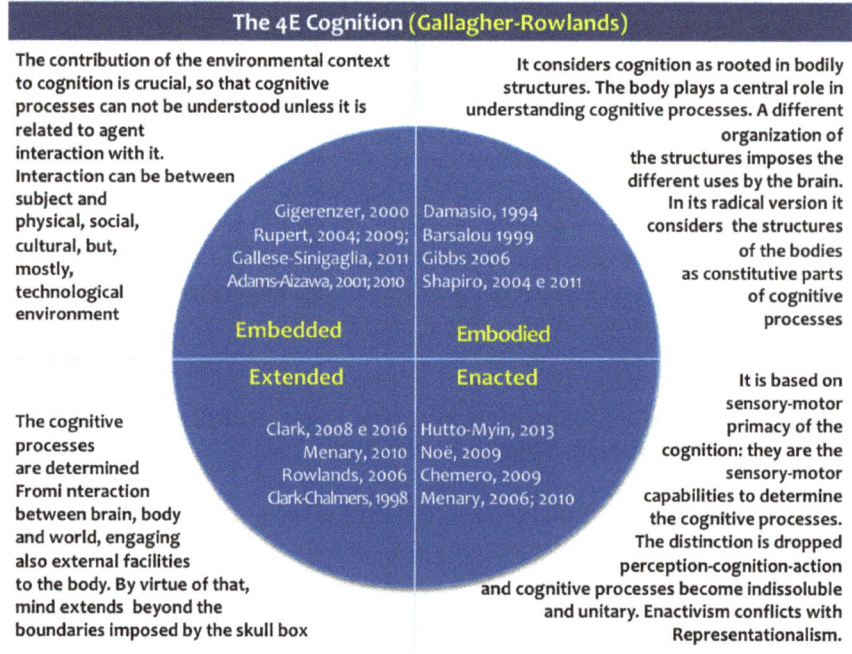

Fig. 2 4e cognition

Enactivism (enacted cognition), for instance, has often focused on the performing resources of biological organisms, their ability to carry out activities, starting with motor activity and also relying on an evolutionary perspective [32–35].

Even the theory of the embodied mind (EC) describes cognition as a function of biological structures. Therefore, as Shapiro attempted to show [4, 28], it is plausible that different body structures can result in different forms of intelligence. The presence or absence of certain structural features can lead to cognitive, mental, and cultural adaptations [6, 33, 34].

The cases of embedded cognition and of the extended mind are less directly addressed to the biological task. In particular, it is not so important to know the constitution of the "bodily technology" [35–37] that makes the development of certain cognitive abilities possible, but, if anything, to understand how digital technologies, developed from certain cognitive abilities, can extend and improve the body's powers [27, 30, 38–43].

All these variants of the embodied mind paradigm, however, can lead to epistemological problems. The first problem is the potential risk of going back to behavioral epistemologies. For instance, the most radical enactivist theses such as Chemero [29] dynamic approach to cognition or the post-artificialist models of Brooks [44, 45] support the idea that the self-organization of systems in perpetual dynamic interaction can account for the entire cognitive process.

The second potential risk is the denial of any form of representationalism as a result of the belief that cognition does not require internal semantic states. Gibson [46], for example, rejects any cognitive function of symbolic processing by attributing to perceptual systems the ability to grasp affordances directly from objects.

The third huge problem is the adoption of a naturalistic perspective centered on the subject and not on the species, which creates a tendency to explain the contribution of corporeality as "weak" or "lower" or "minimal" cognition [29, 47–49].

All these problems represent a price to pay for the EC perspective in cognitive science since they do not meet the fundamental principle of cognitive science: a theory must not describe behaviors, but he has to explain them. It is no accident that Gallagher wonders whether there is any point in conceiving enactivism as a science of mind. Enactivism is not a scientific research program, but a "philosophy of nature" [50, 51], a sort of "comment about the overall image of the natural world made by scientific and non-scientific research" [50] or "a form of naturalism, [which] does not endorse the mechanistic definition of nature" [51].

4 Conclusion

But if with the chiaroscuro of the EC, the critical picture is well defined, the constructive perspectives of a third "neo-naturalistic" phase of the cognitive sciences seem well underway but still need profound rearrangements in order to constitute a basic naturalistic model for the interdisciplinary approach to scientific problems. The abandonment of the mythical strong AI project and its transformation into a new computer science based on the intensive exploitation of big data, as in *deep learning*, has given rise to stunning results in some fields such as visual recognition [16, 52]. Abandoning any idea of simulating the way the human mind works, too different from artificial algorithmic solutions, AI seems to have found a completely adequate position in neo-robotics, in deep mind health, in-home automation, and in all the countless other sectors of "applied intelligence." These developments seem to go precisely in the direction of an unexpected integration between theoretical and applied sciences, demonstrating the rules-free creativity of interdisciplinarity.

On the other hand, contemporary neurosciences, while deepening and perfecting more and more the refinement technologies of the investigation methods for clinical-diagnostic purposes and arriving at the ever-wider diffusion of neural interfaces, seem to have understood the theoretical importance of mitigating the tested internalist approach to embracing also the possibilities of the externalist. In particular, cognitive, affective, and social neurosciences have discovered the fruitfulness of making the relationship between the individual and the collective mind interact both on a psychological and ethological-social level. The extended mind model is no longer limited to assessing the possibilities of cognitive enhancement of individual subjects through the amplification of technological devices. It prolongs, however, to the analysis of the social management of emotions, feelings, primary conditions of aggression, cooperation, for the first-time crossing interaction with evolutionary psychology and with the ethological constraints of evolutionary and developmental biology (Evo-Devo).

Physicalist naturalism, in other words, is increasingly amalgamating in cognitive sciences with what we have previously called the humanistic naturalism of evolutionary biology, according to Ernst Mayr's clear clairvoyant vision (2004) [13]. It is becoming increasingly clear to the entire research field that the decisive leap, which can transform cognitive science into a universal epistemological method, is its definitive metamorphosis into the unitary core-knowledge of the science without borders paradigm.

The evolutionist perspective, contrary to what the most important exponents of twentieth-century physicalism and even of early cognitivism thought, does not use the historical-diachronic axis as an instrument of pure (and useless/idle) description on the origins. On the contrary, the reference to the gradualism of the structures and the constraints of natural selection and the development of form (introduced by Evo-Devo) constitutes *the only form of causalism admissible in natural explanatory processes*. Elsewhere [53, 54], we have called it the principle of "chronological-logic-causalism," meaning by it the indisputable fact that every variation always and obligatorily derives from the previous forms.

In a "disembodied" cognitive science, this type of causalism would be a negligible event. In fact, gradualism operates on structures and not on functions. Functions are never taken for granted in the evolutionary steps; anything can happen, as the numerous cases of exaptation demonstrate [55] and, of all, that of language, decisive for human cognitive ethology [6, 54].

But if we use, as we have tried to show in this essay, a naturalistic-biological perspective, there can be no doubt about the importance of the chronological causalism of the structures. The cognitive science of the new millennium must take for granted that applying a naturalistic perspective means, first of all, always basing cognitive evolution on morphological (anatomical-physiological) evolution that cannot undergo leaps. While the functional jump can produce transformations that we call "emergencies," the evolution of structures does not; all matter is transformed on the basis of its previous forms. A cognitively emergent state is reached, however, only at the conclusion of a very slow anatomical-physiological revolution. Chronological causalism and populational thought explain for themselves the reasons for any change that has proved decisive in the evolutionary history of a species.

The relationship, therefore, no longer between mind and body but between what Alva Nöe and Daniel Hutto call brain–body and non-brain–body (or neural body and non-neural body) is overturned from this neo-naturalistic perspective. As Spinoza had already taught in the seventeenth century, the substance is always one, but it is the non-brain–body that triggers changes in the brain–body. The brain is, as André Leroi-Gourhan happily said, "the tenant of the body" [53, 56]. According to Leroi-Gourhan, in fact, "human evolution did not begin with the brain but with the feet" [57]. With this expression, he summarized the long history of structural transformations in hominids, each derived from the previous one: bipedal mutation —► standing upright —► enlargement of the cranial fan —► vocal tract formation with two 90° intersecting pipes —► possibility of proffering articulated sounds —► language training —► rapid cognitive and cultural evolution Fig. 3

From the first to the last of these stages, seven million years pass. Almost all engaged in slow, gradual transformations of the physio-anatomical structures filtered by selecting the genetic pool. Only the very last part of this path was utterly unpredictable and took place in a brief time (57 thousand years) at the end of an enormously long cycle of structural transformations. If we admit, therefore, that this history of consequential structural transformations has determined the turning points of cognitive evolution that we analyze today with the typical forms of the experimental cognitive method, then we will also have to recognize that the evolutionary perspective is not a useless historical reconstruction without practical applications, but it is the foundation of a new naturalistic causalism on which the idea of science without borders could rest entirely.

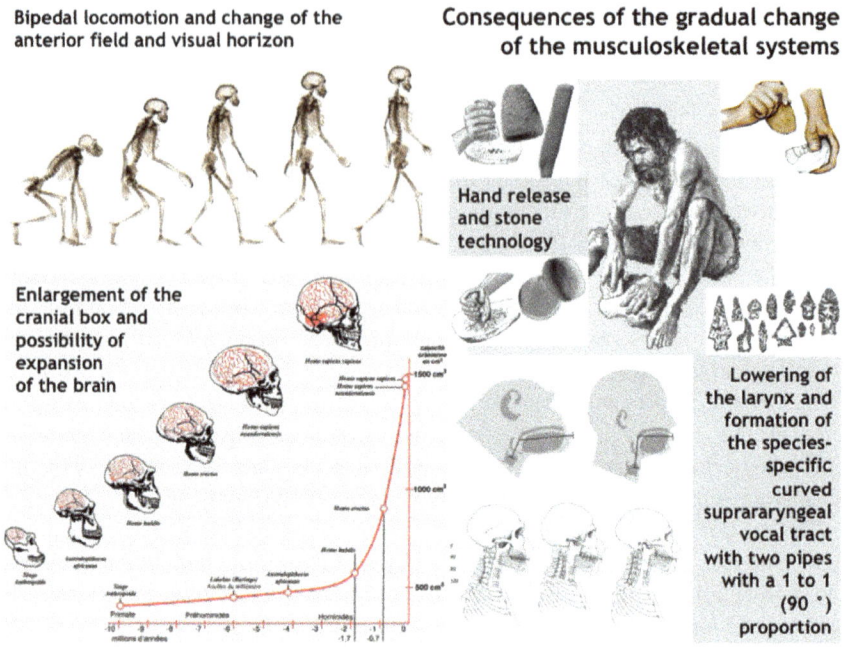

Bipedal locomotion and change of the anterior field and visual horizon

Consequences of the gradual change of the musculoskeletal systems

Hand release and stone technology

Enlargement of the cranial box and possibility of expansion of the brain

Lowering of the larynx and formation of the species-specific curved suprararyngeal vocal tract with two pipes with a 1 to 1 (90 °) proportion

Fig. 3 Consequences of the gradual change of the musculoskeletal system

Core Messages

- Recent developments in cognitive science integrate life sciences, and AI, demonstrating the rules-free creativity of interdisciplinarity.
- Contemporary neuroscience seems to understand the integration of an externalist and an internalist approach to the mind.
- Cognitive neuroscience has discovered the fruitfulness of making the relationship between the individual and the collective.
- The model of the embodied mind is not limited to the cognitive level but at both the psychological and ethological-social levels.
- An evolutionary cognitive perspective is the foundation of a new naturalistic causalism of science without frontiers could rest entirely.

References

1. Hurley S (2001) Perception and action: alternative views. Synth 129(1):3–40
2. Chomsky N (2005) Three factors in language design. Linguist Inquiry 36(1):1–22
3. Berwick RC, Chomsky N (2016) Why only us: Language and evolution. MIT press
4. Shapiro LA (2004) The mind incarnate. MIT Press
5. Putnam H (1981) Reason, truth and history, vol 3. Cambridge University Press
6. Pennisi A, Falzone A (2011) La Scienza della natura e la natura del linguaggio. Mucchi
7. Chomsky N (1959) Verbal behavior by BF Skinner. Bobbs-Merrill
8. Spinoza B (2002) Spinoza: complete works. Hackett Publishing
9. Ramachandaran VS (2003) The emerging mind: the BBC Reith Lectures 2003. Profile Books
10. Chalmers DJ (1996) The conscious mind: in search of a fundamental theory. Oxford Paperbacks
11. Perconti P (2011) Coscienza. Il Mulino
12. Hauser MD, Yang C, Berwick RC, Tattersall I, Ryan MJ, Watumull J et al (2014) The mystery of language evolution. Front Psychol 5:401
13. Mayr E (2004) What makes biology unique?: considerations on the autonomy of a scientific discipline. Cambridge University Press
14. Cavalli-Sforza LL (1996) Geni, popoli e lingue, Adelphi
15. Cavalli-Sforza LL (2004) L'evoluzione della cultura: proposte concrete per studi futuri. Codice
16. Perconti P, Plebe A (2020) Deep learning and cognitive science. Cognition 203:104365
17. Turing AM FRS (1952) The chemical basis of morphogenesis. Philosophical Transactions of the Royal Society of London. Series B, Biological Sciences 237: 37-72
18. Turing AM, Haugeland J (1950) Computing machinery and intelligence. MIT Press
19. Hodges A (1983) Alan Turing: the enigma, Simon & Schuster
20. Pennisi A, Falzone A (2017) Linguaggio, evoluzione e scienze cognitive: un'introduzione. Corisco
21. Graziano M (2019) Mind/Brain and economic behaviour: for a naturalised economics. Axiomathes 29(3):237–264
22. Quine WV (1969) Ontological relativity and other essays. Columbia University Press
23. Goldman AI (1967) A causal theory of knowing. J Philos LXIV N 12 June 22. 357-372
24. Dretske F (1969) Seeing and knowing. University of Chicago press
25. Newberg A, D'Aquili EG (2008) Why God won't go away: brain science and the biology of belief. Ballantine Books
26. Edelman G (1992) Brilliant air, brilliant fire. Basic Books
27. Rowlands MJ (2010) The new science of the mind: from extended mind to embodied phenomenology. MIT Press
28. Shapiro LA (2011) Embodied cognition: lessons from linguistic determinism. Philosophical Topics 39(1):121–140
29. Chemero A (2011) Radical embodied cognitive science. MIT Press
30. Clark A (2008) Supersizing the mind: Embodiment, action, and cognitive extension. Oxford University Press
31. Lakoff G, Johnson M (1999) Philosophy in the flesh: the embodied mind and its challenge to western thought, vol 640. Basic Books
32. Noë A, Noë A (2004) Action in perception. MIT Press
33. Carroll SB (2006) The making of the fittest: DNA and the ultimate forensic record of evolution. WW Norton & Company
34. Falzone A (2014) Structural constraints on language. Reti, Saperi, Linguaggi 1(2):247–266
35. Pennisi A (2013) Per una tecnologia dello speech making: scienze cognitive e specie-specificità del linguaggio umano. In E. Banfi (ed) Sull'origine del linguaggio e delle lingue storico-naturali. Un confronto fra linguisti e non linguisti. Atti del Convegno Interannuale SLI, Bulzoni. 169–183

36. Pennisi A (2014) La tecnologia del linguaggio tra passato e presente. Blitiry II 2:195–220
37. Pennisi A, Parisi F (2013) Corpo, tecnologia, ambiente. Nuove tendenze naturalistiche dell'esperienza estetica. Aisthesis Pratiche, linguaggi e saperi dell'estetico 6(2):235–256
38. Knappett C, Malafouris L (2008) Material agency: towards a non-anthropocentric approach. Springer
39. Menary R (2010) The extended mind. MIT Press
40. Paolucci C (2011) The 'external mind': Semiotics, pragmatism, extended mind and distributed cognition. VS Quaderni di studi semiotici 112–113:69–96
41. Paolucci C (2012) Per una concezione strutturale della cognizione: semiotica e scienze cognitive tra embodiment ed estensione della mente. In Graziano, M e Luverà, C (a cura di) Bioestetica, bioetica, biopolitica, Corisco. 247–276
42. Rowlands M (2003) Externalism: putting mind and world back together again. McGill-Queen's University Press
43. Sheldrake R (2003) The sense of being stared at: and other aspects of the extended mind. Armony
44. Brooks RA (1991) Intelligence without representation. Artificial Intelligence 47(1–3):139–159
45. Brooks RA (2003) Flesh and machines: how robots will change us. Vintage
46. Gibson JJ (1977) The theory of affordances. perceiving, acting, and knowing: toward an ecological psychology. In Shaw R, Bransford J (eds) Perceiving, acting, and Knowing: Toward an ecological psychology Lawrence Erlbaum.127–142
47. Di Paolo E, Rohde M, De Jaegher H (2010) Horizons for the enactive mind: values, social interaction, and play. In J Stewart, O Gapenne, and EA Di Paolo Enaction: towards a new paradigm for cognitive science. MIT Press Scholarship
48. Garzón PC, Keijzer F (2009) Cognition in plants. In: Plant-environment interactions: Behavioral perspective. In Baluŝcka F (ed.)Plant-Environment Interactions: signaling and communication in plants. Springer-Verlag. 247–266
49. Stewart J (2010) Foundational issues in enaction as a paradigm for cognitive science: From the origin of life to consciousness and writing. In J Stewart, O Gapenne, and EA Di Paolo (eds) Enaction: toward a new paradigm for cognitive science. MIT Press Scholarship. 1–31
50. Godfrey-Smith P (2001) On the status and explanatory structure of developmental systems theory. Cycles of contingency: developmental systems and evolution. 283–297
51. Gallagher S (2017) Enactivist interventions: rethinking the mind. Oxford University Press
52. Plebe A, Grasso G (2019) The unbearable shallow understanding of deep learning. Minds and Machines 29(4):515–553
53. Pennisi A (2016) Prospettive evoluzioniste nell'embodied cognition. Il cervello «inquilino del corpo». Reti, saperi, linguaggi 3(1):179–201
54. Pennisi A, Falzone A (2017) Darwinian biolinguistics: Theory and History of a Naturalistic Philosophy of Language and Pragmatics. Springer
55. Gould SJ, Vrba ES (1982) Exaptation—a missing term in the science of form. Paleobiology 8 (1):4–15
56. Pennisi A (2020) Dimensions of the bodily creativity. For an extended theory of performativity. In Pennisi A, Falzone A (eds) The extended theory of cognitive creativity. Springer. 9–40
57. Leroi-Gourhan A (2009) Le Geste et la Parole. Technique et langage, vol 1. Albin Michel

Antonino Pennisi is currently a full professor of Philosophy of Language at the University of Messina (Italy). He was born in Catania (Italy) in 1954. He has been director of the Department of "Cognitive Sciences" University of Messina (2015–2018); Director of the Ph.D. in "Cognitive Science" (the University of Messina and Rome III, since 2000) and President of CRISCAT—International Center for Research on Theoretical and Applied Cognitive Sciences. He discusses naturalistic approaches (evolutionism, ethology) to the philosophy of language and mind. Among his works*Cognitive Dimensions of performativity. Interdisciplinary approaches to the extended theory of bodily creativity Springer*, 2019 (with A. Falzone); *Darwinian Biolinguistics. Theory and history of a naturalistic philosophy of language and pragmatics,* Springer, 2016 (with A. Falzone); *Plato's error. Biopolitics, language and civil rights in times of crisis,* Il Mulino, Bologna, 2014; *What will become of bodies? Spinoza and the Mystery of Embodied Cognition,* Il Mulino, Bologna, 2020.

Donata Chiricò is currently a professor of Ethics of Communication at the University of Calabria (Italy). Born in Soveria Simeri (CZ-Italy). In 1992, she specialized in the École des Hautes Études en Sciences Sociales in Paris. At University "Denis Diderot-Paris VII," she obtained (1993) the Diplóme d'Études Approfondies (DEA) in "Linguistique Théorique et Formelle" and in 1997 the Ph.D. in "Philosophy of language: theory and history" (University of Palermo, Calabria, Rome La Sapienza). In the following years, she studied the relationships between body and language and gained particular interest in cognitive sciences, audiopsychophonology, and the philosophy of linguistic disabilities and cognition. More recently, she has been interested in the philosophical history of sign language and deafness. Has published numerous essays in the journal "Reti, Saperi, Linguaggi. Italian Journal of Cognitive Sciences" and the monograph *When words are things. Language and Enlightenment*, Mimesis, Milano, 2021.

The Aesthetics of Science from the Viewpoint of Neuroscience

3

Hunkoog Jho

"I always regarded mathematics as the method of obtaining the best shapes and dimensions of things; and this meant not only the most useful and economical, but chiefly the most harmonious and the most beautiful."

James Clerk Maxwell

Summary

Whether or not an aesthetic object exists, whether aesthetics is perceived or recognized, or whether the sense of aesthetics can be improved, all are concerned about the ontology of aesthetic reality and epistemology of the aesthetic experience. Such issues have been discussed in aesthetics for a long time. This study intends to revisit them from a neurological viewpoint. Meta-analysis and literature reviews on aesthetics using fMRI have identified brain regions and processes involved in aesthetics perception and judgement. This study tries to explain the difference between aesthetic perception and judgement and discuss how to improve the ability to sense the science of aesthetics based on neuroscience theories such as the emotion circuit, reward system, and cortical pathways of vision.

H. Jho (✉)
Graduate School of Education, Dankook University, Yongin-si, South Korea
e-mail. hjho80@dankook.ac.kr

Room 1304, 107, Guil-ro 8-gil, Guro-gu, Seoul 08323, South Korea

© The Author(s), under exclusive license to Springer Nature Switzerland AG 2022
N. Rezaei (ed.), *Multidisciplinarity and Interdisciplinarity in Health*,
Integrated Science 6, https://doi.org/10.1007/978-3-030-96814-4_3

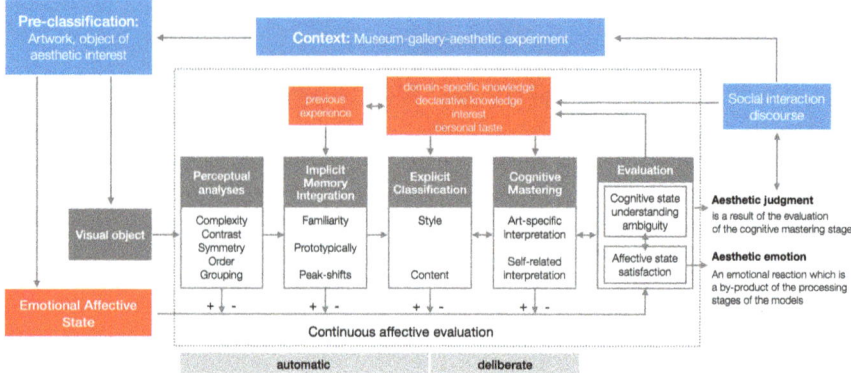

The dynamic model of the aesthetic appreciation [70]

The code of this chapter is *01110000 01110010 01100101 01000001 01100011 01101110 01101001 01101111 01100001 01110100 01101001 01110000.*

Keywords

Aesthetic appreciation · Aesthetics of science · fMRI · Meta-analysis · Neuroscience

1 Introduction

Science is often regarded as a rational discipline distinguished from art and humanities, as C.P. Snow mentioned in his book, The Two Cultures [1]. This image of science, a discipline to find out regularities of the observed data with mathematical and experimental methods, was formed after the scientific revolution in the seventeenth century and is still assumed as a symbol of rational thinking, whereas art is the essence of imagination and creativity. Since the latter half of the twentieth century, science and art historians have shed light on how ingenious scientists and artists accomplished their achievements, and the relationship between science and art seems more reciprocal. For instance, Miller, Root-Bernstein, and others mentioned that modern physicists and artists share similar thinking patterns [2–4]. Both scientists and artists often use observation, depiction, abstraction, and inference as common ways to do creative works. Even scientific ideas offer good material for artworks. A droopy watch in *the persistence of memory* by Dalí stands for the time dilation, and a man putting his arm into the wall in the painting, *universal gravitation* by Magritte, seems to explain the tunnelling effect in quantum mechanics as shown in Griffiths's book, Revolutions in Twentieth-Century Physics [5].

Scientists witness the beauty of science as a crucial objective of science. Poincaré argued that scientists do study not because it is beautiful but because he delights in it. Heisenberg viewed mathematical beauty as the true essence of physical knowledge [6–8]. Kuhn also argued that science's aesthetic preference is one of the compelling reasons for paradigm shift [9]. According to him, the shift from geocentric to heliocentric theory was due to the Copernican theory's simplicity; the heliocentric theory was composed of concentric circles with orbiting planets, whereas the geocentric theory consisted of complex combinations of many epicycles and eccentricities of their orbits.

The aesthetics of science may likely play a significant role in establishing scientific theories. For instance, Thagard reports that the passionate scientists' awe and curiosity towards science evoke research questions and finally lead to science discoveries [10]. Rescher mentions that simplicity plays a significant role in formulating new theories [11]. Even more, the aesthetics of science is categorized into intrinsic (representational) and extrinsic (emotional) aspects [12]. Scientific theories, formulas, and many kinds of visual representations show the innate nature of science: simplicity, symmetry, harmony, order, and others. Such aspects of science can be counted as the value of science in contemporary views on the nature of science, family resemblance approach [13]. On the other hand, those representational forms may evoke subtle feelings such as sublimity, surprise, wonder, and elegance [14]. Such an effective domain may influence to think about how science works. As for aesthetics, both science and art may share some commonalities, and such aspects are crucial in figuring out patterns of the observed data and making conclusions [15].

It is still controversial to interpret developments in scientific theories from the aesthetics of science. Not only symmetry but also symmetry-breaking is regarded as the aesthetics of science [16]. Even uniformity and simplicity are somewhat subjective, and complexity is counted as appropriate in explaining statistical mechanics phenomena [17]. A literature review raises issues about some aspects of the aesthetics as follows;

(i)   the ontological aspect, which is beautiful: nature or science?;
(ii) the epistemological aspect, how can we be aware of beauty, by perception or recognition?; and
(iii) the methodological aspect, is the aesthetic sense of science able to be trained? In other words, is the aesthetic sense of science innate or acquired? [14].

It seems irrelevant to approach such issues scientifically because they are connected to the prolonged philosophical questions since the ancient Greek era. Plato and Aristotle explained the beauty of truth as an *idea* and provided a way to understand beauty. While Plato insisted on the transcendental concept of beauty that cannot be comprehended in an empirical way, apart from the world, Aristotle claimed the secular beauty that a human sense could find. Both thought that beauty is attributed to an object's innate nature and can be sensed by human thoughts or experiences [18]. Over 1500 years, Baumgarten disentangled the aesthetics from philosophy, and Shaftesbury and Kant in the eighteenth century regarded that aesthetic appreciation is perceived by an individual taste of beauty [19]. In the modern era, the aesthetics is not existential but interpretive. Even beauty does not belong to the object but does to an audience's mind. Thus, the questions mentioned above have been aroused in philosophers' minds: *What is truth? Does it really exist? How can we realize the truth?*

The advent of neuroscience may give some clues to explain them scientifically. Neuroimaging studies using, for example, functional magnetic resonance imaging (fMRI) and positron emission tomography (PET) can help understand human thinking patterns by physiological information. Neuroaesthetics is a study of the neural processes underlying aesthetic behaviours [20]. The issues raised before can be tested in neuroscience. Regarding ontology, neuroscience enables us to examine any difference between neurotransmitters and electronic signals when having a look at beautiful objects or not. In terms of epistemology, we can analyse whether the activated parts in the brain are involved in a perceptual mode or a cognitive mode by comparing cognitive neuroscience studies. Last, with respect to the methodology, we may examine any neuroimaging changes before and after the instruction to figure out the consistency of aesthetic preference. This chapter is an effort to tackle the aesthetics of science through theories and studies in neuroscience.

2 Background

Despite the long history of beauty, a neurological perspective on aesthetics has received much attention, especially during the last three decades. As early as 7000 years ago, the man was interested in the nervous system to do medical surgery like boring holes in human skulls. In the Roman Empire, Galen observed human and animal brains and drew their paintings. Over 1500 years, scientists found two different types of brain tissue (grey and white matters), and at the end of the eighteenth century, it was known that the nervous system is constituted of central division (brain and spinal cord) and peripheral division (network of nervous through the body) [21]. In the nineteenth century, Gall, Fritsch, Broca, and others came to know some local brain areas' functions. Circa 1900 scientists were aware of an individual nerve cell, though limited to understanding the brain experimentally. Since the 1980s, the development of molecular biology and atomic physics has led to inventing new ways to detect electrical and physiological processes in the brain: magnetoencephalography (MEG), electroencephalography (EEG), fMRI, and PET. Each measurement has a different range of spatiotemporal resolutions of the imaging, as shown in Fig. 1. Among them, fMRI as a non-invasive measurement makes use of the fact that oxyhemoglobin has a magnetic resonance different from that of deoxyhemoglobin [22]. This blood-oxygen-level-dependent (BOLD) mechanism can detect changes in regional blood flow and metabolism within the brain.

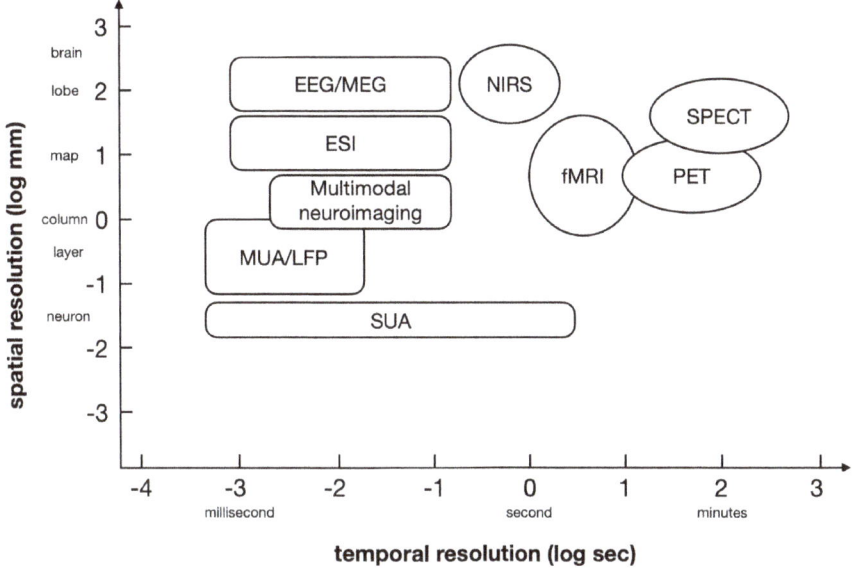

Fig. 1 Spatial and temporal resolution of various neuroimaging techniques [23]

Unlike other sub-disciplines in aesthetics, neuroaesthetics relies upon experimental physiology. From the ancient Greek era, the concept of beauty and the origin of aesthetics was a crucial issue in philosophy. A German philosopher, Alexander Baumgarten, defined aesthetics as a discipline to investigate the sense or taste of beauty [19]. In the nineteenth century, philosophers thought that aesthetic experience is based on creators' and audiences' pleasure and intended to connect aesthetics with physiology [20]. In the twentieth century, some physiologists tried to examine aesthetic behaviours in art. Later in the 1970s, the development of brand-new neuroimaging techniques brought about new approaches to study neural processes related to aesthetic responses. Although visual arts, music, dance, and play are being increasingly investigated in neuroaesthetics, the aesthetics of science remains much less heeded in this realm. For this reason, our current neuroscience of aesthetics relies upon studies on aesthetic perception and judgement in general.

To discuss the issues about aesthetic aspects of science, we need to synthesize the outcomes of neuroaesthetics. Thus, a total of 87 articles dealing with neuroimaging results using fMRI were selected from the Web of Science provided by Clarivate Analytics using the keywords: "beauty", "aesthetics", and "science." The data relating to active brain regions during visual art perception and appreciation were extracted [24]. The positions are illustrated according to MNI (Montreal Neurological Institute) and TAL (Talairach) coordinates. All points were mapped into the MNI coordinate due to the disparity between the two different coordinate systems [25]. The collected data was refined and visualized by R-Studio with oro. nifti and brainR package in the open repository (http://cran.r-project.org/).

3 Active Regions Involved in Aesthetic Behaviour

Various measurements have been adopted to understand the human aesthetic process. In the case of fMRI, the result shows that a total of 245 people (124 male and
121 female) took part in studies included in the review. The majorities of participants
were in their twenties and right-handed (238 people). Their average age was
24.2 years old. Through the literature review, a total of 151 voxels were identified,
as shown in Fig. 2. Parts actively involved in the aesthetic behaviour mainly include
the visual cortex, located on the inside of the occipital lobe. Moreover, Broca's and
Wernicke's areas linked to speech and language are active, and the caudate nucleus,
hippocampus, and cingulate nucleus connected to the limbic system are working.
The left hemisphere responds more rather than the right hemisphere.

Figure 3 shows the cerebral regions actively involved in the aesthetic behaviour
in accordance with the Brodmann area (BA). Motor-sensory areas (6, 8, 32) are
most active because they are involved in visual processing. The retina's visual

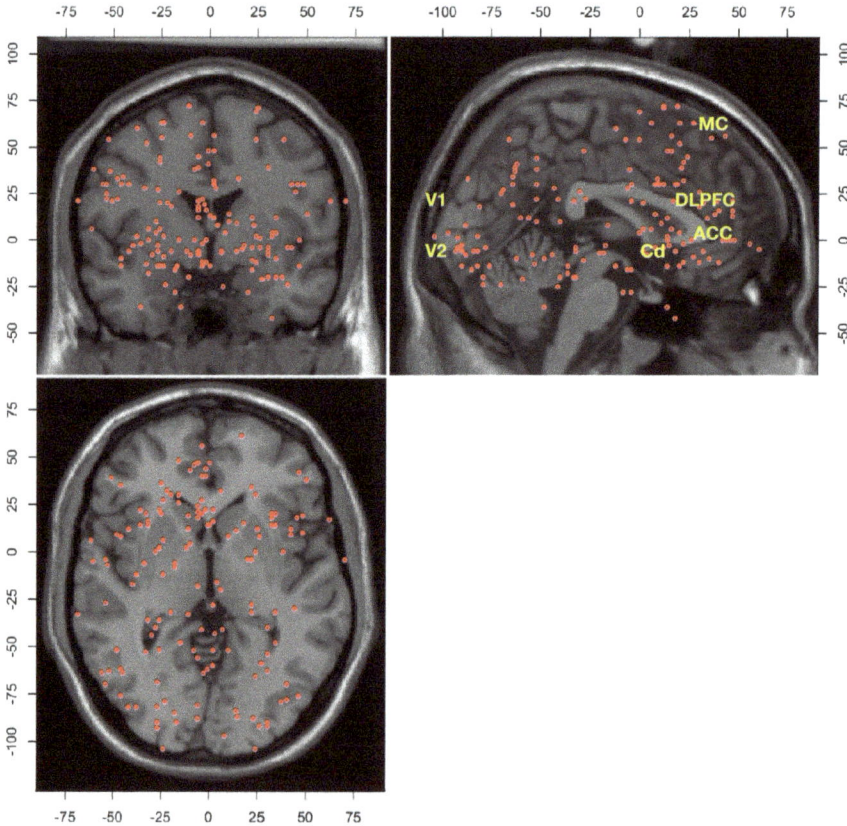

Fig. 2 The voxels related to aesthetic behaviour from the meta-analysis

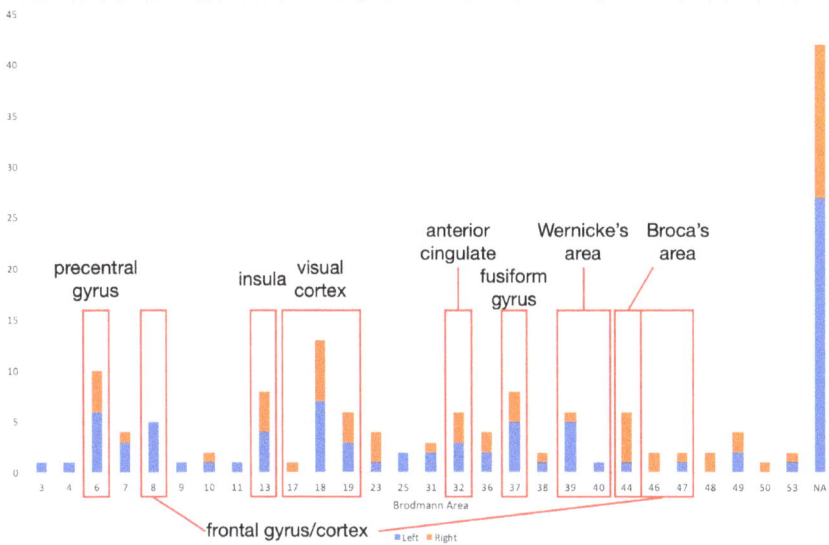

Fig. 3 The distribution of active voxels involved in the aesthetic behaviour according to Brodmann Area. Left: left hemisphere, right: right hemisphere, *y*-axis: frequency

information is delivered to the visual area (17–19) via the lateral geniculate nucleus (LGN). It then flows to two different streams: dorsal stream towards the motor command area in the parietal lobe and ventral stream towards the temporal lobe, known as a face and object recognition area [22]. The dorsal stream is connected to object motions, which is called MT and MST. The area plays mainly three roles: navigation, directing eye movement, and motion perception [21]. The executive area, prefrontal cortex (BA8, frontal eye field), dorsolateral prefrontal cortex (DLPFC, 9, 46), and BA44 (Broca's area) respond actively as well. BA8 functions to control eye movement, and Broca's area is mainly connected to speech production. Recent studies revealed that this area is involved with theory of mind like observation, inference, and projection [26–28]. Intriguing is that the perirhinal cortex is important, involved in visual perception and memory, and mediates stimulus-stimulus associations. It might indicate that cognitive processing gets involved in aesthetic appreciation.

Figure 4 reveals that a variety of regions are connected to aesthetic behaviour. Aligned with the result of Fig. 2, various lobes are involved in aesthetic behaviour. It is interesting to note that the limbic lobe gets involved in aesthetic behaviour. The ring-shaped tissues around the brain stem and corpus callosum participate in the aesthetic behaviour: the amygdala (AB), anterior cingulate cortex (ACC), hippocampus (HPC), and ambient tissues, like the fusiform and lentiform gyrus. The limbic system is known for its role in emotional expression and experience. In particular, the amygdala within the temporal pole undertakes learning some emotions.

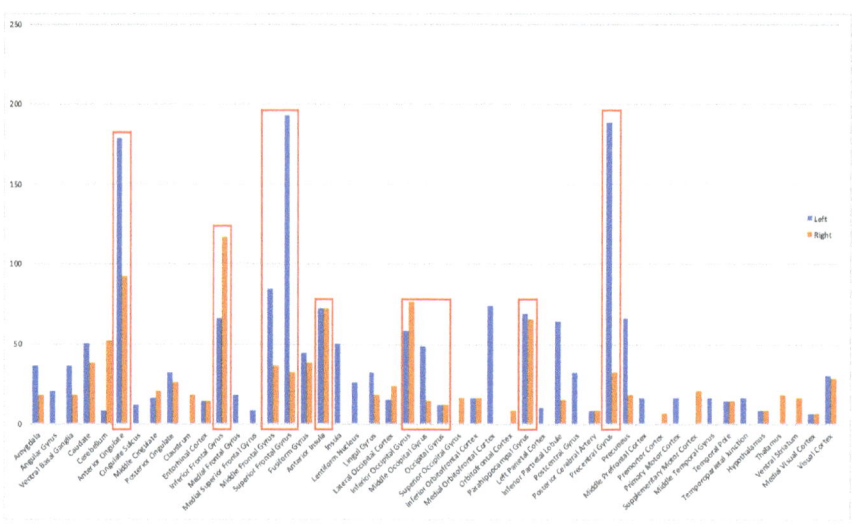

Fig. 4 The distribution of active voxels involved in the aesthetic behaviour Left: left hemisphere, right: right hemisphere, *y*-axis: frequency

Neuroaesthetics assumes that there are specific regions involved in aesthetic behaviour. Studies on aesthetics have dealt with several types of objects: face, artworks, and others. The common areas to make the aesthetic judgement on face and art are superior temporal sulcus (STS), insula, ACC, thalamus, ventral striatum (VS), and orbitofrontal cortex (OFC). The ventral stream in the visual processing carries out facial recognition [29]. As for artworks, both experts and laypeople show active responses on the bilateral medial orbitofrontal cortex and subcallosal cingulate gyrus [30]. While appreciating artworks, mOFC and PFC are activated [31, 32]. The left frontal lobe and bilateral temporal lobes are activated while seeing beauty, and the response lasts from 2000 to 9250 ms (ms) [33]. On the other hand, temporal and occipital lobes are activated in aesthetic judgement [34]. Kesner and others find that ACC and insula are significantly activated in aesthetic judgement [35]. Such regions are linked to visual or linguistic processing, cognition, and empathy. Seeing an artistic image differs from seeing a geometric image in terms of activated brain regions. In modern paintings and geometric images, the left hemisphere likely gets more involved with the precentral gyrus, parietal lobe, and fusiform gyrus, whereas the right hemisphere with ACC and precuneus get more involved in perceiving classical paintings [36–38]. Non-aesthetic stimuli bring about the frontal lobe's activation, whereas beautiful paintings evoke the temporal areas linked to the inference. The aesthetic perception accompanies pleasure, as well as thinking [39]. However, it is not a comprehensive conclusion due to a limited number of studies.

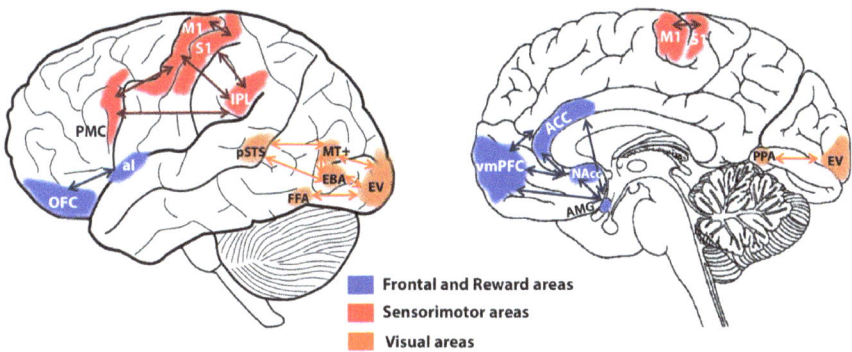

Fig. 5 The schematic representation of the neural circuit implicated in aesthetic judgement tasks [41]

To sum up, the meta-analysis results show that various regions in the brain are involved in the aesthetic behaviour, centring on the regions related to visual processing, and to a lesser extent, on those underlying executive, motor-sensory, and emotional processing. Specific regions are likely involved dependent on objects' nature (representational, geometric, and non-aesthetic). However, this overview is not enough to answer the issues related to the aesthetics of science. By the way, is the aesthetic experience perceptual or cognitive? The process from perception to judgement is as follows: pre-classification, perceptual analysis, implicit memory integration, explicit classification, cognitive mastery, secondary control, and finally, self-awareness and meta-cognitive assessment [40]. Figure 5 represents the neural circuit implicated in aesthetic judgement tasks. The red and orange regions are close to perception, but the blue regions are much closer to cognition. The reward is an outcome of a decision on cost–benefit analysis. After all, the nature of the aesthetic sense is an interplay between seeing and compensating for the beauty.

4 Aesthetic Sense in the Brain, Emotion, or Cognition

Whether or not it is beautiful is a complex decision. When seeing a beautiful painting, joy or pleasure may be evoked from one's mind. However, emotion is based on prior experience, and some criteria interpret it. In the appraisal of artworks, representational paintings like a landscape by Turner and Constable may evoke happiness directly, but abstract painting like cubism by Picasso and geometric abstract painting by Kandinsky requires high-order thinking for the aesthetic experience. The aesthetic evaluation is also often measured by preference decision, e.g. forced-choice preference or Likert scale [42].

In cognitive neuroscience, the aesthetic experience is regarded as cognitive processing. According to Chatterjee and Vartanian, the aesthetic experience is comprised of three domains: knowledge–meaning, sense–action, and emotion–evaluation [43]. The aesthetic judgement can be enhanced by iterative instructions [41, 44]. People are apt to accept the artworks' beauty when they see the paintings more often [45]. With regard to physiology, the aesthetic judgement is determined with 200 ms (ms) as the stimulus is given, and the event-related potential (ERP) result also confirms that the left temporal lobe is activated within 200 ms, which is accepted as rational thinking [46, 47].

On the other hand, the aesthetic experience can be understood as an affective one. Many neurologists report that the limbic system linked to emotional processing, e.g. cingulate gyrus and fusiform gyrus, are activated while feeling the beauty [32, 35, 48]. The VS is mainly activated in the visual processing of artistic pictures but not in the case of non-artistic images [49]. The region is central to the reward circuit [50]. This indicates that aesthetic judgement is an action to pursue the desire or delight from the visual behaviour.

However, it is not easy to distinguish two different aspects of emotion and cognition in the aesthetic experience. First of all, the distinction is ambiguous in perception. When we see an artistic painting, the aesthetic evaluation is primarily linked to emotion and secondarily to unconscious fantasy [51]. As well, aesthetic perception is affected by consideration of the context and prior experience [52]. In aesthetic judgement, different structures in the temporal, occipital, and parietal lobes, such as ACC and insula, become activated and interact with each other [35, 53]. That is to say, the emotional and cognitive regions are involved in the aesthetic

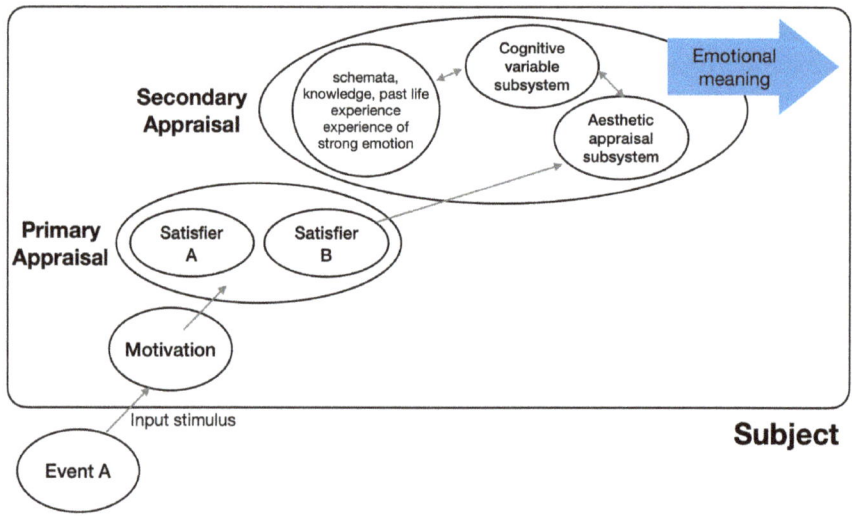

Fig. 6 A depiction of the functional process pertaining to the aesthetic emotion [54]

judgement simultaneously, and it is impossible to discriminate the role of emotion and cognition in the aesthetic experience (Fig. 6).

Concerning the aesthetics of science, let us discuss the perception of symmetry. There is a significant correlation between the perception of symmetry and aesthetic judgement [55]. However, the activated regions during the symmetry perception are different from that of the aesthetic judgement. Symmetry is relatively linked to the left hemisphere (precentral and fusiform gyri), whereas the aesthetic judgement to the right hemisphere (ACC, prefrontal gyrus, and precuneus) [36, 37]. This indicates that science's aesthetic perception/judgement relies more on cognitive aspects than art appreciation. Mead and Loughlin mention that asymmetry stimulates aesthetic preference more [56].

What we should keep in our minds is the significance of context. For example, the paintings seem more beautiful in the artistic context rather than in everyday life [57, 58]. The response to the same painting is different according to the context, in the gallery or on the computer screen [34]. Even the position may influence the aesthetic preference because people tend to prefer the right one rather than the middle one [59].

5 Innate or Acquired Nature of Aesthetics of Science

In the seventh century, philosophers defined art's objective as pursuing pleasure and had a robust debate about whether the aesthetic taste is inborn or learned. From a neurological viewpoint, pleasure or happiness is connected to the reward system. In the brain, there are several dopaminergic pathways pertaining to the reward: ventral tegmental area (VTA) to PFC, VTA to VS, mid-brain to the dorsal striatum (caudate and putamen), hypothalamus to the spinal cord, and others [22]. They are mainly divided into two different ways: mid-brain to the limbic system and mid-brain to cortices. In the latter case, dopamine neurons in VTA have synaptic connections between OFC, and the former neurons are connected to nucleus accumbens in the amygdala's vicinity. The amygdala crucially contributes to mediating and learning emotions and emotion-related experiences. Mechanistically, aesthetic perception is a construct of the brain that links positive feedback of the reward system with pleasure encoded in OFC [29]. This may be contradictory to the canonical theories of aesthetics. Shaftesbury's taste theory and Kant's aesthetic attitudes posit that aesthetic behaviour is blind action to pursue one's pleasure. Their arguments are more concerned about psychological and social behaviours. However, neuroaesthetics intends to unveil the unkenned human structure of emotion. This may sound more comfortable, but the reward system does not reflect the whole processing of the aesthetic experience. It is an occasion that emotional preference does not coincide with the aesthetic appraisal. The medial prefrontal lobe is activated in the appreciation of paintings, whereas the limbic system is more activated in the emotional interest (putamen, caudate nucleus, and precuneus) [60].

The mirror neuron system (MNS) needs to be considered to answer the prolonged question. STS produces higher-order visual images of the observed behaviour. The created representation is transmitted to prefrontal-parietal MNS, coded into design and intention of motion, and finally turns back to STS to compare the predicted outcomes with the observed outcomes. In particular, MNS is related to premotor, motor, and motor-sensory areas. For instance, when we see a painting with dancing women, our prior experience and motor areas react [61]. The parietal lobe involved in the motion is activated, as shown in Figs. 2 and 4. MNS stands for biological sympathy, which means neurons' reaction to the related actions [62, 63]. MT and MST respond more actively in the case of experts when they see a painting. Many researchers report that individual aesthetic preference affects one's perception [64, 65]. The findings propose that the aesthetic experience with visual arts has its roots in a neural network that involves different regions depending on the stimuli' content [66–68]. However, some argue that MNS is not necessary to aesthetic appreciation [69]. Even though we do not know how to make a bronze relief or have no prior experience of the art object, we can feel it beautiful or have a deep impression. As well, it is unclear whether MNS is involved in the aesthetic appreciation of science theories since the beauty of science is more inclined to semantic and symbolic constituents.

6 Conclusion

Whether or not an aesthetic object exists, whether aesthetics is perceived or recognized, whether the sense of aesthetics can be improved, all are concerned about the ontology of aesthetic reality and epistemology of the aesthetic experience. Such issues have been discussed in aesthetics for a long time. This study intends to revisit them from a neurological viewpoint. First of all, it figured out the specific regions involved in the aesthetic appreciation. Various areas in the limbic system and prefrontal, temporal, and occipital lobes are activated during the aesthetic appreciation. In particular, some parts commonly involved in the aesthetic behaviour are ACC, amygdala, caudate, VS, and VTA (the limbic system); PFC and OFC (the frontal lobe); MT and MST (the parietal lobe); Wernicke's area (the temporal lobe); and V1–V3 (the occipital lobe). Different types of objects (face, classical painting, and abstract painting) may activate the different regions in the brain. Some argue that regions involved in aesthetic perception are different from those in aesthetic judgement, but the regions are not sufficiently defined in an empirical way.

Regarding the role of emotion and cognition in the aesthetic experience, aesthetic appreciation is accompanied by cognitive processing. On the one hand, the dopaminergic reward system, known for its role in emotional processing, is also activated during aesthetic appraisal. However, the emotion affected by the reward system, intentional pursuit of pleasure, may not be consistent with traditional aesthetics theories. About the inborn or acquired nature of aesthetic appraisal, the reward system is mainly dependent on the emotion. It can reinforce stronger emotion about the aesthetic experience by increased knowledge and repetitive experiences. Moreover, such experiences may influence MNS, which is distinct according to the individual talents and experiences. This indicates that aesthetic judgement can be improved or changed intentionally, though the aesthetic appraisal of an object is subjective. The results of the studies do not coincide with the attitudes of philosophers towards aesthetics.

With limited studies available, it is hardly possible to propose a comprehensive theory to explain the aesthetic appreciation and the aesthetics of science. Future studies should investigate to understand the aesthetics of science. However, it seems that the aesthetic judgement of science has something in common with that of artistic context and that the aesthetic judgement can be improved. Then, how can we help preprofessional scientists and the public to understand such a nature more? Leder and others propose a model to enhance aesthetic appreciation focussing on emotion [70]. They design the aesthetic experiences step-by-step: "perception, explicit classification, implicit classification, cognitive mastering and evaluation". Each step encourages students to have stronger emotion on the experience and finally bring about satisfaction with the aesthetic objects.

Another possible factor influencing aesthetic appreciation is attention. It is very critical to perceive and cope with the situation. Various brain regions are active during attention processes, while the peripheral nervous system allows the selective perception of stimuli. Lateral inhibition is the reduction of the activity of neighbour neurons by an excited neuron. Lateral inhibition makes a contrast in stimulus, for example, visual illusion and hearing someone's voice in the noisy background. It helps us to focus on one thing we want and to ignore peripheral and trivial information. In the brain, the interplay between emotion and cognition shows a similar work. Intriguing is that working memory is overlapped with attention, and they are controlled by PPC and PFC [22]. This indicates that aesthetic perception and judgement are adjusted by the attention process, which is a conscious behaviour with intention.

The aesthetic appreciation of science is a combination of instantaneous individual preference and cognitive mastering process. Limited studies are insufficient to reveal the prolonged issues of science aesthetics, but dispersed active regions tell us about its complexity. Higher-order thinking requires the response of broader regions in the brain. The more accurate methods such as multiple local field potentials (LPFs) and single-unit activity (SUA) would shed light on the different aesthetic processing sides. How scientists understand nature as a navigator's compass in an unexplored ocean remains to be answered.

Core Messages

- Some brain regions, i.e., DLPFC, ACC, V1–V4, Cd, and MC, are frequently activated in the aesthetic judgement of science and art.
- Brain activation patterns indicate that the role of cognition and emotion in aesthetic appraisal cannot be distinguished.

- The aesthetic emotion evoked in one's mind is likely connected to the dopaminergic reward system.
- The aesthetic behaviour is adjusted by attention.
- Despite the inborn nature of aesthetic appraisal, aesthetic perception and judgement can be improved by structured instruction.

Acknowledgements This work was supported by the National Research Foundation of Korea (NRF) grant funded by the Korean government (MSIT) (No. 2020R1C1C1005754).

References

1. Snow CP (1959) The two cultures. Cambridge University Press, Cambridge
2. Vitz PC, Glimcher AB (1999) Modern art and modern science: the parallel analysis of vision. Praeger, New York
3. Miller AI (2000) Insights of genius: imagery and creativity in science and art. MIT Press, Cambridge, MA
4. Root-Bernstein RS (1999) Sparks of genius: the thirteen thinking tools of the world's most creative people. Houghton MIfflin Co., Boston, MA
5. Griffiths DJ (2013) Revolutions in twentieth-century physics. Cambridge University Press, New York
6. Poincaré H (1946) The foundations of science. Science Press, Lancaster
7. Heisenberg W (1971) Physics and beyond: encounters and conversations. Harper & Row, New York
8. Wilczek F (2016) A beautiful question: finding nature's deep design. Penguin Books, New York
9. Kuhn TS (2012) The structure of scientific revolutions. The University of Chicago Press, Chicago
10. Thagard P (2002) The passionate science: emotion in scientific cognition. In: Carruthers P, Stich S, Siegal M (eds) The cognitive basis of science. Cambridge University Press, Cambridge
11. Rescher N (1990) Aesthetic factors in natural science. University Press of America, New York
12. Girod M (2007) A conceptual overview of the role of beauty and aesthetics in science and science education. Stud Sci Educ 43(1):38–61
13. Erduran S, Dagher Z (2014) Reconceptualizing the nature of science for science education: scientific knowledge, practices and other family categories. Springer, Dordrecht
14. Jho H (2018) Beautiful physics: re-vision of aesthetic features of science through the literature review. J Korean Phys Soc 73(4):401–413
15. Lee J-J (2018) An aesthetic explanation of technoscientific images : focusing on Golgi and Cajal's neuroscientific experimental researches in the 19th century (an aesthetic explanation of technoscientific images: focusing on Golgi and Cajal's neuroscientific experimental researches in the 19th century). J Humanit 39(1):335–363
16. Earman J (2004) Laws, symmetry, and symmetry breaking: invariance, conservation principles, and objectivity. Philos Sci 71(5):1227–1241
17. McAllister JW (1996) Beauty and revolution in science. Cornell University Press, New York
18. Hartmann N (2014) Aesthetics (trans: Kelly E). De Gruyter, Boston

19. Cahn SM, Meskin A (2007) Aesthetics: a comprehensive anthology. Blackwell Publishing, Malden, MA
20. Skov M, Vartanian O (2009) Neuroaesthetics. Routledge, New York
21. Bear MF, Connors BW, Paradiso MA (2016) Neuroscience: exploring the brain. Wolters Kluwer, New York
22. Baars BJ, Gage NM (2010) Cognition, brain, and consciousness. Elsevier, Burlington, MA
23. He B, Liu Z (2008) Multimodal functional neuroimaging: integrating functional MRI and EEG/MEG. IEEE Rev Biomed Eng 5(1):23–40
24. Stocco A (2014) Coordinate-based meta-analysis of fMRI studies with R. R J 6(2):5–15
25. Laird AR, Robinson JL, McMillan KM, Tordesillas-Gutiérrez D, Moran ST, Gonzales SM, Ray KL, Franklin C, Glahn DC, Fox PT, Lancaster JL (2010) Comparison of the disparity between Talairach and MNI coordinates in functional neuroimaging data: validation of the lancaster transform. Neuroimage 51(2):677–683
26. Gweon H, Saxe R (2013) Developmental cognitive neuroscience of theory of mind. In: Rubenstein JLR, Rakic P (eds) Neural circuit development and function in the brain. Elsevier, New York, pp 367–377
27. Mahy CEV, Moses LJ, Pfeifer JH (2014) How and where: theory-of-mind in the brain. Dev Cogn Neurosci 9:68–81
28. Meinhardt-Injac B, Daum MM, Meinhardt G, Persike M (2018) The two-systems about account of theory of mind: tesing the links to social-percpetual and cognitive abilities. Front Hum Neurosci 12:25
29. De Ridder D, Vanneste S (2013) The artful mind: sexual selection and an evolutionary neurobiological approach to aesthetic appreciation. Perspect Biol Med 56(3):327–340. https://doi.org/10.1353/pbm.2013.0029
30. Kirk U, Skov M, Christensen MS, Nygaard N (2009) Brain correlates of aesthetic expertise: a parametric fMRI study. Brain Cogn 69(2):306–315. https://doi.org/10.1016/j.bandc.2008.08.004
31. Kawabata H, Zeki S (2004) Neural correlates of beauty. J Neurophysiol 91(4):1699–1705
32. Vartanian O, Goel V (2004) Neuroanatomical correlates of aesthetic preference for paintings. NeuroReport 15(5):893–897
33. Lengger PG, Fischmeister FPS, Leder H, Bauer H (2007) Functional neuroanatomy of the perception of modern art: A DC-EEG study on the influence of stylistic information on aesthetic experience. Brain Res 1158:93–102. https://doi.org/10.1016/j.brainres.2007.05.001
34. Kirk U, Skov M, Hulme O, Christensen MS, Zeki S (2009) Modulation of aesthetic value by semantic context: an fMRI study. Neuroimage 44(3):1125–1132. https://doi.org/10.1016/j.neuroimage.2008.10.009
35. Kesner L, Grygarová D, Fajnerová I, Lukavskýa J, Nekovářová T, Tintěra J, Zaytseva Y, Horáček J (2018) Perception of direct versus averted Gaze in portrait paintings: an fMRI and eyetracking study. Brain Cogn 125:88–99
36. Jacobsen T, Schubotz RI, Höfel L, van Cramon DY (2006) Brain correlates of aesthetic judgment of beauty. Neuroimage 29(1):276–285. https://doi.org/10.1016/j.neuroimage.2005.07.010
37. de Tommaso M, Pecoraro C, Sardaro M, Serpino C, Lancioni G, Livrea P (2008) Influence of aesthetic perception on visual event-related potentials. Conscious Cogn 17:933–945
38. Lee SB, Jung WH, Son JW, Jo SW (2011) Neural correlates of the aesthetic experience using the fractal images: an fMRI study (Neural correlates of the aesthetic experience using the fractal images: an fMRI study). Sci Emot Sensibility 14(3):403–414
39. Brielmann AA, Pelli DG (2017) Beauty requires thought. Curr Biol 27(10):1506–1513 e1503. https://doi.org/10.1016/j.cub.2017.04.018
40. Pelowski M, Markey PS, Forster M, Gerger G, Leder H (2017) Move me, astonish me... delight my eyes and brain: the Vienna integrated model of top-down and bottom-up processes in art perception (Vimap) and corresponding affective, evaluative andneurophysiological correlates. Phys Life Rev 21:80–125

41. Kirsch LP, Urgesi C, Cross ES (2016) Shaping and reshaping the aesthetic brain: emerging perspectives on the neurobiology of embodied aesthetics. Neurosci Biobehav Rev 62:56–68. https://doi.org/10.1016/j.neubiorev.2015.12.005
42. Huston JP, Nadal M, Mora F, Agnati LF, Cela-Conde CJ (2015) Art, aesthetics and the brain, vol Oxford. University Press, Oxford
43. Chatterjee A, Vartanian O (2014) Neuroaesthetics. Trends Cogn Sci 18(7):370–375. https://doi.org/10.1016/j.tics.2014.03.003
44. Hayn-Leichsenring GU, Kloth N, Schweinberger SR, Redies C (2013) Adaptation effects to attractiveness of face photographs and art portraits are domain-specific. Iperception 4(5):303–316
45. Bohrn IC, Altmann U, Lubrich O, Menninghaus W, Jacobs AM (2013) When we like what we know—a parametric fMRI analysis of beauty and familiarity. Brain Lang 124:1–8
46. Cela-Conde CJ, García-Prieto J, Ramasco JJ, Mirasso CR, Bajo R, Munar E, Flexas A, del-Pozo F, Maestu F (2013) Dynamics of brain networks in the aesthetic appreciation. Proc Natl Acad Sci 110(Supplement 2):10454–10461. https://doi.org/10.1073/pnas.1302855110
47. Munar E, Nadal M, Castellanos NP, Flexas A, Maestu F, Mirasso C, Cela-Conde CJ (2011) Aesthetic appreciation: event-related field and time-frequency analyses. Front Hum Neurosci 5:185. https://doi.org/10.3389/fnhum.2011.00185
48. Yeh Y-c, Lin C-W, Hsu W-C, Kuo W-J, Chan Y-C (2015) Associated and dissociated neural substrates of aesthetic judgment and aesthetic emotion during the appreciation of everyday designed products. Neuropsychologia 73:151–160. https://doi.org/10.1016/j.neuropsychologia.2015.05.010
49. Lacey S, Hagtvedt H, Patrick VM, Anderson A, Stilla R, Deshpande G, Hu X, Sato JR, Reddy S, Sathian K (2011) Art for reward's sake: visual art recruits the ventral striatum. Neuroimage 55:420–433
50. Telzer EH (2016) Dopaminergic reward sensitivity can promote adolescent health: a new perspective on the mechanism of ventral striatum activation. Dev Cogn Neurosci 17:57–67
51. Rose GJ (2004) Aesthetic ambiguity revisited via the artist-model pair and neuroscience. Psychoanal Psychol 21(3):417–427. https://doi.org/10.1037/0736-9735.21.3.417
52. Noguchi Y, Murota M (2013) Temporal dynamics of neural activity in an integration of visual and contextual information in an esthetic preference task. Neuropsychologia 51:1077–1084
53. Zhang W, Lai S, He X, Zhao X, Lai S (2016) Neural correlates for aesthetic appraisal of pictograph and its referent: an fMRI study. Behav Brain Res 305:229–238. https://doi.org/10.1016/j.bbr.2016.02.029
54. Xenakis I, Arnellos A, Darzentas J (2012) The functional role of emotions in aesthetic judgment. New Ideas Psychol 30:212–226
55. Jacobsen T, Höfel L (2003) Descriptive and evaluative judgment processes: behavioral and electrophysiological indices of processing symmetry and aesthetics. Cogn Affect Behav Neurosci 3(4):289–299
56. Mead AM, Loughlin JP (1992) The roles of handedness and stimulus asymmetry in aesthetic preference. Brain Cogn 20(2):300–307
57. Arai S, Kawabata H (2016) Appreciation contexts modulate aesthetic evaluation and perceived duration of pictures. Art Percept 4(3):225–239
58. Bao Y, von Stosch A, Park M, Pöppel E (2017) Complementarity as generative principle: a thought pattern for aesthetic appreciations and cognitive appraisals in general. Front Psychol 8:1–16. https://doi.org/10.3389/fpsyg.2017.00727
59. Beaumont JG (1985) Lateral organization and aesthetic preference: the importance of peripheral visual asymmetries. Neuropsychologia 23(1):103–113
60. Kim J, Shin E-h, Kang H, Kim C-Y (2015) Sad but beautiful; Brain responses to aesthetic judgment and emotion appraisal of visual art. Korean J Cogn Biol Psychol 27(2):231–251
61. Jeffers CS (2010) A still life is really a moving life: the role of mirror neurons and empathy in animating aesthetic response. J Aesthetic Educ 44(2):31–39

62. Freedberg D, Gallese V (2007) Motion, emotion and empathy in esthetic experience. Trends Cogn Sci 11(5):197–203
63. Gallese V (2013) Mirror neurons, embodied simulation and a second-person approach to mindreading. Cortex 49(10):2954–2956. https://doi.org/10.1016/j.cortex.2013.09.008
64. Cattaneo Z, Lega C, Ferrari C, Vecchi T, Cela-Conde CJ, Silvanto J, Nadal M (2015) The role of the lateral occipital cortex in aesthetic appreciation of representational and abstract paintings: a TMS study. Brain Cogn 95:44–53. https://doi.org/10.1016/j.bandc.2015.01.008
65. Boccia M, Barbetti S, Margiotta R, Guariglia C, Ferlazzo F, Giannini AM (2014) Why do you like Arcimboldo's portraits? Effect of perceptual style on aesthetic appreciation of ambiguous artworks. Atten Percept Psychophys 76(6):1516–1521. https://doi.org/10.3758/s13414-014-0739-7
66. Brown S, Gao X, Tisdelle L, Eickhoff SB, Liotti M (2011) Naturalizing aesthetics: brain areas for aesthetic appraisal across sensory modalities. Neuroimage 58(1):250–258
67. Boccia M, Barbetti S, Piccardi L, Guariglia C, Giannini AM (2017) Neuropsychology of aesthetic judgment of ambiguous and non-ambiguous artworks. Behav Sci (Basel) 7(1). https://doi.org/10.3390/bs7010013
68. Boccia M, Barbetti S, Piccardi L, Guariglia C, Ferlazzo F, Giannini AM, Zaidel DW (2016) Where does brain neural activation in aesthetic responses to visual art occur? Meta-analytic evidence from neuroimaging studies. Neurosci Biobehav Rev 60:65–71. https://doi.org/10.1016/j.neubiorev.2015.09.009
69. Casati R, Pignocchi A (2007) Mirror and canonical neurons are not constitutive of aesthetic response. Trends Cogn Sci 11(10):410; author reply 411. https://doi.org/10.1016/j.tics.2007.07.007
70. Leder H, Belke B, Oeberst A, Augustin D (2004) A model of aesthetic appreciation and aesthetic judgments. Br J Psychol 95(4):489–508

Hunkoog Jho is an assistant professor at Dankook University, Republic of Korea. He received his Ph. D. in physics education from Seoul National University in 2012. He has investigated an interdisciplinary approach to science and art from historical and philosophical perspectives, funded by the National Research Foundation of Korea since 2017. He is also interested in applying machine learning to science education.

Neuroscience and Quantum Physics Aspect of Human Brainwaves

4

Zamzuri Idris, Zaitun Zakaria, Faruque Reza,
Abdul Rahman Izaini Ghani, and Jafri Malin Abdullah

*"the mind that opens to a new idea never returns to its original
size."*

Albert Einstein

Summary

The anatomical view is the approach often used to understand the brain. This
view largely relies upon brainwaves, their relation to neuropsychiatric disorders,
namely epilepsy, and their application to brain mapping, brain-computer
interfaces, and consciousness. The research and our knowledge in consciousness
are, however, currently limited. The reason for that perhaps lies in our initial
perception of how the brain works. Thus, to better understand human
consciousness, the alternative view for the brain is all are waves. This view

Z. Idris (✉) · Z. Zakaria · F. Reza · A. R. I. Ghani · J. M. Abdullah
Department of Neurosciences, School of Medical Sciences, Universiti Sains Malaysia,
Kubang Kerian 16150, Kelantan, Malaysia
e-mail: zakariaz@tcd.ie

F. Reza
e-mail: faruque@usm.my

Z. Idris
Integrated Science Association (ISA), Universal Scientific Education and Research Network
(USERN), Kubang Kerian, Malaysia

Z. Idris · Z. Zakaria · A. R. I. Ghani · J. M. Abdullah
Brain and Behaviour Cluster (BBC), School of Medical Sciences, Universiti Sains Malaysia,
Kubang Kerian 16150, Kelantan, Malaysia

Z. Idris · Z. Zakaria · F. Reza · A. R. I. Ghani · J. M. Abdullah
Hospital Universiti Sains Malaysia (HUSM), Universiti Sains Malaysia, Kubang Kerian
16150, Kelantan, Malaysia

has benefited from quantum physics principles that consider a particle, an atom, or a molecule as waves to offer a deeper insight into the brain. The present chapter is an attempt to cover both the neuroscientific and quantum physics aspects of brainwaves. In specific, it comprises the origin of the human brainwaves; microgravity inside the cranium; the seat of the human soul that involves the greater limbic system, consciousness, and death of a person; brain or body transplant; quantum psychiatry; duality of a particle and light; illusionary and borderless universe; infinite shape in our nature and brain; and a shift in energy or dimension.

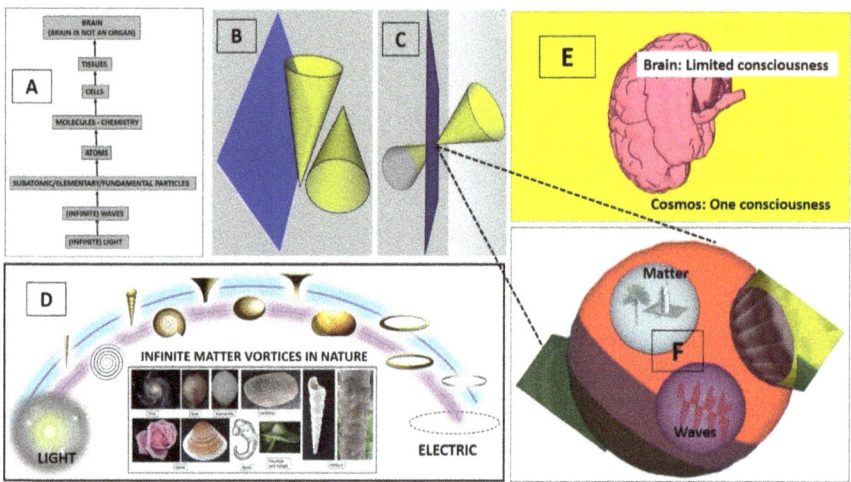

The brain, cosmology, and quantum physics: a, from the lowest quantum to the highest anatomical view for the brain; b-c, Limited projection of reality from infinite dimension or energy (waves) with some waves do intersect (c) while many of them do not (b); d, the intersection of some waves at certain angles with a newly formed physical universe (light or glowing phase after the big bang) create light, matter, and electric vortices, the matter vortices are infinite in their shapes (corresponded to the infinite shape in our nature); e, with the limited projected waves, the brain function is limited, i.e., in the quality of its consciousness; f, with the limited quality for both (physical universe) and the brain (observer), the person cannot see the true nature of the particles which are waves or energy (collapsing the wave function of the particles and duality of them).

The code of this chapter is *01,110,100 01,110,010 01,100,001 01,111,001 01,110,110 01,000,111 01,101,001.*

Keywords

Brain field · Brainwaves · Consciousness · Electromagnetic field · Epilepsy · Infinite dimension · Magnetoencephalography · Psychosis · Quantum field

1 Introduction

The research done on the brain commonly covers gross and microanatomy, cell and tissue histology, brain genetic and chemical molecules, and brainwaves. Brainwaves are the electrical activity of the brain. It denotes the brain's function, and thus, any alteration in it can lead to abnormal clinical or different manifestations. Since a sum of all waves will have a field (of energy) and any electrical field will have the accompanying magnetic field, the well-documented brainwaves are often known as the electromagnetic field (EMF). Lately, the research interest in consciousness has arisen not only among clinicians but also among physicists and mathematicians. They argue the presence of a single quantum field of energy that permeates the whole cosmos [1, 2]. It signifies the brain itself is permeated by and interacts with this energy field. Therefore, among neuroscientists, a shifting interest in studying the brain at a deeper level, i.e., at atomic, subatomic, or fundamental particle levels, is growing. The present chapter follows this scientific argument inspired by earlier scientific work on brain consciousness by an English physicist at the University of Oxford, Sir Roger Penrose, a renowned mathematician, and philosopher. The authors present their own scientific experience and clinical data on the neuroscientific aspect of brainwaves, along with some hypotheses prepared for the quantum physics aspect of human brainwaves. Therefore, this chapter is an example step toward integrating two major scientific fields: neurosciences and physics, which enables us to view the brainwaves' organization and function differently.

2 The Origin of Human Brainwaves

Brainwaves are the most crucial aspect of human brain functions [3]. The definition of a wave itself is referred to as an up and down and thus forms basic energy. Without energy, the human becomes dead, and his or her brainwaves are isoelectric. Interestingly, the dead features of human beings are not only referred to as the isoelectric brainwaves but also the absence of waviness for the heartbeat, respiration, and blood pressure, among others. Therefore, death is commonly being determined by brain death, cardiac death, or circulatory death. To conclude that a person is dead, documenting absence in consciousness (coma state) is one criterion. When discussing consciousness, medical doctors tend to refer it to the brain. The parts of the brain related to consciousness are the diencephalon and brainstem. Any injury to this central and deep structure does affect consciousness, while consciousness remains almost unchanged if the cerebral hemisphere is not present [4–8]. In 1949, Morruzzi and Magoun performed an animal experiment in which stimulation of the brainstem's reticular formation evoked changes in the scalp electroencephalography (EEG) through two pathways, namely the dorsal and ventral pathways. Even though each pathway originated from the brain's central area [9], the brainwaves are different. The dorsal pathway projects to the thalamus and then is diffused to the superficial cortices, while the ventral pathway projects to the hypothalamus and ascends to the basal forebrain amygdala-hippocampus before finally reaching cortices. Therefore, consciousness is in close contact with cortical brainwaves, and both are related to the central area of the brain.

In 1988, Nieuwenhuys et al. [10] studied the central area of the brain and named it a greater limbic system. The structures associated with *the greater limbic system* include the following: (i) classical limbic system; (ii) classical reticular-brainstem system; (iii) basal forebrain; (iv) thalamus; and (v) pineal and pituitary gland. These structures are in a bending position (microgravity) and form a central nervous system's core and para core area [11]. Accordingly, one will agree that the brain's deep and central area is responsible for the whole brain and human function (Fig. 1). Therefore, it is not surprising that ancient philosophers' historical notes had considered this component area of the brain as *the human soul's seat*.

Debates that the seat of the human soul is either in the brain or the heart date back to Plato (424–348 BC), Aristotle (circa 384–383 BC), Galen (circa 200/216 BC), and Leonardo da Vinci era (1452–1519) [12]. Plato and Galen preferred the

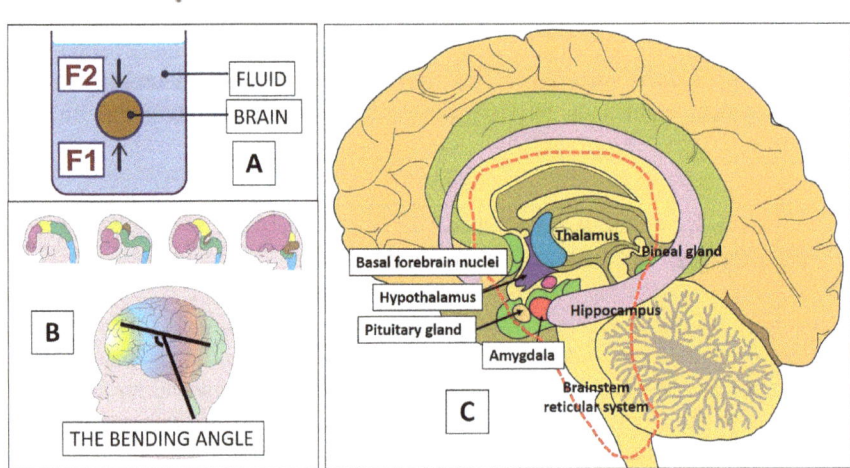

Fig. 1 The greater limbic system and the buoyancy: **a** the anti-gravity of the fluid (weightlessness) and whenever the F1 force equals F2, the object will hardly move; **b** the greater limbic system is in microgravity posture with its bending or curving angle. The central nervous system always requires microgravity since its embryogenesis; **c** the greater limbic system lies deep inside the brain and is at the periventricular areas (dashed line in red)

brain, while Aristotle was the opposite. Later, during the renaissance period in Italy, Leonardo da Vinci described the seat of the human soul in the brain and at the center of the human's skull. He used the Godly or infinity ratio (golden ratio) to draw the lines intercepted at the hypothalamic area, an important area of the greater limbic system. We believe that the soul's seat likely resides at a broader central region of the brain from the above discussion and our previous publications. Indeed, it might be closely related to the autonomic system that carries out consciousness as well as other functions that determine a person is either alive or dead [13–16]. The subsequent discussions are on brainwaves and their correlates with the functions of the brain and human consciousness.

3 The Neuroscience of Human Brainwaves

One of the best ways to visualize brain activity as images are through the study of brainwaves. The brainwaves denote brain function. Therefore, abnormal brainwaves signify an abnormal brain. The methods to study human brainwaves include EEG, electrocorticography (ECoG), depth electrodes, and magnetoencephalography (MEG). EEG and ECoG study the surface or cortical part of the brain, depth electrodes study the deep brain structures, and MEG studies mainly the cortical and subcortical brain regions. EEG allows obtaining brainwaves in a non-invasive manner using electrodes placed on and secured to the scalp. A standard EEG system includes 21 electrodes, an impedance of around 100–5000 Ω, a sensitivity of 5–10 μV/mm, and a low and high-frequency filter at 0.5 and 70 Hz. The low-pass filter typically filters out the galvanic skin response and movement artifact, whereas the high-pass filter removes undesired artifacts, such as electromyographic signals. An additional notch filter with a frequency of 50/60 helps attenuate an artifact caused by the power line frequency. Electrodes are located with reference to the international 10–20 EEG system (10–20% of total length). Electrodes located over the right hemisphere correspond to even numbers, while those on the left hemisphere to odd numbers. Basic kinds of montages are; i, bipolar; ii, common electrode reference (monopolar); iii, average reference; iv, weighted average reference; and v, Laplacian. Advanced quantitative EEG (qEEG) analysis for band power, amplitude, and coherence maps is also possible. In contrast to EEG, ECoG is an invasive way to obtain brainwaves. Open surgery is required to implant the ECoG in the skull. Electrodes can be in the form of either a subdural grid or strip or depth electrodes. Their brainwave signals are regarded as the gold standard because they obtain brainwaves directly from the brain.

Brainwaves can also be studied using a non-invasive gold standard but the expensive method of MEG. MEG records the brainwaves and evoked magnetic fields. To sum up the methodology, a 306-channel Elekta-Neuromag MEG scanner may be used in a magnetically shielded room with a magnetometer and two orthogonal planar gradiometers that occupy each of 102 elements and four head-position indicator (HPI) coils that localize the position of the head. There is a device for digitizing 3D locations of HPI coils at 200 points located on the scalp and fiducial landmarks, including nasion, left pre-auricular point, and right pre-auricular point. MEG data are obtained using the sampling of 1000 Hz. The MaxFilter software (Elekta-Neuromag) is applied as a pre-processing tool to the raw data collected for compensating head movements. The baseline analysis is measured as positivity (P) and negativity (N) in milliseconds (ms). The four analyses commonly used for localization are as follows:

i. motor, during active movement of the index finger that occurs at approximately 50 ms (P50);
ii. somatosensory, following stimulation of median nerve that occurs at approximately 20 ms (N20);
iii. visual, testing each eye separately using a checkerboard that occurs at approximately 75–120 ms (N75-120); and
iv. auditory evoked magnetic fields, testing each ear separately using sounds produced by clicks that occur at approximately 100 ms (P100).

Besides standard MEG waveform analysis, the anatomical magnetic resonance imaging (MRI) data could also be collected and fused with a reconstructed topographic map of the head-model MEG-brainwave data. The modeling analysis is done using brain atlas, brain electrical source analysis (BESA) software, and Matlab-statistical parametric mapping (SPM). Brainwaves fall into epsilon, lower to 0.5 Hz; delta, 0.5–3 Hz; theta, 4–7 Hz; alpha, 8–12 Hz; low-beta, 13–20 Hz; high-beta, 20–30 Hz; low-gamma, 30–60 Hz; high-gamma, 60–100 Hz; hyper-gamma, 100–200 Hz; and lambda, more than 200 Hz. Brainwaves in clinical neuroscience are important in epilepsy management, brain mapping, rehabilitation, brain-computer interfaces, brain death confirmation, and understanding psychiatric disorders [17]. Epilepsy is a brain disorder characterized by aberrant brain activity and seizures when changes in behavior, sensation, and even awareness might occur [18, 19]. Brainwaves suddenly deviating from a normal pattern or rhythm is a characteristic of epilepsy. They include spikes, polyspikes, sharp waves with or without phase reversal, and sharp-and-slow wave complexes. Spikes and sharp waves are interictal epileptiform discharges with spikes that have a pointed peak and duration between 20 and 70 ms, representing the cortical discharges within an area of at least 6 cm^2. Sharp waves also have a pointed peak but last from 70 to 200 ms, representing the discharges of a wider area of the cortex or an area beyond the recording electrode. Sharp-and-slow wave complexes are epileptiform discharges, where sharp waves are associated with slow waves. These can be single or multiple events. Sometimes, before or after the seizures, also known as the pre-ictal or post-ictal phase, slow waves (delta and theta waves) may appear. At the time of ictus, muscle artifacts from the seizures might be noted. Below summarizes the brainwaves the authors could detect in some interesting cases with epilepsy.

3.1 Limbic Epilepsy Associated with Viscero- and Psychosomatic Reaction

An adult man was diagnosed to have a focal brain lesion at the right lower motor cortex. His brain MRI was also suspicious of ipsilateral hippocampal sclerosis. He described his seizures as stomach bloating, hiccup, and abdominal cramp. Also, he had a schizophrenia-like psychotic symptom. The scalp EEG showed only focal abnormality over the lesion. An invasive grid and a depth electrode were implanted into the right lower motor area and the ipsilateral hippocampus. During the surgery, there was an intraaxial lesion, that looked pretty much like normal brain tissues. The monitoring was made in the ward with additional contralateral lower motor strip scalp EEG, and anti-epileptic medications were stopped a few days before brainwave recording. The findings were some epileptic activities over the lesion side, which correspond to his abdominal or visceral seizure semiology. Fascinatingly, on several occasions, some abnormal brainwave activities occurred at the right hippocampal-limbic and contralateral lower motor strip (abdominal viscera) regions. The lesion at the right lower motor strip was removed under awake surgical technique, and the histology revealed a low-grade glioma. The subsequent brainwave monitoring after the surgery showed a much-improved morphology. Both abdominal and psychotic-like seizures were reduced in frequency. Currently, he is still under our regular follow-up. This case highlights the advantages of studying brainwaves in clinical neuroscience (Fig. 2). They can help the clinician localize abnormalities and the involved networks and correlate the semiology with the brainwaves. It implies that the limbic system might contribute to the viscerosomatic and psychosomatic reactions. In 1955, Maclean proposed the limbic system as a visceral brain [20], and our case is an additional proof of that and expands the use of direct cortical and depth electrodes to study the brainwaves.

3.2 Limbic-Vestibular-Insular Epilepsy and the Role of White Matter Tract in Epilepsy

Another case with limbic epilepsy was a 51-year-old lady who presented with short-term memory impairment, heart palpitations associated with hand tremor, and difficulty breathing with laryngeal discomfort. The brain MRI imaging revealed bilateral atrial cysts which lie close to the fornices. She was then subjected to MEG after a normal EEG was taken. The anti-epileptic medications were weaned off a few days before brainwave recordings. The MEG showed grossly abnormal brainwaves with dipole cluster vectors localized near the lesions. The patient was diagnosed to have symptomatic bilateral cystic lesions with a limbic-vestibular-insular seizure network. The semiology was consistent with brainwaves, brain atlas mapping, and diffusion tensor imaging findings (Fig. 3). Chaitanya et al. [21] recently published limbic seizures within the white matter. They demonstrated that local field potential (brainwaves) recorded using the fornix depth electrodes could detect an epileptiform network activity of limbic seizure. Our case echoes a

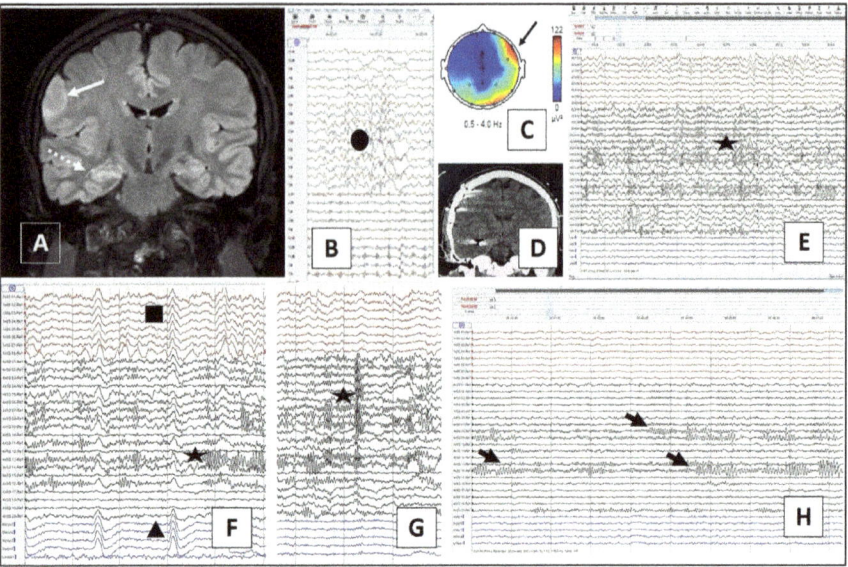

Fig. 2 Limbic epilepsy with viscero- and psychosomatic reaction: **a** the MRI-brain shows the presence of a focal brain lesion at the right visceral motor cortex (white arrow) and suspicious features of ipsilateral hippocampal sclerosis (dashed white arrow); **b** the presence of only focal abnormality over the frontal electrodes (black circle) on the 48 h scalp EEG monitoring; **c** qEEG band power analysis shows the presence of slow waves (delta) over the right frontotemporal area (black arrow); **d** the CT feature of invasive ECoG; **e–g** epileptic activities over the lesion side (black stars) and on several occasions, there were also some abnormal brainwave activities at the right hippocampal-limbic (black rectangle) and contralateral visceral cortex (black triangle); **h** brainwaves improvement after the surgery-note the only presence of *wicket spikes* (black arrowhead brainwaves) after the surgery

similar conclusion that the white matter tract's irritation causes focal seizures related to its network.

3.3 Brain Mapping

MEG can be used for the sensorimotor, visual, and auditory modalities. Figure 4 shows evoked magnetic fields for each modality. For the memory and language mapping, multiple trials of either memory or language tasks are completed in the MEG room, with the average results applied in the mapping (Fig. 5). The MEG mapping for the eloquent or functional brain areas is mostly used in patients with a lesion near those areas requiring neurosurgery. Besides MEG, another commonly used brain mapping technique is direct brain stimulation in awake surgery. Brainwave mappings are commonly used in fundamental and clinical neuroscience [15, 22–24].

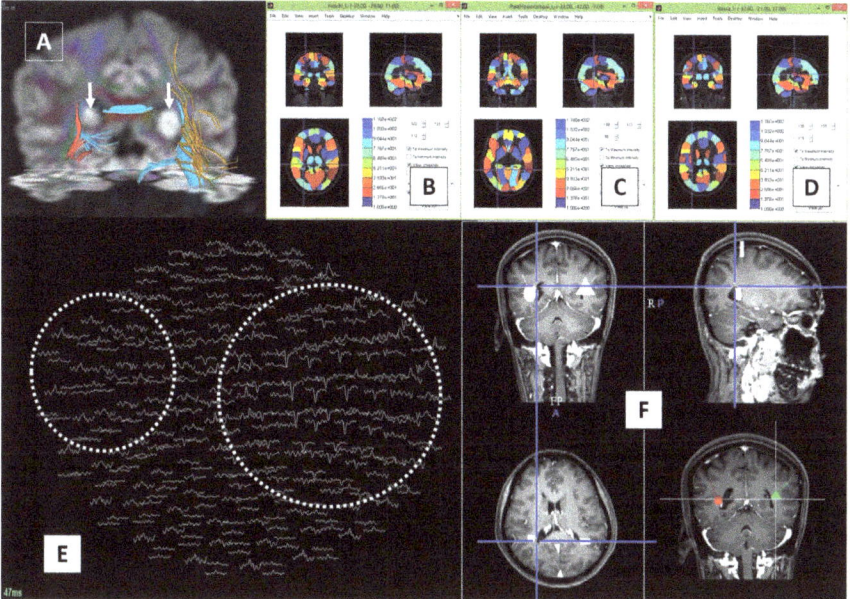

Fig. 3 Limbic-vestibular-insular epilepsy: **a** bilateral atrial brain cysts (white arrows) lie close to the fornices (blue fibers); **b–d** the brain atlas mapping shows each lesion lies adjacent to the Heschl gyrus (vestibular), parahippocampus, and insular; **e** the abnormal sharp waves and sharp-and-slow wave complexes (dashed white circles) on the MEG; **f** the abnormal brainwave vectors (dipole clusters) were localized near the lesions (crossed lines)

3.4   Brainwaves in Psychiatry

The study of human brainwaves in psychiatry is gaining popularity nowadays. Brainwaves are often regarded as visual images of brain activity. Therefore, EEG is commonly employed to study brainwaves. Here we present a few patients with psychiatric disorders monitored with eyes-open resting EEG. The signals were further analyzed using qEEG and found to have some peculiar findings. Figure 6 depicts the loss of hemispheric symmetry in patients with psychiatric disorders. In patients with psychosis, the high-frequency brainwaves (beta and gamma: blue and black) (Fig. 6b) are dominant in one hemisphere (commonly right), and the brainwaves are noted to become nearly symmetrical in treated patients (Fig. 6c). On the other hand, the slow delta (red) brainwaves were dominant in patients suffering from attention disorders (Fig. 6d). Another finding worth mentioning here is that the brainwave power spectrum has double peaks (delta and alpha range) in healthy individuals (Fig. 6a), while in psychosis, there are multiple new peaks, including the gamma peak (gamma band). The double peak has also disappeared for attention deficit disorder, with hemispheric asymmetry and more slow waves dominant in one hemisphere (commonly in the right hemisphere). The findings confirm

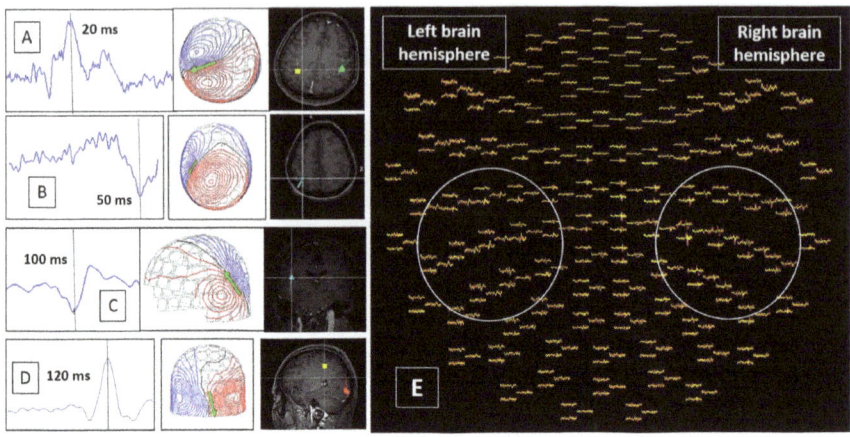

Fig. 4 The MEG brain mapping of the eloquent brain areas: **a** the somatosensory cortical mapping at around N20 (20 ms) with its vector and cortical localization (crossed lines); **b** the cortical motor mapping at 50 ms, which created the magnetic vector over the motor strip (crossed lines); **c** the auditory mapping at 100 ms (P100) with its vector and cortical localization (crossed lines); **d** the visual evoked magnetic fields of N75-120 (75–120 ms) with its vector and cortical visual area (red); **e** the brainwave topography for the somatosensory response (yellow: right-hand sensory stimulation; red: left-hand sensory stimulation) which shows the bilateral hemispheric responses

brainwave asymmetry, gamma-band oscillations, and altered power spectrum in psychiatric patients. Previously, Reilly et al. in 2018 [25], found similar gamma-band oscillations in patients with early psychosis, with some studies reporting brain asymmetry, for example, the frontal lobe asymmetry in patients with attention deficit hyperactivity disorder (ADHD) and a hemispheric asymmetry in patients with psychosis [26, 27].

3.5   Brainwaves of a Blind Person

Brainwaves of a blind person with a large brain tumor compressing the visual cortices were studied. MEG revealed compensatory waves at some regions of the brain (Fig. 7). Supernormal activities were noted, especially in the frontal and temporal lobes. The enhanced brainwaves in these lobes suggest that visual deprivation might trigger reorganization in the brain network to ensure a person's continued survival. Besides enhancing auditory brainwaves, the blind person also seems to have more activation in the amygdala-hippocampal area. It is, thus, easy to conclude that a blind person does have compensatory enhancement not only in hearing but also in memory. Brainwave data may help understanding blindness in a way to create the opportunity to develop brain-computer interfaces and rehabilitative machines for blind people.

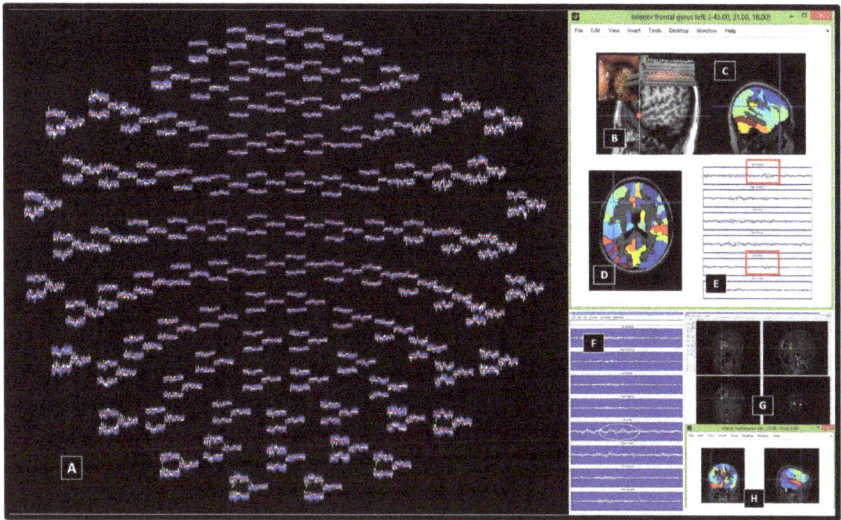

Fig. 5 Memory and language mapping: **a** multiple memory tasks mapping—from easier (red and blue waves) to more complicated (white waves) memory retrieval tasks reveal larger magnetic field (brainwaves) over the right and left occipito-temporo-frontal areas; **b** Intraoperative language mapping using the bipolar neurostimulation which corresponds to the MEG language mapping area (red crossed lines). The dipole cluster was mapped further with the brain atlas, which localized it at the left perisylvian region or left inferior frontal gyrus (**c–d**); **e** the language task (silent word naming) was done in multiple trials in order to produce a language evoked field (red rectangles); **f–h** another patient who had language mapping using the MEG (f, A white circle represents the evoked language field; **g–h** language area was mapped using the MEG and represented onto the inferior frontal gyrus on the brain atlas)

4 Quantum Physics of Brainwaves and Total Brain Field

Quantum physics speaks about tiny objects confined to the scale of atoms and subatomic particles. It is in contrast to classical Newtonian physics, which deals with macroscopic and large-scale objects. The neuroscience of brainwaves regards the brain's fundamental components as particles that form atoms, molecules, cells, tissues, and the brain. The brain tissues consist of neurons capable of producing the action potentials at microscale oscillations. The synchronization of the firing of action potentials from a group of neurons creates synaptic interactions to the neighboring cerebral cortex [28]. At a large-amplitude mesoscale oscillation, the firing pattern can be detected as electromagnetic fields (EMF). Studies on the brain are mostly related to molecules, cells, tissues, the brain, and electromagnetic brainwaves. Two fundamental components of the brain, i.e., particles and atoms, are largely left unexplored by neuroscientists. However, the anatomical brain can also be described in terms of waves [29] according to the principles of quantum physics:

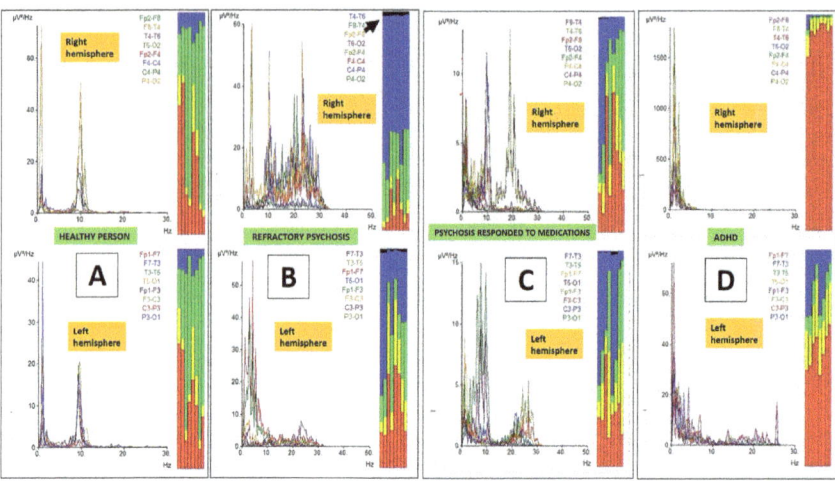

Fig. 6 The psychiatry and the qEEG (notes for the graph and colored spectrum: for the graph—the absolute band power [energy] of the brainwaves in the y-axis and wave frequency in x-axis; for the colored spectrum—it represents the relative band power of brainwaves spectrum, i.e., red for delta, yellow for theta, green for alpha, blue for beta, and black for gamma): **a** a qEEG of a healthy person with (nearly) symmetrical hemispheric brainwaves and a double peak feature on the band power analysis; **b** the qEEG of patient with severe chronic refractory psychosis: A loss of hemispheric symmetry and a dominant presence of higher frequency beta and gamma (black arrowhead) brainwaves; **c** more symmetrical and reappearance of double peak brainwaves were noted in a patient with psychosis who had received adequate medications; **d** the qEEG of ADHD patient with dominant delta and slow theta waves together with marked hemispheric asymmetry. Note that all EEGs of the psychiatric patients presented here have lost the double peak feature (either having single or more than two peaks)

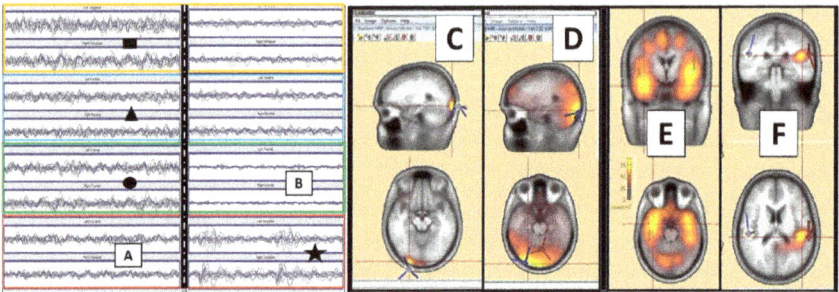

Fig. 7 The blind and normal-healthy person brainwaves: **b–d** the normal-healthy person brainwaves with visual evoked magnetic field in the occipital lobes (black star and crossed lines); **a**, **e**, and **f** the blind person brainwaves which disclose reorganization in the brain networks (black triangle, circle, and rectangle) to ensure the continued survival of a person with marked enhancement in the auditory and limbic (memory, etc.) brainwaves

- *Particle-wave duality for the quantum entity*, an atom or light, can be viewed as either particles or waves; thus, in quantum physics, the brain is viewed at its fundamental level in which only the waves form the essence of the brain. Graphical Abstract (Part A) depicts what the brain consists of, from the lowest quantum level to the highest anatomical level;
- *Coherence and decoherence*, quantum coherence defines waves as having a constant phase difference, while decoherence is a general phenomenon of interference that could counteract the occurrence of coherence;
- *Superposition* is an ensemble of components of a wave. In a wave, there could be many other smaller waves;
- *Quantum tunneling* is referred to as where a quantum particle passes through a barrier;
- *Quantum entanglement* holds multiple subsystems and defines the quantum state of a subsystem in terms of each other's definition, while the individual subsystems may be spatially separated; and
- *Quantum uncertainty or Heisenberg's uncertainty principle* considers the limit one can know about the quantum system, i.e., "The more precise the measurement of position, the more imprecise the measurement of momentum, and vice versa," as quoted by Werner Heisenberg in 1927.

These principles make the physics of a small object not the same as a large object's physics. In summary, the brain can be viewed as; i, classical or Newtonian brain dealt with an anatomical brain and electromagnetic brainwaves; or ii, quantum brain with waves distributed over the whole brain. In the latter view, the anatomical brain comprises waves or energy and therefore corresponds to a quantum field (QF). The quantum field is different from EMF (resulting from neuronal electromagnetic brainwaves described before) and can be defined as a field with discrete energy proportional to the whole wave frequency it represents. Both EMF and quantum field should follow the quantum principles simply because of the wave function. Nonetheless, one might notice some contradictory quantum features for the EMF: i, waves can be localized or determined, which goes against Heisenberg's uncertainty principle (unlocatable, changing, and non-deterministic); and ii, waves have a pilot or directional wave. Therefore, the brain should be perceived as having two types of brainwave or field: electromagnetic brainwaves, which give rise to the brain EMF (electric feature: locatable, focal, and deterministic); and quantum waves, which give rise to the brain-quantum field (light feature: unlocatable, diffuse, and non-deterministic).

From 1925 to 1927, physicists tried to clarify the particle duality (particle or wave) concept in quantum mechanics (physics), the so-called Copenhagen interpretation. Essentially, the interpretation that nothing is real unless it is observed or measured implies that both the EMF and quantum field should be measurable. As we all already know, an EMF is a type of brainwave commonly studied and can be determined and localized. It is likely because of the Bohmian mechanics applied to the EMF and its projected wave feature. The Bohmian mechanics (sometimes called the de Broglie-Bohm theory or pilot-wave theory) is a quantum formulation that describes a genuine position of point particles whose motion is guided by the wave function. Electromagnetic brainwaves are a classical view of the brain regarded as the Newtonian concept that deals with the brain as a large-scale object. Despite having wave properties, they can be studied and determined using Bohmian mechanics. Quantum brainwave or brain-quantum field is a quantum view of the brain referred to as the quantum concept that considers the brain as a small-scale object comprising waves: anatomic-chemical waves or quantum concept for the brain, which should be studied using quantum mechanics. Both perceptions (the electromagnetic physiological and the anatomic-chemical quantum brainwaves whereby the anatomy itself is regarded as energy) give rise to the total brain field (TBF).

The brain QF offers an exciting perspective that all in our universe are waves or vibrations. Waves are interconnected, changing, and diffused. It seems as they have a one-field or oneness concept and are hypothesized by some as closely related to one consciousness [16, 30, 31]. Since all are regarded as waves, then someone should wonder why all these waves do not collide with each other. The likely reason is that they appear to cross the same dimension or space, but in reality, they do not. The medium along which the waves propagate might consist of more than three dimensions (thought of as an infinite dimension). However, we humans live in a three-dimensional (3D) space that presents itself to the eye. Nevertheless, our vision cannot see the waves (energy). The next question should be why we only see a 3D space while the medium has more than three dimensions? It is because our eyes observe a limited projection of reality. As stated in the Copenhagen interpretation, nothing is real unless observed or measured; then, the brain's role is prominent, and a complete comprehension of our universe or brain requires knowledge of quantum physics and neuroscience. Integration of these two fields should be called *Brain-Physics*.

Limited projection of reality might be due to the limited number of waves projected onto our physical universe [32]. Because of limited projection, it may appear as those waves are intersected. In reality, they are not, and they lie in extra dimensions (Graphical Abstract, Part B). Since all are regarded as waves (energy), the extra or infinite dimension may consist of infinite wavy energy. Part of their waves is projected and intersected at certain angles with our newly formed physical universe and form the finite or wavy physical universe (Graphical Abstract, Part C and F). The intersections are hypothesized to have already occurred soon after the big bang when glowing (light-wave) was formed [29]. The consequence of these waves' intersection is the formation of the projected vortices. Here, the word vortex is referred to as projected wave movement. The projected waves or the formed wave vortices cover the electromagnetic or light vortex formed at 0-degree angulation (low dimensional, diffused, and faster); b, the electric vortex formed at 90° angulation (high dimensional, directional, and slower vortex); and c, the matter vortex (in between those degrees) that lies in between light and electric vortices. These matter vortices commonly occur in our nature as having infinite shapes (disk, spiral, bulb, revolute, suprabell, and lentillion) (Graphical Abstract, Part D), and once formed, will follow the physics laws.

In short, the intersection (i.e., infinite waves with finite waves or higher light with lower light) may have created an illusion of physical materials (hologram); they (matter vortex) are actually waves but appear as matter. This argument may explain why atoms, molecules, or light in our physical universe display the duality feature that can exist either in particle (matter is an illusion) or waveform (origin or reality of the matter). Therefore, one may conclude that limited projection of waves or reality is an illusion of physical materials. To observe that reality, we as humans require a brain that has limited consciousness. The concept of one wave field (one field) or one consciousness covers the brain and our cosmos. Therefore, the human brain is thought of as having only partial or limited consciousness. Those limitations cause a collapse of the projected original waves function or energy (the particles' reality/origin) (Graphical Abstract, Part E and F). With this statement, one may notice the opposite-concept is applied to this notion: the reality is not really the reality (referring to matter or photonic light), the reality is actually the waves or energy (referring to the projected waves), and we cannot see with our eyes (brain) that reality (wave or energy). Therefore, the Copenhagen interpretation should be restated as the unreal can only be observed or measured by our brain with limited consciousness (the end of reality). One may also realize that our physical universe is borderless simply because it is formed from the projected infinite waves or part of infinite wavy energy or of infinite dimension. In essence, our illusionary universe is always seen as expanding because of our brain, which can only consider the particle vortices (thus, one will not reach the border).

The first question that should be asked on this argument is there any further relationship between the limited projected waves with our brain? EMF or electromagnetic brainwaves are supposed to have an ideal relationship. It is simply because the EMF is a type of electromagnetic response to our observed world (the physical universe). The projected waves or vortices cause evoked responses inside our brain, theorized here as having only limited consciousness (i.e., limited representation of the true or original waves). Thus, one may say the electrical (electromagnetic) wave brain network is a mirror to the limited projected waves. Given the limitation in both (wave-projection and consciousness), larger evoked electromagnetic fields are naturally observed for seeing, hearing, tasting, and movement of material objects.

As of now, one might have realized the existence of two possible energy fields inside our brain: EMF (via projected waves that have electric features) and quantum field (one-field or oneness or one consciousness or one infinite energy that permeates the whole cosmos, including our brain that has light features). There could be an interaction between the two fields which gives rise to mind-field. Tables 1 and 2 summarize the EMF and quantum field characteristics and brain functions that are related to those fields. In a nutshell, a complete comprehension of our universe or brain does require knowledge in both quantum physics and brain neurosciences. Integration of these two fields should be called Brain-Physics.

Table 1 The general features of the brain electromagnetic field (EMF) and brain quantum field (QF.)

Feature	Electromagnetic field (EMF)	Quantum field (QF)
The origin	The projected waves (physiological waves)	The original waves (anatomic-chemical waves)
Wave pattern	Presence of Pilot/directional wave	Diffused (delocalized) waves
Wave characteristics	High-frequency wave (limited frequency range when alive)	Low-frequency wave (unlimited and cover all frequency range via quantum wave-superposition)
Wavelength	Short wavelength (presence or absence of wave superposition)	Long-wavelength (with absolute presence of wave superposition)
Quantum concept	Deterministic (locatable)	Non-deterministic (unlocatable, changing)
Physics concept	Bohmian mechanics	(Truly) Quantum mechanics
Dimension	High dimension (electric)	Low dimension (light)
System (Energy)	Concentrated organized system (energy) (low entropy)	Widespread diffused random system (energy) (high entropy)
Brain Network	Simple or specific network (few nodes)	Complex, changing or non-specific network (many nodes)
Symmetry	More asymmetry	More symmetry
Evoked response	Large evoked response with few stimuli or trials	Smaller evoked response, need much higher amount of stimuli/trials and quantum wave detector
Neuroplasticity (recovery once treated)	Less likely (because of limited field or focal or less diffused waves)	More likely (because of diffused waves and oneness)
Wholeness/Oneness/One field concept	No (it is a response to the projected/limited field)	Yes (spreading or permeating whole of universe/field)
Related to psychiatry	Less relevant	More relevant (The QF or Mind-field or Quantum-mind is thought of as related to wholeness or reality or one consciousness concept)
Medical concept	Physiological or Electromagnetic Brainwaves (commonly studied waves) Newtonian concept	All are waves: Anatomic-Chemical Quantum Brain Field (still hidden energy; and anatomy is also energy quanta). Quantum concept
Is it related to the seat of human soul? (i.e. why only the brain?)	Yes (because the center part of the brain is the origin for the brainwaves or EMF or all body-functions [seat of soul area/concept])	Yes (cosmos QF can interact meaningfully *with only the brain* which has seat of soul concept resulting in *brain QF or Mind-field or Quantum-mind)*

Table 2 A combination of electromagnetic field (EMF) and quantum field (QF) is responsible for various brain functions.

Brain function	Combination of EMF and QF (2 brain fields/energy or TBF)
Brain function (motor, sensory, vision, sound, touch) and its impairment	Non-cognitive impairment such as stroke affecting motor, sensory, vision, sound, touch EMF is more affected than QF − Measurable − Associated with degree of impairment
Brain function (language, emotion, memory, attention, planning etc.) and its impairment	Cognitive impairment for language, emotion, memory, attention, planning, etc. QF is affected significantly together with EMF − Measurable − Associated with degree of impairment
Brain function and psychosis	Psychotic manifestations such as auditory or visual hallucination, thought insertion, delusions, etc. QF is likely more affected than EMF − Yes or No/presence or absence (not associated with degree of impairment)

4.1 Quantum Features of the Brainwaves

Figure 8a, b provide evidence of quantum entanglement in the brain when the abnormal brainwaves were noted in the abnormal hemisphere and the contralateral brain hemisphere. Figure 8c also shows a similar feature of immediate and bilateral hemispheric response suggesting brain entanglement. Finally, Fig. 8d illustrates the presence of wave coherence and wave-superposition at the time of near-death.

4.2 Death, Meditation, Anesthesia, and Brain Transplant

As mentioned earlier, death is commonly determined by brain death, cardiac death, or circulatory death. The end of the spectrum of cardiac death is brain death. Once the heart stops pumping the blood to the brain, the brainwaves become isoelectric, and brain death is certified. Regarding cardiopulmonary death, if the patient is resuscitated and quickly connected to the ventilator and medications given to make the heart pumping again (re-emergence of cardiopulmonary ups and downs, heartbeat waves, or blood pressure), cardiac death might then be prevented. On the other hand, resuscitating the brainwaves from isoelectric to wavy brainwaves seems like a challenging task, and no man has yet to do so! This argument might suggest that brain death is the actual death of a person. Therefore, one may define death as purely based on brain dead features, e.g., isoelectric brainwaves, fixed and dilated pupils, absence of brainstem reflex, apnea, and comatose, that lie in the seat of the soul brain area. Another point worth noting here is that perhaps the brain is superior to the heart because there is no machine, technology, or medication that can make

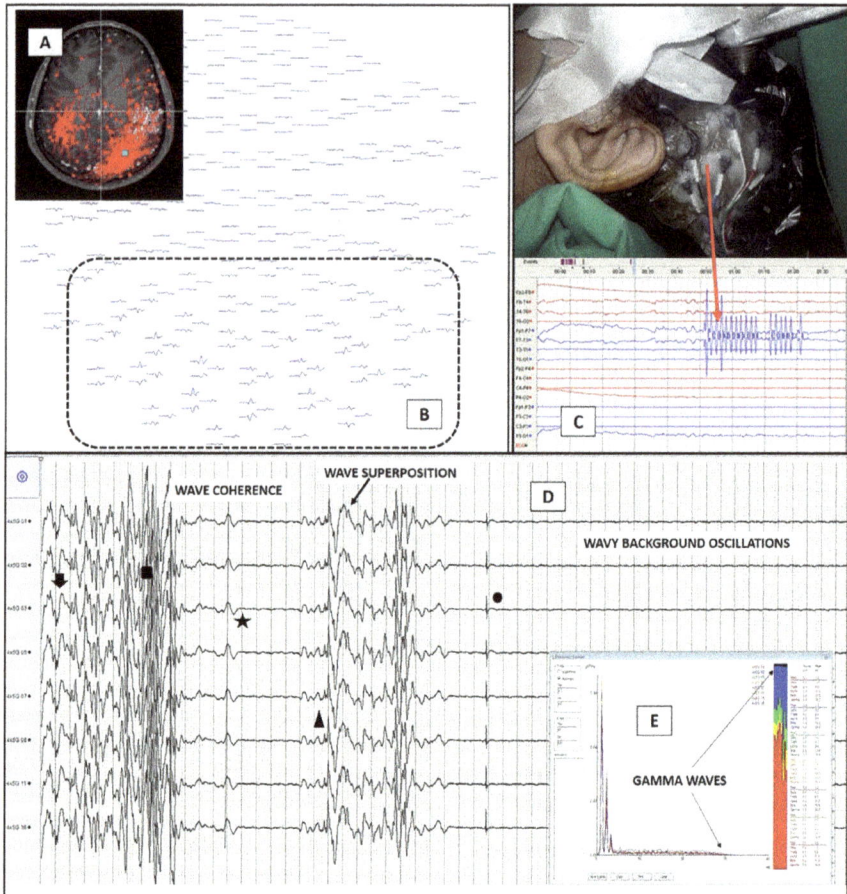

Fig. 8 Quantum features of the brainwaves: **a–c**, immediate and bilateral hemispheric brainwave responses for the abnormal waves near the lesion (a), normal brainwaves for the vision (b), and intraoperative facial cortex stimulation with instantaneous contralateral cortical responses (c). Entanglement may play a role in these three examples; **d** direct cortical brainwaves illustrate the presence of wave coherence, wave-superposition, tiny background oscillations of the flattened-isoelectric waves, and high energy gamma-band/waves (e, an inset for qEEG) at time of near-death. Note the presence of sharp wave (black star), spike (black circle), polyspikes (black rectangle), sharp-and-slow (black arrowhead), and spike-and-slow (black triangle) wave complexes in this illustration (see also the epilepsy session). Abnormal brainwave rhythm or morphology may happen in ischemia (near-death) or any brain-irritation process (e.g., epilepsy)

the isoelectric brainwaves wavy again in patients who have the typical brain dead features mentioned above. Successful in recovering brainwaves when in that particular state, one may be then able to make a dead person become alive again! Brainwaves documented for the dead brain mainly include isoelectric or flattened brainwaves obtained using the scalp EEG. Interestingly, the recorded direct cortical brainwaves (ECoG) of a person transitioning from alive to dead showed a sudden surge in brainwave activities and quickly followed by lower energy waves and the last few spikes before becoming flattened waves at a standard sensitivity-view of 70 uV/cm. However, when viewed at magnified sensitivity of 20 uV/cm, one may still notice the presence of wavy background oscillations (Fig. 8d). Furthermore, brainwaves at that particular time were also noted to become more synchronized or more coherent with the presence of gamma waves (Fig. 8e). Thus, one may conclude that the initial dominance for EMF, reflected in very wavy electrical brainwaves during alive, has now reverted to one consciousness with quantum field dominance (isoelectric on scalp EEG but still the presence of waviness on ECoG). In other words, brainwaves are now getting similar to one-field or one whole cosmos waves (i.e., infinite wavy energy). Notably, the near-death ECoG findings align with the earlier animal study findings made by Borjigin et al. in 2013 and later in 2019 by Zhang *et al.* [33, 34]. Both described a sudden surge of high-frequency and synchronized (highly coherent) brain activities in rats in their near-death state. Therefore, our ECoG description of the near-death state in humans in terms of synchronized and highly coherent activity and the presence of gamma and tiny background oscillations might suggest the brain has shifted to a heightened state of consciousness and support that the brain follows quantum physics and possess quantum field features. Fascinatingly, a shift to a higher state of consciousness is also noted in a person who is doing deep meditation. Studies show brainwaves during meditation can change in two directions, either lower frequency (such as delta and theta bands) or higher frequency bands (such as gamma or hyper-gamma bands) [35–37]. These findings support our ECoG brainwave findings in the near-death state, which correlates with a heightened consciousness level (reverting to one consciousness or one-field or universal quantum field) with the lowest isoelectric brainwaves on scalp EEG (when a person is dead). These findings imply that our current dimension (illusionary physical universe, 1–3 or x, y and z dimension, light-electric-matter vortices universe) is the lowest (energy) when compared to infinite wave-band or infinite dimension (noted that the brain EMF is a mirror to the projected waves that is limited in its frequency bands from the lowest epsilon to lambda bands). In summary, the higher brain frequencies associated with higher consciousness (energy) tend to emerge when brainwaves surpass the beta band oscillations (when a person is still alive) or when an opposite shift happens for brainwaves toward isoelectric EEG, while waviness on ECoG is still present (when a person is dead). Further support on the brain-quantum features comes from the scalp EEG recorded in sedated anesthetized individuals. Scalp EEG findings in anesthesia include sleep spindles (also wicket-like waves) and wave-superposition when the patient was sedated [38]. The wave-superposition is one of the quantum features, and interestingly, the sleep spindles are generated in the thalamo-corticothalamic circuits and thalamic reticular nuclei, i.e., part of the greater limbic system (seat of the soul) that has a role in consciousness [39]. This observation may support the idea that only the brain with a seat of the soul (noted that it is in microgravity, bending, prostration, or sujood posture) can interact with the cosmos-quantum field, leading to the development of the brain-quantum field or mind-field or quantum-mind. Other organs such as the liver, muscles, heart, or bone cannot induce mind-field (brain-quantum field). With this argument, one may also say that the brain should not be regarded as an organ and cannot be transplanted. Brain or head or body transplantation is likely impossible because of the following special features of the brain:

- it has both EMF and the quantum field: the brain should also be perceived as a duality structure, particles and waves (i.e., has two perceptions or considerations);
- the quantum field interactions with the EMF, especially at the seat of the soul (as previously pointed out by Leonardo da Vinci), which control all body functions, including human consciousness;
- the difficulty in separating the interaction effects concerning the brain-quantum field and the cosmos-quantum field;
- the difficulty in preserving the EMF and quantum field component while awaiting to complete the procedure; and
- the presence of autonomic centers in the greater limbic system, which is difficult to preserve their functions (hypothalamus for the parasympathetic and sympathetic system).

5 Conclusion

Brainwaves have ups and downs morphology which correlate with energy features in physics. The energy is required by any system, including the brain, to function normally. Thus, the commonly studied and known electromagnetic brainwaves (EMF) is regarded as brain energy that gives rise to various brain functions. In this chapter, the hypothesis was made that the EMF arises from the projected infinite waves, which could further explain: more significant evoked response noted in EMF; degree of impairment for the EMF; the illusion of the matter; duality feature for the light, particles, atom, molecule, and matter; borderless physical or illusionary universe; our inability to see the waves or energy (collapsing wave function); always the presence of universe expansion; no collision of all waves; and infinite shape in our nature and brain, noted that inside the brain, structures are also infinite in shapes, i.e., the hippocampus (logarithmic spiral), caudate nucleus (Durer spiral), and thalamus (lemniscate). Besides EMF, we also hypothesized the presence but still hidden brain-quantum field for the brain (also arising from the cosmos' energy, i.e., when all is regarded as one consciousness, one field, or one interconnected wave/energy). The quantum field with the oneness concept might explain characteristics of psychiatric disorders, such as delusions and psychosis. Better understanding in this field requires integrating brain sciences and physics to open up a new horizon and innovation (such as quantum wave detectors). In the end, this chapter provided an alternative view and ideas on human brainwaves. Albert Einstein quoted that "the mind that opens to a new idea never returns to its original size;" thus, new ideas are always needed to progress.

Core Messages

- The greater limbic system in microgravity posture is responsible for all body functions and origin for the electromagnetic brainwaves.
- Electromagnetic brainwaves in epilepsy, brain mapping, rehabilitation and brain-computer interface, brain death, and psychiatric disorders.
- The brain has quantum properties, and thus it is not an organ and cannot be transplanted.
- The total brain field combines electromagnetic field (EMF) with quantum field (QF).
- A better comprehension of our universe or brain requires knowledge in quantum physics and neurosciences; integrating these two fields should be called Brain-Physics.

Acknowledgements The authors would like to acknowledge M. Aizam Hakim Zamzuri for assistance in designing figures. The USM Research University Grant supported the work related to ECoG with Grant No: 1001/PPSP/8012261. It was done with ethical approval obtained through the local ethics committee: JEPeM-USM-ethic committee with No. 18010074. The procedures performed in the study were under the ethical standards of the institutional and/or national research committee and with the 1964 Helsinki declaration and its later amendments or comparable ethical standards.

References

1. Hameroff S, Penrose R (2014) Consciousness in the universe: a review of the 'orch or' theory. Phys Life Rev 11(1):39–78
2. Keppler J (2018) The role of the brain in conscious processes: a new way of looking at the neural correlates of consciousness. Front Psychol 9(1346):1–8
3. Idris Z (2014) Searching for the origin through central nervous system: a review and thought which related to microgravity, evolution, big bang theory and universes, soul and brainwaves, greater limbic system and seat of the soul. Malays J Med Sci 21(4):4–11
4. Guldenmund P, Soddu A, Baquero K, Vanhaudenhuyse A, Bruno MA, Gosseries O, Laureys S, Gómez F (2016) Structural brain injury in patients with disorders of consciousness: a voxel-based morphometry study. Brain Inj 30(3):343–352
5. Jang SH, Yeo SS (2016) Injury of the lower ascending reticular activating system in patients with pontine hemorrhage: diffusion tensor imaging study. Medicine (Baltimore) 95(50):e5527
6. Lutkenhoff ES, Chiang J, Tshibanda L, Kamau E, Kirsch M, Pickard JD, Laureys S, Owen AM, Monti MM (2015) Thalamic and extrathalamic mechanisms of consciousness after severe brain injury. Ann Neurol 78(1):68–76
7. Merker B (2007) Consciousness without a cerebral cortex: a challenge for neuroscience and medicine. Behav Brain Sci 30(1):63–81; discussion 81–134

8. Jang SH, Chang CH, Jung YJ, Kim JH, Kwon YH (2019) Relationship between impaired consciousness and injury of ascending reticular activating system in patients with intracerebral hemorrhage. Stroke 50(8):2234–2237

9. Moruzzi G, Magoun HW (1949) Brain stem reticular formation and activation of the EEG. Electroencephalogr Clin Neurophysiol 1(4):455–473

10. Nieuwenhuys R, Veening JG, van Domburg P (1988) Core and paracores; some new chemoarchitectural entities in the mammalian neuraxis. Acta Morphol Neerl Scand 26(2–3):131–163

11. Nieuwenhuys R, Voogd J, van Huijzen C (2008) Greater limbic system. In: Nieuwenhuys R, Voogd J, van Huijzen C (eds) The human central nervous system: a synopsis and atlas. Springer, Berlin, Germany, pp 917–946

12. Dolan B (2007) Soul searching: a brief history of the mind/body debate in the neurosciences. Neurosurg Focus 23(1):E2

13. Idris Z, Mustapha M, Ghani R, Idris B, Kandasamy R, Abdullah JM (2014) Principles, anatomical origin and applications of brainwaves: a review, our experience and hypothesis related to microgravity and the question on soul. J Biomed Sci Eng 7(8):435–445

14. Idris Z, Kandasamy R, Reza F, Abdullah JM (2014) Neural oscillation, network, eloquent cortex and epileptogenic zone revealed by magnetoencephalography and awake craniotomy. Asian J Neurosurg 9(3):144–152

15. Idris Z, Reza F, Abdullah JM (2017) Human brain anatomy: prospective, microgravity, hemispheric brain specialisation and death of a person. In: Sisu AM (ed) Human anatomy—reviews and medical advances. IntechOpen, London, pp 61–88

16. Idris Z (2019) An infinite-light and infinite-frequency in cosmology and neurosciences. Open J Phil 9(2):236–251

17. Richardson MP (2012) Large scale brain models of epilepsy: dynamics meets connectomics. J Neurol Neurosurg Psychiatry 83(12):1238–1248

18. Lopes da Silva F, Blanes W, Kalitzin SN, Parra J, Suffczynski P, Velis DN (2003) Epilepsies as dynamical diseases of brain systems: basic models of the transition between normal and epileptic activity. Epilepsia 44(Suppl 12):72–83

19. Maillard L, Ramantani G (2018) Epilepsy surgery for polymicrogyria: a challenge to be undertaken. Epileptic Disord 20(5):319–338

20. Maclean PD (1955) The limbic system ("visceral brain") and emotional behavior. AMA Arch Neurol Psychiatry 73(2):130–134

21. Chaitanya G, Toth E, Ilyas A, Pati S (2020) Sensing limbic seizures within the fornical white matter: a technical report. Clin Neurophysiol 131(6):1320–1322

22. Amunts K, Zilles K (2015) Architectonic mapping of the human brain beyond brodmann. Neuron 88(6):1086–1107

23. Idris Z, Wan Hassan WMN, Mustapha M, Idris B, Ghani ARI, Abdullah JM (2013) Functional MRI, diffusion tensor imaging, magnetic source imaging and intraoperative neuromonitoring guided brain tumor resection in awake and under general anaesthesia. In: Lichtor T (ed) Clinical management and evolving novel therapeutic strategies for patients with brain tumors. IntechOpen, London, pp 17–54

24. Lim LH, Idris Z, Reza F, Wan Hassan WMN, Mukmin LA, Abdullah JM (2018) Language mapping in awake surgery: report of two cases with review of language networks. Asian J Neurosurg 13(2):507–513

25. Reilly TJ, Nottage JF, Studerus E, Rutigliano G, Micheli AI, Fusar-Poli P, McGuire P (2018) Gamma band oscillations in the early phase of psychosis: A systematic review. Neurosci Biobehav Rev 90:381–399

26. Keune PM, Wiedemann E, Schneidt A, Schönenberg M (2015) Frontal brain asymmetry in adult attention-deficit/hyperactivity disorder (ADHD): extending the motivational dysfunction hypothesis. Clin Neurophysiol 126(4):711–720

27. Begić D, Popović-Knapić V, Grubišin J, Kosanović-Rajačić B, Filipčić I, Telarović I, Jakovljević M (2011) Quantitative electroencephalography in schizophrenia and depression. Psychiatr Danub 23(4):355–362
28. Kalitzin S, Petkov G, Suffczynski P, Grigorovsky V, Bardakjian BL, Lopes da Silva F, Carlen PL (2019) Epilepsy as a manifestation of a multistate network of oscillatory systems. Neurobiol Dis 130:104488
29. Idris Z, Zakaria Z, Yee AS, Fitzrol DN, Ghani ARI, Abdullah JM, Wan Hassan WMN, Hassan MH, Manaf AA, Chong Heng RO (2021) Quantum and electromagnetic fields in our universe and brain: a new perspective to comprehend brain function. Brain Sci 11(5):558
30. Idris Z (2020) Quantum physics perspective on electromagnetic and quantum fields inside the brain. Malays J Med Sci 27(1):1–5
31. Meijer DKF, Raggett S (2014) Quantum physics in consciousness studies. The quantum mind extended. In: Raggett S (eds) Quantum mind UK. Quantum Mind UK, UK, pp 146–157
32. Venis PA (2017) Waves and dimensions. http://www.infinity-theory.com/en/science/Main_pages/Waves_and_Dimensions. Accessed 14th November 2019
33. Borjigin J, Lee U, Liu T, Pal D, Huff S, Klarr D, Sloboda J, Hernandez J, Wang MM, Mashour GA (2013) Surge of neurophysiological coherence and connectivity in the dying brain. Proc Natl Acad Sci USA 110(35):14432–14437
34. Zhang Y, Li Z, Zhang J, Zhao Z, Zhang H, Vreugdenhil M, Lu C (2019) Near-death high-frequency hyper-synchronization in the rat hippocampus. Front Neurosci 13(800):1–9
35. Hinterberger T, Schmidt S, Kamei T, Walach H (2014) Decreased electrophysiological activity represents the conscious state of emptiness in meditation. Front Psychol 5(99):1–14
36. Vialatte FB, Bakardjian H, Prasad R, Cichocki A (2009) EEG paroxysmal gamma waves during bhramari pranayama: a yoga breathing technique. Conscious Cogn 18(4):977–988
37. Wong WP, Camfield DA, Woods W, Sarris J, Pipingas A (2015) Spectral power and functional connectivity changes during mindfulness meditation with eyes open: a magnetoencephalography (MEG) study in long-term meditators. Int J Psychophysiol 98(1):95–111
38. Hagihira S (2017) Brain mechanisms during course of anesthesia: what we know from EEG changes during induction and recovery. Front Syst Neurosci 11(39):1–5
39. Steriade M, McCormick DA, Sejnowski TJ (1993) Thalamocortical oscillations in the sleeping and aroused brain. Science 262(5134):679–685

Zamzuri Idris had graduated from the University of Wales College of Medicine (UWCM) Cardiff UK (1994), then pursued his postgraduate career in Neurosurgery locally at Universiti Sains Malaysia (2005). In 2007, he completed his fellowship program in Belgium for Neuroendoscopy, Functional and Epilepsy Neurosurgery. He was further trained in Neurosurgery in Toronto, Canada, Switzerland, Germany, Italy, the United States, Japan, and the Netherlands. He is currently the Head of Neuroscience Department in USM, Head for the Malaysia Neurosurgical Programme, Chief for the Malaysian Qualification Agency (MQA) for Neurosurgery Programme, and the Executive Committee for the Neurosurgical Association of Malaysia (NAM). He is also professionally associated with the Belgian Neurological Surgery Society and European Society for Stereotactic and Functional Neurosurgery

Jafri Malin Abdullah studied in Sekolah Menengah Sains Bukit Mertajan and SMS Kelantan, Pengkalan Chepa and Leederville Technical College, Perth, Western Australia and the University of Western Australia. After receiving an M.D. from the School of Medical Sciences, Universiti Sains Malaysia (USM) in 1986, he was awarded the Diplomate Certification of Specialization in Neurosurgery from the University of Ghent, Belgium in 1994 and a Ph.D. (Magna Cum Laude) in 1995. He is a Fellow of the Academy of Science of Malaysia, American College of Surgeons, the Royal College of Surgeons of Edinburgh, International College of Surgeons (USA), and the Royal Society of Medicine (UK). He is currently a Grade A Professor of Neurosciences and the Chairman of the Brain Behaviour Cluster of the USM

Modern Psychiatry: Confluence of Mind, Science, and Society

5

Veeraraghavan J. Iyer

"For a behavior to be termed a psychiatric disorder it has to be regularly accompanied by subjective distress and/or some generalized impairment in social effectiveness of functioning."

Robert Spitzer

Summary

Psychiatry has grown by leaps and bounds from Hippocrates' times through Freud's times and now in the age of progressing neuroscience. This chapter gives a brief overview of the history and progress through the ages. The chapter also provides a fresh view of psychiatric disorders and recent directions in psychiatry. More importantly, it highlights the aftermath of psychiatric disorders and the large impact on social learning via the hypothalamic–pituitary–adrenal axis. This chapter addresses the impact of stigma on mental health. Psychiatry, thus, has been presented as a confluence of mind, science, and society.

V. J. Iyer (✉)
Institute of Living/Hartford Hospital, Hartford, CT, USA

Integrated Science Association (ISA), Universal Scientific Education
and Research Network (USERN), Springfield, MA, USA

Baystate Med Center, CAPTU, 759 Chestnut St, Springfield, MA 01199, USA

© The Author(s), under exclusive license to Springer Nature Switzerland AG 2022
N. Rezaei (ed.), *Multidisciplinarity and Interdisciplinarity in Health*,
Integrated Science 6, https://doi.org/10.1007/978-3-030-96814-4_5

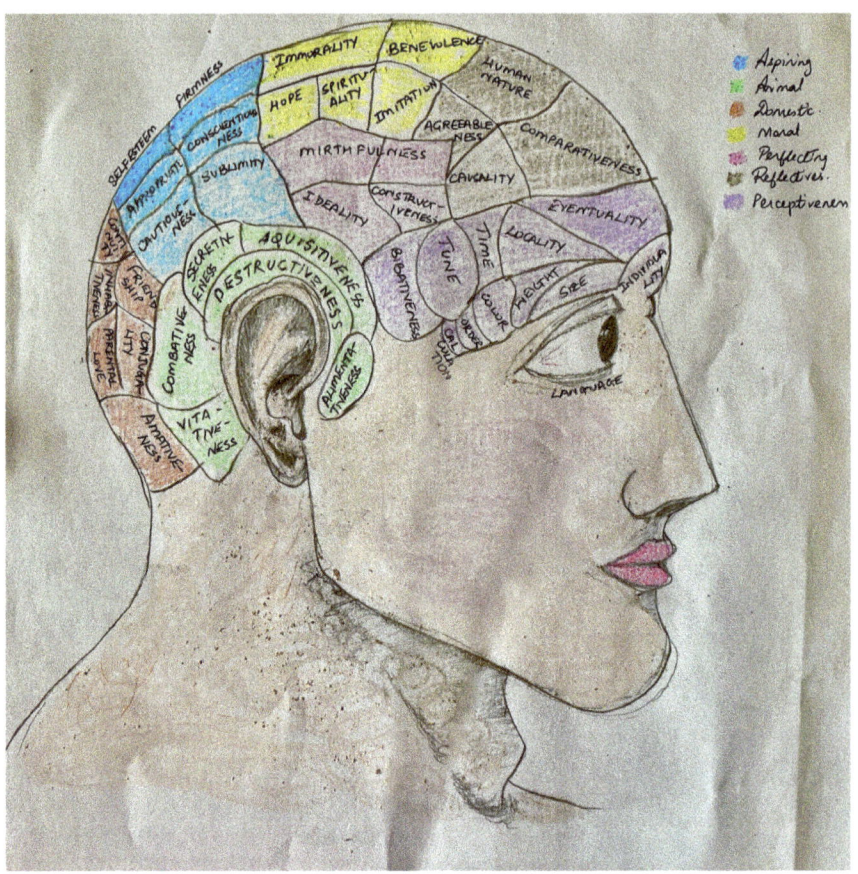

Franz Gall's illustration of the different functional areas of the brain

The code of this chapter is *01101001 01100001 01110100 01010011 01100111 01101101*.

Keywords

Confluence · HPA axis · Mind · Psychiatry · Science · Society · Stigma

1   Introduction

Human thinking, behavior, and brain function have been inspiring awe since time immemorial. Our brain is an organ weighing a mere three pounds but is hailed unanimously as the seat of thinking, reasoning, and intelligence. For this, it has ever been a topic of provocative curiosity, inspiring centuries of scientific and analytic scrutiny.

As early as the fifth century AD, Hippocrates stated how madness resulted from moistness in the mind in his astute observational brilliance. This followed Rene Descartes, who put forth the mind–body dualism principle in the seventeenth century. This dualism suggested that the brain controls perceptions, passions, memory, and motor functions, whereas the mind accounts for maintaining conscious awareness through the brain's machinery mediated by the pineal gland [1].

In the nineteenth Century, Charles Darwin proposed evolution theories, including the revolutionary idea of animal behaviors serving as models for understanding human behaviors. These theories of evolution laid the foundations for ethology and experimental psychology. Before Darwin, psychological and brain studies were limited to radical ideas proposed from brain dissections and linking anatomical findings to behaviors [1].

In 1800, a Viennese physician and neuroanatomist, Franz Joseph Gall, suggested a physiognomy-based theory. This theory put forth two radical ideas. First, that the physical brain functions as the organ of the mind and that all mental functions emanate from the brain. Second, that particular regions of the brain perform specific functions (graphical abstract). His methods lacked scientific validation and investigation.

Gall's hypotheses found validation in the mid-nineteenth century by neurologists Paul Pierre Broca, Carl Wernicke, Hughlings Jackson, and their scientific demonstrations. The works of Wernicke and Jules Dejerine were particularly informative in linking behaviors to interconnected brain regions. This school of thought propagated the knowledge of biology in the pursuit of understanding the ways of the brain [1].

Around the same time, another school of thought emerged with two neurologists, Sigmund Freud and Pierre Janet. This school of thought was the psychoanalytic one, which promoted that a person's unique life experiences manifest as psychiatric symptoms which could be remedied by *conscious processing and corrective experiences*. Freud and Janet were both under the supervision of Jean-Martin Charcot, who thought of hysteria as a result of traumatic events and physiological causes. Janet added that traumatic memories often manifested as physical symptoms, while in Freud's opinion, traumatic memories were mental events restricted to exist in the unconscious thought. Janet used hypnosis, and Freud promoted interview techniques to bring traumatic repressed memories to consciousness.

2 The Semantics of Psychiatric Diagnoses

Diagnosis is a core component of medicine that has multi-dimensional implications. The correct diagnosis helps guide treatment optimally, ascertain prognosis, and explain causal relationships between disorders and symptoms. For instance, appendicitis is an infection of the appendix from different causes (Cause). It presents clinically as nausea, vomiting, abdominal pain, and diarrhea (symptoms). Some complications of appendicitis are perforation and peritonitis (effect). Thus, medical diagnoses can be clearly described in terms of cause, symptoms, and effects.

Psychiatric diagnoses have been treated differently [2]. For instance, depression is a decreased mood state affecting functioning in terms of sleep, appetite, general motivation, and energy (effect). It is also described to cause a cluster of symptoms such as poor sleep, appetite, poor energy, and poor motivation (cause and symptoms). This categorization of psychiatric diagnoses is treated both as cause and effect. This is a philosophical quandary because causation is distinct from effect in that the former is the reason and, the latter, the consequence. With time and the advent of investigative modalities, this distinctive relationship is continually evaluated by the scientific community. This elucidation will become clearer as neural underpinnings and valid research data become robust.

This chapter explores literature in neurobiology, epigenetics, and other psychiatry aspects to better understand such semantics.

3 Concept of Epigenetics

Epigenetics is the influence of environmental factors on genes [3]. For instance, a child with phenylketonuria (PKU) will have neuropsychiatric manifestations like intellectual disability if the necessary dietary correction (low-phenylalanine diet) is not prescribed [4]. Thus, an environmental factor (dietary supplement) can influence the genetic component to manifest a certain trait (intellectual disability). This concept that the environment has a significant bearing on the genes, and hence, a person's traits are particularly important in psychiatry. We will delve into more disorder-specific instances and research in this regard in subsequent sections.

4 Bio-Psycho-Social Profile

When we consider a person's mental constitution, we consider the person's physical/medical health, genetic profile (biological vulnerability), social predicament, inherent temperament, and coping styles. This form of conceptualizing a person's psychological makeup is called a bio-psycho-social formulation [5]. This recognizes and validates that psychiatric disorders are not standalone conditions but have contributions from the above sectors of life in their making.

For instance, consider an 18-year-old man with a strong family history of schizophrenia who presents to the hospital with auditory hallucinations. After thoroughly investigating and ruling out possible medical explanations such as brain lesions or delirious/infectious processes, a preliminary diagnosis of schizophrenia is considered. He is admitted to the inpatient psychiatric unit and stabilized on a regimen of psychotropic medications. He is discharged to an uncertain social predicament (for instance, a temporary shelter). It is also determined that he has a stubborn inclination for using illicit substances (maladaptive coping), and his motivation to follow up with recommendations are pre-contemplative at best. He has no family involved in his care, no sustainable source of income because of the disabling disease process, legal troubles because he has resorted to burglary or panhandling to make ends meet. To add to his misery, he had no means to transport himself to outpatient appointments. Altogether, even if the correct diagnosis is identified and optimal treatment is prescribed, the likelihood for success is meager when his predicament is conceptualized from a bio-psycho-social perspective.

The bio-psycho-social profile nudges the good clinician to consider other dimensions of care, not just to remedy his/her symptoms in the interim. After all, if there were a way to improve his financial and social distress, he would presumably not be involved in a crime life. He would be able to transport himself to appointments and be self-sufficient. This paradigm shift has resulted in sociological and administrative research, which has led to improved social services delivery and better access to care, such as the exemplary assertive community treatment teams (ACT) [6].

5 Concepts for Child Mental Health

Eighteenth-century Empiricists believed in the concept of *tabula rasa* (blank slate) [7]. It means human brains are akin to a blank slate's receptiveness, shaped over time by subjective sensory experiences. This evolved into the idea that inherent qualities contribute to the perception of the experiences.

The most influential theories in child psychiatry came from Jean Piaget, Lev Vygotsky, and John Bowlby. Jean Piaget proposed that children's thinking is different from adults', and it evolves biologically as the child matures with age. Vygotsky proposed the importance of social interactions in learning. John Bowlby wrote prolifically about parent–child attachments and the impact of deprivation and institutionalization on children. This had reverberating influences on child care policies worldwide. Ericson's proposed psychosocial development and Kohlberg's moral development were remarkable additions to the field, which helped understand and implement specific psychotherapy forms using these stage-specific guidelines (Table 1).

One important concept is temperament. Temperament is understood as a complex of innate characteristics of a person that are not dependent upon social learning or parenting (Table 2) [8]. Depending upon the temperamental characteristics, a child could be loosely categorized as easy, slow to warm up, shy, or difficult/challenging. Temperament often has implications for parental responses. For instance, an easy-going infant might receive pleasant responses from parents, reinforcing the child to respond pleasantly. A parent might respond with anger and frustration to a "challenging child," which might trickle down into a child's frustrating experience. This also helps gauge and integrate appropriate response framing and supports for the parent as part of treatment planning.

Table 1 Comparison of developmental stages: Piaget's cognitive, Freud's psychosexual, Ericson's psychosocial, and Kohlberg's moral stages

Age	Piaget cognitive	Freud psychosexual	Erickson psychosocial	Kohlberg moral
6 months	**Sensorimotor** – Experience the world through senses and interaction – Stranger anxiety	**Oral** Sucking and rooting	Trust versus mistrust	**Preconventional** **1st stage (pre-moral):** Bad behavior is punished and good not **2nd stage**
12 months (1 yr)		**Anal** Toilet training		**(Hedonism):** Good is pleasant and desirable, and bad behavior is undesirable
18 months (1.5 yrs)				
24 months (2 yrs)	**Preoperational** – Symbolism and representational thinking – Pretend play – Egocentric thinking (unable to take other's perspective)	**Phallic** – Genital exploration – Preference for the opposite-sex parent	**Shame and self-doubt versus independence**	
36 months (3 yrs)				
48 months (4 yrs)				
60 months (5 yrs)			**Initiative versus guilt**	**Conventional:** **1st stage** – Good behavior wins approval from parents, peers, and teachers – Praise = moral **2nd stage** – Moral = conforming to rules and norms – Obeys rules and authority figures
72 months (6 yrs)				
84 months (7 yrs)		**Latency** – Resolution of parental preference – Plays with the same gender	**Industry versus inferiority**	
8 yrs	**Concrete operational** – Concrete binary concepts – Abstraction – Basic math operations – Begin internalizing thoughts and reflections			
9 yrs				
10 yrs				
11 yrs				
12 yrs	**Formal operations** – Full abilities for abstraction – Non-binary concepts	**Genital** – Sexual curiosity and preference formation – Pursues romantic relationships	**Identity and role diffusion versus role confusion**	
13 yrs				
14 yrs				
15 yrs				
Adolescent				**Post conventional** – Judges own moral behavior – Principles of conscience
Young adult			**Intimation versus isolation**	
Adulthood			**Generativity versus stagnation**	
Mature			**Integrity versus despair**	

Table 2 Temperamental characteristics [7]

Activity level	The level of physical activity, motion, restlessness or fidgety behavior that a child demonstrates in daily activities (and which also may affect sleep)
Rhythmicity or regularity	The presence or absence of a regular pattern for basic physical functions such as appetite, sleep and bowel habits
Approach and withdrawal	The way a child initially responds to a new stimulus (rapid and bold or slow and hesitant), whether it be people, situations, places, foods, changes in routines or other transitions
Adaptability	The degree of ease or difficulty with which a child adjusts to change or a new situation, and how well the youngster can modify his reaction
Intensity	The energy level with which a child responds to a situation, whether positive or negative
Mood	The mood, positive or negative, or degree of pleasantness or unfriendliness in a child's words and behaviors
Attention span	The ability to concentrate or stay with a task, with or without distraction
Distractibility	The ease with which a child can be distracted from a task by environmental (usually visual or auditory) stimuli
Sensory threshold	The amount of stimulation required for a child to respond. Some children respond to the slightest stimulation, and others require intense amounts

6   The Equation of Conceptualization

G X E
  Gene-by-environment interaction (Epigenetics)

6.1   Gene

Genes are considered the fundamental units of heredity, both physical and functional. What constitutes these units are deoxyribonucleotides (DNAs) that code for proteins, which, in turn, affect various functional aspects of the body. Proteins are required for energy formation, brain responses, and learning through second messenger mechanisms. Proteins and second messengers are also necessary for programmed cell death (apoptosis), which mediates neural plasticity (a mechanism mediating the pruning of those neural circuits which are not employed actively and frequently by the brain; thus, siphoning necessary growth energy into strengthening circuits that are employed frequently). Thus, fundamentally, genes mediate learning and adaptation.

6.2 Environment

The environment in psychiatry is considered to be any external factor that modulates behavior [9]. The commonest illustration could be family accommodation in school avoidance. Consider the case of Sam, who adamantly refuses to go to school because of social anxiety, awkwardness, and fear of humiliation. To further his cause, Sam frequently complained of unrelenting stomach pain. As much as his parents suspect this to be a functional pain to avoid school, if they give in to his demand, he learns that persisting with his made-up symptom will somehow tire his parents into giving in; this external factor of familial accommodation results in reinforcement of his school avoidance behavior.

6.3 Modulators

Factors that could affect genetics could be spontaneous mutations or other genetic disorders which could manifest as neuropsychiatric disorders. Physical trauma could reshape the course of pregnancy. For instance, an abdominal injury to a mother may induce a head injury to the fetus or reduce blood flow to the brain, causing injury to vital brain tissues at a tenderly formative age. This could also manifest as an intellectual and developmental disability. Familial trauma plays a significant role in this. The detailed mechanism of genetic trauma will be discussed in the subsequent section.

The philosopher Aristotle once commented that *poverty is the parent of crime*. Poverty is a direct result of poor education and unemployment. It has far-reaching repercussions by affecting the basic needs in Maslow's hierarchy [10]. It affects housing, shelter, food, and nutrition, resulting in insurmountable psychological distress and social uncertainty. Other factors that contribute to this unpalatable distress are perceived discrimination based on socioeconomic status, which further affects future employment prospects. The social rift invariably leads to turmoil. As history has taught us, this culminates into theft, crime, and other unlawful activities. Inadvertently, this leads to experiential trauma and learned helplessness. Thus, poverty has the scope to induce psychiatric comorbidities.

Through the formative years, neglect, trauma, and nutrition weigh significantly in psychological endurance. Poor nutrition affects not only the virility of the mind but in doing so, it robs a person of viable psychological skills. A person striving for a meal a day could not afford the opportunity to analyze and process his stressful predicament.

Recent researches have given a biological basis for the aftermath of enduring experiences. There is evidence that trauma results in histone deacetylation and methylation. This is a process by which certain genes are silenced, resulting in an inability to produce proteins necessary for brain mechanisms affecting neural plasticity, learning, and coping.

7 Adverse Childhood Experiences Study (ACES)

The ACES study by Dube et al. [11] reported that children exposed to early life trauma, neglect, and other forms of toxic stress have a 4–12-fold risk of alcohol use, substance use, depression, suicides, and sexually transmitted diseases [11]. The ACES study, to some length, validates at least two important perspectives to consider, the neural plasticity model and the maladaptive learning model.

7.1 The Neural Plasticity Model

As development progresses, brain growth becomes more selective. This is an important evolutionary context. Only the most frequently employed circuitry or neural mechanisms are reinforced. From an evolutionary perspective, this affords at least a few advantages. First, these reinforced mechanisms become easily accessible by such preferential development. Second, because such mechanisms are rehearsed, sharpened, and learned effortlessly, they become second nature over time. Third, over generations, these reinforced mechanisms are further strengthened by focusing the genetic energy in amplifying proteins required to conduct these functions leading to silencing of energy devoted to less-used neural mechanisms.

This epigenetic model introduces the concept of genetic trauma. The idea is that trauma induces genetic changes that are carried and preserved in the traumatized individual's offspring. Yehuda and Lehrner (2018) review seminal works in the realm and discuss animal models that have succeeded in explaining genetic trauma [12].

7.2 Maladaptive Learning

When faced with inescapable stress, physiologically, there are specific brain and body changes to prepare for the proverbial apocalypse. The brain siphons psychic energy mediated by the fear complex (amygdala) and the thalamic gateway into the pituitary. This hasty short-circuiting and vigorous overriding *escape mechanism* is learned by the memory complex (hippocampus). This mechanism is refined to near perfection by linking the interoceptive cues to fear-inducing phenomena to reduce the insular system's subsequent response times. Finally, as if in this tumultuous dance sequel's climax, the posterior pituitary releases precursors to catecholamines (steroids). These precursors prime and stimulate adrenal glands sitting majestically above the kidneys to form catecholamines such as adrenaline and norepinephrine that bring to fruition the hasty end. The result is an increase in heart rate and blood perfusion in vital organs and skeletal muscles; therefore, fleeing is a viable option; breaking down glycogen to make glucose and energy available. It, in turn, would increase brain activity in the breathing center so the lungs could support the body in the intense response to the stressor (fight, flight, flee, and freeze). This survival sequel starts with stress perception and ends with the catecholamines' activation called the hypothalamic-pituitary-adrenal axis (HPA). The unwavering focus is self-preservation. The brain's thinking or analytical part is abdicated from its executive throne to maintain the prime focus on escaping conflict. The abilities to learn and analyze are relinquished temporarily. Thus, in a state of panic, learning is merely a forsaken luxury.

Adverse childhood experiences result in an adept short-circuiting mechanism such that the analytic part of the brain is left hopelessly impotent in mastering trust, love, resilience, self-regulation, and self-adequacy. The results are stunted maturity and proneness to highly stimulating and habit-forming tendencies such as substance use and other impulsive avenues to seek remedy. The following sections will find how different psychiatric disorders fit into the above-described models. This will shed more light on the confluence of mind, science, and society, which is the primary premise of this chapter.

8 Disorders

8.1 Anxiety

When faced with a test, we all have some form of reaction in anticipation of the test. This could be general restlessness, increased sweating, urge and frequency to use the restroom, palpitation, and shortness of breath, to name some. The test in itself might not be the subject of anxiety, but the uncertainty of the outcome or validity of performance is. Thus, the anxiety-inducing event may be vague. It is deemed a disorder if it permeates into an individual's day-to-day functioning; for instance, if it affects a person's eating or sleep habits, ability to function at work or school, or relationships such that they, knowingly or unknowingly, relinquish everyday responsibilities. In most instances, there might be a tendency to avoid stressful situations to avoid being overwhelmed.

Fear, on the contrary, is a physiological response to a threatening situation. Thus, the most distinct separation between fear and anxiety is the definition of the threat. Fear responds to a specific threat and anxiety to a vague, non-specific threat [13].

There are two schools of thought:

(i)   The first thinks of the amygdala as where anxiety originates and relays experiences to the prefrontal cortex; and
(ii) The second proposes the two-system model where behavioral and physiological changes are mediated by the subcortical circuitry, particularly the amygdala, and the conscious feeling states mediated by the cortical circuitry.

More recent neuroimaging studies have shown that the bed nucleus stria terminalis (BNST) bed nucleus is the primary origin of signals when faced with an unknown threat (anxiety). Perceptive signals (visual, tactile, olfactory, and auditory from their respective centers in the brain) are input into the BNST, which then communicates with the amygdala. The central amygdala feeds into the thalamic gateway, which relays these communications to the HPA axis leading, ultimately, to catecholamines production, as mentioned above.

8.1.1 Role of Breathing

The brainstem begins regulating breathing in utero at 11–13 weeks of gestational age. While breathing is not a conscious process, the breathing process requires the diaphragm, which can be activated using accessory muscles such as intercostals, neck muscles, and pharyngeal and facial muscles. Thus, the unconscious act of breathing can be mediated using voluntary muscles of the neck, face, and chest.

The pacemakers of respiration, ventral and dorsal respiratory groups of neurons, are located in the medulla. The dorsal group of neurons responds to hypoxia and lung inflation. The ventral group coordinates respiratory motor output. The rostral part of the ventral group near the nucleus ambiguous, called the pre-Botzinger complex (Pre-BotC), is critical for producing normal ventilatory rhythm. Serotonergic neurons in this Pre-BotC may play at least two roles in maintaining the ventilatory rhythm. Firstly, they possibly increase the firing rate in response to an increase in pH (decrease in carbon dioxide, CO_2). Secondly, they are closely associated with large arteries in the ventral medulla, monitoring CO_2 changes. Inputs descending from the limbic system mimic a fight or flight response anticipating a higher need for oxygen. This 'presumed' hypoxia may activate the ventilatory neurons to cause hyperventilation and choking sensation, as seen in anxiety and a panic attack.

The reverse can also be therapeutic. Using the muscles of the neck, chest, and face to maintain breathing could regulate this perception of hypoxia in the brain. These regulatory signals are sent to the HPA axis, which may help in mitigating a panic state.

8.2 Bipolar Disorder

According to the Diagnostic and Statistical Manual (DSM) 5, bipolar disorders are referred to as distinct mood states consisting of signs and symptoms of depression persisting for two or more weeks and mania consisting of elevated mood state for at least a week (between 3 days to a week of manic symptoms with reduced severity are called hypomania) [14].

Overall, research reveals bipolar disorders as communicative pathology. Studies using magnetic resonance imaging-diffuse tensor imaging (MRI-DTI), which studies the brain on a functional level, have provided evidence that the gray matter that connects different parts of the brain functions less efficiently in bipolar disorders [15]. There is also evidence that, as in schizophrenia, there may be abnormal development of brain areas in bipolar disorders.

8.2.1 Brain Tissue Migration

Brain tissues undergo complex changes during embryonic stages to form the highly complicated structure of nerves. The brain forms from the outer part of the embryo (the ectoderm). The process starts when a portion of the ectoderm folds onto itself to form the neural crest and, eventually, the neural tube. This neural tube is a tube that runs from the head to the end of what would be the spinal cord. The head end of the neural tube forms specialized vesicles that eventually divide and proliferate further to form the forebrain, the hindbrain, and the hindbrain. The forebrain forms the cerebral cortex, the thalamus, and the hypothalamus.

Another process subsists on a similar microcellular level. Near what would become the brain's ventricles holds neural progenitor cells (ancestor cells that differentiate to form specialized neural cells). This zone is called the ventricular zone (VZ). Through complex processes, the cells move in an inside-out fashion. They differentiate and move outwardly, forming six layers (I–VI), with layer VI being the deepest part of the dorsolateral prefrontal cortex (DLPFC). The specific protein that instructs migratory cells where to stop and align themselves is called Reelin. Reelin abnormalities result in premature migration arrest, reduced volume of specific regions of the cerebral cortex, poor maturation, and abnormal functioning [16].

Neural tissue studies in bipolar disorders have evidenced a reduction in layers III through VI in the DlPFC and hippocampal tissues.

8.2.2 Other Bipolar Related Findings

Other studies have observed reduced synaptic receptors that help synaptic vesicle migration from the cell body into the synapse resulting in a proposed reduction in presynaptic neurotransmitter release. Another consistent finding in bipolar disorder seems to be reduced fractional anisotropy (FA). FA is a direct measure of white matter integrity and is studied using MRI-DTI. Reduced FA indicates a decreased connectivity of white matter between brain regions.

Moreover, genetic association studies have associated bipolar disorders with single nucleotide polymorphisms (SNPs) in the genes encoding calcium channel proteins that drive the second messenger pathways. Second messengers are specialized pathways involving chemicals such as Inositol Phosphate 3, cyclic Adenosine Mono-phosphate (cAMP), and other molecules. These second messenger pathways bring about genetic changes and help with producing proteins with specific purposes. Some examples of neurotransmitter productions using second messengers are corticotropin-releasing hormone (CRH) signaling, phospholipase C (PLC) signaling, and glutamate receptor signaling, among many others [17].

8.3 Depression

As mentioned earlier, depression is a mood state characterized by a persistent reduction in general morale [18]. Depression could be severe enough to result in anhedonia, hopelessness, helplessness, and suicidal thoughts or actions. Severe depression could also result in perceptual problems such as auditory hallucinations and delusional thinking. Persistence is defined by an arbitrary period of two weeks of symptoms.

Studies in depression have observed a reduction of brain volumes, especially in the hippocampus [19]. There are also reductions in anterior cingulate gyrus volumes. Studies have reported increased blood flow to the ventromedial prefrontal cortex (VMPFC) and reduction to DLPFC. VMPFC mediates pain, aggression, and other drives, whereas DLPFC mediates executive functioning, task-specific attention, and higher thought functioning. The findings of functional MRI (fMRI) translate to reduced drives and decreased cognitive abilities. Other fMRI studies have reported reduced connectivity between the amygdala and the anterior cingulate cortex (ACC). ACC is hypothesized to have two functions: the anterior portion is thought to mediate motivation, and the posterior is thought to be a part of the cognitive network in collaboration with DLPFC. Reduced functioning of ACC affects both motivation and cognition.

Kindling is a phenomenon that reduces the threshold for stress with symptom recurrence. Kindling is closely related to depression. Chronic stress can often lead to overactivation of the HPA axis, as discussed previously. Increased corticosteroid activation, as a result, may directly cause toxic damage to the hippocampus. Kindling, thus, mediates HPA overactivity-induced brain damage with subsequent episodes of depression. Brain-derived neurotrophic factor (BDNF) is considered a major protective factor in this toxic damage over time. BDNF is produced as a result of second messenger systems. Serotonin reuptake inhibition and norepinephrine reuptake inhibition in the synaptic systems result in increased cAMP and increased activity of protein kinase A, which increases BDNF production in the brain via the second messenger system.

8.4 Schizophrenia-Spectrum Disorders

Schizophrenia is a complex psychiatric condition characterized by hallucinations, delusions, disorganized thinking, and disorganized behaviors. Eugene Bleuler first coined the term schizophrenia. *Schizo* means 'split' and *phrene* means 'mind.' This was considered an apt description of the split thinking states of the afflicted.

Schizophrenia was divided into multiple subtypes: disorganized, catatonic, paranoid, residual, and undifferentiated in previous DSM versions. This was consolidated into one diagnosis with universally described symptom profiles as positive, negative, and cognitive dysfunction.

Positive symptoms are considered the presence of hallucinations, delusions, and disorganized and bizarre thinking and behavior. Negative symptoms are regarded as 4 A's: Apathy, Avolition, Alogia, and Asociality. Multiple phenomenological papers and discourses have resulted in the more acceptable 4A's for negative symptoms. Particular importance is given to avolition in differentiating this from anhedonia. Cross-sectional studies have observed the distinction that people with schizophrenia have preserved interests and do not typically exhibit anhedonia [20].

Schizophrenia research is heterogeneous [21]. Multiple genetic linkages, copy number variants, SNPs, and other genetic mutations have been correlated without replicability. Likewise, fMRI studies implicate *hypofrontality* or poor prefrontal cortex functioning, reduced gray matter volumes, decreased connectivity, and decreased blood flow to vital brain structures.

One hypothesis for such non-uniform findings is the concept of epigenetics [22]. DNA methylation and histone deacetylation are two largely investigated mechanisms in most psychiatric disorders. There is new evidence of increased histone deacetylase (HDAC 1) in the post-mortem brains of people with schizophrenia. Other findings include oxytocin receptor changes and mitochondrial ribonucleic acid (MiRNA) methylation.

Receptor theories have gained large-scale recognition in schizophrenia. Dopamine theory has been pre-eminent since the 1960s since the discovery of Chlorpromazine. Subsequently, 5-Hydroxytryptamine (Serotonin) theory arose from similarities in presentation between a model of psychosis induced by lysergic acid diethylamide (LSD), a 5HT agonist, and schizophrenia [23]. The similarities with phencyclidine (PCP)-induced psychosis gave rise to the N-methyl-D-aspartate (NMDA) theory. Also, gamma-aminobutyric acid (GABA) receptors are being investigated actively as new therapeutic targets for schizophrenia.

9 Society and Its Role in Psychiatry

Some studies report stigma regarding mental health. Evidence observing the correlation between stigma, suboptimal acceptance of services, and poor functioning is overwhelming. This introduces the concept of societal accommodation.

As noted previously, accommodation is a process of enabling and reinforcing maladaptive mechanisms. This often worsens behavior many folds by perpetuating patterns. We have reviewed, so far, the neurobiological basis of psychiatric conditions. After examining scientific evidence, it is clear that without interventions, there is little scope for remediation.

Applying two perspectives of the ACEs as described above to all the described disorders, we can surmise that suboptimal treatment could result in not just dysfunction of an individual but could impact generations by genetic means. Maladaptive learning due to cognitive dysfunction and activation of the HPA axis is not particularly helpful in prosperous living either. Thus, the implications of unaddressed psychiatric conditions are far more distressing and complex enough to derange multiple levels of functioning. Societal intervention by spreading awareness is the call of the day.

One factor highly under-looked remains caregiver burnout [24]. While the implications of caregiver burnout are widely studied in the literature, very little is offered to remedy said burnout, and little is known about large-scale interventions to improve caregiver burnout.

9.1   Shame and Self-accommodation

The shame and embarrassment of being *different* from others are all-consuming. Being the butt of all societal and familial jokes compounds it many times over. This would exacerbate anxiety. In turn, the added anxiety fans the HPA flames vigorously, ultimately worsening the overall prognosis. The cycle of shame and self-accommodation, thus, is a self-perpetuating one.

9.2   Treatment

Treatment of psychiatric conditions generally involves medically addressing the brain changes, socially empowering individuals to break old patterns, learn better adaptive skills, and educate and spread awareness among families and societies. The medical and psychotherapeutic descriptions are beyond this chapter's scope but are certainly worth the reader's attention.

9.3   System of Dispensing Care

As pointed out previously, poverty is of grave concern on multiple levels. Having a system of care in a place that governs the equitable distribution of psychiatric services regardless of a person's ability to afford would improve the quality of care and, therefore, functioning at a societal level [25].

A few generalizations could be derived regarding health care management trends. Firstly, increased access to health insurance promotes the use of health services. There is also consensus that health management trends derive direction primarily from profit-driven principles. Secondly, the disparities in access to insurance coverage are well-established. Lower socioeconomic groups and families with lower levels of education are largely affected. This makes information dissemination about the availability of access all the more important. Thirdly, a compelling argument can be made that preventive primary care visits and school-based screening systems are the best strategies for early screening, referral, and thereby efficient means of access to mental health. Perhaps, an additional measure for consideration through such screening processes could be to identify families from lower socioeconomic levels and provide necessary information to obtain optimal health care access.

10 Conclusion

"Imagine a society that subjects people to conditions that make them terribly unhappy then gives them the drugs to take away their unhappiness. Instead of removing the conditions that make people depressed modern society gives them antidepressant drugs. In effect antidepressants are a means of modifying an individual's internal state in such a way as to enable him to tolerate social conditions that one would otherwise find intolerable."

A Critic

Modern psychiatry, as a science, has truly emerged into a confluence of mind, science, and society. As reviewed in detail, the scientific evidence overwhelmingly supports this confluence theory where each domain feeds off the other. It is also evident that psychiatric problems incapacitate learning by hijacking the HPA axis and leave an evolutionary disadvantage by genetic expression of maladaptive patterns and trauma. Every psychiatric disorder fits into this theoretical framework one way or another. New research will further elucidate the fundamental nature of psychiatric conditions to help the philosophical questions of cause and effect.

Core Messages

- Psychiatry is no longer an abstract theoretical discipline of medicine.
- Research supports the neurochemical and biological basis of mental health conditions and treatments.
- There is an intimate interplay between science, society (environment), genes, and the mind.
- Social learning is affected by overstimulation of the HPA axis resulting in poor outcomes and maladaptive behaviors.
- Stigma and poor awareness regarding psychiatry would only add to society's detriment.

References

1. Kandel E, Schwartz JH, Jessell T, Siegelbaum SA, Hudspeth AJ Overall perspective. In: Principles of neural science, fifth. McGraw-Hill Medical
2. Maung HH (2016) To what do psychiatric diagnoses refer? A two-dimensional semantic analysis of diagnostic terms. Stud Hist Philos Biol Biomed Sci 55:1–10. https://doi.org/10.1016/j.shpsc.2015.10.001
3. Egger G, Liang G, Aparicio A, Jones PA (2004) Epigenetics in human disease and prospects for epigenetic therapy. Nature 429:457–463
4. Kandel E, Schwartz JH, Jessell T, Siegelbaum SA, Hudspeth AJ Genes and behavior. In: Principles of neural science, 5th edn. McGraw-Hill Medical

5. Rey JM, Assumpção Jr FB, Bernad CA, Çetin Çuhadaroğlu F, Evans B, Fung D, Harper G, Loidreau L, Ono Y, Pūras D, Remschmidt H, Robertson B, Rusakoskaya OA, Schleimer K, History SK, Professor of Psychiatry F, Assumpção Jr Professor FB, Çetin Çuhadaroğlu Professor of Child F, Psychiatry A (2015) History of child psychiatry. In: IACAPAP textbook of child and adolescent mental health
6. Schöttle D, Schimmelmann BG, Ruppelt F, Bussopulos A, Frieling M, Nika E, Nawara LA, Golks D, Kerstan A, Lange M, Schödlbauer M, Daubmann A, Wegscheider K, Rohenkohl A, Sarikaya G, Sengutta M, Luedecke D, Wittmann L, Ohm G, Meigel-Schleiff C, Gallinat J, Wiedemann K, Bock T, Karow A, Lambert M (2018) Effectiveness of integrated care including therapeutic assertive community treatment in severe schizophrenia-spectrum and bipolar I disorders: four-year follow-up of the ACCESS II study. PLoS ONE 13:e0192929. https://doi.org/10.1371/journal.pone.0192929
7. How to Understand Your Child's Temperament—HealthyChildren.org. https://www. healthychildren.org/English/ages-stages/gradeschool/Pages/How-to-Understand-Your-Childs-Temperament.aspx. Accessed 4 July 2020
8. What is a gene?—Genetics Home Reference—NIH. https://ghr.nlm.nih.gov/primer/basics/gene. Accessed 22 July 2020
9. Compton MT, Shim RS (2015) Social determinants of mental health
10. Wahba MA, Bridwell LG (1973) Maslow reconsidered: a review of research on the need hierarchy theory. Acad Manag Proc 1973:514–520. https://doi.org/10.5465/ambpp.1973.4981593
11. Dube SR, Anda RF, Felitti VJ, Edwards VJ, Croft JB (2002) Adverse childhood experiences and personal alcohol abuse as an adult. Addict Behav 27(5):713–725
12. Yehuda R, Lehrner A, Peters Bronx JJ (2018) Intergenerational transmission of trauma effects: putative role of epigenetic mechanisms The origin of studies of intergenerational trauma effects
13. LeDoux JE, Pine DS (2016) Using neuroscience to help understand fear and anxiety: a two-system framework. Am J Psychiatry 173:1083–1093
14. What are bipolar disorders? https://www.psychiatry.org/patients-families/bipolar-disorders/what-are-bipolar-disorders. Accessed 25 July 2020
15. Mahon K, Russo M, Perez-Rodriguez MM, Burdick KE (2015) The neurobiology of bipolar disorder: neuroimaging and genetics update. Focus (Madison) 13:3–11. https://doi.org/10.1176/appi.focus.130111
16. Stiles J, Jernigan TL The basics of brain development. https://doi.org/10.1007/s11065-010-9148-4
17. Martinowich K, Schloesser RJ, Manji HK (2009) Bipolar disorder: from genes to behavior pathways. J Clin Invest 119:726–736
18. What is depression? https://www.psychiatry.org/patients-families/depression/what-is-depression. Accessed 25 July 2020
19. Maletic V, Robinson M, Oakes T, Iyengar S, Ball SG, Russell J Neurobiology of depression: an integrated view of key findings. https://doi.org/10.1111/j.1742-1241.2007.01602.x
20. Foussias G Remington G negative symptoms in schizophrenia: avolition and occam's razor. https://doi.org/10.1093/schbul/sbn094
21. Aetiology of schizophrenia: the neurobiology of schizophrenia|eNetMD. http://www.enetmd.com/content/aetiology-schizophrenia-neurobiology-schizophrenia. Accessed 26 July 2020
22. Cariaga-Martinez A, Saiz-Ruiz J, Alelú-Paz R (2016) From linkage studies to epigenetics: what we know and what we need to know in the neurobiology of schizophrenia. Front Neurosci 10:202

23. Sato M, Numachi Y, Hamamura T (1992) Relapse of paranoid psychotic state in methamphetamine model of schizophrenia. Schizophr Bull 18:115–122
24. Iyer VJ (2014) Effect of family dynamics on mental health: a case report. Bhuvana 3:400705. https://doi.org/10.1001/archpsyc
25. Iyer VJ (2019) An analysis of services for children with special health care needs in connecticut. NADD Bull 22

Veeraraghavan J. Iyer is a psychiatrist in Massachussetts, USA. He completed his general training from Rutgers NJMS and his child psychiatry training from the Institute of Living, Hartford Hospital. He holds special interests in psychiatric care systems, autism spectrum disorders, motivational interviewing, and psychopharmacology. To this length, he has published some papers. When he is not practicing psychiatry, he could be seen honing his photography skills, being a doting husband, and playing with his infant daughter.

Schizophrenia: A Disorder of Timing and Sensorimotor Integration During Decision-Making

6

Juliana Bittencourt, Bruna Velasques, Silmar Teixeira, Danielle Aprígio, Mariana Gongora, Mauricio Cagy, Thayaná Fernandes, Pedro Ribeiro, and Victor Marinho

"Space and time, in fact, are sides of the same coin. The time is relative and cannot be measured in exactly the same way everywhere, thus, the time passes differently for each individual shows a curvature in the fabric of space-time."

Albert Einstein

Summary

Sensorimotor integration and decision-making essentially contribute to time perception by integrating different sources of sensory stimuli and using them for performance in cognition and adaptation to the environment. Schizophrenia is a

J. Bittencourt · B. Velasques · D. Aprígio · M. Gongora · M. Cagy · P. Ribeiro
Brain Mapping and Sensory Motor Integration, Federal University of Rio de Janeiro, Rio de Janeiro—RJ, Brazil

J. Bittencourt
Veiga de Almeida University, Rio de Janeiro, Brazil

S. Teixeira · T. Fernandes · V. Marinho (✉)
Federal University of Parnaíba Delta, Parnaíba 64202-020, Brazil
e-mail: silmarteixeira@ufpi.edu.br

S. Teixeira · V. Marinho
The Northeast Biotechnology Network (RENORBIO), Federal University of Piauí, Teresina, Brazil

V. Marinho
Integrated Science Association (ISA), Universal Scientific Education and Research Network (USERN), Parnaíba-PI, Brazil

Neuro-Innovation Technology and Brain Mapping Laboratory, Laboratory, Federal University of Delta Do Parnaíba, Parnaíba 64202-020, Brazil neurodevelopmental disorder where brain connectivity, and therefore, the integration of cognitive, motor and sensory information is disturbed. The present chapter discusses changes in sensorimotor integration and decision-making, such as in time interval judgment in schizophrenic patients. The results highlight that schizophrenic patients cannot build appropriate and accurate representations of the environment to perform tasks. Some defend schizophrenia as a disorder associated with abnormal neural inputs causing difficulty for these individuals to deal with decision-making feedback. Moreover, the hypothesis that these failures may be due to a dysconnectivity among different brain areas during the sensorimotor integration, and decision-making became attractive.

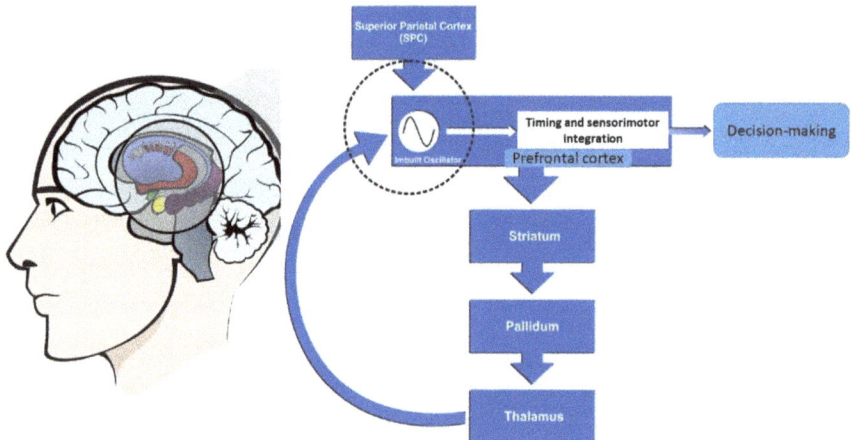

Connections between the superior parietal cortex and prefrontal cortex during timing and sensorimotor integration. The proposed inbuilt modular clock in the prefrontal cortex is responsible for the decision-making task. The superior parietal cortex does not function properly when there is reduced interference in the role of striatal neural oscillations in the representation of time intervals in the brain during decision-making.

The code of this chapter is *01110011 01,101,100 01,100,011 01,001,111 01,110,100 01,101,001 01,101,100 01,101,001 01,101,110 01,100,001 01,101,111.*

Keywords

Decision-making · Executive functions · Neurobiological domains · Schizophrenia · Time perception

1   Introduction

The subjectivity in the time perception and decision-making during time judgment passes like a train that runs into the future at a constant rhythm. The process of decision-making, therefore, essentially ensures survival. It is, however, dependent on the environment and will be dynamically affected by environmental changes. This explains why individuals are skilled in temporal information processing [1, 2]. The executive functions and cognition are part of a fundamental adaptation process to the environment among animal species [2, 3]. It reflects the ability to make choices and weigh the expected reward and potential positive or negative consequences [4]. In particular, decision-making and sensorimotor integration occur through neural circuit synchronization (i.e., frontoparietal network and striatum-tactical and striatum-orbitofrontal loops) [5, 6].

Thus, calculation and the ability to predict cost/benefit are inherent to human beings. However, neuropsychiatric disorders, especially schizophrenia, promote changes in sensorimotor integration and timing for decision-making [2, 7, 8]. Erroneous decision-making, as well as the impaired capacity to judge timing, is behavioral characteristics in people with schizophrenia, along with changes in neurotransmission in specific cortical areas, such as in the frontal cortex, that underpin planning, sensorimotor integration, and decision-making (Fig. 1) [9].

The central nervous system (CNS) needs to be intact and unchanged (i.e., no lesions and/or deficiencies in the ineffective dopamine recruitment in the mesolimbic pathway) to estimate the time intervals with accuracy during executive functions [10, 11]. Schizophrenic patients present a disintegration in thought processes and emotional response to decision-making, auditory hallucinations, delusions, and paranoia [12–14]. Also, they do not accurately perform cognitive functions that require time judgment, even daily activities involving the motor response time, like preparing food or waiting for a green light [1, 15, 16].

The CNS organizes sensory processing in different orders, from milliseconds (ms) for motor coordination and seconds (s) to minutes (min) for time perception to hours for circadian processes (Fig. 2) [2, 17]. Those processes allow for the acquisition and refinement of many behaviors (i.e., motor control, learning, and decision-making) [18, 19]. Therefore, to develop perception and execution, it is necessary to create an internal model that is ready to reorganize CNS contents when facing and integrating sensory stimuli from the environment, individual, and task [20]. Thus, cognitive processes are critically integrated into executive functions, and patterning of this integration is repeated in behavioral features that characterize a healthy brain [21]. Studies have tried to seek relations among time perception, executive functions, and psychiatric disorders [1, 20, 22, 23]. Particularly, in patients with schizophrenia who cannot construct appropriate and accurate environment representations to execute a task [24], features related to the judgment of time intervals and sensorimotor integration are changed (Table 1) [25]. The current evidence is, however, not conclusive. Therefore, this chapter will discuss the association between time perception and sensorimotor integration during decision-

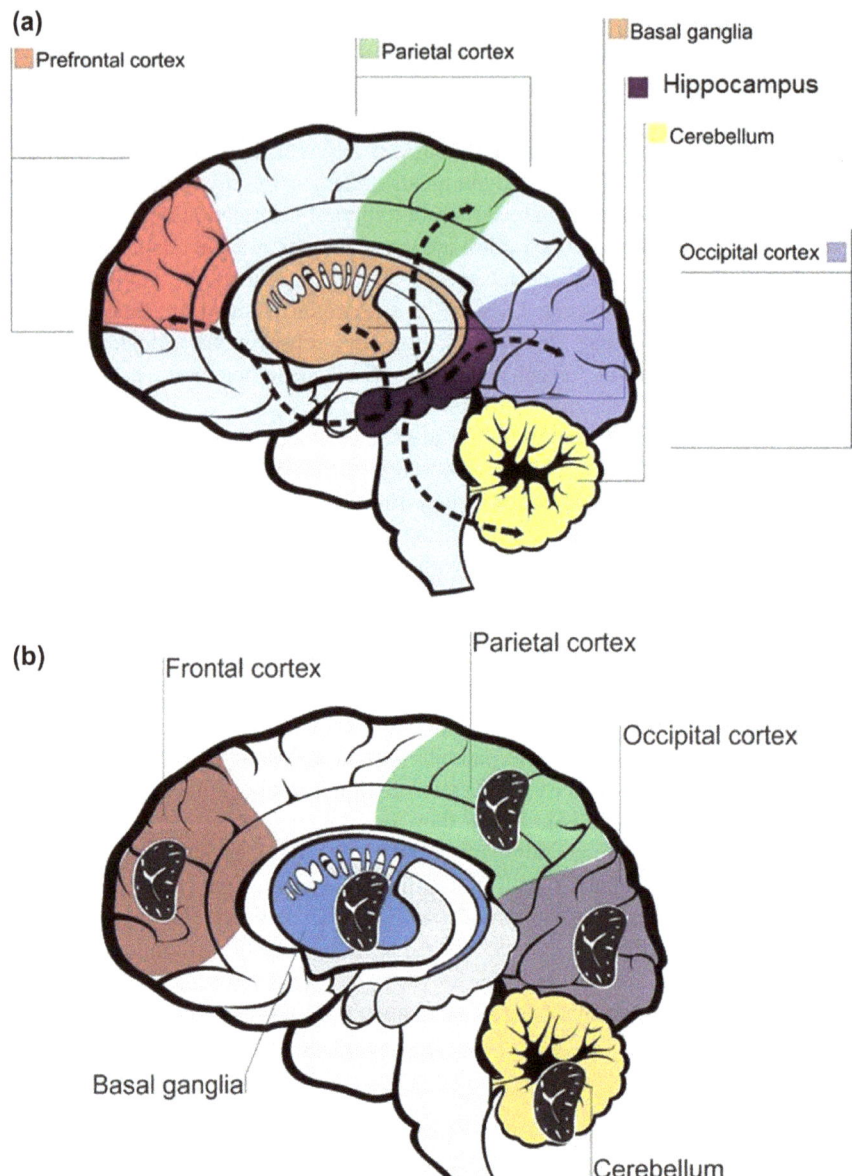

Fig. 1 Neurotransmission in specific cortical areas that underpin planning, sensorimotor integration, and decision-making. Decision-making and the brain: **a** The neurochemical and neural network differences in the function of cortical and subcortical areas predispose the individuals to different behavioral phenotypes; this can be associated with inadequate recruitment, reduction, or increase of the neurochemistry levels resulting from differentiated protein expression; **b** brain structures act as a dedicated system in the timing and decision-making. Lesions in these brain areas cause timing deficits in decision-making, perceptual capacity, and sensorimotor integration

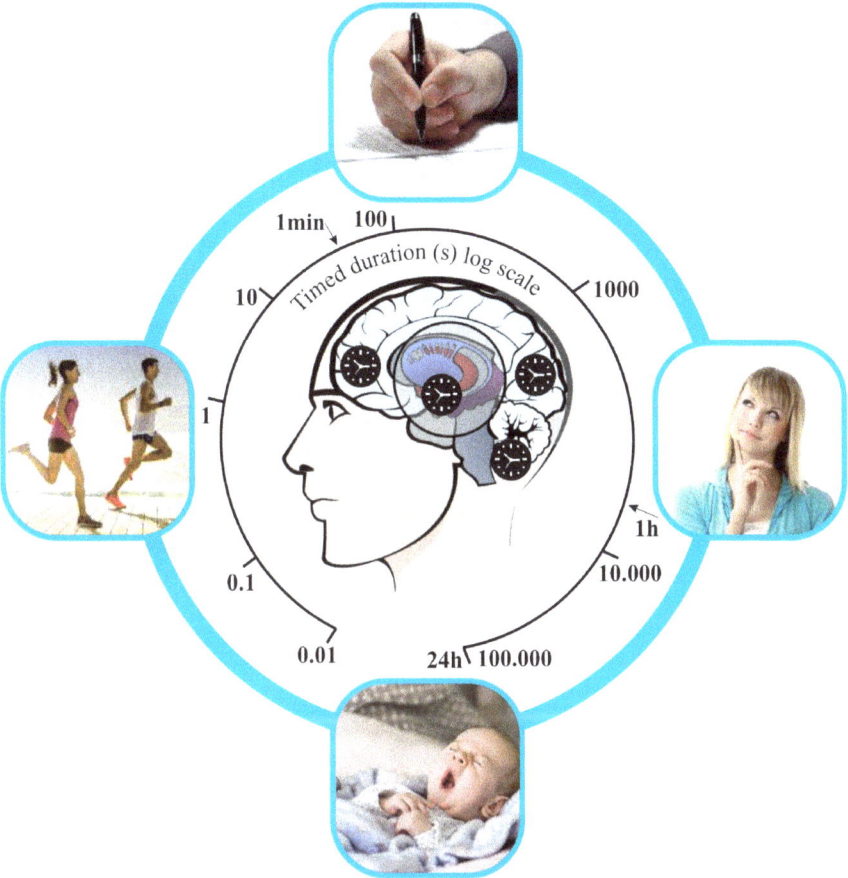

Fig. 2 The CNS processes sensory stimuli and develops a key role in perception, an important brain executive function. Neural inputs in cortical areas form neurotransmission that underpins time perception in a vast range, from milliseconds to seconds to longer daily rhythms (motor control, decision-making, learning, and wake cycle, respectively)

making in schizophrenic patients. Our hypothesis is that the difficulty to process sensory information from the environment will lead to failures in the sensorimotor integration process and the production of motor acts in these patients. Accordingly, a neurobiological perspective is required to explore how sensorimotor integration occurs in schizophrenic patients (e.g., perception, memory, attention, and motor control) when carrying out cognitive tasks. Finally, the chapter presents a state-of-the-art sensorimotor integration during decision-making and judgment of time intervals in schizophrenia.

Table 1 Summary of studies investigating the impact of decision-making and/or time perception in patients with schizophrenia

Study	Study group: N	Task stimulus	Standard duration	Study findings
Papageorgiou et al. (2013)	Schizophrenia: 60 Control: 60	Discrimination task (using EPQ questionnaire) Auditory	50 ms, 1 s, 2 s	Patients with schizophrenia showed incorrect responses in the discrimination of time intervals, except in the standard conditions
Davalos et al. (2011)	Schizophrenia: 16 Control: 18	Discrimination tasks Auditory	200 ms, 70 ms, 100 ms, 300 ms, 500 ms	Two groups showed no difference in temporal processing of "easy" and "difficult" tasks According to fMRI, people with schizophrenia had reduced activation in SMA, insula/opercular cortex, and DLPFC
Carrol et al. (2008)	Schizophrenia: 23 Control: 22	Time estimation task Auditory, visual	300 ms, 600 ms, 350 ms, 400 ms, 450 ms, 500 ms, 550 ms	The schizophrenia group exhibited overestimation regarding both auditory and visual duration
Penney et al. (2005)	HrSz: 17 HrAff: 16 high-risk Control: 34	Time estimation task Auditory, visual	3 s, 3.67 s, 3.78 s, 4.24 s, 4.76 s, 5.34 s, 6 s	HrSz group showed more difference between auditory and visual in timing task than in the control or HrAff groups The HrSz participants underestimated the time
Davalos et al. (2003)	Schizophrenia: 15 Control: 16	Time estimation task Auditory	500 ms, 1 s, 3 s	Patients with schizophrenia overestimated the time

EPQ, Eysenck Personality Questionnaire; fMRI, functional magnetic resonance imaging; SMA, supplementary motor area; DLPFC, dorsolateral prefrontal cortex; HrSz, high-risk for schizophrenia; HrAff, high-risk for major affective disorder

2 The Attentional Level and Time Perception in Patients with Schizophrenia

Attention is an important cognitive process both in the decision-making process and motor execution [7, 26]. In patients with schizophrenia, there is the disintegration of thought and emotional responses, as seen in auditory hallucinations, paranoia, and delirium, which consequently cause changes in attention level and response to environmental stimuli. Therefore, perception and attention are compromised, making integration between internal and external environments difficult [20, 27]. Deficient synchronization of sensory and motor outputs is partly due to attention changes [28–30]. This explains cognitive deficits observed in people with schizophrenia contribute to impairments in executive function [29–31].

The tasks that involve saccadic eye movement are useful for assessing attention level and sensorimotor integration during decision-making [8, 32]. Some studies have detected flaws in the preattentive stage of visual search in patients with schizophrenia [33], but there is evidence that this stage remains intact in these patients [34]. Elahipanah et al. [35] found that schizophrenic patients are less efficient in detecting and making decisions on target hits, especially when they are placed among some distractors. At the neuronal level, distracting events cause dorsolateral prefrontal cortex connections prepared for inhibitory control of irrelevant stimuli. Such connectivity is reduced in schizophrenic patients, leading to longer reaction times and lowered attentional levels. Moreover, schizophrenic patients appear unable to process the stimulus exposed to the visual field periphery. This explains a reduced visual range that might lie in impairments related to the frontal eye field (FEF), inferior frontal gyrus, magnocellular system, and dorsal visual pathway [35]. These specific brain areas are involved in movement and space perception [35–39].

A functional magnetic resonance imaging (fMRI) study of visual tasks observed the integrity of centers related to attention in patients with schizophrenia at different points of time from before to after treatment [8]. The study showed [8] that non-medicated schizophrenics had reduced activity in frontal, parietal, and cerebellar areas—brain regions involved in eye movements. Treatment with antipsychotics led to an improvement in neural activity in the cerebellum as well as in the frontal and parietal areas. Consistent with this, patients showed better attention and performance during decision-making with visual stimuli.

According to neuroimaging studies of attention, the cortical activity in schizophrenic patients is different from that in healthy individuals [40], with modulated activation in the frontal and parietal areas [41]. These results may vary according to the methodology and task applied. In this sense, a study compared the cortical activity in healthy subjects with schizophrenic patients during tasks requiring sustained attention and transient attention. The same regions were activated in both task types, but schizophrenic patients showed more voxels during the sustained task and a smaller number of voxels during the transient attention task than control subjects. There might be a difficulty in modulating attention in response to different task types. More clearly, patients with schizophrenia seem to fail at adaptation to new stimuli and thus produce responses that do not address episodic information [29].

Although some experiments have shown failures of attentional processes and decreased accuracy in selective attention tasks in patients with schizophrenia [8, 29, 30], there is no consensus regarding this matter. For example, Minzenberg et al. [28] evaluated functional connectivity in the frontoparietal network during decision-making tasks and showed no significant difference between patients with schizophrenia and controls.

Electrophysiological and neuroimaging studies have shown deficient perception of motion and sensory processing problems in schizophrenia. Changes in motion processing occur at different levels, including the basic cortical circuitry [42]. Particularly, evidence indicates dysfunction of the magnocellular system, one of the two main routes from the retina into the visual cortex, linked to contrast sensitivity in the transmission of visual information at low spatial frequencies and rapid and transient movements in response to high temporal frequency stimulation. This dysfunction might also contribute to poor performance in tasks that involve visual masking, decision-making based on the target speed, movement coherence, eye movements, and processing of the direction of motion of visual stimuli [42, 43]. Taken together, perception is the fundamental basis of the cognitive system; improving perceptual skills offers a potential cognitive and therefore behavioral intervention for schizophrenic patients. In healthy people, movement perception can be improved through perceptual learning. However, the neurobiological aspects pertinent to this perceptual plasticity remain unclear in schizophrenic patients [14, 25, 42].

Distortions in the time perception happen to different neuropsychiatric conditions, interestingly schizophrenia (Table 1) [2, 14]. Carrol et al. [43] assessed time perception in schizophrenic patients through discrimination between auditory and visual signals. There was a lower temporal hearing accuracy in patients with schizophrenia compared to the control group. It implies that temporal coordination that is essential to perception, cognition, motor, and consciousness processes is altered to develop thought disorder, disorganization, and improper behavior, which are classic symptoms of schizophrenia [1, 7]. Consistent with this, the impairment of numerous brain regions and neurotransmitter systems contributing to time perception is well appreciated in schizophrenia [43].

Deficient synchronization of time intervals in patients with schizophrenia might arise due to genomic and proteomic alterations that alter dopamine levels [44–47] and therefore the interaction between the nigrostriatal pathway and prefrontal cortex (Fig. 1). Ward [48] proposed hypodopaminergic activity over the prefrontal cortex in schizophrenia is concurrent with a hyperdopaminergic activity within neurons of the mesolimbic pathway. Moreover, oscillation impairments explain positive symptoms, e.g., disorganized behavior and paranoid delusions, in schizophrenia [49, 50]. Generally, patients with schizophrenia tend to overestimate time [51]. Uhlhaas et al. [52] used anatomical and EEG correlates and showed aberrant beta and gamma frequency oscillations in patients with schizophrenia. It indicates that oscillation impairments might contribute to the cognitive deficit and other classical symptoms such as hallucinations and social withdrawal in schizophrenia. Oscillation impairments may be related to aberrant neural connectivity in the mesolimbic pathway. Davalos et al. [53] carried out an fMRI study comparing patients with clinically stable schizophrenia and healthy participants when making decisions at two levels of difficulty implemented with an objective interval of an ms. The result showed no significant difference between the groups, although reaction time correlated with higher categorization errors at both levels of difficulty in the schizophrenic group. Also, patients with schizophrenia had lower neural activation, though they often made decisions in the proposed time interval [50].

Using temporal discrimination tasks, Penney et al. [54] investigated neural activity in people at high genetic risk for schizophrenia in response to auditory and visual stimuli when the interstimulus time was 3 s to 6 s. In people genetically at risk for schizophrenia, psychometric functions for visual stimuli shifted more to the right than controls. Brain locations that showed activity in response to auditory stimuli did not differ between the groups. This result is in line with neurochemical and genetic association studies [56, 57] that corroborate the role the internal biological clock plays in time perception [55]. Decision-making studies in long time intervals found that schizophrenic patients were less able to discriminate the tested interval and less accurate to reproducing and replicating sound sequences [43, 58]. When investigating cognitive alterations in schizophrenic patients, Seeman [59] suggested that a decrease in the temporal information processing for decision-making might lie in the prefrontal cortex's altered functional neuroanatomy hippocampus, as if the schizophrenic clock runs slow relative to the standard one [14].

3 Memory Processing, Recall, and Recruitment in Patients with Schizophrenia

Schizophrenia is associated with dopaminergic modulation of the mesolimbic pathway and cortical areas [15, 48], which are involved in consolidating performance measures in executive functions [45, 60, 61]. It interferes with the dopaminergic system's neural integration in the hippocampus during memory formation [61, 62]. The ability to store, transmit, retain, and evoke information relevant to short-term and long-term memory is affected [63–65]. People with schizophrenia have limitations in processing, interpreting, and consequently consolidating the information in memory. Ranganath et al. [61] point out that the ability to retain information, for example, time information [66], is based on an internal model that can be modified or have its speed adjusted through pharmacological treatment by antipsychotics. The corticohippocampal circuit seems to be able to retain information, and its disorder would lead to difficulty in building internal representations [67].

Schizophrenic patients suffer from memory impairment, especially in working memory [60]. Working memory can take information from short-term memory and store that in long-term memory [68]. Besides, it requires a time sequence of successful encodings between events to be selected, based on the importance the event represents to the individual (i.e., emotionally charged situations) or the frequency with which the event occurs (i.e., task-dependent conditioning) [69]. Disturbed working memory in patients with schizophrenia might be due to a compromised connection between the dorsolateral prefrontal cortex and hippocampus, which triggers problems during information processing, thus interfering with information consolidation [61, 68, 70]. Interestingly, the presence of working memory impairments in unaffected relatives of patients with schizophrenia supports this characteristic as an endophenotype of schizophrenia [31, 71]. For example, an experiment showed that patients with schizophrenia and their siblings made a decision over prolonged durations when performing sustained attention tasks. However, siblings performed better than patients when processing working memory and executive functioning tasks [31, 71].

Decompensated oscillations might be mentioned as contributing to impaired memory retention in patients with schizophrenia [52]. Other memory subtypes are also compromised in these patients, e.g., long-term memory and predictive memory [67, 71–73]. The deficit observed in long-term memory in non-medicated patients with first-episode schizophrenia is similar to that in previously treated patients. Predictive memory allows access to past experiences bringing a degree of automatism to interpret and perceive the event [73]. The study by Kraus and Keefe [74] suggests that individuals with schizophrenia deal with disrupted ability to evoke memory, which may underlie the development of disillusions and hallucinations. In this situation, patients with schizophrenia are unable to interpret the complete images of everyday objects. Also, episodic memory disorders have been reported and appear to be a consequence of the altered association between the hippocampus, prefrontal, and medial temporal cortex, important structures for memory consolidation and recovery. The episodic memory associated with the long term allows selecting social behavior options suitable for environmental contexts. Failures in this process lead to deficits in action planning, action generation, and problem-solving, which characterize the classical schizophrenia-related behavioral phenotype [50]. Moreover, Landgraf et al. [73] evaluated short-term memory using a list of words in people with schizophrenia compared to healthy individuals. Patients had a lower ability to process and retain information than healthy individuals. However, the implicit memory seems to be intact in these patients [75].

Some striate-frontal and striate-temporal connections or disconnections seem to be related to mnemonic failures in people with schizophrenia [61, 76], in the same way, that the dorsolateral prefrontal cortex dysfunction interferes with the ability to maintain active targets in working memory (for review see [61]). Dysfunctional structures proposed to mediate memory deficits in schizophrenia are similar to those of Alzheimer's disease. In particular, the role of the prefrontal cortex and hippocampus and the interactions between these regions in consolidating and evoking memories are proposed based on stimulus timescale and transformation of temporal sequences into a set of codes that can be consolidated. This is supported by adaptive learning models, which are based on the power of synchronization in the hippocampus and its modulation in learning every day or conditioned events [77], as well as behavioral studies in neuropsychiatric diseases (i.e., schizophrenia, Alzheimer) [78, 79]. Therefore, the importance of temporal contiguity in memory organization is prominent [80, 81].

4 Sensorimotor Integration During Decision-Making in Schizophrenia

Changes in neural networks in schizophrenia are linked to deficient coordination and temporal processing of information, which affects perceptual, cognitive, and motor decision-making [7, 43, 82, 83]. Cognitive dysfunction may be present before the first schizophrenia episode [84]. To better understand physiological processes associated with motor preparation and decision-making, it is extremely important to evaluate brain activity.

Weisbrod et al. [85] used EEG to compare 16 schizophrenic patients with 16 healthy individuals during an auditory task based on Go/NoGo decision-making. EEG by monitoring changes in cortical activity caused by decision-making in the sensorimotor task execution allows the analysis of long-range temporal correlations (LRTC) [86, 87]. Schizophrenic patients showed a deficit in inhibition response in the No/Go condition. A dysfunction was also identified in the left frontal region, where cognitive functions required for performing the task are moderated. Also, the striatum-tactical loop circuits contribute to the synchronization of relevant information to perceive and integrate decision-making in all-time task domains [88]. Studies report that this cortex area is involved in executive control, consequently activating the inhibition process [87, 89].

Studies using the P300 tool revealed changes in information processing and decision-making in schizophrenia [90–92]. Decreased P300 amplitude might reflect a decrease in temporal lobe volume, albeit with inefficient attention in sensorimotor tasks [93, 94]. Bestelmeyer [95] evaluated 21 schizophrenic patients with the oddball paradigm associated with P300 amplitude and showed that these patients were more sensitive to distractor stimuli and showed reduced inhibitory control by decreasing P300 amplitude [95]. Another study that used the same paradigm demonstrated that age and disease duration influenced the change in P300 latency [93].

Saccadic reaction times guided by visual stimuli are considered an objective correlate of attentional and sensorimotor integration. The tasks involve seeking visual stimuli with specific colors, while the stimuli are static or dynamic. Patients with schizophrenia presented impaired inhibitory control in stimulus detection when distractor components were present, especially in the task requiring decision-making to hit the target stimulus associated with movement [96].

Recently, the research studied electroencephalographic responses during transcranial magnetic stimulation (TMS) in the motor cortex of 16 schizophrenic patients and 16 healthy individuals. Compared to healthy individuals, there was an increase in the gamma-band cortical activation in the front-temporal and central regions in patients with schizophrenia. Increased gamma-band activation correlates with classic symptoms (i.e., delusions and hallucinations) in schizophrenia. Thus, it is suggested that excessive cortical activation in this area in response to TMS might lead to abnormal signal propagation, which consequently results in deficient information processing [97].

Seok et al. [98] used PET scans during a sustained attention task to investigate cerebral blood flow (CBF) and brain metabolism in patients with schizophrenia, patients with major depression, and healthy individuals. Compared to healthy individuals, schizophrenic patients showed a decreased CBF in the lower-left frontal gyrus, left cuneus, and right upper parietal lobe, and an increased CBF in the right upper frontal gyrus and right cuneus. Patients with depression symptoms were not different from controls. The findings imply a frontoparietal network dysfunction in schizophrenia.

Data from fMRI studies confirm dopaminergic dysfunction in the prefrontal cortex and the basal ganglia [99]. Functional abnormalities in the nigrostriatal pathway during sensorimotor tasks that involve working memory put forward the hypothesis of a common neurobiological domain linking the pathogenesis of schizophrenia to cognitive deficits and psychosis [100]. However, further research is necessary for assessing decision-making and sensorimotor integration in schizophrenia.

5   Conclusion

The chapter reviewed decision-making neglect in patients with schizophrenia, including impaired timing and sensorimotor synchronism and long-term cognitive impairment. Conversely, the approach to decision-making for time intervals and motor task judgment is useful in neuroscience. It helps to characterize behavioral phenotypes and neural circuits involved in schizophrenia and possibly in other neuropsychiatric disorders. There is increasing interest in therapeutic and rehabilitation methods to minimize the effects of schizophrenia-related phenotype on the life quality of patients and their caregivers. In parallel with conventional therapies, the importance of cognitive tasks is increasingly prominent, mainly due to its non-invasive nature, making it a suitable option for a long time [101].

Core Messages

- Psychiatric disorders alter the decision-making, and consequently, modulate the mechanisms of adaptation and response to the environment.
- Patients with schizophrenia show lower temporal accuracy in time perception and sensorimotor integration than healthy subjects.
- Assessing decision-making for time intervals helps to understand the schizophrenia-related behavioral phenotypes and neural circuits.

References

1. Matthews WJ, Meck WH (2014) Time perception: the bad news and the good. Wiley Interdiscip Rev Cogn Sci 5(4):429–446
2. Marinho V, Oliveira T, Bandeira J, Pinto GR, Gomes A, Lima V, Magalhães F, Rocha K, Ayres C, Carvalho V, Velasques B, Ribeiro P, Orsini M, Bastos VH, Gupta D, Teixeira S (2018) Genetic influence alters the brain synchronism in perception and timing. J Biomed Sci 25(1):61
3. Starcke K, Brand M (2012) Decision making under stress: a selective review. Neurosci Biobehav Rev 36(4):1228–1248
4. Orsini CA, Moorman DE, Young JW, Setlow B, Floresco SB (2015) Neural mechanisms regulating different forms of risk-related decision-making: Insights from animal models. Neurosci Biobehav 58:147–167
5. Frank MJ, Claus ED (2006) Anatomy of a decision: striato-orbito-frontal interactions in reinforcement learning, decision making, and reversal. Psychol 113(2):300–326
6. Bechara A, Damasio H, Damasio AR, Damasio (2000) Emotion, decision making and the orbitofrontal cortex. Cereb Cortex 10(3):295–307
7. Fontes R, Ribeiro J, Gupta DS, Machado D, Lopes-Júnior F, Magalhães F, Bastos VH, Rocha K, Marinho V, Lim G, Velasques B, Ribeiro P, Orsini M, Pessoa B, Leite MA, Teixeira S (2016) Time perception mechanisms at central nervous system. Neurol Int 8 (1):5939
8. Keedy SK, Rosen C, Khine T, Rajarethinam R, Janicak PG, Sweeney JA (2009) An fMRI study of visual attention and sensorimotor function before and after antipsychotic treatment in first-episode schizophrenia. Psychiatry Res 172(1):16–23
9. Harrington DL, Zimbelman JL, Hinton SC, Rao SM (2010) Neural modulation of temporal encoding, maintenance, and decision processes. Cereb Cortex 20(6):1274–1285
10. Assadi SM, Yücel M, Pantelis C (2009) Dopamine modulates neural networks involved in effort-based decision-making. Neurosci Biobehav Rev 33(3):383–393
11. Diamond A (2013) Executive functions. Annu Rev Psychol 64:135–168
12. Lee KH, Bhaker RS, Mysore A, Parks RW, Birkett PB, Woodruff PW (2009) Time perception and its neuropsychological correlates in patients with schizophrenia and in healthy volunteers. Psychiatry Res 166(2–3):174–183
13. Fan J, Semenzin E, Meng W, Giubilato E, Zhang Y, Critto A, Zabeo A, Zhou Y, Ding S, Wan J, He M, Lin C (2015) Ecological status classification of the Taizi River Basin, China: a comparison of integrated risk assessment approaches. Environ Sci Pollut Res Int 22 (19):14738–14754
14. Teixeira S, Machado S, Paes F, Velasques B, Silva JG, Sanfim AL, Minc D, Anghinah R, Menegaldo LL, Salama M, Cagy M, Nardi AE, Pöppel E, Bao Y, Szelag E, Ribeiro P,

Arias-Carrión O (2013) Time perception distortion in neuropsychiatric and neurological disorders. CNS and Neurol Disorders—Drug Targets 12:567–582

15. Gómez J, Marín-Méndez JJ, Molero P, Atakan Z, Ortuño F (2014) Time perception networks and cognition in schizophrenia: Aa review and a proposal. Psychiatry Res 220:737–744

16. Elvevåg B, Brown GD, McCormack T, Vousden JI, Goldberg TE (2004) Identification of tone duration, line length, and letter position: an experimental approach to timing and working memory deficits in schizophrenia. J Abnorm Psychol 113(4):509–521

17. Bandeira J, Teixeira S, Rebouças Pinto G, Figueiredo R, Martins FC, Marinho V (2019) Association of SLC6A4 5-HTTLPR and 5HTR2A T102C in the neurobiological domains associated with time perception: genetic and behavioral correlates. jneuropsychiatry.org 9 (6):2476–2484

18. Dorris MC, Olivier E, Munoz DP (2007) Competitive integration of visual and preparatory signals in the superior colliculus during saccadic programming. J Neurosci 27(19):5053–5062

19. Klaes C, Schneegans S, Schöner G, Gail A (2012) Sensorimotor learning biases choice behavior: a learning neural field model for decision making. PLoS Comput Biol 8(11): e1002774

20. Velasques B, Machado S, Paes F, Cunha M, Sanfim A, Budde H, Cagy M, Anghinah R, Basile LF, Piedade R, Ribeiro P (2011) Sensorimotor integration and psychopathology: motor control abnormalities related to psychiatric disorders. World J Biol Psychiatry 12 (8):560–573

21. Hassiotis A, Brown E, Harris J, Helm D, Munir K, Salvador-Carulla L, Bertelli M, Baghdadli A, Wieland J, Novell-Alsina R, Cid J, Vergés L, Martínez-Leal R, Mutluer T, Ismayilov F, Emerson E (2019) Association of borderline intellectual functioning and adverse childhood experience with adult psychiatric morbidity. Findings from a British birth cohort. BMC Psychiatry 19(1):387

22. Damaraju E, Allen EA, Belger A, Ford JM, McEwen S, Mathalon DH, Mueller BA, Pearlson GD, Potkin SG, Preda A, Turner JA, Vaidya JG, van Erp TG, Calhoun VD (2014) Dynamic functional connectivity analysis reveals transient states of dysconnectivity in Schizophrenia. Neuroimage Clin 5:298–308

23. Kent M, Bardi M, Hazelgrove A, Sewell K, Kirk E, Thompson B, Trexler K, Terhune-Cotter B, Lambert K (2017) Profiling coping strategies in male and female rats: potential neurobehavioral markers of increased resilience to depressive symptoms. Horm Behav 95:33–43

24. Rowland LM, Spieker EA, Francis A, Barker PB, Carpenter WT, Buchanan RW (2009) White matter alterations in deficit schizophrenia. Neuropsychopharmacology 34(6):1514–1522

25. McTeague LM, Huemer J, Carreon DM, Jiang Y, Eickhoff SB, Etkin A (2017) Identification of common neural circuit disruptions in cognitive control across psychiatric disorders. Am J Psychiatry 174(7):676–685

26. Mathis KI, Wynn JK, Breitmeyer B, Nuechterlein KH, Green MF (2011) The attentional blink in schizophrenia: isolating the perception/attention interface. J Psychiatr Res 45 (10):1346–1351

27. Sweeney DJ (1999) Servo motor driven fill system. U.S. Patent No 5.865.226

28. Minzenberg MJ, Firl AJ, Yoon JH, Gomes GC, Reinking C, Carter CS (2010) Gamma oscillatory power is impaired during cognitive control independent of medication status in first-episode schizophrenia. Neuropsychopharmacology 35(13):2590–2599

29. Ravizza SM, Moua KC, Long D, Carter CS (2010) The impact of context processing deficits on task-switching performance in Schizophrenia. Schizophr Res 116(2–3):274–279

30. Lalanne L, Dufour A, Després O, Giersch A (2012) Attention and masking in schizophrenia. Biol Psychiatry 71(2):162–168

31. Giakoumaki SG, Roussos P, Pallis EG, Bitsios P (2011) Sustained attention and working memory deficits follow a familial pattern in schizophrenia. Arch Clin Neuropsychol 26 (7):687–695
32. Bittencourt J, Velasques B, Teixeira S, Basile LF, Salles JI, Nardi AE, Budde H, Cagy M, Piedade R, Ribeiro P (2013) Saccadic eye movement applications for psychiatric disorders. Neuropsychiatr Dis Treat 9:1393–1409
33. Lieb K, Merklin G, Rieth C, Schüttler R, Hess R (1994) Preattentive information processing in schizophrenia. Schizophr Res 14(1):47–56
34. Gould RA, Mueser KT, Bolton E, Mays V, Goff D (2001) Cognitive therapy for psychosis in schizophrenia: an effect size analysis. Schizophr Res 48(2–3):335–342
35. Elahipanah A, Christensen BK, Reingold EM (2011) Controlling the spotlight of attention: visual span size and flexibility in schizophrenia. Neuropsychologia 49(12):3370–3376
36. Brittain PJ, Surguladze S, McKendrick AM, Ffytche DH (2010) Backward and forward visual masking in schizophrenia and its relation to global motion and global form perception. Schizophr Res 124(1–3):134–141
37. Ortuño F, Guillén-Grima F, López-Garcia P, Gómez J, Pla J (2011) Functional neural networks of time perception: challenge and opportunity for schizophrenia research. Schizophr Res 125:129–135
38. Leszczyńska A (2015) Facial emotion perception and schizophrenia symptoms. Psychiatr Pol 49(6):1159–1168
39. Su L, Wyble B, Zhou LQ, Wang K, Wang YN, Cheung EF, Chan RC (2015) Temporal perception deficits in schizophrenia: integration is the problem, not deployment of attentions. Sci Rep 5:9745
40. Alústiza I, Radua J, Albajes-Eizagirre A, Domínguez M, Aubá E, Ortuño F (2016) Meta-analysis of functional neuroimaging and cognitive control studies in schizophrenia: preliminary elucidation of a core dysfunctional timing network. Front Psychol 7:192
41. Voegler R, Becker MP, Nitsch A, Miltner WH, Straube T (2016) Aberrant network connectivity during error processing in patients with schizophrenia. J Psychiatry Neurosc JPN 41(2):E3–E12
42. Potvin S, Marchand S (2008) Hypoalgesia in schizophrenia is independent of antipsychotic drugs: a systematic quantitative review of experimental studies. Pain 138(1):70–78
43. Carroll CA, Boggs J, O'Donnell BF, Shekhar A, Hetrick WP (2008) Temporal processing dysfunction in schizophrenia. Brain Cogn 67(2):150–161
44. Monakhov M, Golimbet V, Abramova L, Kaleda V, Karpov V (2008) Association study of three polymorphisms in the dopamine D2 receptor gene and schizophrenia in the Russian population. Schizophr Res 100(1–3):302–307
45. Eisenberg DP, Berman KF (2010) Executive function, neural circuitry, and genetic mechanisms in Schizophrenia. Neuropsychopharmacol Rev 35:258–277
46. Kunii Y, Miura I, Matsumoto J, Hino M, Wada A, Niwa S, Nawa H, Sakai M, Someya T, Takahashi H, Kakita A, Yabe H (2014) Elevated postmortem striatal t-DARPP expression in schizophrenia and associations with DRD2/ANKK1 polymorphism. Prog Neuropsychopharmacol Biol Psychiatry 53:123–128
47. Fehér Á, Juhász A, Pákáski M, Kálmán J, Janka Z (2014) Association between the 9 repeat allele of the dopamine transporter 40bp variable tandem repeat polymorphism and Alzheimer's disease. Psychiatry Res 220(1–2):730–731
48. Ward RD, Kellendonk C, Kandel ER, Balsam PD (2012) Timing as a window on cognition in Schizophrenia. Neuropharmacology 62(3):1175–1181
49. Cordeiro Q, Vallada H (2014) Association study between the Taq1A (rs1800497) polymorphism and schizophrenia in a Brazilian sample. Arq Neuropsiquiatr 72(8):582–586
50. Bonnot O, de Montalembert M, Kermarrec S, Botbol M, Walter M, Coulon N (2011) Are impairments of time perception in schizophrenia a neglected phenomenon? J Physiol Paris 105:164–169

51. Zalla T, Verlut I, Franck N, Puzenat D, Sirigu A (2004) Perception of dynamic action in patients with schizophrenia. Psychiatry Res 128(1):39–51
52. Uhlhaas PJ, Singer W (2010) Abnormal neural oscillations and synchrony in schizophrenia. Nat Rev Neurosci 11:100–113
53. Davalos DB, Rojas DC, Tregellas JR (2011) Temporal processing in schizophrenia: effects of task-difficulty on behavioural discrimination and neuronal responses. Schizophrenia Res 127:123–130
54. Penney TB, Meck WH, Roberts SA, Gibbon J, Erlenmeyer-Kimling L (2005) Interval-timing deficits in individuals at high risk for schizophrenia. Brain Cogn 58 (1):109–118
55. Roy M, Grondin S, Roy MA (2012) Time perception disorders are related to working memory impairments in schizophrenia. Psychiatry Res 200:159–166
56. Bertolino A, Fazio L, Caforio G, Blasi G, Rampino A, Romano R, Di Giorgio A, Taurisano P, Papp A, Pinsonneault J, Wang D, Nardini M, Popolizio T, Sadee W (2009) Functional variants of the dopamine receptor D2 gene modulate prefronto-striatal phenotypes in schizophrenia. Brain: A J Neurol 132(Pt 2):417–425.
57. Yao J, Pan YQ, Ding M, Pang H, Wang BJ (2015) Association between DRD2 (rs1799732 and rs1801028) and ANKK1 (rs1800497) polymorphisms and schizophrenia: a meta-analysis. American journal of medical genetics. Part B, Neuropsychiatric genetics: the official publication of the International Society of Psychiatric Genetics 168B(1):1–13
58. Papageorgiou C, Karanasiou IS, Kapsali F, Stachtea X, Kyprianou M, Tsianaka EI, Karakatsanis NA, Rabavilas AD, Uzunoglu NK, Papadimitriou GN (2013) Temporal processing dysfunction in schizophrenia as measured by time interval discrimination and tempo reproduction tasks. Prog Neuropsychopharmacol Biol Psychiatry 40:173–179
59. Seeman P (2013) Schizophrenia and dopamine receptors. Eur Neuropsychopharmacol 23 (9):999–1009
60. Chan RC, Xu T, Heinrichs RW, Yu Y, Wang Y (2010) Neurological soft signs in schizophrenia: a meta-analysis. Schizophr Bull 36(6):1089–1104
61. Ranganath C, Minzenberg MJ, Ragland JD (2008) The cognitive neuroscience of memory function and dysfunction in schizophrenia. Biol Psychiatry 64(1):18–25
62. Wiltgen BJ, Zhou M, Cai Y, Balaji J, Karlsson MG, Parivash SN, Li W, Silva AJ (2010) The hippocampus plays a selective role in the retrieval of detailed contextual memories. Curr Biol 20(15):1336–1344
63. Hölscher C (2003) Time, space and hippocampal functions. Rev Neurosci 14:253–284
64. Tubridy S, Davachi L (2011) Medial temporal lobe contributions to episodic sequence encoding. Cereb Cortex 21:272–280
65. Eichenbaum H (2014) Time cells in the hippocampus: a new dimension for mapping memories. Nat Rev Neurosci 15(11):732–744
66. Buhusi CV, Meck WH (2002) Differential effects of methamphetamine and haloperidol on the control of an internal clock. Behav Neurosci 116(2):291–297
67. Giersch A, van Assche M, Huron C, Luck D (2011) Visuo-perceptual organization and working memory in patients with schizophrenia. Neuropsychologia 49(3):435–443
68. Mayer JS, Park S (2012) Working memory encoding and false memory in schizophrenia and bipolar disorder in a spatial delayed response task. J Abnorm Psychol 121(3):784–794
69. MacDonald CJ (2014) Prospective and retrospective duration memory in the hippocampus: is time in the foreground or background? philosophical transactions of the royal society of London. Series B Biolog Sci 369(1637):20120463
70. Nielson DM, Smith TA, Sreekumar V, Dennis S, Sederberg PB (2015) Human hippocampus represents space and time during retrieval of real-world memories. Proc Natl Acad Sci USA 112(35):11078–11083
71. Olsen EK, Bjorkquist OA, Bodapati AS, Shankman SA, Herbener ES (2015) Associations between trait anhedonia and emotional memory deficits in females with schizophrenia versus major depression. Psychiatry Res 230(2):323–330

72. Rannikko I, Murray GK, Juola P, Salo H, Haapea M, Miettunen J, Veijola J, Barnett JH, Husa AP, Jones PB, Järvelin MR, Isohanni M, Jääskeläinen E (2015) Poor premorbid school performance, but not severity of illness, predicts cognitive decline in schizophrenia in midlife. Schizophrenia Res Cognition 2(3):120–126
73. Landgraf S, Steingen J, Eppert Y, Niedermeyer U, van der Meer E, Krueger F (2011) Temporal information processing in short- and long-term memory of patients with schizophrenia. PloS one 6(10):e26140
74. Kraus MS, Keefe RS, Krishnan RK (2009) Memory-prediction errors and their consequences in schizophrenia. Neuropsychol Rev 19(3):336–352
75. Rass O, Schacht RL, Buckheit K, Johnson MW, Strain EC, Mintzer MZ (2015) A randomized controlled trial of the effects of working memory training in methadone maintenance patients. Drug Alcohol Depend 156:38–46
76. Deserno L, Sterzer P, Wüstenberg T, Heinz A, Schlagenhauf F (2012) Reduced prefrontal-parietal effective connectivity and working memory deficits in schizophrenia. J Neurosci 32(1):12–20
77. Modi KK, Jana A, Ghosh S, Watson R, Pahan K (2017) Correction: a physically-modified saline suppresses neuronal apoptosis, attenuates tau phosphorylation and protects memory in an animal model of alzheimer's disease. PLoS One 12(6):e0180602
78. Gill PR, Mizumori SJY, Smith DM (2011) Hippocampal episode fields develop with learning. Hippocampus 21:1240–1249
79. Kluge M, Schacht A, Himmerich H, Rummel-Kluge C, Wehmeier PM, Dalal M, Hinze-Selch D, Kraus T, Dittmann RW, Pollmächer T, Schuld A (2014) Olanzapine and clozapine differently affect sleep in patients with schizophrenia: results from a double-blind, polysomnographic study and review of the literature. Schizophr Res 152(1):255–260
80. Yin B, Troger AB (2011) Exploring the 4th dimension: hippocampus, time, and memory revisited. Front Integr Neurosci 5:36
81. Jacobs NS, Allen TA, Nguyen N, Fortin NJ (2013) Critical role of the hippocampus in memory for elapsed time. J Neurosci 33:13888–13893
82. Bressler SL (2003) Cortical coordination dynamics and the disorganization syndrome in schizophrenia. Neuropsychopharmacology 28(Suppl 1):S35–S39
83. Enriquez-Geppert S, Konrad C, Pantev C, Huster RJ (2010) Conflict and inhibition differentially affect the N200/P300 complex in a combined go/nogo and stop-signal task. Neuroimage 51(2):877–887
84. Vaz-Serra A, Palha A, Figueira ML, Bessa-Peixoto A, Brissos S, Casquinha P, Damas-Reis F, Ferreira L, Gago J, Jara J, Relvas J, Marques-Teixeira J (2010) Cognição, cognição social e funcionalidade na esquizofrenia [Cognition, social cognition and functioning in schizophrenia]. Acta Med Port 23(6):1043–1058
85. Weisbrod M, Kiefer M, Marzinzik F, Spitzer M (2000) Executive control is disturbed in schizophrenia: evidence from event-related potentials in a Go/NoGo task. Biol Psychiatry 47 (1):51–60
86. Beggs JM, Plenz D (2003) Neuronal avalanches in neocortical circuits. J Neurosci 23 (35):11167–11177
87. Ertekin E, Üçok A, Keskin-Ergen Y, Devrim-Üçok M (2017) Deficits in Go and NoGo P3 potentials in patients with schizophrenia. Psychiatry Res 254:126–132
88. Allman MJ, Meck WH (2012) Pathophysiological distortions in time perception and timed performance. Brain 135(Pt 3):656–677
89. D'Esposito M, Detre JA, Alsop DC, Shin RK, Atlas S, Grossman M (1995) The neural basis of the central executive system of working memory. Nature 378(6554):279–281
90. Sassi FC, Matas CG, de Mendonça LI, de Andrade CR (2011) Stuttering treatment control using P300 event-related potentials. J Fluency Disord 36(2):130–138
91. Almeida PR, Vieira JB, Silveira C, Ferreira-Santos F, Chaves PL, Barbosa F, Marques-Teixeira J (2011) Exploring the dynamics of P300 amplitude in patients with schizophrenia. Int J Psychophysiol 81(3):159–168

92. Gaspar PA, Ruiz S, Zamorano F, Altayó M, Pérez C, Bosman CA, Aboitiz F (2011) P300 amplitude is insensitive to working memory load in schizophrenia. BMC Psychiatry 11:29
93. Yeon YW, Polich J (2003) Meta-analysis of P300 and schizophrenia: patients, paradigms, and practical implications. Psychophysiology 40(5):684–701
94. Petersen SE, Posner MI (2012) The attention system of the human brain: 20 years after. Annu Rev Neurosci 35:73–89
95. Bestelmeyer PE (2012) The visual P3a in schizophrenia and bipolar disorder: effects of target and distractor stimuli on the P300. Psychiatry Res 197(1–2):140–144
96. Curtin A, Sun J, Zhao Q, Onaral B, Wang J, Tong S, Ayaz H (2019) Visuospatial task-related prefrontal activity is correlated with negative symptoms in schizophrenia. Sci Rep 9(1):9575
97. Frantseva M, Cui J, Farzan F, Chinta LV, Perez Velazquez JL, Daskalakis ZJ (2014) Disrupted cortical conductivity in schizophrenia: TMS-EEG study. Cereb Cortex 24(1):211–221
98. Seok JH, Park HJ, Lee JD, Kim HS, Chun JW, Son SJ, Oh MK, Ku J, Lee H, Kim JJ (2012) Regional cerebral blood flow changes and performance deficit during a sustained attention task in schizophrenia: (15) O-water positron emission tomography. Psychiatry Clin Neurosci 66(7):564–572
99. Yoon YS, Lee HS (2013) Projections from melanin-concentrating hormone (MCH) neurons to the dorsal raphe or the nuclear core of the locus coeruleus in the rat. Brain Res 1490:72–82
100. Williams SN, Undieh AS (2016) Dopamine-sensitive signaling mediators modulate psychostimulant-induced ultrasonic vocalization behavior in rats. Behav Brain Res 296:1–6
101. Magalhães F, Rocha K, Marinho V, Ribeiro J, Oliveira T, Ayres C, Bento T, Leite F, Gupta D, Bastos VH, Velasques B, Ribeiro P, Orsini M, Teixeira S (2018) Neurochemical changes in basal ganglia affect time perception in parkinsonians. J Biomed Sci 25(1):26

Juliana Bittencourt is a physiotherapy graduate from the Serra dos Órgãos University Center, Teresópolis (UNIFESO)—RJ. She has a master's degree and a Ph.D. in mental health from the Federal University of Rio de Janeiro—UFRJ. Juliana is researching at the Laboratory of Brain Mapping and Sensorimotor Integration and the Laboratory of Neurophysiology and Neuropsychology of Attention—UFRJ and the Institute of Applied Neurosciences—INA. She serves as professor of the undergraduate course in physiotherapy at the University Veiga de Almeida—UVA and the postgraduate course in applied neurosciences at the Federal University of Rio de Janeiro—UFRJ. She is a delegate to the Brazilian Association of Neurofunctional Physiotherapy (ABRAFIN).

Victor Marinho is a biomedicine graduate from the Federal University of Piauí, with clinical analysis and clinical pathology qualifications. He has a master's degree and Ph.D. in Biotechnology from the Federal University of Piauí—UFPI. He has a background in higher education management at the International Faculty of Delta and microbiology at Faculdade Única de Ipatinga—MG. He is researching neuroscience and genetics at the Laboratory of Technological Neuro-Innovation and Brain Mapping (NitLab) and Laboratory of Genetics and Molecular Biology. He serves as a clinical analysis professor and biomedical scientist.

Getting to Know Ourselves Through Recognizing Ourselves in Others: Neuroanatomy of Empathy in a Social Neuroscientific Model

Roberto E. Mercadillo and Daniel Atilano-Barbosa

"Inside us there is something that has no name, that something is what we are."

José Saramago, Blindness

Summary

This chapter revisits the intricate notion of 'human being' elaborated by Duns Scotus to argue that empathy is a broad concept overlapping emotional, cognitive, and social components, allowing people to recognize another's mental and physical states and to motivate prosocial behaviors. Also, an analytical review is offered and initiated with the German term *einfühlung* used in esthetics and phenomenology. It was then translated as *empathy*, with meanings used in the contemporary cognitive sciences and evolutionary perspectives. The analytical review guided a critical revision of neuroimaging studies to provide

R. E. Mercadillo (✉)
National Council for Science and Technology, Unidad Iztapalapa. San Rafael Atlixco No. 186, Col. Vicentina, Del. Iztapalapa, C.P. 09340 Ciudad de México, Mexico
e-mail: remercadilloca@conacyt.mx

Area of Neurosciences, Department of Biology of Reproduction, Universidad Autónoma Metropolitana, Iztapalapa Unit, Mexico City, Mexico

D. Atilano-Barbosa
Institute of Neurobiology, Universidad Nacional Autónoma de México, Mexico City, Mexico

143
N. Rezaei (ed.), *Multidisciplinarity and Interdisciplinarity in Health*,
Integrated Science 6, https://doi.org/10.1007/978-3-030-96814-4_7

a neuroanatomical mapping of empathy useful for mental health issues and the analysis of social behaviors in response to others' needs. The mapping involves overlapped neural functions related to emotional and motor experiences (insula and anterior cingulate cortex), emotional contagion and recognition (amygdala, inferior frontal gyrus, inferior parietal lobe, and premotor cortex), cognitive perspective-taking and mentalizing (medial and dorsolateral prefrontal cortex, posterior cingulate cortex, superior temporal sulcus, temporal pole, temporoparietal junction, and precuneus), inference of emotional states (ventromedial prefrontal cortex), and the distinction between the self and the other (dorsomedial and ventromedial regions of the prefrontal cortex and inferior parietal lobe). The neuroanatomical mapping is interpreted under the *social neuroscience* proposal, which examines neurobiological processes related to social interactions. The chapter concludes with a pictorial representation to illustrate and explain empathic processes expressed in the human social world. An interdisciplinary model to comprehend empathy through neurosciences, experimental biology, ethology, sociology, anthropology, history, philosophy, and psychology is finally depicted.

'Empathic Being'
[Made by Roberto E. Mercadillo, 2019; Photography by Cinthia Montiel].
The code of this chapter is *01100101 01110010 01101101 01001101 01111001 01101111.*

Keywords

Brain function · Complexity · Empathy · Interdisciplinary · Neuroanatomy ·
Neuroimaging · Social neuroscience · Transdisciplinary

1 Introduction

Perhaps, empathy represents one of the most complex concepts and processes
embracing essential affective qualities of human relations. Its complexity is illustrated in the artwork titled Empathic Being that opens this chapter. This creation is
made with colored yarn on beeswax, which emulates an artisanal technique used by
the Huichol or Wixáritari indigenous communities that live in Northern Mexico.
The shamans and visionaries of these communities use the colors and shapes of the
yarn to express their knowledge, which emerges after taking peyote, through
dreams or under states of profound introspection and reflection. These pieces are
both ornament and knowledge because they represent their world view, as well as
the origins, composition, and well-being of their culture and members [1].

In Empathic Being, the author presents an allegory of the notion 'human being'
elaborated in the Middle Ages by the Franciscan philosopher-theologian John Duns
Scotus (1266–1308) [2]. For this philosopher, the mind and body make up a single
psychophysical unit that cannot be understood separately. Duns Scotus is represented at the bottom of the artwork, dressed in his Franciscan monk habit that
symbolizes the beliefs he decided to represent and permeates his world vision. His
face cannot be distinguished because it is covered with a Greek theatrical mask that
grants human beings a personality and vibrancy of expression that allows them to
be heard by the theater's audience, in other words, those other people who hear and
observe the person. Although he is wearing his habit, his heart and lungs are visible
since, according to Scotus, both organs allow a person to be understood not as an
abstraction but as a being with a body, whose blood and breath grant him feelings,
life, and movement.

Scotus' concept of 'human being' is shown in the four representations above his
head. His notion of the 'human' necessarily has a relational quality; one single
isolated individual cannot represent a person; others are always part of its construction. However, the notion of 'the other' is preceded by a notion of 'oneself' so
that an individual may discern their own unit and then explore and reflect on
themselves and others. This capacity to reflect is possible due to the human being's
intellectual nature. However, Scotus is careful not to reduce the 'intellect' to 'rationality' or the capacity to observe the world outside of the body to make reality
objective. He alludes to the Latin root of intellectus, which refers to the capacity of
making what happens inside and outside of the body readable, to intelligence,
comprehension, and discernment, and to the ability to understand, discover, and
appreciate the world both through the body and the imagination.

Scotus' notion of a human being is shown sitting down face forward since the philosopher argued that the ability to understand the other comes from the capacity and intent to lean forward and put oneself at the level of those being observed to distinguish their mutual expressions through close inspection, face to face. Expressions are distinguished through a person's sensitive and intellectual qualities, which in contemporary terms are interpreted as sensory and cognitive qualities based on the primordial central and autonomous nervous systems illustrated in the representations of the piece. Paradoxically, the ability to lean toward another allows for building the self and self-affirmation. This implies Ultima Solitudo, the individual's quality which, according to Scotus, refers to a differentiation from others but always in relation to them, emulating an open and closed system. This dual quality allows for solidarity and openness as any identity depends on others. The narrative of life and behavior is what transform a person's identity, suggests Scotus. Thus, he equips human beings with another quality; the Homo Viator, a traveler who constantly walks under the sun, as shown in the artwork. Scotus' Viator requires a travel itinerary throughout his life to configure himself because his beliefs, attributes, and thoughts about himself, and others can be modified as his existence moves forward.

According to the artwork's author, the medieval notion of a human person provided by Scotus is complex and deeply empathic since it is based on relational properties and cognitive and affective dynamics, allowing human beings to think about their own and others´ feelings. Since contemporary scientists have new theoretical schemes and more sophisticated tools and instruments than in medieval times to analyze our human constitution, this chapter revisits Scotus's empathic notion of 'human being' with contemporary premises postulated by social neuroscience. We also use neuroimaging techniques that have allowed scientists to perform non-invasive studies to elucidate the neuroanatomical and neurofunctional bases of cognition, affective experiences, decision-making, memory, and language. Most of these studies are task-based, with people performing the task while scanning their brain activity.

Several experimental studies and reviews based on neuroimaging findings have been made concerning empathy [3]. Nevertheless, most research has focused on emotional recognition abilities since these constitute the more salient empathic elements and have not integrated the wide variety of multilevel aspects regarding the other's cognitive, affective, and social attributes. This chapter reviews concepts and theories concerning empathy and neuroimaging studies that show different elements related to empathy and connected to human–social interactions. Our analytical revision involves different crucial cognitive and affective processes embraced in Scotus's proposal [2], which must be defined separately, but interpreted jointly to configure the neuroanatomical basis of empathy: emotional experiences affecting empathy, emotional contagion, cognitive simulation, effects of personal distress, emotional autoregulation, mentalizing, and the distinction between self and other. Finally, a set of neuroanatomical regions was made into a map illustrating the cognitive, affective, and social processes involving empathy. Also, we provide two models to understand human empathy from a multilevel perspective and emphasize the interdisciplinary work needed to study and comprehend different aspects of empathic experiences and expressions.

2 Complex Meanings of Empathy: Connections with Mental Health and Social Neuroscience

The current concept of empathy can be traced to the German term einfühlung, minted by the philosopher Robert Vischer at the end of the nineteenth century, which means appreciating beauty through an affective bond set between an observer and an esthetic object. The philosopher and psychologist, Theodore Lipps, expanded this term to explain that esthetic experiences involve an inherent imitation of expressions perceived in the objects, which evoke firsthand feelings in the observers; those expressions represent affections and attributions that are projected onto others. Therefore, einfühlung refers to a connection between a sentimental self and an expressive self present in the observer [4, 5].

In his phenomenological approach developed at the beginning of the twentieth century, Edmund Husserl also used the term einfühlung, not in the field of esthetics, but to refer to an intentional apprehension allowing for the other's body to be perceived as equivalent or analogous to the perceiver's own body through a motivation encompassing similarities between the other's and the perceiver's own behaviors. Thus, otherness is not a mere duplication of the observer but an intentional adaption of the self [6]. In 1917, the philosopher and Carmelite nun Edith Stein expanded Husserl's proposal considering einfühlung as a conscience experience of the other obtained through a similar experience of their feelings. According to Stein, finding this similar experience is possible through remembering one's own flow of experiences identifying equivalent situations lived during the observer's life. During this mnemonic and reflective process, one's body and the other's body are transposed, so experiences are felt and accomplished by mimicking the other's bodily expressions [7].

The English American psychologist Edward B. Titchener was interested in introspection as a process and method to assess one's own feelings, impulses, and thoughts to build identity. To conceptualize this process, in 1909, Titchener translated the term einfühlung into the English word empathy, which included the original einfühlung attributions, and also incorporated the etymological Greek meaning of empatheia (ἐμπάθεια) referring to 'en' (in) and 'pathos' (feeling, suffering) [8]. Therefore, empathy refers to how a person can feel or share another's emotional states, so their feelings, particularly suffering, can be felt by the perceiver as their own and kinesthetically expressed.

More recently, the term empathy was disseminated and operationalized in psychological studies, particularly by the cognitivism movement during the second half of the twentieth century. Currently, empathy is generally used as a concept to comprehend mental and behavioral processes involving the representation of others, the inference of others' inner and physical states, emotional contagion, and the self's projection onto others' situations [9, 10]. The study of empathy was promptly adopted by several biological–behavioral approaches to understand social relations. For example, from an evolutionary perspective, empathy is proposed as a phylogenetically ancient mechanism, probably presented in mammals and birds, to motivate altruistic behaviors in response to another's pain, need, or distress by matching emotional states between the perceivers and the observed [11]. The study of this conceivably evolved empathic mechanism includes ethological reports in a variety of non-human primate species in captive and free environments, such as stressed chimpanzees in zoos or disabled Japanese macaques in natural parks. These studies aim to identify empathy when expressing anguish or self-consciousness, transmitting information between members of a group, and manipulating social relations. This ethological approach could be particularly interesting to understand evolutionary aspects of empathy since human and non-human primates share certain brain regions and functions associated with recognizing another's needs and feelings to help them effectively [12].

Concerning mental health, the empathic framework has been useful to understand some psychiatric issues related to violence, offense, addiction, and the consequences of some neurological diseases. For example, people showing psychopathic qualities manifest no or reduced understanding of another's distress and empathy when negative emotional stimuli or punishments are performed [13], while the abuse of psychoactive substances is related to low detection and a limited experience of others' emotional states [14]. Related to neurological diseases, frontotemporal dementia (a neurodegenerative disease that implies atrophy of the frontal and/or anterior temporal lobes) is characterized by a deficit in empathy and social behaviors causing relationship dissolutions [15], while patients with spinocerebellar ataxia type 2 (a genetic disease causing degeneration in Purkinje cells) manifest lack of social and emotional interests on others associated with frontal, temporal, and cerebellar reduced volumes [16].

When working with patients with autism, a neurological disorder involving impairments in communication and social interaction, neurologist Oliver Sacks wrote: 'I observed these patients closely, felt for them, and tried, as a physician, to bring out their positive potentials. I tried to engage them, whenever possible, in the morally neutral real to play' [17, p. 210]. Sack's quote represents the relational–emotional perspective of empathy and suggests its relevance in nursing and caring dynamics involving mental health. Empathic attitudes and positions are necessary to mediate physician–patient relations to understand the disease better and look for treatments and elaborate educational training that improves students' and specialists' attitudes toward people in need, as recently claimed by several health institutions [18, 19].

In cultural domains, the empathic background allows scientists to understand fundamental human phenomena and their relation to brain function, such as religious or spiritual practices based on empathy and otherness. For example, the Franciscan Catholic tradition understands human beings 'as relational properties where the self and others are interdependent spiritual and corporal entities', so Franciscans focus on alleviating human suffering with solidarity and acceptance of different conditions [20]. This practice can be seen in some religious texts used for self-support groups dealing with drug abuse. Reading them evokes brain activations in motor, parietal, and cognitive regions involved in empathic attitudes and cooperative behaviors, favoring a collectivistic dynamic to overcome addiction [21]. Another example is the loving-kindness meditation practiced in the Theravada Buddhist tradition to cultivate compassion, empathy, and positive feelings toward oneself and others. This practice continuum elicits an increment of brain volume in the angular and parahippocampal gyrus associated with affective regulation and empathic responses [22].

Moreover, empathy may represent a powerful impulse for social actions, particularly in situations that cause suffering or conditions that could be considered unfair. This is possible because empathy is considered a basic disposition to experience so-called moral emotions, such as compassion or indignation that 'are elicited by perceived transgressions that do not affect the perceiver's well-being directly but do affect the well-being of others and motivate prosocial decisions to restore the transgression' [23, 24]. Consequently, empathy is linked to the morality of acting when people consider something is bad or good, right or wrong. An example of this link is illustrated by brain functions related to feeling indignation and empathy activated when people watch scenes representing victims' cases during the apartheid period in South Africa [25]. Another example is seen in educational programs focused on generating empathic attitudes that favor multi-cultural co-existence and mediation in aggressive environments at schools shaped by multi-origin migrant teenagers [26]. Empathy is at the basis of some international initiatives to develop 'a culture of peace and non-violence', as is the Manifesto 2000 proposed by UNSECO and signed by 75 million people around the world to respect life and dignity without prejudice, to listen for understanding, to create new forms of civic solidarity, to share time and material resources with generosity, and to guard freedom of expression and cultural diversity.

As can be seen above, and similar to Scotus' proposal, the complexity of empathy implies the notion of otherness linked to cognitive and affective processes involving social life. The neuroscience of empathy may not only contribute to understanding certain psychiatric and neurological disorders, but also, if we consider the brain as the biological organ that regulates behavior and cognition, it may be possible to understand the way different elements of empathy are assembled in the brain anatomy and functions, and how these functions are exhibited in the human social world. Accordingly, the neurobiological basis of empathy must be defined and interpreted while considering the complexity of social domains. One way of approaching this is to link the neuroanatomy of empathy with the framework of what has been termed 'social neuroscience', which constitutes the inter-disciplinary study of neurobiological processes related to interactions in the social world.

The social neuroscience proposal was developed from behavioral, functional, and systemic neuroscience traditions using neurophysiological and neuroanatomical methods to explore mechanisms behind complex functions and the notion of social cognition applied in social psychology and cognitive neuroscience [27]. Social neuroscience denotes an integrative multilevel approach (from the neural level to the social level) proposed by John Cacioppo and Gary Berntson in 1992 to guide the interpretation of neurobiological findings based on three analytical determinants:

i. Multiplicity, an event (e.g., empathy) belonging to a level of organization (e.g., neural), can take multiple determinants through multiple levels of organization related to biological, psychological, or social domains;

ii. Non-additive, the reductionist approximation of the phenomenon does not always explain its totality until an analysis of the phenomenon is performed through a multilevel perspective; and

iii. Reciprocity, there is evidence of influences and relationships between social and biological factors when determining behaviors [28].

In the following section, we present theoretical proposals and neuroimaging studies exploring affective and cognitive processes related to empathy, framed in the social neuroscientific perspective.

3 Empathic Processes and Their Neuroanatomical Correlations

3.1 Emotional Experience

In 1872, the naturalist Charles Darwin revealed the relevance of communicating inner mental states through emotional expressions, which could be similar among young and adult people of different origins and also have an equivalent between humans and animals. This evolutionary analysis of emotions, complemented with transcultural studies performed with Papua New Guinea tribe members, was the basis of Paul Ekman´s proposal about universal emotional expressions [29]. According to Ekman, these expressions are not culturally dependent but biologically originated to denote fear, happiness, anger, sadness, disgust, contempt, and surprise. This universalistic viewpoint was then refined to classify emotions in a negative category triggering avoidance behaviors and uncomfortable experiences and a positive category related to wellness experiences that may be intentionally reinforced [30]. Although the universalistic view assumes that social learning may generate certain individual or collective variances in emotional expressions; it also suggests that different emotional experiences are attached to differentiated neuronal and physiological functions [31]. Coincidentally, functional magnetic resonance imaging (fMRI) has evidenced how the induction of positive emotions, such as happiness, elicits brain activation in the right dorsolateral prefrontal cortex, left dorsal-posterior portion of the cingulate cortex, and bilateral cerebellum. In contrast, the induction of negative emotions, such as sadness, relates to activations in the ventrolateral prefrontal cortex bilaterally, left anterior cingulate cortex, and superior temporal cortex bilaterally [32]. Nevertheless, studies have also shown that both positive and negative experiences share a common brain activation in limbic and paralimbic regions [32, 33]. Therefore, common brain limbic activation may imply a basic neural function underlying any emotional experience, while differential brain activation in the cortex may reflect positive or negative emotional properties triggering different related behaviors and experiences, such as wellness or avoidance.

Even the universalistic viewpoint of emotions may explain brain function differences as related to basic or universal feelings, differences can also be explained by the diversity of personal histories and experiences that individuals may have throughout their life, and which represent a variety of positive and negative valences given to a multiplicity of circumstances [34]. This perspective is focused on individual differences assuming that interoceptive and exteroceptive processes shape emotional experiences. Interoception regards homeostasis and conscious or non-conscious representations of physiological states, such as heartbeat, respiratory, or thermal sensations, while exteroception implies sensitive systems, such as vision, sound, or touch, allowing for people's bodily awareness and movements. When facing any circumstance involving emotions, both processes generate a brain map of embodied states that are translated into action programs guided by the function of several brain regions, such as the brainstem, hypothalamus, thalamus, and amygdala. These brain regions converge to generate mental images set on the somatosensorial cortex and insula, engaging the conscious experience of body states and subsequent evocation as feelings [35–37]. Since an individual's body exists in social spaces with continued interpersonal interactions, these somatic maps are unique and individually represented but modulated by social interfaces. Thus, emotional experiences are infused with beliefs and objects related to particular social spaces. Individual social experiences may serve as emotion regulation strategies and action programs to deal with psychological and social circumstances [38, 39]. Such strategies involve complex neural and cognitive functions set on phylogenically recent evolved cortical brain regions, including the dorsomedial prefrontal cortex, orbitofrontal cortex, lateral prefrontal cortex, and anterior cingulate cortex related to decision-making, planning, and goals [40]. Particularly, the ventromedial prefrontal cortex may underlie such strategies since this brain region is considered a neural center for conceptual and mnemonic representations, decoding the affective meaning of social situations, and allowing people to change their physiological and emotional responses processed in subcortical brain structures [41]. Hence, brain functions involving emotional experiences reflect not only biological adaptions to communicate inner states but also social adjustments needed to recognize our own reactions and to know and interact with others in particular spaces.

The individual–collective and biological–social dichotomist properties of emotional experiences proposed above are embedded in empathy since, as evolutionary perspectives suggest, it entails a biological mechanism to favor social interactions; but also, it implies individual and collective experiences, as phenomenological traditions indicate. Nonetheless, empathy may emphasize a particular kind of emotional experience related to the perception of others' suffering [42]. Accordingly, fMRI findings of empathy reveal bilateral activation in the supplementary motor area when people perceive various emotional scenarios involving otherness; the left middle temporal gyrus is activated when reading emotional situations, while the right amygdala and right dorsal lateral prefrontal cortex are activated when watching emotional facial expressions. Interestingly, when perceiving others' physical pain, vast brain activations are seen in the bilateral inferior parietal lobule, left middle cingulate cortex, left middle occipital gyrus, bilateral fusiform gyrus, and left anterior insula [43].

Therefore, empathic experiences may relate to brain functions involving motor processes, linguistic domains, and facial expression recognition. Further, the perception of another's pain may be salient as inferred by the brain functions related to sensory integration (inferior parietal lobule), firsthand pain perception (cingulate cortex), visual association (occipital cortex), facial recognition (fusiform gyrus), and interoceptive processes (insula). These findings may denote a human bias to process negative feelings, as suggested by the vast vocabulary, terms, and cultural representations referring to negative experiences impacting memory, learning, attention, and moral judgments [44]. This bias may be essential to infer another's needs and motivate cooperation in social dynamics.

Though there is possible bias, empathy cannot be reduced to negative experiences, as showed in some spiritual practices favoring others' wellness as illustrated by the Franciscan or the Theravada traditions mentioned earlier [21, 22]. In this sense, positive emotions involving wellness experiences and concerning empathy have also shown complex neurobiological functions. For example, pleasant tactile stimulation either directly received or observed in another person activates the medial orbitofrontal cortex bilaterally [45]. The mere intention to empathize with positive situations elicits the activation of the septal area and the ventromedial prefrontal cortex involving the reward brain system and prosocial motivations [46]. Moreover, a common brain activation involving the medial prefrontal cortex, insula, and inferior frontal cortex has been observed when people read emotional statements representing either positive or negative situations affecting them or others [47]. Hence, a general brain function involving interoception and memory is involved when experiencing empathy. As Stein proposed, different brain functions are involved besides the mnemonic and auto-reflexive processes when positive or negative situations are faced.

3.2 Emotional Contagion and Simulation

Some authors argue empathy implies an emotional contagion of another's emotional states, which influences empathic accuracy or the ability to infer their feelings and thoughts satisfactorily [48]. Some authors understand this contagion as a human tendency to imitate and synchronize with others' expressions, vocalizations, postures, and movements, affecting their own emotional experience expressed in spaces sharing certain common cultural traits, such as ethnicity or language

[49–51]. Similarly, when people perceive sadness in those who belong to their ingroup identity, an asymmetric electroencephalographic activity in the right prefrontal cortex is presented and associated with negative emotional experiences. However, this brain functional asymmetry is not manifested when the observer presents prejudiced and racist attitudes toward people from an outgroup who express sadness [52]. Controversially, empathic accuracy manifested by equivalent physiological responses (such as cardiac beat or skin conductance level) measured in people from different ethnic origins (African–American, European–American, Mexican–American, or Chinese–American) attributed to others when experiencing emotions has been reported independently of ethnic origin. In this last case, empathic accuracy may be not greater or influenced by their ingroup identity or by the perception of ethnic differences in others [53]. Both studies above may imply that, though emotional contagion may be an automatic process favoring empathic accuracy, this process may be inhibited or mediated by certain learned cultural traits that are attributed to others (based on aspects such as prejudice or racism).

About emotional contagion, simulation theory proposes that people make pre-reflective, intuitive, and non-conscious representations of the other's mental states and actions through a mental simulation, in which perceptual and action patterns are generated in the perceiver's own mind and is represented as equivalent to the other's states and actions [54, 55]. One of the actual empirical bases of simulation theory is the mirror neurons, which exist in the F5 frontal area of non-human primate brains and project to the inferior parietal lobe. These projections shape a putative sensorial-motor frontoparietal network to pre-reflectively, pre-linguistically, and implicitly understand another's action, internally represent another's actions, and generate goal-directed behaviors [56, 57]. In humans, the proposed putative network involves ventral and dorsal portions of the premotor area (Brodmann areas 6 and 44), supplementary motor area, inferior and superior parietal cortex, medial temporal cortex, medial portions of the cingulate cortex, somatosensory areas embracing BA 1, BA 2, and BA 3, and cerebellum [58].

Some authors have suggested that representations of motor actions in the putative mirror neuron system play a causal role in emotional contagion [59]. A recent fMRI study suggests that the human mirror neuron system participates in discriminating emotional facial expressions since activations in the fusiform gyrus, superior temporal sulcus, amygdala, insula, inferior parietal lobe, and frontal regions embracing BA 44 are activated when performing these types of tasks [60]. However, the fundamental role of the human mirror neuron system in empathy is controversial because the function of this system can vary according to social learning and is influenced by different types of visual and social stimuli. Furthermore, neuroanatomical equivalences of the mirror neurons system between humans and non-human primates have been difficult to shape. The F5 area in non-human primate brains has been suggested as homologous to the Broca's area (BA 44) located inside the inferior frontal gyrus of the human brain. However, this brain region is marginally involved in watching actions that involve grasping objects or communicative mouth movements. Therefore, motor outputs from said area to control hand or phonoarticulatory movements have not been demonstrated [61, 62].

Despite the controversies, the mirror neurons paradigm and the simulation theory have allowed some authors to develop proposals regarding automatic mental representations to understand others' emotional states. One of these proposals is the perception–action model [63], which considers empathy as a process in which internal representations are generated from perceiving another's state to activate somatic and visceral responses automatically. This process allows subjects to understand or infer the other's internal states through representations distributed in their own brain function to facilitate the observer's visceral responses and actions. In line with this model, empathic concern for others' affliction is predicted by the functional connectivity of a network comprised of the inferior frontal gyrus, inferior parietal lobule, superior temporal sulcus, insular cortex, and amygdala favoring visceral functions, emotional and painful experiences, and motor behaviors [64]. Additionally, empathy considered by this meaning has also contemplated the inclusion of certain social–emotional situations. For example, empathy toward victims who express fear and avoid an aversive situation provokes similar experiences in the observers associated with activations in the right insula, bilateral inferior parietal lobe (BA 40), and right fusiform gyrus (BA 20/37) whose functions relate to the identification of facial expressions [65]. Moreover, both the experiences of physical pain and the perception of pain felt by a loved one are correlated with the activation of a similar neural network involving the bilateral anterior insular cortex, anterior cingulate cortex, lateral cerebellum, and the brainstem [66]. The anterior insular cortex and the dorsal portion of the anterior cingulate cortex are proposed as a network involved in firsthand pain experiences and the perception of pain inflicted on others [67].

An interesting finding reveals that the functional connectivity between the left amygdala and the ventromedial prefrontal cortex relates to regulating emotions and understanding others' mental states. This connectivity is present when people perceive pain in others, but only when the pain is intentionally caused by others [68]. Another finding indicates that when drug users are considered responsible for getting acquired immunodeficiency syndrome (AIDS), there is a reduced empathic disposition for their pain. This is associated with lower responses in the anterior cingulate cortex involved in the pain network based on firsthand and another's pain [69]. Therefore, the brain function related to emotional contagion, simulation, and empathy for another's pain is not necessarily automatic but can be modulated by complex moral traits and the intentionality attributed to their actions.

3.3 Personal Distress and Emotional Regulation

Personal distress is considered an empathic dimension related to high and automatic contagion of another's experiences, particularly negative ones. Unlike the emotional contagion mentioned in the section above, personal distress involves states of alarm, anguish, or disturbances. These states reduce self-regulation and one's own reflection on the aversive experience [70, 71]. Some experimental findings show the complex consequences of this phenomenon. For example, when people watched videos showing similar aversive experiences to those lived by the observers, lower recognition of the other's emotions, difficulties to imagine the other's reactions, and reduced empathic accuracy were presented [72]. Therefore, sharing similar negative experiences between the person who is perceiving and the one who is being perceived promotes high personal distress and, paradoxically, this affects the ability to understand the other person's emotions. Consequently, personal distress and self-regulation may represent a dualistic process needed to empathize. fMRI studies have shown that different brain regions regulate this double process. When watching pictures that elicit aversive experiences and personal distress, lateralized regions are activated in the left hemisphere, including the medial prefrontal cortex (BA 9/10), the posterior cingulate (BA 23/30), and the amygdala, related to the amplification of affective labels denoting an increase in emotional intensity. In contrast, when participants performed self-regulation to evaluate their own and another's emotions presented in the pictures, activations were presented in the dorsolateral (BA 9/8) and orbital (BA 44) prefrontal regions of the right hemisphere associated with episodic memory and judgment as well as a decrease in the activation of the bilateral amygdala, which is typically involved in the elicitation of negative emotional expressions. The joint activation of these last three regions is proposed as self-regulative functions since they may mediate affective responses and cognitive control [73]. Moreover, it seems that the self-regulation needed to reduce personal distress implies different brain functions depending on whether the circumstances being faced represent physical or emotional pain. For example, when reading narratives denoting emotional pain, activation in the right lateral prefrontal cortex and a decreased activation in the bilateral amygdala has been observed. However, when the narratives denoted physical pain, activations were observed in the anterior insula and the right frontal operculum, together with a decreased activation in the motor and somatosensorial cortices. Furthermore, brain functional connectivity associated with emotional pain indicates an inhibitory effect of the right prefrontal cortex over the amygdala, while for physical pain, the superior temporal sulcus, associated with mentalizing, has an inhibitory effect on the amygdala [74]. The process is neuroanatomically complex. fMRI studies on functional brain connectivity have shown that when people intentionally decide to empathize while perceiving another's feelings, strengthening of the functional connectivity is presented in a large neural network that involves the inferior, middle, and superior frontal gyri, premotor cortex, cingulate cortex, primary and secondary somatosensorial cortices, the inferior parietal and the right posterior inferior temporal cortices. Cognitive functions associated with this network embrace embodied cognition, motor and perceptual bodily aspects, reasoning and categories of concepts, and analysis of the environment [75].

Culture has been mentioned as a crucial element to be considered for personal distress and self-regulation since living in collectivist or individualistic societies may influence both processes [76]. Experimental research in this sense has shown that people belonging to collectivistic cultures with strong social interdependence manifest a greater activation in the left dorsolateral prefrontal cortex (BA 9), related to inhibitory processes and emotional regulation when intentionally empathizing with the other's anger. In contrast, people belonging to individualistic cultures that emphasize individual independence manifest activations in brain areas involved in perspective-taking and emotional understanding, such as the right temporoparietal junction (BA 40), right superior and inferior temporal gyrus (BA 20/22), and right middle insula (BA 13) [77].

3.4 Cognitive Components of Empathy and Theory of Mind

As proposed by Edith Stein and as inferred from Scotus' proposal about human beings, empathy does involve not only immediate automatic reactions and contagion but also self-reflective cognitive mechanisms to understand others' intentions, beliefs, and desires allowing the perceiver to control their experience and learn about himself and others through perceived actions. Social psychology had previously exposed this mechanism as attribution [78, 79], which the current cognitive trend calls the theory of mind (ToM) or mentalizing. These processes are considered cognitive components of empathy; they allow the person to attribute or ascribe certain mental states to others, such as feelings, ideas, or thoughts, through a causal inferential process based on the perception of their behaviors and expressions, knowing that these belong to the other but not to oneself [80–84].

Mental state attribution has been correlated to activations in two crucial brain regions: the superior temporal sulcus and medial prefrontal cortex [85]. However, attribution or ToM concerns a variety of elements regulated by different brain regions. For example, the superior temporal sulcus and the temporoparietal junction functions are involved in incorporating others' perspectives and predicting their mental states; the temporal pole and medial prefrontal cortex are associated with storage and interpretation of information related to people based on context, and precuneus is related to mental imagery [86, 87]. The right temporoparietal junction also has a prominent role since its function has been related to generating, testing, and correcting predictions based on external sensory events attributed to others [88, 89].

Though ToM can be considered part of a broad concept of empathy, different brain functions have been associated with either attribution of mental states or typical empathy (inferring others' feelings). For example, when attributing intentions and empathizing with people with schizophrenia, both processes share activations in the medial prefrontal cortex, temporoparietal junction, and temporal poles. Differentially, ToM is associated with activations in the lateral orbitofrontal cortex, middle frontal gyrus, cuneus, and superior temporal gyrus, while emotional attribution has been associated with activations in the paracingulate, anterior and posterior cingulate, and amygdala [90].

Experimental studies about the neural correlates of ToM and emotional attribution using fMRI are vast; thus, we present two of the most relevant for our study. The first focused on how participants reacted to descriptions by Corradi-Dell'Acqua, Hofstetter, and Vuilleumier (2014). In their study, a group of participants were asked to read stories describing people in various scenarios. Then, they were asked to attribute either beliefs or emotional states or somatic pain states to the participants. When attributing both beliefs (non-emotional mental states) and emotional states with no pain, activations were presented in the temporoparietal junction, the medial temporal gyrus, and the dorsolateral prefrontal cortex, while for stimuli presenting painful somatic attributions, activations were observed in the supramarginal gyrus, middle cingulate cortex, and middle insular cortex [91]. In the other study, Tholen, Trautwein, Böckler, Singer, and Kanske (2020) asked a group of participants to react to videos showing people in different emotional and non-emotional situations and then rated their own feelings assuming that this exercise implied both empathic and ToM processes, attributing feelings and/or non-emotional states to the people in the videos. When processes were assumed as empathic, brain activations were presented in the anterior insula, anterior cingulate cortex, dorsomedial prefrontal cortex, inferior frontal gyrus, dorsal temporoparietal junction, and supramarginal gyrus. However, when assuming ToM processes, activations in the ventral temporoparietal junction, superior temporal gyrus, superior temporal sulcus, and temporal poles were observed [92]. These results suggest a broad concept of empathy involved in ToM processes, which includes processing perceptual, emotional, and social information that is then integrated into higher-level processing centers. According to Mitchell and Phillips (2015), these centers include the medial prefrontal cortex, amygdala, dorsolateral prefrontal cortex, temporoparietal junction, and superior temporal sulcus [93].

Further, studies suggest that ToM and the inference of others' emotional states imply self-projection or mental processes projecting the self into another time, place, and another's perspective. These processes involve the function of frontal regions related to planning and the regulation of changes in perspective, but also temporoparietal regions related to episodic memory, which allows predictions based on own past experiences [94]. Language, considered a fundamental cultural trait, has an important role in self-projection processes since stimuli with semantic content describing internal states of other people preferentially activate a neural system involved in ToM compared to visual stimuli with no linguistic properties [95].

3.5 The Notion of the Self and the Other

Social psychology has argued that humans have an egocentric bias toward others' behaviors, feelings, opinions, and characteristics [96]. For some authors, this last idea is evident because it is always from the self that motivations emerge to preserve and stabilize one's own representations, so others' impressions are made under a self-constructed worldview [97]. Therefore, as marked by Duns Scotus, the distinction between the self and the other is crucial for empathy because although this process may imply imposing one's own reflections or projections onto others, even a kind of experiential fusion, the other is always the one who is interpreted. This complexity is also reflected in neuroimaging studies. Although, certain brain regions, e.g., the medial prefrontal cortex, posterior cingulate cortex, precuneus, and the superior temporal gyrus, are activated when inferring one's own and others' emotional responses, the ventrolateral region of the prefrontal cortex is activated when attributing emotional states to others. However, the dorsomedial portion is associated with the emotional distinction of the self and the other [73]. Particularly, the dorsomedial prefrontal cortex also involves monitoring a possible conflict between one's perspective and the other's, as identified when watching video clips and being required to infer a person's false belief regarding the location of an object [98].

The notion of the 'self' in empathy implies a neural network involving recently evolved phylogenetical brain regions, such as the prefrontal cortex, where functions relate to high cognitive processes needed to elaborate concepts and abstractions. However, the self is not only an abstract and conceptual attribute because this network also involves processing bodily representations concerning the inferior parietal cortex, which coordinates and contrasts sensorial-motor information to distinguish the self and the other cognitively [99, 100]. This kind of embodied cognition is relevant for this interpretation. Some experimental studies support this last idea in relation to empathy. The repetitive transcranial magnetic stimulation disrupts the cortical activity in the right inferior parietal lobe, causing a deficit that allows for the differentiation between one's face and another's [101]. An experiment using positron emission tomography was made to identify brain function when participants adopt their own perspective or their mothers' perspective in situations that cause social emotions such as embarrassment, pride, shame, guilt, or admiration. When processing social emotions, activation in the amygdala was present from either their own perspective or another's perspective. But, when adopting the other's perspective (mothers), activations were presented in the medial prefrontal cortex, left temporoparietal junction, left temporal pole, and right inferior parietal cortex, while when adopting one's own point of view, postcentral gyrus in the somatosensory cortex was observed [102].

The concept of self is complex because it involves different kinds of cognitive processes. Two types of selves have been proposed: a 'phenomenal self', shaped by the present and immediate experience allowing for a person's self-awareness, and a 'referential self', constituting a state of reflective consciousness that extends subjectivity into a temporal flow where the self is recreated based on past experiences [103]. For the phenomenal self, various brain regions are implicated: insula and somatosensory cortex relate to interoceptive and exteroceptive processing, ventromedial prefrontal and amygdala relate to affective evaluations, and posterior parietal cortex relates to conscious access referring to the body and contextual space. The referential self involves a transition process between pre-reflective experiences represented in somatic markers and temporal, autobiographical, and self-reflective processes related to the function of the medial prefrontal cortex, orbital medial prefrontal cortex, anterior cingulate cortex, medial parietal cortex, posterior cingulate cortex, and the retrosplenial cortex [104, 105].

4 Conclusion

The Empathic Being, shown at the beginning of this chapter, presents an allegory of Duns Scotus' notion of a 'human being'. His proposal comes from the Middle Ages and does not specifically mention empathy, but the attributes he associates with humans coincide with theoretical proposals and contemporary empathy studies. According to Scotus, the individual is defined as intellectual; the self is a unique and individual identity, but it is built based on the interactions, thoughts, and feelings shared with other people. Additionally, the individual has a mutable existence; change is triggered by lived experiences throughout their life. Although indeed, Duns Scotus does not mention culture and social life as a part of these experiences; we are currently aware of how these aspects necessarily include beliefs, education, and memories located in the cultural world the individual moves in. The human being also self-reflects and projects, which coincides with einfühlung and Vischer and Stein's phenomenology. Scotus argues that this reflection also allows for solidarity and action in the face of other people's pain, which is consistent with Titchener's notion of empathy and is taken up by contemporary cognitive sciences. Additionally, according to Scotus, the individual is not an abstraction but has qualities based on a biological body that is also evidenced in the interceptive, motor, and sensory functions that make up part of the current concept of empathy.

Empathic being was made with yarn over beeswax following an artisanal technique used by shamans and visionaries from the Huichol or Wixáritari communities. The purpose behind this exercise was to show that the role scientists play across diverse human communities in the twenty-first century might not be too different from the role of shamans and visionaries. After arduous research and reflection, scientists are in charge of putting together the various aspects that make up our world's reality. Just as in Huichol art pieces, the knowledge poured from the scientific community functions as a guide for recognizing what makes us human beings and guides our behaviors. Therefore, scientific labor is a commitment to our community. Maybe this labor and commitment contribute to elucidating empathy as that which the writer José Saramago refers to in Blindness: 'Inside us there is something that has no name, that something is what we are'.

As evolutionary perspectives suggest, empathy may imply a sense of cooperation among members of a species. This theory explains the biological bases of neural mechanisms present in the body to infer others' emotional states, emphasizing needs, affliction, and well-being [106]. In human beings, neurobiology and cultural traits, including norms, beliefs, identities, and social learning, underlie empathy [107], as neuroimaging studies have verified. Therefore, to identify the neuroanatomical bases of empathy and have their cognitive and affective correlated functions be useful for mental health, neurological diseases, and general comprehension of human behavior, empathy should be analyzed considering biological–social relations based on interdisciplinary and transdisciplinary approaches. We consider it is particularly crucial to focus on the interdisciplinary nature of social neuroscience. This relatively new neuroscientific area focuses on the elucidation of neurobiological mechanisms involved in social interaction, behavior, and cognition, taking into account how affects, motivations, and perceptions can be modeled by social contexts [27, 108, 109]

Some authors state that social psychology is interested in how and when empathic attributions emerge, while social neuroscience attempts to describe where (in the brain) these processes are elaborated [84]. Nevertheless, as shown in this chapter, neuroimaging findings do not solely indicate a brain location. Rather, neuroanatomy allows researchers to infer the cognitive, affective, and social processes and elements that make up empathic experiences. These inferences may be used to contrast and build new theoretical approaches concerning empathy. However, experimental neuroimaging studies have applied various cognitive designs, populations, and phenomena that may complicate consensual results and cross-cultural comparisons. A serious discussion should be held regarding how social neuroscience may offer directions to contribute to this experimental field to have comparable findings across different populations throughout the world.

We argue that human empathy is a broad concept involving complex and overlapping emotional, cognitive, and social components that allow people to recognize another's mental and physical states and motivate prosocial behaviors [110–113]. This complexity and its overlapping nature are not only present in the concept of empathy itself but also among the neuroanatomical connections evidenced in this chapter. Therefore, the neuroanatomical bases of empathy involve brain regions for emotional and motor experiences related to simulating and mimicking the physical or psychological state of the other (e.g., insula and anterior cingulate cortex), emotional contagion and recognition (e.g., inferior parietal lobe, premotor cortex, inferior frontal gyrus, and amygdala), cognitive perspective-taking and mentalizing (e.g., temporoparietal junction, superior temporal sulcus, temporal pole, medial and dorsolateral prefrontal cortex, precuneus, posterior cingulate cortex), or inference of emotional states (e.g., ventromedial prefrontal cortex) [65, 111, 114]. Finally, these neuroanatomical bases of empathy involve brain regions that distinguish between cognitive and emotional states of the self and the other (e.g., dorsomedial and ventromedial regions of the prefrontal cortex, inferior parietal lobe, and frontal pole) [115].

A particularly distinctive brain region we argue is involved in empathy is the cerebellum. Although the cerebellum is typically related to motor processes, its functional relevance for empathy may be explained as a multiple domain region involved in the mirror neuron system and mentalizing [58, 116, 117], the perception of others' physical pain [66, 118], the experience and regulation of emotions [119], identifying one's own perspective and that of others [120, 121], and in understanding social contexts [116, 122].

In conclusion, we offer a neuroanatomical map showing the brain regions that participate in different components or processes related to empathy described throughout this chapter (Table 1; Fig. 1). Based on social neuroscience, this map also includes information about brain correlates of social interaction, the associations between context and social stimuli, and the modulation of empathic experiences aimed at guiding socially oriented responses (e.g., anterior cingulate cortex, dorsolateral prefrontal cortex, medial temporal lobe, orbitofrontal cortex, anterior insula, and frontal pole) [110, 123–125]. Also, we provide a pictorial representation to illustrate and explain empathic processes in the human social world (Figs. 2 and 3) and a model involving different disciplines that contribute to understanding empathy from the social neuroscience approach (Fig. 4).

Table 1 Brain regions involved in empathic components and processes

Empathic components and processes	Related brain regions
Emotional experience	– Right amygdala – Septal area – Left middle cingulate cortex – Left anterior insula – Right dorsal lateral prefrontal cortex – Bilateral supplementary motor area – Medial orbitofrontal cortex – Ventromedial prefrontal corTex – Medial prefrontal cortex – Inferior frontal gyrus – Inferior parietal lobule – Cerebellum
Emotional contagion and simulation	– Amygdala – Anterior insula – Anterior cingulate cortex – Inferior frontal gyrus – Right prefrontal cortex – Superior temporal sulcus – Inferior parietal lobe – Cerebellum
Personal distress	– Left amygdala – Left posterior cingulate cortex – Left medial prefrontal cortex
Emotional regulation	– Anterior insula – Dorsolateral prefrontal cortex – Right lateral prefrontal cortex – Right orbitofrontal cortex – Right frontal operculum – Superior temporal sulcus – Cerebellum
Mentalizing	– Amygdala – Anterior cingulate cortex – Posterior cingulate cortex – Medial/dorsolateral prefrontal Cortex – Right temporo-parietal junction – Superior temporal sulcus – Middle temporal gyrus – Temporal pole – Precuneus – Cerebellum

(continued)

Table 1 (continued)

Empathic components and processes	Related brain regions
Self and other distinction	– Posterior cingulate cortex
	– Dorsomedial prefrontal cortex
	– Ventromedial prefrontal cortex
	– Frontal pole
	– Superior temporal gyrus
	– Left temporal pole
	– Left temporoparietal junction
	– Precuneus
	– Inferior parietal lobe
	– Cerebellum
Social context	– Anterior insula
	– Anterior cingulate cortex
	– Frontal pole
	– Dorsolateral prefrontal cortex
	– Orbitofrontal cortex
	– Medial temporal lobe
	– Cerebellum

Core Messages

- Empathy as a response to another's pain or need is the expression of biological properties embracing our species' natural history and cultural scenarios built throughout human civilizations.
- The cognition that guides our empathic responses to a particular moment necessarily involves our bodily and affective individual history stored in our memory.
- To profoundly comprehend the complex neurobiology of empathy, neuroscientists should listen and engage with theories and discussions focused on other elements besides the brain.
- Empathy may be one of the most sublime ways humans have to know ourselves by recognizing ourselves in others.

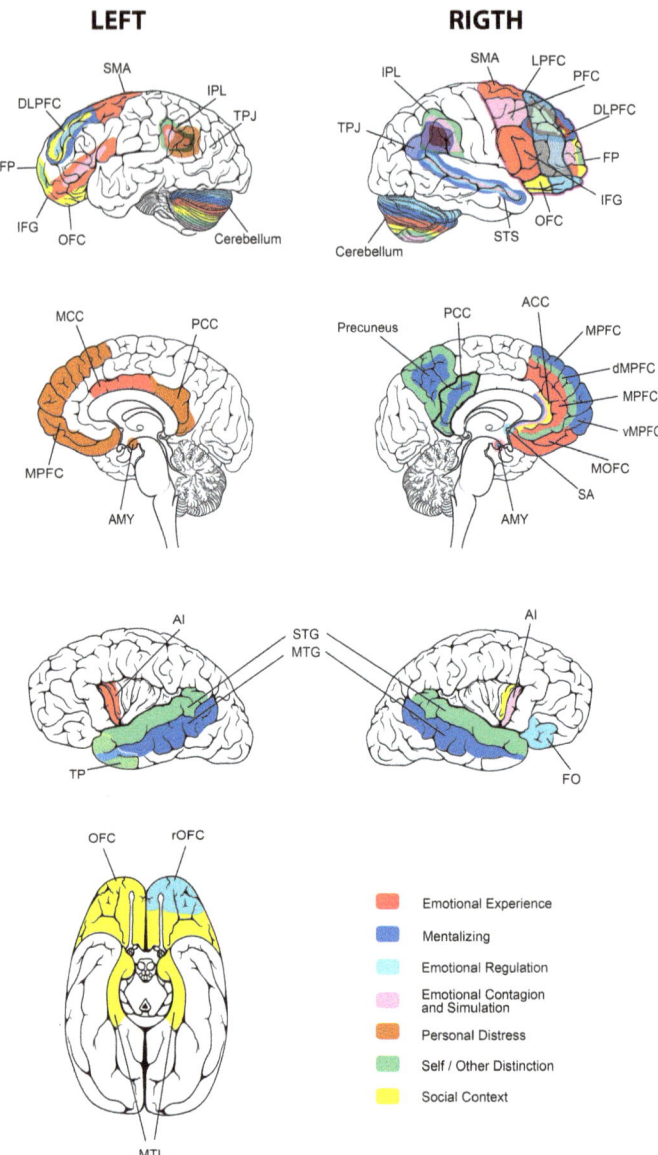

Fig. 1 Neuroanatomical map showing brain regions identified by neuroimaging studies and which functions have been related to empathic components and processes. OFC, orbitofrontal cortex; rOFC, right orbitofrontal cortex; LPFC, lateral prefrontal cortex; FO, frontal operculum; AI, anterior insula; AMY, amygdala; MCC, middle cingulate cortex; DLPFC, dorsolateral prefrontal cortex; PFC, prefrontal cortex; SMA, supplementary motor area; MOFC, medial orbitofrontal cortex; SA, septal area; vMPFC, ventromedial prefrontal cortex; MPFC, medial prefrontal cortex; dMPFC, dorsomedial prefrontal cortex; IFG, inferior frontal gyrus; ACC, anterior cingulate cortex; PCC, posterior cingulate cortex, TP, temporal pole; STS, superior temporal sulcus; STG, superior temporal gyrus; TPJ, temporoparietal junction; IPL, inferior parietal lobe; FP, frontal pole; MTL, medial temporal lobe; MTG, medial temporal gyrus

Fig. 2 Quotidian empathic scene. A landscape in Mexico City shows various people and institutions, making an urban common social scene in which two main characters are exhibited: a man (empathic perceiver) who leans forward at a homeless person (other) who is perceived in a vulnerable social circumstance

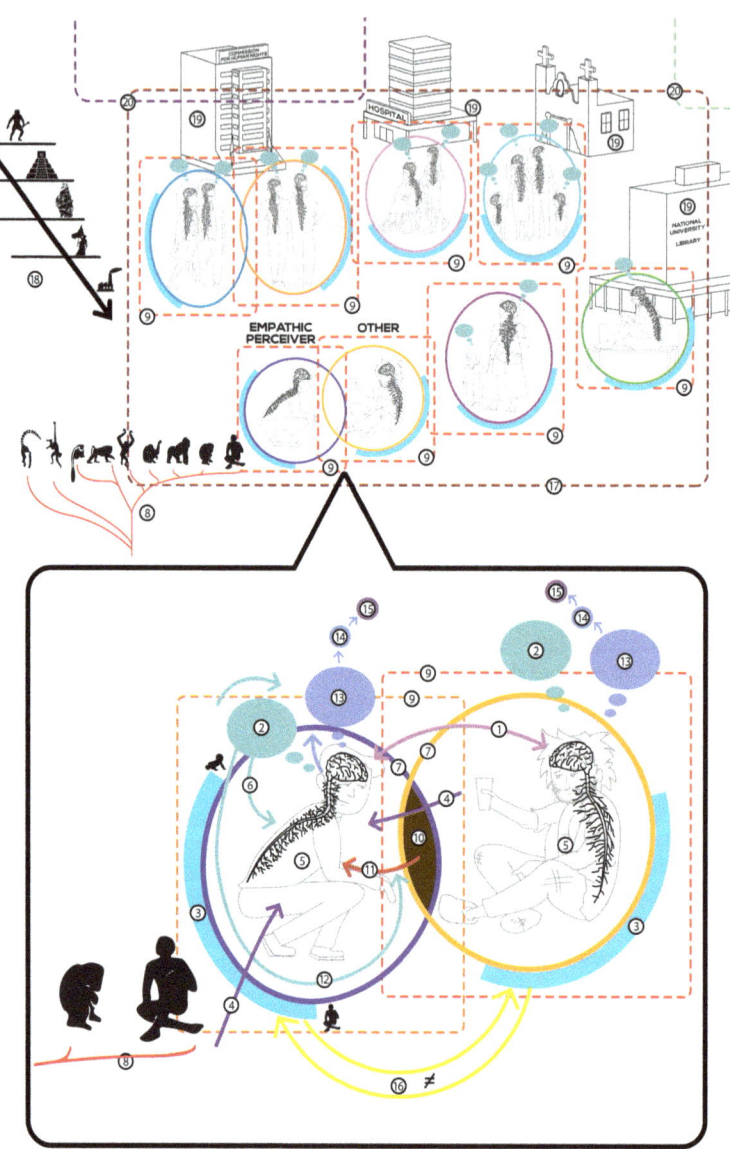

◄ **Fig. 3** Pictorial exemplification of empathic processes expressed in a quotidian human social scene. The central nervous system is enhanced into the body of each person in the scene. This represents the neuroanatomical route to perceive and to move in the social space, as well as to empathize with other persons. The two main characters are closely illustrated in the amplified frame below the scene. By leaning forward to be face to face, the empathic perceiver and the homeless person mutually recognize their facial and bodily emotional expressions, for example, sadness or affliction (1). Their mutual emotional recognition arouses memories in the empathic perceiver and in the homeless person, which may be linked to similar past experiences, for example, those representing helping behaviors or actions in response to people in vulnerability. These mnemonic processes are also performed for all the persons in the scene when moving in their quotidian social space (2). Memories accomplish the own personal experiences guiding emotional recognition and representing the individual's ontogeny and history. Individuals are presented for all the persons in the scene, so expressions and actions in response to another's vulnerability may differ from one person to another (3). When empathizing, exteroception is presented to process visual, olfactory, auditory, or tactile sensory information allowing the empathic perceiver to bodily awareness and movements toward the homeless person (4). Simultaneously, interoception regarding homeostasis and physiological states is presented in the empathic perceiver (for example, heartbeat, respiratory, or thermal sensations). Interoception may be presented in the homeless person too when attending to the other's closeness and may elicit his expressions influencing the empathic perceiver's reactions (5). The quality of interoception is influenced by memories associated with the actual perceived person and context (6). Memory, exteroception, and interoception configure the quality of the emotional experience outlining the empathic perceiver when facing the homeless person's circumstance. Emotional experiences are bordered by the individual's history and also outline each person in the scene but influenced by the people and particular circumstances perceived in their quotidian social moment (7). Emotional experience and expressions concerning the empathic perceiver when facing the homeless person are partially defined by the natural history of the human species, which have configured evolved cooperative mechanisms in response to another's needs (8). Besides the influence of those evolved mechanisms, the empathic perceiver's emotional experience and expressions are framed by his personal learning of cultural traits representing his community values, beliefs, and knowledge, for example, to consider helping a person in vulnerability as a correct/incorrect or a good/bad action. However, each person and group in the social scene keep their particular frame, representing a variety or learned of cultural traits defining the diversity of social relations (9). The empathic perceiver may experience an emotional contagion of the homeless person's state since both persons share certain evolved physiological mechanisms and personal learned cultural traits configuring their respective experiences (10). This emotional contagion felt by the empathic perceiver could elicit personal distress and/or simulation of the other's expressions influencing both, interoceptive processes and the accuracy to infer the homeless person's expressions (11). Personal distress and/or simulation could be intentionally regulated by memories and learned affective and cognitive strategies, attributions, and values (12). Emotional experience outlining both the empathic perceiver and the homeless person involves individuals' history, memories, interoception, exteroception, emotional contagion, learned values, and beliefs influencing Theory of Mind or mentalizing processes (13), which induce attributions toward the other (14), and consequent decisions-making or actions considered as congruent with the experience (15). For example, if memories, emotional contagion, empathic accuracy, and attribution of vulnerability are linked to certain moral values and beliefs reinforcing cooperative attitudes, so the empathic perceiver may execute helping behaviors toward the homeless person in such circumstances, but his behavior may be different if the elements present a different shape. As well, if the homeless person understands the empathic perceiver's emotional expressions as linked to memories indicating helping behaviors, so his expressions communicating needs may be performed, but his expressions may be inhibited if the perceiver's expressions are linked to discrimination. Even the empathic perceiver's experience may imply emotional contagion and shared traits, and a self-other differentiation must be performed to identify own experiences and decisions (16). So, empathy when facing others' circumstances may allow for self-recognition and knowledge about oneself. The affective and cognitive processes exemplified through the empathic perceiver in the illustration are necessarily framed in a wide cultural setting that involves all the persons belonging to a community or a society (17). This cultural setting represents dynamic pieces of knowledge derived from human history (18) and are changeable, transmitted, and expressed through cultural devices, such as educational, political, civil, religious, or health institutions (19). The described empathic processes and their cultural determinants illustrate a particular cultural setting in Mexico City. It is true that human beings may share certain evolved mechanisms and global historical knowledge, which could elicit similar empathic experiences. Nevertheless, differences in empathy may be presented according to different cultural settings which intersections must be considered to understand cross-cultural concerns (20)

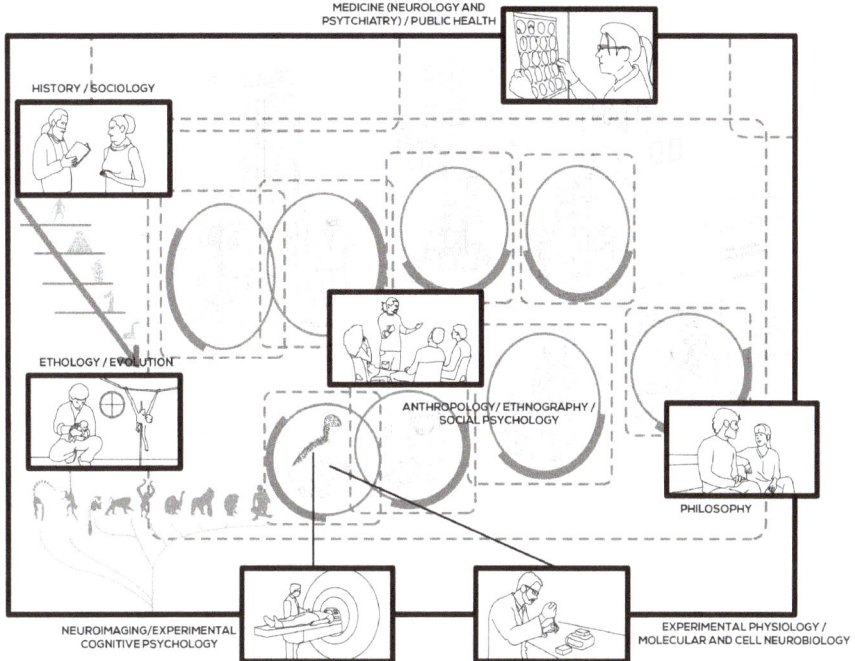

Fig. 4 Pictorial representation of disciplines contributing to the comprehension of empathy from a social neuroscience perspective in a quotidian human scene. It involves ineludibly interdisciplinary and transdisciplinary interactions where each discipline contributes to elucidate particular properties conforming to a wide empathic process. For example, neuroimaging and experimental cognitive psychology allow neuroscientists to identify brain anatomy and functions related to empathic processes by designing cognitive and affective particular tasks performed in functional magnetic resonance imaging. An extensive interpretation of these neural functions requires the analysis of emotional expressions and motivational evolved mechanisms embracing the natural history human species. Ethology and evolution may contribute to this line by performing comparative behavioral studies across primate species. In addition, experimental physiology and molecular and cell neurobiology allows for describing hormonal and/or neurochemical properties embracing the brain function identified by neuroimaging and related to evolved behaviors. Anthropology, ethnography, and social psychology may provide micro-social descriptions and community interventions to analyze when and how empathy is experienced and expressed in particular social and cultural domains revealing the quotidian people's life. So, cognitive tasks could be designed by using stimuli representing social circumstances to be significant for the participants performing the tasks. Knowledge about brain anatomy and functions, physiology, and cultural traits embracing empathy can be used in medicine to comprehend certain mental issues related, such as antisocial behaviors or affective disorders, but also to propose Public Health programs in which empathy could guide physician–patient relations and health education. Neurobiological, psychological, anthropological, and medical knowledge concerning empathy can be discussed from history and sociology to understand this knowledge as part of wider social structures defined and built during human history. All the elements can be integrated by philosophical approaches allowing scientists to elaborate new proposals not only about what empathy means but also concerning some related issues, such as human brain–mind relations, affective–cognitive dichotomies, or nature–nurture properties

Acknowledgements The authors thank Jessica González and Paulina Barrios from Acento Traducciones for revising the text, Cinthia Montiel for the art photography, and Salvador Nava for the design of Figs. 1–4.

References

1. Neurath J (2013) La vida de las imágenes. In: Arte huichol. CONACULTA, México
2. Mercadillo RE, García-Fuentes DO (2018) La persona de acuerdo con Duns Escoto: Una lectura desde la perspectiva neuroética. Ludus vitalis 26(49):183–206
3. Nomi JS, Scherfeld D, Friederichs S et al (2008) On the neural networks of empathy: a principal component analysis of an fMRI study. Behav Brain Funct 4:41. https://doi.org/10.1186/1744-9081-4-41
4. Bridge H (2011) Empathy theory and Heinrich Wölfflin: a reconsideration. J Eur Stud 41(1):3–22. https://doi.org/10.1177/0047244110391033
5. Zahavi D (2010) Empathy, embodiment and interpersonal understanding: from Lipps to Schutz. Inquiry 53(3):285–306. https://doi.org/10.1080/00201741003784663
6. Walton R (2001) Fenomenología de la empatía. Philosophica 24–25:24–25
7. Stein E (1994) Sobre el problema de la empatía. Universidad Iberoamericana, México
8. Davis MH (1996) Empathy: a social psychological approach. Westview Press, Boulder
9. Wispé L (1986) The distinction between sympathy and empathy: to call forth a concept, a word is needed. J Pers Soc Psychol 50(2):314. https://psycnet.apa.org/, https://doi.org/10.1037/0022-3514.50.2.314
10. Batson CD (2009) These things called empathy: eight related but distinct phenomena. In: Decety J, Ickes W (eds) The social neuroscience of empathy. MIT Press, Cambridge, pp 3–15
11. de Waal F (2008) Putting the altruism back into altruism: the evolution of empathy. Annu Rev Psychol 59:279–300. https://doi.org/10.1146/annurev.psych.59.103006.093625
12. de Waal F (1997) Bien natural. In: Los orígenes del bien y del mal en los humanos y otros animales. Herder, Barcelona
13. Viding E, McCrory E, Seara-Cardoso A (2014) Psychopathy. Curr Biol 24(18):R871–R874. https://doi.org/10.1016/j.cub.2014.06.055
14. Ferrari V, Smeraldi E, Bottero G, Politi E (2014) Addiction and empathy: a preliminary analysis. Neurol Sci 35(6):855–859. https://doi.org/10.1007/s10072-013-1611-6
15. Takeda A, Sturm VE, Rankin KP et al (2019) Relationship turmoil and emotional empathy in frontotemporal dementia. Alzheimer Dis Assoc Disord 33(3):260–265. https://doi.org/10.1097/WAD.0000000000000317
16. Mercadillo RE, Galvez V, Díaz R et al (2015) Social and cultural elements associated with neurocognitive dysfunctions in spinocerebellar ataxia type 2 patients. Front Psychiatry 6:90. https://doi.org/10.3389/fpsyt.2015.00090
17. Sacks O (2015) On the move. a life. Penguin Random House, U.S.A.
18. Richardson C, Percy M, Hughes J (2015) Nursing therapeutics: teaching student nurses care, compassion and empathy. Nurse Educ Today 35(5):e1-5. https://doi.org/10.1016/j.nedt.2015.01.016
19. Gholamzadeh S, Khastavaneh M, Khademian Z et al (2018) The effects of empathy skills training on nursing students' empathy and attitudes toward elderly people. BMC Med Educ 18(1):198. https://doi.org/10.1186/s12909-018-1297-9
20. Brice J, Kourie C (2006) Contemplation and compassion: the heart of a Franciscan spirituality of clinical pastoral supervision. J Pastoral Care Counsel 60:109–116. https://doi.org/10.1177/154230500606000111
21. Mercadillo RE, Fernandez-Ruiz J, Cadena O et al (2017) The Franciscan prayer elicits empathic and cooperative intentions in atheists: a neurocognitive and phenomenological enquiry. Front Sociol 2:1–18. https://doi.org/10.3389/fsoc.2017.00022

22. Leung MK, Chan CC, Yin J et al (2013) Increased gray matter volume in the right angular and posterior parahippocampal gyri in loving-kindness meditators. Soc Cogn Affect Neurosci 8(1):34–39. https://doi.org/10.1093/scan/nss076

23. Haidt J (2003) The moral emotions. In: Davidson RJ, Scherer KR, Goldsmith HH (eds) Series in affective science. Handbook of affective sciences. Oxford University Press, U.K., pp 852–870

24. Mercadillo RE, Barrios FA, Díaz JL (2007) Neurobiología de las emociones morales. Salud Mental 30(3):1–11

25. Fourie MM, Stein DJ, Solms M et al (2017) Empathy and moral emotions in post-apartheid South Africa: an fMRI investigation. Soc Cogn Affect Neurosci 12(6):881–892. https://doi.org/10.1093/scan/nsx019

26. Pagani C, Robustelli F (2010) Young people, multiculturalism, and educational interventions for the development of empathy. Int Soc Sci J 61(200–201):247–261. https://doi.org/10.1111/j.1468-2451.2011.01761.x

27. Grande-García I (2009) Neurociencia social: el maridaje entre la psicología social y las neurociencias cognitivas. revisión e introducción a una nueva disciplina. An Psicol-Spain 25(1):1–20

28. Cacioppo JT, Berntson GG (1992) Social psychological contributions to the decade of the brain: doctrine of multilevel analysis. Am Psychol 47(8):1019–1028. https://doi.org/10.1037/0003-066X.47.8.1019

29. Ekman P (1992) An argument for basic emotions. Cogn Emot 6(3–4):169–200. https://doi.org/10.1080/02699939208411068

30. Cacioppo JT, Gardner WL (1999) Emotion. Annu Rev Psychol 50(1):191–214. https://doi.org/10.1146/annurev.psych.50.1.191

31. Ekman P, Cordaro D (2011) What is meant by calling emotions basic. Emot Rev 3(4):364–370. https://doi.org/10.1177/1754073911410740

32. Habel U, Klein M, Kellermann T et al (2005) Same or different? neural correlates of happy and sad mood in healthy males. Neuroimage 26(1):206–214. https://doi.org/10.1016/j.neuroimage.2005.01.014

33. Lindquist KA, Satpute AB, Wager TD et al (2015) The brain basis of positive and negative affect: evidence from a meta-analysis of the human neuroimaging literature. Cereb Cortex 26(5):1910–1922. https://doi.org/10.1093/cercor/bhv001

34. Norris CJ, Larsen JT, Crawford LE, Cacioppo JT (2011) Better (or worse) for some than others: individual differences in the positivity offset and negativity bias. J Res Pers 45(1):100–111. https://doi.org/10.1016/j.jrp.2010.12.001

35. Damasio A (2010) Y el cerebro creó al hombre. Destino, Barcelona

36. Damasio A, Carvalho GB (2013) The nature of feelings: evolutionary and neurobiological origins. Nat Rev Neurosci 14(2):143–152. https://doi.org/10.1038/nrn3403

37. Damasio AR, Grabowski TJ, Bechara A et al (2000) Subcortical and cortical brain activity during the feeling of self-generated emotions. Nat Neurosci 3(10):1049–1056. https://doi.org/10.1038/79871

38. Lazarus RS (1991) Cognition and motivation in emotion. Am Psychol 46(4):352. https://doi.org/10.1037/0003-066X.46.4.352

39. Lazarus RS (1998) From psychological stress to the emotions: a history of changing outlooks. Annu Rev Psychol 44:1–21. https://doi.org/10.1146/annurev.ps.44.020193.000245

40. Kim S, Hamann SB (2007) Neural correlates of positive and negative emotion regulation. J Cogn Neurosci 19(5):776–798. https://doi.org/10.1162/jocn.2007.19.5.776

41. Roy M, Shohamy D, Wager TD (2012) Ventromedial prefrontal-subcortical systems and the generation of affective meaning. Trends Cogn Sci 16(3):147–156. https://doi.org/10.1016/j.tics.2012.01.005
42. Hoffman ML (2001) Empathy and moral development: implications for caring and justice. Cambridge University Press, Cambridge
43. Ding R, Ren J, Li S et al (2020) Domain-general and domain-preferential neural correlates underlying empathy towards physical pain, emotional situation and emotional faces: an ALE meta-analysis. Neuropsychologia 137:107286. https://doi.org/10.1016/j.neuropsychologia.2019.107286
44. Rozin P, Royzman EB (2001) Negativity bias, negativity dominance, and contagion. Pers Soc Psychol Rev 5(4):296–320. https://doi.org/10.1207%2FS15327957PSPR0504_2
45. Lamm C, Silani G, Singer T (2015) Distinct neural networks underlying empathy for pleasant and unpleasant touch. Cortex 70(9):79–89. https://doi.org/10.1016/j.cortex.2015.01.021
46. Morelli SA, Rameson LT, Lieberman MD (2014) The neural components of empathy: predicting daily prosocial behavior. Soc Cogn Affect Neurosci 9(1):39–47. https://doi.org/10.1093/scan/nss088
47. Perry D, Hendler T, Shamay-Tsoory SG (2012) Can we share the joy of others? Empathic neural responses to distress vs joy. Soc Cogn Affect Neurosci 7(8):909–916. https://doi.org/10.1093/scan/nsr073
48. Ickes W (2009) Empathic accuracy: its links to clinical, cognitive, developmental, social, and physiological psychology. In: Decety J, Ickes W (eds) The social neuroscience of empathy. MIT Press, Cambridge, pp 57–70
49. Hatfield E, Cacioppo JT, Rapson RL (1993) Emotional contagion. Curr Dir Psychol Sci 2(3):96–100. https://doi.org/10.1111/1467-8721.ep10770953
50. Hatfield E, Rapson RL, Yen-Chi L (2009) Emotional contagion and empathy. In: Decety J, Ickes W (eds) The social neuroscience of empathy. MIT Press, Cambridge, pp 19–30
51. Elfenbein HA, Ambady N (2002) On the universality and cultural specificity of emotion recognition: a meta-analysis. Psychol Bull 128(2):203–235. https://doi.org/10.1037/0033-2909.128.2.203
52. Gutsell JN, Inzlicht M (2012) Intergroup differences in the sharing of emotive states: neural evidence of an empathy gap. Soc Cogn Affect Neurosci 7(5):596–603. https://doi.org/10.1093/scan/nsr035
53. Soto JA, Levenson RW (2009) Emotion recognition across cultures: the influence of ethnicity on empathic accuracy and physiological linkage. Emotion 9(6):874–884. https://doi.org/10.1037/a0017399
54. Cruz J, Gordon RM (2006) Simulation theory. In: Nadel L (ed) Encyclopedia of cognitive science. Wiley, New York, pp 9–14
55. Keysers C, Gazzola V (2007) Integrating simulation and theory of mind: from self to social cognition. Trends Cogn Sci 11(5):194–196. https://doi.org/10.1016/j.tics.2007.02.002
56. Rizzolatti G, Craighero L (2004) The mirror-neuron system. Annu Rev Neurosci 1:169–192. https://doi.org/10.1146/annurev.neuro.27.070203.144230
57. Gallese V (2003) The roots of empathy: the shared manifold hypothesis and the neural basis of intersubjectivity. Psychopathology 36(4):171–180. https://doi.org/10.1016/S1364-6613(99)01417-5
58. Gazzola V, Keysers C (2009) The observation and execution of actions share motor and somatosensory voxels in all tested subjects: single-subject analyses of unsmoothed fMRI data. Cereb Cortex 19:1239–1255. https://doi.org/10.1093/cercor/bhn181
59. Jabbi M, Keysers C (2008) Inferior frontal gyrus activity triggers anterior insula response to emotional facial expressions. Emotion 8(6):775. https://doi.org/10.1037/a0014194
60. Schmidt SNL, Sojer CA, Hass J et al (2020) fMRI adaptation reveals: the human mirror neuron system discriminates emotional valence. Cortex. https://doi.org/10.1016/j.cortex.2020.03.026

61. Lamm C, Majdandžić J (2015) The role of shared neural activations, mirror neurons, and morality in empathy–a critical comment. J Neurosci Res 90:15–24. https://doi.org/10.1016/j.neures.2014.10.008
62. Cerri G, Cabinio M, Blasi V et al (2015) The mirror neuron system and the strange case of Broca's area. Hum Brain Mapp 36(3):1010–1027. https://doi.org/10.1002/hbm.22682
63. Preston SD, De Waal F (2002) Empathy: its ultimate and proximate bases. Behav Brain Sci 25(01):1–20. https://doi.org/10.1017/S0140525X02000018
64. Christov-Moore L, Reggente N, Douglas PK et al (2020) Predicting empathy from resting state brain connectivity: a multivariate approach. Front Integr Neurosci 14:3. https://doi.org/10.3389/fnint.2020.00003
65. Nummenmaa L, Hirvonen J, Parkkola R, Hietanen JK (2008) Is emotional contagion special? an fMRI study on neural systems for affective and cognitive empathy. Neuroimage 43(3):571–580. https://doi.org/10.1016/j.neuroimage.2008.08.014
66. Singer T, Seymour B, O'Doherty J et al (2004) Empathy for pain involves the affective but not sensory components of pain. Science 303(5661):1157–1162. https://doi.org/10.1126/science.1093535
67. Lamm C, Decety J, Singer T (2011) Meta-analytic evidence for common and distinct neural networks associated with directly experienced pain and empathy for pain. Neuroimage 54(3):2492–2502. https://doi.org/10.1016/j.neuroimage.2010.10.014
68. Akitsuki Y, Decety J (2009) Social context and perceived agency affects empathy for pain: an event-related fmri investigation. Neuroimage 47(2):722–734. https://doi.org/10.1016/j.neuroimage.2009.04.091
69. Decety J, Echols S, Correll J (2010) The blame game: the effect of responsibility and social stigma on empathy for pain. J Cogn Neurosci 22(5):985–997. https://doi.org/10.1162/jocn.2009.21266
70. Batson CD, Fultz J, Schoenrade PA (1987) Distress and empathy: two qualitatively distinct vicarious emotions with different motivational consequences. J Pers 55(1):19–39. https://doi.org/10.1111/j.1467-6494.1987.tb00426.x
71. Eisenberg N, Eggum ND (2009) Empathic responding: sympathy and personal distress. In: Decety J, Ickes W (eds) The social neuroscience of empathy. MIT Press, Cambridge, p 71–83. https://doi.org/10.7551/mitpress/9780262012973.003.0007
72. Israelashvili J, Sauter DA, Fischer AH (2020) Different faces of empathy: feelings of similarity disrupt recognition of negative emotions. J Exp Soc Psychol 87:103912. https://doi.org/10.1016/j.jesp.2019.103912
73. Ochsner KN, Knierim K, Ludlow DH et al (2004) Reflecting upon feelings: an fMRI study of neural systems supporting the attribution of emotion to self and other. J Cogn Neurosci 16(10):1746–1772. https://doi.org/10.1162/0898929042947829
74. Bruneau EG, Jacoby N, Saxe R (2015) Empathic control through coordinated interaction of amygdala, theory of mind and extended pain matrix brain regions. Neuroimage 114:105–119. https://doi.org/10.1016/j.neuroimage.2015.04.034
75. Borja Jimenez KC, Abdelgabar AR, de Angelis L et al (2020) Changes in brain activity following the voluntary control of empathy. Neuroimage 116529. https://doi.org/10.1016/j.neuroimage.2020.116529
76. Chiao JY (2011) Towards a cultural neuroscience of empathy and prosociality. Emot Rev 3:111–112. https://doi.org/10.1177/1754073910384159
77. de Greck M, Shi Z, Wang G et al (2012) Culture modulates brain activity during empathy with anger. Neuroimage 59(3):2871–2882. https://doi.org/10.1016/j.neuroimage.2011.09.052
78. Malle BF (2011) Attribution theories: how people make sense of behavior. In: Chadee D (ed) Theories in social psychology. Wiley Blackwell, London, pp 72 95
79. Kelley HH, Michela JL (1980) Attribution theory and research. Annu Rev Psychol 31(1):457–501. https://doi.org/10.1146/annurev.ps.31.020180.002325

80. Frith D, Frith U (2008) Implicit and explicit processes in social cognition. Neuron 60 (3):503–510. https://doi.org/10.1016/j.neuron.2008.10.032
81. Brüne M, Brüne-Cohrs U (2006) Theory of mind—evolution, ontogeny, brain mechanisms and psychopathology. Neurosci Biobehav Rev 30(4):437–455. https://doi.org/10.1016/j. neubiorev.2005.08.001
82. Premack D, Woodruff G (1978) Does the chimpanzee have a theory of mind? Behav Brain Sci 1(04):515–526. https://doi.org/10.1017/S0140525X00076512
83. Schaafsma SM, Pfaff DW, Spunt RP et al (2015) Deconstructing and reconstructing theory of mind. Trends Cogn Sci 19(2):65–72. https://doi.org/10.1016/j.tics.2014.11.007
84. Mahy CE, Moses LJ, Pfeifer JH (2014) How and where: theory-of-mind in the brain. Dev Cogn Neurosci 9:68–81. https://doi.org/10.1016/j.dcn.2014.01.002
85. Harris LT, Todorov A, Fiske ST (2005) Attributions on the brain: neuro-imaging dispositional inferences, beyond theory of mind. Neuroimage 28(4):763–769. https://doi. org/10.1016/j.neuroimage.2005.05.021
86. Frith D, Frith U (2006) The neural basis of mentalizing. Neuron 50(4):531–534. https://doi. org/10.1016/j.neuron.2006.05.001
87. Schurz M, Radua J, Aichhorn M et al (2014) Fractionating theory of mind: a meta-analysis of functional brain imaging studies. Neurosci Biobehav Rev 42:9–34. https://doi.org/10. 1016/j.neubiorev.2014.01.009
88. Decety J, Lamm C (2007) The role of the right temporoparietal junction in social interaction: how low-level computational processes contribute to meta-cognition. Neuroscientist 13(6): 580–593. https://doi.org/10.1177/1073858407304654
89. Scholz J, Triantafyllou C, Whitfield-Gabrieli S et al (2009) Distinct regions of right temporo-parietal junction are selective for theory of mind and exogenous attention. PLoS ONE 4(3):e4869. https://doi.org/10.1371/journal.pone.0004869
90. Völlm BA, Taylor AN, Richardson P et al (2006) Neuronal correlates of theory of mind and empathy: a functional magnetic resonance imaging study in a nonverbal task. Neuroimage 29(1):90–98. https://doi.org/10.1016/j.neuroimage.2005.07.022
91. Corradi-Dell'Acqua C, Hofstetter C, Vuilleumier P (2014) Cognitive and affective theory of mind share the same local patterns of activity in posterior temporal but not medial prefrontal cortex. Soc Cogn Affect Neurosci 9(8):1175–1184. https://doi.org/10.1093/scan/nst097
92. Tholen MG, Trautwein F, Böckler A et al (2020) Functional magnetic resonance imaging (fMRI) item analysis of empathy and theory of mind. Hum Brain Mapp 1–18. https://doi.org/ 10.1002/hbm.24966
93. Mitchell RL, Phillips LH (2015) The overlapping relationship between emotion perception and theory of mind. Neuropsychologia 70:1–10. https://doi.org/10.1016/j.neuropsychologia. 2015.02.018
94. Buckner RL, Carroll DC (2007) Self-projection and the brain. Trends Cogn Sci 11(2):49–57. https://doi.org/10.1016/j.tics.2006.11.004
95. Zaki J, Hennigan K, Weber J et al (2010) Social cognitive conflict resolution: contributions of domain-general and domain-specific neural systems. J Neurosci 30(25):8481–8488. https://doi.org/10.1523/JNEUROSCI.0382-10.2010
96. Ross L, Greene D, House P (1977) The "false consensus effect": an egocentric bias in social perception and attribution processes. J Exp Soc Psychol 13(3):279–330. https://doi.org/10. 1016/0022-1031(77)90049-X
97. Green JD, Sedikides C (2001) When do self-schemas shape social perception?: the role of descriptive ambiguity. Motiv Emot 25(1):67–83. https://doi.org/10.1023/A:1010611922816
98. Van der Meer L, Groenewold NA, Nolen WA et al (2011) Inhibit yourself and understand the other: neural basis of distinct processes underlying theory of mind. Neuroimage 56(4): 2364–2374. https://doi.org/10.1016/j.neuroimage.2011.03.053
99. Decety J, Chaminade T (2003) When the self represents the other: a new cognitive neuroscience view on psychological identification. Conscious Cogn 12(4):577–596. https:// doi.org/10.1016/S1053-8100(03)00076-X

100. Decety J, Sommerville JA (2003) Shared representations between self and other: a social cognitive neuroscience view. Trends Cogn Sci 7(12):527–533. https://doi.org/10.1016/j.tics.2003.10.004

101. Uddin LQ, Molnar-Szakacs I, Zaidel E et al (2006) rTMS to the right inferior parietal lobule disrupts self–other discrimination. Soc Cogn Affect Neurosci 1(1):65–71. https://doi.org/10.1093/scan/nsl003

102. Ruby P, Decety J (2004) How would you feel versus how do you think she would feel? a neuroimaging study of perspective-taking with social emotions. J Cogn Neurosci 16(6):988–999. https://doi.org/10.1162/0898929041502661

103. Gallager S (2000) Philosophical conceptions of the self: implications for cognitive science. Trends Cogn Sci 4(1):14–21. https://doi.org/10.1016/S1364-6613(99)01417-5

104. Kim H (2012) A dual-subsystem model of the brain's default network: self-referential processing, memory retrieval processes, and autobiographical memory retrieval. Neuroimage 61(4):966–977. https://doi.org/10.1016/j.neuroimage.2012.03.025

105. Tagini A, Raffone A (2010) The 'I' and the 'Me' in self-referential awareness: a neurocognitive hypothesis. Cogn Process 11(1):9–20. https://doi.org/10.1007/s10339-09-0336-1

106. Gazzaniga SM (2010) ¿Qué nos hace humanos?: la explicación científica de nuestra singularidad como especie. Paidós, Madrid

107. Hollan D, Throop JC (2008) Whatever happened to empathy?: introduction. Ethos 36(4):385–401. https://doi.org/10.1111/j.1548-1352.2008.00023.x

108. Cacioppo JT, Berntson GG, Decety J (2010) Social neuroscience and its relationship to social psychology. Soc Cogn 28(6):675–685. https://doi.org/10.1521/soco.2010.28.6.675

109. Todorov A, Harris LT, Fiske ST (2006) Toward socially inspired social neuroscience. Brain Res 1079(1):76–85. https://doi.org/10.1016/j.brainres.2005.12.114

110. Decety J, Jackson PL (2004) The functional architecture of human empathy. Behav Cogn Neurosci Rev 3(2):71–100

111. Perry A, Shamay-Tsoory S (2013) Understanding emotional and cognitive empathy. In: Baron-Cohen S, Lombardo M, Tager-Flusberg H, Cohen D (eds) Understanding other minds: perspectives from developmental social neuroscience. Oxford University Press, Oxford, pp 178–194

112. Smith A (2006) Cognitive empathy and emotional empathy in human behavior and evolution. Psychol Rec 56(1):3–21. https://doi.org/10.1007/BF03395534

113. Zaki J, Ochsner KN (2012) The neuroscience of empathy: progress, pitfalls and promise. Nat Neurosci 15(5):675–680. https://doi.org/10.1038/nn.3085

114. Kanske P, Böckler A, Trautwein FM et al (2015) Dissecting the social brain: introducing the empatom to reveal distinct neural networks and brain–behavior relations for empathy and theory of mind. Neuroimage 122:6–19. https://doi.org/10.1016/j.neuroimage.2015.07.082

115. Decety J, Jackson PL (2006) A social-neuroscience perspective on empathy. Curr Dir Psychol Sci 15(2):54–58. https://doi.org/10.1111/j.0963-7214.2006.00406.x

116. van Overwalle F, Baetens K, Mariën P et al (2014) Social cognition and the cerebellum: a meta-analysis of over 350 fMRI studies. Neuroimage 86:554–572. https://doi.org/10.1016/j.neuroimage.2013.09.033

117. van Overwalle F, Mariën P (2016) Functional connectivity between the cerebrum and cerebellum in social cognition: a multi-study analysis. Neuroimage 124:248–255. https://doi.org/10.1016/j.neuroimage.2015.09.001

118. Christov-Moore L, Iacoboni M (2016) Self-other resonance, its control and prosocial inclinations: brain-behavior relationships. Hum Brain Mapp 37(4):1544–1558. https://doi.org/10.1002/hbm.23119

119. Clausi S, Iacobacci C, Lupo M et al (2017) The role of the cerebellum in unconscious and conscious processing of emotions: a review. Appl Sci 7(5):521. https://doi.org/10.3390/app7050521
120. David N, Bewernick BH, Cohen MX et al (2006) Neural representations of self-versus other: visual-spatial perspective taking and agency in a virtual ball-tossing game. J Cogn Neurosci 18(6):898–910. https://doi.org/10.1162/jocn.2006.18.6.898
121. Macuga KL, Frey SH (2011) Selective responses in right inferior frontal and supramarginal gyri differentiate between observed movements of oneself vs. another. Neuropsychologia 49 (5):1202–1207. https://doi.org/10.1016/j.neuropsychologia.2011.01.005
122. Heleven E, van Dun K, van Overwalle F (2019) The posterior cerebellum is involved in constructing social action sequences: an fMRI study. Sci Rep 9(1):1–11. https://doi.org/10.1038/s41598-019-46962-7
123. Bernhardt BC, Singer T (2012) The neural basis of empathy. Annu Rev Neurosci 35:1–23. https://doi.org/10.1146/annurev-neuro-062111-150536
124. Decety J (2011) Dissecting the neural mechanisms mediating empathy. Emot Rev 3(1):92–108. https://doi.org/10.1177/1754073910374662
125. Melloni M, Lopez V, Ibanez A (2014) Empathy and contextual social cognition. Cogn Affect Behav Neurosci 14(1):407–425. https://doi.org/10.3758/s13415-013-0205-3

Roberto E. Mercadillo obtained his doctorate degree in biomedical sciences at the Neurobiology Institute, Universidad Nacional Autónoma de México. He is a fellow of the National Council for Science and Technology, Mexico, in the Neurosciences Area of the Universidad Autónoma Metropolitana, Iztapalapa, and a Professor in the National School of Anthropology and History. His scientific research interests include transdisciplinary approaches to study drug consumption, moral emotions, and social cognition in both basic models and human populations in vulnerability. He is the author of more than 30 peer-reviewed articles and author and editor of six books regarding social cognition on empathy and compassion, neuropsychiatric diseases including attentional disorders and spinocerebellar ataxia, human behavioral evolution, neuroethics, the culture of peace, and ethnographic studies on war and conflicts. He is the Mexican representative of International Colloquia on Brain and Aggression.

Daniel Atilano-Barbosa is a psychologist and master in sciences student at the Neurobiology Institute, Universidad Nacional Autónoma de México, with experience as a research analyst at the Legal Research Institute in the same university. His research interests include fMRI analysis techniques, psychometric and neuropsychological assessment of empathy and social cognition, and neuropathologies presented in inhalants users and youth people living on the streets of Mexico City.

Why People Make Irrational Choices About Their Health?

8

Jakub Šrol and Vladimíra Čavojová

"The essence of the independent mind lies not in what it thinks, but in how it thinks."

Christopher Hitchens

Summary

When it comes to health, people often hold beliefs that are either unsupported or directly opposed by scientific evidence, i.e., epistemically suspect beliefs (ESB). They prefer alternatives (e.g., homeopathy, healing by crystals, magnets, or herbs) over standard medical treatments. Even in the case of severe diseases, they oppose vaccinations or believe that it is harmful to eat genetically modified foods, while the oral use of disinfectants is beneficial for their overall health. Such beliefs are widespread for many reasons, including people forming their views based on circumstantial evidence, adopting causal interpretations for spuriously correlated or non-contingent events, or rejecting scientific evidence because of the tendency to endorse conspiracy theories. This chapter focuses specifically on one of the factors, scientific reasoning ability, which can help people debunk ESB in the domain of health and, ultimately, lead them to better health choices. In the first part of the chapter, we mention various health-related

J. Šrol (✉) · V. Čavojová
Institute of Experimental Psychology, Centre of Social and Psychological Sciences, Slovak Academy of Sciences, Dúbravská cesta 9, 841 04 Bratislava, Slovakia
e-mail: jakub.srol@savba.sk

Integrated Science Association (ISA), Universal Scientific Education and Research Network (USERN), Bratislava, Slovakia

N. Rezaei (ed.), *Multidisciplinarity and Interdisciplinarity in Health*,
Integrated Science 6, https://doi.org/10.1007/978-3-030-96814-4_8

177

ESB and outline some reasons why these beliefs are so prevalent. The second part of the chapter defines scientific reasoning and explains its relation to health-related beliefs and behaviors.

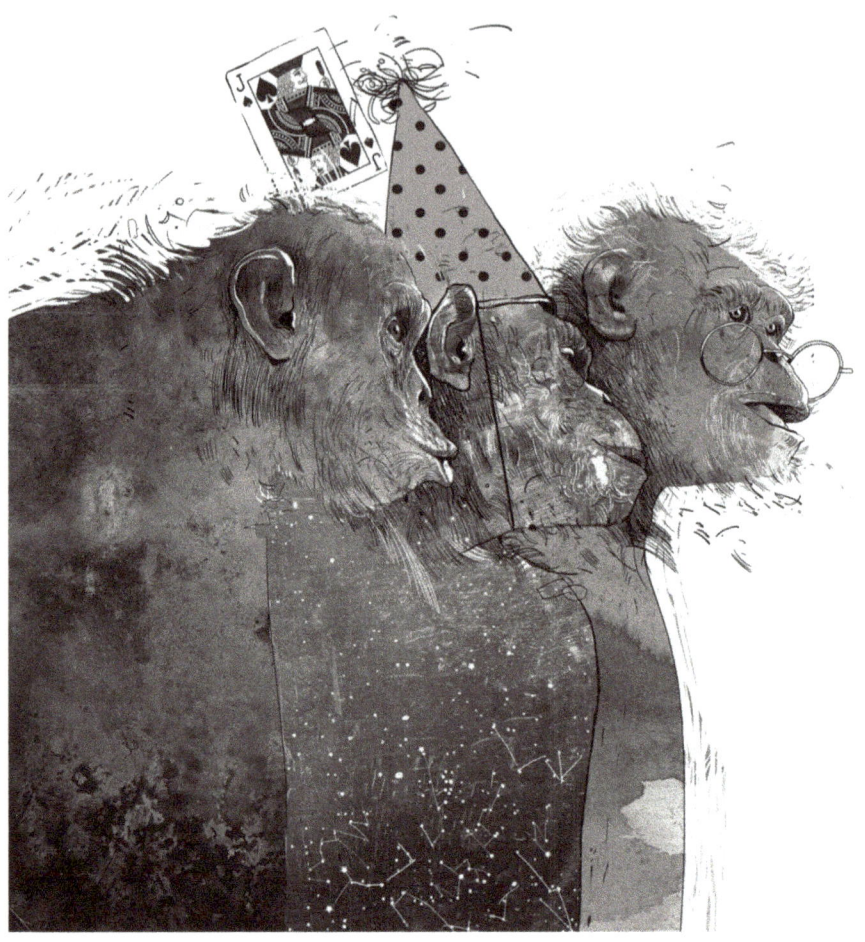

The Forer's effect
[Made by Ján Kurinec, https://www.jankurinec.com/].
The code of this chapter is *01101100 01100101 01001000 01100001 01110100 01101000.*

Keywords

Alternative medicine · Anti-vaccination attitudes · Epistemically suspect beliefs · ESB · Illusory pattern perception · Scientific reasoning

1   Introduction

The year 2020 will probably be remembered as the year of the coronavirus disease 2019 (COVID-19) pandemic. As people worldwide watched COVID-19 rapidly spread from mainland China to other countries, many "guaranteed" treatments and theories about its origin spread along with it. Most of these treatments are built on traditional herbal remedies, such as recommendations to gargle vinegar and rose-water, drink mint or white willow beverages, ingest spices like saffron, turmeric, and cinnamon, or chew garlic and ginger. Of course, the need to do something to protect one's health, especially during disease outbreaks, is natural and under-standable. However, holding unfounded and false beliefs, such as those about the effectiveness of remedies as mentioned above to treat or protect oneself from the coronavirus, can lead to endangering the health of individuals and groups. Just look at the advice given by the American president, Donald Trump, who suggested that people could try to inject disinfectant or expose themselves to powerful doses of ultra-violet light.[1] Such self-medication may have fatal consequences.

It may seem that these beliefs about possible remedies and preventive measures for the COVID-19 provide merely circumstantial evidence that people fall prey to various misinformation. As such, they are not truly indicative of some widespread tendency to hold unsubstantiated beliefs. Consider, however, the results of surveys in various stages of early coronavirus outbreaks around the world. In March 2020, around one quarter (23%) of American respondents believed that the COVID-19 was intentionally created in a laboratory rather than that it came around naturally [1]. At the end of February and early March, a similar proportion (25.6% United States, 29.6% of United Kingdom) of people thought it would be best to avoid eating at Chinese restaurants [2]. In our research about people's early responses to the coro-navirus outbreak in Slovakia, we have found that at least 20% of people believed that the COVID-19 pandemic might have been stopped at the beginning of the outbreak, had it not been for the big pharma companies that wanted to make the profit out of it, and that severe acute respiratory syndrome coronavirus 2 (SARS-CoV-2) was a biological weapon designed to eliminate overpopulated human race [3].

However, there was another trend that emerged along with the spread of various unsubstantiated beliefs. People seemed to have realized that they needed to rely on the advice of doctors and scientists to help tackle the COVID-19 outbreak. Despite the proliferation of various suggested natural remedies and other possible treat-ments, people accepted that in crises such as the COVID-19 pandemic, they need scientists to help them understand the origin of the virus and not act on unverified measures. However, why does it take a crisis for people to start listening to sci-entists for advice about their health? Why are people's beliefs regarding their health often based on unfounded information?

There is a line of research that shows what makes people susceptible to various unfounded beliefs is that they are relatively incapable of applying basic principles of scientific reasoning, such as objectively evaluating information and evidence.

---

[1] https://www.bbc.com/news/world-us-canada-52407177.

In this chapter, we will attempt to examine the link between such scientific reasoning skills and holding various unfounded beliefs, which may subsequently lead people to make suboptimal choices about health. The chapter contains two parts. In the first part, we outline some commonly held unfounded beliefs, explain several reasons for why they are so prevalent among people, and consider the consequences of such beliefs for choices and behaviors in the domain of health. Secondly, we show how scientific reasoning, referred to as the "ability to understand methods and principles of scientific inquiry," may help people reason better about information and evidence around them and how it can help identify and debunk misinformation about health and thus contribute to better choices and beliefs in this domain.

2 Health-Related Epistemically Suspect Beliefs (ESB)

We already mentioned several questionable beliefs related to the outbreak of COVID-19 in 2020, but it would be a mistake to think that unfounded beliefs occur only in times of crises or pandemics. Various conspiracy and pseudoscientific beliefs about health were surprisingly widespread long before COVID-19, and there is no reason to suspect they will go away when it recedes. Across psychological studies and sociological surveys, some of the most prevalent unfounded beliefs around are related to people's health. Beliefs that the medical industry and pharmaceutical industry work together to create new diseases intentionally, that childhood vaccines might contribute to autism spectrum disorder (ASD) that acquired immunodeficiency syndrome (AIDS) was created intentionally to infect certain groups of people (mostly black and gay men), or that homeopathy and alternative treatments are just as effective for serious diseases (such as cancer) are being held by 10–40% of participants across samples from various countries all around the world [4–9].

These beliefs share a common ground that is either unsupported or directly contradicted by most current scientific evidence. For this reason, researchers use the term "epistemically suspect beliefs" or "epistemically unwarranted beliefs" to signify that such beliefs are not in line with the knowledge accumulated across various scientific disciplines [5, 7, 10–12]. The most commonly recognized types of ESB are the belief in paranormal phenomena, the endorsement of conspiracy theories, and the acceptance of pseudoscience. Interestingly, the topic of health and medicine is at the center of many of these popular ESB and covers all three aforementioned main types of such beliefs, especially conspiracy and pseudoscientific beliefs. These beliefs are not harmless either; they may have dire consequences, such as negative attitudes toward vaccination and a lower tendency to take vaccines [3, 13], high-risk sexual behavior [9], or preference of alternative medicine instead of standard treatments even in cases of severe diseases [14]. Despite such negative associations, the prevalence of ESB does not seem to be dropping over the last few years, begging the question, where do these beliefs come from, and why are they so prevalent?

3   Why Do People Adopt Epistemically Suspect Beliefs?

One thing we need to make clear right from the beginning is that people do not hold unsubstantiated beliefs just because they are gullible, stupid, or because they do not have access to relevant information and evidence [15]. Though this might be true in some cases, in general, people are equipped with a powerful brain that enables them to process large amounts of information with incredible accuracy. Many irrational beliefs result from the over-application of otherwise effective and valid information processing [15]. Just consider the example of Kary Mullis, 1993 winner of the Nobel prize for chemistry. He believes that climate change and the connection between human immunodeficiency viruses (HIV) and AIDS are the results of a conspiracy of environmentalists, government agencies, and scientists trying to salvage their careers and earn money. Similarly, Linus Pauling, the two-time Nobelist, held to his belief that mega doses of vitamin C help you to cure a cold, cancer, and schizophrenia even in the face of contradictory evidence. After being diagnosed with prostate and rectal cancer, he underwent two surgeries and refused other standard medical treatment favoring mega doses of vitamin C, but he lived only three years after diagnosis [16]. As should be clear from these cases, some people are exceptionally intelligent, and yet, they are not immune to various types of unfounded information. If it is not a lack of intelligence or relevant information, what then are the drivers of ESB?

3.1   Illusory Pattern Perception

Evidence suggests that it may be exactly our well-evolved mental faculties, making people prone to adopt some unfounded beliefs. Specifically, humans have a pronounced ability to see patterns in their environment, an ability that helped our ancestors to quickly adapt to changing conditions, spot danger, and navigate life in social groups [17]. It appears that this pattern of perception may be so pronounced that people sometimes see patterns where there are none, which leads them to misperceive and misinterpret random data and, ultimately, to dubious judgments and beliefs. Many studies, for example, demonstrated that people have wrong intuitions about what random sequences look like [18, 19] and thus can develop suspicions about some hidden connection or intention. For example, people expect that when tossing the coin, the head and tail will alternate more frequently than is the case. Because just by random tossing, there are less of these alternations than what our intuition tells us, even truly random sequences can look too orderly or "chunky." When people cannot correctly recognize randomness in various events, it often leads to the belief that something truly random, chaotic, and illusory is rather systematic, orderly, and intentional.

Take, for example, that you wish you did not have to go to your mother-in-law's birthday party, and later that day, you come down with a high fever. It may seem there is a connection between the two events, and you may start to believe that you have caused the illness by your wish (or that it is a punishment for your wish sent by your all-knowing witch-like mother-in-law). This tendency to identify meaningful relationships among random or unrelated events, sometimes called the illusory pattern perception, has been argued to constitute the basis of many paranormal and conspiracy beliefs [20–22].

3.2 Sense of Control

Illusory pattern perception is presumably a result of the tendency to impose structure or control over random and unpredictable events [21]. People need predictability and feeling of control to feel good, and if they cannot gain the sense of control objectively, they try to achieve it at least perceptually. Whitson and Galinsky tested whether a lack of control would lead to an increased perception of illusory patterns in a set of random or unconnected stimuli [23]. Participants with a lower feeling of control were more likely to see illusory patterns and hidden figures in "noisy" pictures; they created illusory correlations from stock market information, preferred conspiracy explanations, and formed superstitious beliefs. These findings were corroborated by recent research on bullshit reception where people who perceived meaningless statements, for example, "Good health imparts reality to subtle creativity," as profound more frequently could see meaningful patterns in random shapes, such as clouds [24].

Crisis and stress situations, as we see during pandemics, natural disasters, and stock market crashes, greatly contribute to beliefs in conspiracy theories associated with anxiety and loss of control [25, 26]. Similarly, when anxiety and helplessness are experimentally induced, it increases the acceptance of conspiracy theories [25, 27]. The recent COVID-19 pandemic has enabled scientists to study this consequence of heightened anxiety and lack of control in real life. Harsh social distancing measures affected in some way almost everyone and the majority of people did not encounter crises of such magnitude in their lifetime. That has naturally led to uncertainty about the future and increased anxiety. In our study [28], we collected data when the first cases of COVID-19 were diagnosed in our country. The results, indeed, showed that people who had more feelings of anxiety and loss of control as a result of the ongoing pandemic of COVID-19 had more conspiracy explanations about the origins, spread, and treatment of the coronavirus. For example, they were more likely to believe that the coronavirus was created intentionally as "a biological weapon to eliminate the overcrowded human population" or that the cure is already available but is being kept secret. Moreover, conspiracy beliefs about the coronavirus were also strongly correlated with various other generic ESB. This highlights the generality of belief in unwarranted claims. It means that people who are prone to have belief in a specific type of suspect beliefs are also much more susceptible to hold various other ESB as well.

3.3 Illusory Correlations and Causal Illusion

As mentioned above, situations inducing anxiety and lack of control can increase people's need to control their environment by perceiving meaningful patterns in random or unrelated stimuli or events, which can ultimately lead to an increased tendency to endorse conspiracy explanations of events taking place. However, the tendency to see patterns in random information around us is also exacerbated by the fact that people often do not realize that when they monitor a large number of events, they can expect to find a few strong—even if random—correlations between some of them. Then, once people have a suspicion that a phenomenon exists, they have no difficulties in explaining why it exists and what does it mean [15].

If you are not already familiar with it, please take a moment to check the webpage created by Tyler Vigen,[2] where you can find examples of some interesting spurious correlations between random variables. If you do so, you can find that there is a strong relationship between the number of films Nicolas Cage appeared in and the number of deaths by drowning in a pool; or almost a perfect relationship between per capita cheese consumption and the number of deaths by strangling in bedsheets; as well as between the divorce rate in state Maine and margarine consumption. When people are asked to think about the relationship between cheese consumption and strangling in one's bed sheets, they can come up with various explanations: eating cheese before going to bed may disturb your digestion and lead to violent dreams that, in turn, cause a person to get strangled in their sheets and die; or, substances contained in cheese might interact with sleeping pills or alcohol ingested before sleep; or, the relationship is because people in rich countries not only consume more cheese, but they also have bedsheets they can get strangled in the first place, while people in poor countries consume less cheese and do not have bedsheets. If you came up with a similar (or even more unlikely) scenario, you could see how people are exceptionally good at finding meaningful relationships even when none exists. As we can see, our brains are quick and efficient in explaining things that we believe in [29, 30].

So, not only are people very adept at seeing correlations (patterns, relationships) where none exist, they even further tend to explain the (non-existent) relationship by a causal link. This tendency to develop causal explanations for events, which are not contingent on each other, is called "causal illusion" or "causal bias," and it has been shown to associate with the endorsement of ESB [31, 32]. The research on this issue used a contingency learning task where people are usually asked to judge the effectiveness of a fictitious medicine based on several rounds of case studies where fictitious patients suffer some symptoms, whether they were given the medicine or not, and subsequently were cured or not. Although the cases were intentionally manipulated so that the cure is not effective (the same amount of patients are cured with the medicine as without medicine), people develop causal illusions, i.e., they believe that the medicine helped cure the symptom. As shown by Blanco et al. [31], people with paranormal beliefs developed more such causal illusions, mainly

---

[2] www.tylervigen.com

because they exposed themselves to biased information, they focused more on cases where the medicine was administered, and they were less interested in cases when the medicine was not administered.

Recently, Torres et al. [32] used a similar task. They presented participants with cases where people suffered from a headache. Some were given an herb from Amazon; others were not. The proportion of people who got rid of the headache was the same regardless of whether they got the herb or not. Still, people thought the herb worked; they believed there was a causal link between getting a herb and curing headaches. This extent of such causal illusion (the belief in the effectiveness of the herb) was positively (although modestly) correlated with a range of pseudoscientific beliefs, including the belief that the positive attitude helps prevent cancer, that the Wi-Fi causes headaches, or belief in the effectiveness of homeopathy, natural remedies, acupuncture, nutritional supplements, magnetic healing, and detox therapy. It should be mentioned that the results of Torres et al. [32] showed no relationship between the strength of causal illusion and the endorsement of superstitious beliefs, contrary to what was observed by Blanco et al. [31], and even the link between participants' causal illusion and the endorsement of pseudoscientific beliefs was relatively modest. Still, their results are interesting because the task to measure causal illusion resembled real-life learning of (illusory) causal links.

3.4 Confirmation Bias and Myside Bias

Another factor that can contribute to cementing dubious beliefs is that once the people "reveal" the cause of some event, i.e., when they form a causal illusion between non-contingent events, they might start to actively collect the information that will further confirm this causal link at the expense of other alternative explanations. This is because of a very well-documented tendency to search for, interpret, and evaluate information to confirm their previous beliefs, a tendency called confirmation bias [33] or myside bias [34, 35]. Imagine that we knew some parents whose child was diagnosed with ASD in a short time after receiving the measles, mumps, and rubella (MMR) vaccine. The age when an ASD is often diagnosed coincides with the age when the MMR vaccine is usually administered, and this temporal coincidence has been misinterpreted as the causal relationship in which MMR vaccines cause ASD [36]. Once we form a belief that vaccines may be hurtful, it is easy to search specifically for evidence that vaccines may be associated with negative health consequences. If we take circumstantial evidence presented on blogs and anti-vax web pages and ignore the scientific evidence on this issue, we would quickly infer that vaccines are dangerous. When we do not look for evidence that counters our beliefs or if we actively ignore such information, we can easily conclude that what we believe is the only reasonable account of that particular issue.

3.5 The Interrelationships Among Various Types of ESB

Lastly, one reason why people might be more prone to adopt certain ESB about their health is simply that they already believe in some other unwarranted claims in the domain of paranormal phenomena and superstitions, conspiracy theories, or pseudo-science. It is well documented that people who endorse one type of ESB, for example, those who believe that the medical industry works with the pharmaceutical industry to create diseases for profit, are much more likely also to have various other unwarranted beliefs as well [5, 11, 37]. As was shown by Lobato et al. [7], the endorsement of one type of suspect belief is best predicted by the endorsement of other types of ESB, over and above various personality dimensions and cognitive factors.

Interestingly, this link does not only exist among conceptually related beliefs. It is quite understandable that someone who believes a conspiracy theory that United States agencies created AIDS to eliminate gay and black men will also be more likely to think that the scientific fact that HIV causes AIDS is wrong. However, the link between the endorsement of various ESB seems to be strong even if the beliefs are not, on their surface, related. This is perhaps best documented in the domain of conspiracy beliefs, where it was shown that some people are so prone to believe conspiracy explanations of certain events that they are likely to endorse even mutually contradictory conspiracy theories [38]. As was further shown in two studies by Lewandowsky and his colleagues [8, 39], conspiracist ideation is linked with a higher rejection of established scientific findings, such as the link between smoking and cancer, the safety of currently used genetically modified foods, and the beneficial role of vaccinations.

As we demonstrated in this chapter, there are many reasons people commonly tend to adopt various ESB regarding their health. This is also why there is much research focusing on how these beliefs can be amended and how to stop the spread of misinformation, which forms the basis of such beliefs. Without trying to review all of the factors which are associated with the endorsement of health-related ESB, we would like to mention one specific factor which showed up to be an important determinant of those beliefs in our recent research, i.e., scientific reasoning ability. In the next section, we outline what scientific reasoning is and how it can help people make more optimal choices about their health.

4 The Role of Scientific Reasoning in Our Beliefs About Health

4.1 What is Scientific Reasoning?

People who do not make science for a living often underappreciate the extent of scientific reasoning needed in daily life. Suppose you are a recreational runner and have some concerns about whether you should continue with your habit during a coronavirus outbreak when officials recommend people to stay put in their homes.

In a morning paper, you read about a study that reports a negative correlation between doing some outdoor physical exercise and being diagnosed with COVID-19. Should you conclude, same as the reporter seems to do, that outdoor activity decreases the chances of contracting a COVID-19? If despite your clear wish for the reporter's conclusion to be true, you start to question it because you know that you cannot make causal inferences based on correlations, or if you wonder about the design of the study and the number of participants it was based on, or if you think about an alternative explanation of the results (e.g., that sick people tend to stay at home and not engage in outdoor physical exercise, etc.), you just engaged in scientific reasoning.

The example above illustrates the notion that being scientifically literate helps you not just to understand the scientific supplement of your favorite magazine but also to be a more informed citizen and function properly in society [40]. This is the case, particularly when we need to quickly distinguish reliable information from disinformation during the recent COVID-19 crisis.

At this point, we should stress that by scientific reasoning, we do not mean simply having more scientific knowledge. Just knowing that paracetamol is acetaminophen, which "consists of a benzene ring core, substituted by one hydroxyl group and the nitrogen atom of an amide group in the para pattern" [41] and it is used to treat fever does not mean that person can reason scientifically. Rather, we use the term scientific reasoning as "the ability to understand methods and principles of scientific inquiry and the acquisition of the skills required to formulate, test, and revise theories and reflect the process of generating evidence" [3, p. 2]; or, more specifically, as the "application of the methods or principles of scientific inquiry to reasoning or problem-solving situations" [42, p. 173] and see also [43–45]. Defined thusly, it is obvious that scientific reasoning is not the type of thinking reserved exclusively for scientists. Rather, we see it as a part of skills that are necessary when people contemplate any complex issue in their lives, such as issues related to health, personal finance, etc. Using the abovementioned example with paracetamol, scientific reasoning would imply that you understand how the evidence about its efficacy for fever was obtained and how to evaluate whether it is reliable (using the control group, excluding confounding variables, using random assignments to groups, double-blinding, etc.). To measure the understanding of science, some previous studies relied on participants' agreement with statements on which there is a scientific consensus [46]. Alternatively, several studies used various knowledge-based tests [47–52] to measure the knowledge of basic scientific concepts. However, there was a limitation to these measures, as the knowledge of scientific concepts, as indicated by these measures, may not have adequately reflected a true understanding of those concepts [47] and, in some cases (as with items regarding politically controversial scientific topics), may have reflected people's attitudes more than their actual understanding of science [48, 53].

Knowledge of scientific facts and the ability to draw conclusions based on them are not the same thing, as was also shown by our research. In 2017, we tested the general population's scientific literacy by using a test of science knowledge as well as a measure of scientific reasoning (using similar scenarios as the example above with

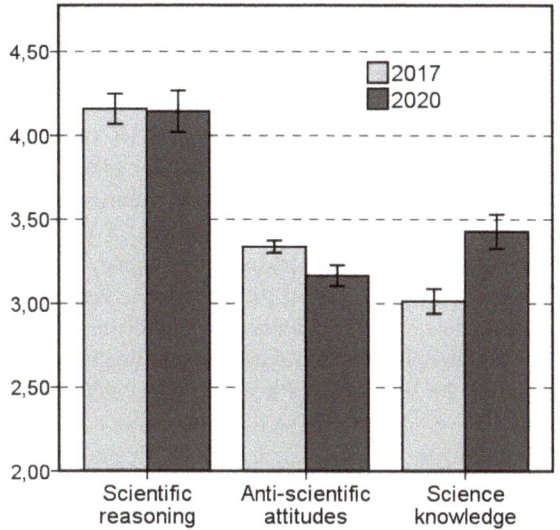

Fig. 1 Comparison of scientific reasoning, anti-scientific attitudes, and science knowledge in two Slovak samples between 2017 and 2020 [54]. The error bars represent a 95% confidence interval

the runner) and anti-scientific attitudes. Three years later, we repeated the survey after the COVID-19 outbreak in Slovakia. We found out that while scientific knowledge slightly but significantly increased, mainly due to people's increased understanding of the questions related to differences between viruses and bacteria and the efficiency of using antibiotics for treating viral diseases, scientific reasoning stayed virtually the same (Fig. 1) [54]. Although people had more accurate information about specific scientific issues that presumably have been brought to their attention during the coronavirus outbreak, this did not mean that they had a greater understanding of how the knowledge is produced and what evidence is valid and reliable.

4.2 Scientific Reasoning as a Factor Contributing to Fewer ESB

Understanding science by having more knowledge and better scientific reasoning skills is also reflected in more positive attitudes toward science and less trust for intuitively appealing or magical explanations of events and endorsing less ESB in general [5, 10, 11].

There are various reasons why scientific reasoning ability is an important predictor of whether people believe certain epistemically suspect beliefs. Firstly, the most obvious reason is that at the basis of various such beliefs are claims that were already directly disproven by scientific research [55]. Examples of this are various pseudoscientific treatments, such as homeopathy, healing by prayer, or palm reading. The second reason has to do with the fact that while scientific theories are judged according to strict criteria, such as those regarding falsifiability and consistency, pseudoscientific accounts and conspiracy theories are not held to the same level of scrutiny [8]. This is why the explanatory potential of these unsubstantiated accounts is higher, and so is their potential to be understood and spread further by the general public. For example, people may believe that COVID-19 was developed as a biological weapon (despite evidence of the contrary) and use it to explain unrelated events (lack of facial masks in the drugstores after the coronavirus outbreak). Lastly, people who endorse ESB often adopt arguments that resemble scientific arguments [56]. Also, they might "base their claims on theories or methods that seem scientific but lack empirical evidence collected by the scientific method" [57]. For example, proponents of dubious therapies often use scientific language (measuring the frequency of energy wavelengths coming from the body). However, the efficiency of these therapies was neither established by double-blind clinical trials, nor the evidence of inefficiency is ignored.

Thus, people who can reason scientifically and know at least some basic principles of the scientific method are better able to recognize the flaws in argumentation and evidence provided by the proponents of ESB claims and are also less susceptible to adopting such beliefs. Besides, it has been shown that scientific reasoning predicted having more beliefs that are consistent with the scientific consensus even after education, political and religious affiliation, and scientific knowledge were taken into account [58].

Several studies examined the link between science understanding and unfounded beliefs. Earlier research used various proxy measures to disentangle the effects of science understanding, such as comparing science students with students from other fields or using formal education as a proxy for better scientific reasoning skills. For example, Aarnio and Lindeman [59] compared Finnish university students who were assumed to have better science knowledge than students from vocational schools. As expected, university students scored lower on paranormal beliefs than their vocational counterparts, mainly because they prefer analytical thinking. Moreover, by comparing the study fields of students, they found out that future medical doctors and psychologists held relatively few paranormal beliefs, while future teachers and theologians held the most paranormal beliefs. Thus, the amount of formal science education was more important than the length of the study.

Similarly, Grimmer and White [60] examined the prevalence of ESB among Australian students of medicine, science, and art. Based on the factor analysis, they extracted one factor they labeled "pseudo practitioners" involving belief in practices such as astrology, numerology, iridology, homeopathy, biofeedback, telepathy in plants, and water divining. The authors found that medical students had the fewest such beliefs, and they explained it by their most extensive science education since high school. Moreover, they also argue that "it is not simply gaining the content knowledge that mediates belief but also the socialization that accompanies education within a particular discipline" [60, p.526]. Also, other studies showed the relationship between various dubious beliefs, such as questionable health practices and pseudoscience, and the amount of education (or attending science courses). However, the results are not always conclusive [61–64].

Other studies examined scientific understanding and its link to ESB by more specific measures, usually factual knowledge of science [52, 65] and scales measuring trust in science. For example, Fasce and Picó [66] found that both factual knowledge of science and trust in science and scientists correlated negatively with the endorsement of pseudoscientific and paranormal beliefs. However, trust in science and scientists correlated negatively with conspiracy theories. There was no relationship with the factual knowledge of science. Also, other studies found a (weak) negative association between pseudoscientific beliefs and knowledge of science facts [47], but no correlation between pseudoscientific beliefs and understanding of scientific concepts. Similarly, Majima [67] found that science understanding predicted paranormal beliefs better than cognitive ability, even after analytic cognitive style and intuitive cognitive style were introduced into the model. On the other hand, Lundström and Jakobsson [68] examined pseudoscientific and superstitious ideas and their relationship with ideas and knowledge of the human body and health. Besides finding that many students believed pseudoscientific claims, such as telepathy and the influence of the Moon on people's health, they found no correlation between scientific knowledge (about the human body) and pseudoscientific beliefs.

These mixed findings might lie in the factual knowledge of science facts, although moderately correlated with the scientific reasoning ability [10, 54, 58], does not necessarily reflect a proper understanding of scientific concepts or the ability to reason scientifically. It is supported by the fact that while the above-mentioned evidence on the link between scientific knowledge and the endorsement of various ESB is mixed, research on scientific reasoning ability consistently shows that it is associated with fewer unfounded beliefs in the domains of paranormal beliefs, conspiracy theories, or pseudoscientific claims specifically related to health. For example, in our studies, scientific reasoning predicted more strongly having fewer ESB than cognitive ability and thinking dispositions [11], or religious faith and political orientation [5], all of which were previously shown to be associated with the endorsement of epistemically suspect beliefs.

In the context of the recent COVID-19 outbreak, we examined whether scientific reasoning would protect people against various dubious health-related beliefs [3]. Interestingly, people with higher scientific reasoning ability had not only fewer generic pseudoscientific and conspiracy beliefs related to health but also endorsed fewer conspiracy beliefs related specifically to COVID-19. Not only do our findings support the results of previous studies that the best predictor of accepting some new ESB is already having some other suspect beliefs, but they also explain why misinformation after a disaster forms and spread so quickly [7, 11]. The study also showed that the more scientifically sophisticated people could better navigate the constantly updating information about the coronavirus. Scientific reasoning enabled them to distinguish between reliable and false information and acquire more accurate knowledge about the new coronavirus. However, the relationship was relatively weak [3].

4.3 Scientific Reasoning and Suboptimal Health Choices

We illustrated how scientific reasoning could protect us against any claims that exploit the tendency of people for magical and superstitious thinking that is behind many ESB. Now let us look at why these kinds of beliefs can be detrimental to our health and how scientific reasoning can help us to avoid making poor health choices. Examples of fatal consequences of poor health choices can be found when parents refuse vaccination of and/or antibiotics for their children [69, 70], or when people refuse conventional treatment in favor of an alternative one until it is too late [71, 72] or until the alternative treatment harms them [73, 74]. In this section, we will review studies that examined the connection between scientific reasoning and the belief in the efficacy of complementary and alternative medicine, vaccination attitudes and vaccination uptake, and preventive behaviors during recent COVID-19 pandemics.

Considering the proponents of various alternatives to the standard medical treatments, we have found that having general critical thinking skills was not sufficient; it was rather scientific reasoning associated with less trust and use of complementary and alternative medicine [14]. However, it seems that preference for complementary and alternative medicine is driven mostly by spiritual beliefs. In other words, people often choose alternative treatments because they believe that they are more "natural" (so-called naturalistic fallacy) and more in line with their holistic health beliefs. Similarly, religiousness is often a significant predictor of medical and political conspiracies [75], which are, in turn, associated with refusing vaccination and conventional medical health care [76].

The more obvious connection between scientific reasoning and health-related behavior is visible when we look at vaccination as one of the prime examples of health-related behavior. People with higher scientific reasoning have been shown to have more positive attitudes toward vaccination [3, 77]. Moreover, as was shown in our later research, anti-vaccination attitudes are associated with conspiracy thinking and having more ESB in general [3]. However, besides anti-vaccination attitudes, we also asked participants if they would be willing to get the vaccine against COVID-19 had it been available at the time (the study was run in March 2020) and whether they got or intended to get vaccinated against seasonal flu that year. While the refusal of potential vaccination against COVID-19 was associated with anti-vaccination attitudes, coronavirus-specific conspiracy theories, and generic health-related ESB, there was no correlation between the refusal of a potential vaccine for the coronavirus and scientific reasoning [3]. Neither did we find a relation between scientific reasoning and intention to get a shot against seasonal flu that year, though, again, the flu-vaccination intention was negatively related to anti-vaccination attitudes and refusal of potential vaccination against COVID-19 [3]. These results suggest that people who have strong anti-vaccination attitudes that are associated with holding various other ESB would refuse any vaccination—existing as well as potential new ones—regardless of the information they get or knowledge about the disease they have.

Similarly, Sarathchandra et al. [77, p. 5] concluded that while general vaccine acceptance is "substantially eroded by conspirational thinking," it is also modestly reduced by political conservativism. It seems obvious that public opinions on vaccination became influenced by deliberate misinformation and political polarization, similar to climate change, nuclear energy, and genetically modified organisms (GMOs). It might explain why "neither educational attainment nor scientific literacy consistently increases vaccine acceptance" [77, p. 5]. Research shows that people do not search for evidence and interpret information in an objective way but do so in a biased manner which allows them to confirm their strongly held beliefs [33, 35, 78]. Therefore, among people with strong anti-vaccination attitudes, higher scientific reasoning may actually contribute to supporting their own worldview even though it is not in line with the scientific consensus.[3]

Moreover, in a study from the beginning of the coronavirus outbreak in Europe (middle of March 2020), we were interested in whether scientific reasoning might help people distinguish accurate information and misinformation about COVID-19 and adopt responsible preventive behaviors to slow down the spread of the coronavirus [3]. We asked people if they engage in various preventive behaviors, some of which were recommended by official authorities (wearing a facemask, washing hands, avoiding crowded spaces, etc.). In contrast, others were advocated by laypeople on social media, but their validity as preventive measures was questionable or downright false (e.g., drinking strong alcohol, spraying one's surroundings with alcohol or chlorine). We observed that people who were better at scientific reasoning had a higher ability to recognize false information about the coronavirus, such as that garlic or ginger are effective coronavirus treatments or that it is possible to kill the virus using the hot-air hand drier. However, the findings were much more confusing when we looked at the preventive behaviors that people reported undertaking in the first days of the outbreak of COVID-19.

However, we found no significant relationships between scientific reasoning and more reasonable health practices. There was no difference between people with better scientific reasoning and lower scientific reasoning in most generally recommended behaviors, such as washing hands and staying at home after general lockdown. We observed more frequent behaviors that could be labeled as "momisms" [78]—things told by your mom but unsubstantiated by any evidence—in people with lower scientific reasoning. Moreover, people with higher scientific reasoning also engaged in behavior that was either not completely reasonable (spraying around chlorine and/or alcohol) or recommended by authorities (making supplies of food and water). They also reported more generally advisable behavior (sporting actively).

Although this pattern of behavior was confusing, at first sight, we identified several possible explanations. First, there were contextual factors, such as high uncertainty, ambivalent information in media, and experts' disagreement on the best way to proceed. In such a confusing situation, people with higher scientific reasoning may not have been better off deciding the best course of action than people with less pronounced scientific reasoning skills. Secondly, in a related study, Erceg et al. [79] found that people who reported engaging in generally recommended preventive behaviors (social distancing and hygienic measures during coughing/sneezing) did not have more accurate information about COVID-19. However, they had fewer unfounded beliefs (conspiracy beliefs and beliefs about pseudoscientific treatments for the coronavirus) and were more scientifically curious (followed scientific news, etc.). It should be mentioned that the strongest correlate of preventive behaviors in

---

[3] More on this is included in the section "Downfalls of scientific reasoning" below.

their study was to worry about COVID-19. As could be expected, the more worried people were about the coronavirus, the more they engaged in the aforementioned preventive behaviors. Thus, the feeling of anxiety and threat may be a better predictor of any preventive behavior—reasonable or questionable—than cognitive factors. Lastly, another explanation could be provided by the intention–behavior gap that mostly occurs, particularly regarding health-related behavior [80–82]. Although people with higher scientific reasoning appeared to be more likely to have beliefs in line with the scientific consensus on various issues [10, 58], this does not automatically mean that they are more willing to act on those beliefs.

4.4 Downfalls of Scientific Reasoning

We have highlighted many positive effects of scientific reasoning on health-related beliefs and behaviors in the previous sections. However, unfortunately, having better scientific reasoning skills does not guarantee that a person will have no unfounded beliefs or always behave optimally regarding their health. It has been shown that for beliefs that concern topics that have been heavily politically polarized (climate change, GMOs, and gun control), higher education or scientific reasoning may not help people reason better but somewhat further increases the polarization among people with different views and opinions [53, 83]. This is because people use "motivated cognition:" they oppose the findings or evidence, which threatens their core beliefs or worldview, but do not question the information aligned with their views [84, 85]. Since people with higher scientific reasoning are more adept at finding limitations or alternative explanations of presented evidence, they can reject findings against their prior beliefs. For example, Kahan et al. [53] found that not only does scientific literacy not reduce false beliefs about climate change, but it even leads to greater polarization between people with a hierarchical-individual and egalitarian-communitarian value orientation. They argue that public division over climate change is not caused by public incomprehension of science but by the reluctance of people to examine (and change) beliefs that are central to their identity. In other words, some beliefs are at the core of the cultural identity of a group of people or individuals who will use cognitive skills to defend them rather than to examine them critically.

The support for this interpretation can also be found in the literature on motivated reasoning and myside bias [34, 35, 85, 86]. It makes it possible for people of higher scientific reasoning and cognitive skills, in general, who are only more skeptical of ESB to go against their beliefs but not of those that correspond more to their political or religious affiliations. For example, if someone is confident that vaccinations can cause autism or other developmental problems for their child, simply educating their scientific reasoning ability might not make them adopt more beliefs in line with the scientific consensus. Rather, it could help the person to identify the limitations of existing research more easily and increase the confidence in their anti-vaccination attitudes.

5 Conclusion

As shown in this chapter, people hold many dubious and potentially dangerous beliefs related to health. We also outlined several processes responsible for our susceptibility to falling for these kinds of beliefs. To conclude the chapter, we formulated three recommendations for the education of scientific reasoning. First, we have to start teaching scientific reasoning early. We have shown that the endorsement of an ESB is a strong predictor of believing in other unfounded claims as well [7, 28]. Thus, scientific reasoning helps by acquiring less ESB that can potentially endanger people's health and have tools to evaluate evidence for any newly encountered claims. This is important because research suggests that the effect of scientific reasoning may be limited when it comes to stressful events (crises, threats) and reasoning about one's core beliefs, so the first step would be to help children to learn the tools that will help them to acquire beliefs that are in line with the current scientific knowledge. Second, for people to be able to reason scientifically, they need to realize why it is important. Scientific reasoning is a way to test our intuitions about the world, and we have shown in the chapter that our intuitions are often wrong; we see patterns in random stimuli, draw conclusions on incomplete data, and tend to see causality where there is none. It is important to be aware of the limits of our reasoning and circumstances when these limits are most likely to occur, such as when we will feel threatened or out of control. Lastly, we should change the way we teach science to foster scientific reasoning. We could encourage children and adults to focus not only on what we know but also on how we know it, the evidence for and against it, and how reliable it is. By doing so, we help children and adults distinguish between more and less reliable evidence and help them be better equipped to deal with misinformation that accompanies modern life.

> *"...people who can reason scientifically and know at least some basic principles of the scientific method, are better able to recognize the flaws in argumentation and evidence provided by the proponents of ESB claims and are also less susceptible to adopt such beliefs."*

Jakub Šrol, Vladimíra Čavojová.

Core Messages

- People hold epistemically suspect beliefs (ESBs) related to health and health care.
- ESBs are common because they stem from natural human cognitive processes.
- The endorsement of conspiracy theories and pseudoscientific beliefs is associated with suboptimal health behavior.
- People with better scientific reasoning can evaluate evidence.
- Scientific reasoning is less useful for stressful events (crises, threats); it should be considered when educating it.

Acknowledgements The research was supported by the Slovak Research and Development Agency as part of the research project APVV-20-0335: "Reducing the spread of disinformation, pseudoscience and bullshit" and by the scientific grant agency of the Ministry ofEducation, Science, Research and Sport of the Slovak Republic as part of the project VEGA 2/0053/21: "Examining unfoundedbeliefs about controversial social issues".

References

1. Schaeffer K (2020) Nearly three-in-ten Americans believe COVID-19 was made in a lab. https://www.pewresearch.org/fact-tank/2020/04/08/nearly-three-in-ten-americans-believe-covid-19-was-made-in-a-lab/. Accessed 21 July 2020
2. Geldsetzer P (2020) Use of rapid online surveys to assess people's perceptions during infectious disease outbreaks: a Cross-sectional Survey on COVID-19. J Med Internet Res 22:1–13. https://doi.org/10.2196/18790
3. Čavojová V, Šrol J, Ballová Mikušková E (2020) How scientific reasoning correlates with health-related beliefs and behaviors during the COVID-19 pandemic? J Health Psychol. https://doi.org/10.1177/1359105320962266
4. Jensen T (2013) Democrats and Republicans differ on conspiracy theory beliefs. https://www.publicpolicypolling.com/polls/democrats-and-republicans-differ-on-conspiracy-theory-beliefs/. Accessed 21 July 2020
5. Šrol J (2021) Individual differences in epistemically suspect beliefs: the role of analytic thinking and susceptibility to cognitive biases. Think Reason. https://doi.org/10.1080/13546783.2021.1938220
6. Mancosu M, Vassallo S, Vezzoni C (2017) Believing in conspiracy theories: evidence from an exploratory analysis of Italian survey data. South Eur Soc Polit 22:327–344. https://doi.org/10.1080/13608746.2017.1359894
7. Lobato E, Mendoza J, Sims V, Chin M (2014) Examining the relationship between conspiracy theories, paranormal beliefs, and pseudoscience acceptance among a university population. Appl Cogn Psychol 28:617–625. https://doi.org/10.1002/acp.3042
8. Lewandowsky S, Gignac GE, Oberauer K (2013) The role of conspiracist ideation and worldviews in predicting rejection of science. PLoS One 8:e75637. https://doi.org/10.1371/journal.pone.0075637
9. Grebe E, Nattrass N (2012) AIDS conspiracy beliefs and unsafe sex in Cape Town. AIDS Behav 16:761–773. https://doi.org/10.1007/s10461-011-9958-2
10. Bašnáková J, Čavojová V, Šrol J (2021) Does concrete content help people to reason scientifically? Adaptation of scientific reasoning scale. Sci Educ 30:809–826. https://doi.org/10.1007/s11191-021-00207-0
11. Čavojová V, Šrol J, Jurkovič M (2020) Why should we try to think like scientists? Scientific reasoning and susceptibility to epistemically suspect beliefs and cognitive biases. Appl Cogn Psychol 34:85–95. https://doi.org/10.1002/acp.3595
12. Pennycook G, Fugelsang JA, Koehler DJ (2015) Everyday consequences of analytic thinking. Curr Dir Psychol Sci 24:425–432. https://doi.org/10.1177/0963721415604610
13. Browne M, Thomson P, Rockloff MJ, Pennycook G (2015) Going against the herd: Psychological and cultural factors underlying the "vaccination confidence gap." PLoS ONE 10:1–14. https://doi.org/10.1371/journal.pone.0132562
14. Čavojová V, Ersoy S (2019) The role of scientific reasoning and religious beliefs in use of complementary and alternative medicine. J Public Health. https://doi.org/10.1093/pubmed/fdz120
15. Gilovich T (1991) How we know what isn't so: the fallibility of human reason in everyday life. The Free Press, New York

16. Lilienfeld SO, Basterfield C, Bowes SM, Costello TH (2020) Nobelists gone wild: case studies in the domain specificity of critical thinking. In: Sternberg RJ, Halpern DF (eds) Critical thinking in psychology. Cambridge University Press, New York, pp 10–38

17. van Prooijen J-W, van Vugt M (2018) Conspiracy theories: evolved functions and psychological mechanisms. Perspect Psychol Sci 13:770–788. https://doi.org/10.1177/1745691618774270

18. Bar-Hillel M, Wagenaar WA (1991) The perception of randomness. Adv Appl Math 12:428–454. https://doi.org/10.1016/0196-8858(91)90029-I

19. Tversky A, Kahneman D (1974) Judgment under uncertainty: heuristics and biases. Science 185:1124–1131. https://doi.org/10.1126/science.185.4157.1124

20. Dagnall N, Parker A, Munley G (2007) Paranormal belief and reasoning. Pers Individ Dif 43:1406–1415. https://doi.org/10.1016/j.paid.2007.04.017

21. van der Wal RC, Sutton RM, Lange J, Braga JPN (2018) Suspicious binds: conspiracy thinking and tenuous perceptions of causal connections between co-occurring and spuriously correlated events. Eur J Soc Psychol 48:970–989. https://doi.org/10.1002/ejsp.2507

22. van Prooijen J-W, Douglas KM, De IC (2018) Connecting the dots: illusory pattern perception predicts belief in conspiracies and the supernatural. Eur J Soc Psychol 48:320–335. https://doi.org/10.1002/ejsp.2331

23. Whitson JA, Galinsky AD (2008) Lacking control increases illusory pattern perception. Science 322:115–117. https://doi.org/10.1126/science.1159845

24. Walker AC, Turpin MH, Stolz JA et al (2019) Finding meaning in the clouds: illusory pattern perception predicts receptivity to pseudo-profound bullshit. Judgement Decis Mak 14:109–119

25. van Prooijen J-W (2019) An existential threat model of conspiracy theories. Eur Psychol 25:16–25. https://doi.org/10.1027/1016-9040/a000381

26. van Prooijen J-W, Douglas KM (2017) Conspiracy theories as part of history: the role of societal crisis situations. Mem Stud 10:323–333. https://doi.org/10.1177/1750698017701615

27. van Prooijen JW, Acker M (2015) The influence of control on belief in conspiracy theories: conceptual and applied extensions. Appl Cogn Psychol 29:753–761. https://doi.org/10.1002/acp.3161

28. Šrol J, Ballová Mikušková E, Čavojová V (2020) When we are worried, what are we thinking? Anxiety, lack of control, and conspiracy beliefs amidst the COVID-19 pandemic. Appl Cogn Psychol 35:720–729. https://doi.org/10.1002/acp.3798

29. Mercier H, Sperber D (2017) The enigma of reason. Pinguin Books, London, A new theory of human understanding

30. Shermer M (2011) The believing brain: from ghosts and gods to politics and conspiracies. Holt, New York

31. Blanco F, Barberia I, Matute H (2015) Individuals who believe in the paranormal expose themselves to biased information and develop more causal illusions than nonbelievers in the laboratory. PLoS ONE 10:1–16. https://doi.org/10.1371/journal.pone.0131378

32. Torres MN, Barberia I, Rodríguez-Ferreiro J (2020) Causal illusion as a cognitive basis of pseudoscientific beliefs. Br J Psychol. https://doi.org/10.1111/bjop.12441

33. Nickerson RS (1998) Confirmation bias: a uniquitous phenomenon in many guises. Rev Gen Psychol 2:175–220. https://doi.org/10.1037/1089-2680.2.2.175

34. Baron J (1995) Myside bias in thinking about abortion. Think Reason 1:221–235. https://doi.org/10.1080/13546789508256909

35. Čavojová V, Šrol J, Adamus M (2018) My point is valid, yours is not: myside bias in reasoning about abortion. J Cogn Psychol 30:656–669. https://doi.org/10.1080/20445911.2018.1518961

36. Li N, Stroud NJ, Jamieson KH (2017) Overcoming false causal attribution: debunking the MMR–autism association. In: Jamieson KH, Kahan DM, Scheufele DA (eds) The Oxford handbook of the science of science communication. Oxford University Press, New York, pp 433–444

37. Fasce A, Picó A (2019) Conceptual foundations and validation of the pseudoscientific belief scale. Appl Cogn Psychol 33:617–628. https://doi.org/10.1002/acp.3501

38. Wood MJ, Douglas KM, Sutton RM (2012) Dead and alive: beliefs in contradictory conspiracy theories. Soc Psychol Personal Sci 3:767–773. https://doi.org/10.1177/1948550611434786

39. Lewandowsky S, Oberauer K, Gignac GE (2013) NASA faked the moon landing-therefore, (climate) science is a Hoax: an anatomy of the motivated rejection of science. Psychol Sci 24:622–633. https://doi.org/10.1177/0956797612457686

40. Trefil J (2008) Why science? https://books.google.sk/books/about/Why_science.html?id=EMXaAAAAMAAJ&redir_esc=y. Accessed 21 July 2020

41. World of Molecules (2021) Paracetamol molecule—Tylenol. https://www.worldofmolecules.com/3D/tylenol_3d.htm. Accessed 9 Nov 2021

42. Zimmerman C (2007) The development of scientific thinking skills in elementary and middle school. Dev Rev 27:172–223. https://doi.org/10.1016/j.dr.2006.12.001

43. Wilkening F, Sodian B (2005) Scientific reasoning in young children: introduction. Swiss J Psychol 64:137–139. https://doi.org/10.1024/1421-0185.64.3.137

44. Koslowski B (1996) Theory and evidence: the development of scientific reasoning. The MIT Press, Cambridge

45. Kuhn D, Franklin S (2006) The Second Decade: What Develops (and How). In: Kuhn D, Siegler RS (eds) Handbook of Child Psychology, vol 2. Cognition, perception and language. John Wiley & Sons, Hoboken, pp 953–993

46. Lobato EJC, Zimmerman C (2019) Examining how people reason about controversial scientific topics. Think Reason 25:231–255. https://doi.org/10.1007/BF00376456

47. Johnson M, Pigliucci M (2004) Is knowledge of science associated with higher skepticism of pseudoscientific claims? Am Biol Teach 66:536–548. https://doi.org/10.2307/4451737

48. Kahan DM (2017) "Ordinary science intelligence": a science comprehension measure for use in the study of risk perception and science communication. J Risk Res 20:995–1016. https://doi.org/10.1080/13669877.2016.1148067

49. McCloskey M (1983) Intuitive physics. Sci Am 248:122–130. https://doi.org/10.1038/scientificamerican0483-122

50. Miller JD (1983) Scientific literacy: a conceptual and empirical review. Daedalus 112:29–48. https://doi.org/10.2307/20024852

51. Shtulman A, Valcarcel J (2012) Scientific knowledge suppresses but does not supplant earlier intuitions. Cognition 124:209–215. https://doi.org/10.1016/j.cognition.2012.04.005

52. Miller JD (1998) The measurement of civic scientific literacy. Public Underst Sci 7:203–223. https://doi.org/10.1088/0963-6625/7/3/001

53. Kahan DM, Peters E, Wittlin M et al (2012) The polarizing impact of science literacy and numeracy on perceived climate change risks. Nat Clim Chang 2:732–735. https://doi.org/10.1038/nclimate1547

54. Čavojová V, Šrol J (2020) COVID-19 pandemics changed our attitudes to science, but not our ability to reason scientifically. Institute of Experimental Psychology, CSPS, SAS, Bratislava

55. Hines T (2002) Pseudoscience and the paranormal. Prometheus Books, New York

56. Harambam J, Aupers S (2015) Contesting epistemic authority: Conspiracy theories on the boundaries of science. Public Underst Sci 24:466–480. https://doi.org/10.1177/0963662514559891

57. Tsai CY, Shein PP, Jack BM et al (2012) Effects of exposure to pseudoscientific television programs upon Taiwanese Citizens' pseudoscientific beliefs. Int J Sci Educ 2:175–194. https://doi.org/10.1080/21548455.2011.6101324

58. Drummond C, Fischhoff B (2017) Development and validation of the scientific reasoning scale. J Behav Decis Mak 30:26–38. https://doi.org/10.1002/bdm.1906

59. Aarnio K, Lindeman M (2005) Paranormal beliefs, education, and thinking styles. Pers Individ Dif 39:1227–1236. https://doi.org/10.1016/j.paid.2005.04.009

60. Grimmer MR, White KD (1992) Nonconventional beliefs among Australian science and nonscience students. J Psychol Interdiscip Appl 126:521–528. https://doi.org/10.1080/00223980.1992.10543385

61. Tobacyk J (1984) Paranormal belief and college grade point average. Psychol Rep 54:217–218. https://doi.org/10.2466/pr0.1984.54.1.217
62. Warburton FW (1956) Beliefs concerning human nature among students in a university department of education. Br J Educ Psychol 26:156–162. https://doi.org/10.1111/j.2044-8279.1956.tb01375.x
63. Jahoda G (1968) Scientific training and the persistence of traditional beliefs among West African university students. Nature 220:1356. https://doi.org/10.1038/2201356a0
64. Pasachoff JM, Cohen RJ, Pasachoff NW (1970) Belief in the supernatural among Harvard and West African university students. Nature 227:971–972. https://doi.org/10.1038/227971a0
65. National Science Board (2010) Science and engineering indicators 2010. National Science Foundation, Arlington
66. Fasce A, Picó A (2019) Science as a vaccine. Sci Educ 28:109–125. https://doi.org/10.1007/s11191-018-00022-0
67. Majima Y (2015) Belief in pseudoscience, cognitive style and science literacy. Appl Cogn Psychol 29:552–559. https://doi.org/10.1002/acp.3136
68. Lundström M, Jakobsson A (2009) Students' ideas regarding science and pseudo-science in relation to the human body and health. Nor Dina 5:3–17. https://doi.org/10.5617/nordina.279
69. Saint-Victor DS, Omer SB (2013) Vaccine refusal and the endgame: walking the last mile first. Philos Trans R Soc Lond B Biol Sci 368:20120148. https://doi.org/10.1098/rstb.2012.0148
70. Liliana Barbacariu C (2014) Parents´ refusal to vaccinate their children: an increasing social phenomenon which threatens public health. Procedia Soc Behav Sci 149:84–91. https://doi.org/10.1016/j.sbspro.2014.08.165
71. Boström H, Rössner S (1990) Quality of alternative medicine–complications and avoidable deaths. Qual Assur Heal care Off J Int Soc Qual Assur Heal Care 2:111–117. https://doi.org/10.1093/intqhc/2.2.111
72. Johnson SB, Park HS, Gross CP, Yu JB (2018) Use of alternative medicine for cancer and its impact on survival. JNCI J Natl Cancer Inst 110:121–124. https://doi.org/10.1093/jnci/djx145
73. Robbins M (2010) The man who encourages the sick and dying to drink industrial bleach. https://www.theguardian.com/science/2010/sep/15/miracle-mineral-solutions-mms-bleach. Accessed 21 July 2020
74. Farrington R, Musgrave I, Nash C, Byard RW (2018) Potential forensic issues in overseas travellers exposed to local herbal products. J Forensic Leg Med 60:1–2. https://doi.org/10.1016/J.JFLM.2018.08.003
75. Galliford N, Furnham A (2017) Individual difference factors and beliefs in medical and political conspiracy theories. Scand J Psychol 58:422–428. https://doi.org/10.1111/sjop.12382
76. Hornsey MJ, Harris EA, Fielding KS (2018) The psychological roots of anti-vaccination attitudes: a 24-nation investigation. Heal Psychol 37:307–315. https://doi.org/10.1037/hea0000586
77. Sarathchandra D, Navin MC, Largent MA, McCright AM (2018) A survey instrument for measuring vaccine acceptance. Prev Med (Baltim) 109:1–7. https://doi.org/10.1016/j.ypmed.2018.01.006
78. Heinzen TE, Lilienfeld SO, Nolan SA (2015) Horse that won't go away: clever hans, facilitated communication, and the need for clear thinking. Worth Publishers, New York
79. Erceg N, Ružojčić M, Galic Z (2020) Misbehaving in the corona crisis: the role of anxiety and unfounded beliefs. https://doi.org/10.31234/OSF.IO/CGJW8. Accessed 21 July 2020
80. Sheeran P, Webb TL (2016) The intention-behavior gap. Soc Personal Psychol Compass 10:503–518. https://doi.org/10.1111/spc3.12265
81. Cobb Leonard K, Scott-Jones D (2010) A belief-behavior gap? Exploring religiosity and sexual activity among high school seniors. J Adolesc Res 25:578–600. https://doi.org/10.1177/0743558409357732

82. Baumann S, Gaertner B, Schnuerer I et al (2015) Belief incongruence and the intention-behavior gap in persons with at-risk alcohol use. Addict Behav 48:5–11. https://doi.org/10.1016/j.addbeh.2015.04.007
83. Drummond C, Fischhoff B (2017) Individuals with greater science literacy and education have more polarized beliefs on controversial science topics. Proc Natl Acad Sci 114:9587–9592. https://doi.org/10.1073/pnas.1704882114
84. Lewandowsky S, Oberauer K (2016) Motivated rejection of science. Curr Dir Psychol Sci 25:217–222. https://doi.org/10.1177/0963721416654436
85. Klaczynski PA, Gordon DH (1996) Self-serving influences on adolescents' evaluations of belief-relevant evidence. J Exp Child Psychol 62:317–339. https://doi.org/10.1006/jecp.1996.0033
86. Stanovich KE, West RF (2008) On the failure of cognitive ability to predict myside and one-sided thinking biases. Think Reason 14:129–167. https://doi.org/10.1080/13546780701679764

Jakub Šrol is a postdoc researcher at the Institute of Experimental Psychology of the Centre of Social and Psychological Sciences at the Slovak Academy of Sciences. His professional interests include the study of individual differences in the susceptibility to cognitive biases and epistemically suspect beliefs. Lately, his research is focused on the negative consequences of the spread of misinformation in society, especially in the domain of health behavior and social prejudice and discrimination. He received several scientific awards for his work at the Slovak Academy of Sciences.

Vladimíra Čavojová is a senior researcher and the deputy director at the Institute of Experimental Psychology of the Centre of Social and Psychological Sciences at the Slovak Academy of Sciences. She also serves as an Editor-in-Chief of the international scientific journal Studia Psychologica. In her research, she focuses on examining various kinds of misinformation, scientific reasoning, and efforts to improve scientific communication to reduce the spread of misinformation. Besides her numerous scientific accomplishments, she is also passionate about the popularization of science. She received an Award from the Slovak Academy of Sciences; she wrote or edited several well-received popular science books and organized lectures and workshops for the general public on intuitive thinking and the spread of misinformation.

Adherence to Treatment: At the Interface of Biological, Medical, and Social Sciences

9

Veronica K. Emmerich, Esther A. Balogh, and Steven R. Feldman

"Drugs don't work in patients who don't take them."

Charles Everett Koop

Summary

Medication adherence is an important topic in clinical medicine and a crucial factor in patient outcomes. Poor medication adherence is responsible for more than 100,000 deaths every year in the U.S. alone. By improving adherence, the

V. K. Emmerich · E. A. Balogh · S. R. Feldman (✉)
Center for Dermatology Research, Department of Dermatology, Wake Forest School of Medicine, Winston-Salem, North Carolina, USA
e-mail: sfeldman@wakehealth.edu

V. K. Emmerich
e-mail: vemmeric@wakehealth.edu

E. A. Balogh
e-mail: ebalogh@wakehealth.edu

V. K. Emmerich · E. A. Balogh · S. R. Feldman
Integrated Science Association (ISA), Universal Scientific Education and Research Network (USERN), Winston-Salem, NC, USA

S. R. Feldman
Department of Pathology, Wake Forest School of Medicine, Winston-Salem, NC, USA

Department of Social Sciences & Health Policy, Wake Forest School of Medicine, Winston-Salem, NC, USA

Department of Dermatology, University of Southern Denmark, Odense, Denmark

Department of Dermatology, Wake Forest School of Medicine Medical Center Boulevard, Winston-Salem, NC 27157-1071, USA

© The Author(s), under exclusive license to Springer Nature Switzerland AG 2022
N. Rezaei (ed.), *Multidisciplinarity and Interdisciplinarity in Health*,
Integrated Science 6, https://doi.org/10.1007/978-3-030-96814-4_9

effect on health status in the whole population can be far more significant than any specific medical therapy or procedure. Poor adherence is often blamed on patient shortcomings, with little attention paid to physicians and healthcare systems' roles. The physician's role has traditionally been prescribing medicines and interventions, but modern medicine demands that the physician also attends to adherence behaviors. Medication adherence can be bolstered with a variety of interventions. Basic interventions include building trust, creating an inviting clinic environment, and increasing feelings of accountability. These are the "foundation" of improving adherence. More advanced interventions build upon this foundation and borrow from the fields of psychology and behavioral economics. These advanced interventions use human behavior and cognitive biases, e.g., framing, anchoring, loss aversion, saliency, inertia, motivational interviewing, and anecdotes to overcome common barriers to adherence. Discussion of these tactics must occur in the context of medical ethics. It requires us to respect the patients' autonomy, beneficence, nonmaleficence, and justice. These requirements vary from individual to individual; they are met by careful ethical consideration of each case.

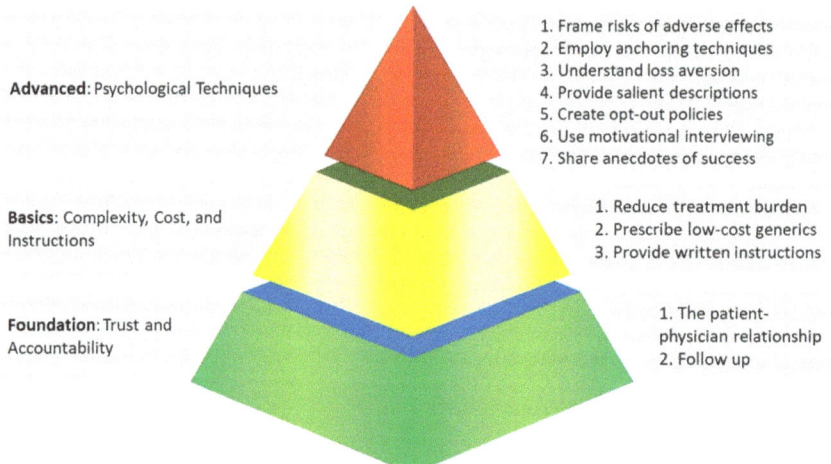

Adherence intervention pyramid: A visual representation of possible interventions

[Adapted with permission from [11]*].*

The code of this chapter is *01110110 01100001 01100101 01101000 01000010 01101001 01110010 01101111.*

Keywords

Adherence · Behavioral economics · Dermatology · Ethics · Health · Interventions · Medicine · Psychology

1 Introduction

Adherence, or the degree the patient considers medical advice, is one of the most important clinical medicine topics and is a crucial factor in determining patient outcomes. Adherence is such an important issue that in 2003 the World Health Organization (WHO) wrote that "increasing the effectiveness of adherence interventions may have a far greater impact on the health of a population than any improvement in specific medical treatments" [1, 2]. Beyond its health implications, adherence also carries a tremendous economic impact: morbidity and mortality from nonadherence are estimated to incur $300 billion in costs annually in the U.S. alone [3]. It is not an issue to take lightly.

Before tackling the problem of poor adherence, it is helpful to have a framework for thinking about it in place. Adherence can be viewed as an interface between biology, medicine, and social science in a linear model [2] as follows:

Treatment (medicine) → Adherence (social science) → Outcome (biology).

In this system, treatment is an input, adherence is a variable, and health outcomes, both on an individual and population level, are an output. By adhering to treatment, the biology and pathophysiology of a disease state can be altered. For an individual, this might look like regularly taking medicine to control high blood pressure. For a population, following the recommended vaccination schedules prevents outbreaks of infectious disease. To take an example from dermatology, a physician might prescribe a topical corticosteroid to control a patient's psoriasis. The patient might fill the prescription and then might use the topical corticosteroid as prescribed. The patient's outcome here is dependent on both the appropriateness of the treatment and the quality of adherence. For topical corticosteroids, the outcomes are changes in gene transcription leading to a reduction in inflammation, which leads to better disease control. These cellular and molecular changes ultimately contribute to a better individual health outcome. Similar analogies can be drawn in the realm of public health, where adhering to vaccination schedules and wearing a mask in the setting of a pandemic, for example, contribute to a population's health as a whole.

Of course, reality never subscribes perfectly to models, especially not linear ones. These three factors, medicine, social science, and biology, are more intricately intertwined in actual practice. A good outcome achieved by a highly adherent patient may warrant a subsequent change in treatment as the disease regresses. Better disease control may improve a patient's life quality in a way that affects a positive feedback loop of increased adherence. Adherence behavior itself is a variable that is affected by other variables. Some patients are naturally (biologically) inclined to be more adherent, some medications are more difficult or unpleasant to apply than others, and the strength of a patient's trust in and relationship physician may all inform adherence behavior. Thus, biology and social factors both play a role in adherence behavior.

2   Why Does Adherence Matter?

Adherence is widely studied because it is universally terrible and recognized as an area with great potential for improving health. Adherence is typically examined in two ways: initiation of treatment versus quality of execution. Primary adherence is concerned with initiating treatment, and it asks the questions: "did they fill the prescription? Did they start using the medication?" Secondary adherence is concerned with the quality of execution and asks: "if they are using the medication after filling the prescription, are they using the medication as it was prescribed?" [4]. Rates of both adherence types are terrible: prescription filling data show that medications are often not even filled, and fill rates are worse when the treatment is more complex. In a study of 143 patients diagnosed with acne, patients were given a treatment consisting of one, two, or three medications. Patients failed to fill these prescriptions 9%, 40%, and 31% of the time, respectively [5]; even a modest treatment plan consisting of two medications is significantly less likely to be filled than a plan containing only one medication. Patients who never fill their prescription or fill the prescription but never subsequently use the medication are considered primarily nonadherent.

Secondary nonadherence can be assessed objectively through electronic monitoring. For example, Medication Event Monitoring System (MEMS) caps are specialized medication packages that electronically record dosing events. Every time the packaging is opened, one event is recorded. In a MEMS cap study of adolescent patients prescribed topical benzoyl peroxide gel, mean adherence declined drastically from 82 to 45% over 6 weeks [6]. One might think that doses of topical treatments are particularly prone to be skipped due to the effort of application, but poor adherence transcends medical specialties and treatment modalities. MEMS cap studies have also revealed that approximately 50% of patients discontinue entirely their antihypertensive medications after one year. Of those who continue to engage with their treatment regimen, almost half have at least one "drug holiday" (when drug dose is missed for three or more consecutive days) per year [4]. These patients are considered secondarily nonadherent.

Although prescription filling records and MEMS caps give us data on patients' adherence behaviors (or lack thereof), they do not offer any insight on their reasons for nonadherence. Nonadherence is often ascribed to some shortcomings of the patient: forgetfulness, poor motivation, perceived burden of treatment, fear of medication, and lack of trust; all these have been suggested as contributors to poor adherence. Additionally, if it is true that some people are naturally (biologically) inclined to be more adherent to medication, it is also true that some people are inclined to be less adherent. However, these factors place full blame on the patient and ignore "physicians" and health care systems' considerable role in low adherence rates. Indeed, many of these issues can be reframed as problems that the physician or health care system should address. Behind every patient who does not understand how to take their medication is a physician who has not adequately transmitted information or instructions on how to do so. A patient who does not believe that treatment will work has a physician who has not paid enough attention to developing a trusting relationship with their patient. An unmotivated patient suffering from depression will continue to be unmotivated until their underlying depression is adequately addressed [7] (Table 1).

The consequences of poor adherence cannot be understated. On an individual level, poor adherence at best leads to therapeutic failure and disappointment. Frequently, it leads to disease progression and the development of complications. For example, elevated blood pressure correlates with a high risk of heart failure, stroke, and acute coronary syndromes. Poor adherence causes an estimated 125,000 deaths annually in the U.S. alone. Moreover, nonadherence accounts for nearly two-thirds

Table 1 Why are patients nonadherent?

Problem	The explanation that faults the patient	Physician or health system intervention
Poor motivation	The patient may not be particularly bothered to use their medication	Address underlying reasons for poor motivation, such as unclear or complicated treatment regimen, depression
Secondary gain	The patient is seeking disability or another gain	Explore patient's motivations and social situation by fostering trust and improving the patient-physician relationship
Lack of trust in the doctor	Poor patient-physician relationship	Attend to improving trust with the patient
Fear of medication	Founded or unfounded fear of treatment or associated side effects	Address the patient's fear of medication, correct unfounded fears, or explore other treatment options
Do not know what to do	Patients may not remember oral instructions or cannot understand written instructions	Provide simple written instructions and have patients teach-back orally to ensure they understand their treatment plan
Burden of treatment	Sometimes the treatment is worse than the disease	Explore other treatment options, simplify the number of medications or dosing frequency
Perceived burden	Sometimes the treatment seems worse than the disease	Educate patients about the need for treatment of mild or asymptomatic disease
Passing responsibility	With multiple caregivers at home, no one may take responsibility for the patient's outcome	Address the patient's support system (or lack thereof)
Forgetfulness	"Pavlov's dog" problem	Put reminder system in place to help patients remember to take their medication
Resignation	Some patients have just given up	Address patient's beliefs about their condition, and address potential comorbid conditions such as depression

of medication-related hospital admissions [8, 9]. These hospital admissions cost patients thousands of dollars in medical care annually, and the morbidity of worsening disease costs patients' life quality. At a population level, poor adherence manifests as outbreaks of infectious disease, decreased life expectancy, and increased chronic disease burden. Despite the mountain of evidence highlighting the poor outcomes linked to nonadherence, getting patients to follow medical advice remains challenging. Offering a treatment alone and expecting adherence is, in most cases, not realistic.

3 Factors in Adherence

The model described in the introduction is not perfectly linear. Besides prescribing treatment, a physician can also improve adherence to behavioral interventions. The job is not done when he or she prescribes a treatment; he or she must also encourage the patient to adhere to it. Often, we think of treatment as a purely medical approach; in actual practice, getting people to use medication depends on several factors.

The first factor is biology. Patients will have different natural tendencies and habits regarding using medication. Some will use their medication religiously every day within the same five-minute window. Others will skip doses for days or weeks before remembering they are supposed to be using anything at all. Additionally, not all medications will work equally well for all patients. Genetic variation has a significant influence on individual responses to a drug, and this variation can explain why some people report more adverse effects while others have a purely therapeutic response [3]. Patients who experience adverse effects or, more rarely, do not respond to medication due to these genetic variants are less likely to continue using the medication than patients with therapeutic responses. Generally, this is not a modifiable factor.

The next factor is the medication itself. The complexity and cost of treatment will affect patients' use and, therefore, their outcomes. More complicated and costly treatment regimens are associated, unsurprisingly, with worse adherence. In the research in patients with acne vulgaris, some participants received a clindamycin-tretinoin combination product for one application daily, and other participants used separate clindamycin and tretinoin products for a total of two applications daily. Patients enrolled in the combination product group showed higher median adherence than patients in the separate products group (88% vs. 61%) [10]. The best treatment is the one the patient will use, which is often the most straightforward treatment. Fortunately, this is often a modifiable factor.

The final and perhaps most important component is social context. A strong patient-physician relationship is critical to establishing trust and improving medication use. A patient who trusts that his or her physician has prescribed appropriate therapy, that his or her physician does not have an ulterior motive to prescribe medication, and that his or her physician will be available if any problems arise during treatment is more likely to be adherent than a patient who distrusts his or her physician. Family, environment, and other social support systems also influence patients' adherence behaviors. This applies not only to medical therapies but also to habits like smoking, alcohol consumption, and diet.

In short, adherence behavior is multifactorial, with "nature" and "nurture" both playing a role. Little can be done to alter the "nature" aspect, although adaptive behaviors can help compensate for a natural inclination toward forgetfulness. On the "nurture" side, one of the most potent ways a physician can encourage adherence is to cultivate a trusting patient-physician relationship. A strong relationship becomes more of a partnership in which both parties are invested in treatment success.

4   Improving Adherence: The Basics

The basics of improving adherence are simple. They can be applied to any transactional relationship and include building trust, attending to context and appearance, and increasing feelings of accountability.

If interventions for improving adherence were a pyramid, building trust would be its foundation [11]. Patient trust and satisfaction are highly correlated and are strong predictors of a patient's intent to stay with their physician, thus contributing to continuous and consistent care and recommending the physician to others [12]. Therefore, it can be assumed that a high level of patient satisfaction translates to a high level of trust. In a study of over 20,000 patients, patient satisfaction with physicians was most highly correlated with a "caring, friendly attitude" followed distantly by "time spent with physician" [13]. This implies that a caring, friendly attitude contributes to the development of trust. Therefore, attention to empathy is a worthwhile investment for physicians aiming to improve patient adherence.

Physicians, as a group, generally care about their patients and are invested in improving patients' lives. There is, however, a subtle distinction between caring about your patients and *showing* that you care about your patients. There are many ways to show that you care. Beyond expressing empathy through spoken words, providing a handwritten personal phone number or offering tailored instructions are powerful tools for demonstrating empathy. These acts are effective because they give patients the sense that they are getting something exclusive and not offered to everyone; a pre-printed set of instructions or a business card does not convey the same sense of specialness. Even if every set of instructions is tailored exactly the same way, termed "placebo tailoring," the suggestion of personalization increases patients' intent to adhere to treatment. A study of patients with eczema assigned some participants to a standardized treatment plan and other participants to a treatment plan with placebo tailoring; participants exposed to the placebo tailored plan expressed greater willingness to follow treatment recommendations [14] (Table 2).

Table 2 Placebo Tailoring. An example of the same treatment plan presented two different ways. Patients express a greater willingness to follow treatment plans that appear to be personalized *[Adapted with permission from* [14]*]*

Your treatment plan	Standard treatment plan	Treatment plan with Placebo Tailoring		
Moisturizers	Emollient	Humectant	*Emollient*	Occlusive
Topical Corticosteroid	Fluocinonide	Triamcinolone	*Fluocinonide*	Hydrocortisone
Topical Corticosteroid Frequency	2x/day	1x/day	*2x/day*	4x/day
Bath	Regular lukewarm	Diluted bleach	Oatmeal	*Regular lukewarm*
Diet	Anti-inflammatory	*Anti-inflammatory*	Dairy-free	*Regular diet*
Other	Liquid paraffin	*Liquid paraffin*	Cool mist vaporizer	
Additional details	**Oatmeal bath:** Combine 2–3 cups colloidal oatmeal (oats ground into a fine powder) in a bathtub full of water. Soak for 10 min. Rinse completely with warm tap water **Anti-inflammatory diet:** Avoid foods high in saturated fats, refined grains, processed meat. Consume foods high in omega-3 fatty acids (fish), probiotics (yogurt with live cultures), flavonoids (colorful fruits and vegetables)	**Diluted bleach bath:** Combine 1/4 cup household bleach (6.15% sodium hypochlorite) in a bathtub half-full of water. Soak for 5–10 min. Rinse completely with warm tap water **Oatmeal bath:** Combine 2–3 cups colloidal oatmeal (oats ground into a fine powder) in a bathtub full of water. Soak for 10 min. Rinse completely with warm tap water **Anti-inflammatory diet:** Avoid foods high in saturated fats, refined grains, processed meat. Consume foods high in omega-3 fatty acids (fish), probiotics (yogurt with live cultures), flavonoids (colorful fruits and vegetables) **Dairy-free diet:** Avoid ingredients such as lactose, whey, casein, cheese powder, lactalbumin. Liquid paraffin: Use liquid paraffin spray as needed for hard-to-reach areas		

The context and appearance of the physical setting also matter. An open, organized front desk with welcoming and friendly signs invites patients' trust far more than a cluttered waiting room with signs insisting on immediate payment, "cash or check only." One of these scenarios says, "I care about you." The other says, "I care about your money." Priorities matter, but so does how those priorities are perceived.

After attending to appearance and setting, another simple intervention that improves adherence is increasing the patient's feelings of accountability. MEMS cap studies show that, in the treatment of psoriasis, adherence to topical therapy rapidly declines to an abysmal level even among patients who are informed that their medication use will be monitored (though how this monitoring is accomplished is not divulged). In a MEMS cap study measuring adherence to topical salicylic acid gel for psoriasis, overall adherence to treatment declined from 84.6 to 51% over eight weeks [15]. However, there were statistically significant upticks in adherence at regular intervals that coincided with follow-up visits. This increase in adherence around office visits is termed the "white coat effect." This effect can be combined with frequent follow-up visits to promote adherence as a habit. Adding a return visit at one week improves adherence markedly [15].

What about what *not* to do? The thing you should not do is blame the patient. Blame will not inspire long-term adherence, will not improve trust, and will damage the patient-physician relationship. When it comes to patient adherence, it is better to be loved than feared.

5 A More Advanced Approach

Beyond the basics of increasing trust and accountability, more sophisticated methods of improving adherence rely on behavioral science principles. These more advanced interventions use psychology and behavioral economics to build upon the foundation of a strong and trusting patient-physician relationship.

5.1 Framing

Framing is the principle that our choices are influenced by how they are presented to us. This principle applies even in scenarios that do not require a conscious choice. Human nature is to value a sure gain over a possible gain and a possible loss over a sure loss. For example, when a physician delivers bad news to a patient, she may state, "your condition will worsen," or, alternatively, she may state that "your condition is not likely to improve." Both statements contain negative framing, but the latter will be perceived more positively by patients because it suggests a possibility of worsening rather than an inevitable worsening. Even if the latter statement is revised to the more certain "your condition will not improve," it will still be perceived more positively. This suggests that word choice and negations play a substantial role in mitigating the impact of bad news. In contrast, with positive statements, it is better to state the good news directly. A patient awakening from surgery would rather hear "you are going to live" than "you are not going to die." Such framing affects a patient's evaluation of the physician, the message, prospects of living with a disease, and medical adherence intentions [16].

Framing can be applied to almost all aspects of patient interactions. If a patient is reluctant to use medication because it is said that it causes serious side effects in one of every 100 patients, this can be reframed by telling the patient that 99 out of 100 patients never develop the serious side effect. A therapy that does not work for 5% of the population *does* work for 95% of the population. These examples emphasize a high probability of success over a slim probability of failure. Careful framing affects not only people's decisions but also how they feel about those decisions and the person who offered them.

5.2 Anchoring

Anchoring refers to our tendency to rely upon an initial piece of information when making decisions. Humans are subjective creatures who are constantly making value judgments, and anchoring makes use of these judgments. Anchoring is used frequently in negotiations and in the retail world; the entire concept of a "sale" relies upon customers anchoring to an original price, which is always crossed out yet still visible. This initial "anchor" price serves as the reference point against which potential perceptions and decisions are weighed.

In medicine, anchoring can be used to increase patients' willingness to try a therapy that is difficult or otherwise unpleasant to use. When starting a new therapy, the initial anchor is always whatever the patient tried before for their condition, which is often nothing. This initial anchor of "nothing" can be a barrier to adhering to treatment because treatment will always require more effort than doing nothing. One potential solution to this is to reset the patient's anchor.

In a study of patients with psoriasis, patients were far more willing to take a monthly injection of medication if initially presented with the option of a daily injection [17]. The offering of a daily injection altered the initial reference point (or anchor) of no injections. In other words, a monthly injection sounds terrible relative to no injections, but a monthly injection sounds significantly better compared to daily injections.

5.3 Loss Aversion

The principle of loss aversion comes from behavioral economics. People are inherently averse to losing, and most people will prefer avoiding a loss over acquiring an equivalent gain. For example, people would rather not lose $20 than acquire $20 [18].

A few studies have examined loss aversion as an adherence intervention in a very literal sense. A small randomized controlled trial of glaucoma patients concluded that patients are significantly more adherent when motivated by an adherence-contingent rebate. In this study, monetary rebates on treatment and medication costs were offered in proportion to how successful a patient was in adhering to their prescribed therapy. Adherence was monitored with electronic packaging, and patients in the rebate group were, on average, 12% more adherent than patients in the control group at six-month follow-up [19]. Similar models have been developed as phone apps that offer a monthly rebate and penalize the user for missing medication doses. Utilizing behavioral economics to establish a habit is a powerful way to improve adherence and outcomes.

Although loss aversion is closely related to behavioral economics, it does not require a monetary penalty. The penalty can be anything of value, including the intangible. In a dermatology clinic, this might look like counseling patients to use sunscreen because they risk losing the youthful look of their skin and acquiring wrinkles rather than telling them to use sunscreen to maintain their youthful look. The message is essentially the same, but the impact is greater when it is framed as a loss.

5.4   Saliency

Saliency is the quality of being prominent or having high visibility, and as a cognitive bias, it refers to the human tendency to pay more attention to things or events that are more noticeable. In medicine, this might mean making perceived consequences more personal or more vivid to improve adherence. Combined with loss aversion, increasing the salience of adverse health consequences can increase adherence. For example, telling a patient that "not wearing a hat increases your risk of skin cancer" does not have the same weight as "if you don't wear a hat, you could end up with a golf ball-sized skin cancer that would require surgical removal of your nose and the placement of a rubber prosthesis." A large-scale randomized controlled trial conducted in the U.K. supports this claim. The study's authors hypothesized that increasing the risk salience would increase adherence. The study included over 16,000 patients divided into four possible intervention groups. Interventions consisted of a signed pledge, a commitment to take medication as prescribed, on a sticker attached to patients' medication packaging. Some stickers contained the pledge only, some emphasized societal costs and the NHS's potential bankruptcy, and some emphasized personal health risks. The study groups were no commitment, commitment only, commitment plus emphasis on societal costs of nonadherence, or commitment plus emphasis on personal costs of nonadherence. After nine months, only those assigned to the "personal health risks" group were more adherent to their medication. The authors deemed that patients were not motivated to be more adherent to avoid societal costs or by pledges alone [20]. This is not to say that all patients do not care about costs to society, some certainly do, but personal costs are almost always more persuasive in terms of improving adherence behaviors as they are more self-relevant.

A risk described at a population level, a statistic, is abstract, while a risk described at an individual level becomes much more concrete. Thus, highlighting the personal risks of nonadherence may be an effective tool in improving adherence.

5.5   Inertia

Humans tend to be psychologically inert. Topics that are not at the forefront of consciousness are permitted to follow the path of least resistance. This means that the general flow of human behavior will follow a "default," which will always be the behavior that requires the least amount of intervention. What happens if the default is changed?

The creation of "opt-out" policies takes advantage of this inertia to increase beneficial health behaviors such as vaccination and organ donation. These opt-out policies make beneficial behaviors the default by requiring effort *not* to partake. In a study of influenza vaccination rates, an opt-out policy increased vaccination rates by 1.4-fold in a primary care setting that traditionally offered flu shots on request and at a weekly flu shot clinic. This increase did not consider that the vaccine supply was interrupted for more than a week during the 26-week opt-out period [21]. Similarly, participation in colorectal cancer screening dramatically increases when an opt-out policy is implemented [22]. These policies create a barrier to unwise medical decisions. These thoughtful policies increase beneficial decision-making by creating a "choice architecture" that takes advantage of human psychology [23]. Ironically, this good decision-making is increased by removing the need to make decisions at all. Both adherence and outcomes improve when valuable treatments and policies become the paths of least resistance.

5.6 Motivational Interviewing

Motivational interviewing is a method widely used in counseling that attempts to enhance patients' intrinsic motivations. It involves a conversational style designed to change patients' behavior by helping them self-discover their motivations and goals and identifying what they perceive as barriers to those goals. It works by finding discrepancies between patients' behaviors and stated goals, helping them reflect on psychological roadblocks, and asking open-ended questions about overcoming those roadblocks. Clinical psychologists originally developed motivational interviewing to help patients with substance use disorders, but it has been adopted in many other areas that require behavioral change [24]. Used in counseling, it is a complete and structured process that requires significant expertise, but elements of the process can still be used in other areas where behavior change is desired. Multiple meta-analyses have found that motivational interviewing is effective in increasing patient adherence [25]. For example, a lifelong smoker who wants to quit but feels they do not have the strength to do so might be asked, "what would increase your confidence in your ability to quit?" A reckless driver with an addiction to alcohol might be asked about their long-term goals and how they see their behaviors fitting with those goals. Similarly, a patient who reliably forgets to take their medication could be asked, "what would help you remember to take your medication every day?"

5.7 Anecdotes

Storytelling is more effective than statistics; the power of a moving narrative is almost always greater than that of pure data. It was the power of anecdotes that propelled the spread of the anti-vaccine movement, with concerned parents sharing personal encounters and photos on social media documenting their children's harrowing experiences with vaccines. Data and scientific consensus did not matter; the fear generated by these personal horror stories outweighed any rational argument [26]. Social media did not help, but the primary tool of the movement was the anecdote.

Why are anecdotes so much more effective than data? Several factors contribute to anecdotes' power. First, they are relatable. It is much easier for people to imagine themselves as the protagonist in a story than it is for them to interpret data into something meaningful. For example, a new parent reading a story about a young child who has autism after vaccination is susceptible to thinking, "this could happen to me if I vaccinate my child." Even though scientific evidence does not support this belief, the anecdote can produce a knee-jerk reaction in those unaware that it is patently false. It is even easier to relate when the storyteller is a trusted friend or family member. This relatability makes the anecdote more salient.

Second, people prefer narratives over statistics. Children ask for bedtime stories, not bedtime clinical trial data. The average person's scientific knowledge is obtained via television or social media, not data or peer-reviewed papers [26]. These media forms are based on narrative transmission of information; they are a modern oral tradition. Even a narrative that is not relatable is more engaging than a set of numbers or a table. Stories tend to be simple to follow, but analysis of data requires complex and abstract thought.

Finally, there may be a biological reason that people respond more powerfully to anecdotes than to data. An anecdote is more personal, more visceral, and perhaps elicits a more significant response in the amygdala, the brain area that plays a central role in responding to external threats. However, anecdotes do not have to rely on fear to be effective.

Regardless of the reason, anecdotes are more powerful than data. A study assessing caregivers' willingness to use a topical corticosteroid to treat children with atopic dermatitis found that anecdotal reassurance was more effective than clinical trial data in increasing willingness [27]. Despite clinical trials being one of the most robust scientific inquiry tools, they do not have the same effect as anecdotes in convincing and persuading people to accept an idea. To increase adherence, this does not mean that clinical trial evidence is not useful; it does positively affect increasing patients' willingness to use therapies. It is just that the effect of anecdotes is more substantial in influencing patients' adherence behaviors (Table 3).

5.8 Ethics

These behavioral interventions cannot be discussed without a concomitant discussion of ethics. After all, there is an element of psychological manipulation inherent in each intervention. Before examining the ethical implications of these interventions, it is important to understand the ethics themselves. All medical ethics principles are based on four fundamental tenets: autonomy, beneficence, nonmaleficence, and justice.

Table 3 Summary of Adherence Interventions

Intervention	Example
Building trust	Provide a handwritten phone number or (pseudo) personalized instructions for treatment
Context and appearance	Have an organized front desk with signs thanking patients for their trust and referrals, which allow the practice to grow, as opposed to signs demanding payment
Increasing accountability	Decrease intervals between follow-up appointments
Framing	Emphasize that the treatment works for a vast majority of people, rather than that it does not work for only a few
Anchoring	Offer less pleasant treatment options first, or explain treatment using a more unpleasant example (e.g. "this treatment is similar to insulin in that you will need to give it as an injection, but unlike daily insulin injections, you only need to use this once a month")
Loss Aversion	Emphasize what the patient stands to lose from not adhering to treatment, rather than what they stand to gain from adhering to treatment
Saliency	Emphasize the personal consequences of not adhering to treatment by painting a vivid picture
Inertia	Create opt-out policies that make beneficial health decisions the default choice and require no effort from patients
Motivational Interviewing	Help patients explore what would increase their motivation to use a treatment, what psychological roadblocks they are facing, and discrepancies between their goals and current behavior
Anecdotes	Offer a story about a specific person or situation, instead of (or in addition to) offering clinical trial data and statistics

Autonomy, or self-rule, dictates that patients must make their own decisions based on informed consent. It is the physician's responsibility to convey complete and accurate information regarding the risks and benefits of potential decisions, but ultimately it is up to the patient to make his or her own healthcare decisions. Making an informed decision demands complete honesty between physician and patient because a decision made from incomplete or incorrect information cannot be considered informed. However, the scope of autonomy extends far beyond informed consent. Respect for autonomy includes maintaining confidentiality, keeping promises, and avoiding deception in communications. People organize aspects of their lives around the assumption that these principles are upheld, so maintaining their trust contributes to autonomy [28].

Beneficence and nonmaleficence are closely related and mandate that a physician provides a net benefit to patients while minimizing harm. This requires knowledge not only of the risks and benefits of interventions, including lack of intervention, but also their corresponding statistical probabilities. The best way to assess these qualities is through rigorous medical research, so medical research is necessary. A physician who wishes to practice beneficence must also consider patient autonomy because a patient's priorities may not always align with what is medically indicated.

Justice, or fairness, is concerned with the equitable distribution of limited resources, respect for rights, and respect of laws [28]. This tenet is not particularly relevant for the adherence interventions discussed here. However, the inability to afford medication or to visit a physician to prescribe the medication in the first place (which hints at an inequitable distribution of resources) certainly plays a role in preventing primary adherence. We are more concerned with the first three tenets, i.e., autonomy, beneficence, and nonmaleficence, although justice does play a minor role.

The central conflict arises between respecting patients' autonomy and recognizing a professional responsibility to practice beneficence. One could argue that altering patients' perspectives to increase their likelihood of accepting a treatment does not respect autonomy by possibly swaying patients to do something they otherwise would not do. On the other hand, not attending to adherence leads to worse outcomes in patients who might otherwise have used their medication, which neglects beneficence [29].

The principle of autonomy demands communication free of deception. None of the adherence interventions discussed involve deception, but they do involve some manipulation. Despite the negative connotations associated with manipulation, it is frequently used with good intentions. A therapist helping a patient with addiction or a teacher encouraging students to do homework are both inarguably using manipulation for good. A physician who shares anecdotal stories of patients' success with a drug to improve adherence is, in the same vein, using the human perspective's subjective nature to promote beneficial health behavior. However, physicians are held in a position of great trust by patients, and that trust requires careful maintenance. Even understanding that it was done with their best interest in mind, a patient who feels manipulated may feel that their trust toward that physician has been eroded.

The principles of beneficence and nonmaleficence demand that physicians maximize benefit while minimizing harm. Patients coming to see physicians usually have a complaint and seek expert opinions on how to resolve it. The physician has to resolve the complaint by prescribing treatment and encouraging the patient to *use* that treatment. A treatment prescribed but never used has zero benefits at best and deleterious health consequences at worst. Either way, this fails to resolve the complaint; benefit is not maximized and harm is not minimized.

Failure of treatment due to poor adherence also does not respect the tenet of justice. This failure is worse than not seeking treatment at all. The health outcomes of failing to adhere and not seeking treatment are the same, but failure to adhere has the additional insult of wasting finite resources such as office resources, physician time and effort, and insurance payments. In other words, using adherence interventions promotes justice by minimizing the waste of resources.

Ultimately, the question is whether the ends justify the means. Is it acceptable to potentially undermine a patient's trust in order to achieve better health outcomes? Should the use of these interventions depend on the magnitude of possible harm from nonadherence? The disappointment of a poor outcome is also capable of damaging the patient-physician relationship, and encouraging patients to use medicine that will control severe hypertension is clearly different from enticing patients to purchase cosmetic services. Even though patients often understand that these interventions are done with their best interest in mind, these techniques should be used judiciously, with careful ethical consideration of each unique situation.

6 Conclusion

Poor adherence is a public health crisis in its own genre. It carries both high personal and societal costs, yet adherence interventions carry a meager cost. It is not enough for physicians to make the correct diagnosis and prescribe the correct treatment: adherence is a fundamental part of a successful outcome, so physicians should aim to improve adherence through building trust and behavioral interventions. These interventions draw upon concepts from psychology and behavioral economics and have been proven, time after time, to be effective in increasing patient adherence.

To understand how adherence fits into healthcare overall, it helps understand the treatment $\rightarrow$ adherence $\rightarrow$ outcome model, although this is a simplified representation of a complex interface. Adherence is a major variable between treatment and health outcomes, and it has traditionally been considered entirely patient-dependent. We know that the reality is not this simple; improved health outcomes can encourage improved adherence and cause subsequent changes in treatment, and better health outcomes can also improve social outcomes, which change people's interactions with others. Regardless of complexity, the success of patient outcomes is dependent on both the appropriateness of treatment and the quality of patient adherence.

Every individual case should take the ethics of adherence interventions into account. Medical ethics demand that physicians respect the tenets of autonomy, beneficence, nonmaleficence, and justice. Adherence interventions maximize health benefits and minimize the harms of complications and disease progression, which satisfies both beneficence and nonmaleficence. Additionally, improved adherence decreases the waste of finite resources, which respects justice. However, patients can potentially feel manipulated, which may damage the patient-physician relationship and potentially disrespect patient autonomy. Every case warrants careful consideration of these tenets, with the level of intervention commensurate with the potential benefit of treatment or potential harm of nontreatment, and always with the individual patient's goals in mind.

Core Messages

- Poor adherence to medication is an enormous clinical problem with high costs, both financially and in quality of life.
- Adherence behavior is a complex product of biological, social, and medical influences.

- Prescribing a treatment and expecting adherence is usually not enough.
- The foundation of improving adherence rests on building a trusting patient-physician relationship and increasing feelings of accountability.
- Adherence interventions are proven to increase patients' adherence to therapy, when used judiciously and with careful consideration of medical ethics in each case.

References

1. De Geest S, Sabate E (2003) Adherence to long-term therapies: evidence for action. Eur J Cardiovasc Nurs 2(4):323. https://doi.org/10.1016/S1474-5151(03)00091-4
2. Brown MT, Bussell JK (2011) Medication adherence: WHO cares? Mayo Clin Proc 86(4): 304–314. https://doi.org/10.4065/mcp.2010.0575
3. Haga SB, LaPointe NM (2013) The potential impact of pharmacogenetic testing on medication adherence. Pharmacogenomics J 13(6):481–483. https://doi.org/10.1038/tpj.2013.33
4. Vrijens B, Vincze G, Kristanto P, Urquhart J, Burnier M (2008) Adherence to prescribed antihypertensive drug treatments: longitudinal study of electronically compiled dosing histories. BMJ 336(7653):1114–1117. https://doi.org/10.1136/bmj.39553.670231.25
5. Anderson KL, Dothard EH, Huang KE, Feldman SR (2015) Frequency of primary nonadherence to acne treatment. JAMA Dermatol 151(6):623–626. https://doi.org/10.1001/jamadermatol.2014.5254
6. Yentzer BA, Alikhan A, Teuschler H, Williams LL, Tusa M, Fleischer AB Jr, Kaur M, Balkrishnan R, Feldman SR (2009) An exploratory study of adherence to topical benzoyl peroxide in patients with acne vulgaris. J Am Acad Dermatol 60(5):879–880. https://doi.org/10.1016/j.jaad.2008.11.019
7. Devine F, Edwards T, Feldman SR (2018) Barriers to treatment: describing them from a different perspective. Patient Prefer Adherence 12:129–133. https://doi.org/10.2147/PPA.S147420
8. McCarthy R (1998) The price you pay for the drug not taken. Bus Health 16(10):27–28, 30, 32–23
9. Osterberg L, Blaschke T (2005) Adherence to medication. N Engl J Med 353(5):487–497. https://doi.org/10.1056/NEJMra050100
10. Yentzer BA, Ade RA, Fountain JM, Clark AR, Taylor SL, Fleischer AB Jr, Feldman SR (2010) Simplifying regimens promotes greater adherence and outcomes with topical acne medications: a randomized controlled trial. Cutis 86(2):103–108
11. Lewis DJ, Feldman SR (2017) Practical ways to improve patient adherence. CreateSpace Independent Publishing Platform
12. Platonova EA, Kennedy KN, Shewchuk RM (2008) Understanding patient satisfaction, trust, and loyalty to primary care physicians. Med Care Res Rev 65(6):696–712. https://doi.org/10.1177/1077558708322863
13. Uhas AA, Camacho FT, Feldman SR, Balkrishnan R (2008) The relationship between physician friendliness and caring, and patient satisfaction: findings from an internet-based survey. Patient 1(2):91–96. https://doi.org/10.2165/01312067-200801020-00004
14. Bashyam AM, Cuellar-Barboza A, Ghamrawi RI, Feldman SR (2020) Placebo tailoring improves patient satisfaction of treatment plans in atopic dermatitis. J Am Acad Dermatol. https://doi.org/10.1016/j.jaad.2020.01.044
15. Carroll CL, Feldman SR, Camacho FT, Manuel JC, Balkrishnan R (2004) Adherence to topical therapy decreases during the course of an 8-week psoriasis clinical trial: commonly

used methods of measuring adherence to topical therapy overestimate actual use. J Am Acad Dermatol 51(2):212–216. https://doi.org/10.1016/j.jaad.2004.01.052

16. Burgers C, Beukeboom CJ, Sparks L (2012) How the doc should (not) talk: when breaking bad news with negations influences patients' immediate responses and medical adherence intentions. Patient Educ Couns 89(2):267–273. https://doi.org/10.1016/j.pec.2012.08.008

17. Oussedik E, Cardwell LA, Patel NU, Onikoyi O, Feldman SR (2017) An anchoring-based intervention to increase patient willingness to use injectable medication in psoriasis. JAMA Dermatol 153(9):932–934. https://doi.org/10.1001/jamadermatol.2017.1271

18. Kokmotou K, Cook S, Xie Y, Wright H, Soto V, Fallon N, Giesbrecht T, Pantelous A, Stancak A (2017) Effects of loss aversion on neural responses to loss outcomes: an event-related potential study. Biol Psychol 126:30–40. https://doi.org/10.1016/j.biopsycho.2017.04.005

19. Bilger M, Wong TT, Lee JY, Howard KL, Bundoc FG, Lamoureux EL, Finkelstein EA (2019) Using adherence-contingent rebates on chronic disease treatment costs to promote medication adherence: results from a randomized controlled trial. Appl Health Econ Health Policy 17(6):841–855. https://doi.org/10.1007/s40258-019-00497-0

20. Jachimowicz JM, Gladstone JJ, Berry DAN, Kirkdale CL, Thornley T, Galinsky AD (2019) Making medications stick: improving medication adherence by highlighting the personal health costs of non-compliance. Behav Public Policy:1–21. https://doi.org/10.1017/bpp.2019.1

21. Logue E, Dudley P, Imhoff T, Smucker W, Stapin J, DiSabato J, Schueller C (2011) An opt-out influenza vaccination policy improves immunization rates in primary care. J Health Care Poor Underserved 22(1):232–242. https://doi.org/10.1353/hpu.2011.0009

22. Mehta SJ, Khan T, Guerra C, Reitz C, McAuliffe T, Volpp KG, Asch DA, Doubeni CA (2018) A randomized controlled trial of opt-in versus opt-out colorectal cancer screening outreach. Am J Gastroenterol 113(12):1848–1854. https://doi.org/10.1038/s41395-018-0151-3

23. Thaler RH, Sunstein CR Nudge: improving decisions about health, wealth, and happiness

24. Allsop S (2007) What is this thing called motivational interviewing? Addiction 102(3): 343–345. https://doi.org/10.1111/j.1360-0443.2006.01712.x

25. Palacio A, Garay D, Langer B, Taylor J, Wood BA, Tamariz L (2016) Motivational interviewing improves medication adherence: a systematic review and meta-analysis. J Gen Intern Med 31(8):929–940. https://doi.org/10.1007/s11606-016-3685-3

26. Shelby A, Ernst K (2013) Story and science: how providers and parents can utilize storytelling to combat anti-vaccine misinformation. Hum Vaccin Immunother 9(8):1795–1801. https://doi.org/10.4161/hv.24828

27. Johnson MC, Pona A, Adler-Neal AL, Kesty C, Cline A, Feldman SR (2020) Assessing the effect of clinical trial evidence and anecdote on caregivers' willingness to use corticosteroids: a randomized controlled trial [formula: see text]. J Cutan Med Surg 24(1):17–22. https://doi.org/10.1177/1203475419871050

28. Gillon R (1994) Medical ethics: four principles plus attention to scope. BMJ 309(6948): 184–188. https://doi.org/10.1136/bmj.309.6948.184

29. Oussedik E, Feldman SR (2019) Manipulating mindsets to improve patient outcomes: is it ethical? Can it be avoided? J Eur Acad Dermatol Venereol 33(2):e79–e81. https://doi.org/10.1111/jdv.15224

Veronica Emmerich is a fourth-year medical student at Wake Forest School of Medicine. She studied political science at the University of California, Santa Barbara, and biochemistry at North Carolina State University before deciding to pursue a career in dermatology. In her spare time, she enjoys learning languages, traveling, and building tiny homes. Her dream is to own an apartment in southern Italy and share it with friends.

Steven Feldman serves as a professor of Dermatology, Pathology, and Social Sciences & Health Policy at Wake Forest. A protein chemist, Feldman was hired in 1991 by Joe Jorizzo to run a test tube research lab, a career that failed miserably. He got sucked into doing more clinical care, for which he was, arguably, ill-suited. Because of his natural lack of interpersonal skills, his patients were particularly poorly adherent to treatment, leading to his research studies into patients' adherence to topical treatments that transformed how physicians understand and manipulate patients' use of treatment for chronic skin diseases. Dr. Feldman was the founder of www. DrScore.com, an online doctor rating/patient feedback website, and Causa Research, an adherence company. With the help of his many minions, Steven has authored over 1,000 peer-reviewed publications. He has been rated by ExpertScape.com as the #1 expert in the world on psoriasis and the #2 expert on treatment adherence.

Social Cognition and Food Decisions in Obesity

10

Hélio A. Tonelli and Luisa de Siqueira Rotenberg

"And emotions often have something to add to or subtract from decisions one would have expected to be purely rational."

Antonio Damasio

Summary

This chapter discusses the role that social cognition (SC) plays in food decisions in individuals with obesity. It begins with an overview of neuropsychological decision-making mechanisms focusing on eating decisions, followed by a conceptualization of SC. We argue that some social cognitive domains may be affected by the toxic effects of obesity on the brain. We also review the few studies, to date, showing SC deficits in obesity and discuss the role of brain areas regarding decision-making and SC. Such areas comprise the ventromedial prefrontal cortex, insula, and anterior cingulate cortex, recruited in emotion recognition and theory of mind, whose proper function is central for a healthy social life. Impairments of those areas' functions impact social regulation mechanisms of emotions, particularly the aversive ones, leading to dysregulation of eating behaviors, e.g., binges, emotional eating, and food addiction. We conclude by examining some of the toxic effects of obesity on the brain, such as inflammation, gut microbiota changes, and insulin resistance, frequently described in individuals suffering from obesity. Understanding how SC affects

H. A. Tonelli (✉)
Department of Neuropsychology, and Caetano Marchesini Clinic, FAE Business School, Avenida Cândido de Abreu, 526, cj. 311 Torre B, Curitiba, Paraná, Brasil

L. de Siqueira Rotenberg
Bipolar Disorders Research Program, Department of Psychiatry, São Paulo Medical School, São Paulo, Brazil

© The Author(s), under exclusive license to Springer Nature Switzerland AG 2022   219
N. Rezaei (ed.), *Multidisciplinarity and Interdisciplinarity in Health*,
Integrated Science 6, https://doi.org/10.1007/978-3-030-96814-4_10

food decisions in people with obesity may help develop novel strategies to treat and prevent obesity.

FOOD DECISIONS

A Quick Breakdown

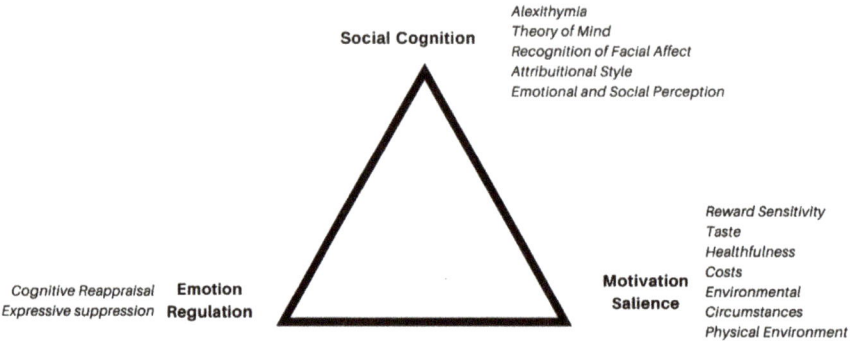

Psychological variables that come into play in dietary decision-making. In addition to those usually considered decisive factors in consumption choices, such as those related to motivational reward/pleasure/salience, preferences, health, and environment, social cognitive skills are fundamental for a healthy social life. The ability to properly regulate unpleasant emotions (many of them stem from unsatisfactory interpersonal relationships) protects against the possibility that they might be regulated not only by palatable foods but also by alcohol and psychoactive substances. Any damage to the factors illustrated above is capable of compromising food decisions.

The code of this chapter is *01101101 01,101,110 01,100,110 01,101,100 01,100,001 01,101,110 01,100,001 01,101,101 01,110,100 01,101,001 01,001,001 01,101,111*.

Keywords

Decision-making · Emotion recognition · Obesity · Social cognition · Theory of mind · Drives · Motivations

1 Introduction

Obesity is a chronic, complex, and heterogeneous condition associated with several other disorders that may disturb the whole organism [1]. All countries, including developing ones, currently face an obesity epidemic [2, 3]. It calls for a more comprehensive understanding of the mechanisms involved in food decisions, paving the way to develop new strategies to prevent and treat obesity.

Eating behavior and food choices are mediated by an intricate network of feedback loops involving the hypothalamus, the brain stem, and other upper brain centers, as well as organs such as the stomach, liver, pancreas, intestine, thyroid, and adipose tissue [4]. Communication within this network is damaged in obesity due to numerous pathophysiological mechanisms resulting in behavior changes. Such changes affect eating experience and food decisions, causing morbid changes in satiety and taste experience [5, 6], uncontrolled and impulsive behaviors toward caloric foods [7], and even cognitive problems [8].

Decision-making comprises cognitive operations through which the brain transforms sensations into actions. However, it must be taken into account that not all "decisions" involve reflective and deliberative processes [9]. Even if temporarily remote, sensory stimuli interact with cognitive, motivational, emotional, and autonomic systems to generate behavior. In decision-making, the brain/mind chooses, consciously and, to some extent, in a flexible and modifiable fashion, one among several possible behaviors [10, 11].

Food choices involve diverse, multifaceted decision types, including whether, what, where, when, with whom, how, and how much to eat [12]. Moreover, decision-making processes are studied through various theoretical perspectives [12] and methodologies, resulting in heterogeneous discussions. In this sense, our framework assumes that food decisions of individuals with obesity undergo multiple interferences, among them, from social cognitive processing, including those related to identifying one's own emotions and those of other people. Social cognition (SC) directly influences emotion regulation processes since emotions are essentially social phenomena besides being predominantly regulated in social contexts [13]. Emotion regulation impairments seem to influence food decisions, not exclusively, but remarkably, in people with obesity. Dysfunctional eating behaviors, such as binge eating [14], emotional eating [15], loss of control over eating [15], and addition to food [16], frequently occur in patients with obesity, being habitually triggered and perpetuated by emotions such as anger and loneliness, so frequent in interpersonal contexts [17, 18].

SC is a growing area of science concerned with neuropsychological processes of social living [19]. It involves several domains, namely facial affect recognition [20], understanding emotions in body postures, gestures [21], and attitudes of others [22], as well as the ability to infer others' mental states, also known as "theory of mind" or ToM processing [23]. SC deficits are common to people with mental disorders, in particular, autism [24], schizophrenia [25], mood disorders [26], and eating disorders [27], leading to social dysfunctionality, which may worsen these conditions. Findings pointing that SC deficits may lie behind people's problematic eating behaviors with eating disorders have instigated the interest in investigating similar deficits in individuals with obesity. Indeed, this population mostly shows binges, emotional eating, addiction to food, and loss of control over eating, isolated or in combination [28]. Pathological eating behaviors, in turn, would result from deficits in emotional regulation, which, at least in part, derive from impairments in SC processing. Patients with eating disorders and obesity seem to display higher scores on instruments measuring alexithymia [29, 30], a transdiagnostic condition characterized by difficulties in identifying and describing their own emotional states [31]. Moreover, alexithymic individuals may suffer from impairments in interoceptive processing that interfere with their ability to distinguish specific interoceptive signals, thus having a hard time discerning basic signals of hunger (which is a motivational state) from emotions [32, 33]. Such a scenario favors eating in order to mitigate aversive emotional experiences. Although traditionally neglected by science, emotions are now recognized as essential components of high-level cognitive function, including decision-making [32].

SC theories based on simulation propose that the ability to understand others' emotional states is built upon the ability to recognize our own emotions [34, 35]. As a result, people who fail to process their own emotions appropriately, as seen in patients with alexithymia, might have impairments in their ability to perceive and interpret others' emotional states. Such deficits would ultimately compromise social life, leading to mental suffering and isolation and, consequently, bad food decisions. These decisions would not serve primarily as a nutritional purpose but would act as regulators of aversive emotions.

Despite being scarce, studies on SC in obesity point to impairments of recognition and interpretation of others' emotional states in this population and how they will affect food decisions. The insights provided can also contribute to developing prevention and treatment strategies based on SC aimed at obesity.

2 Food Decisions and Emotion Regulation

Emotions are affective responses triggered by stimuli, situations, or environmental events with reinforcing potential [36] and have different motivational functions affecting all eating behavior, including inclination toward eating: what motivates a food choice, how one emotionally responds to food digestion, the amount and speed that goes into eating, as well as metabolism and digestion [37].

Although closely connected, the nature of the connection between emotions and eating behavior is not yet fully understood. For instance, hunger is a potent emotional modulator since hungry animals and humans tend to be more alert and irritable. Likewise, different stimuli elicit diverse emotional responses in satiated and hungry individuals [38]. Brain areas traditionally linked to the processing of the reward associated with food, such as the mesencephalic ventral tegmental area (mVTA), the nucleus accumbens (NA), the anterior cingulate cortex (ACC), and the ventromedial prefrontal cortex (vmPFC), are dysfunctional in several emotional disorders such as depression, bipolar disorder, and even schizophrenia [39].

Making decisions, including food decisions, is executive processes encompassing a wide range of inputs, such as sensory information, implicit and explicit memories, and emotions, which need to be associated with expectations and possible outcomes in order for an adaptive decision to be made [40]. Rational decision-making models like those used in economics and classical decision theory, in which emotions are not considered central cognitive elements, but undesirable noises, do not predict precisely most decision-making processes, including those concerning food decisions. The neuroscience of decision-making has shown that, in general, our behavior is much less rational than we would like to admit because it is constantly subject to the influence of emotions and other implicit and automatic cognitive processes.

Emotion regulation is a multidimensional construct that comprises the ability to respond depending on environmental requirements by managing emotions and proper behaviors with accepted sociality and flexibility. It might also be referred to as the ability to control or restrain automatic responses when required [41]. It incorporates intrinsic and extrinsic psychological processes such as monitoring, appreciating, and changing the magnitude of emotional reactions themselves [41]. One of the most studied emotion regulation models encompasses two mechanisms, cognitive reappraisal and expressive suppression [42]. The former is more adaptive and involves the cognitive effort of modifying a given condition's emotional potential, redefining it in non-emotional terms, while the latter covers the modulation of the emotional response. Both strategies require some ability to perceive and reflect on one's own emotions, a capacity that is not evenly distributed among the general population [29], as discussed above.

Recent findings show that pathological food decisions behind binges and restrictive eating behaviors present in anorexia nervosa result from dysfunctional alternatives to regulate or suppress unpleasant emotions [42]. In the same way as individuals with eating disorders, individuals with obesity seem to have greater difficulty in identifying and describing their own feelings, besides presenting externally oriented thinking, which is a style of perceiving and thinking disconnected from emotions [43] and is a characteristic of alexithymia. As discussed above, alexithymia stems from interoceptive deficits; however, such deficits are not restricted to affective signs but involve impairments of the appropriate interpretation of signs of hunger, proprioception, tiredness, and temperature [44]. Thus, as discussed above, alexithymic people would find it difficult to differentiate anger from tiredness, hunger, or fever. Impairments of one's emotion identification present in alexithymia interfere with the accomplishment of emotional regulation strategies such as cognitive reappraisal and affective suppression, leading to emotion dysregulation concerning food, addictive substances, gambling, shopping, or pornography, for instance. Such impairments also disrupt the adequate processing of social information, damaging the social regulation of emotion. As will be discussed below, most studies examining SC in individuals with obesity show that this population displays deficits in facial affect recognition and ToM processing. However, the relationship of such subdomains of SC with alexithymia remains controversial.

3 Social Cognition in Obesity

Food decisions may be impaired in individuals with obesity due to deficits in emotion regulation. Consequently, these individuals may relieve unpleasant emotions through dysfunctional eating behaviors, such as binges. Deficits of emotion regulation, in turn, may result from difficulties with one's own emotion identification, therefore interfering with the recognition of others' mental states and emotions and the social regulation of emotions.

It is important to notice that social cognitive impairments in obesity, as in other psychiatric conditions, might result from cognitive deficits in general [45], despite the heterogeneity of the opinions regarding the independence of social cognitive domains from general cognitive domains, extensively discussed in the literature [46]. Studies that investigate both cognition and affect are needed to comprehend these complex relationships better. Following this demand, three recent studies have taken this into account and addressed the association between general and social cognitive impairments by investigating tasks assessing general cognitive domains and emotion recognition tasks [47–49]. Manderino et al. [47] showed that a correlation between performance on neuropsychological tests might be a significant predictor of emotion recognition accuracy and speed in bariatric surgery candidates. Other authors [48] concluded that slower processing speed in overweight children, as well as in children with obesity, may explain their impairments in the accuracy of emotional recognition, whereas [49] found ToM deficits in individuals with obesity, although the performance on a general cognition task did not differ from that of controls. Most studies on SC in obesity are not concerned with the impact of general cognitive effects on SC, which must be highlighted.

Two studies approached the putative developmental impacts of SC impairments in mothers with obesity on their children's performance on tasks assessing SC [50, 51]. Baldaro et al. [50] showed that mothers of children with severe early-onset obesity displayed greater difficulty in recognizing facial emotions than mothers of children slightly overweight. Moreover, children with severe obesity made more mistakes when identifying facial affect. More recently, Bergman et al. [51] did not document differences in facial emotion recognition abilities between mothers with and without obesity; nevertheless, the findings of these authors show that mothers with obesity have less emotional availability when interacting with their children, possibly due to impairments in their ability to understand their children's emotional states. These authors confirm, at least in part, an old hypothesis [52] proposing that, due to difficulties in properly interpreting emotional signs of their children, mothers with obesity are prone to confuse further emotional demands with hunger, which leads them to train their offspring to regulate emotions with food involuntarily.

Whether deficiencies in identifying emotions observed in people with obesity are a by-product of alexithymic traits (or, conversely, whether alexithymia is a by-product of deficits in SC processing in this population) has been a matter of debate, possibly a consequence of the great amount of research on alexithymia in people suffering from obesity [53]. Three studies addressed this issue [54–56] with contradictory conclusions. Surcinelli et al. [54] investigated alexithymia and impairments of emotional recognition in children and adolescents with obesity and conjectured that emotional recognition and expression deficits might lead to alexithymic traits. However, they have not documented significant differences in recognizing facial emotions between normal-weight participants and participants with obesity, despite the differences found in the ability to describe emotions, one dimension of alexithymia. Other authors [55] found that regardless of depressive symptoms, men and women with obesity have greater impairments in emotion perception. These findings indicate that men and women with obesity scored higher in alexithymia when compared to normal-weight men and women. Hence, these results suggest an association between obesity, mood, cognition, and eating behavior. Establishing a causal relationship between alexithymia and impaired recognition of emotion in obesity remains to be proven and studied. Aloi et al. [56] studied ToM in participants with obesity with or without a binge-eating disorder. They showed that, although all participants displayed similar abilities to attribute emotional states to other people, only those with binges exhibited impairments on their own emotions' recognition, possibly reflecting alexithymic traits. Other authors focused on the role of psychiatric conditions or symptoms on SC impairments of people with obesity. Cserjési et al. [57] found that participants with obesity showed difficulties in paying attention to schematic faces depicting negative emotions such as sadness or anger, regardless of self-reported depression and anxiety. Similarly, Koch and Polatos [48] compared children with obesity and normal-weight children regarding their ability to recognize facial emotions, thus indicating that such impairments may connect with emotional eating. They showed that overweight children might display lower categorization accuracy and slower reaction times for angry facial expressions, which, in turn, did not correlate with emotional eating. Turan et al. [58] also showed social cognitive impairments, specifically ToM deficits, in adolescents with obesity, independently of the presence of binge eating.

Although not all the studies reviewed above approached specifically the relationship between SC and eating behaviors of people with obesity, they provided a series of insights about the impact that deficits in this cognitive domain can have on this population's food decisions. Moreover, such insights may also support the development of preventive and therapeutic strategies based on SC which may help patients suffering from obesity improve their food decisions. Some of such strategies are briefly discussed below.

4 Neuroanatomy of Food Decisions

To understand what motivates and instigates an individual's food choices, many variables must be considered, such as the activity of specific brain areas allegedly involved with food decisions. Interestingly, many of the areas related to dietary decisions are also recruited experimentally by social cognitive tasks, such as brain circuits associated with reward processing, vmPFC, orbitofrontal cortex (OFC), ACC, and insula. This is made clear by functional neuroimaging (fMRI) studies, which employ an important and frequently used tool in investigating and comprehending neural activity behind food decisions and SC.

fMRI studies focusing on pathway activation in response to food images show greater activation of hedonic pathways comprising not only mATV but also subcortical (NA and amygdala) and cortical areas (OFC). Hedonic eating is called eating in the absence of hunger, i.e., when metabolic feedback does not regulate food intake. This eating behavior comprises reward, emotion, and motivation and is regulated through cognitive strategies [59]. Greater activation of hedonic pathways has been reported in studies with satiety and food consumption [60] and prediction of short-term weight loss [61] or weight gain [62]. However, cross-sectional studies that included obesity and/or binge-eating disorders show inconsistent findings regarding activation areas in response to food-related stimuli [63]. This was made evident by a recent meta-analysis that examined the neural correlates of the processing of food visual cues, which established that the concurrence of studies was moderate, with only 41% contributing to a significant activation cluster [64]). In this sense, recent findings suggest that the neural activity involved in response to food stimuli comprises several networks behind a food choice. Sensory, motivational, emotional, and cognitive variables (including SC ones) play an important role [65].

Hence, the mesocorticolimbic system described above is better characterized as a motivational salience network instead of purely reward-related. This is reiterated by recent fMRI studies showing that the activity within these areas is related to the processing of motivation, rather than specifically to the hedonic value of food [65]. Understanding the neurobiological mechanisms of reward, motivation, and cognitive control that lead to food decisions is crucial in generating new knowledge about how and why overeating occurs and potential therapeutic interventions. Noteworthy is that social interactions also recruit mesocorticolimbic circuits responsible for processing pleasure, reward, and motivation in association with the environment, which undoubtedly interferes with food choices.

Recent reviews on neuroimaging suggest that other brain areas appear to be affected by obesity. For instance, in addition to hyperactivity in circuits of pleasure, reward, and motivation, represented by NA, amygdala, putamen, and OFC, obesity is related to hypoactivity within prefrontal areas, such as vmPFC, ACC, and prefrontal cortex (PFC) [66], which are also inactive in social contexts, as will be discussed later. The insula is related to the processing of interoceptive signals and social cognitive variables, including morality and empathy [67]). A recent pooled meta-analysis of gray matter changes in obesity revealed a large cluster of reduced gray matter volume in patients with obesity in the right inferior frontal gyrus (IFG), as well as the insula region. Reduced gray matter volume found in obese patients in the insula region may indicate deficient sensory and social cognitive processing [68]. IFG is related to emotional regulation processes [68] and participates in inhibitory control mechanisms [69]. Thereby, gray matter volume reductions found in the insula and the IFG play an important role in the processing of deficits of food cues and/or food intake, attention toward satiety, as well as deficient cognitive control, emotional regulation abilities, and SC. Also, the insula has been activated during cravings, motivational salience in addiction [67], and food cravings [70]. A recent genome-wide association study found that the expression of BMI/obesity susceptibility genes was strongly enhanced in specific brain areas such as the insula and substantia nigra, which suggests that specific genes may influence food decisions, which, in turn, may lead to food addiction and emotional eating [71].

The OFC integrates sensory modalities such as taste, smell, and vision through its dense reciprocal projections into thalamic, midbrain, and striatal regions that are active in decision-making processes [72]. Thus, this brain structure is anatomically and functionally crucial when it comes to comprehending feeding behavior. The medial and lateral portions of OFC focus on pleasure since they were largely equally responsive to stimuli representing positive and negative affect. The OFC is crucial to the decision-making process, for it estimates the probability of a specific outcome to guide future responses, as supported by lesion and inactivation studies [73]. Studies investigating this brain region in people with obesity have found structural and volumetric differences, such as decreased gray matter volume of OFC [74] and decreased total OFC volume [75], leading to implications in altered executive functioning and decision-making [74]. These findings suggest that the OFC activation in response to food cues may be crucial in understanding overeating and dieting difficulties regardless of satiety. Since the OFC integrates sensory and limbic cues that help guide behavior, anatomic and functional abnormalities might help explain response inhibition deficits. Such findings support the notion that the OFC is critically implicated in obesity.

Thus, fMRI findings of activation of the same brain areas both by SC tasks and during food choices may help understand the role of socio-cognitive variables in our food decisions. Hence, food choices are constantly mediated by social encounters, perceptions, and interactions. Many of these areas receive peripheral signals, fundamental for the configuration of embodied aspects of cognition, which, among others, comprise emotion recognition and sensorimotor mirroring processes abundantly described in studies on the ability to understand the other's mental states [76–79].

Real-time fMRI neurofeedback (rt-fMRI-NF) is an important tool for comprehending brain activation in people with obesity [80]. The mechanism behind rt-fMRI-Nf consists of measuring the blood oxygen level-dependent signal (BOLD) and subsequently presenting it to the subject in an MRI scanner. Thus, the subject can learn self-control and regulate brain activity through visual and task-related cues, translating to the corresponding mental state [80] (Fig. 1).

Investigations in this field are still relatively new, although studies involving the anterior insula (AI) show promising results [81]. In these studies, neurofeedback promoted upregulation in healthy people and patients with schizophrenia, and AI downregulation was successful in phobia patients [81]. These results might indicate a promising path for future studies since people with obesity show increased BOLD activity in brain regions that encode cognitive control and reward [82]. Current studies have started to investigate rt-fMRI-NF as an intervention tool for people with obesity. Kohl and colleagues (2019) [83] tested rt-fMRI-NF neurofeedback to up-regulate dorsolateral prefrontal cortex activity, a region responsible for cognitive control. Another study intended to appraise the effect of near-infrared hemoencephalography in neurofeedback training on appetite control, weight, and food-

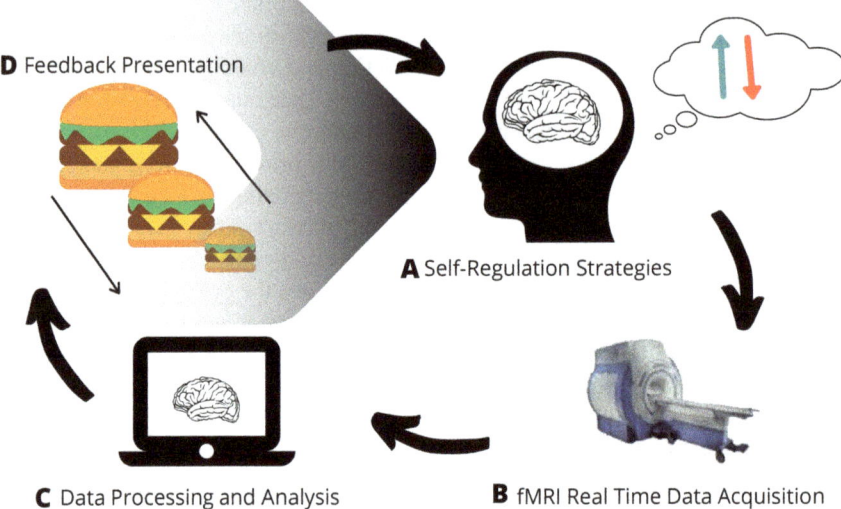

D Feedback Presentation

A Self-Regulation Strategies

C Data Processing and Analysis

B fMRI Real Time Data Acquisition

Fig. 1 Real-time neurofeedback can be an adequate tool for the follow-up of people with obesity and dysfunctional eating behaviors since it allows developing a more efficient regulation of their brain activity

related brain activity [84]. This study focused on OFC, demonstrating a positive trend of increased self-control and inhibition of feeding behaviors. Hence, this study may indicate that neurofeedback training can aid as a therapeutic strategy focusing on self-control in obesity, although further investigation regarding specific mechanisms is needed [83].

Another proposed tool based on these neuroscientific findings is behavioral interventions that positively affect those susceptible to cue-based eating [85]. One of the main behavioral interventions in people with obesity disorder is cognitive regulation, in which strategies of thought suppression and reappraisal are used as effective tools for reducing self-reported cravings [65]. A randomized controlled trial (RCT) used mindfulness training, in conjunction with diet and exercise, to reduce reward-based eating [86]. Cognitive neuroscience may be a way to provide valuable endophenotypic markers to the heterogeneous pathways to obesity and cater to specific target interventions, such as the ones cited above. However, long-term and larger studies are warranted to confirm and gage the effect of behavioral interventions based on neural pathways.

With the increasing use of fMRI as the main investigation tool into neural activity behind food decisions, some methodological errors have become evident. Since fMRI is able to measure regional responses and inter-regional relationships, a need for accurate and precise behavioral, metabolic, genetic, and cognitive profiling is a must [87]. More complex task-dependent measurements should be considered to achieve this than those currently used in studies with participants with obesity. The lack of attention toward subtle processes appears to lie behind the striking inconsistency across studies. In this sense, we highlight the importance of a clear mental task for participants that considers behavioral, temporal, metabolic, genetic, and cognitive individual profiling processes. Future studies should employ paradigms that ensure a narrow control over the mental process of interest to interpret neural findings and clearly ensure consistency in this field. We also highly recommend using a multivariate approach for data analysis to avoid simple cross-sectional measures such as body mass index (BMI), which does not translate the complexity of this disorder. Obesity is a highly heterogeneous disorder, and therefore, the use of only one model might lead as a deterrent rather than guide future research [87].

5 How Obesity Affects the Brain and May Impair SC and Food Decisions

Obesity affects brain function through different processes, including chronic low-grade inflammation, changes in the intestinal microbiome [88], and insulin resistance [89]. These processes interfere with the proper functioning of areas related to mood regulation, motivation, consummatory and anticipatory rewards associated with food and its value, as well as to cognitive control, and even to the configuration of food memories and mental representations [89, 90]. For instance, not only the early evidence of immunological suppression and increased inflammatory activity in depressed patients [91] but also more recent findings of greater activity of mesolimbic reward systems triggered by food cues in obese patients with high inflammatory markers support the notion that a complex network of bidirectional dimensions between the immune system and the brain may lie behind several behavioral disorders in humans. Moreover, inflammation due to obesity induces both changes in intestinal bacteria and insulin resistance [88].

Studies on obesity-related changes in gut microbiota show that individuals with chocolate cravings may have different microbial metabolites than those presented by individuals indifferent to chocolate, as well as that butyrate, a short-chain fatty acid produced by intestinal tract bacteria that can affect mood and behavior. Such findings highlight the importance of the gut microbiome in food decisions [92].

Insulin is an anorexigenic peptide whose receptors are distributed over several brain regions. Malfunctions to these receptors caused by obesity-induced inflammation disturb the activity of several brain areas, including those responsible for the processing of hedonic aspects of eating, cognitive control, and the generation of food memories and images, distorting food decisions [89].

Nonetheless, the extent to which food decisions of individuals with obesity may also result from dysfunction induced by the processes described above in brain areas processing SC is a point to be investigated. Some controlled studies approached the influences of inflammation on SC processing in participants without obesity, even though their results are contradictory. For instance, Salmonella Typhi vaccination (STV)-induced systemic inflammation diminished the performance of male participants on a ToM task [93], The Reading the Mind in the Eyes Test (RMET). Likewise, an endotoxin administration derived from *Escherichia coli* impaired male and female participants' performance on the RMET [94]. However, the acute inflammation induced by the injection of a bacterial lipopolysaccharide not only did not affect the performance of male participants on the RMET but also increased the responses of brain areas commonly associated with SC, as showed by fMRI [95].

To our knowledge, at the time of writing this chapter, no studies exist addressing the direct impact of inflammation, gut microbiota changes, and insulin resistance on SC in individuals with obesity, specifically. However, studies focusing on other mental conditions have shown some significant findings. For instance, neuroinflammation plays an important role in the onset and progression of Alzheimer's disease-related cognitive deficits [96], including impairments of SC. Likewise, ToM and emotion recognition impairments in patients with psychosis and individuals with ultra-high risk of psychosis correlate to levels of interleukin-4 (IL-4), a biomarker of inflammatory activity. Negative biases in social and emotional processing in depressed patients may be triggered by inflammation since there is evidence that experimentally induced inflammation can affect not only CS but also other cognitive domains, such as attention, executive functioning, and memory [97]. There is evidence that higher levels of inflammatory mediators, e.g., cytokines, are also implicated in eating disorders (ED), even though studies on the impact of neuroinflammation on the SC processing in patients with ED are not available so far. Individuals with anorexia nervosa (AN), binge-eating disorder, and night-eating syndrome may display altered concentrations of interleukin-1, interleukin-10, epidermal growth factor, and interferon-gamma [98]. Interestingly, a recent meta-analysis found that, compared to healthy controls, individuals with AN, unlike individuals with bulimia nervosa, show higher tumor necrosis factor-alfa (TNF-α) and IL-6 [99]. Obesity seems to mediate inflammation and dysfunctional eating behaviors, such as binges, since individuals with obesity and binges have a worse inflammatory profile, exhibiting significantly higher levels of high-sensitive C-reactive protein and higher insulin resistance [100].

It is commonly accepted that neuroinflammation due to obesity causes synaptic remodeling and neurodegeneration, mainly at the hypothalamus, amygdala, and reward processing areas, leading to cognitive deficits [101] and, consequently, to inappropriate food decisions. Human studies on inflammation secondary to obesity that show affected brain areas are commonly associated with SC processing, such as the ACC, vmPFC, insula, superior temporal sulcus (STS), and fusiform area (FA), though scarce might shed light on how SC impairments might influence eating behavior. These studies will explain how biomarkers of inflammation impact neurotransmitter systems located within these and other areas with which they connect, originating SC deficits behind bad food decisions. Inflammatory processes affect ACC and insula, among other areas, in patients suffering from anxiety, mood disorders, and post-traumatic stress disorder [102, 103]. Such results can be extrapolated to the inflammation induced by obesity, disrupting the neuronal function within these areas. Several social cognitive processes can be disrupted by inflammation, for instance, error processing, as well as benefits and costs during social interactions (ACC) [104]; emotion recognition and ToM abilities (vmPFC) [105]; interoceptive processing, crucial to understanding one's and other's emotions (insula) [106]; empathy (insula and ACC) [107]; morality (insula) [108]; and face perception (STS and FA) [109, 110]. As discussed at the beginning of this chapter, problems in such SC domains generate difficulties in interpersonal relationships, which, in turn, cause psychosocial stress and negative emotions that may be mitigated with food.

6   Conclusion

Contrary to our common-sense knowledge, food decisions cannot be considered purely rational processes. Food choices are strongly influenced by emotions, being even more pronounced in those who suffer from obesity. Toxic processes derived from obesity affect the brain, impairing the proper functioning of multiple areas related to emotional processing and regulation. Many of these areas relate to SC processing and, when damaged, may disrupt multiple cognitive domains that are essential for satisfactory social interaction. Social cognitive deficits are increasingly described in individuals with obesity and may lie behind their inadequate food decisions, often attributed to an alleged "lack of willpower" to control their eating behavior. Such a false belief, commonly shared by the general population and even by some health professionals, in addition to not adding to weight loss, has contributed to increased feelings of demoralization and stigma among those who suffer from obesity. Understanding how social cognitive deficits can affect emotional regulation and, therefore, influence food decisions of individuals with obesity can be of immense value for developing prevention and treatment strategies for their dysfunctional eating behaviors.

In summary, obesity has toxic effects on the brain. These effects can affect areas responsible for properly processing social information, such as identifying facial emotions and inferring others' mental states. Impairments to the functioning of these areas can create further difficulties for people with obesity, who commonly suffer from stigmatization, prejudice, and social exclusion. Such difficulties include problems in properly "deciphering" other people, generating aversive emotions that can be regulated through dysfunctional emotion regulation strategies, such as binges, emotional eating, and food addiction.

Core Messages

- Social cognition is crucial for interpersonal relationships, and deficits in this domain have been described in individuals with obesity. We propose that these impairments may play an important role in inadequate food decisions and behaviors.

- Obesity has toxic effects on the brain that may be responsible for properly processing social cues, such as identifying facial emotions and inferring other people's intentions (also known as the theory of mind).
- The relationship between social cognition and obesity is complex and seems to appear in a cyclic pattern. Impairments in SC may lead to psychosocial stress and difficulties in dealing with negative emotions that tend to be mitigated with food and thus resulting in the toxicity consequences of obesity that may impair even more social processing.

References

1. Srivastava G, Apovian CM (2018) Current pharmacotherapy for obesity. Nat Rev Endocrinol 14(1):12–24. https://doi.org/10.1038/nrendo.2017.122
2. Pontzer H, Raichlen DA, Wood BM, Mabulla AZ, Racette SB, Marlowe FW (2012) Hunter-gatherer energetics and human obesity. PLoS One 7(7) https://doi.org/10.1371/journal.pone.0040503
3. Pandit R, de Jong JW, Vanderschuren LJ, Adan RA (2011) Neurobiology of overeating and obesity: the role of melanocortins and beyond. Eur J Pharmacol 660(1):28–42
4. Sweeting AN, Hocking SL, Markovic TP (2015) Pharmacotherapy for the treatment of obesity. Mol Cell Endocrinol 418(2):173–183. https://doi.org/10.1016/j.mce.2015.09.005
5. Berridge KC, Ho CY, Richard JM, DiFeliceantonio AG (2010) The tempted brain eats: pleasure and desire circuits in obesity and eating disorders. Brain Res 1350:43–64. https://doi.org/10.1016/j.brainres.2010.04.003
6. Goldstone AP, Miras AD, Scholtz S, Jackson S, Neff KJ, Pénicaud L, Geoghegan J, Chhina N, Durighel G, Bell JD, Meillon S, le Roux CW (2016) Link between increased satiety gut hormones and reduced food reward after gastric bypass surgery for obesity. J Clin Endocrinol Metab 101(2):599–609. https://doi.org/10.1210/jc.2015-2665
7. Brockmeyer T, Hamze Sinno M, Skunde M, Wu M, Woehning A, Rudofsky G, Friederich HC (2016) Inhibitory control and hedonic response towards food interactively predict success in a weight loss programme for adults with obesity. Obes Facts 9(5):299–309. https://doi.org/10.1159/000447492
8. Pearce AL, Mackey E, Cherry JBC, Olson A, You X, Magge SN, Mietus-Snyder M, Nadler EP, Vaidya CJ (2017) Effect of adolescent bariatric surgery on the brain and cognition: a pilot study. Obesity (Silver Spring) 25(11):1852–1860. https://doi.org/10.1002/oby.22013
9. de Araujo IE, Simon AS (2009) The gustatory cortex and multisensory integration. Int J Obesity (London) 33(2):S34-43. https://doi.org/10.1038/ijo.2009.70
10. Lee D (2013) Decision making: from neuroscience to psychiatry. Neuron 78(2):233–248. https://doi.org/10.1016/j.neuron.2013.04.008
11. Rangel A, Camerer C, Montague PR (2008) A framework for studying the neurobiology of value-based decision making. Nat Rev Neurosci 9(7):545–556. https://doi.org/10.1038/nrn2357
12. Sobal J, Bisogni CA (2009) Constructing food decisions. Ann Behav Med 38(1):S37–S46. https://doi.org/10.1007/s12160-009-9124-5
13. Reeck C, Ames DR, Ochsner KN (2016) The social regulation of emotion: an integrative, cross- disciplinary model. Trends Cognitive Sci 20(1):47–63. https://doi.org/10.1016/j.tics.2015.09.003

14. Berg KC, Crosby RD, Cao L et al (2015) Negative affect prior to and following overeating-only, loss of control eating-only, and binge eating episodes in obese adults. Int J Eating Disorders 48(6):641–53. https://doi.org/10.1002/eat.22401
15. Wiedemann AA, Ivezaj V, Barnes RD (2018) Characterizing emotional overeating among patients with and without binge-eating disorder in primary care. Gen Hosp Psychiatry 55:38–43. https://doi.org/10.1016/j.genhosppsych.2018.09.003
16. Ivezaj V, Wiedemann AA, Grilo CM (2017) Food addiction and bariatric surgery: a systematic review of the literature. Obes Rev 18(12):1386–1397. https://doi.org/10.1111/obr.12600
17. Gianini LM, White MA, Masheb RM (2013) Eating pathology, emotion regulation, and emotional overeating in obese adults with binge eating disorder. Eat Behav 14(3):309–313. https://doi.org/10.1016/j.eatbeh.2013.05.008
18. Reas DL, Grilo CM (2014) Current and emerging drug treatments for binge eating disorder. Expert Opin Emerg Drugs 19(1):99–142. https://doi.org/10.1517/14728214.2014.879291
19. Carlston DE (2013) On the nature of social cognition: my defining moment. In: Carlston DE (ed) The Oxford Handbook of Social Cognition, 1st edn. Oxford University Press, New York, NY, pp 3–15
20. Hugenberg K, Wilson JP (2013) Faces are central to social cognition. In: Carlston DE (ed) The oxford handbook of social cognition, 1st edn. Oxford University Press, New York, NY, pp 167–193
21. Cunningham WA, Haas IJ, Jahn A (2015) Attitudes. In: Decety J, Cacioppo JT (eds) The Oxford handbook of social neurosciences, 1st edn. Oxford University Press, New York, NY, pp 212–226
22. Ames DL, Fiske ST, Todorov AT (2015) Impression formation: a focus on other's intents. In: Decety J, Cacioppo JT (eds) The Oxford handbook of social neurosciences, 1st edn. Oxford University Press, New York, NY, pp 419–433
23. Frith CD, Frith U (2000) The physiological basis of theory of mind. In: Baron-Cohen S, Tager-Flusberg H, Cohen D (eds) Understanding other minds: perspective from developmental social neuroscience. Oxford Library of Psychology, Oxford, pp 335–356
24. Isaksson J, Van't Westeinde A, Cauvet É, Kuja-Halkola R, Lundin K, Neufeld J, Willfors C, Bölte S (2019) Social cognition in autism and other neurodevelopmental disorders: a co-twin control study. J Autism Dev Dis 49(7), 2838–2848. https://doi.org/10.1007/s10803-019-04001-4
25. Javed A, Charles A (2018) The importance of social cognition in improving functional outcomes in schizophrenia. Front Psych 9:157. https://doi.org/10.3389/fpsyt.2018.00157
26. Ospina L, Nitzburg G, Shanahan M et al (2018) Social cognition moderates the relationship between neurocognition and community functioning in bipolar disorder. J Affect Disord 235:7–14. https://doi.org/10.1016/j.jad.2018.03.013
27. Cardi V, Corfield F, Leppanen J et al (2015) Emotional processing, recognition, empathy and evoked facial expression in eating disorders: an experimental study to map deficits in social cognition. PLoS ONE 10(8). https://doi.org/10.1371/journal.pone.0133827
28. Meany G, Conceição E, Mitchell JE (2014) Binge eating, binge eating disorder and loss of control eating: effects on weight outcomes after bariatric surgery. Eur Eat Disord Rev 22(2):87–91. https://doi.org/10.1002/erv.2273
29. Nowakowski ME, McFarlane T, Cassin S (2013) Alexithymia and eating disorders: a critical review of the literature. J Eat Disord 1:21. https://doi.org/10.1186/2050-2974-1-21
30. van Strien T (2018) Causes of emotional eating and matched treatment of obesity. Curr DiabRep 18(6):35. https://doi.org/10.1007/s11892-018-1000-x
31. Coffey E, Berenbaum H, Kerns J (2003) Brief Report. Cogn Emot 17(4):671–679
32. Murphy J, Brewer R, Catmur C, Bird G (2017) Interoception and psychopathology: a developmental neuroscience perspective. Dev Cogn Neurosci 23:45–56. https://doi.org/10.1016/j.dcn.2016.12.006

33. Damasio A (2018) The strange order of things. Life, feeling, and the making of cultures. Pantheon, New York, NY
34. Blakemore SJ, Decety J (2001) From the perception of action to the understanding of intention. Nat Rev Neurosci 2(8):561–567. https://doi.org/10.1038/35086023
35. Goldman AI (2006) Simulating minds: the philosophy, psychology, and neuroscience of mindreading. Oxford University Press, New York, NY
36. Baumeister RF, Bratslavsky E, Muraven M, Tice DM (1998) Ego depletion: is the active self a limited resource? J Pers Soc Psychol 74(5):1252–1265. https://doi.org/10.1037//0022-3514.74.5.1252
37. Macht M (2008) How emotions affect eating: a five-way model. Appetite 50(1):1–11. https://doi.org/10.1016/j.appet.2007.07.002
38. Garcia-Burgos D, Maglieri S, Vögele C, Munsch S (2018) How does food taste in Anorexia and Bulimia Nervosa? A protocol for a quasi-experimental, cross-sectional design to investigate taste aversion or increased hedonic valence of food in eating disorders. Front Psychol 9:264. https://doi.org/10.3389/fpsyg.2018.00264
39. Der-Avakian A, Markou A (2012) The neurobiology of anhedonia and other reward-related deficits. Trends Neurosci 35(1):68–77. https://doi.org/10.1016/j.tins.2011.11.005
40. Gutnik LA, Hakimzada AF, Yoskowitz NA, Patel VL (2006) The role of emotion in decision-making: a cognitive neuroeconomic approach towards understanding sexual risk behavior. J Biomed Inform 39(6):720–736. https://doi.org/10.1016/j.jbi.2006.03.002
41. Casagrande M, Boncompagni I, Forte G, Guarino A, Favieri F (2019) Emotion and overeating behavior: effects of alexithymia and emotional regulation on overweight and obesity. Eating and weight disorders: EWD. Advance online publication. https://doi.org/10.1007/s40519-019-00767-9
42. Berking M, Wupperman P (2012) Emotion regulation and mental health: recent findings, current challenges, and future directions. Curr Opin Psychiatry 25(2):128–134. https://doi.org/10.1097/YCO.0b013e3283503669
43. Li X, Lu J, Li B, Li H, Jin L, Qiu J (2017) The role of ventromedial prefrontal cortex volume in the association of expressive suppression and externally oriented thinking. J Affect Disord 222:112–119. https://doi.org/10.1016/j.jad.2017.06.054
44. Brewer R, Cook R, Bird G (2016) Alexithymia: a general deficit of interoception. Royal Soc Open Sci 3(10):150664. https://doi.org/10.1098/rsos.150664
45. Sargénius HL, Lydersen S, Hestad K (2017) Neuropsychological function in individuals with morbid obesity: a cross-sectional study. BMC Obesity 4:6. https://doi.org/10.1186/s40608-017-0143-7
46. Tonelli HA (2009) Processamento cognitivo 'Teoria da Mente' no transtorno bipolar [Cognitive 'Theory of Mind' processing in bipolar disorder]. Revista brasileira de psiquiatria (Sao Paulo, Brazil : 1999), 31(4):369–374. https://doi.org/10.1590/s1516-44462009000400015
47. Manderino L, Spitznagel MB, Strain G, Devlin M, Cohen R, Crosby RD, Mitchell JE, Gunstad J (2015) Cognitive dysfunction predicts poorer emotion recognition in bariatric surgery candidates. Obes Sci Pract 1(2):97–103. https://doi.org/10.1002/osp4.9
48. Koch A, Pollatos O (2015) Reduced facial emotion recognition in overweight and obese children. J Psychosom Res 79(6):635–639. https://doi.org/10.1016/j.jpsychores.2015.06.005
49. Caldú X, Ottino-González J, Sánchez-Garre C, Hernan I, Tor E, Sender-Palacios MJ, Dreher JC, Garolera M, Jurado MÁ (2019) Effect of the catechol-O-methyltransferase Val 158 Met polymorphism on theory of mind in obesity. Eur Eating Disorders Rev: The J Eating Disorders Assoc 27(4):401–409. https://doi.org/10.1002/erv.2665
50. Baldaro B, Balsamo A, Caterina R, Fabbrici C, Cacciari E, Trombini G (1996) Decoding difficulties of facial expression of emotions in mothers of children suffering from developmental obesity. Psychother Psychosom 65(5):258–261. https://doi.org/10.1159/000289085
51. Bergmann S, von Klitzing K, Keitel-Korndörfer A, Wendt V, Grube M, Herpertz S, Schütz A, Klein AM (2016) Emotional availability, understanding emotions, and recognition of

facial emotions in obese mothers with young children. J Psychosom Res 80:44–52. https://doi.org/10.1016/j.jpsychores.2015.11.005

52. Bruch H (1973) Obesity, anorexia nervosa and the person within. Basic Books, New York, NY

53. Fernandes J, Ferreira-Santos F, Miller K, Torres S (2018) Emotional processing in obesity: a systematic review and exploratory meta-analysis. Obes Rev 19(1):111–120. https://doi.org/10.1111/obr.12607

54. Surcinelli P, Baldaro B, Balsamo A, Bolzani R, Gennari M, Rossi NC (2007) Emotion recognition and expression in young obese participants: preliminary study. Percept Mot Skills 105(2):477–482. https://doi.org/10.2466/pms.105.2.477-482

55. Giel KE, Hartmann A, Zeeck A, Jux A, Vuck A, Gierthmuehlen PC, Wetzler-Burmeister E, Sandholz A, Marjanovic G, Joos A (2016) Decreased emotional perception in obesity. Eur Eat Disord Rev 24(4):341–346. https://doi.org/10.1002/erv.2444

56. Aloi M, Rania M, Caroleo M, De Fazio P, Segura-García C (2017) Social cognition and emotional functioning in patients with binge eating disorder. Eur Eat Disord Rev 25(3):172–178. https://doi.org/10.1002/erv.2504

57. Cserjési R, Vermeulen N, Lénárd L, Luminet O (2011) Reduced capacity in automatic processing of facial expression in restrictive anorexia nervosa and obesity. Psychiatry Res 188(2):253–257. https://doi.org/10.1016/j.psychres.2010.12.008

58. Turan S, Özyurt G, Çatlı G, Öztürk Y, Abacı A, Akay AP (2019) Social cognition and emotion regulation may be impaired in adolescents with obesity independent of the presence of binge eating disorder: a two-center study. Psychiatry Clin Psychopharmacol 29(4):887–894. https://doi.org/10.1080/24750573.2019.1693727

59. Berthoud HR (2011) Metabolic and hedonic drives in the neural control of appetite: who is the boss? Curr Opin Neurobiol 21(6):888–896. https://doi.org/10.1016/j.conb.2011.09.004

60. Finlayson G (2017) Food addiction and obesity: unnecessary medicalization of hedonic overeating. Nat Rev Endocrinol 13(8):493–498. https://doi.org/10.1038/nrendo.2017.61

61. Murdaugh DL, Cox JE, Cook EW 3rd, Weller RE (2012) fMRI reactivity to high-calorie food pictures predicts short- and long-term outcome in a weight-loss program. Neuroimage 59(3):2709–2721. https://doi.org/10.1016/j.neuroimage.2011.10.071

62. Demos KE, Heatherton TF, Kelley WM (2012) Individual differences in nucleus accumbens activity to food and sexual images predict weight gain and sexual behavior. J Neurosci 32 (16):5549–5552. https://doi.org/10.1523/JNEUROSCI.5958-11.2012

63. Lee PC, Dixon JB (2017) Food for thought: reward mechanisms and hedonic overeating in obesity. Curr Obes Rep 6(4):353–361. https://doi.org/10.1007/s13679-017-0280-9

64. van der Laan LN, de Ridder DT, Viergever MA, Smeets PA (2011) The first taste is always with the eyes: a meta-analysis on the neural correlates of processing visual food cues. Neuroimage 55(1):296–303. https://doi.org/10.1016/j.neuroimage.2010.11.055

65. Roefs A, Franssen S, Jansen A (2018) The dynamic nature of food reward processing in the brain. Curr Opin Clin Nutr Metab Care 21(6):444–448. https://doi.org/10.1097/MCO.0000000000000504

66. Herrmann MJ, Tesar AK, Beier J, Berg M, Warrings B (2019) Grey matter alterations in obesity: a meta-analysis of whole-brain studies. Obes Rev 20(3):464–471. https://doi.org/10.1111/obr.12799

67. Naqvi NH, Bechara A (2009) The hidden island of addiction: the insula. Trends Neurosci 32 (1):56–67. https://doi.org/10.1016/j.tins.2008.09.009

68. Herrmann MJ, Beier JS, Simons B, Polak T (2016) Transcranial Direct Current Stimulation (tDCS) of the right inferior frontal gyrus attenuates skin conductance responses to unpredictable threat conditions. Front Hum Neurosci 10:352. https://doi.org/10.3389/fnhum.2016.00352

69. Swann N, Tandon N, Canolty R, Ellmore TM, McEvoy LK, Dreyer S, DiSano M, Aron AR (2009) Intracranial EEG reveals a time- and frequency-specific role for the right inferior

frontal gyrus and primary motor cortex in stopping initiated responses. J Neurosci 29 (40):12675–12685. https://doi.org/10.1523/JNEUROSCI.3359-09.2009

70. Carnell S, Gibson C, Benson L, Ochner CN, Geliebter A (2012) Neuroimaging and obesity: current knowledge and future directions. Obes Rev 13(1):43–56. https://doi.org/10.1111/j.1467-789X.2011.00927.x

71. Ndiaye FK, Huyvaert M, Ortalli A, Canouil M, Lecoeur C, Verbanck M, Lobbens S, Khamis A, Marselli L, Marchetti P, Kerr-Conte J, Pattou F, Marre M, Roussel R, Balkau B, Froguel P, Bonnefond A (2020) The expression of genes in top obesity-associated loci is enriched in insula and substantia nigra brain regions involved in addiction and reward. Int J Obes 44(2):539–543. https://doi.org/10.1038/s41366-019-0428-7

72. Seabrook LT, Borgland SL (2020) The orbitofrontal cortex, food intake and obesity. J Ppsychiatry Nneurosc: JPN 45(3):190163. https://doi.org/10.1503/jpn.190163

73. Izquierdo A (2017) Functional heterogeneity within rat orbitofrontal cortex in reward learning and decision making. J Neurosci 37(44):10529–10540. https://doi.org/10.1523/JNEUROSCI.1678-17.2017

74. Walther K, Birdsill AC, Glisky EL, Ryan L (2010) Structural brain differences and cognitive functioning related to body mass index in older females. Hum Brain Mapp 31(7):1052–1064. https://doi.org/10.1002/hbm.20916

75. Cazettes F, Cohen JI, Yau PL, Talbot H, Convit A (2011) Obesity-mediated inflammation may damage the brain circuit that regulates food intake. Brain Res 1373:101–109. https://doi.org/10.1016/j.brainres.2010.12.008

76. Singer T, Seymour B, O'doherty J, Kaube H, Dolan RJ, Frith CD (2004) Empathy for pain involves the affective but not sensory components of pain. Science 303(5661):1157–1162

77. Wicker B, Keysers C, Plailly J, Royet J-P, Gallese V, Rizzolatti G (2003) Both of us disgusted in My insula: the common neural basis of seeing and feeling disgust. Neuron 40 (3):655–664. https://doi.org/10.1016/s0896-6273(03)00679-2

78. Carr L, Iacoboni M, Dubeau M-C, Mazziotta JC, Lenzi GL (2003) Neural mechanisms of empathy in humans: a relay from neural systems for imitation to limbic areas. Proc Natl Acad Sci 100(9):5497–5502. https://doi.org/10.1073/pnas.0935845100

79. Von Der Heide RJ, Skipper LM, Klobusicky E, Olson IR (2013) Dissecting the uncinate fasciculus: disorders, controversies and a hypothesis. Brain 136(6):1692–1707. https://doi.org/10.1093/brain/awt094

80. Linhartová P, Látalová A, Kóša B, Kašpárek T, Schmahl C, Paret C (2019) fMRI neurofeedback in emotion regulation: a literature review. Neuroimage 193:75–92. https://doi.org/10.1016/j.neuroimage.2019.03.011

81. Zilverstand A, Sorger B, Sarkheil P, Goebel R (2015) fMRI neurofeedback facilitates anxiety regulation in females with spider phobia. Front Behav Neurosci 9:148. https://doi.org/10.3389/fnbeh.2015.00148

82. Lowe CJ, Reichelt AC, Hall PA (2019) The prefrontal cortex and obesity: a health neuroscience perspective. Trends Cogn Sci 23(4):349–361. https://doi.org/10.1016/j.tics.2019.01.005

83. Kohl SH, Veit R, Spetter MS, Günther A, Rina A, Lührs M, Birbaumer N, Preissl H, Hallschmid M (2019) Real-time fMRI neurofeedback training to improve eating behavior by self-regulation of the dorsolateral prefrontal cortex: a randomized controlled trial in overweight and obese subjects. Neuroimage 191:596–609. https://doi.org/10.1016/j.neuroimage.2019.02.033

84. Percik R, Cina J, Even B, Gitler A, Geva D, Seluk L, Livny A (2019) A pilot study of a novel therapeutic approach to obesity: CNS modification by N.I.R. H.E.G. neurofeedback. Clinical nutrition (Edinburgh, Scotland), 38(1):258–263. https://doi.org/10.1016/j.clnu.2018.01.023

85. Forman EM, Butryn ML (2015) A new look at the science of weight control: how acceptance and commitment strategies can address the challenge of self-regulation. Appetite 84:171–180. https://doi.org/10.1016/j.appet.2014.10.004

86. Mason AE, Epel ES, Aschbacher K, Lustig RH, Acree M, Kristeller J, Cohn M, Dallman M, Moran PJ, Bacchetti P, Laraia B, Hecht FM, Daubenmier J (2016) Reduced reward-driven eating accounts for the impact of a mindfulness-based diet and exercise intervention on weight loss: Data from the SHINE randomized controlled trial. Appetite 100:86–93. https://doi.org/10.1016/j.appet.2016.02.009

87. Ziauddeen H, Farooqi IS, Fletcher PC (2012) Obesity and the brain: how convincing is the addiction model? Nat Rev Neurosci 13(4):279–286. https://doi.org/10.1038/nrn3212

88. Cox AJ, West NP, Cripps AW (2015) Obesity, inflammation, and the gut microbiota. Lancet Diabetes Endocrinol 3(3):207–215. https://doi.org/10.1016/S2213-8587(14)70134-2

89. Kullmann S, Heni M, Hallschmid M, Fritsche A, Preissl H, Häring HU (2016) Brain insulin resistance at the crossroads of metabolic and cognitive disorders in humans. Physiol Rev 96 (4):1169–1209. https://doi.org/10.1152/physrev.00032.2015

90. Mraz M, Haluzik M (2014) The role of adipose tissue immune cells in obesity and low-grade inflammation. J Endocrinol 222(3):R113–R127. https://doi.org/10.1530/JOE-14-0283

91. Leonard BE (2018) Inflammation and depression: a causal or coincidental link to the pathophysiology? Acta neuropsychiatrica 30(1):1–16. https://doi.org/10.1017/neu.2016.69

92. John GK, Mullin GE (2016) The gut microbiome and obesity. Curr Oncol Rep 18(7):45. https://doi.org/10.1007/s11912-016-0528-7

93. Balter L, Hulsken S, Aldred S, Drayson MT, Higgs S, Veldhuijzen van Zanten J, Raymond JE, Bosch JA (2018) Low-grade inflammation decreases emotion recognition—evidence from the vaccination model of inflammation. Brain Behav Immun 73:216–221. https://doi.org/10.1016/j.bbi.2018.05.006

94. Moieni M, Irwin MR, Jevtic I, Breen EC, Eisenberger NI (2015) Inflammation impairs social cognitive processing: a randomized controlled trial of endotoxin. Brain Behav Immun 48:132–138. https://doi.org/10.1016/j.bbi.2015.03.002

95. Kullmann JS, Grigoleit JS, Wolf OT, Engler H, Oberbeck R, Elsenbruch S, Forsting M, Schedlowski M, Gizewski ER (2014) Experimental human endotoxemia enhances brain activity during social cognition. Soc Ccognit Aaffect Nneurosc 9(6):786–793. https://doi.org/10.1093/scan/nst049

96. Liu Y, Zhang Y, Zheng X, Fang T, Yang X, Luo X, Guo A, Newell KA, Huang XF, Yu Y (2018) Galantamine improves cognition, hippocampal inflammation, and synaptic plasticity impairments induced by lipopolysaccharide in mice. J Neuroinflammation 15(1):112. https://doi.org/10.1186/s12974-018-1141-5

97. Bollen J, Trick L, Llewellyn D, Dickens C (2017) The effects of acute inflammation on cognitive functioning and emotional processing in humans: a systematic review of experimental studies. J Psychosom Res 94:47–55. https://doi.org/10.1016/j.jpsychores.2017.01.002

98. Caroleo M, Carbone EA, Greco M, Corigliano DM, Arcidiacono B, Fazia G, Rania M, Aloi M, Gallelli L, Segura-Garcia C, Foti DP, Brunetti A (2019) Brain-behavior-immune interaction: serum cytokines and growth factors in patients with eating disorders at extremes of the body mass index (BMI) spectrum. Nutrients 11(9):1995. https://doi.org/10.3390/nu11091995

99. Dalton B, Bartholdy S, Robinson L, Solmi M, Ibrahim M, Breen G, Schmidt U, Himmerich H (2018) A meta-analysis of cytokine concentrations in eating disorders. J Psychiatr Res 103:252–264. https://doi.org/10.1016/j.jpsychires.2018.06.002

100. Succurro E, Segura-Garcia C, Ruffo M, Caroleo M, Rania M, Aloi M, De Fazio P, Sesti G, Arturi F (2015) Obese patients with a binge eating disorder have an unfavorable metabolic and inflammatory profile. Medicine 94(52):e2098. https://doi.org/10.1097/MD.0000000000002098

101. Miller AA, Spencer SJ (2014) Obesity and neuroinflammation: a pathway to cognitive impairment. Brain Behav Immun 42:10–21. https://doi.org/10.1016/j.bbi.2014.04.001

102. Felger JC (2018) Imaging the role of inflammation in mood and anxiety-related disorders. Curr Neuropharmacol 16(5):533–558. https://doi.org/10.2174/1570159X15666171123201142

103. Attwells S, Setiawan E, Wilson AA, Rusjan PM, Mizrahi R, Miler L, Xu C, Richter MA, Kahn A, Kish SJ, Houle S, Ravindran L, Meyer JH (2017) Inflammation in the neurocircuitry of obsessive-compulsive disorder. JAMA Psychiat 74(8):833–840. https://doi.org/10.1001/jamapsychiatry.2017.1567
104. Apps MA, Rushworth MF, Chang SW (2016) The anterior cingulate gyrus and social cognition: tracking the motivation of others. Neuron 90(4):692–707. https://doi.org/10.1016/j.neuron.2016.04.018
105. Hiser J, Koenigs M (2018) The multifaceted role of the ventromedial prefrontal cortex in emotion, decision making, social cognition, and psychopathology. Biol Psychiat 83(8):638–647. https://doi.org/10.1016/j.biopsych.2017.10.030
106. Gasquoine PG (2014) Contributions of the insula to cognition and emotion. Neuropsychol Rev 24(2):77–87. https://doi.org/10.1007/s11065-014-9246-9
107. Bernhardt BC, Singer T (2012) The neural basis of empathy. Annu Rev Neurosci 35:1–23. https://doi.org/10.1146/annurev-neuro-062111-150536
108. Ying X, Luo J, Chiu CY, Wu Y, Xu Y, Fan J (2018) Functional dissociation of the posterior and anterior insula in moral disgust. Front Psychol 9:860. https://doi.org/10.3389/fpsyg.2018.00860
109. Deen B, Koldewyn K, Kanwisher N, Saxe R (2015) Functional organization of social perception and cognition in the superior temporal sulcus. Cereb Cortex 25(11):4596–4609. https://doi.org/10.1093/cercor/bhv111
110. Schultz RT, Grelotti DJ, Klin A, Kleinman J, Van der Gaag C, Marois R, Skudlarski P (2003) The role of the fusiform face area in social cognition: implications for the pathobiology of autism. Philosophical transactions of the Royal Society of London. Series B, Biological sciences 358(1430):415–427. https://doi.org/10.1098/rstb.2002.1208

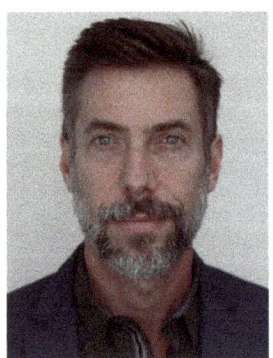

Hélio Tonelli, M.D. obtained his title of specialist in Psychiatry at the University of São Paulo, São Paulo. He attained his specialist and master's degrees in Pharmacology at the Federal University of Parana. He is also a guest professor of postgraduate studies in Neuropsychology at FAE Business School, in Curitiba, Parana, where he is in charge of Social Cognition and Neuroscience and Psychotahology disciplines. He also works as an intern consultant psychiatrist and researcher on the obesity-behavior interface at Caetano Marchesini's bariatric surgery service in Curitiba, Parana. He is also a member of the Brazilian Society of Bariatric and Metabolic Surgery (SBCBM) and the International Federation of Surgery of Obesity (IFSO).

Luisa de Siqueira Rotenberg is a psychologist at the University of São Paulo`s Psychiatry Institute (IPq-FMUSP). She is currently a Ph.D. student under the orientation of Dr. Beny Lafer, focusing on social cognition rehabilitation of patients with bipolar disorder. Her main interest areas are Social Cognition, Bipolar Disorder, Neuropsychology, and Emotion Regulation. She also works as a clinical psychologist in her private office in São Paulo, Brazil.

Nuclear Medicine: A Transdisciplinary Field to Integrate Formal, Physical, Biological, and Medical Sciences

11

Sergio Baldari, Fabio Minutoli, and Riccardo Laudicella

> *"We shall not cease from exploration and the end of all our esploring will be to arrive where we started and know the place for the first time."*
>
> T. S. Eliot

Summary

Nuclear medicine is an emerging field of imaging, deeply grounded on basic sciences, namely physics and biology. It is a unique discipline able to supply precise answers to unsolved clinical questions. Also, it offers an opportunity to treat inoperable patients with minimal/absent collateral effects. Notably, nuclear medicine physicians can use a single radiopharmaceutical to treat and map step by step the illness through the theragnostic approach. Further, nuclear medicine equipment is evolving fast, and the main upcoming advancements are scanner upgrading, new therapeutic probes, and artificial intelligence approaches. Nuclear medicine scanners mainly include SPECT/CT and PET/CT, but PET/MR has raised interest as a state-of-the-art approach to the soft-tissue assessment and the determination of metabolic information.

S. Baldari · F. Minutoli · R. Laudicella (✉)
Nuclear Medicine Unit, Department of Biomedical and Dental Sciences and Morpho-Functional Imaging, University of Messina, Messina, Italy
e-mail: sbaldari@unime.it

F. Minutoli
e-mail: fminutoli@unime.it

© The Author(s), under exclusive license to Springer Nature Switzerland AG 2022   241
N. Rezaei (ed.), *Multidisciplinarity and Interdisciplinarity in Health*,
Integrated Science 6, https://doi.org/10.1007/978-3-030-96814-4_11

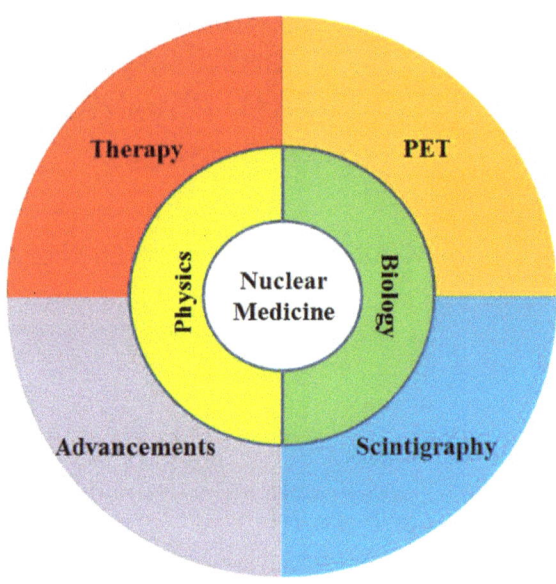

Nuclear medicine

The code of this chapter is *01110010 01101111 01100011 01101110 01101001 01010000 01100101 01110011 01101001*.

Keywords

Artificial intelligence · Nuclear medicine · PET · Scintigraphy · Therapy

1 Introduction

Henri Becquerel discovered radioactivity in 1896: Following the first X-ray capture made by Wilhelm Röntgen, Becquerel enfolded a fluorescent material (potassium uranyl sulfate) in photographic plates and obscure material during an experiment. In 1903, he was awarded the Nobel Prize in Physics (shared with the Curies) for his work on nuclear radiation. The Becquerel (Bq) was subsequently established as the International System (SI) unit of radioactivity, while before that, a non-SI unit Curie (Ci) existed for use. Nowadays, the isotope radioactivity can be measured as disintegrations per second, and one disintegration/second is equivalent to 1 Bq (37 MBq = 1 mCi). In the early 1900s, isotopes were initially used for physiological and blood flow measurements. Then, Enrico Fermi divided the atom, and reactors were established, with 131iodine (131I) being the most used product to discover thyroid tissue through Geiger counters and treat thyroid diseases. Then, in the 1960s, rectilinear scanners became available, and Anger developed the single-view gamma camera. However, the discovery of technetium at the Berkeley cyclotron and the announcement of the first single-photon emission computed tomography (SPECT) in 1968 pushed the discipline forward over the following three decades to the development of the modern multi-headed SPECT camera, able to acquire not only static or dynamic images, but also obtaining time activity curves (TACs) in a certain region of interest (ROI). The renogram is a remarkable application of this device (Fig. 1), along with tomographic images that display a proper radiopharmaceutical distribution in a body section. Then, positron emission tomography (PET) was discovered in 1976 and achieved another milestone in nuclear medicine [1]. Thanks to its incorporation with computed tomography (CT) in 2001 by Townsend and Cherry [2] and more recently with magnetic resonance (MR), PET has been increasingly important in the 2000s for the evaluation of several pathologies, mainly in the oncology field [3]. Hybrid SPECT/CT or PET/CT notably magnified the diagnostic accuracy of nuclear medicine examinations: The combination of anatomical data (CT) and PET (PET/CT) or SPECT information (SPECT/CT) in a sequential, registered, integrated system allows obtaining

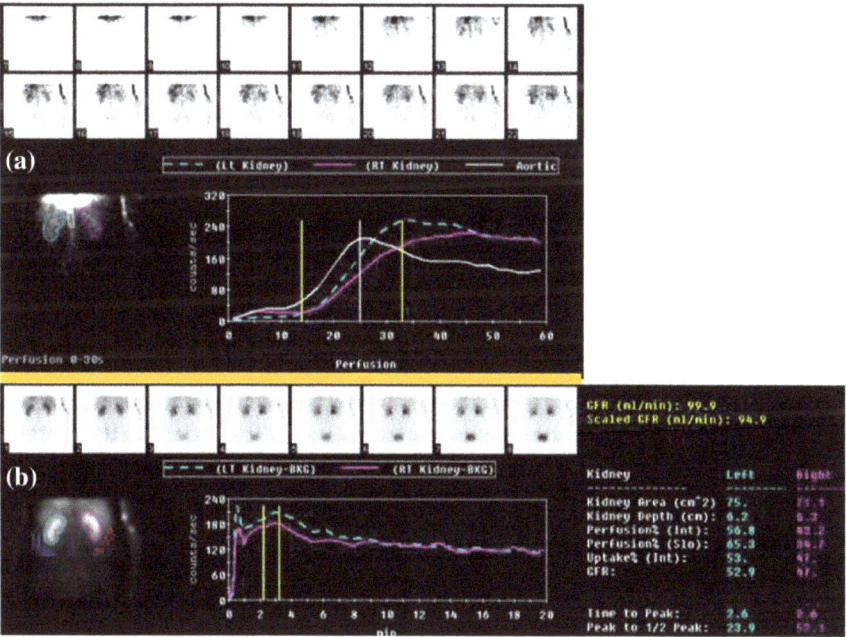

Fig. 1 **a** Angioscintigraphic phase acquired during the first minute after the administration of 99mTc-DTPA able to assess the radiopharmaceutical's distribution in both kidneys and in the aorta; **b** Renogram representative of the single kidney's functions; semiquantitative data of each single kidney and glomerular fraction rate value. Images from the Nuclear Medicine Department of the University of Messina

precise and unique results (anatomical, functional, and metabolic) thus reducing the false positive rates and improving the specificity and the overall diagnostic accuracy. More recently, PET/MR was developed as a combined machine to simultaneously provide state-of-the-art soft-tissue resolution and functional information from MR in relation to the metabolic data from PET in a one-stop-shop exam. Despite some disadvantages (i.e., cost and availability), MR imaging helps determine functional imaging parameters and improves motion correction and patients' radiation exposure due to CT [4, 5].

2 Physical Basis

In the presence of a certain (imbalanced) group of neutrons and protons, the atom's nucleus might be unstable, and this causes an emission process named radioactive decay. The three main expressions of radioactive decay are alpha (α), beta (β), and gamma (γ). Radioactive decay follows an exponential behavior, and the decay time needed for half of the atoms is called the half-life ($T_{1/2}$). α-emission of two protons and two neutrons (a helium nucleus) expelled from the atom is typical of the periodic table's heavy elements with great applicability to nuclear medicine therapy. B-emission can occur into two forms of decay: β-(negative) with therapeutic potential, in which a neutron is transformed into an electron and proton, followed by the ejection of an electron with a neutrino (a particle without masses or electric charge); β+ (positive) with diagnostic potential, in which a proton is transformed into a positron and a neutron, followed by the emission of a positron with a neutrino. Further, γ-emission has diagnostic utilities and can occur in several pathways due to different transitions and energy among the three possible atom's states: the ground state (stable); the excited state (unstable); the metastable state (an unstable state with a lifetime higher than 10^{-12} s). However, even an α- or β-emission with a metastable daughter nucleus may determine a final γ-emission [6]. When describing the interactions between matter and radiation, it is essential to assess the X-ray window wavelength. When an electron is upraised to a higher energetic level (excited atom) and the vacancy space takes on an electron in the outer shell, the energy difference is emitted through electromagnetic radiation (X-ray). However, it is important to distinguish between directly ionizing charged particles (e.g., α, β, and protons) and indirectly ionizing particles (photons and neutrons). Ionizing radiation can be described as an energy transfer as particles or electromagnetic waves (wavelength $\leq$ 100 nm) able to produce ions directly or indirectly [7]. We refer to the energy unit, i.e., electron volt (eV), as the energy quantity gained by an electron when moving through a potential difference (voltage) of one volt. Below 13.6 eV, no ionizing radiation is available. The energy of X-rays and γ-rays mainly determines three kinds of interactions in nuclear medicine:

(i)   Photoelectric absorption (PA): A given atom engrosses the whole energy of the incident photon, and this energy is used to expel one photoelectron (orbital electron);

(ii) Compton scattering (CS): The incident photon is transported with a loosely bound external shell electron, and the photon will be deflected with a scattering angle depending on the incident photon energy;

(iii) Pair production: A higher energetic photon (at least 1.02 meV) cooperates with a charged particle's electric field, photon disappears, and its energy is used to develop a negative–positive electron pair.

Both CS and PA do not determine ionization directly, but the expulsion of orbital electrons and the formation of positive–negative electrons will determine ionization and radiobiological effects. In the real imaging scenario, both PA and CS lead to missing/misplaced information because the signal is lessened ("attenuated"), and the ultimate image needs to be corrected accordingly. The total attenuation of a specific tissue varies depending on the incident photon's energy, tissues' thickness, density, and mean atomic number. For SPECT techniques, the photon energy mainly results in CS. SPECT artifacts are partially solved by attenuation correction using SPECT/CT devices in which the CT images are used, including co-registration, resolution, and energy scaling. In a PET device, the detection of anticoincidence gamma rays (resulting from positron annihilation) is also influenced by random and scattered events. Scattering can change the γ-photon emission, and therefore, the coincidence may be allocated to a wrong line of response (LOR). Random events are due to the non-infinitely small-time window of coincident photon acceptance (5–10 ns), which determine a non-negligible probability that two photons from diverse annihilation events may be detected. CS and random event rectifications are achieved in the PET images' raw data. Attenuation, CS, and random events contribute to a distorted PET signal and the images' noise (more evident in the inner body). Therefore, corrections are essential, and attenuation can be measured with different approaches (μ-map for CT and Dixon sequence for MR). However, it is important to say that all improvements will supply the reconstructed images' global noise features, while the average image pixel values (unbiased) will be more similar to the true signal.

3   Biological Basis

Radiation exposure can damage DNA, and consequently, single-strand breaks (SSBs), double-strand breaks (DSBs, whose repair is more complex), base damage, and protein–protein/DNA cross-links occur. When tissue is irradiated, the response is mainly determined by the absorbed dose (D), namely the mean amount of energy imparted per unit mass: For an absorbed radiation dose of 1 Gy, there are approximately 1000 SSBs, 1000 base damage, and 40 DSBs. If DNA repair fails, the cellular injury may manifest as mutation, chromosome aberration, transformation, reproductive failure, or cell death. High rates of cell proliferation and tissue growth result in increased radiosensitivity (i.e., a fetus is more radiosensitive than a child to radiation exposure). Another variable that must be considered is the linear energy transfer (LET, KeV/μm), namely the density of the matter at which energy is deposited through radiation. Photons and electrons (i.e., X-, γ-, and β-ray) are low-LET radiations (LET values between 0.2 and 10 keV/μm) with predominant indirect actions. Otherwise, Auger electrons, α-particles, protons, and neutrons are high-LET radiations (LET values between 10 and 100 keV/μm) whose direct action is mainly represented by ionizing radiation. Higher LET radiation will deposit more energy than necessary to cause damage ("overkill effect"). A LET of 100 keV/μm (Auger electron and α-particles) is optimal in producing biological effects. At this density, the mean separation in ionizing events is equal to the DNA double helix diameter, which corresponds to the highest probability of DSBs.

The biological effects of ionizing radiations can be divided into stochastic effects and deterministic effects. Effects whose probability increases with the dose are referred to as stochastic effects. The most representatives are cancer and genetic effects. We might refer to the International Commission on Radiological Protection (ICRP) that "the detriment adjusted nominal risk coefficient for cancer for the whole population after exposure to radiation at low dose rate is 5.5% per Sievert (Sv) effective dose." Furthermore, as above-mentioned, exposure at an early age results in higher nominal risk factors, and females are slightly more susceptible. According to experimental studies, the ICRP sets nominal risk coefficients at 0.2% per Sv effective dose to account for genetic effects in the whole population. Differently, deterministic effects are characterized by an incidence and severity that increase above a certain threshold with increasing dose. Early tissue reactions may appear within a few weeks (inflammatory effects associated with release or loss of cell factors), while late tissue reactions can occur over the years (directly resulting in the formation of a specific tissue or from early tissue reactions). Such reactions can appear after partial body (cataract, erythema) or whole-body exposure (acute radiation syndrome) over a specific threshold dose stated by ICRP, and for doses <100 mGy, no tissue expresses relevant functional impairment. LET, dose rate, dose fractionation, oxygen, radionuclide decay, and biological clearance can affect cellular survival, particularly in radionuclide therapy. For these reasons, the biologically effective dose (BED) was introduced to compare different types of treatment. Particular attention must be paid to the embryonic and fetal state because of a higher radiosensitivity, especially in the early fetal period (organogenesis) with the potential occurrence of stochastic and deterministic effects such as malformations, growth and intelligence quotient (IQ) decrease, and neonatal death. Table 1 describes the main types of interactions.

Table 1 Main types of interactions

Type of interaction	Description	Products
Ionization	Energy deposition higher than the binding energy of an orbital electron, that leaves the atom and produce ion pairs	Free radicals with very reactive unpaired electrons
Excitation	Energy deposition determines the elevation of the orbital electrons to a higher energy shell, as well as the atom, is raised to a higher energy level	Free radicals with very reactive unpaired electrons
Radiolysis	Energy deposition in cellular water determining a complex series of chemical changes	Free hydroxyl and highly reactive radicals

4 Scintigraphy

SPECT and PET are the most common nuclear medicine imaging techniques based on γ-radiation detection. Namely, in SPECT radiotracers, γ-emission is obtained through the de-excitation of one short-lived excited nuclei emitting one γ-photon per nuclear decay, detected by one to three rotating detectors (mainly two fixed detectors at 90°/180° on a rotating gantry). Both SPECT and PET detectors are scintillators (able to convert γ-rays into visible scintillation light) coupled through a light guide to a photodetector, photomultiplier tubes (PMT), and then converted into an electrical pulse. PMT are organized in an entrance photo-cathode able to detect the photons (γ-radiation) from the scintillator; such photons are then emitted by the photoelectric effect to the dynodes in a vacuum tube, being accelerated and amplified by a rising high voltage, to a final photo-anode at the end of the tube. To precisely determine the photons' direction, high-density physical collimators are required for SPECT imaging, often made by tungsten or lead, to adsorb γ-photons from different directions. SPECT and PET scintillators have different physical properties which differentiate cameras and detectors. The atomic number and the detectors' constituent element's density determine the efficiency (sensitivity), and the radionuclides' decay time affects the luminescence signal decay rate after an excitation. The light yield determines the energy resolution and the time, representing the number of optical photons generated from any γ-interaction. The photodetector and its quantum efficiency depend on the rate of converted photons into electrons. However, the precise localization and the distribution of the radio-pharmaceutical uptake in each tissue or organ cannot be easily determined. Anatomical and morphological information is of primary importance, also for magnifying the diagnostic value of nuclear medicine techniques. As for SPECT and PET imaging, the common method is to use and acquire CT simultaneously [8]. State-of-the-art SPECT/CT is now equipped with high-resolution and fast diagnostic CT to improve the accuracy, operators' confidence, and patients' comfort (Fig. 2). Also, CT may be useful in scatter and attenuation correction by photon counting and determining photon direction in the alignment. Furthermore,

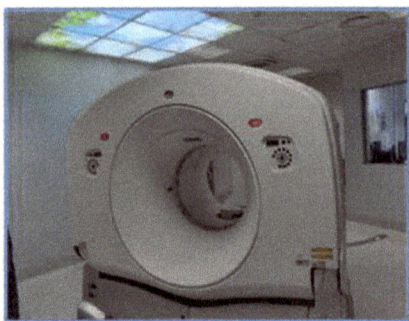

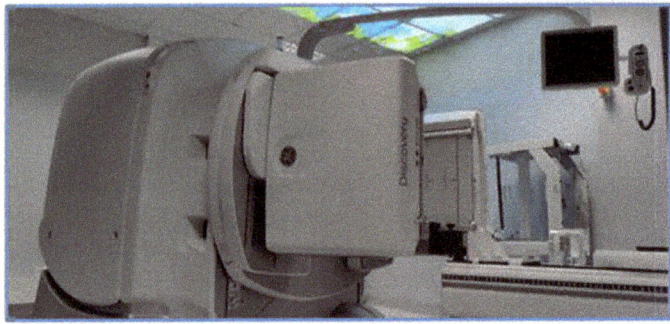

Fig. 2 GE healthcare discovery SPECT/CT in use in the Nuclear Medicine Department of the University of Messina

SPECT/CT improves dosimetry procedures and their precision and reliability [9]. There are several applications of different SPECT imaging radiopharmaceuticals in both non-oncological and oncological settings. ^{99m}Tc is the most widely used radionuclide in diagnosis, applied to more than 90% of SPECT scans and almost 80% of all nuclear medicine tests [10]. One of SPECT and SPECT/CT's main oncological applications with a dramatic impact on patient management is the sentinel lymph node mapping. After the interstitial radiocolloid injection at the primary tumor's site, this procedure can identify the first lymph node that drains primary cancer and select it for eventual biopsy. It is widely used for breast cancer and melanoma, but it also has the potentiality for head/neck cancer, gynecological, and urological malignancies [11]. Bone scintigraphy is a widely used scintigraphic technique. Metastable technetium (99mTc) labeled with a diphosphonate given through an intravenous route can detect metastases when bone formation is reactively increased (osteoblastic activity), with great sensitivity but a low specificity mainly for lytic lesions in which CT might be beneficial [12] (Fig. 3). Also, for thyroid metabolism assessment, scintigraphy is essential for the characterization of "hot" (usually benign) and "cold" areas (potentially malignant) through the administration of $[^{99m}Tc]O^{4-}$, which can assess the NIS sodium iodide symporter (NIS) expression on thyroid tissue [13] (Fig. 4). Also, labeling the 99mTc with sestamibi (MIBI) makes it possible to assess the parathyroid glands' status [14].

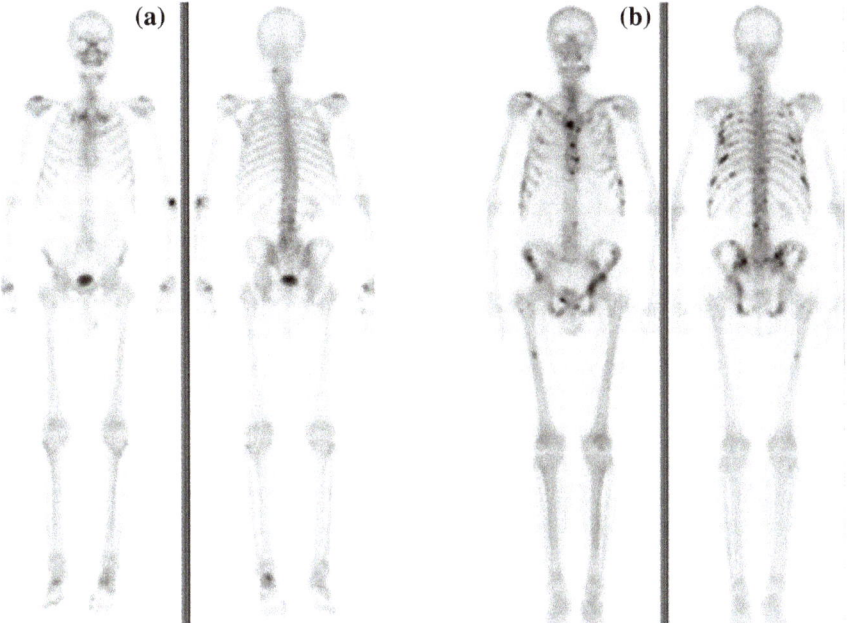

Fig. 3   **a** Physiologic distribution of 99mTc-MDP; **b** Pathological distribution with several bone lesions. Images from the Nuclear Medicine Department of the University of Messina

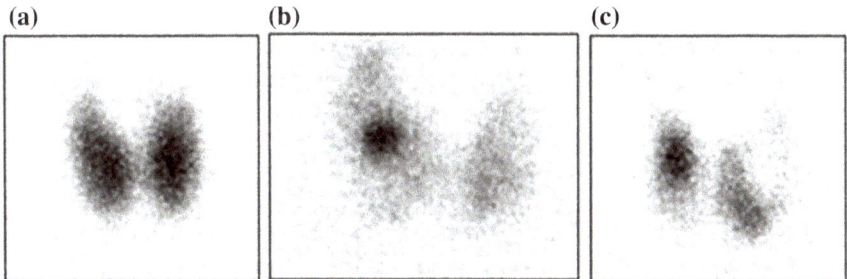

Fig. 4   **a** Anterior view of the normal distribution of ^{99m}Tc-O$_4$ in the thyroid gland; **b** Anterior view of a focal hot area in the right thyroid middle lobe; **c** Anterior view of a focal cold area in the left thyroid lobe (middle and upper third of the gland). Images from the Nuclear Medicine Department of the University of Messina

Other non-oncological applications include nuclear cardiology, in which the cardiac stress/rest SPECT exam has a high predictive value. It is validated for diagnosis, risk stratification, and prognosis of coronary artery disease with a high accuracy thanks to the CT attenuation correction and the possibility of the calcium score

Table 2 Half-lives and emission energies of ^{99m}Tc, ^{123}I, and ^{111}In

Radionuclide	Emission energy (KeV)	Half-life (hours)
^{99m}Tc	141	6.02
^{123}I	159	13.22
^{111}In	171 and 245	67.2

calculation [15] and for cardiac amyloidosis [16]. Regarding neurology, SPECT assists with perfusion evaluation, mainly in dementia and epilepsy through [^{99m}TC] HMPAO or ECD [17, 18] and also with the dopaminergic system quantification through [123I]ioflupane that can be used in the case of parkinsonism [19]. SPECT can also help detect the bleeding source in the gastrointestinal tract, diagnose Meckel's diverticulum, and monitor inflammatory bowel disease with red blood cells labeled with 99mTc [20–22]. In the lung, ventilation (99mTc-technegas) and perfusion (99mTc-macroaggregate of albumin) SPECT can be used, whereas pulmonary embolism is suspected or for preoperative quantification of lung function and for the assessment of the regional changes in chronic disease [23]. Physical half-lives and emission energies of ^{99m}Tc, ^{123}I, and ^{111}In are described in Table 2.

5 Positron Emitting Tomography

For PET imaging, the rapid radioactive decay of positron-emitting isotopes is used: after their production in cyclotrons (e.g., 18F) or elution from generators (e.g., 68 Ga), isotopes are inserted through chemical reactions into high biological-relevant molecules (e.g., FDG or Choline for metabolism, DOTA-peptide for receptors expression). The so-obtained radiopharmaceuticals are administered intravenously in patients, and positrons are emitted; their rapid annihilation with electrons produces two antiparallel photons (γ-radiation), with an energy of 511 keV for each, detected by a ring system detector. Table 3 reports the main aspects of the most used PET radionuclides. These high-energy antiparallels (emitted at an angle of approximately 180°) and coincident (within a few nanoseconds) photons, originating from the same point of the body, will be then revealed by the PET detector ring made of peculiar material (e.g., lutetium/gadolinium oxyorthosilicate) able to convert high-energy photons into low-energy photons, detected and amplified by PMT, followed by the so-called electronic collimation. However, a fraction of annihilation photons is diverted (scattered) by the patient's body, potentially discriminated due to lower energy or being undetected by PET detectors. Therefore, CT by providing an attenuation map is beneficial, using the information about the tissue electron density to adjust photons considering their energy. However, misalignment (misregistration between emission and transmission) and metal devices in the patient's body may affect the PET tracer's quantification, and reviewing the non-attenuation corrected images is essential in this scenario. In summary, a typical PET system hardware mainly consists of "scintillators, photodetectors, front-end readout electronics, back-end

Table 3 Main aspects of the most used PET radionuclides

Radionuclide	Positron energy max (MeV)	Positron max penetration range (mm)	Half-life (minutes)
^{18}F	0.65	2.5	110.0
^{68}Ga	1.90	9.0	67.7
^{11}C	0.96	4.2	20.4
^{13}N	1.73	8.4	10.0
^{15}O	1.20	5.5	2.0

data acquisition system, and a computer workstation to control the acquisition and image reconstruction." Scintillators convert high-energy annihilation photons (511 keV, not visible) to low-energy photons (1–10 eV) belonging to the visible light range (ultra-violet), detectable by photodetectors that further convert them into electric current (EC). Then, EC is forwarded to the "front-end readout electronics and data acquisition system, which amplify the signal and extract the energy, time (for coincidence detection), and position information from the signal. The energy, time, and position information are then sent to a workstation for image reconstruction to generate the PET images." A typical PET workflow is depicted in Fig. 5, while in Fig. 6, we reported the PET/CT scanner of our department. Furthermore, PET hybrid imaging provides semiquantitative ROI metrics, namely standardized uptake values (SUV), indicating the radiopharmaceuticals' metabolic amount, being considered as a biomarker of response and outcome [24]. Clinical applications of different PET imaging radiopharmaceuticals are summarized in Table 4. In the oncological setting, PET helps differentiate benign from malignant lesions, search for an unknown primary tumor (especially useful in case of a paraneoplastic syndrome or metastatic disease), determine tumor stage, monitor treatment response, detect tumor recurrence in high-risk and suspicious cases (i.e., increased tumor markers), diagnose between recurrence and post-treatment changes (i.e., fibrosis, necrosis), select biopsy site to increase the diagnostic accuracy, and plan radiation therapy [25–27]. As described in Tables 1 and 2, several radiopharmaceuticals are available for different purposes. Fluorine-18 (18F) is a short half-life (110 min) cyclotron-produced radioisotope that emits positrons. It can be labeled with numerous molecular tracers that can be acquired for imaging several hours after administration. The first clinical fluorine-labeled PET radiopharmaceutical (1982, by the group of Di Chiro et al.) remains the most widely used: 18F-2-deoxy-2-fluoro-D-glucose (^{18}F-FDG) acts as a radioactive glucose form, and it is transported through the cellular membrane by carrier proteins, the so-called glucose transporter molecules (GLUT). Upon intracellular availability, the phosphorylation rate of FDG-6 Phosphatase (FDG-6P) relative to the rate of dephosphorylation determines the tracer's accumulation in the cell. Body tissues that possess high amounts of GLUT, hyperperfusion, or high hexokinase activity might reveal upgraded uptake of [^{18}F]FDG described as increased uptake foci in scans. Such criteria are typical of several cancers, although they have been reported in

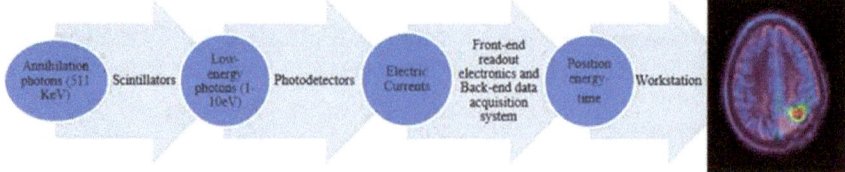

Fig. 5 Typical PET workflow with a FET-fused PET/MR indicative of pathological recurrence from astrocytic glioma in left parietal brain lobe

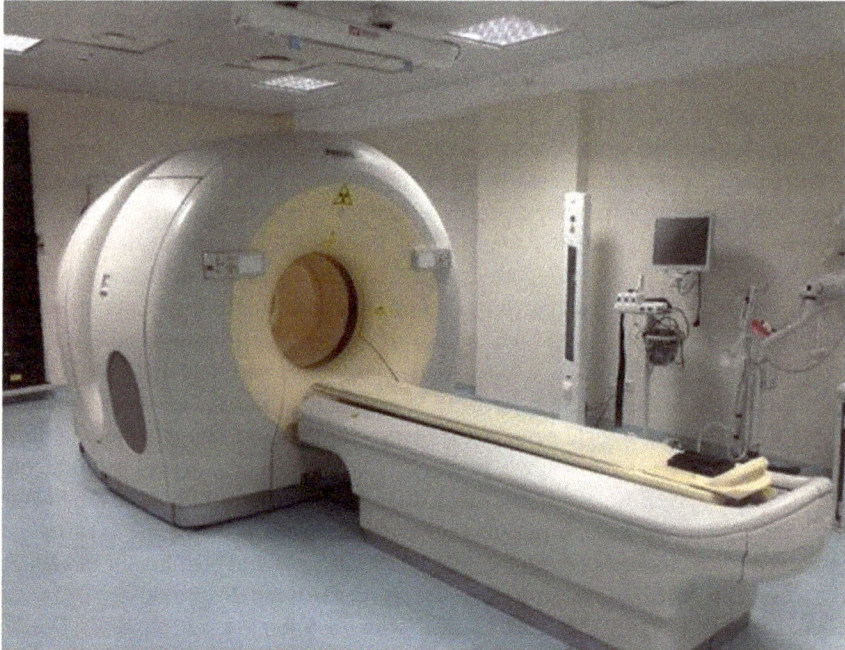

Fig. 6 Philips TOF Gemini PECT/CT in use in the Nuclear Medicine Department of the University of Messina

other physiological and pathological conditions [28]. The appropriateness criteria for PET in oncology are evolving [29]: The main applications of whole-body (skull base to mid-thigh) FDG-PET in this setting are connected to high-glucose avid neoplasms, such as lymphoma (Hodgkin and non-Hodgkin), especially for nodal involvement. FDG-PET might upstage the disease in about 30% of cases, while its accuracy remains high compared to conventional imaging techniques such as contrast-enhanced CT [30] (Fig. 7). Also, endorsing the Deauville and Lugano criteria, PET offers a valuable tool for assessing treatment response in lymphoma

Table 4   Application of different PET imaging radiopharmaceuticals

Biological/Biochemical process	Radiopharmaceutical
Proliferation	FLT
Aminoacid transport	MET, FET, fluciclovine
Hypoxia	FMISO, FAZA, ATSM
Receptors expression	DOTA-peptides, FES, galacto-RGD, bombesin
Glucose metabolism	FDG
Dopamine metabolism	FDOPA
Lipid and fatty acid metabolism	CH, FCH, acetate
Antigen	PSMA
Fibroblast protein	FAPI
Calcium analog	NaF

[31, 32]. FDG-PET is also beneficial in the characterization of suspicious lung nodules and lung cancer staging, in which PET was able to detect occult metastases in about 30% of patients [33], particularly bone metastases [34]. Also, PET could predict and evaluate treatment response and its correlates with the metabolic response [35], as shown in Fig. 8. Another big killer that has benefited from FDG-PET is metastatic breast cancer, thanks to accurate metabolic information [36]. In advanced-stage melanoma patients, [^{18}F]FDG-PET endorses a significant role in monitoring and determining the extent of the disease [4]. Another scenario in which PET has reached consensus over the years is prostate cancer with several radiopharmaceuticals such as choline, fluciclovine, bombesin, and the most impacting prostate-specific membrane antigen (PSMA) [37–41]. Further, in neuro-oncology, several steps have been taken in molecular imaging. Advanced PET tracers of great application to brain tumors are fluoro-ethyl-tyrosine (FET, Fig. 5), whose TACs estimation is a useful instrument [42]; fluorothymidine (FLT) useful in high-grade glioma; hypoxic tracers such as fluoromisonidazole (FMISO) able to assess tumor hypoxia and treatment-related modifications; DOTA-peptides useful in meningiomas, and under-survey immuno-PET tracers with antibodies as labeled tracers [43, 44]. The specific PET radiopharmaceutical for cortical bone evaluation is [^{18}F]NaF, introduced in 1962. This radiopharmaceutical accumulation depends on the blood flow and the degree of osteoblastic activity [45], while for bone and soft-tissue tumors evaluation, FDG is still the best choice [46]. Another big chapter of PET application was opened for neuroendocrine tumors, whose somatostatin receptor expression is well-documented by radiolabeled DOTA-peptides PET imaging [47]. Non-oncological PET studies are also present in several clinical scenarios such as cardiovascular [48–50], neurological [51–53], and infectious disease [54–56].

(a) (b)

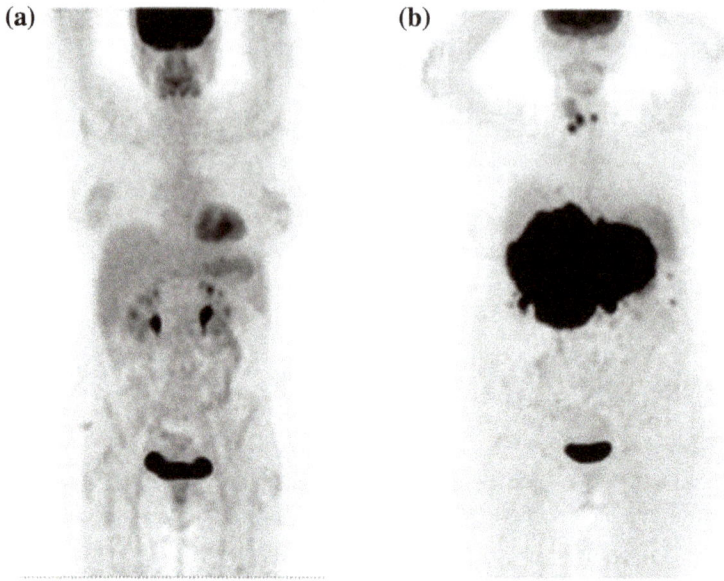

Fig. 7 **a** Anterior view of the normal biodistribution of FDG with physiological uptake in the brain, hearth, urinary tract, and bladder; **b** Anterior view of the extensive mediastinal FDG uptake due to a large B-cell lymphoma. Images from the Nuclear Medicine Department of the University of Messina

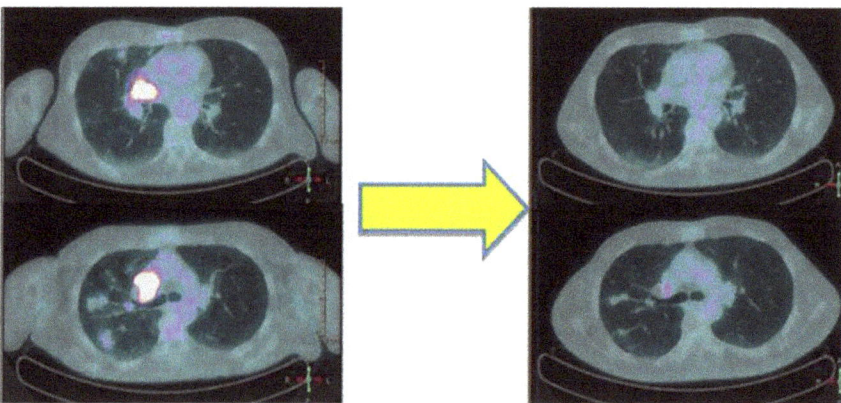

Fig. 8 Axial view of a partial response to chemotherapy (right panels) in adenocarcinoma of the right lung described in the Nuclear Medicine Department of the University of Messina

6 Therapy

Historically, the initial focuses of nuclear medicine were toward therapy. Phosphorus (isotope 32P, produced by cyclotron) and iodine (isotope 131I, produced by nuclear reactor) were the first radiopharmaceuticals proposed for nuclear medicine therapy in the late 1930s [57, 58]. Further, radioiodine was the first "theragnostic" agent. Theragnostic (THERApeutic plus diaGNOSTIC) potentially enables, at the same time and with the same radiopharmaceutical administration, the detection and targeting of a pathologic process (in this case, neoplastic thyroid tissue). Theragnostic uses the same/very similar tracer (molecules), either labeled with different nuclides or isotopes, administered in a different dosage to identify, diagnose (low activity), and treat (high activity) a specific disease. Notably, through the theragnostic approach, it is possible to label the same agent with γ (i.e., 111In-pentetreotide) or $\beta+$ (68 Ga-DOTA-peptides) radionuclide for imaging, as well as an α or $\beta-$ (177Lu-DOTA-peptides) radionuclide for therapy, making nuclear medicine an appropriate discipline for theragnostic toward personalized medicine (Fig. 9). As mentioned, β-emitters enhance cytotoxicity to tumors due to the long-range penetration of electrons to increase the average delivered dose. However, on the opposite, such penetration's capability also involves the nearby healthy tissue. Also, the low ionization electrons' capability (low LET) of β-radiation cannot produce a fatal dose single-focused tumor cell [59]. Differently, α-emitters have a shorter range of penetration and are highly cytotoxic (higher LET). In other words, α-emitters can target tumor cells over a short range (diameters of numerous cells) and cause complex DNA lesions [60, 61]. Therefore, α-radiations, due to their short-range effects and high energies, are better suitable for small-volume disease treatment, being more efficient and specific, reducing distant radiation burden to normal cells [62]. As regards iodine, it is trapped by the thyroid gland's follicular cells, stored in thyroglobulin, and released as the hormones T3 and T4. Therefore, the thyroid is the physiological seat of iodine storage in the body, and radioactive iodine is used to treat hyperthyroidism [63] and thyroid cancer with high success rates [64]. Namely, post-surgery iodine is orally administered (radioactive iodine pills) to ablate/find tissue residues remaining from thyroidectomy; a higher dose is reserved to treat/re-treat recurrences or metastatic patients, with rare adverse effects [65]. Another challenge for nuclear medicine therapy is neuroblastoma which is a neoplasm that develops from immature nerve cells in different parts of the body. It can be studied with a diagnostic scan of low-dose $[^{131}I]$meta-iodobenzylguanidine (MIBG) and treated with higher doses of the same radionuclide (Fig. 10) [66]. Radioimmunotherapy is another application of nuclear medicine therapy in which a monoclonal antibody is labeled by a β-emitting radionuclide and used for therapy of non-Hodgkin's lymphoma (NHL) that overexpresses CD20 antigen. In clinical oncology, anti-CD20, such as Rituximab, is used to treat NHL, while refractory/relapsed NHL may be treated with Rituximab labeled with 90Y- or $[^{131}I]$antiCD20Mab [67, 68]. Metastatic bone

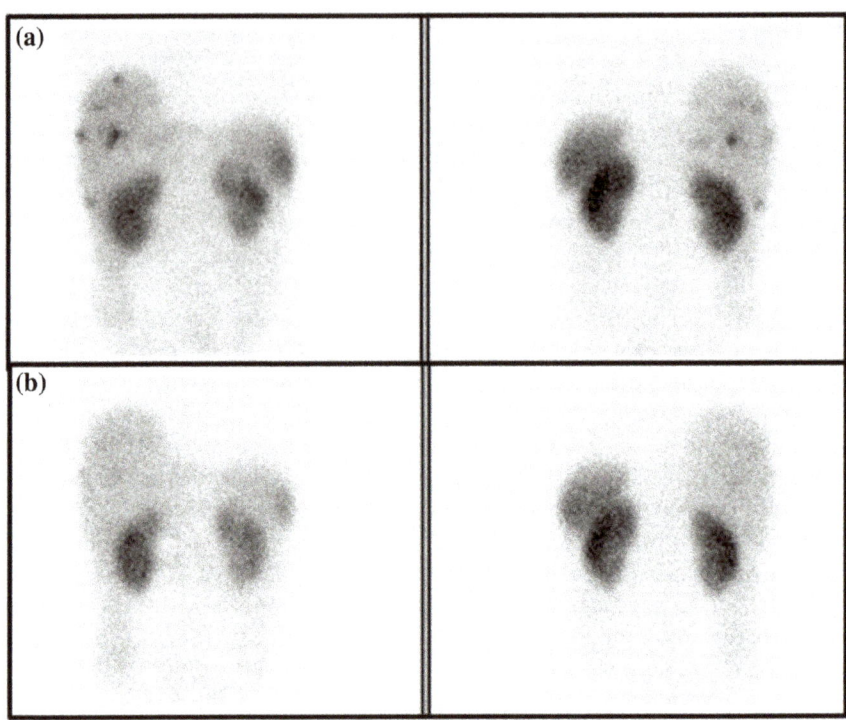

Fig. 9 Response to peptide receptor radionuclide therapy (PRRT) in a GEP-NET patient with liver metastases, treated in the Nuclear Medicine Department of the University of Messina. **a** Anterior (left) and posterior (right) view of post-treatment imaging after I cycle of PRRT with 177Lu-DOTATOC; **b** Anterior (left) and posterior (right) view of post-treatment imaging after II cycle of PRRT with 177Lu-DOTATOC

pain can also benefit from nuclear medicine's efforts. Prostate, breast, and lung tumors that account for about 50% of cancers have a high probability of bone metastases that can be assessed through a bone scan, as already mentioned. Several radionuclides can be used for metastatic bone pain relief, such as [^{188}Re]HEDP, [^{177}Lu]EDTMP, [153Sm]EDTMP (more recently), and the α-emitting 223RaCl2 specific for prostate cancer bone metastases [69, 70]. Also, liver tumors and metastases can be treated using selective internal radiotherapy (SIRT) and trans-arterial radioembolization (TARE), exploiting the peculiar arterial feeding of tumoral lesions through the intra-arterial administration of radionuclides such as 90Y-labeled microspheres, with minimal damage to normal hepatocytes [71]. Theragnostic [72, 73] is also used for neuroendocrine neoplasms, as demonstrated with Lutathera [74–77] and prostate cancer with [^{177}Lu]PSMA [78]. Novel targets and theragnostic radiopharmaceuticals are, for example, fibroblast activation protein inhibitors (FAPI), neurotensin receptors, glycoprotein B7, CA19.9 antibody

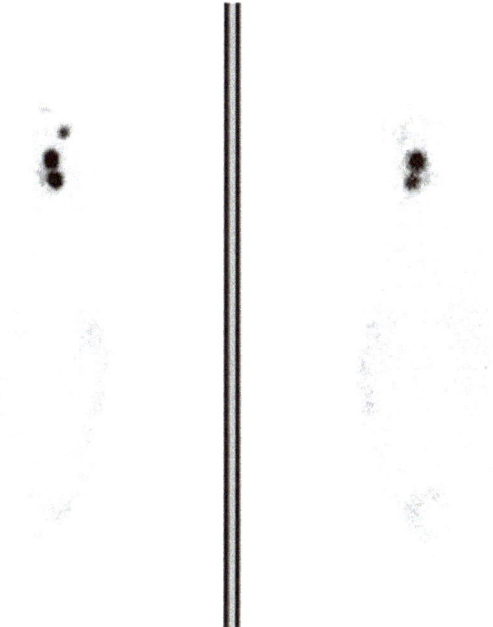

Fig. 10 Anterior (left) and posterior (right) view of post-dose imaging after ablation metabolic radiotherapy with 131I in differentiated thyroid cancer, treated in the Nuclear Medicine Department of the University of Messina

Table 5 Characteristics of the most used radiopharmaceuticals for therapy

Radionuclide	Half-life	Emission	Penetration range	Main application
^{111}In	67.9 h	γ (173 and 247 keV)	0.002–0.5 mm	Imaging and therapy
^{90}Y	64 h	β- (2288 keV)	4–8 mm	Therapy
^{177}Lu	6.7 days	β- (500 keV) γ (113 and 208 keV)	1–2 mm	Therapy
^{131}I	8.02 days	β- (606 keV) γ (364 keV)	0.6–2 mm	Imaging and therapy
^{223}Ra	11.4 days	α (5–7.5 meV)	<10 μm	Therapy

[72, 79], and the paramagnetic 166Holmium [80]. Promising results have also been obtained with α-emitters such as 225Actinium and 213Bismuth [81, 82]. Table 5 describes the characteristics of the most commonly used radiopharmaceuticals for therapy.

7   Technological Insights and Future Perspectives

Nuclear medicine applications are also expanding through technological innovations in hardware and software. Several steps and improvements have been reached in hybrid PET scanners in the last two decades, enabling improved spatial resolution, image reconstruction quality, diagnostic accuracy, disease staging, and therapeutical assessment. Specifically, time of flight (TOF) technology consents to the radiopharmaceutical's dose reduction. It does not decrease image quality but also provides a higher spatial resolution and reduces the scan time [83]. Since 2015, a new generation of integrated PET/CT has been available. Digital PET/CT scanners are equipped with digital photon counter (higher detection efficiency and counting rate), high-performance digital detectors, and implemented photodiodes. Several studies highlighted the improved accuracy and the better image quality provided with these digital scanners. Also, the new generation of silicon photomultiplier can directly convert photons into a digital signal, reducing/excluding signal noise/loss for a more accurate and efficient scanner [84–86] in terms of spatial/contrast resolution (smaller lesions detection) and semiquantitative measurements, also with larger field of view (FOV). Further, new scanners incorporate the latest diagnostic CT innovations enabling dose reduction and an improved assessment of potential artifacts (i.e., metal implants) while saving valuable time and therefore improving the daily workflow. Another significant innovation in nuclear medicine was the introduction of simultaneous PET/MR that holds great promise because it combines in a one-stop-shop technique the state-of-the-art soft-tissue contrast/functional information from MR (perfusion and diffusivity) with the metabolic PET information. Apart from the synchronous evaluation of anatomical, microstructural, functional, and metabolic data, the MR technique is free of radiation (of great importance in the pediatric population and cases where frequent exams are required). Also, PET/MR enables PET partial-volume correction, real-time motion correction, and pharmacokinetic studies through simultaneous dynamic acquisitions; thanks to the increased sensitivity of avalanche photodiodes, it allows radioactivity dose reduction without reducing the diagnostic quality [44, 87–90]. Another arrow in the quiver is represented by artificial intelligence (AI) algorithms that could be helpful in medical imaging and nuclear medicine. AI makes a system able to *interpret* data and *learn* from them, obtaining awareness to accomplish specific tasks through flexible adaptation. AI includes machine learning (ML) and deep learning (DL) techniques. ML algorithms can assist with the prediction of diagnostic and prognostic outcomes by using human-made features. Namely, DL can be described as the application of neural networks organized in *subsequent, progressive, and multiple nonlinear processes (layers) related to each other*. When input images are registered, DL algorithms simultaneously and automatically learn, discovering respective features and models, without the so-called features' engineering. Therefore, DL is particularly well suited for large, diverse, complex datasets, and challenging tasks such as classification and image segmentation. Furthermore, radiomics is an application of standard ML to medical images "assuming that, any smallest part (i.e., any pixel) of a medical image of different origin may contain characteristics ('features') of tumor phenotypes that can reflect the underlying pathophysiological process, potentially related to outcome and treatment response." Several promising applications of such techniques are available in the literature [91–93]; however, AI suffers from significant challenges because it requires large quantities of labeled data that must be as standardized as possible. Creating large datasets is under investigation and warranted with the goal of using AI models in clinical settings [94], along with the development of more promising molecules and radionuclides. Initial and interesting results are described for immuno-PET (i.e., ^{89}Zr-labeled monoclonal antibody), enabling the non-invasive assessment of tumor-related antigens expression in vivo [95–97], discovering new biochemical aspects of pathologies, assessing their behavior [98–100], as well as improving the efficacy of available probes [101–107].

8 Conclusion

Nuclear medicine imaging provides functional information of human tissues and organs thanks to the administration of radiopharmaceuticals. Nuclear medicine applications are expanding as new therapeutic agents and hybrid techniques come to emerge, mainly SPECT/CT, PET/CT, and PET/MR that have raised interest as a combined machine. [^{18}F] FDG-PET still represents the nuclear medicine main actor, but several other radiopharmaceuticals are available (i.e., PSMA, FET) or under survey (i.e., FAPI). Hardware and software advancements (digital PET) further improved the quality of images obtained by PET, enabling the reduction of scan time and radiopharmaceutical's dose required for detecting small lesions that might critically determine the disease assessment. Despite several standardization issues, AI applications could further enhance the diagnostic accuracy and non-invasive molecular in vivo grading of several pathologies. However, new and more specific radionuclides are necessary as well as hybrid physicians with mixed backgrounds (nuclear medicine and radiology/oncology) toward personalized treatments and diagnosis; we suppose and forecast further and noteworthy improvements in nuclear medicine, making this a thrilling moment to stay in molecular imaging, always headed to precision medicine.

Core Messages

- Nuclear medicine is a transdisciplinary field whose basis is represented by physics and biology.
- SPECT and PET are the most common nuclear medicine imaging techniques, both based on γ-radiation detection.
- Hybrid SPECT/CT or PET/CT–PET/MR scanners have enhanced the diagnostic precision of nuclear medicine.

- The combination of morphology (CT/MR) and SPECT/PET functional and metabolic information allows unique consideration.
- Artificial intelligence, new radionuclides, technical improvements are paving the way for precision nuclear medicine.

References

1. Alavi A, Reivich M (2002) Guest editorial: the conception of FDG-PET imaging. Semin Nucl Med 32:2–5
2. Townsend DW, Cherry SR (2001) Combining anatomy and function: the path to true image fusion. Eur Radiol 11:1968–1974
3. Tagliabue L, Del Sole A (2014) Appropriate use of positron emission tomography with [(18)F] fluorodeoxyglucose for staging of oncology patients. Eur J Intern Med 25:6–11
4. Laudicella R, Baratto L, Minutoli F, Baldari S, Iagaru A (2020) Malignant cutaneous melanoma: updates in PET imaging. Curr Radiopharm 13:14–23
5. Laudicella R, Davidzon G, Vasanawala BS, Iagaru A (2019) 18F-FDG PET/MR refines evaluation in newly diagnosed metastatic urethral adenocarcinoma. Nucl Med Mol Imaging 53:296–299
6. Cherry SR, Sorenson JA, Phelps ME (2003) Physics in nuclear medicine, 3rd edn. Saunders, Philadelphia
7. Council Directive 2013/59/Euratom of 5 December (2013)
8. Seo Y, Mari C, Hasegawa BH (2008) Technological development and advances in single-photon emission computed tomography/computed tomography. Semin Nucl Med 38:177–198
9. Gleisner KS, Ljungberg M (2012) Patient-specific whole-body attenuation correction maps from a CT system for conjugate-view-based activity quantification: method development and evaluation. Canc Biother Radiopharm 27:10
10. Vaz SC, Oliveira F, Herrmann K, Veit-Haibach P (2020) Nuclear medicine and molecular imaging advances in the 21st century. Br J Radiol 93:20200095
11. Israel O, Pellet O, Biassoni L, De Palma D, Estrada-Lobato E, Gnanasegaran G, Kuwert T, la Fougere C, Mariani G, Massalha S, Paez D, Giammarile F (2019) Two decades of SPECT/CT—the coming of age of a technology: an updated review of literature evidence. Eur J Nucl Med Mol Imaging 46:1990–2012
12. Davila D, Antoniou A, Chaudhry MA (2015) Evaluation of osseous metastasis in bone scintigraphy. Semin Nucl Med 45:3–15
13. Giovanella L, D'Aurizio F, Campennì A, Ruggeri RM, Baldari S, Verburg FA, Trimboli P, Ceriani L (2016) Searching for the most effective thyrotropin (TSH) threshold to rule-out autonomously functioning thyroid nodules in iodine deficient regions. Endocrine 54: 757–761
14. Campenni' A, Giovinazzo S, Pignata SA, Di Mauro F, Santoro D, Curtò L, Trimarchi F, Ruggeri RM, Baldari S (2017) Association of parathyroid carcinoma and thyroid disorders: a clinicale review. Endocrine 56:19–26
15. Mordi IR, Badar AA, Irving RJ, Weir-McCall JR, Houston JG, Lang CC (2017) Efficacy of non-invasive cardiac imaging tests in diagnosis and management of stable coronary artery disease. Vasc Health Risk Manag 13:427–437

16. Minutoli F, Di Bella G, Mazzeo A, Laudicella R, Gentile L, Russo M, Vita G, Baldari S (2019) Serial scanning with 99mTc-3, 3-diphosphono-1, 2-propanodicarboxylic acid (99mTc-DPD) for early detection of cardiac amyloid deposition and prediction of clinical worsening in subjects carrying a transthyretin gene mutation. J Nucl Cardiol

17. Koulibaly PM, Nobili F, Migneco O, Vitali P, Robert PH, Girtler N, Darcourt J, Rodriguez G (2003) 99mTc-HMPAO and 99mTc-ECD perform differently in typically hypoperfused areas in Alzheimer's disease. Eur J Nucl Med Mol Imaging 30:1009–1013

18. Henry TR, Van Heertum RL (2003) Positron emission tomography and single photon emission computed tomography in epilepsy care. Semin Nucl Med 33:88–104

19. Seckin ZI, Whitwell JL, Utianski RL (2020) Ioflupane 123I (DAT scan) SPECT identifies dopamine receptor dysfunction early in the disease course in progressive apraxia of speech. J Neurol

20. Schillaci O, Spanu A, Tagliabue L, Filippi L, Danieli R, Palumbo B, Del Sole A, Madeddu G (2009) SPECT/CT with a hybrid imaging system in the study of lower gastrointestinal bleeding with technetium-99m red blood cells. Q J Nucl Med Mol Imaging 53:281–289

21. Parisi MT, Otjen JP, Stanescu AL, Shulkin BL (2018) Radionuclide imaging of infection and inflammation in children: a review. Semin Nucl Med 48:148–165

22. Caobelli F, Evangelista L, Quartuccio N, Familiari D, Altini C, Castello A, Cucinotta M, Di Dato R, Ferrari C, Kokomani A, Laghai I, Laudicella R, Migliari S, Orsini F, Pignata SA, Popescu C, Puta E, Ricci M, Seghezzi S, Sindoni A, Sollini M, Sturial L, Svyridenka A, Vergura V, Alongi P (2016) Role of molecular imaging in the management of patients affected by inflammatory bowel disease: state-of-the-art. World J Radiol 8:829–845

23. Roach PJ, Schembri GP, Baile L (2013) V/Q scanning using SPECT and SPECT/CT. J Nucl Med 54:1588–1596

24. Weber WA (2006) Positron emission tomography as an imaging biomarker. J Clin Oncol 24:3282±92

25. Boellaard R, Delgado-Bolton R, Oyen WJ, Giammarile F, Tatsch K, Eschner W, Verzijlbergen FJ, Barrington SF, Pike LC, Weber WA, Stroobants S, Delbeke D, Donohoe KJ, Holbrook S, Graham MM, Testanera G, Hoekstra OS, Zijlstra J, Visser E, Hoekstra CJ, Pruim J, Willemsen A, Arends B, Kotzerke J, Bockisch A, Beyer T, Chiti A, Krause BJ (2015) European association of nuclear medicine (EANM). FDG PET/CT: EANM procedure guidelines for tumour imaging: version 2.0. Eur J Nucl Med Mol Imaging 42:328–354

26. Alongi P, Laudicella R, Desideri I, Chiaravalloti A, Borghetti P, Queartuccio N, Fiore M, Evangelista L, Marino L, Caobelli F, Tuscano C, Mapelli P, Lancellotta V, Annunziata S, Ricci M, Ciurlia E, Fiorentino A (2019) Positron emission tomography with computed tomography imaging (PET/CT) for the radiotherapy planning definition of the biological target volume: PART 1. Crit Rev Oncol Hematol 140:74–79

27. Fiorentino A, Laudicella R, Ciurlia E, Annunziata S, Lancellotta V, Mapelli P, Tuscano C, Caobelli F, Evangelista L, Marino L, Quartuccio N, Fiore M, Borghetti P, Chiaravalloti A, Ricci M, Desideri I, Alongi P (2019) Positron emission tomography with computed tomography imaging (PET/CT) for the radiotherapy planning definition of the biological target volume: PART 2. Crit Rev Oncol Hematol 139:117–124

28. Kim JW, Dang CV (2006) Cancer's molecular sweet tooth and the Warburg effect. Cancer Res 66:8927–8930

29. Agrawal A, Rangarajan V (2015) Appropriateness criteria of FDG PET/CT in oncology. Indian J Radiol Imaging 25:88–101

30. Kostakoglu L, Cheson BD (2014) Current role of FDG PET/CT in lymphoma. Eur J Nucl Med Mol Imaging 41:1004–1027

31. Lopci E, Meignan M (2020) Current evidence on PET response assessment to immunotherapy in lymphomas. PET Clin 15:23–34

32. Albano D, Laudicella R, Ferro P, Allocca M, Abenavoli E, Buschiazzo A, Castellino A, Chiaravalloti A, Cuccaro A, Cuppari L, Durmo R, Evangelista L, Frantellizzi V, Kovalchuk S, Linguanti F, Santo G, Bauckneht M, Annunziata S (2019) The Role of 18F-FDG PET/CT in staging and prognostication of mantle cell lymphoma: an Italian multicentric study. Cancers (Basel) 11:1831

33. Quartuccio N, Evangelista L, Alongi P, Caobelli F, Altini C, Cistaro A, Lambertini A, Schiorlin I, Popescu CE, Linguanti F, Laudicella R, Scalorbi F, Di Pierro G, Asabella AN, Cuppari L, Margotti S, Lima GM, Scalisi S, Pacella S, Kokomani A, Ciaccio A, Sturiale L, Vento A, Cardile D, Baldari S, Panareo S, Fanti S, Rubini G, Schillaci O, Chiaravalloti A (2019) Prognostic and diagnostic value of [18F]FDG-PET/CT in restaging patients with small cell lung carcinoma: an Italian multicenter study. Nucl Med Commun 40:808–814

34. Qu X, Huang X, Yan W, Wu L, Dai K (2012) A meta-analysis of ^{18}FDG-PET-CT, ^{18}FDG-PET, MRI and bone scintigraphy for diagnosis of bone metastases in patients with lung cancer. Eur J Radiol 81:1007–1015

35. Mac Manus MP, Hicks RJ, Matthews JP, McKenzie A, Rischin D, Salminen EK, Ball DL (2003) Positron emission tomography is superior to computed tomography scanning for response-assessment after radical radiotherapy or chemoradiotherapy in patients with non-small-cell lung cancer. J Clin Oncol 21:1285–1292

36. Groheux D, Espié M, Giacchetti S, Hindié E (2013) Performance of FDG PET/CT in the clinical management of breast cancer. Radiology 266:388–405

37. Mapelli P, Picchio M (2015) Initial prostate cancer diagnosis and disease staging–the role of choline-PET-CT. Nat Rev Urol 12:510–518

38. Laudicella R, Albano D, Alongi P, Argiroffi G, Bauckneht M, Baldari S, Bertagna F, Boero M, De Vincentis G, Sole AD, Rubini G, Fantechi L, Frantellizzi V, Ganduscio G, Guglielmo P, Nappi AG, Evangelista L (2019) (18)F-Facbc in prostate cancer: a systematic review and meta-analysis. Cancers (Basel) 11:1348

39. Baratto L, Duan H, Laudicella R, Toriihara A, Hatami N, Ferri V, Iagaru A (2020) Physiological 68Ga-RM2 uptake in patients with biochemically recurrent prostate cancer: an atlas of semiquantitative measurements. Eur J Nucl Med Mol Imaging 47:115–122

40. Song H, Harrison C, Duan H, Guja K, Hatami N, Franc BL, Moradi F, Aparici CM, Davidzon GA, Iagaru A (2020) Prospective evaluation of 18F-DCFPyL PET/CT in biochemically recurrent prostate cancer in an academic center: a focus on disease localization and changes in management. J Nucl Med 61:546–551

41. Müller J, Ferraro DA, Muehlematter UJ, Schuler HIG, Kedzia S, Eberli D, Guckenberger M, Kroeze SGC, Sulser T, Schmid DM, Omlin A, Muller A, Zilli T, John H, Kranzbuelher H, Kaufmann P, von Schulthess GK, Burger IA (2019) Clinical impact of 68Ga-PSMA-11 PET on patient management and outcome, including all patients referred for an increase in PSA level during the first year after its clinical introduction. Eur J Nucl Med Mol Imaging 46:889–900

42. Abdalla G, Hammam A, Anjari M, D'arco F, Bisdas S (2020) Glioma surveillance imaging: current strategies, shortcomings, challenges and outlook. Br J Radiol

43. Laudicella R, Albano D, Annunziata S, Calabrò D, Argiroffi G, Abenavoli E, Linguanti F, Albano D, Vento A, Bruno A, Alongi P, Bauckneht M (2019) Theragnostic use of radiolabelled dota-peptides in meningioma: from clinical demand to future applications. Cancers (Basel) 11:1412

44. Laudicella R, Iagaru A, Minutoli F, Gaeta M, Baldari S, Bisdas S (2020) PET/MR in neuro-oncology: is it ready for prime-time? Clin Transl Imaging

45. Czernin J, Satyamurthy N, Schiepers C (2010) Molecular mechanisms of bone 18F-NaF deposition. J Nucl Med 51:1826–1829

46. Costelloe CM, Chuang HH, Madewell JE (2014) FDG PET/CT of primary bone tumors. AJR Am J Roentgenol 202:W521–W531

47. Moradi F, Jamali M, Barkhodari A, Schneider B, Chin F, Quon A, Mittra ES, Iagaru A (2016) Spectrum of 68Ga-DOTA TATE uptake in patients with neuroendocrine tumors. Clin Nucl Med 41:e281–e287
48. Sarikaya I (2015) Cardiac applications of PET. Nucl Med Commun 36:971–985
49. Minutoli F, Di Bella G, Vita G, Laudicella R, Bogaert J, Baldari S (2017) Non-invasive cardiac imaging in patients with systemic amyloidosis: a practical approach with emphasis on clinical contribution of bone-seeking radiotracers. Clin Transl Imaging 5:545–559
50. Laudicella R, Minutoli F, Baldari S (2019) Prognostic insights of molecular imaging in cardiac sarcoidosis. J Nucl Cardiol
51. Hellwig S, Domschke K, Meyer PT (2019) Update on PET in neurodegenerative and neuroinflammatory disorders manifesting on a behavioural level: imaging for differential diagnosis. Curr Opin Neurol 32:548–556
52. Xia C, Dickerson BC (2017) Multimodal PET imaging of amyloid and Tau pathology in Alzheimer disease and non-Alzheimer disease dementias. PET Clin 12:351–359
53. Rice L, Bisdas S (2017) The diagnostic value of FDG and amyloid PET in Alzheimer's disease—a systematic review. Eur J Radiol 94:16–24
54. Sconfienza LM, Signore A, Cassar-Pullicino V, Cataldo MA, Gheysens O, Borens O, Trampuz A, Wortler K, Petrosillo N, Winkler H, Vanhoenacker FMHM, Jutte PC, Glaudemans AWJM (2019) Diagnosis of peripheral bone and prosthetic joint infections: overview on the consensus documents by the EANM, EBJIS, and ESR (with ESCMID endorsement). Eur Radiol 29:6425–6438
55. Slart RHJA (2018) FDG-PET/CT(A) imaging in large vessel vasculitis and polymyalgia rheumatica: joint procedural recommendation of the EANM, SNMMI, and the PET Interest Group (PIG), and endorsed by the ASNC. Eur J Nucl Med Mol Imaging 45:1250–1269
56. Jamar F, Buscombe J, Chiti A, Christian PE, Delbeke D, Donohoe KJ, Israel O, Martin-Comin J, Signore A (2013) EANM/SNMMI guideline for 18F-FDG use in inflammation and infection. J Nucl Med 54:647–658
57. Seidlin SM, Marinelli LD, Oshry E (1946) Radioactive iodine therapy; effect on functioning metastases of adenocarcinoma of the thyroid. J Am Med Assoc 132:838–847
58. Doan CA, Wiseman BK (1947) Radioactive phosphorus, p32; a six-year clinical evaluation of internal radiation therapy. J Lab Clin Med 32:943–69 (For this there is no more info about other authors on Pubmed)
59. Marcu LG, Bezak E, Filip SM (2012) The role of PET imaging in overcoming radiobiological challenges in the treatment of advanced head and neck cancer. Cancer Treat Rev 38:185–193
60. Mulford DA, Scheinberg DA, Jurcic JG (2005) The promise of targeted {alpha}-particle therapy. J Nucl Med 1:199S-204S
61. Pouget JP, Lozza C, Deshayes E, Boudousq V, Navarro-Teulon I (2015) Introduction to radiobiology of targeted radionuclide therapy. Front Med (Lausanne) 2:12
62. Allen GS, Frank J (2007) Structural insights on the translation initiation complex: ghosts of a universal initiation complex. Mol Microbiol 63:941–950
63. Amato E, Campennì A, Leotta S, Ruggeri RM, Baldari S (2016) Treatment of hyperthyroidism with radioiodine targeted activity: a comparison between two dosimetric methods. Phys Med 32:847–853
64. Campennì A, Giovanella L, Pignata SA, Violi MA, Siracusa M, Alibrandi A, Moleti M, Amato E, Ruggeri RM, Vermiglio F, Baldari S (2015) Thyroid remnant ablation in differentiated thyroid cancer: searching for the most effective radioiodine activity and stimulation strategy in a real-life scenario. Nucl Med Commun 36:1100–1106
65. Campennì A, Amato E, Laudicella R, Alibrandi A, Cardile D, Pignata SA, Trimarchi F, Ruggeri RM, Auditore L, Baldari S (2019) Recombinant human thyrotropin (rhTSH) versus Levo thyroxine withdrawal in radioiodine therapy of differentiated thyroid cancer patients: differences in abdominal absorbed dose. Endocrine 65:132–137

66. Iagaru A, Peterson D, Quon A, Dutta S, Twist C, Daghighian F, Gambhir SS, Albanese C (2008) 123I MIBG mapping with intraoperative gamma probe for recurrent neuroblastoma. Mol Imaging Biol 10:19–23

67. Iagaru A, Mittra ES, Ganjoo K, Knox SJ, Goris ML (2010) 131I-Tositumomab (Bexxar) versus 90Y-Ibritumomab (Zevalin) therapy of low-grade refractory/relapsed non-Hodgkin lymphoma. Mol Imaging Biol 12:198–203

68. Iagaru A, Gambhir SS, Goris ML (2008) 90Y-ibritumomab therapy in refractory non-Hodgkin's lymphoma: observations from 111In-ibritumomab pretreatment imaging. J Nucl Med 49:1809–1812

69. Silberstein EB, Eugene L, Saenger SR (2001) Painful osteoblastic metastases: the role of nuclear medicine. Oncology (Williston Park) 15:157–163

70. Picciotto M, Franchina T, Russo A, Ricciardi GRR, Provazza G, Sava S, Baldari S, Caffo O, Adamo V (2017) Emerging role of Radium-223 in the growing therapeutic armamentarium of metastatic castration-resistant prostate cancer. Expert Opin Pharmacother 18:899–908

71. Wu L, Shen F, Xia Y, Yang YF (2016) Evolving role of radiopharmaceuticals in hepatocellular carcinoma treatment. Anticancer Agents Med Chem 16:1155–1165

72. Langbein T, Weber WA, Eiber M (2019) Future of theranostics: an outlook on precision oncology in nuclear medicine. J Nucl Med 60:13S-19S

73. Sherman M, Levine R (2019) Nuclear medicine and wall street: an evolving relationship. J Nucl Med 60:20S-24S

74. Minutoli F, Amato E, Sindoni A, Cardile D, Conti A, Herberg A, Baldari S (2014) Peptide receptor radionuclide therapy in patients with inoperable meningiomas: our experience and review of the literature. Cancer Biother Radiopharm 29:193–199

75. Bodei L, Kidd M, Prasad V, Modlin IM (2015) Peptide receptor radionuclide therapy of neuroendocrine tumors. Front Horm Res 44:198–215

76. Giuffrida G, Ferraù F, Laudicella R, Cotta OR, Messina E, Granata F, Angileri FF, Vento A, Alibrandi A, Baldari S, Cannavò S (2019) Peptide receptor radionuclide therapy for aggressive pituitary tumors: a monocentric experience. Endocr Connect 8:528–535

77. Prasad V, Srirajaskanthan R, Toumpanakis C, Grana CM, Baldari S, Shah T, Lamarca A, Courbon F, Scheidhauer K, Baudin E, Thanh XMT, Houchard A, Dromain C, Bodei L (2020) Lessons from a multicentre retrospective study of peptide receptor radionuclide therapy combined with lanreotide for neuroendocrine tumours: a need for standardised practice. Eur J Nucl Med Mol Imaging

78. Kratochwil C, Fendler WP, Eiber M, Baum R, Bozkurt MF, Czernin J, Delgado Bolton RC, Ezzidin S, Forrer F, Hicks RJ, Hope TA, Kabasakal L, Konijnenberg M, Kopka K, Lassmann M, Mottaghy FM, Oyen W, Rahbar K, Schoder H, Virgolini I, Wester HJ, Bodei L, Fanti S, Haberkorn U, Herrmann K (2019) EANM procedure guidelines for radionuclide therapy with 177Lu-labelled PSMA-ligands (177Lu-PSMA-RLT). Eur J Nucl Med Mol Imaging 46:2536–2544

79. Lindner T, Loktev A, Altmann A, Giesel F, Kratochwil C, Debus J, Jager D, Mier W, Haberkorn U (2018) Development of quinoline-based theranostic ligands for the targeting of fibroblast activation protein. J Nucl Med 59:1415–1422

80. Reinders MTM, Smits MLJ, van Roekel C, Braat AJAT (2019) Holmium-166 microsphere radioembolization of hepatic malignancies. Semin Nucl Med 49:237–243

81. Kratochwil C, Bruchertseifer F, Giesel FL, Weis M, Verburg FA, Mottaghy F, Kopka K, Apostolidis C, Haberkorn U, Morgenstern A (2016) 225Ac-PSMA-617 for PSMA-targeted α-radiation therapy of metastatic castration-resistant prostate cancer. J Nucl Med 57:1941–1944

82. Morgenstern A, Apostolidis C, Kratochwil C, Sathekge M, Krolicki L, Bruchertseifer F (2018) An overview of targeted alpha therapy with 225Actinium and 213Bismuth. Curr Radiopharm 11:200–208

83. El Fakhri G, Surti S, Trott CM, Scheurmann J, Karp JS (2011) Improvement in lesion detection with whole-body oncologic time-of-flight PET. J Nucl Med 52:347–353

84. Nguyen NC, Vercher-Conejero JL, Sattar A, Miller MA, Maniawski PJ, Jordan DW, Muzic RF Jr, Su KH, O'Donnell JK, Faulhaber PF (2015) Image quality and diagnostic performance of a digital PET prototype in patients with oncologic diseases: initial experience and comparison with analog PET. J Nucl Med 56:1378–1385

85. Baratto L, Park SY, Hatami N, Davidzon G, Srinivas S, Gambhir SS, Iagaru A (2017) 18F-FDG silicon photomultiplier PET/CT: a pilot study comparing semiquantitative measurements with standard PET/CT. PLoS One 12:e0178936

86. Hsu DF, Ilan E, Peterson WT, Uribe J, Lubberink M, Levin CS (2017) Studies of a next generation silicon-photomultiplier-based time-of-flight PET/CT system. J Nucl Med 58:1511–1518

87. Kjær A, Torigian DA (2016) Clinical PET/MR imaging in oncology: future perspectives. PET Clin 11:489–493

88. Kwatra NS, Lim R, Gee MS, States LJ, Vossough A, Lee EY (2019) PET/MR imaging: current updates on pediatric applications. Magn Reson Imaging Clin N Am 27:387–407

89. Sałyga A, Guzikowska-Ruszkowska I, Czepczyński R, Ruchala M (2016) PET/MR—a rapidly growing technique of imaging in oncology and neurology. Nucl Med Rev Cent East Eur 19:37–41

90. Wehrl HF, Sauter AW, Divine MR, Pichler BJ (2015) Combined PET/MR: a technology becomes mature. J Nucl Med 56:165–168

91. Aktolun C (2019) Artificial intelligence and radiomics in nuclear medicine: potentials and challenges. Eur J Nucl Med Mol Imaging 46:2731–2736

92. Bisdas S, Shen H, Thust S, Katsaros V, Stranjalis G, Boskos C, Brandner S, Zhang J (2018) Texture analysis- and support vector machine-assisted diffusional kurtosis imaging may allow in vivo gliomas grading and IDH-mutation status prediction: a preliminary study. Sci Rep 8:6108

93. Alongi P, Laudicella R, Stefano A, Caobelli F, Comelli A, Vento A, Sardina D, Ganduscio G, Toia P, Ceci F, Mapelli P, Picchio M, Midiri M, Baldari S, Lagalla R, Russo G (2020) Choline PET/CT features to predict survival outcome in high risk prostate cancer restaging: a preliminary machine-learning radiomics study QJNM

94. Laudicella R, Comelli A, Stefano A, Szostek M, Croce L, Vento A, Spataro A, Comis AD, La Torre F, Gaeta M, Baldari S, Alongi P (2020) Artificial neural networks in cardiovascular diseases and its potential for clinical application in molecular imaging. Curr Radioph

95. Quartuccio N, Laudicella R, Mapelli P, Guglielmo P, Pizzuto DA, Boero M, Arnone G, Picchio M (2020) Hypoxia PET imaging beyond 18F-FMISO in patients with high-grade glioma: 18F-FAZA and other hypoxia radiotracers. Clin Transl Imaging 8:11–20

96. Shooli H, Dadgar H, Wáng YJ, Vafaee MS, Kashuk SR, Nemati R, Jafari E, Nabipour I, Gholamrezanezhad A, Assadi M, Larvie M (2019) An update on PET-based molecular imaging in neuro-oncology: challenges and implementation for a precision medicine approach in cancer care. Quant Imaging Med Surg 9:1597–1610

97. Waaijer SJ, Giesen D, Ishiguro T, Sano Y, Sugaya N, Schroder CP, de Vries EG, Lub-de Hooge MN (2020) Preclinical PET imaging of bispecific antibody ERY974 targeting CD3 and glypican 3 reveals that tumor uptake correlates to T cell infiltrate. J Immunother Cancer 8:e000548

98. Beinat C, Patel CB, Haywood T, Shen B, Naya L, Gandhi H, Holley D, Khalighi M, Iagaru A, Davidzon G, Gambhir SS (2020) Human biodistribution and radiation dosimetry of [18F] DASA-23, a PET probe targeting pyruvate kinase M2. Eur J Nucl Med Mol Imaging

99. Koole M, Lohith TG, Valentine JL, Bennacef I, Declercq R, Reynders T, Riffel K, Celen S, Serdons K, Bormans G, Ferry-Martin S, Laroque P, Walji A, Hostetler ED, Briscoe RJ, de Hoon J, Sur C, Van Laere K, Struyk A (2020) Preclinical safety evaluation and human dosimetry of [18F]MK-6240, a novel PET tracer for imaging neurofibrillary tangles. Mol Imaging Biol 22:173–180

100. Cicone F, Denoël T, Gnesin S, Riggi N, Irving M, Jakka G, Schaefer N, Viertl D, Coukos G, Prior JO (2020). Preclinical evaluation and dosimetry of [111In] CHX-DTPA-scFv78-Fc targeting endosialin/tumor endothelial marker 1 (TEM1). Mol Imaging Biol
101. Ilhan H, Lindner S, Todica A, Cyran CC, Tiling R, Auernhammer CJ, Spitzweg C, Boeck S, Unterrainer M, Gildehaus FJ, Boning G, Jurkschat K, Wangler C, Wangler B, Schirrmacher R, Bartenstein P (2020) Biodistribution and first clinical results of 18F-SiFAlin-TATE PET: a novel 18F-labeled somatostatin analog for imaging of neuroendocrine tumors. Eur J Nucl Med Mol Imaging 47:870–880
102. Carlos Dos Santos J, Beijer B, Bauder-Wüst U, Schafer M, Leotta K, Eder M, Benesova M, Kleist C, Giesel F, Kratochwil C, Kopka K, Haberkorn U, Mier W (2020) Development of novel PSMA ligands for imaging and therapy with copper isotopes. J Nucl Med 2020 (61):70–79
103. Werner RA, Kircher S, Higuchi T, Kircher M, Schirbel A, Wester HJ, Buck AK, Pomper MG, Rowe SP, Lapa C (2019) CXCR4-directed imaging in solid tumors. Front Oncol 9:770
104. Frantellizzi V, Conte M, Pontico M, Pani A, Pani R, De Vincentis G (2020) New frontiers in molecular imaging with superparamagnetic iron oxide nanoparticles (SPIONs): efficacy, toxicity, and future applications. Nucl Med Mol Imaging 54:65–80
105. Sanders VA, Cutler CS (2020) Radioarsenic: a promising theragnostic candidate for nuclear medicine. Nucl Med Biol S0969–8051(20):30063–30069
106. Rahmati S, Shojaei F, Shojaeian A, Rezakhani L, Dehkordi MB (2020) An overview of current knowledge in biological functions and potential theragnostic applications of exosomes. Chem Phys Lipids 226:104836
107. Zhou YY, Zhang PP, Lin RT, Chen TS, Liu XY, Liu WJ, Liang YN, Chen S, Pan X, Ni GY, Wang TF, Liu XS, Yuan JW (2020) Investigating the theragnostic potential of 131I-caerin peptide in thyroid cancer. Hell, J Nucl Med 23:27–33

Sergio Baldari is a Full Professor of nuclear medicine and the Head of the Department of Biomedical and Dental Sciences and Morphofunctional Imaging at the University of Messina. He is also the Head of the Unit of Nuclear Medicine at A.O.U. "Policlinico G. Martino" in Messina and has been the Head of the Training School in nuclear medicine for 20 years. His main current scientific interests include diagnosing and treating neuroendocrine tumors and treating bone metastases with beta- and alfa-emitters. Further, he has been the Head of the Italian Association of Nuclear Medicine (AIMN) Educational Center of excellence "Diagnosis and Therapy of Neuroendocrine Tumors" of Messina. Sergio serves as an expert for research project evaluation in the Italian Ministry of Health and the Italian Ministry of University and Research.

Riccardo Laudicella is a Nuclear Medicine Consultant at the Fondazione Istituto G. Giglio in Cefalù. He is also a Ph. D. student in "Bioengineering Applied to Medical Sciences" at the University of Messina, and he is a fellow in Zurich (USZ and ETH) in "hybrid imaging of genito-urinary cancer, internal radiotherapy, and glucose metabolism." He is graduated from the University of Palermo, where he spent six months at the Ovidius University in Constanta (Romania) for the Erasmus project. He completed an internship in the Radiology Department of Palermo with a bachelor thesis about "Morphovolumetric Studies in the Neuroradiological Field." Then, he completed the 4-year Nuclear Medicine Residency Program at the University of Messina with a bachelor thesis about "Applications of Radiomics on 18F-Choline PET/CT Images". During his Residency, he spent four months as Visiting Trainee and Researcher in the Division of Nuclear Medicine and Molecular Imaging at Stanford University (2018) and in the Nuclear Medicine Unit of the Fondazione Istituto G. Giglio, Cefalù (2019). Riccardo is also the Head of the Italian Association of Young Nuclear Medicine Physician/Residents, and his main interests are artificial intelligence applications, nuclear medicine therapy, and PET/MR.

Beyond the Borders of Dentistry: Interprofessional and Interdisciplinary Approach to Oral Health Promotion

12

Mohammad R. Khami, Morenike Oluwatoyin Folayan, Armando E. Soto-Rojas, Heikki Murtomaa, Prathip Phantumvanit, and Farid Farrokhi

> *"Some tortures are physical and some are mental, but the one that is both is dental."*
>
> Ogden Nash

Summary

Oral health is an important but usually neglected health issue. Such as other aspects of health, it is under the influence of several factors at various levels. Thus, promoting the population's oral health requires an interdisciplinary and multi-dimensional approach. This chapter defines oral health, and its relation to general health will be presented. It reviews disciplines that contribute to oral health promotion. Then, the concepts of oral health promotion will be reviewed, and finally, the teamwork approach in the oral healthcare delivery system will be emphasized.

M. R. Khami (✉)
Research Center for Caries Prevention, Dentistry Research Institute, Tehran University of Medical Sciences, No. 1417614411, Quds Street, Tehran, Iran
e-mail: mkhami@tums.ac.ir

Integrated Science Association (ISA), Universal Scientific Education and Research Network (USERN), Tehran, Iran

M. O. Folayan
Department of Child Dental Health, Obafemi Awolowo University, Ile-Ife, Nigeria
e-mail: toyinukpong@oauife.edu.ng

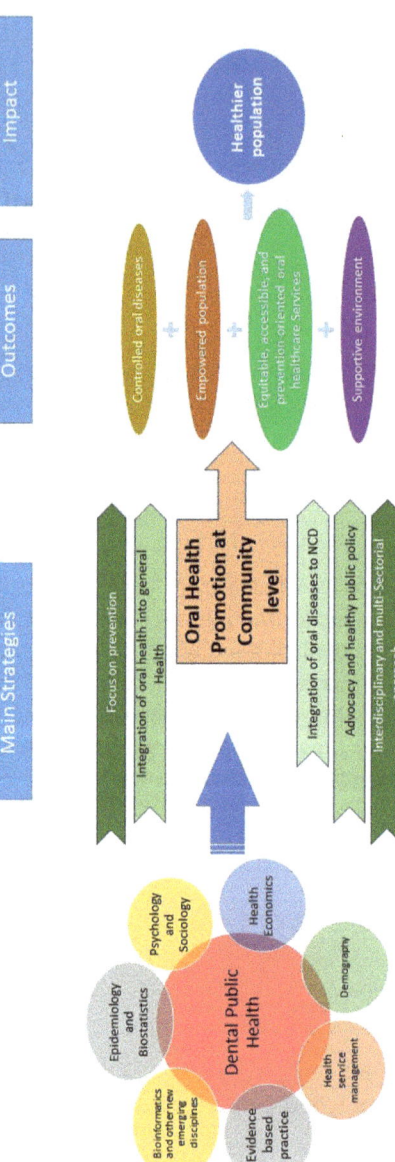

A schematic illustration to show how the implementation of interdisciplinary and interprofessional approaches will help to promote the population's oral and general health.

The code of this chapter is *01110010 01,101,101 01,101,110 01,101,111 01,101,111 01,101,001 01,101,111 01,110,100 01,010,000.*

Keywords

Oral health care · Oral health promotion · Interdisciplinary approach · Social determinants of health · Interprofessional training · Preventive dentistry

1 Introduction

Ideally, health should be the "complete physical, mental and social well-being and not merely the absence of disease" [1]. It is, however, practically defined as "the extent to which an individual or group is able to realize aspirations and satisfy needs and to change or cope with the environment. It is a resource for everyday life, not the objective of living; it is a positive concept, emphasizing social and personal resources, as well as physical capacities" [2]. That multi-dimensional concept includes physical, mental, social, emotional, spiritual, and environmental health [3].

The health and disease models have jettisoned the compartmentalized biomedical approach and embraced the integrated public health approach [4]. As acknowledged by the integrated public health approach, oral health is crucial to general health, well-being, and life quality. Oral health is a concept that embraces different dimensions, including our capability to *speak, smile, touch, smell, taste, chew, and swallow* and express feelings by facial expressions without pain and discomfort [5]. Our mouth can show signs of nutritional deficiencies and infections. Some systemic diseases can first be diagnosed in the oral cavity [6]. Not surprisingly, the Iranian traditional medicine described the tongue as the "mirror of the body."

A. E. Soto-Rojas
Cariology, Operative Dentistry and Dental Public Health Department, Indiana University School of Dentistry, Indianapolis, US
e-mail: arsoto@iu.edu

H. Murtomaa
Oral Public Health, University of Helsinki, Helsinki, Finland
e-mail: heikki.murtomaa@helsinki.fi

P. Phantumvanit
Faculty of Dentistry, Thammasat University, Bangkok, Thailand
e-mail: prathprathip@tu.ac.th

F. Farrokhi
Community Oral Health Department, Tehran University of Medical Sciences, Tehran, Iran
e-mail: f-farrokhi@razi.tums.ac.ir

There is a heavy connection between oral and general health. The connection of oral and non-communicable diseases through inflammation can be seen in different pieces of literature. The low-grade chronic inflammation resulting from poor oral hygiene is associated with chronic oxidative stress causing molecular mimicry that triggers cancers, diabetes, osteoporosis, and neurodegenerative disease [7] by inducing the production of inflammatory mediators, importantly tumor necrosis factor-alpha (TNF-α), interleukin-1 (IL-1), and interleukin-6 (IL-6) [8]. The inflammatory pathway partly explains the bidirectional relationship between chronic periodontitis, insulin resistance, and type two diabetes [9]. Chronic periodontitis also raises the circulating inflammatory cytokines and changes the serum albumin and c-reactive protein levels. It might cause hypoalbuminemia, altering the mechanism of the dialysis treatment, and worsening kidney diseases [10]. Increased concentrations of pro-inflammatory cytokines are also a result of dental caries [11] that activate a set of inflammatory processes [12–14] connected with damaged growth [15, 18] and mechanisms that cause fatness [12, 19]. There are other links between oral and general health. For example, tooth loss is associated with nutritional deficiencies, obesity, and nutritional intake deficiency in adults and elderlies, while it correlates with malnutrition and growth retardation in children [20, 22].

There are links recognized in periodontal disease with self-reported hypertension [23, 24], poor oral hygiene with low sperm count [25], and chronic periodontitis with a longer time to conception [26, 27]. But some of these links are hard to describe [28]. Periodontal microbiota increases the carotid intima-media thickness, leading to subclinical atherosclerosis and cardiovascular disease [29].

Oral health status can also be considered a sign of poor health and mortality in middle-aged men [30, 31]. The risk of cause-specific mortality and all-cause mortality was higher in men with a low level of oral hygiene, which was worse for edentulous people [31]. Hence, we can consider oral health status an early sign of lifestyle, risk-taking, and health-compromising behaviors related to all-cause mortality. Less attention is paid to evaluating oral health in patients with systemic diseases, which will result in underestimating oral health's impact on general health [32].

On the other hand, there are common risk factors identified for oral and systemic disorders. For children, high sugar consumption is a common risk factor for obesity and caries. Children with caries also have higher concentrations of pro-inflammatory cytokines, which are mediators of inflammation, infection, and immunological processes. The pro-inflammatory cytokines produce free radicals followed by the release of peroxides, prostaglandin E2, IL-6, TNF-α, and cysteine leukotrienes, which correlate with a higher risk for obesity, an inflammatory disease [33]. Furthermore, caries-related bacteria may increase the concentration of TNF-α and lipopolysaccharides, reduce the concentration of adiponectin, and induce insulin resistance, thereby increasing the risk for obesity. Cytokines produced by activated monocytes and macrophages in obese patients play a role in infection and inflammatory diseases and correlate with caries experience [34].

For young people, oral, mental, sexual, and conceptive well-being are connected. Mental issues are among the main reasons for disability in youths. Depression is one of the top three reasons for years lived with disability in adolescents [35]. Mental issues such as anxiety, stress, and depression correlate with decreased saliva pH and xerostomia, contributing to increased caries risk, poor dental hygiene, and gum and periodontal diseases [36]. Poor mental health also increases the risk for substance abuse associated with increased risk for oral cancer and dental health issues [37]. In addition, the prevalence of sexual violence faced by adolescents, especially female adolescents, result in posttraumatic stress disorders associated with low self-esteem and a high risk of caries [38]. Also, the increasing independence of adolescents paves the way for high free sugar intakes that correlate with increased risk of caries as well as mental health issues, e.g., attention deficit hyperactivity disorder (ADHD) [39] and suicidal ideation [40, 41]. Orientation toward suicide is connected to psychological and mental issues in adolescents. Such issues often lie in poor living conditions and poverty. These conditions increase the risk for caries and sexual and reproductive health challenges, too [42, 43]. For adults, tobacco and alcohol are common risk factors for oral cancer as well as lung and oropharyngeal cancer. The changes induced by tobacco use and alcohol in oral microflora correspond to developing and/or progressing dental caries [44]. Periodontal pockets, attrition, and loss of attachment are increased with the duration of tobacco use [45].

2 Social Determinants of Health

"I wrote a song about dental floss, but did anyone's teeth get cleaner?"

Frank Zappa

Oral health socially depends on factors related to the conditions "where people are born, grow, live, work, and age" [46]. These factors determine the profile of oral health. The relationship between these factors and oral health is not simple. It is important for global health as they affect how individuals access and utilize oral healthcare services. They also affect an individual's oral health and, therefore, his or her life quality [47]. The social determinants of oral health influence the socioeconomic status of people in society and determine exposure to everyday life circumstances known to have health advantages/disadvantages [48]. Of concern is the possible cumulative impact of the social determinants of oral health on the accumulated lifetime trajectories of health across the life course, which can be transferred across generations [49]. Many oral diseases are in relation to socioeconomic status, educational fulfillment, job status, household income, housing, and physical/mental health. In poorer areas, living in affluent neighborhoods has been associated with general health—and oral health—promoting effects in children and young people [50].

More recently, empirical evidence has shown the huge impact of social and structural factors on oral health at the macro-level. A direct association was found between the risk of caries in preschool children and an increased proportion of residents living in slums and below the poverty line [51], a higher percentage of females participating in decisions related to visiting family, relatives, and friends, and more women living under 50% of median income [52]. The risk of caries in preschool children is also associated with the governance and political systems and structures. The latter is, in turn, represented in the gross national income per capita for women and their work status, importantly, the employment rate of women, their chance to take part as leaders, legislators, top officials, and managers, and their ability to take time from work to care for their neonates [53]. A higher prevalence of caries in infants and toddlers has been observed in countries with a higher percentage of females participating in their own healthcare decisions [52].

The prevalence of early childhood caries (ECC) is also associated with access to universal health coverage (UHC) [54]. In addition, the amount of governance, employment ratio, gross domestic product (GDP) per capita, human development index, income inequality, type of welfare regime, type of political regime, government expenditure on health, and out-of-pocket (private) health expenditure by citizens of any country correlates with oral health in children [55]. The environment also poses a risk for oral health. The risk for caries in preschool children may be higher with more economic growth, industrialization, and urbanization. On the other hand, a vital economic ecosystem can make more rational decisions considering resources and lifestyle behaviors to promote health and prevent ECC as well [56].

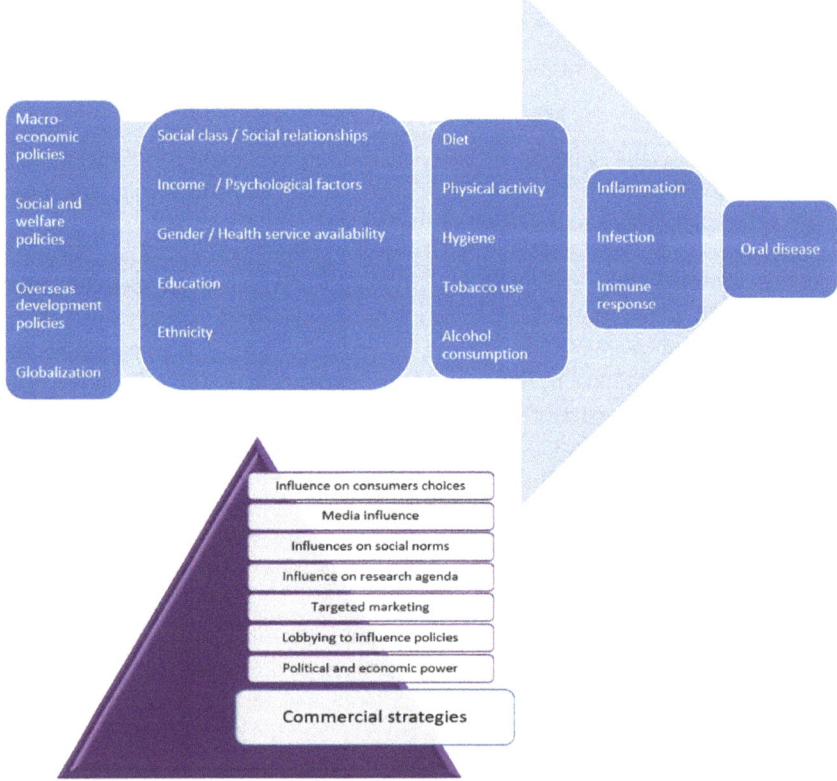

Fig. 1 Conceptual framework merging the social and commercial determinants of oral health [58, 59]

Commercial activities can also affect oral health through four ways by marketing tactics to:

i. Improve product acceptability;
ii. Design desired public health policies;
iii. Increase producer's acceptability through sponsorship of public events, like sports, and initiation of creative healthcare practices; and
iv. Extend supply chains internationally [57].

Figure 1 is a framework of oral health that shows the convergence of social and commercial factors and their interactions [58, 59].

The possible impact of social determinants on oral health has been recognized for some time. Still, there is a need to accelerate policymaking and application to deal with these determinants [60] by evaluating the impact of modifying these social factors like education, urban planning, community development, housing, income, and occupation on oral health [61]. The concept of social determinants of health clearly shows that oral health promotion cannot be confined to dentistry borders. The interdisciplinary approach to oral healthcare management is important to make the needed changes.

3 Multidisciplinary Approach to Oral Health Care

Dental public health has been known and established as an independent discipline. However, it is a wide-ranging topic that looks to intensify the focus of the dental profession on some factors that affect oral health and are influential in preventing and treating oral health problems [4]. Some of the sciences and disciplines implicated in dental public health are discussed below.

3.1 Epidemiology and Oral Health

The dentistry practice needs to be multidisciplinary in its approach to address the multiple etiological factors and the multi-dimensional impact of oral diseases. Information about the epidemiology of oral disease helps policymakers and oral health programmers understand the history, etiology, effects, and impact of oral health diseases. The evidence generated through epidemiological studies is also important for optimal clinical decisions about diagnosing and treating oral diseases [4, 62]. In their way toward oral health promotion, oral epidemiology and biostatistics walk hand in hand, as biostatistics provides the possibility of interpretation of the data as accurately as possible.

3.2 Psychology and Oral Health

The role and importance of psychology in the manifestation of diseases have expanded over the years. Individuals' mental health affects the adoption of safe oral health behaviors. Several studies reveal the place and role of psychosocial factors in the etiology of oral diseases. For instance, an individual's social support, defined as the "network of family, friends, neighbors, and community members that is available in times of need to give psychological, physical, and financial help" [63], enhances "resilience to stress, help protect against developing trauma-related psychopathology, decrease the functional consequences of trauma-induced disorders such as posttraumatic stress disorder, and reduce medical morbidity and mortality" [63]. Individuals with rich social networks are less likely to be engaged in risky unhealthy behaviors, have negative appraisals, and fail to adhere to treatment. Low levels of social support are associated with increased disease-related morbidity and mortality [64]. It has been shown that access to social support from their neighboring environment of immigrants in the form of structural or functional support plays an important role in promoting their oral health [65]. On the other hand, parental access to social support does not significantly help improve children's oral health [66].

Psychosocial factors are associated with oral health outcomes. More precisely, depression is associated with reduced energy and motivation, causing oral health neglect and a higher risk of dental caries and periodontal disease [67]. The use of antidepressants is also associated with xerostomia, known to increase the risk of caries [68]. Another important factor is stress, which has a major influence on mood, well-being, behavior, and health [69]. Stress increases the risk for teeth grinding, periodontal disease, xerostomia, and candida infection [70]. Parenting stress is also associated with an increased risk for maternal caries [71]. General anxiety and dental anxiety are also associated with oral health problems [72, 73]. Moreover, adverse childhood experiences [74–76] are associated with an increased risk for poor oral health in children and adolescents.

Other important psychosocial factors that act as moderators or mediators of factors associated with oral health include having a high internal locus of control [77, 78] associated with more frequent daily tooth brushing and regular dental check-ups [79]. It is an important factor in designing personalized oral health improvement programs [80]. Other factors are a sense of coherence, which is a resource for coping with stressors that improve the quality of life [81, 82] and oral health [83–86]. The executive function is described as a group of hierarchical mental processes needed when an individual needs concentration, instinct, or intuition [87] and reduces the risk for caries [88]. Self-esteem is an important health and oral health promotional factor, especially in adolescents [89–91], which is also associated with an increased risk for poor dental hygiene and caries [38]. Resilience, a multi-dimensional construct of individual characteristics, a process or a result [92], also influences oral health [38].

3.3 Sociology and Oral Health

Oral health is an essential matter for physical and social functioning. The mouth is remarked as "the boundary between the internal body and the external sources of pollution" [93], by which we could have food and drink, and is a tool through which we express ourselves [94, 95]. Oral health, the mouth, and teeth appearance play a role in an individual's social interactions [96]. The position of the mouth in the face makes it highly visible; therefore, it impacts the outer appearance, to which other individuals react to us [97–100]. Especially, the appearance of teeth is essential and may affect the judgment about people [101, 102]; although such factors will shape these judgments as cultural, social, and historical factors [102].

One of the most potent influences of sociology on oral health is the study of cultures and oral health. Culture, an experience that is learned, shared, and transmitted, can influence diet, oral self-care habits, and understandings of risk from oral diseases. Culture influences the concept of disease etiology and cure. For many, the cause of a disease is physical and/or spiritual (breach of a taboo, evil eye, past sin, spiritual intrusion, the wrath of God) [103]. Cultural influences on oral health are a complex phenomenon that is difficult to disentangle. Culture is in association with oral health literacy, personal experience, and socioeconomic status in complex ways [104]. Culture can shape the beliefs about teeth and the oral cavity, help-seeking and preventive care behavior, oral self-care practices, and the use of folk remedies [103]. Culture can also influence the perception of the need to use dental services. Thus, even when there is no financial barrier and services are accessible, learned behaviors can determine health-seeking behavior [105].

Culture influences body modification in ways to enhance beauty or as preventive or therapeutic remedies. The alteration of human dentition and oral soft tissues is common in many societies, both historically and in current contexts. Tooth modification practices include changing the shape of teeth, dyeing and lacquering the teeth, tooth avulsion, tooth bud enucleation, and inlays and/or onlays decoration [103, 106, 110]. Some of these modifications were done to relieve toothache [111] or other forms of childhood illnesses like diarrhea [112, 113]. There is also non-therapeutic extraction or filling of teeth [111, 114], and dyeing of teeth [102] for reasons ranging from initiation and rites ceremonies to mimicking the appearance of certain animals, improving personal appearance, providing a form of tribal and intra-tribal class identification, a sign of mourning and improved mastication and speech [115, 117]. Though the prevalence of tooth modification has decreased significantly over the years, its prevalence in certain forms remains at significant levels even today [103]. Understanding these contexts is important for forensic odontology.

Anthropologists have also contributed to the understanding of the evolution of tooth eruption. Biological anthropology has contributed significantly to identifying population-specific growth standards and aging where birth records do not exist. Forensic dentistry works on data of dental development, eruption, and morphometries, bearing in mind that the incidence of mass deaths and disasters is increasing globally [118].

Dental development is also important for studying developmental patterns, e.g., age at sexual maturation of adolescents, age at first reproduction, number of child (ren), interpregnancy intervals, amount of parental investment, aging, and death [118, 119].

Exploration and juxtaposition of developmental patterns between species, especially in closely related great apes, had caused queries to be raised about the relationship between weaning and mandibular first molar emergence [120, 122]. Studies on the correlation between age of first mandibular molar emergence, age at completion of tooth emergence, and brain size in hominids are ongoing [123] because of the strong correlation between brain size and dental development in primates [124].

3.4 Health Economics and Oral Health

Economics has been described by Haycox [125] as the science of limitation. Economics investigates how decisions about limited goods and services have to be made and prioritized by individuals to maximize benefit [125].

Even though progress in health is seen in most countries, the cost of health care is also rising. In the USA, for instance, healthcare costs comprise 15% of gross domestic product (GDP), and the corresponding figure in the UK is 17% [126]. Such factors as the price of materials, personnel salaries, and the growing use of more advanced technology may play a role. However, the evidence shows that increased spending has not led to better health [127].

This is now more than obvious that the resources for health care are finite, while the demand for health care seems to be infinite [128]. The economic analysis enables us to systematically analyze and respond to questions regarding the use of these finite and limited health resources. It helps us find solutions to some healthcare problems [126]. Thus, health economics is studying how we can use economic theory in decision-making about health and health care [126, 129].

While it is easy to apply economic laws to various sectors, dentistry and medicine are substantially different from the market for cars, fruits, or vegetables, for instance. An appointment with a dentist includes factors that are not economic and thus can be very different from a normal consumer deal or settlement. However, the economy certainly influences people's health. At last, when we study this relationship, the influence of health itself on the economy should not be ignored. The individual's health condition has an important influence on their performance, affecting their productivity at work. Even insignificant disorders such as toothache and the common cold can cause hours of being absent at work, while more severe problems like major surgery may result in months of absence from productivity [130].

The economy has its main roots in limitation, choice, and opportunity cost. Limitation happens when the accessible resources are less than those needed for whatever it is needed to do. Thus, we have to choose how to use available resources. It should also be mentioned that these choices are not easy. What is the opportunity cost? It will be called the opportunity cost, "if the benefits generated from the way we choose to use resources exceed the benefits generated by using the same resources in their most productive alternative uses" [131]. Thus, with the help of the term opportunity cost, we can choose how to redirect supplementary resources and manage their consumption for oral health promotion.

Dental treatment costs too much! Estimations show that globally in 2015, the direct and indirect costs of dental diseases have been US$356·80 billion and US$187·61 billion, respectively [132]. When the costs of different diseases were compared in 2015 among the 28 EU members, dental diseases (€90 billion) graded third after diabetes (€119 billion) and cardiovascular diseases (€111 billion) [58]. Besides, indirect costs such as productivity loss for people with oral diseases have to be considered [133].

The traditional curative model of oral health care is shown to be too expensive, in both human and financial resources, in a situation where we have an increase in demand worldwide. Expensive programs focusing only on oral diseases are not effective options for low and middle-income countries (LMICs), where resources are limited, and the effect of untreated caries is mostly high. A common representative of health systems in LMICs is that they do their job under unpredictable and changing conditions involving different areas such as financial resources, human resources, and supplies [134]. Families in LMICs that pay for dental care spend a large share of their income and have a higher risk of poverty, comparing those who do not pay for dental care [135].

The causes of dental diseases are identified, and effective preventive actions have been recognized. Yet, treatment services still control oral health systems worldwide [136]. To overcome this problem, the most important and effective job that could be done is prevention. Preventing oral disease is reachable. There are evidence-based, cost-effective, and simple preventive methods, but they have to be meticulously implemented [137].

One of the strategies is integrating oral health with universal health coverage (UHC), which is supported by evidence [138]. UHC, which covers oral health, can help people in low- and middle-income countries who are financially weak by the inclusion of: a, necessary oral health services and the basic oral care as a package; and b, dental care coverage in health insurances and also financial protection [139].

3.5 Demography and Oral Health

The relationship between aging and health has always been a challenge, but it is clear more than ever due to demographic changes. We have been progressing, from average life expectancy at birth of 35–50 years in the last 70–80 years to community averages of 55–80 years. This process is not finished, or its terminal point is known if there is actually a limit. Additionally, it is unknown that how long a dentition suffered from caries can bear normal life stresses or what other oral diseases will need to be cared for in whole communities at different ages. We must also remember the effect of the trends in common oral diseases on quality of life [140].

One of the important demographic trends, especially in developing countries, is the aging of the population. Effect on dental practice, which results from these growing numbers of older people, has become well determined [140]. Another important demographic pattern is the matter of health inequalities, which evidently affect minority groups. These groups in the USA mainly include African Americans, Hispanics, American Indians, and Alaska Natives. Individuals in these groups suffer from the unequal burden of disease and disability. The health inequalities will result in decreased quality of life, low life expectancy, loss of economic opportunities, and observation of injustice" [142].

3.6 Bioinformatics and Oral Health

Nowadays, we can observe medical informatics importance as a multidisciplinary science developed on the interaction of information sciences with medicine and health care in harmony with the specified level of information technology. Today's healthcare settings use electronic versions of health records that are shared between different computer systems and may be spread over many places and between organizations in order to deliver information to internal consumers, to payers, and also to respond to external demands. Patient data is accumulating in different places, with increasing movement of people, but it has to be available in a systematized way on a national and even international measure [143].

Health informatics is a mixture of data from medical areas like pre-clinical, clinical, post-clinical, and healthcare administration management. The developments in information technology gave this opportunity to the scientists to develop the public health, healthcare system, and biomedical field. Delivering complete information of the patient's health to healthcare professionals is the healthcare informatics' main goal. This data makes it easy to decide at the right time concerning the treatment. Also, the patient from the rural area can get the opinion from the best healthcare professionals by using healthcare informatics [144, 145]. Bioinformatics combines software engineering, computer science statistics, informatics, and engineering [146].

4 Oral Health Promotion

Good health is a resource for social, personal, and economic progress. Political, economic, social, cultural, environmental, behavioral, and biological factors can improve or damage health. Health promotion enables people to improve their health; thus, it intends to make these conditions beneficial to health. Accordingly, it is believed that health promotion goes further than the healthcare system. It tries to put health on the policymakers' agenda, pointing them to be aware of the impact of their decisions on health and forcing them to accept health responsibilities. Health promotion policy combines different approaches, including financial measures, taxation, legislation, and organizational change. Health promotion is done through specific and effective community actions to prioritize, make decisions, plan strategies, and perform them to gain better health [147].

Health promotion has three important essential fundamentals: first, it concentrates on addressing the determinants of health and inequalities; second, it focuses on the coalition with a range of agencies and sectors; and third, it adopts a strategic approach utilizing a complementary range of actions for health promotion of the population [4].

Health promotion tackles the wider range of health determinants and intends to reduce risks through organized actions and policies. Health promotion in the environment where people live, work, and learn is the most cost-effective way to improve oral health and quality of life. To have a level of physical, mental, and social well-being, people must identify desires, fulfill needs, and change with the environment. The health sector is not just responsible for health promotion, but it has to be in our lifestyle [147].

Ottawa Charter for Health Promotion [148] determines five areas for health promotion activities. The five items of the Ottawa Charter, with an oral health example in each, are mentioned below:

i. Building healthy public policy

- Banning delivery of unhealthy and sugary food in school buffets; no sugar added to food and drink in the first two years of age

- Policymaking to reduce the advertisement of unhealthy food and beverages to children; sugar tax in beverage/soda

- Developing and implementing community fluorides such as water, salt, or milk

ii. Creating supportive environments

- Providing fluoridated kinds of toothpaste at a reduced cost, especially for low-income groups
- Social marketing and mass media campaigns to enhance oral health

iii. Strengthening community action

- Developing target-oriented oral health promotion programs based on community cultures
- Engaging and encouraging communities to be involved in oral health programs

iv. Developing personal skills

- Supervised tooth brushing programs in schools
- Smoking cessation activities guided by oral health professionals

v. Re orientating health services

- Integrating oral health to primary healthcare system; health through oral health activities
- Focusing on preventive services in the oral healthcare system
- Facilitating the practice of early caries detection by primary healthcare professionals [147, 148].

Both general and oral health will be influenced by poor hygiene. It is often the case for developing countries to execute hygiene educational programs aimed at promoting healthy lifestyle behaviors for children and their relatives. Communities need to actively contribute to these programs and help develop school healthcare and appropriate water hygiene. These programs are ideally expected to incorporate different methods of oral health activities that meet challenges posed by urbanization and demographic factors. Progress in oral health is unlikely to be achieved by isolated interventions. The most effective interventions are a combination of social policy and individual responsibility by which we can be sure that healthy behaviors are adopted. WHO contribution is crucial for integrating oral health promotion into general health promotion. Countries can use local experiences, encourage communities to cooperate for their healthy future, and enable community empowerment [149]. WHO has released some recommendations for oral health promotion as summarized in Table 1 [150]. Also, Fig. 2 shows a representative of oral health promotion program [151].

Table 1 WHO recommendations for oral health promotion [149]

Main purpose	Proposed line(s)
Disease burden confinement	Oral disease load and disability, especially in poor populations, should be reduced
Promotion of healthy behaviors	Healthy lifestyles need to be promoted while decreasing risk factors of oral disease, e.g., environmental, economic, social factors
Development of competent health systems	Competent health systems should be built to improve oral health outcomes, respond to people's appropriate demands, and be financially reasonable
Revising health policies	Health policies should be included in national and community health programs, and promoted as a practical aspect of the developmental policy of society

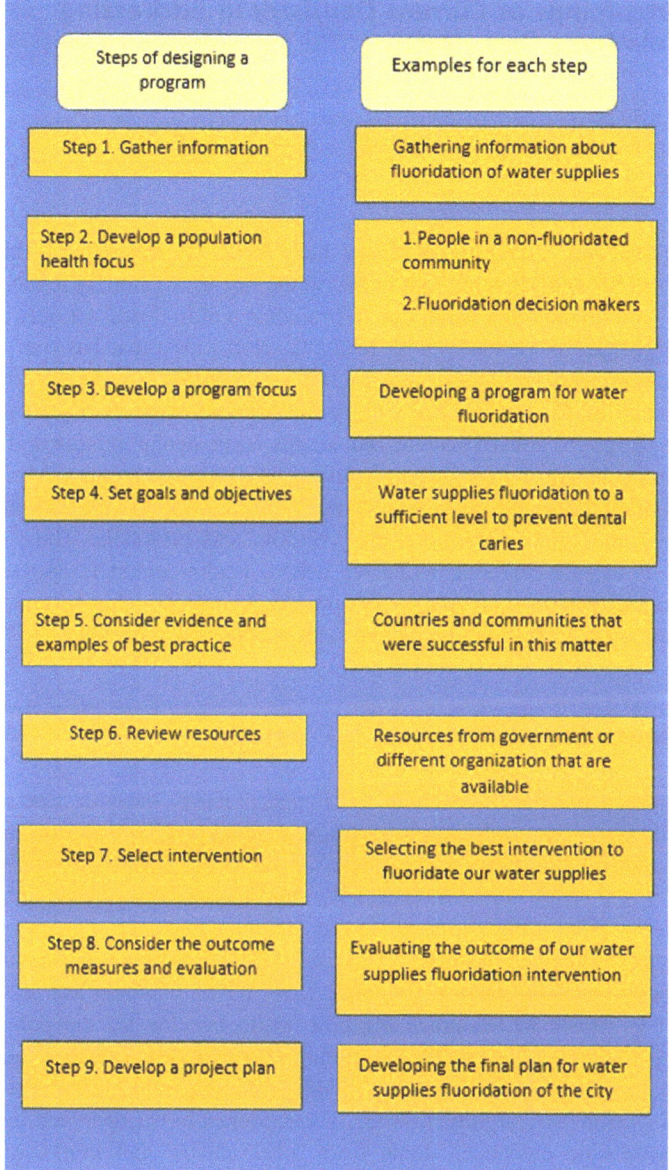

Fig. 2   Steps to design an oral health promotion program with an example of water supplies [150]

5 Shortcomings of Current Dentistry in Addressing Population's Oral Health

"Solving problems of disease is not the same thing as creating health and happiness. Health and happiness are the expression of the manner in which the individual responds and adapts to the challenges that he meets in everyday life."

Rene Dubos

Oral diseases and, therefore, dentistry have been a challenge of this century worldwide [152–154]. It is thought to be, on the one hand, directly related to dental clinicians, and on the other hand, due to problems within dental services and dental care. The practice of dentistry with technical philosophy that has the patterns of dental disease from 80 years ago is still treatment-focused [155, 156]. This approach highlights a biomedical side of disease and assumes that treatment and technology might be able to manage oral health. Basic dental training and education have been the same for a long time, focusing on certain techniques. Moreover, the paradigm of oral surgery continues to make oral surgeons prepared for surgical treatments rather than preventive interventions and protocols. Thus, such preparedness is disease-centered, while we need a health-centered approach then to expect general health—and oral health-promoting effects. This approach in dentistry is complex. Yet, it can be started with a universal movement [157].

6 Evidence-Based Oral Health Promotion

Evidence-based practice is essential to clinical practice, meaning that all clinical decisions need to take a thorough assessment of the available scientific evidence into account [4]. Recently, more emphasis has been placed on evidence-based practice for oral health sciences, especially after many advancements in health care and medicine [158, 159]. No need to say, without evidence, clinical practice is experimental and subjective [160]. The use of evidence-based decision-making is not confined to clinical practice. Evidence-based health planning and services are a practical mechanism to recognize the best health benefits for the population we work for [161, 162]. The nature and quality of evidence, especially the evidence pertaining to general health and oral health-promoting effects, has also been the subject of considerable argument [163]. The evidence-based approach lies in high-quality, accessible evidence, stems from local values, and thinks of resource availability and accessibility as crucial to the practice of evidence-based oral health promotion [164]. Implementation of an evidence-based approach in oral health promotion programs faces such obstacles and difficulties as an absence or scarcity of quality evidence on health promotion; inadequate attention on the development of strategies to evaluating the efficacy and effectiveness of health-promoting actions; improper generalization of the characteristics of randomized controlled trials (RCTs) to population-based cohort studies; and lack of enough information of current evidence-based approaches among healthcare personnel [163, 165].

7   Dental Auxiliaries, Teamwork, and Their Potential Role in Oral Health Promotion

The application of a professional team in dentistry is not well-configured. It might be a heritage of the oral surgeon paradigm, wherein the oral surgeon is referred to as the only person that accepts the responsibility of disease management from diagnosis to treatment. Many dental schools worldwide teach their students to be dentist who works and treats separately instead of training dental care professionals with diverse skills that complement each other. They can make interventions for their patients and local populations to effectively resolve their oral care needs. The need can be from preventive treatments to complex ones. These treatments' delivery, efficiency, and effectiveness can be heightened when being covered by oral health workforces with proper and different sets of skills. Also, mid-level practitioners can help enhance access to oral health care in groups under special support and remote population groups. Training dental care that is community-based more than that is in usual dentistry program might help resolve the lack of workforce in many settings, especially LMICs. Dental care professionals are different between countries in terms of the types and roles each type plays. However, they often consist of dental hygienists, dental therapists, denturists, dental assistants, dental nurses, and dental technicians. Like other professional fields, to which degree different dental professionals and respective thoughts/practices are dependent on/independent of each other is a complex question to ask. The negotiation over which it is decided that these professionals can do what, and under what conditions (including supervision), is often decided as an agreement between the involved professionals rather than community needs [157].

8   Conclusion

Oral health is crucial to general health, well-being, and quality of life. The current approach in dentistry is complex yet inadequate in reducing the global oral disease burden. Expensive programs merely focused on oral diseases are not effective options. To overcome this problem, the most important and effective job that could be done is prevention. Oral health is maintained and ensured if the conditions are provided; for instance, if proper health policy, health economics, health literacy, and other health components are in harmony with related sciences for oral health. Although oral health is believed to be an independent discipline, it is closely interrelated to other disciplines and sciences—sciences that their entrance and inclusion in Dental Public Health intensify the dental profession's focus on some of the factors that affect oral health, and are influential in preventing and treating oral health problems.

Core Messages

- Oral health is crucial to general health.
- "Health through oral health" and "No UHC without oral health" concepts should be advocated.
- An interdisciplinary and multi-sectorial approach is needed to promote the oral health of the population.
- To effectively manage the oral health problems of the community, the dentistry profession should go beyond its borders.
- Oral health promotion programs need to benefit from an evidence-based approach.

References

1. World Health Organization (2005) Constitution of the World Health Organization. In: World Health Organization, 45th ed. Basic documents, Geneva
2. World Health Organization. Regional Office for Europe (1984) Health promotion: a discussion document on the concept and principles: summary report of the Working Group on Concept and Principles of Health Promotion, Copenhagen, pp 9–13
3. Ewles L, Simnet I (1992) Promoting health, 2nd edn. Scutari Press, London
4. Daly B, Batchelor P, Treasure E T, Watt RG (2013) Essential dental public health, 2nd ed. Oxford University Press, pp 99–174, 231–238
5. Glick M, Williams DM, Kleinman DV, Vujicic M, Watt RG, Weyant RJ (2016) A new definition for oral health developed by the FDI World Dental Federation opens the door to a universal definition of oral health. Int Dent J 66:322–324
6. American Dental Association (2019). https://www.ada.org/en/member-center/oral-health-topics/oral-systemic-health. Accessed on 23 September 2019
7. Alibek K, Grechanyi L, Klimenko T, Pashkova A (2008) Fifth revolution in medicine: on the role of infections in pathogenesis of aging and chronic diseases. Lik Sprava 3–30
8. Cardoso EM, Reis C, Manzanares-Cespedes MC (2018) Chronic periodontitis, inflammatory cytokines, and interrelationship with other chronic diseases. Postgrad Med 130:98–104
9. Sanz M, Ceriello A, Buysschaert M, Chapple I, Demmer RT, Graziani F, Herrera D, Jepsen S, Lione L, Madianos P, Mathur M, Montanya E, Shapira L, Tonetti M, Vegh D (2018) Scientific evidence on the links between periodontal diseases and diabetes: consensus report and guidelines of the joint workshop on periodontal diseases and diabetes by the International Diabetes Federation and the European Federation of Periodontology. J Clin Periodontol 45:138–149
10. Wahid A, Chaudhry S, Ehsan A, Butt S, Ali Khan A (2013) Bidirectional relationship between chronic kidney disease and periodontal disease. Pak J Med Sci 29:211–215
11. Lima GQT, Brondani MA, da Silva AAM, do Carmo CDS, da Silva RA, Ribeiro CCC (2018) Serum levels of pro-inflammatory cytokines are high in early childhood caries. Cytokine 111:490–495
12. Werner H, Katz J (2004) The emerging role of the insulin-like growth factors in oral biology. J Dent Res 83(11):832–836

13. Ribeiro CCC, Pachêco CDJB, Costa EL, Ladeira LLC, Costa JF, da Silva RA, Carmo CDS (2018) Pro-inflammatory cytokines in early childhood caries: salivary analysis in the mother/children pair. Cytokine 107:113–117
14. Zehnder M, Delaleu N, Bickel M (2003) Cytokine gene expression-part of host defence in pulpitis. Cytokine 22(3–4):84–88
15. Haluk D, Mehmet A, Sekeroğlu MR, Tarakçioğlu M, Noyan T, Cesur Y, Balahoroğlu R (2002) Pro-inflammatory cytokines in Turkish children with protein-energy malnutrition. Mediat Inflamm 11(6):363–365
16. Santetti D, de Albuquerque Wilasco MI, Dornelles CT, Werlang ICR, Fontella FU, Kieling CO, dos Santos JL, Gonçalves Vieira SM, Sueno Goldani HA (2015) Serum pro-inflammatory cytokines and nutritional status in pediatric chronic liver disease. World J Gastroenterol 21(29):8927–8934
17. Mehta NM, Corkins MR, Lyman B, Malone A, Goday PS, Carney LN, Jessica L, Monczka SWP, Schwenk WF (2013) Defining pediatric malnutrition: a paradigm shift toward etiology-related definitions. American Society for Parenteral and Enteral Nutrition Board of Directors. J Parenter Enteral Nutr 37(4):460–481
18. Zoico E, Roubenoff R (2002) The role of cytokines in regulating protein metabolism and muscle function. Nutr Rev 60(2):39–51
19. Cregger RA, Langworthy KL, Salako NO, Streckfus C (2017) Relationship between salivary cytokines, and caries experience in children with different body mass indices. J Dent Oral Health 3(5):075
20. Alkarimi HA, Watt RG, Pikhart H (2014) Dental caries and growth in school-age children. Pediatrics 133(3):e616–e623
21. Lee S, Sabbah W (2018) Association between number of teeth, use of dentures and musculoskeletal frailty among older adults. Geriatr Gerontol Int 18(4):592–598
22. Gaewkhiew P, Sabbah W, Bernabé E (2017) Does tooth loss affect dietary intake and nutritional status? A systematic review of longitudinal studies. J Dent 67:1–8
23. Arowojolu MO, Oladapo O, Opeodu OI, Nwhator SO (2016) An evaluation of the possible relationship between chronic periodontitis and hypertension. J West Afr Coll Surg 6(2):20–38
24. Umeizudike KA, Ayanbadejo PO, Onajole AT, Umeizudike TI, Alade GO (2016) Periodontal status and its association with self-reported hypertension in non-medical staff in a university teaching hospital in Nigeria. Odontostomatol Trop 39:47–55
25. Nwhator SO, Umeizudike KA, Po A et al (2014) Another reason for impeccable oral hygiene: oral hygiene-sperm count link. J Contemp Dent Pract 15(3):352–358
26. Nwhator S, Opeodu O, Ayanbadejo P, Opeodu OI, Olamijulo JA, Sorsa T (2014) Could periodontitis affect time to conception? Ann Med Health Sci Res 4(5):817–822
27. Hart R, Doherty DA, Pennell CE et al (2012) Periodontal disease: a potential modifiable risk factor limiting conception. Hum Reprod 27:1332–1342
28. Dorfer C, Benz C, Aida J, Campard G (2017) The relationship of oral health with general health and NCDs: a brief review. Int Dent J 67(2):14–18
29. Desvarieux M, Demmer RT, Rundek T, Boden-albalal B, Jacobs DRJR, Sacco RL, Papapanou PN (2005) Periodontal microbiota and carotid intima-media thickness: the oral infections and vascular disease epidemiology study (INVEST). Circulation 111:576–582
30. Fedele S, Sabbah W, Donos N, Porter S, D'Aiuto F (2011) Common oral mucosal diseases, systemic inflammation, and cardiovascular diseases in a large cross-sectional US survey. Am Heart J 161(2):344–350
31. Sabbah W, Mortensen LH, Sheiham A, Batty GD (2014) Oral health as a risk factor for mortality in middle-aged men: the role of socioeconomic position and health behaviors. J Epidemiol Community Health 68(2):191
32. Morita I, Inagaki K, Nakamura F, Noguchi T, Matsubara T, Yoshii S, Nakagaki H, Mizuno K, Sheiham A, Sabbah W (2012) Relationship between periodontal status and levels of glycated hemoglobin. J Dent Res 91(2):161–166

33. Ramos-Nino ME (2013) The role of chronic inflammation in obesity-associated cancers. ISRN Oncol 697521. https://doi.org/10.1155/2013/697521

34. Folayan MO, El Tantawi M, Ramos-Gomez F, Sabbah W Sugar consumption, exclusive breastfeeding and its association with early childhood caries and overweight: an ecological study. Peer J 40360

35. World Health Organization (2017) Depression and other common mental disorders: global health estimates. Geneva, License: CC BY-NC-SA 3.0 IGO

36. Kisely S, Sawyer E, Siskind D, Lalloo R (2016) The oral health of people with anxiety and depressive disorders—a systematic review and meta-analysis. J Affect Disord 200:119–132

37. Folayan MO, Adeniyi AA, Oziegbe EO, Fatusi AO, Harrison A (2016) Integrated oral, mental and sexual health management for adolescents: a call for professional collaboration. Int J Adolesc Med Health 30(3). https://doi.org/10.1515/ijamh-2016-0060

38. Folayan MO, Oginni O, Arowolo O, El Tantawi M (2020) Association between adverse childhood experiences, bullying, self-esteem, resilience, social support, caries and oral hygiene in children and adolescents in sub-urban Nigeria. BMC Oral Health 20(1):202

39. Del-Ponte B, Quinte GC, Cruz S, Grellert M, Santos IS (2019) Dietary patterns and attention deficit/hyperactivity disorder (ADHD): a systematic review and meta-analysis. J Affect Disord 252:160–173

40. Pan X, Zhang C, Shi Z (2011) Soft drink and sweet food consumption and suicidal behaviours among Chinese adolescents. Acta Paediatr 100(11):215–222

41. Jacob L, Stubbs B, Koyanagi A (2020) Consumption of carbonated soft drinks and suicide attempts among 105,061 adolescents aged 12–15 years from 6 high-income, 22 middle-income, -and 4 low-income countries. Clin Nutr 39(3):886–892

42. Joury E (2019) Syria profile of the epidemiology and management of early childhood caries before and during the time of crisis. Front Public Health 7:271

43. Tunçalp Ö, Fall IS, Phillips SJ, Williams I, Sacko M, Touré OB, Thomas LJ, Say L (2015) Conflict, displacement and sexual and reproductive health services in mali: analysis of 2013 health resources availability mapping system (HeRAMS) survey. Confl Health 9:28

44. Rooban T, Vidya K, Joshua E, Rao A, Ranganathan S, Rao UK, Ranganathan K (2011) Tooth decay in alcohol and tobacco abusers. JOMFP 15(1):14–21

45. Mahapatra S, Chaly PE, Mohapatra SC, Madhumitha M (2018) Influence of tobacco chewing on oral health: a hospital-based cross-sectional study in Odisha. Indian J Public Health 62(4):282–286

46. Tellez M, Zini A, Estupiñan-Day S (2014) Social determinants and oral health: an update. Curr Oral Health Rep 1:148–152

47. Dean HD, Williams KM, Fenton KA (2013) From theory to action: applying social determinants of health to public health practice. Public Health Rep 128(3):1–4

48. Lee JY, Divaris K (2013) The ethical imperative of addressing oral health disparities: a unifying framework. J Dent Res 4:1–7

49. Russell E, Johnson B, Larsen H, Novilla MLB, van Olmen J, Swanson RC (2013) Health systems in context: a systematic review of the integration of the social determinants of health within health systems frameworks. Rev Panam Salud Publica 34(6):461–467

50. Williams DM, Sheiham A, Watt RG (2013) Oral health professionals and social determinants. Br Dent J 214(9):427

51. Folayan MO, El Tantawi M, Aly NM, Al-Batayneh OB, Schroth JR, Castillo JL, Virtanen JI, Gaffar BO, Amalia R, Kemoli A, Vulkovic A, Feldens CA, the ECCAG (2020) Association between early childhood caries and poverty in low and middle income countries. BMC Oral Health 20(1):8

52. Folayan MO, El Tantawi M, Vukovic A, Schroth R, Gaffar B, Al-Batayneh OB, Amalia R, Arheiam A, Obiyan M, Daryanavard H, Early Childhood Caries Advocacy Group (2020) Women's economic empowerment, participation in decision-making and exposure to violence as risk indicators for early childhood caries. BMC Oral Health 20(1):54

53. Folayan MO, Tantawi M, Vukovic A, Schroth RJ, Alade M, Mohebbi SZ, Al-Batayneh OB, Arheiam A, Amalia R, Gaffar B, Onyejaka NK, Daryanavard H, Kemoli A, Díaz ACM, Grewal N, on behalf of the Global Early Childhood Caries Research Group (2020) Governance, maternal well-being and early childhood caries in 3–5-year-old children. BMC Oral Health 20(1):166

54. Tantawi M, Folayan MO, Mehaina M, Vukovic A, Castillo JL, Gaffar BO, Arheiam A, Al-Batayneh OB, Kemoli AM, Schroth RJ, Lee GHM (2018) Prevalence and data availability of early childhood caries in 193 United Nations Countries, 2007–2017. Am J Public Health 108(8):1066–1072

55. Baker SR, Foster Page L, Thomson WM, Broomhead T, Bekes K, Benson PE, Diaz FA, Do L, Hirsch C, Marshman Z, McGrath C, Mohamed A, Robinson PG, Traebert J, Turton B, Gibson BJ (2018) Structural determinants and children's oral health: a cross-national study. J Dent Res 97(10):1129–1136

56. Folayan MO, El Tantawi M, Schroth RJ, Kemoli AM, Gaffar B, Amalia R, Feldens C, ECCAG (2020) Association between environmental health, ecosystem vitality, and early childhood caries. Front Pediatr 8:196

57. Kickbusch I, Allen L, Franz C (2016) The commercial determinants of health. Lancet Glob Health 4:895–896

58. Watt RG, Sheiham A (2012) Integrating the common risk factor approach into a social determinants framework. Community Dent Oral Epidemiol 40:289–296

59. Solar O, Irwin A (2010) A conceptual framework for action on the social determinants of health. In: Social Determinants of Health Discussion Paper 2 (Policy and Practice). World Health Organization, Geneva

60. Peres MA, Macpherson LMD, Weyant RJ, Daly B, Venturelli R, Mathur MR, Listl S, Celeste RK, Guarnizo-Herreño CC, Kearns C, Benzian H, Allison P, Watt RG (2019) Oral diseases: a global public health challenge. Lancet 394:249–260

61. Thornton RLJ, Glover CM, Cené CW, Glik DC, Henderson JA, Williams DR (2016) Evaluating strategies for reducing health disparities by addressing the social determinants of health. Health Aff 35(8):1416–1423

62. Chattopadhyay A (2011) Oral health epidemiology: principles and practice. Jones & Bartlett Learning, pp 3–22

63. Ozbay F, Johnson DC, Dimoulas E, Morgan CA, Charney D, Southwick S (2007) Social support and resilience to stress: from neurobiology to clinical practice. Psychiatry (Edgmont) 4(5):35

64. Strom JL, Egede LE (2012) The impact of social support on outcomes in adult patients with type 2 diabetes: a systematic review. Curr Diab Rep 12(6):769–781

65. Dahlan R, Ghazal E, Saltaji H, Salami B, Amin M (2019) Impact of social support on oral health among immigrants and ethnic minorities: a systematic review. PLoS One 14(6): e0218678

66. Qiu RM, Tao Y, Zhou Y, Tao Y, Lin HC (2016) The relationship between children's oral health-related behaviors and their caregiver's social support. BMC Oral Health 16(1):86

67. Park SJ, Ko KD, Shin SI, Ha YJ, Kim GY, Kim HA (2014) Association of oral health behaviors and status with depression: results from the Korean National Health and Nutrition Examination Survey, 2010. J Public Health Dent 74(2):127–138

68. Niklander S, Veas L, Barrera C, Fuentes F, Chiappini G, Marshall M (2017) Risk factors, hypo salivation and impact of xerostomia on oral health-related quality of life. Braz Oral Res 31:e14

69. Schneiderman N, Ironson G, Siegel SD (2005) Stress and health: psychological, behavioral, and biological determinants. Ann Rev Clin Psychol 1:607–628

70. Vasiliou A, Shankardass K, Nisenbaum R, Quiñonez C (2016) Current stress and poor oral health. BMC Oral Health 16(1):88

71. Folayan MO, El Tantawi M, Oginni A, Adeniyi A, Alade M, Finlayson TL (2020) Psychosocial, education, economic factors, decision-making ability, and caries status of mothers of children younger than 6 years in suburban Nigeria. BMC Oral Health 20(1):131
72. Zinke A, Hannig C, Berth H (2018) Comparing oral health in patients with different levels of dental anxiety. Head Face Med 14(1):25
73. Okoro CA, Strine TW, Eke PI, Dhingra SS, Balluz LS (2012) The association between depression and anxiety and use of oral health services and tooth loss. Community Dent Oral Epidemiol 40(2):134–144
74. Bright MA, Alford SM, Hinojosa MS, Knapp C, Fernandez-Baca DE (2015) Adverse childhood experiences and dental health in children and adolescents. Community Dent Oral Epidemiol 43(3):193–199
75. Bosch J, Weaver TL, Arnold LD (2019) Impact of adverse childhood experiences on oral health among women in the United States: findings from the behavioral risk factor surveillance system. J Interpers Violence. https://doi.org/10.1177/0886260519883872
76. Crouch E, Nelson J, Radcliff E, Martin A (2019) Exploring associations between adverse childhood experiences and oral health among children and adolescents. J Public Health Dent 79(4):352–360
77. Albino J, Tiwari T, Henderson WG, Thomas JF, Braun PA, Batliner TS (2018) Parental psychosocial factors and childhood caries prevention: data from an American Indian population. Community Dent Oral Epidemiol 46(4):360–368
78. Padmaja P, Kulkarni S, Doshi D, Reddy S, Reddy P, Reddy KS (2018) Impact of health locus of control on oral health status among a cohort of IT professionals. Oral Health Prev Dent 16(3):259–264
79. Peker K, Bermek G (2011) Oral health: locus of control, health behavior, self-rated oral health and socio-demographic factors in Istanbul adults. Acta Odontol Scand 69(1):54–64
80. Kent GG, Matthews RM, White FH (1984) Locus of control and oral health. J Am Dent Assoc 109(1):67–69
81. Eriksson M, Lindström B (2005) Validity of Antonovsky's sense of coherence scale: a systematic review. J Epidemiol Community Health 59(6):460–466
82. Eriksson M, Lindström B (2007) Antonovsky's sense of coherence scale and its relation with quality of life: a systematic review. J Epidemiol Community Health 61(11):938–944
83. Davoglio RS, Abegg C, Fontanive VN, de Oliviera MMC, de Castro Aerts DRG, Cavalheiro CH (2016) Relationship between sense of coherence and oral health in adults and elderly Brazilians. Braz Oral Res 30(1):e56
84. Savolainen J, Suominen-Taipale AL, Hausen H, Harju P, Uutela A, Martelin T, Knuuttila M (2005) Sense of coherence as a determinant of the oral health-related quality of life: a national study in Finnish adults. Eur J Oral Sci 113(2):121–127
85. Savolainen J, Suominen-Taipale A, Uutela A, Aromaa A, Härkänen T, Knuuttila M (2009) Sense of coherence associates with oral and general health behaviours. Community Dent Health 26(4):197–203
86. Elyasi M, Abreu LG, Badri P, Saltaji H, Flores-Mir C, Amin M (2015) Impact of sense of coherence on oral health behaviors: a systematic review. PLoS One 10(8):e0133918
87. Diamond A (2013) Executive functions. Annu Rev Psychol 64:135–168
88. Dourado MR, Andrade PM, Ramos-Jorge ML Moreira RN, Oliveira-Ferreira F (2013) Association between executive/attentional functions and caries in children with cerebral palsy. Res Dev Disabil 34(9):2493–2499
89. Agou S, Locker D, Muirhead V, Tompson B, Streiner DL (2011) Does psychological well-being influence oral-health-related quality of life reports in children receiving orthodontic treatment? Am J Orthod Dentofacial Orthop 139(3):369–377
90. Bandeira CDM, Hutz CS (2010) As implicações do bullying na auto-estima de adolescentes. Psicologia Escolar e Educacional 14(1):131–138

91. Foster Page LA, Thomson WM, Ukra A (2013) Clinical status in adolescents: is its impact on oral health-related quality of life influenced by psychological characteristics? Eur J Oral Sci 121(3pt1):182–187

92. Lee TY, Cheung CK, Kwong WM (2012) Resilience as a positive youth development construct: a conceptual review. Sci World J 12:1–9

93. Nettleton S (1988) Protecting a vulnerable margin: towards an analysis of how the mouth came to be separated from the body. Sociol Health Illn 10(2):156–169

94. Thorogood N (2000) Mouth rules and the construction of sexual identities. Sexualities 3(2): 165–182

95. Gibson B (2008) Cultural history of the mouth and teeth. In: Pitts-Taylor V (ed) [1963] Cultural encyclopaedia of the body. Greenwood Press, Westport

96. McGrath C, Bedi R (1998) A study of the impact of oral health on the quality of life of older people in the UK-findings from a National Survey. Gerodontology 15(2):93–98

97. Goffman E (2009) Stigma—notes on the management of spoiled identity. Simon and Schuster, London

98. Shilling C (1993) The body and social theory. Sage Publications Ltd., London

99. Giddens A (1991) Modernity and self-identity: self and society in late modern age. Basil Blackwell Ltd, Oxford

100. Featherstone M, Hepworth M, Turner BS (1991) The body. Social process and cultural theory. Sage, London

101. Alkhatib N, Holt R, Bedi R (2005) Age and perception of dental appearance and tooth colour. Gerodontology 22:32–36

102. Gregory J, Gibson B, Robinson PG (2005) Variation and change in the meaning of oral health related quality of life: a 'grounded' systems approach. Soc Sci Med 60:1859–1868

103. Chopra M, Marya CM, Nagpal R (2015) Culture and oral health. GRIN Verlag, Munich

104. Baskaradoss JK (2018) Relationship between oral health literacy and oral health status. BMC Oral Health 18(1):172

105. Latunji OO, Akinyemi OO (2018) Factors influencing health-seeking behaviour among civil servants in Ibadan, Nigeria. Ann Ib Postgrad Med 16(1):52–60

106. Tayanin GL, Bratthall D (2006) Black teeth: beauty or caries prevention? Practice and beliefs of the Kammu people. Community Dent Oral Epidemiol 34(2):81-86

107. Pindborg JJ (1982) Painting of teeth black in Asia. Tandlaegebladet 86(7):235–236

108. Guarda L, Mason PN (1989) Mutilazioni dentarie, un enigma etnologico. Dentista modern 6:65–71

109. Zumbroich TJ (2009) "When Black Teeth Were Beautiful"—the history and ethnography of dental modifications in Luzon, Philippines. Int J Asian Stud 10:125–169

110. Zumbroich TJ (2009) 'Teeth as black as a bumble bee's wings': the ethnobotany of teeth blackening in Southeast Asia. Ethnobotany Res Appl 7:381–398

111. Vukovic A, Bajsman A, Zukic S (2009) Cosmetic dentistry in ancient times—a short review. Bull Int Assoc Paleodont 3(2):9–13

112. Musinguzi N, Kemoli A, Okullo I (2019) Prevalence and dental effects of infant oral mutilation or Ebiino among 3–5 year-old children from a rural district in Uganda. BMC Oral Health 19(1):204

113. Edwards PC, Levering N, Wetzel E, Saini T (2008) Extirpation of the primary canine tooth follicles: a form of infant oral mutilation. J Am Dent Assoc 139(4):442–450

114. Pinchi V, Barbieri P, Pradella F, Focardi M, Bartolini V, Norelli GA (2015) Dental ritual mutilations and forensic odontologist practice: a review of the literature. Acta Stomatol Croat 49(1):3–13

115. Morinis A (1985) The ritual experience: pain and the transformation of consciousness in ordeals of initiation. Ethos 13(2):150–174

116. van Wyk CW (1976) Oral lesions caused by habits. Forensic Sci 7(1):41–49

117. Ozaki K (1971) Custom of dental extraction in Australian natives. Nihon Shika Ishikai Zasshi 24(4):368

118. Esan TA, Schepartz LA (2019) Dental development: anthropological perspectives. In: Folayan MO (ed) A compendium on oral health of children around the world: early childhood caries. Nova Science Publishers Inc, NY
119. Stearns SC (2000) Life history evolution: successes, limitations, and prospects. Naturwissenschaften 87(11):476–486
120. Dirks W, Bowman JE (2007) Life history theory and dental development in four species of catarrhine primates. J Hum Evol 53(3):309–320
121. Robson SL, Wood B (2008) Hominin life history: reconstruction and evolution. J Anat 212(4):394–425
122. Humphrey LT (2010) Weaning behavior in human evolution. Semin Cell Dev Biol 21:453–461
123. De Castro JB, Rozzi FR, Martinón-Torres (2003) Patterns of dental development in Lower and Middle Pleistocene hominins from Atapuerca (Spain). In: Thompson JL, Krovitz G, Nelson A (eds) Patterns of growth and development in the genus homo. Cambridge University Press, Cambridge, pp 246—270
124. Smith TM (2013) Teeth and human life-history evolution. Ann Rev Anthropol 42:191–208
125. Haycox A (2009) What is health economics? http://www.whatisseries.co.uk/whatis/pdfs/What_is_health_econ. Accessed Sept 2012
126. Morris S, Devlin N, Parkin D (2007) Economic analysis in health care. Wiley, London
127. Abel-Smith B (1996) The escalation of health care costs: how did we get there? Health care reform. The will to change. OECD, Paris, pp 17–30
128. Cohen D (2008) In: Naidoo J, Wills J (eds) Health economics in health studies: an introduction, 2nd ed.
129. Mooney G (2003) Economics, medicine and healthcare. Pearson Education, London
130. Tuominen R (2018) Health economics in dentistry, 2nd ed. Cambridge Scholars
131. Listl S, Grytten JI, Birch S (2019) What is health economics? Community Dent Health 36(4):262–274
132. Righolt AJ, Jevdjevic M, Marcenes W, Listl S (2018) Global-, regional-, and country-level economic impacts of dental diseases in 2015. J Dent Res 97:501–507
133. Glick M (2012) FDI Vision 2020: shaping the future of oral health. Int Dent J 62:278–291
134. Fejerskov O, Nyvad B, Kidd E (2015) Dental caries: the disease and its clinical management. Wiley
135. Bernabé E, Masood M, Vujicic M (2017) The impact of out-of-pocket payments for dental care on household finances in low and middle income countries. BMC Public Health 17:109
136. Blinkhorn A (1998) Dental health education: what lessons have we ignored. Br Dent J 184:58–59
137. Lancet (2009) Oral health: prevention is key. Lancet 373(9657):1
138. Mathur MR, Williams DM, Reddy KS, Watt RG (2015) Universal health coverage a unique policy opportunity for oral health. J Dent Res 94(3):3S-5S
139. Fisher J, Selikowitz H-S, Mathur M, Varenne B (2018) Strengthening oral health for universal health coverage. The Lancet 392(10151):899–901
140. Barmes DE (2000) Public policy on oral health and old age: a global view. J Public Health Dent 60(4):335–371
141. Garcia RI, Cadoret C, Henshaw M (2008) Multicultural issues in oral health. Dent Clin N Am 52:319–332
142. Agency for Healthcare Research and Quality (2003) National Healthcare Quality and Disparities Report. Rockville, MD, US, Pub. No. 04-0035
143. Berka P, Rauch J, Zighed DA (2009) Data mining and medical knowledge management: cases and applications. IGI Global, pp 1–37
144. Nadri H, Rahimi B, Timpka T, Sedghi S (2017) The top 100 articles in the medical informatics: a bibliometric analysis. J Med Syst 41:150
145. Ohno-Machado L (2013) Data science and informatics: when it comes to biomedical data, is there a real distinction? JAMIA 20(6):1009

146. Majhi V, Paul S, Jain R (2019) Bioinformatics for healthcare applications. In: AICAI, pp 204—207
147. Ottawa Charter for Health Promotion (1986) Health Promotion Int 1(4):405–405
148. Niranjan VR, Kathuria V, Salve A (2017) Oral health promotion: evidences and strategies. Insights into various aspects of oral health. https://doi.org/10.5772/intechopen.69330
149. World Health Organization (2020). https://www.who.int/oral_health/strategies/hp/en. 'Accessed' on 2020
150. The World Health Organization (2003) Oral health promotion: an essential element of a health promoting school. WHO Information Series on School Health. https://apps.who.int/iris/handle/10665/70207. Accessed on 2020
151. Ministry of Health (2008) Promoting oral health: a toolkit to assist the development, planning, implementation and evaluation of oral health promotion in New Zealand. Ministry of Health, Wellington. Accessed on 02 February 2008
152. Kassebaum NJ, Smith AGC, Bernabé E, Fleming T D, Reynolds AE, Vos T, Murray CJL, Marcenes W, GBD 2015 Oral Health Collaborators (2017) Global, regional, and national prevalence, incidence, and disability-adjusted life years for oral conditions for 195 countries, 1990–2015: a systematic analysis for the global burden of diseases, injuries, and risk factors. J Dent Res 96:380–387
153. Cohen L, Dahlen G, Escobar A (2017) Dentistry in crisis: time to change. La Cascada Declaration. Aust Dent J 62:258–260
154. Vujicic M (2018) Our dental care system is stuck. J Am Dent Assoc 149:167–169
155. Fejerskov O, Escobar G, Jøssing M, Baelum V (2013) A functional natural dentition for all —and for life? The oral healthcare system needs revision. J Oral Rehabil 40:707–722
156. Baelum V, Helderman P van V, Hugoson A, Yee R, Fejerskov O (2007) A global perspective on changes in the burden of caries and periodontitis: implications for dentistry. J Oral Rehabil 34:872–906
157. Watt RG, Daly B, Allison P, Macpherson LMD, Venturelli R, Listl S, Weyant RJ, Mathur MR, Guarnizo-Herreño CC, Celeste RK, Peres MA, Kearns C, Benzian H (2019) Ending the neglect of global oral health—time for radical action. Lancet 394:262–272
158. Kusiak J, Somerman M (2016) Data science at the National Institute of Dental and Craniofacial Research. J Am Dent Assoc 147(8):597–599
159. Stohler C (2015) Editorial: orofacial pain and the prospects of precision medicine. J Oral Facial Pain Headache 29(4):321
160. Finkelstein J, Zhang F, Levitin BSSA, Cappelli D (2020) Using big data to promote precision oral health in the context of a learning healthcare system. J Public Health Dent 80: S43–S58
161. Gray JAM (1997) Evidence-based health care. how to make health policy and management decisions. Churchill Livingstone, London
162. Richards D, Lawrence A (1998) Evidence based dentistry. Evid-Based Dent 1(1):7–10
163. Wiggers J, Sanson-Fisher R (1998) Evidence-based health promotion. In: Scott D, Weston R (eds) Evaluating health promotion. Stanley Thornes Publishers, Cheltenham, UK, pp 127–145
164. Petersen PE (2004) Challenges to improvement of oral health in the 21st century—the approach of the WHO Global Oral Health Program. Int Dent J 54(6 suppl 1):329–343
165. Raphael D (2000) The question of evidence in health promotion. Health Promot Int 15(4): 355–367

Mohammad Reza Khami born in 1978, received his Doctor of Dental Surgery (DDS) degree from Tehran University of Medical Sciences (TUMS) in 2002 and his Ph.D. in Community Oral Health from University of Helsinki and Shahid Beheshti University of Medical Sciences in 2007. He is currently a full professor in the Community Oral Health Department, TUMS School of Dentistry, and the head of Research Center for Caries Prevention, Dentistry Research Institute of TUMS. He also is the Chief Consultant of Vice-chancellor for International Affairs, TUMS. He is an Associate Editor in BMC Oral Health journal (part of Springer-Nature) and Editorial Board Member of the Current Dentistry journal (Bentham Science). His research interests are oral health promotion and dental education. He was selected as a distinguished young researcher in the Razi National Research Festival in 2009 and as a distinguished educator in TUMS Avicenna Education and Research Festival in 2012. He won the WHO Patient Safety research grant in 2008.

Farid Farrokhi born in 1989, received his Doctor of Dental Surgery (DDS) degree from Yerevan State Medical University (YSMU) in 2012 and Babol University of Medical Sciences in 2015. He is currently a Ph.D. student at the Community Oral Health Department, School of Dentistry, Tehran University of Medical Sciences, and the Chief Resident of the department. He is working on his thesis about the complications of providing oral health services by Health Care Providers and their interactive learning in the field of oral health. He was awarded certificates for presentation in the first scientific conference of Iranian medical students in Armenia in 2010, a presentation in the first Conference of Dental Students' scientific club at YSMU in 2012, and a poster presentation entitled "Data mining and its implications in health care system research" in the 4th National Community Oral Health congress in 2019. His recent project was a study about COVID-19 and dental education.

Drug Discovery in Big Pharma: Where "Birds" and "Fish" Collaborate to Find New Medicines

13

Donald R. Kirsch

"We live in a society exquisitely dependent on science and technology, in which hardly anyone knows anything about science and technology."

Carl Sagan (American Astronomer)

Summary

Biomedical scientists discover new drugs. But the high cost of this research, estimated to be $1.4 billion per Food and Drug Administration (FDA)-approved drug, means that drug discovery scientists must collaborate with business people to obtain financial support for their research. These are vastly different groups in terms of both their professional goals and restraints. Not surprisingly, these differences commonly lead to ineffective collaboration and project loss. This problem will be illustrated by three drugs: captopril (Capoten), atorvastatin (Lipitor), and sirolimus (Rapamune), each of which was initially opposed by commercial management and later went on to become a clinically and commercially successful drug. Captopril was the founding member of the angiotensin-converting enzyme (ACE) inhibitor class of antihypertensive therapy that today is a clinical standard of care. Atorvastatin is the leading drug for the treatment of hyperlipidemia and is the most profitable drug ever developed. Sirolimus is a clinically highly important drug for organ transplantation.

D. R. Kirsch (✉)
Department of Biology, Columbia University, 1130 Amsterdam Ave, New York, NY 10027, USA
e-mail: dk3148@columbia.edu

© The Author(s), under exclusive license to Springer Nature Switzerland AG 2022
N. Rezaei (ed.), *Multidisciplinarity and Interdisciplinarity in Health*,
Integrated Science 6, https://doi.org/10.1007/978-3-030-96814-4_13

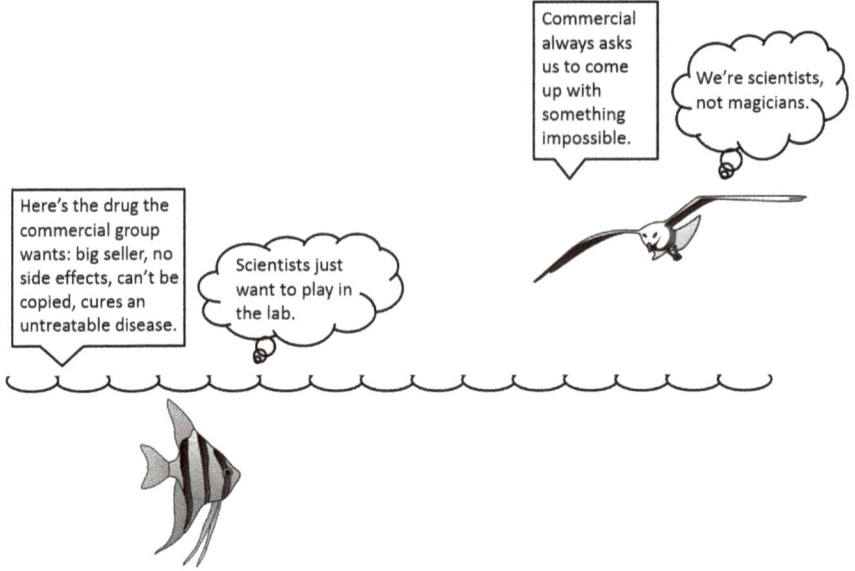

A commercial management "fish" miscommunicating with a scientist "bird"
The code of this chapter is *01100101 01100001 01101101 01101110
01001101 01110100 01100001 01100111 01100101 01101110.*

Keywords

Collaboration · Drug discovery · Pharmaceutical management

1   Introduction

"Do you know the only thing that gives me pleasure? It's to see my dividends coming in."

John D. Rockefeller (founder of Standard Oil)

Drug discovery is, in its essence, a collaborative activity. The most recent drug discovered by a single individual was diethyl ether for anesthesia in 1846 [1]. Since then, all drugs have been discovered by teams. The discovery of conventional "small molecule" drugs (think aspirin or hydrocortisone) requires the collaboration of medicinal chemists, pharmacologists, and animal scientists. For biologics, also called "large molecule" drugs, such as Humira (adalimumab), Eliquis (apixaban), or Herceptin (trastuzumab), you substitute monoclonal antibody scientists and protein engineers for the medicinal chemists. On top of that, you add in scientists specific for the therapeutic area. For an antibiotic, you also need microbiologists, for a cardiovascular drug, physiologists, for an enzyme inhibiting drug, enzymologists, for a cancer drug, cell biologists, etc.

Although these researchers come from diverse disciplines, they are all scientists and operate under similar professional goals and constraints: producing new knowledge, improving the human condition, adding to and protecting the integrity of the scientific record, ensuring that all data are of the highest quality, are statistically significant, and can be reproduced by unaffiliated laboratories. Scientists lose sleep worrying about making a significant discovery, something which is probably ten years away.

That drug discovery is most commonly pursued by commercial entities introduces an extra additional complication. Industrial scientists need to work productively with their commercial colleagues. These business people work under a different set of professional goals and constraints: maximizing return on investment, minimizing risk, optimizing the return on invested capital, producing consistent and growing positive cash flows, identifying and exploiting the most attractive markets, and maximizing profit margins and market penetration. Business people lose sleep worrying about financial performance in the current financial quarter.

It is in many regards astonishing that such a collaboration could be workable. Consider the metaphor of a bird and a fish. It would seem unreasonable to expect a bird and a fish to collaborate effectively. First of all, if they wanted to work together, where would these inhabitants of vastly different fluid environments find "common ground" to meet? Secondly, what would their common language and experiences be, their means of effective communication? And thirdly, what common professional interests and goals would they have to form the basis for a productive work relationship? However, the business and scientific birds and fish somehow manage to work together and, improbable as it may seem, regularly produce successful drug discovery projects.

2 Birds and Fish

Conflicts between science and business occur commonly and at a pretty fundamental level in the pharmaceutical industry. Comparing two different imaginary biopharmaceutical companies has previously been used to illustrate issues in science/business collaboration [2]. This author wrote:

> "NewBio develops a new drug. They are fortunate to find investors to fund the development of the drug through phase 3 clinical trials and the FDA approves the drug for marketing. However, the patient population receiving the drug is small and payer issues limit reimbursement. Although the drug provides significant clinical value to patients, earnings are limited and they ultimately never repay the R&D costs to discover and develop the drug."

> "NewPharma also develops a new drug. Their drug shows very positive results in its phase 2 clinical trial, and the drug is then purchased by a large pharmaceutical company for a high upfront payment, greatly more than the costs incurred by NewPharma, to bring it through phase 2. The large pharmaceutical company then takes the drug through phase 3 clinical trials where it unfortunately does not reach significance in achieving the desired outcome measures. As a result, the large pharmaceutical company drops the project."

"The scientists at NewBio are pleased that their project was a success while the scientists at NewPharma are disappointed by their failure. On the other side, the business people at NewBio feel that they failed while the business people at NewPharma are delighted by their success. With such a basic disagreement regarding the definition of success and failure, it should be of little surprise that this fundamental difference in perspective leads to conflict between the technical and commercial groups within biopharmaceutical industry. And pressures from each side of the organization can bring the other side into undesired territory."

Real-world examples of joint technical and commercial success are evident and easy to point out, as joint technical and commercial failures are too. However, there are also many cases of technical failure and commercial success. Sirtris Pharmaceuticals provides a good example of this [3]. Sirtris was launched in August 2004 to discover therapeutics that would extend longevity and treat diseases of old age based on a five-stage $102.9 million venture capital investment. Sirtis's initial public offering (IPO) in 2007 achieved a market capitalization of $278 million. A year later, Sirtris was purchased by Glaxo SmithKline for $720 million [4].

Glaxo SmithKline shut down Sirtris just five years later [5]. It currently appears that Sirtris science has yielded no new therapies. The initial investors who held their Sirtris investment for four years realized a sixfold profit on their investment, producing an annual rate of return of over 60%, not a bad profit from a technology that never worked.

3 Cost of Developing New Drugs

It has been estimated that the cost of development for a single new drug is approximately $1.4 billion [6]. The overall process takes 14–15 years from conception to regulatory approval, with an overall success rate in the 1% range [7–11]. Moreover, many marketed drugs never repay their R&D costs [12]. Thus, drug discovery success depends upon a high level of risk tolerance and the long-term availability of substantial funding.

In drug discovery, scientists conceive of and promote ideas for the new therapeutic agents, and commercial management supports the pursuit of ideas with the potential for future economic gain. Moreover, given their extremely different backgrounds and perspectives, negotiations regarding what projects should be pursued are often acrimonious with much disagreement [13].

In the early 2000s, Pedro Cuatrecases found himself in a somewhat unique position to assess how scientists and commercial management work together in large pharmaceutical companies. In 1975, after an almost 25-year academic medical research career at Washington University, the NIH, and Johns Hopkins University, Cuatrecasas moved to industry, taking a position at Burroughs Wellcome Co. the head of R&D. Following the merger between Burroughs Wellcome and Glaxo, he then became head of R&D at Glaxo. In 1989, he moved to Warner–Lambert/Park Davis until his retirement in 1997. Thus, he had an extensive academic career

Table 1 Drugs not supported by commercial management [13]	Company	Drug
	Burroughs Wellcome	Acyclovir (Zovirax) Azidothymidine (Retrovir) Bupropion (Wellbutrin)
	Park Davis/Warner Lambert	Gabapentin (Neurontin) Atorvastatin (Lipitor) Troglitazone (Rezulin) Fosphenytoin (Cerebyx) Pregabalin (Lyrica)
	AstraZeneca	Propranolol (Inderal)
	SmithKline Beecham	Cimetidine (Tagamet)
	Squibb	Captopril (Capoten)
	Merck	Lovastatin (Mevacor)
	Pfizer	Sildenafil (Viagra)

followed by 22 years of research management experience in what could be argued to be four culturally different pharmaceutical companies.

Through his extensive insider knowledge, Cuatrecases compiled a list of 13 regulatory approved and commercially successful drugs opposed by commercial management (Table 1). Many of these drugs were breakthrough therapies at the time of their clinical introduction, and all were profitable. From Cuatrecases' work and my own experiences in the pharmaceutical industry, I will illustrate the scientific/commercial dynamic using three major drugs, each of which initially was not supported by commercial management, each of which was developed despite this, each of which received regulatory approval, and each of which went on to become a leader in their respective therapeutic area.

4 Captopril (Capoten)

When I was a student in the 1970s, some textbooks still referred to what we today call idiopathic hypertension as essential hypertension, essential in the sense that it was necessary for the health of the patient. John Hay, who served as a medical professor at Liverpool University, wrote that "there is some truth in the saying that the greatest danger to a man with a high blood pressure lies in its discovery because then some fool is certain to try and reduce it" in 1931. And as recently as the 1950s, many physicians considered elevated blood pressure as required for adequate tissue perfusion and for vital organs to function appropriately and was thus dangerous to lower [14].

Studies published in the 1960s changed that view. According to the long-term longitudinal Framingham Heart Study, high blood pressure is associated with an increased risk of myocardial infarction, congestive heart failure, kidney disease, and stroke [15]. Moreover, a Veterans Administration study published in 1967 showed that drug treatment of blood pressure in severely hypertensive patients lowered the risk of congestive heart failure, kidney disease, and stroke [16]. The investigators in the Veterans Administration study faced extreme experimental design challenges. There were no highly effective, well-tolerated drugs for treating hypertension, especially for patients with the most extreme levels of increased high blood pressure.

All of the drugs available at that time, such as reserpine and hydralazine, produced severe, intolerable side effects. The only reasonably well-tolerated drug was chlorothiazide (Diuril), but chlorothiazide worked poorly in extreme hypertension patients [14]. The new understanding of high blood pressure created a high level of interest in the pharmaceutical industry. High blood pressure was a common condition, and patients taking antihypertensive therapy would be taking their drugs daily for the rest of their lives: years if not decades. Thus, the sales potential for antihypertensive medications was enormous.

However, patients who have high blood pressure feel perfectly fine and will not continue taking drugs that produce severe side effects. What was needed were drugs that reduced high blood pressure without producing adverse effects. The first of such compounds to be discovered were beta-blockers, which inhibit a specific subset of the receptors for epinephrine (adrenaline) and norepinephrine (noradrenaline).

Injected epinephrine or norepinephrine increases heart rate (tachycardia) and blood pressure through actions on two different types of receptors. Some researchers thought it should be possible to lower blood pressure without negatively affecting heart rate by selectively blocking only those receptors that modulate blood pressure and not those that increase heart rate. The first compound of this type, propranolol, was discovered by the British physician James Black in the early 1960s while working at ICI Pharmaceuticals [17, 18]. Soon other companies produced almost a dozen of their own versions of propranolol. While all of these drugs were a vast improvement over previous therapies, beta-blockers were not perfect compounds because they produced an array of unpleasant side effects, e.g., gastrointestinal side effects, pulmonary side effects, cardiac side effects, and in men sexual side effects. Thus, there was keen interest in finding better approaches for the control of blood pressure.

Researchers at Squibb Pharmaceuticals decided to focus their efforts on a different and, at the time, poorly characterized system for blood pressure control: the renin/angiotensin system [19]. At this time, the renin/angiotensin system was thought to be of only minor importance in regulating blood pressure. Most scientists predicted that drugs acting on this system would have small and likely not clinically valuable blood pressure control effects.

The Squibb scientists felt otherwise. Squibb had hired the British scientist John Vane as a consultant, and Vane promoted the idea of targeting a particular component of the renin/angiotensin system, an angiotensin-converting enzyme called ACE for short. However, without a selective ACE inhibitor, it was not possible to conclusively prove whether or not ACE inhibition would produce a substantial lowering of blood pressure. Vane suggested using the venom from the Brazilian pit viper snake, *Bothrops jararaca*, for this purpose. Vane knew of an unpublished study on the venom's effects on the renin/angiotensin system and suggested that purifying the ACE inhibiting active ingredient from the venom could produce the needed reagent.

Squibb scientists soon isolated teprotide from the snake venom, which they showed to be an effective antihypertensive drug, proving that ACE inhibition was a good target for antihypertensive therapy. Teprotide was not a clinically useful drug because it was costly to synthesize and lacked oral activity. However, a different molecule with teprotide's ACE inhibitory activity could theoretically become a clinically important drug [20].

However, the teprotide project was discontinued as a clinical candidate because of a lack of commercial interest. The chemistry team leader, Miguel Ondetti, was reassigned to antibiotic projects, and the pharmacology team leader, Dave Cushman, to a prostaglandin project. Fortunately, two key R&D managers, Zola Horowitz (pharmacology department director) and Arnold Welch (head of research) were willing to take the risk of promoting the continuation of the ACE inhibitor project in the absence of commercial support. But how would the desired clinically useful ACE inhibitor be found? The team tested 2000 diverse chemical structures for ACE inhibition with no hint of useful activity.

On March 13, 1974, the ACE inhibition team discussed a paper on an enzyme similar to ACE called carboxypeptidase A [21]. This paper described the design of an inhibitor of carboxypeptidase A based on the substrate's structure for the enzyme and its catalytic action. The Squibb team realized that they could take an analogous approach to design an inhibitor of ACE.

A year later, the team identified an orally active ACE inhibitor, and after an additional half-year of structural optimization, they discovered captopril. It was lightning-fast work for a small team, and the approach they took is now hailed as the model for the rational design strategy in drug discovery. Cushman and Ondetti won the 1999 Lasker Medical Research Award (often called America's Nobel) for this discovery.

The problem was that commercial management was not interested. Clinical studies are much more expensive than preclinical laboratory studies, and commercial management did not want to spend a lot of R&D money on what they considered to be a risky project. First, they were wary of pursuing an unproven approach that many highly respected scientists in the field claimed was unlikely to work. Secondly, they also worried about their own "me too" beta-blocker, nadolol (Corgard). Nadolol had decent sales, and it was argued that if patients taking nadolol switched to captopril, there would be no net increase in sales. The project was over.

Despite the disinterest from commercial management, Cushman and Ondetti were proud of their work and wanted to publish their findings for the scientific community. Since patent applications had already been filed, they argued that there was no risk in publishing their work should commercial management change their minds. Commercial management allowed them to publish their work, but with the restriction that they spend full time on their new projects and only work on their captopril manuscripts at night and on weekends. Over the next few years, Cushman and Ondetti completed three papers and published them in highly respected scientific journals [22–24].

After these papers appeared, Squibb was contacted by several top American medical schools expressing interest in participating in the clinical trials for captopril. However, no such trials had been planned! It made commercial management rethink their decision. They had been worried that Cushman and Ondetti had not been objective in promoting their pet project and discounted their arguments. However, the scientists at top-tier medical schools had no reason to show a bias in supporting captopril. The medical school professors were only interested in helping their patients.

As a result, commercial management reversed their decision and decided to fund the clinical development of captopril. Although there were some problems with early trials due to overconfidence in the drug's safety, later studies showed that the drug was a safe, well-tolerated, and highly effective treatment for hypertension [25]. Captopril was approved by the US FDA on April 6, 1981, and soon went on to become a blockbuster drug (annual sales greater than $1 billion).

Cushman and Ondetti's strategy for promoting their project to Squibb management was to let good science speak for itself. And that is pretty much what happened. Once commercial management saw the enthusiasm for captopril expressed by outside biomedical scientists, they decided to commercialize and were richly rewarded for their efforts.

Captopril was wildly successful financially, earning Squibb more money than even the most optimistic supporters of the drug expected. Soon Bristol-Myers realized that the Squibb stock price did not reflect the financial prospects for captopril. They, therefore, successfully launched a hostile takeover of Squibb that resulted in the creation of Bristol-Myers Squibb in 1989 [26]. The Squibb stock price rocketed from $24.75 per share to $112.50 per share following the $12 billion Bristol-Myers merger offer. Thus ironically, captopril's commercial success led to the demise of Squibb as an independent pharmaceutical company.

Lastly, while the overall project was highly profitable, the lack of good teamwork between R&D and commercial management resulted in significant missed opportunities. Following the commercial launch of captopril, both sides were mentally and emotionally exhausted. Captopril was not a perfect drug because it had to be dosed 2–3 times a day and had some side effects, such as producing a metallic taste, because captopril is was a sulfhydryl compound. Long-term dosed drugs will be more successful if they are clean of side effects and are easy to take. This liability was appreciated by Merck, who, after reading the Cushman and Ondetti papers published in 1977–1978, started their own programs to come up with improved versions of captopril, something Squibb could have easily done but chose not to do.

Merck quickly designed compounds that had the desired improved dosing and tolerability properties, including enalapril (Vasotec) and lisinopril (Prinivil). Lisinopril has now become a standard of care for the management of hypertension [27]. Squibb made much money but lost significant market share to competing compounds that could be quickly discovered based on the Squibb publications. Squibb could have made a lot more if they had delayed publication of their work, accelerated the development of captopril, and continued their research efforts to discover improved follow-on compounds. But commercial management was suspicious that their own scientists had lost objectivity on their own work and decided to support the project only after outside key opinion leaders expressed support for Squibb's published work on captopril. It was a scientific and commercial success but also a missed opportunity.

5 Atorvastatin (Lipitor)

Like hypertension, high serum cholesterol in producing disease risk is well-documented [28]. In the later 1940s, a highly regarded biochemistry textbook stated that "[t]here is no satisfactory evidence that the incidence of atherosclerosis bears any relationship to the concentration of cholesterol in the blood" [28]. Then, similar to its role in showing the relationship between hypertension and disease, the Framingham Heart Study showed that the risk of developing coronary artery disease was directly related to serum cholesterol levels [29]. Also, similar to the treatment of hypertension, when the relationship between high serum cholesterol and coronary artery disease was demonstrated, no drugs existed that effectively lowered cholesterol levels at well-tolerated doses. For example, niacin will lower serum cholesterol when dosed at high levels, but this commonly produced flushing and dyspepsia that most patients found to be intolerable. Moreover, similar to hypertension, high serum cholesterol, which is called hyperlipidemia, is common and would be treated as a chronic condition for years, if not decades. Also, similar to hypertension, the sales potential for new medications to treat hyperlipidemia was enormous.

Drug discovery scientists took a biochemical approach to find drugs to treat hyperlipidemia. Cells synthesize cholesterol through a long, thirty-step pathway [30]. Since 1960, 3-hydroxy-3-methylglutaryl coenzyme A (HMG-CoA) reductase had been shown as the key, rate-limiting enzyme for cholesterol synthesis [31]. Thus, the drug discovery strategy was to find a selective inhibitor of this enzyme for hyperlipidemia treatment. The tactic first used to find such an inhibitor was to randomly screen natural product chemicals produced by soil microorganisms. The first HMG-CoA-reductase inhibitor, compactin, produced by the soil fungus, *Penicillium citrinum*, was discovered by scientists working for Sankyo Pharmaceuticals in 1976 [32]. Shortly after that, scientists at Merck reported their own chemically very similar inhibitor lovastatin (Mevacor), which is produced by the edible fungus *Pleurotus ostreatus* [33].

Merck aggressively pursued the development of lovastatin, beginning clinical trials in 1980 [34]. A few months later, rumors circulated that Sankyo had abandoned the development of its very similar compound, compactin, because it was carcinogenic. As a result, the lovastatin clinical development project was terminated.

Fortunately, Merck at that time was unusual within the pharmaceutical industry in that R&D was given a powerful voice in decision-making, in no small part because the Merck CEO at the time, Roy Vagelos, was a physician-scientist. So, Merck scientists continued to study lovastatin. Their carefully designed studies showed that lovastatin was not carcinogenic and studies in a small group of highly ill patients with extraordinarily high serum cholesterol levels showed that the drug was highly effective and well-tolerated [35].

The rumors of compactin carcinogenicity were never substantiated, although Sankyo never developed the drug and instead substituted a second cholesterol drug they had discovered, pravastatin (Pravachol) [30]. Squibb soon licensed pravastatin for development in the USA using their newfound riches from sales of captopril. In 1987, lovastatin was the first drug of its type to be approved for sale by the FDA, followed by three additional HMG-CoA reductase inhibitors during the next seven years.

The first four reported HMG-CoA reductase inhibitors were natural products. Companies needed to have a dedicated natural products group to pursue this type of research, and companies that lacked such a resource were thought to be at a significant disadvantage for the development of drugs of this type. Merck had pursued a parallel standard chemical effort to produce an HMG-CoA reductase inhibitor by conventional organic synthesis but never found a compound that was superior to the two natural product-based drugs they had found [36].

However, despite the lack of a natural products group, several companies pursued their own HMG-CoA reductase inhibitors via conventional chemical strategies. One of these was Warner–Lambert. Scientists at Warner–Lambert produced atorvastatin (Lipitor) in the mid-1980s following a conventional chemical synthesis approach. The problem was that there would likely be four other HMG-CoA reductase inhibitors on the market by the time atorvastatin was approved [30].

Warner–Lambert commercial management felt that being the fifth entry to the market with a drug that appeared to have no distinct advantages over compounds already on or soon to be on the market was a flawed idea. Commercial management thought that atorvastatin was going to be a financial loser. However, their arguments were based on a central paradox of pharmaceutical drug discovery. The only way to see if your compound is clinically superior is to perform clinical trials. However, clinical trials are costly, and it is hard to justify spending that money if a marketing advantage is not already apparent.

Atorvastatin was much more potent than competing drugs in laboratory studies, and it was arguable that patients would likely need to take less drug substance to achieve the same therapeutic effect. It was arguable that by taking less drug patients were likely to see fewer side effects. However, commercial management was close to terminating the project anyway. Bruce Roth, the inventor of atorvastatin, argued for one last chance to test atorvastatin in a human clinical trial and was supported by the Warner–Lambert head of research, Ronald Cresswell.

The Warner–Lambert scientists won commercial management approval for a small clinical trial including only 24 subjects [37]. This trial showed that the earlier animal studies had greatly underestimated the effects of atorvastatin. Dosed at just ten milligrams, it reduced low-density lipoprotein (LDL), also known as the bad cholesterol, more effectively than rival drugs dosed at their far higher FDA-recommended doses. It was clear from this simple clinical experiment that atorvastatin was a superior drug.

Warner–Lambert needed a large marketing organization to maximize profits from their late market entry. So, in 1996, they signed an agreement with Pfizer, who had no cholesterol-lowering drug of their own but an enormous sales force, to co-market atorvastatin. Between the initial drug launch in 1997 and the patent expiration, atorvastatin (Lipitor) generated more than $80 billion in sales, the world's best-selling blockbuster drug [38].

What was most important in this instance was that: (i) commercial management at Warner–Lambert was willing to compromise with R&D to support a limited clinical trial; and (ii) commercial management fully accepted the results of that trial and took the decision to aggressively move forward with atorvastatin (Lipitor) development and commercialization.

6 Sirolimus/Rapamycin (Rapamune)

American Home Products (AHP) was founded as a holding company in 1926, and the holding company philosophy was central to their business strategy [39]. AHP had no allegiance to any particular industry or business sector. They bought and sold consumer goods companies based solely upon their financial performance. In the last half of the twentieth century, AHP products included Dristan nasal spray, Anacin headache remedy, Preparation H for hemorrhoids, Chef Boyardee spaghetti, Mama Leone's pasta, Gulden's mustard, Wrigley's chewing gum, Woolite laundry detergent, Black Flag insecticide, Old English furniture polish, PAM cooking spray, and a line of aluminum pots and pans. They also owned two pharmaceutical companies.

In 1931, AHP purchased Wyeth, an old-line nineteenth-century pharmaceutical company. In 1943, AHP acquired the Canadian pharmaceutical company Ayerst, McKenna and Harrison, Ltd. AHP was so invested in their holding company philosophy that they did not merge their two pharmaceutical companies until 1987, waiting almost 45 years despite the obvious economies of scale cost savings that would result. It is said that the Wyeth-Ayerst merger was one of the most contentious in the pharmaceutical industry. For four and a half decades, the two companies were operated as bitter rivals competing for corporate resources and had built up a long history of ill will and animosity.

In the early 1970s, Ayerst started a project to discover a new antifungal agent, which resulted in the discovery of sirolimus [40]. Sirolimus is an antibiotic produced by a soil streptomycete bacterium that Ayerst scientists had collected on Easter Island, and which they originally named rapamycin after the Polynesian name for Easter Island, Rapa Nui. Sirolimus is a potent antifungal drug, but further testing showed that it also has potent immunosuppressive properties. Antibiotics need to work in concert with the immune system to cure infections, so sirolimus would not be a clinically useful antifungal agent.

Suren Sehgal, the sirolimus project leader, soon became intrigued by the immunosuppressive properties of the drug. He thought that it might be useful as a treatment for organ transplant patients. Organ transplant patients need to take drugs that suppress their immune systems not to reject the transplanted organ. AHP had no interest in organ transplant drugs, so commercial management terminated the project.

It is a common misapprehension that new drugs are designed like mobile phones or electric cars. However, this is not true [1]. In reality, drugs are discovered, and as a result, you find what you find, which might not be what you wanted. This is not because scientists are will fully disregarding the corporate business plan. The challenge is to see whether you can come up with a clinical use for whatever you find and a way to make it economically viable.

In the late 1970s, Sandoz Pharmaceuticals was supporting the development of an immunosuppressant therapy called cyclosporine (Sandimmune) [41]. Their drug eventually proved to be an important therapy for organ transplant and helpful in treating autoimmune diseases [42]. AHP business management seemed to feel differently. Because of their holding company strategy, AHP was not very supportive of long-term R&D. Long-term R&D investments do not pay off when subsidiaries are bought and sold for short-term gain. Some managers argued against supporting any R&D at all, saying that if they needed new products to support their businesses, they could negotiate good terms to buy the products they needed from other companies. Commercial management knew what they wanted and did not want to endure the risks and business circumnavigations produced by an R&D organization.

Sehgal was told to terminate the project and, to make sure that this happened, he was told to kill the soil microbe that produced sirolimus by autoclave sterilization. Once the soil microbe that made sirolimus was dead, there would be no way to make the compound and thus no further argument about pursuing the project. Although Sehgal was an extremely honest man, this time, he did something a little devious. He sterilized the culture but also brought a replicate culture home with him that he stored in his basement freezer [43].

After that, there was the Wyeth merger and many management changes. Each time commercial management changed, Sehgal pitched his project to a new group, hoping to finally find someone interested in supporting the development of an immunosuppressant therapy.

Fortunately for Sehgal and AHP, in 1987, a similar drug, tacrolimus, was discovered by Fujisawa Pharmaceuticals. Fujisawa decided to advance the drug into clinical development as an immunosuppressant therapeutic [44]. Since Fujisawa saw a financial opportunity with their drug, AHP decided that it might be worthwhile to pursue something similar. So Sehgal finally found support for his project, and, almost 30 years after its discovery, sirolimus was finally approved in 1999 for use as an immunosuppressive drug in renal transplantation [44]. Sirolimus is also used as a coating for coronary stents to prevent restenosis [45]. More recently, sirolimus has been approved to treat lymphangioleiomyomatosis, a rare progressive neoplastic disease associated with thin-walled cysts in the lungs and angiomyolipomas in the kidneys [46]. Beyond its clinical use, studies of the sirolimus mechanism of action enabled the elucidation of a major new cell regulatory pathway that connects cellular metabolism and growth [47].

7 Conclusion

The clinical utility of a new drug can only be determined through hugely expensive clinical trials. As a result, we do not know how many valuable drugs were discovered and then discarded due to financial constraints before their clinical utility was ever determined. At the very least, it is commonly said that every successful drug has at some point in its development a near-death experience. Fortunately, the three drugs described here were able to proceed to clinical and commercial success, each through their own unique route.

In the case of captopril, Cushman and Ondetti relied on their belief that good science would speak for itself and eventually lead to success. As they hoped, following the publication of Cushman and Ondetti's research, disinterested scientists from major medical schools came forward to show their support for the compound, convincing commercial management to advance captopril into successful clinical trials.

Atorvastatin was moved into full clinical development due to an agreement worked out between the scientists and commercial management at Warner–Lambert to run a very small clinical trial to determine whether the compound showed superior performance in patients. The trial results were positive, and both scientists and commercial management trusted these results, leading to the aggressive support of atorvastatin for commercial development.

The success of sirolimus was based upon Suren Sehgal's long-term cautious, gentlemanly championing of the compound, keeping the project alive until a competitor decided to commercially advance a similar compound, which convinced Wyeth commercial management to proceed.

Core Messages

- Discovering new drugs is costly and requires a very high level of technical skill.
- Drug discovery depends on the collaboration between scientists and commercial management.
- There is no single best practice to discover a new drug.

References

1. Kirsch DR, Ogas O (2016) The drug hunters: the improbable quest to discover new medicines. Arcade Publishing, New York
2. Kirsch DR (2020) Therapeutic drug development and human clinical trials in biotechnology entrepreneurship: starting, managing, and leading biotech companies (Shimasaki C ed) 2nd edn. Elsevier, Amsterdam
3. Stuart T, Kiron D (2008) Sirtris pharmaceuticals: living healthier, longer. Harvard Business School case study N9-808-112
4. Herper (2008) Why Glaxo Bought Sirtris. https://www.forbes.com/2008/04/23/pharmacuticals-sirtris-glaxosmithkline-biz-healthcare-cx_mh_0424glaxo.html#5695e7011762. Retrieved 8 Aug 2019
5. Ledford (2013) GSK absorbs controversial 'longevity' company: news blog. http://blogs.nature.com/news/2013/03/gsk-absorbs-controversial-longevity-company.html. Retrieved 8 Aug 2018
6. DiMassi JA et al (2016) Innovation in the pharmaceutical industry: new estimates of R&D costs. J Health Econ 47:20–33
7. Brown D, Superti-Furga G (2003) Rediscovering the sweet spot in drug discovery. Drug Discovery Today 8:1067–1077
8. Nwaka S, Ridley RG (2003) Virtual drug discovery and development for neglected diseases through public–private partnerships. Nature Rev Drug Discov 2:919–928
9. Koppal T (2004) Wyeth's internal revolution. Drug Discov Devel 7:24–28
10. Reichert JM (2003) Trends in development and approval times for new drugs in the United States. Nature Rev Drug Discov 2:695–702
11. Drews J (2000) Drug discovery: a historical perspective. Science 287:1960–1964
12. Garnier J-P (2008) Rebuilding the R&D engine in big pharma. Harvard Business Review
13. Cuatrecasas P (2006) Drug discovery in jeopardy. J Clin Invest 116:2837–2842
14. Block (2006) Historical perspectives on the management of hypertension. J Clin Hypertens 8(supp2):15–20
15. Kannel WB et al (1969) Blood pressure and risk of coronary heart disease: the Framingham study. Dis Chest 56:43–52
16. VA Cooperative Study Group (1967) Effects of treatment on morbidity in hypertension. Results in patients with diastolic blood pressures averaging 115 through 129 mm Hg. JAMA 202(11):1028–1034
17. Stapleton MP (1997) Sir James black and propranolol. Texas Heart Inst J 24:336–342
18. Black J (1989) Drugs from emasculated hormones: the principle of syntopic antagonism. In Vitro Cell Devel Bio 25:311–320
19. Cushman DW, Ondetti MA (1991) History of the design of captopril and related inhibitors of angiotensin converting enzyme. Hypertension 17:589–592
20. Cushman DW, Ondetti MA (1999) Design of angiotensin converting enzyme inhibitors. Nat Med 5:1110–1112
21. Byers LD, Wolfensen R (1973) Binding of the by-product analog benzylsuccinic acid by carboxypeptidase A. Biochemistry 12:2070–2078
22. Cushman DW et al (1977) Design of potent competitive inhibitors of angiotensin-converting enzyme. Carboxyalkanoyl Mercaptoalkanoyl Amino Acids Biochem 16:5484–5490
23. Ondetti MA et al (1978) Design of specific inhibitors of angiotensin-converting enzyme: a new class of orally active antihypertensive agents. Science 96:441–444
24. Cushman DW et al (1978) Design of new antihypertensive drugs: potent and specific inhibitors of angiotensin-converting enzyme. Progress Cadiovascular Dis 21:176–182
25. Frohlich ED et al (1984) Review of the overall experience of captopril in hypertension. Arch Intern Med 144:1441–1444
26. Brooks (1989) Los Angeles Times. https://www.latimes.com/archives/la-xpm-1989-07-28-fi-295-story.html. Retrieved 24 April 2020

27. Zaman MA et al (2002) Drugs targeting the renin-angiotensin system. Nat Rev Drug Disc 1:621–626
28. Steinberg D, Gotto AM Jr (1999) Preventing coronary heart disease by lowering cholesterol levels. JAMA 282:2043–2050
29. Castelli WP (1984) Epidemiology of coronary heart disease: the Framingham study. Amer J Med 76(sup2A):4–12
30. Tobert JA (2003) Lovastatin and beyond: the history of the HMG-CoA reductase inhibitors. Nat Rev Drug Discov 2:517–526
31. Bucher NLR et al (1960) β-hydroxy-β-methylglutaryl coenzyme a reductase, cleavage and condensing enzymes in relation to cholersterol formation in rat liver. BBA 40:491–501
32. Endo A et al (1976) ML-236A, ML-236B and ML-236C, new inhibitors of cholesterolgenesis produced by Penicillium citrinum. J Antibiotics 29:1346–1348
33. Alberts AW et al (1980) Mevinolin: a highly potent competitive inhibitor of hydroxymethylglutaryl-coenzyme A reductase and a cholesterol-lowering agent. PNAS 77:3957–3961
34. Vagelos PR (1991) Are prescription prices high? Science 252:1080–1084
35. Illingworth DR, Sexton GJ (1984) Hypocholesterolemic effects of mevinolin in patients with heterozygous familial hypercholesterolemia. J Clin Invest 74:1972–1978
36. Roth BD (2002) The discovery and development of atorvastatin, a potent novel hypolipidemic agent. Progress Med Chem 40:1–22
37. Simmons J (2003) The $10 billion pill, Fortune, 20 Jan 2003. https://archive.fortune.com/magazines/fortune/fortune_archive/2003/01/20/335643/index.htm. Retrieved 29 April 2020
38. Jack A (2009) The last of the blockbusters. Financial Times 216–219
39. Chandler (2005) New learning at American home products. Harvard Business School Working Knowledge, April 25. https://hbswk.hbs.edu/item/new-learning-at-american-home-products. Retrieved 1 May 2020
40. Sehgal SN (2003) Sirolimus: its discovery, biological properties, and mechanism of action. Transpl Proc 35(Suppl 3A):7S-14S
41. Borel JF et al (1976) Biological effects of cyclosporin A: a new antilymphocytic agent. Agents Actions 6:468–475
42. Laupacis A et al (1982) Cyclosporin A: a powerful immunosuppressant. Can Med Assoc J 129:1041–1046
43. Loria K (2015) A rogue doctor saved a potential miracle drug by storing samples in his home after being told to throw them away. Business Insider, February 12. https://www.businessinsider.com/suren-sehgal-saved-rapamycin-anti-aging-drug-2015-2. Retrieved 4 May 2010
44. Napoli KL, Taylor PJ (2001) From beach to bedside: history of the development of sirolimus. Ther Drug Monit 23:559–586
45. Moses JW et al (2002) Perspectives of drug-eluting stents the next revolution. Am J Cardiovasc Drugs 2:163–172
46. Xu F-F et al (2020) Lymphangioleiomyomatosis. Semin Respir Crit Care Med 41:256–268
47. Kim J, Guan K-L (2019) mTOR as a central hub of nutrient signalling and cell growth. Nat Cell Biol 21:63–71

Donald R. Kirsch is a bio/pharmaceutical industry consultant with more than thirty-five years of industrial research experience, including stints at Bristol-Myers Squibb, Cyanamid/Lederle, American Home Products, and Genetics Institute/Wyeth. He was most recently the chief scientific officer at a Cambridge Kendall Square biotech startup, Cambria Pharmaceuticals. He received his bachelor's degree from Rutgers College and his master's degree and doctorate from Princeton University. He served as an instructor in the Pharmacology Department of Robert Wood Johnson Medical School before embarking on his industrial career. He received a certificate in technical management from Stanford Business School in 1991. He is the author of over fifty scientific research papers and review articles and an inventor on twenty-seven patents and patent applications. He currently has an adjunct faculty position at Columbia University, where he teaches a drug discovery course.

A Phenomenological Analysis of the Pandemic: Philosophy and Life

14

Juan José Garrido Periñán

"Der Mensch ist in seinem Wesen das Gedächtnis des Seins..."

Martin Heidegger

Summary

The chapter is about generating an opportunity to reconfigure a sense of the philosophical question from the context initiated by the current pandemic. Understanding philosophy only in terms of philosophizing, the author shows how philosophy can only be born from disruptive, extreme, and insecure situations. In this respect, an uncertain time like the present COVID-19 pandemic means an opportunity to develop the philosophy. Finally, the project of an ontology of human life based on the tension between proximity and distance is outlined.

J. J. Garrido Periñán (✉)
Faculty of Philosophy, Department of Aesthetics and History of Philosophy,
University of Seville, Calle Camilo José Cela S/N, 41018 Sevilla, Spain
e-mail: jjgarper@us.es

Integrated Science Association (ISA), Universal Scientific Education and Research Network
(USERN), Sevilla, Spain

N. Rezaei (ed.), *Multidisciplinarity and Interdisciplinarity in Health*,
Integrated Science 6, https://doi.org/10.1007/978-3-030-96814-4_14

311

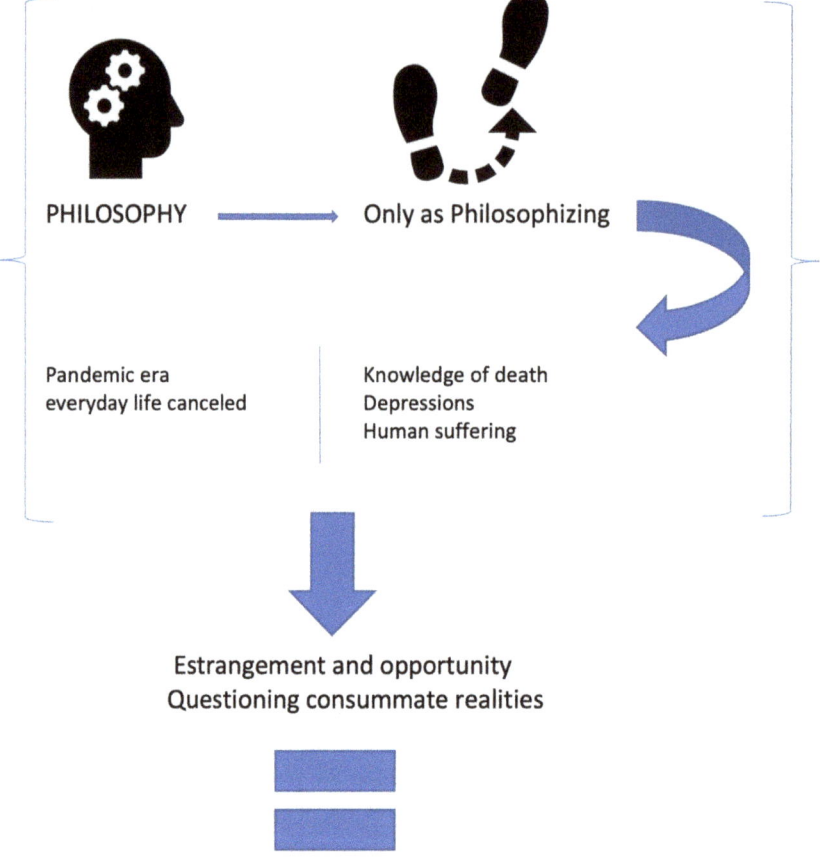

Philosophy: a rehabilitation exercise in favor of life

The code of this chapter is *01101001 01,101,110 01,100,111 01,101,001 01,000,010 01,100,100 01,101,110*.

Keywords

Estrangement · Existence · Pandemic · Philosophizing

1 Introduction

Philosophy is an existential exercise ever-evolved in the course of a here and now, through which the human being, almost by imperative, must face up to questioning realities made and consummated. Philosophy and crisis are constitutive elements of the same being, especially if we understand crisis with respect to its Greek heritage (*krisis*) as a "decision." In the etymological dictionary of the Spanish language, which is cited here because Spanish is the author's mother tongue, Joan Corominas says for "crisis": a "serious change that occurs in an illness, for improvement or worsening," "a decisive moment in a matter of importance"[1] [1]. If one reads between the lines, one can perfectly understand that by crisis, one is alluding to a fundamental moment that would augur a radical transformation,[2] either of a medical matter or in relation to a person, a human life. This sense, philosophy understood as a matter of practical conditions, is not new but has a long tradition dating back to ancient Greece. Crisis at the same time is related to a decisive, vital moment so transcendental that once it has happened, it will give birth to a new reality, a new existence. Philosophy would then be that knowledge capable of carrying out the decision that feeds on the critical moments. The wide range of these critical moments is known to all: from the news of a loved one's death to the elusive circumstances we have recently experienced thanks to coronavirus disease (COVID-19). These are the moments that call for the possibility of opening up philosophizing because they allow for the weakening of consecrated realities, making possible the establishment of other ways of questioning, which are usually silent and omitted in the course of normal, daily life. In this exceptional state, with lives enveloped in a viral pandemic, where human individuals have necessarily had to modify their modes of existence, where convictions and routines are turned upside down, philosophy has an opportunity, a propitious occasion to flourish. The sciences study concrete realities under the optic of a specific method, where mathematics is, in general, very valuable. Philosophy, more than studying and observing reality, gets involved in it, or better said: it immerses itself in it in order to, if possible, emerge, cast its gaze about and decide, dedicate itself to resolve a situation or circumstance, but with *phronesis* that is, with prudence.[3] Philosophy is very much in need of these critical moments; it is deeply involved in them and cannot take refuge in icy objectivity or the unwavering vision of an impartial observer.[4]

---

[1] My translation.

[2] "Philosophy itself can only be achieved by a reversal in the path, but not by a simple reversal, so that knowledge would be directed only to other objects, but, in a more radical way, by "an authentic transformation" (*eine eigentliche Umwandlung*)" [2]. My translation.

[3] See what Heidegger says about that: "Philosophy cannot be defined in the usual way, nor can it be characterized through an ordering in a «thematic relationship» (*Sachzusammenhang*), as when it is said: chemistry is a science and painting is an art. An attempt has also been made to put philosophy in its place through a «conceptual system» (*Regriffssystem*), saying that it dealt with a certain object in a certain way. But here, too, the scientific conception of philosophy is being inoculated. The principles of thinking and knowledge remain constantly unexplained..." [2]. My translation.

[4] "Philosophy cannot be science; it may not lapse into attitudinal determination. Philosophizing lies before the turn into attitude and before the shaping of experience into the tasks of theoretical research (...) «All worldview philosophy spoils the primordial motive of all philosophizing» (*Alle Weltanschauungsphilosophie verdirbt das Ursprüngliche Motiv alles Philosophierens*) [3].

In what follows, the chapter reconstructs the conditions under which philoso-phizing[5] can grow by looking at the face of existence and emerging from it with the idea of rooting philosophical questions in the specific and current ground. Such soil is aerated in the circumstances and situations typical of a pandemic era, like ours, evoked by COVID-19. The chapter will not deal with the pandemic under any guiding idea that seeks to capture it under a definition or model, nor will it try to give a concrete definition of the so-called "COVID-19" virus. Rather, it seeks *to reconfigure an omitted possibility that the current situation of pandemic has made possible* and which would be related to the resurgence of an existential way of understanding philosophy; that is, philosophy as philosophizing, nothing more.

2 Philosophy in the Face of Human Suffering

Even though COVID-19 is today spreading over the Earth even more strongly than a few months ago, it is still not yet possible to compare the effects that such a viral condition will leave in human beings without referring exclusively to the medical effects related to the human physiological condition. Nor is it possible to imagine the existential effects, including the so-called "diseases of the soul," that this pandemic will bring about for millions of people's lives. In strictly philosophical terms, mental health is about the human condition facing its existence with a minimum of pleasure and to the detriment of moments that lead us to suffer, which, as we know, is a source of displeasure. The situation before the pandemic was not at all encouraging, as depression was already the most widespread existential disease.

It can be seen how the number of cases of depression has gradually increased, paradoxically, at the same time that human living conditions have materially improved, including a decrease in poverty. One does not have to be very perceptive to know that COVID-19 will not improve these data but will instead aggravate them. During several months, people have been forced, through the policies of some governments, to live cloistered for some months, with a clear and forceful limitation to their social lives [4]. The social life allows, in general, the unfolding of extroverted behavior and the human coexistence, where affective bonds, without a doubt very important, such as friendship and love, are formed.

What can philosophy say about this? Can it provide us with any remedy for this situation? Honestly, the author regrets having to answer the second question with a no, which is not categorical, though it is unequivocal: philosophy is not here to serve as self-help, to summon up lenitive attitudes before a reality that engages us more and more urgently. However, according to this author's criterion, neither can philosophy become an auxiliary of scientific research. Many academics imagine it

---

[5] I will intentionally use the expressions "philosophy" and "philosophizing" as identical, even if I only understand that philosophy is to "philosophize". The term "philosophy" would refer to an erudite and theoretical knowledge, of historical study, "philosophizing" to the action of questioning the world here and now. The reason why I will use both expressions is because of style, so as not to repeat the word "philosophizing" too often within the body of the text.

to be under the rather unclear argument that philosophy is *knowledge of a second degree* that is, a complement of the same sciences. In this sense, the philosopher should be aware of the novelty of scientific discoveries to advance them. Not sharing this idea, the author will say what philosophy can do if it is understood not as a theoretical discipline, but as an existential possibility whose undertaking is to transform human life lucidly: *philosophy can make it possible for a human being, always stressed by the affective dispositions that surround and govern us, to learn from his condition of suffering being, of being-pathic*. In relation to depression, as a problem, philosophizing cannot even start from the definition of an "illness" because it would also have to define "health" and because it does not have the competencies extracted from a certain scientific method to position itself [5]. However, this does not mean that philosophy should be silenced and become the usufruct of an erudite attitude in which only dead knowledge is involved. Philosophy incorporates, from the root, knowledge about *pathos* that is: philosophy is in the first instance a pathology, a science of suffering, an attempt to clarify what we suffer because what we suffer is always already experienced and through this prism understood, assumed or interiorized in a certain way. An example will be sufficient for understanding what said: a human subject who is mourning the loss of a loved one does not only suffer because of one condition or another (emotional ties, memories, etc.); his suffering is also because he cannot dissociate suffering from what he is, from his existence; a human individual incorporates suffering because he experiences it; he makes it his own, whether he wants it or not. A brief mention of the work of two psychiatrists will suffice for understanding the imposition of this pathetic condition of existence: Emil Kraepelin (1856–1926), a pioneer in the field of modern psychiatry, linked "mental illnesses," understood as "psychic disturbances" and as "biological alterations," based on the materiality of the body. A little later, Kurt Schneider (1887–1967), the great theoretician of the study of schizophrenia, related so-called "mental illnesses" to learning processes linked to social *praxis*. Schneider's step was a giant one, in so far as he made it possible to expand on mental illness causes, taking them out of the biological realm and incorporating them into social processes.

As mentioned above, philosophy cannot define what "disease" is; it has no competence to do so. Philosophizing cannot define pathological behavior because it cannot precisely differentiate a usual behavior from an unusual one. What it can do is to accept human life as it is, understanding that everything that can be considered as a "psychic illness" is a way of existing derived from *being-in-the-world*, where the task to be done is not resolved, and it demands that we do something that we understand it, or that we assume it in a certain way. Schneider's gesture of associating illness and social praxis is precious in that it implies a regression in which philosophizing can be at ease: every psychic illness is an existential position that an individual assumes, whether he or she takes responsibility for it or not. Rooted in the human pathetical condition, philosophy is situated in a condition in which it abstains from judging in the face of what is healthy and sick: life, which is existence, is always already suffering; what is at stake in it are the ways, our guises, as phenomena of assumption and self-understanding in the face of what happens to us.

Is philosophy so blind as to not realize that there are harmful impulses within human behavior? Is philosophy incapable of pronouncing itself when it observes that human subjects suffer from pathological and neurotic behavior, which often leads to self-destruction? The answer, the author, does believe, is obvious: philosophy knows but refrains, at first glance, from a definition.[6] By considering that every possible psychic illness is a way of being-in-the-world, to follow an expression of the philosopher Martin Heidegger [6], one does not want to deny the variety of ways susceptible to human behavior. This is the second philosophical sense that can be given to the coinage of "pathology": philosophizing, starting from a generous attitude before what is given and what happens, would have to glimpse the sense by which a human life faints, renounces its being, depowers, denies itself [7]. Here a philosophical designation can already appear regarding what can be considered as "sick,"[7] which is not opposed to "health," but to life itself that is, to what prevents us from being and growing in the development of our existence. Having said that human life is nurtured by ways of existence that bear the sign of self-understanding, the "pathological" would be understood as all that bombards us to destroy our course in the world. To exist is to resolve oneself as a power of being; *the behavior labeled as "sickly" would be those that insert their respective powerlessness into this power of existing.*

The following section gives philosophical meaning to this powerlessness, helping elucidate what the philosopher Martin Heidegger understands as *being-toward-death.*

3   Philosophy and Knowledge of Death

Being-toward-death is an expression that does not sound good, even in German: *Sein-zum-Tode.* It seems that the German philosopher only wanted to express something very well known: that human beings must die that death is the true destiny. Philosophy is often a refutation of common sense and would therefore be redundant if Heidegger, in this respect, were merely repeating what is already known. The treatment of being toward death is a complex issue that requires a lengthy explanation.[8] Suffice it to say that for Heidegger, being-toward-death *is an expression that attempts to rescue the original meaning of human temporality* [10]. But what does it have to do with what was said before about pathology, about psychic illness? Precisely, in the phenomenological analysis that Heidegger carries out in his 1927 work *Being and Time*, being-toward-death serves so that a human

---

[6] Let it be clear that I consider that philosophizing does not abstain from existing, but from performing, at first, a reflexive act, such as the one that summons the conceptual definition of something.
[7] I will also ambivalently use the words "illness" and "sickness". Both are synonymous. Although in relation to the "psychic" realm the term "mental illness" is often used [8].
[8] For a detailed development, which I cannot do here, I would like to quote my own article [9], where I try to make this interpretation explicit from my reading of *Being and Time*.

being *gains an understanding of the radical powerlessness that embraces and constitutes him*. Death is understood there not as the fatal moment of ceasing to be to become a corpse, but as the constitutive shadow of all powerlessness which, in turn—and this is the paradox—allows the power of being in favor of human life. When it is said that a human being carries death behind him, what is meant is that his power to exist is only opened from what powerlessness allows him; hence all human actions are intertwined with the sign of expiration and contingency [11]. This, according to the Platonic tradition gathered in the dialogue *Phaedo* (63e8-69e4), means that philosophy is linked to the knowledge of death: knowing about death means having the task, often turned into a feat, of knowing what to do with death, knowing that we are going to die one day, and incorporating it into our daily actions in pursuit of existential transformation.[9] Consequently, the practical vocation of this eminently philosophical knowledge is already understood; its difficulty, therefore, lies in modifying behavior that are rooted in routine processes of fixation, most of the time against what this "knowledge" opens up to us: contingency and finitude. The routine processes of life, attending to the analysis of the one [6], concentrate on mechanisms that tend to assure a certain regularity that grant security, to cover possible deficiencies and fickleness to which a certain conscience of our mortal condition follows. In this respect, it does not matter whether the individual who lives his life on a daily basis is aware or not of this clandestine background, made up and omitted, of the knowledge of death.

If one pays attention to this kind of knowledge of our death and its relationship to the lifestyles generated around a pandemic situation, one soon realizes that the link between the two factors is quite evident. However, this relationship is not based exclusively on the fact that pandemics often take our fellow citizens' lives and that by force majeure, we see others die. The author maintains the hypothesis that even in a health emergency, such as the one that recently occurred in Spain, people reinforce this everlasting tendency to lead their lives under routine processes that are capable of imprinting the criteria of regularity and security on their lives, as well as peace of mind, at all costs. Determining the psychological causes for this attitude is not the matter at hand here, but rather to emphasize the impossibility that philosophizing can grow and develop in a situation of normality. Thus, in a pandemic situation like the one generated by COVID-19 human life becomes an opportunity to the extent that the human subject is forced to do something. This pandemic imperative is observed today with changes in human lifestyles, and these changes have the face of impossibility: not being able to play sports, not being able to go to the university to teach, not being able to see your loved ones or a grandfather who is in a residential care home. This powerlessness (impotence), relating as always to the basis of human existence, which is always a "potentiality-for-being" (*Seinkönnen*)

---

[9] I will avoid an explicit treatment that relates knowledge of death and illness, because of the spatial limitations that this writing must respect. It is evident that there would be a link between certain sufferings, which turn into behaviours labelled as pathological, and this knowledge of death, most of the time, remains omitted in the life of human beings, following the Heideggerian expression "I already know it, but I don't think about it" (*Ich weiß schon, aber ich denke nicht daran*) [12]. My translation.

[6], is not inert but inexorably means an *opportunity*. Because powerlessness is a testimony in favor of human contingency, its finiteness, and teaching that demands something from us; this type of experience can be termed, almost without fear of error, as *estrangement*. As far as it can be affirmed, estrangement is an eminently human experience based on the capacity of those who exist to summon a *distance* from what they have lived. Precisely, when an individual is living his life through the protection of a daily routine, it can be affirmed that this subject is attached to his life and is in constant closeness with that which surrounds him. When a state of emergency occurs, provoked, for example, by COVID-19, a distancing in the way of living is generated, in the way in which the world that surrounds us, from its local contexts of familiarity, becomes strange, weird, even inhospitable. Philosophy, if it is clearly understood as philosophizing, as an act that takes place in the bosom of a circumstance, the here and now, and not as erudite knowledge nor as a complement of the current sciences, makes the strangeness, the conflictive relation between closeness and distance, necessary. In its "between" is born the possibility of a question that questions the world, the power of political institutions, and makes existence vulnerable by recognizing itself as contingent.

4 Philosophy and the Binding Force of Human Contingency

Philosophical questioning is born from a fugue, from the friction between closeness and distance; it is an expression of human finitude and needs a reality that is not leaden that is not assured. The person who asks whether he wants to or not is questioning a given reality as if it were consummated. This type of human life, based on questioning, is eminently the root of all science, which is founded on the aspiration to know, and which in turn is grounded in a non-conformity with those around us. When, as tradition has it, Socrates was asking questions in his polis, Athens, generating unease among his fellow citizens, he did so because he did not have enough of the answers already given in his time. The end of Socrates, narrated with sublime beauty in Plato's Apology, gives an account of the destiny of a Socrates accused of corruption, when, as can be seen in the Platonic dialogues, he dedicated himself to a life committed to learning, to taking nothing for granted. This has already been implicitly stated when dealing with the knowledge of death: while we are alive, we always exist in a possibility; death is the accomplished fact. Then the individual who asks, calling into question the belonging to his own reality, knows himself to be contingent, finite; he attends to this type of experience that here was named as an eminently philosophical experience,[10] but not exclusively because

---

[10] "We call this the «fundamental philosophical experience» (*philosophische Grunderfahrung*) (this is the proof of this motive). This is not a special enlightenment, but it is possible in every «concrete existence» (*konkreten Dasein*), where the worried concern returns to the present existence (...) and from there the entire «terminology of philosophy» (*Begrifflichkeit der Philosophie*) can be understood and determined. It is from this that the original purpose of

as the author does believe, it can also serve as a basis for the development of any scientific activity. Now, if the human being who philosophizes lacks this experience, his knowledge will be orphaned from the soil. In this sense, moreover, philosophy, in the author's opinion, does find a differentiating element concerning the sciences: having been derived from a type of experience which can vary from anxiety to amazement, philosophers cannot abandon such questionability to accommodate themselves to consummate answers.[11] Since we are always involved in ways of questioning the reality that embraces us, philosophizing has to involve itself with what it is trying to clarify, making the modern ideal of absolutely objective science, or, like Descartes, among others, presented, the *Mathesis Universalis*, impossible. From here, it is possible to understand what is called the COVID-19 pandemic without having to become a famous virologist.[12] COVID-19 is a disruptive phenomenon that has resulted in the cancellation of our familiar lifestyles, bringing us closer, almost by force, to questions that have brought us into contact with the experience of our vulnerability, based on the finiteness and contingency of human life. That philosophy finds itself at home in a state of exception and a health emergency, such as that generated in the shadow of COVID-19, should call our attention to the value of a type of knowledge, inherited by tradition from at least Plato, which finds itself more emaciated and diminished in the curricula of European countries with every day that passes. The author does not wish to affirm here that philosophy is exhausted in this self-exploration of finitude, but more that it has its home there. Then a multitude of ways to understand, assume and finally transform what we do with that knowledge arise. It can be said without hesitation of a human being who has abdicated himself, not questioning the reality around him that: *he has died in life*. Perhaps the COVID-19 pandemic, even though it can generate more depressive lives, becomes an opportunity for the indelible generation of new forms of questioning by which the human subject is reconciled with that emotional, experiential ground that critical situations are capable of generating. *Not only does one learn from suffering, but suffering itself is already a learning*. Philosophy is, as can be seen in Socrates's case, a real nuisance regarding everyday states of life, vain in terms of their security, but necessary in moments of emergency when our life really comes to the fore. This is the moment of truth, of the unveiling of those who constitute us. And philosophy is its future (Fig. 1).

---

philosophy itself acquires its meaning. «The rigour of philosophy is more original than all the scientific rigour» (*Die Strenge der Philosophie ist ursprünglicher als alle wissenschaftliche Strenge*). It is an explanation beyond all scientific rigour to elevate the «worried-being» (*Bekümmertsein*) in its constant renewal in the facticity of existence «and to make present existence finally insecure» (*and the current Dasein letztlich unsicher zu machen*)" [13]. My translation.

[11] Scientific knowledge is not consummated either, because it changes and progresses. But philosophical knowledge would be constantly challenged by its own foundations and, consequently, it would not be possible to "progress".

[12] It is necessary to say that my words should not be read as a pejorative criticism of the virologist's knowledge, or of science in general. I am striving to give a philosophical and legitimate meaning to COVID-19.

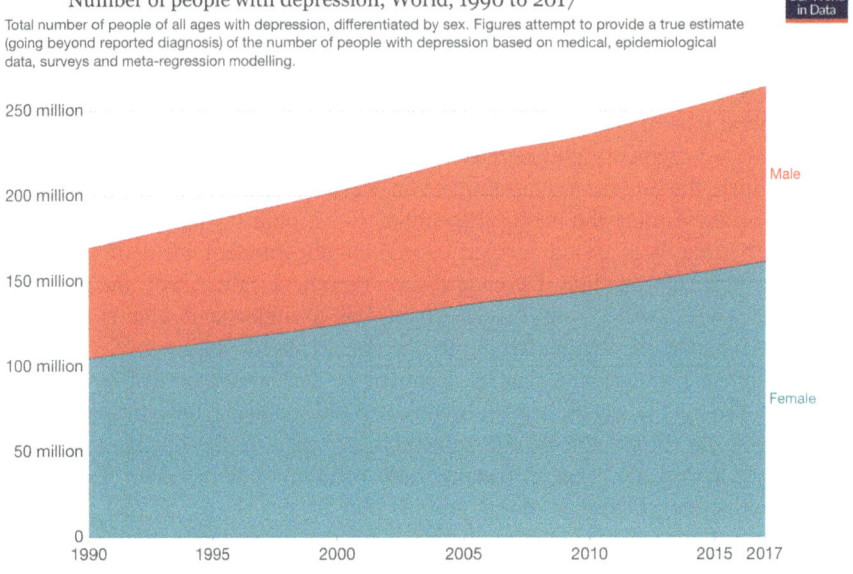

Source: IHME, Global Burden of Disease

Fig. 1 The number of people with depression in the world between 1990 and 2017 [Adapted from IHME, Global Burden of Disease, https://ourworldindata.org/grapher/number-of-people-with-depression]

5 Conclusion

What has been presented thus far is a mere sketch and should pave the way for research. The demarcation produced on the phenomenon called "philosophy," understood exclusively as the act of philosophizing here and now, has allowed a return so that we leave the daily traffic of daily transactions[13] and can *reconfigure a positioning*. The position, as almost always, is very important because, as mentioned above, we walk blinded, lost, or seduced by the idioms of lives imprisoned in routines. This is not meant to "condemn" human life in view of its attachment to regular processes, which provide it with security. *The more security human existence demands, the fewer possibilities it will have to rehabilitate, from an affective state, the possibility of questioning that leads it to free itself from the world to become what it is.* This is precisely the task of all high philosophy: "gnothi seauton," i.e., "know thyself," collecting the legacy, according to tradition, which remained as an inscription on the façade of the temple in honor of Apollo at Delphi. But to know oneself, as you can imagine, is the most challenging thing a human being can achieve because this is exacting knowledge that demands from us a

---

[13] Just because you get out of the traffic of everyday life does not mean you cease to exist.

constant struggle against all our prejudices as well as the need for transformation regarding our habits, the way we usually behave in the world. Once positioning has been established, which in philosophical terms would be a "hermeneutic situation," an ontology of human life would be required that would provide us with the basis through which the transformation, emanating from this type of philosophical knowledge, could have a direction in sight. Because it is not worth any transformation of human life: human life must be transformed to become what it is. Even if we can only briefly mention this ontology, which can be taken implicitly from this text, we would not like to finish without at least opening a small thematic sketch. Human life appears, seen from this type of ontology, from a description of human life's ontological characters, as *existence*. And existence means that human life, while it is, never ends; it presents an intolerance to what is consummated and finished; it is maintained in the "between," pressed between what we have called the characters of proximity and distance. This means that while clinging to our traditions or history, to a country, a nation or a flag, a human being always has the power to miss himself, to become a pariah or a stranger to himself. *Estrangement is very important because it allows us to reconcile ourselves with what makes us familiar and close*. Otherwise, we would not know how to measure the value of the familiar, but we would always be running away from what we are. But why? Because, as has been seen, what is close, if it comes from every day of routine processes, of equal existences, does not put us in contact with disruptive elements, extremes, in which powerlessness is transparent, as does the knowledge of death. Although it may seem abstract, this kind of philosophical thinking suddenly has practical benefits: if the human condition is to be a "between," to be always in a controversial relationship between closeness and distance, then what value do nations, homelands, and nationalisms have in this world? Should not we recognize ourselves in all the statelessness, in the refugees that swarm around Europe today? Are not we always in constant exile concerning who we are? These last questions are only a tiny sample of what philosophical questioning makes possible. In this sense, the pandemic caused by COVID-19 has given us an opportunity so that the existing one can return to itself by rethinking its reality, its way of life. In this return to himself, the human being will have a chance to become more lucid.

Core Messages

- Philosophy only means to awaken in you to philosophizing.
- The COVID-19 pandemic phenomenon is an opportunity to rethink our modes of existence.
- Philosophizing since ancient times has always been committed to the question of how to live, always dwelling in the insecurity of all experience.
- The human being is constituted by his temporality, which only opens from the tragic bursting of the knowledge of death.

Acknowledgements This chapter of the book benefited from the support of the R&D project: "Dynamics of care and the disturbing. Figures of the disturbing in the contemporary phenomenological debate and the possibilities of a Philosophical Orientation. Theoretical and methodological configuration" (FFI2017-83770-P), financed by the Ministry of Science, Innovation, and Universities of the Kingdom of Spain. It was also carried out through a grant provided by the VI Research and Transfer Plan of the University of Seville.

References

1. Corominas J (1987) Breve diccionario etimológico de la lengua castellana. Editorial Gredos, Madrid
2. Heidegger M (1995) Phänomenologie des religiösen Lebens. Frankfurt a. M.: Vittorio Klostermann
3. Heidegger M (2010) Phenomenology of intuition and expression. Continuum, New York
4. España BOdRd (2020) Real Decreto 463/2020, de 14 de marzo, por el que se declara el estado de alarma para la gestión de la situación de crisis sanitaria ocasionada por el COVID-19.: https://www.boe.es/eli/es/rd/2020/03/14/463
5. La OJ (2016) Orientación Filosófica (OrFi): una aplicación de la filosofía en la intervención terapéutica. In: Garrido-Periñán JJ, De Bravo C, Ordóñez J (eds) La Filosofía como terapia en la sociedad actual: Desafíos filosóficos de nuestro, tiempo. Fénix Editora, Sevilla, pp 109–126
6. Heidegger M (1996) Being and time (trans: Stambaugh). State University of New York, Albany
7. Sáez L (2011) Enfermedades de occidente. Patologías actuales del vacío desde el nexo entre filosofía y psicopatología. In: Sáez L, Pérez P, Hoyos I (eds) Occidente enfermo. Filosofía y patologías de civilización ed. München: Grin; pp 71–92
8. Dictionaries OLs. London: https://www.oxfordlearnersdictionaries.com/definition/english/illness?q=illness
9. Garrido-Periñán JJ (2019) La aportación no-apofántica de la disposición afectiva y la mismidad del Dasein: análisis fenomenológico a parrir del momento estructural ser-en. Pensamiento: Revista de investigación e información filosófica. pp 877–911
10. Garrido-Periñán JJ (2019) Vinculabilidad entre cuidado y mismidad en los §§ 39-42 de Ser y Tiempo. Alpha: Revista de artes, letras y filosofía. pp 159–75
11. Garrido-Periñán JJ (2019) La pregunta por el quién del ser-con: Heidegger en su Ser y Tiempo. Anales del Seminario de Historia de la Filosofía, pp 175–200
12. Heidegger M. Der Begriff der Zeit. Frankfurt a. M.: Vittorio Klotermann; 2004.
13. Heidegger M (1993) Grundprobleme der Phänomenologie. Frankfurt a. M.: Vittorio Klostermann

Juan José Garrido Periñán (Seville, 1988) earned his Ph.D. from the University of Seville and graduated in Philosophy with the qualification of Extraordinary Award. His research revolves around a critical interpretation of Heidegger's work and the possibilities of developing a self's phenomenology. He is a founding member of the Martin Heidegger Archive of the University of Seville (delegation Martin Heidegger Meßkirch Archive). With Dr. Denker, he has conducted two summer schools on Heidegger in collaboration with M. Heidegger Meßkirch Archive. He is also the secretary of "Differenz" (https://revistascientificas.us.es/index.php/Differenz), an international journal specialized in the diffusion of the German philosopher's thought. Juan is currently an assistant professor in the Department of Aesthetics and History of Philosophy at the University of Seville.

Engineering, Environment, and Health: Why Interdisciplinarity Matters?

15

Kaushik Sarkar, Monica Lakhanpaul, and Priti Parikh

"The future of research is interdisciplinary and will quickly take us into areas that today we cannot even foresee."

Michael Tanner

* This author is the principal investigator of the project.
# These authors equally contributed to the project.

K. Sarkar
Malaria No More, Seattle, USA

Aceso Global Health Consultants Limited, New Delhi, India

K. Sarkar · M. Lakhanpaul (✉) · P. Parikh
Childhood Infection and Pollution Consortium, London, UK
e-mail: m.lakhanpaul@ucl.ac.uk

P. Parikh
e-mail: priti.parikh@ucl.ac.uk

K. Sarkar · M. Lakhanpaul · P. Parikh
Integrated Science Association (ISA), Universal Scientific Education and Research Network (USERN), New Delhi, India

M. Lakhanpaul
UCL Great Ormond Street Institute of Child Health, UCL, 30 Guilford Street,
London WC1N 1EH, UK

Whittington Hospital NHS Trust, London, UK

P. Parikh
Engineering for International Development (EFID) Research Centre,
University College of London, London, UK

© The Author(s), under exclusive license to Springer Nature Switzerland AG 2022
N. Rezaei (ed.), *Multidisciplinarity and Interdisciplinarity in Health*,
Integrated Science 6, https://doi.org/10.1007/978-3-030-96814-4_15

Summary

Urbanization, climate change, and pandemics have put humanity's future at a crossroads with complex challenges and overlapping vulnerabilities. The convolutional nature of the challenges warrants urgent actions by experts from multiple disciplines to co-develop knowledge, integrated tools, and methodologies to solve global challenges. More than ever, there is a dire need to set and standardize interdisciplinary strategies, actions, and tools to combat the emerging challenges of the twenty-first century. This chapter describes the importance of an integrated interdisciplinary approach to achieve sustainable development; highlights the scope of integration of various disciplines, focusing on health, education, engineering, and environment (HEEE); outlines the challenges in interdisciplinary applications; and provides the tools to apply interdisciplinary approaches.

Interdisciplinary connections of public health research

The code of this chapter is *01101001 01100011 01001100 01110010 01100101 01100001 01111001 01110100*.

Keywords

Engineering · Environmental health · Global health · Interdisciplinary science · Intersectoral convergence · Participatory research · Public health · Sustainable development

1 Introduction

1.1 Interdisciplinary Approach for Sustainable Development

The 2030 agenda for sustainable development provides a unifying framework to develop holistic interventions to achieve human well-being and development. The 17 goals and 169 targets encompass social, environmental, economic, and institutional dimensions of well-being which are integral to health, engineering, social, and environmental interventions. While each of these disciplines have their own niche in creating and contributing to the development of the planet and its people, each also requires concerted action by the other disciplines to make the development sustainable. For example, the most important transformative vision of the 2030 agenda is the elimination of poverty, which requires direct economic interventions to overcome macro- and micro-economic challenges across the world. However, poverty is widely recognized as a socio-economic outcome of living conditions, diseases, and disparities, and a crucial determinant of health, well-being, and built environment. These linkages often form a vicious cycle that results in the poor and marginalized suffering from ill health, poor living conditions, lack of food security, and access to nutrition, education, and welfare services. Modern research, therefore, typically focuses on the interlinkages of these multitudes of problems, which otherwise cannot be tackled with the expertise of one discipline. For example, a Kenyan case study in 2009 categorized poverty and food security as a "twin problem" emerging from disproportionate strain on natural resources due to population increase, climate change, and environmental degradation [1]. A more pro-active approach from experts of individual disciplines in recent times has been the identification of synergies between the corresponding discipline and the SDG targets. For example, Fuso et al. [2] identified synergies between energy systems and 143 out of the 169 SDG targets [2]. Similarly, Parikh et al. [3] identified synergies between sanitation and 130 out of the 169 SDG targets demonstrating the wide-ranging benefits of investing in engineering infrastructure solutions [3]. These articles highlight the scope of leveraging the knowledge, tools, and strategies of one discipline in solving problems, which were otherwise considered earlier as isolated problems of health, nutrition, environment, or economy. For example, the impact of engineering solutions goes beyond the sector to encompass health, education, poverty reduction, equity, and partnerships. Simple engineering solutions like handwashing stations improve the health and well-being of young children by reducing the incidence of diarrhea, malnutrition, and enteric diseases [4, 5]. The role of environmental interventions, like the integration of urban green and blue spaces to fuel economic growth, conservation of natural ecosystems, and public health, was proven even earlier in the late twentieth century [6]. These cases of application of one discipline to overcome the cross-disciplinary challenges have formed the stepping stone for fostering interdisciplinary partnerships, harmonizing policies, and investments in cross-disciplinary interventions. In 2019, Professor Robert Chambers highlighted the limitations in perspectives and scopes of one discipline as "Blind Spots." He advanced the opportunity of using participatory research and innovative interdisciplinary partnerships to bridge the knowledge gap [7].

1.2 The Urgency for Integration of Health, Engineering, and Environment

The concept of an interdisciplinary approach is not new. In fact, since its inception, public health has always remained closely aligned with engineering. Engineers and entomologists worked hand-in-hand during World War II to make the western world, especially the US, free of one of the longest-standing public health scourges, malaria. Following the elimination of malaria in the US in 1951, as the focus of the US center for disease control expanded to all communicable diseases, the discipline of epidemiology gained currency for communicable disease investigation, and at the same time, many of the works of the so-called sanitarians (the erstwhile sanitation/public health engineers) were integrated to the core domain of public health. In the mid-1960s, the tasks previously considered as the responsibility of a public health engineer were included in the discipline of environmental health [8]. In 2017, the environmental experts prioritized 10 grand challenges for environmental engineering to ensure a sustainable future for developing regions or areas [9]. These challenges highlighted the need to prioritize linkages in health outcomes and ecosystem management; water-energy systems; water, sanitation, and hygiene (WASH), food security, resource recovery; green economy; culture, perceptions, and behavior with science and technology; and climate and pollution effects. These challenges clearly need expertise beyond the scope of one single discipline. Cross-disciplinary collaborations have been fostered by the University College London (UCL) grand challenges for the last decade. These collaborations are planned to prepare solutions that address the most important social issues. The six areas included in the UCL's cross-disciplinary research strategy are "global health, sustainable cities, cultural understanding, human well-being, justice and equality, and transformative technology" [10]. One of the recognized works funded to apply the cross-disciplinary approach was health, education, engineering, and environment (HEEE) interlinkages to improve nutritional outcomes for children and young people in India [11–13]. (Table 1) Given the diversity and complexity of the modern world's challenges and the lack of unisectoral expertise to address the problems, policy interventions across the world are currently focusing on "concerted" multisectoral actions. The concept of concerted action is more holistic than the simpler interdisciplinary approach, as the former requires integration along with multisectoral action.

Multisectoral action in Global Health gained momentum in the 1990s. By this time, many countries worldwide already had the preparation with the overarching focus on primary health care. Besides, shifting the paradigm from vertical to horizontal programming started getting attention. In the Africa region, horizontal programming's pragmatism was under discussion in the 1980s [14]. In the health

Table 1 PANChSHEEEL project

Participatory approach for nutrition in children: strengthening health, education engineering and environment linkages (PANChSHEEEL)

PANChSHEEEL project*, funded by the Medical research Council UK, focusses on infant and young child feeding and nutrition in Rural Banswara, Rajasathan, India. It provides a unique example of IDR overcoming the epistemological, managerial, and governance challenges. The project was designed by a highly collaborative and community-centric research partnership between University College of London, Save The Children, Jawaharlal Nehru University, Indian Institute of Technology, Delhi, and other partners. Three disciplines Health, Education and Engineering came together to design solutions for IYCF issues to inform India's POSHAN Mission. The formative research, using research approaches recognized by all the disciplines uncovered detailed information about factors across the Health, Education, Engineering and Environment (HEEE) domains impacting on WASH, nutrition and infection and tied them back to IYCF to determine priority areas for intervention from a socio-ecological viewpoint (individual, inter-personal, organizational, community and governmental levels). Working with community champions and using a highly participatory approach a socio-culturally appropriate, tailored, integrated and interdisciplinary HEEE package was developed and assessed for its acceptability and feasibility of the intervention through engagement and co-design with the community using the MARKS framework (Motivation, Awareness, Resources, Knowledge, and Skills).

sector, this paradigm shift meant a change from disease-specific focus to overall health focus and prioritizing health promotion through primary care over higher care levels. The actualization of this, however, happened post-2000 in the developing world. For example, India changed their approach by launching the National Rural Health Mission in 2005, targeting all health programs under one umbrella mission. In the African region, horizontal interventions came even later.

In the post-2000 era, with the launch of the Millennium Development Goals, many developed countries began focusing on integrated intersectoral action at policy levels to promote health and well-being. Integration is an add-on paradigm over the multisectoral action that focuses on the deduplication of efforts and resources. For example, public health engineering focusing on improved sanitation and hand hygiene promotion through behavioral improvements can be exemplified as a multisectoral action against childhood diarrhea. However, they are not integrated until the strategies, program planning, and delivery operations are not integrated at each level. Therefore, multisectoral interventions still have the window of parallel and duplicated efforts, while integration typically focuses on synergy and deduplication. However, integration in multisectoral interventions was still cursory even in the developed countries before the beginning of the current decade.

Countries in the WHO European region applied multisectoral action in various areas of health, and many of these countries, e.g., Austria, Czech Republic, Denmark, Estonia, applied the approach within their national strategic plan for health promotion. Like Azerbaijan, Belgium, Bosnia, and Croatia, other countries applied multisectoral action in specific health areas, like mental health, infectious diseases, and non-communicable diseases. The primary reasons behind taking up multisectoral actions included the inability of the health sector to manage and address health and well-being challenges alone, "improve coherence" through multisectoral action, and increased resources through better resource mobilization [15].

In Asia, the Indian Government recognized the need for revamping the nutrition program for mothers and children, prioritizing multisectoral action in 2014. Recognizing that malnutrition is a problem that is "multi-dimensional, multifaceted and inter-generational in nature and linked to an inter-related set of factors," the Indian Government called for coordinated action by different sectors: "agriculture including horticulture, food, health, rural development, biotechnology, water and sanitation, education, information and broadcasting, women, and child development" [16]. However, the multisectoral actions mostly remain limited to the policy-setting, establishing multisectoral committees, and laying out multisectoral action plans. The grassroots level integration continues to be cursory because of myriads of reasons (described later).

In 2017, The Institute for Global Environmental Strategies used the social network analysis technique to advance a novel analytical framework for SDG inter-linkages across nine countries, working with 51 indicators with trackable data [17]. The SDG indicators were found as a densely packed network, in which six targets were identified to have the most significant influence (based on their in-degree and out-degree centrality) on other targets, namely double agriculture productivity, sustainable food production systems, global access to safe water for drinking, sanitation, and hygiene, and energy, and development of resilient infrastructure.

The above targets require a direct focus on public health and hygiene, resource conservation for environmental sustainability, and technology from various engineering and agriculture branches. The integration of the three disciplines, i.e., health, engineering, and environment, is therefore critical for the achievement of the SDGs across the globe. One such example is the Childhood Infection and Pollution (CHIP) Research that has been operationalizing a "One Health" approach (described later) to reduce the burden of infections and antimicrobial resistance among urban slum children. Using the combined expertise of health, engineering, and environment, CHIP identifies and addresses the linkages between biological, micro- and macro-environmental factors and behavioral and vector-related factors to break the chain of transmission of infection [18]. A similar yet more policy and technology-focused initiative, Forecasting Healthy Futures (FHF), was launched in 2020 at the intersection of climate and malaria. The policy arm of this initiative provides a unique opportunity to address the inequality and injustices of the impact of climate change, often driven by the disproportionately low allocation of climate resources to the marginalized settings, which has been widely contested at the outset of the 26th Conference of the Parties (COP26). FHF's Malaria Planning and Prediction Tool uses artificial intelligence (AI) to integrate and model hyperlocal weather and earth observation data with different health and demographic information to provide early warning against malaria outbreaks and decision support for program delivery [19]. The unique common strength of both CHIP and FHF lies in the respective communities of practices formed by a network of researchers collaborating across countries and continents to promote interdisciplinary solutions.

1.3 Understanding the Scope of Cross-Disciplinary Overlap for Integration

1.3.1 Public Health, Epidemiology, and Integrated Science

Health is referred to as a "state of complete physical, mental, and social well-being and not merely the absence of disease or infirmity" by the World Health Organization (WHO). Over time the scope of disease as a niche domain within health has expanded, so has the International Classification of Diseases (ICD) become progressively inclusive in classifying different health conditions as diseases.

Beyond the small pie of diseases within the entire spectrum of health, many factors determine the well-being of a human being. Therefore, the entire spectrum of health is broadly described in five dimensions: emotional, mental, physical, social, and spiritual; and at any point in time, an individual's health is as a single moving point within the hyperplane created by the continuum of all these dimensions. At the core of this hyperplane is human health's biomedical status, which comprises biophysical, biochemical, and physiological state variables (SV). According to the "biomedical relativity concept," none of these SVs reflect biomedical health in isolation, and they together form a system of health [20]. All these SVs are also functioning of various factors, including but not limited to social, economic, and environmental factors, which are interlinked and have a complexly nested relationship.

Epidemiology is a scientific discipline that focuses on controlling health problems by studying the distribution and determinants of health-related states or events. An advanced model of the triangle of epidemiology has been developed to explain this multifactorial causation. In this model, the term "agent" has been replaced by "causative factors" to include two or more causes of a disease; the term "host" has been substituted by "groups or populations and their characteristics" to include both individual and population-level factors; and "Behavior," "Culture," "Physiological factors," "Ecological factors" have been added to the term "environment" [21]. Multifactorial origins of diseases or health conditions have increased manifold following industrialization. The understanding of multifactorial syndromes has also become more popular. However, the conceptual inability to formulate a causal hypothesis regarding the multifactorial causation of different health conditions has led to a perceived difficulty in "integrated scientific understanding" [22].

MacMohan and Pugh advanced the "web of causation" approach to illustrate a multifactorial condition's causation through the linkages between several predisposing factors in the form of a spider web [23]. However, even this concept failed to explain the conditions in which the outcome is not static and may vary across a spectrum. In the 1980s, an extended version of an older model, known as the framework of potential outcomes, gained currency as the Rubin Causal Model.

Between 1980 and 2010, this model has been used in medicine, statistics, economics, social and political sciences, and law. The original model included two potential outcomes for each unit based on the treatment status, treated or untreated. The difference between these two potential outcomes was considered as the causal effect of the treatment. More recently, this concept has been advanced in observational studies. There were potential outcomes considered for each level of exposure as well [24]. Because of its more extensive use across disciplines, this model's understanding has brought several disciplines closer in explaining one or multiple potential outcomes due to multiple factors.

The advancement in epidemiology and the realization of complex theories have led to better integration of various disciplines to explain health conditions across a spectrum. Epidemiological methods are critical to finding causal and non-causal relationships and linkages and are vital for identifying, designing, and testing health interventions.

Health interventions are planned "to assess, improve, maintain, promote or modify health, functioning or health conditions" by acting "with or on behalf of a person" [25]. According to the International Classification of health interventions, these interventions are related to the following domains:

  i. body systems and functions;
 ii. activities and participation;
iii. environment; and
 iv. health-related behaviors.

Population health interventions, therefore, typically require an integrated approach involving various disciplines [26]. While many of these disciplines belong to the core scientific domains, the others typically belong to an overlapping domain of art and science. The latter attribution has gained momentum with the paralleling growth in art-science collaboration.

1.3.2 Influence of Other Disciplines on Health

Various attributes of human society as well as individual and population behavior are intricately related to human and population health. Social conditions like poverty, hunger, and factors like built environment, access to safe water and improved sanitation, politics, economy have widespread impact on the overall human health as well as in various diseases. For example, poverty and hunger can lead to pathological conditions like malnutrition, overcrowding, and a poorly built environment. Moreover, lack of access to safe water and basic sanitation can lead to diarrheal diseases. While social and behavioral studies used to be considered within the domain of humanity, many of these disciplines have been included under the domain of science with the increasing application of various scientific methods in these disciplines.

Socio-Cultural System

Two systems are important within the domain of Anthropology of health and medicine: (i) the cultural system of health in Medical Anthropology focuses on "the understanding of health and includes knowledge, perceptions, and cognitions used to define, classify, perceive and explain disease;" and (ii) the social system of health, on the other hand, focuses on health institutions, organization of health specialist roles, power relations, and interactions [27].

Culture is a deep-rooted social phenomenon that influences attitudes, values, and practices toward health. Every population group or community develops its own culture, which eventually influences the group members' perspective toward health. Therefore, the concept of hygiene is embedded within human societies' culture as each society has its way of categorizing things as clean or dirty. The diversity of the human perspective leads to the specific practices of hygiene in different communities.

Beyond cleanliness, divinity is another aspect that is also important in terms of determining health behavior. For example, animal blood can be considered dirty in some cultures, while it can be used for religious practices in other cultures. Such usage may lead to exposure of the community to various infections that are transmitted from animals. While cleanliness, divinity, and religious practices are closer to a value system, habit is another aspect that can increase or decrease health hazards. For example, the habit of handwashing can reduce the risk of various diarrheal illnesses and enteric diseases, while the habit of drinking milk directly from the udder of an animal can lead to increased health risk for various zoonotic diseases.

In the context of social systems, the organization of health, health providers, institutions, and power relations play an important role in determining health-seeking behavior and practices. For example, the reliance on faith-based and traditional healing systems may influence families to seek treatment from unskilled practitioners. This is a prominent system in marginalized tribal communities, in which proximity to a health provider and the preference toward traditional healing techniques play an important role in determining the treatment-seeking behavior. Similarly, different power relations leave a large impact, especially concerning decision-making related to treatment. For example, in traditional communities, the head of the community often influences the other families with regard to accepting a vaccine or treatment.

Gender

Women bear the burden of poor infrastructure access, which has a knock-on impact on well-being and health [28]. As caregivers, they are responsible for collecting water, cleaning, and feeding activities, leaving them with limited time to invest in children's education, appropriate feeding and hygiene practices for children, limiting the gains that could be achieved through traditional sectoral health programs. There is a need for multisectoral and appropriate interventions for women, especially in resource-challenged settings where environmental conditions are insufficient and pre-existing health vulnerabilities are high.

The gender difference in culture and societal gender norms also play a crucial role in determining health risks. For example, menstruation is often considered unholy, and in different parts of the world, women having periods have various implications on health status and health interventions. Women sleeping under a long-lasting insecticidal nets (LLINs) to prevent malaria are compelled by social norms in some communities to wash their LLINs, right after using them during menstruation. These LLINs typically last for three years or 20 washes. Washing LLINs after every cycle of menstruation means 24 washes in two years. The practice, therefore, affects the longevity of LLINs, and the real world effectiveness of this tool for the prevention of malaria.

Economy

"The biggest enemy of health in the developing world is poverty."

Kofi Annan.

The three most important and long-standing infectious diseases, tuberculosis, malaria, and HIV/AIDS, are primarily known as poverty diseases. More than 90% of the victims of these three diseases live in developing countries [29]. The glaring impact of poverty on infectious diseases made BioMed Central and the China CDC launch a new open-access journal, "Infectious Diseases of Poverty," in 2012 [30].

Several environmental determinants of health are also directly linked to poverty. Lack of access to safe water affordable for drinking, sanitation, and hygiene can cause myriads of water-borne illnesses, including cholera, enteric diseases, soil, and water transmitted helminthic infections, to name a few. Deaths in under-five children due to these diseases are most common in society's most deprived quintile [31]. Contrary to the popular belief that poverty only increases the risk of infections, poverty, and lack of access to safe water have resulted in several non-communicable diseases (NCDs) and health problems. Particularly in women, daily collecting and carrying water for long distances result in chronic back pain, abortion, and uterine prolapse, especially in rural areas [31]. Conversely, NCDs also force millions of people to live in poverty in lower-income countries [32]. 75% of global GDP is lost due to the loss of output from major NCDs [33].

Other Disciplines

Social and behavioral sciences play an overlapping role in determining health and health interventions. In particular, political decisions play a vital role in health financing, health care delivery, access to care, payer policies, and public health programs. In the UK, the Brexit decision received some criticism due to the risk of reduction in the health workforce and access to pharmaceuticals, technology, and transplant organs [34]. In India, the policy and focus on Digital India have revolutionized the access to and the use of data to improve public health. Reversal of pro-health decisions has been witnessed in Italy and Australia as they abandoned the mandatory child health immunization and smoking ban decisions of their predecessors, respectively [35]. Political will has also been widely recognized as an important driver of disease elimination programs. For example, the Indian prime minister's pledge to eliminate malaria followed a substantial increase in the National Vector Borne Disease Control Program's domestic budgetary allocation. It enabled the Indian Government to apply life-saving commodities like long-lasting insecticidal nets in most high-endemic regions. Paralleling this intervention, India witnessed the highest malaria burden reduction compared to other high-endemic countries in 2018. Besides, in the developed countries, value-based care is gaining currency over the fee-for-services system as the former focuses on the quality of care and pay-for-performance models. The Affordable Care Act (ACA), also known popularly as the Obama Care, is a remarkable example of political will influencing health sector reform through better accountability setting, quality improvement, transparency, and dollar-return in terms of value [36].

1.4 Engineering: Interdisciplinarity and Applications

Engineering as a discipline focuses on the "creative application of scientific principles to design or develop structures, machines, apparatus, or manufacturing processes, or works" [37]. Interdisciplinary engineering can relate to various scientific domains, including computer science, energy, health, logistics, manufacturing, mobility, natural sciences, psychology, and sociology [38]. Engineering as a discipline has its application in various sustainable development areas, including but is not limited to improving quality of life, trade growth, resource conservation, green sources of energy, safe water, sanitation, food security, and climate change [39]. Engineering research brings advanced technologies to mitigate challenges, improve access, and increase value for each of the above challenges. One of the most important engineering applications in the SDG era is the provision of cost-effective solutions for various challenges. In order to achieve cost-effectiveness, engineering techniques and principles can be applied in various disciplines. One such case is the application of genetic engineering technology to develop cheap and safe vaccines. The cost-efficacy data is receiving paramount importance in managed care systems in the developed world [40]. In the developing world, engineering technologies enable countries and systems to leapfrog toward quality-focused care.

Civil engineering and information technology play a critical role in environmental sustainability, improving access and quality of care. One of the most popular civil engineering applications toward sustainability is the "Jubilee River," built in the UK in 2001. It enabled flood water to bypass lands to be protected through flood relief channels and create new habitats using the same [41]. The technology has profound significance on land and water conservation, protection of natural resources, environmental sustainability, and of course, the life and health of the population. One of the early-stage civil engineering technology applications for health protection was removing breeding places for mosquitoes to prevent malaria. Engineering technologies have also been applied to improve access to safe water and sanitation to prevent water-borne illnesses. However, advancement in WASH technology has been less researched. In a 2017 systematic analysis, researchers found no evidence to support WASH intervention for disease outbreaks with regard to "water trucking, well rehabilitation, bucket chlorination, latrine building, hand-washing, and household spraying." The quality of evidence on environmental hygiene was very low, while the low-quality evidence was found in the WASH package and home water treatment [42]. Information technology and mobile devices have revolutionized the data management information system, disease surveillance, and care delivery. Mobile-based platforms are now also used world-wide for health promotion, behavior change communication, and information exchange.

2 Challenges in Interdisciplinary Approach

The most important barriers to adopting an interdisciplinary approach are epistemological incompatibility, gaps in common understanding, gaps in tools and methodology co-developed, and management and data governance—the difficulty in cross-learning results from the differences in understanding among experts from different backgrounds. The outlook toward different situations and contexts will vary among experts. The key barrier in cross-learning is linguistic differences. The same terminologies are used with different meanings in different disciplines. For example, the terminology "sanitation" has a very different meaning in infectious diseases, food science, and engineering. Sanitation has been used to mean "safe disposal of human excreta" in the context of health [43]. In food science, sanitation is used to mean "the application of a science to provide wholesome food processed, prepared, merchandised, and sold in a clean environment by healthy workers" [44]. With learnings from the debate between biologists and anthropologists regarding biological diversity and economic development in central Africa, researchers have documented disciplinary differences in terms of "facts," "rigor," "causal explanation," and "research goals" [45].

The tools and techniques of different disciplines are also poorly understood by the experts of other disciplines. At times they are also considered competing. For example, big data technologies threaten the importance of the use of core statistics in health models. Although modern-day computations are based on mathematics and statistics principles, the derivation of machine learning techniques from these basic sciences is at times less understood by biomedical experts. While the problems in cross-learning, grounds of commonality, and scarcity of tools can be overcome with time as more interdisciplinary works are conducted and regulated, one challenge that requires appropriate strategizing, even before attempting an interdisciplinary work, is the overall management and data governance. Conflicting interests and competition of power among different disciplines often prevent effective collaboration and accountability settings. For the same reason, data sharing and governance become difficult. Sophisticated management and operation techniques, including tailored log frames and data regulations, are the need of the hour to promote interdisciplinary science.

3 Tools and Techniques of Interdisciplinary Research

"Interdisciplinary research is a mode of research by teams or individuals that integrates information, data, techniques, tools, perspectives, concepts, and/or theories from two or more disciplines or bodies of specialized knowledge to advance fundamental understanding or to solve problems whose solutions are beyond the scope of a single discipline or area of research practice."

Committee on Facilitating Interdisciplinary Research [46]

3.1 Convergence in Interdisciplinary Agenda

Convergence is the central tenet of interdisciplinary research (IDR). Convergent research focuses on improving life and the economy by applying knowledge of various disciplines. The convergence can be natural or spontaneous, or influenced by mutual interest. Spontaneous convergence can be seen in cases where no single discipline can adequately solve the research problems. For example, during the COVID-19 pandemic, AI-powered medical imaging techniques have become very popular. AI is provided by AI engineers, while specialists in radiodiagnosis provide domain expertise to design the diagnostic technology. Specialists in HTA research would then help test the efficacy and effectiveness of this tool. In this case, all the domains enjoy equal stake, and the participation is spontaneous. However, in many other contexts, the primary stake lies in one discipline. For example, in food production, food science and agricultural engineering are the primary disciplines that derive the research, while the product technology is applied in health and nutrition context to improve the nutritional status of the population. In the latter example, true integration seldom occurs; instead, translation or technology transfer drives the initiative. Based on the nature and scope of integration, the tools for IDR will also vary, e.g., in the case of true integration, a single project team comprising experts of various domains is quintessential. A prominent example of such collaboration for a national-level project is the Children in Homeless Accommodations Managing Pandemic Invisibility or Non-inclusive Strategies (CHAMPIONS) project, in which researchers of different disciplines including health, engineering, social sciences, and arts are collaborating to co-develop with communities and service-providers recommendations critical to mitigating the impact of the COVID-19 on children under-5 living in temporary accommodations [47]. However, in the case of technology transfer, the project is typically laid out by the recipient discipline, while the source discipline provides technical expertise for capacity building. This can be achieved through various co-development approaches. Below are the usual steps to follow:

   i. identifying and defining the problem by recipient discipline;
  ii. sourcing solutions from other disciplines (either serendipity or deliberate consultation);
 iii. extensive consultative discussions on technology transfer;

iv. co-development of research design;
v. capacity building of recipient discipline by source discipline; and
vi. technology transfer and test in the real world.

3.2 Catalytic Framework

Interdisciplinary theoretical frameworks are usually laid out at a level higher than program convergence (typically at strategy or policy convergence levels). In this case, the nature of the problem policymakers or strategic planners face is so complex that even the problem domain and the factors determining and resulting from the problems are not understood with input from one discipline. In the health system, zoonotic diseases were considered one such problem where the transmission of zoonotic infections could not be explained alone by either medical or veterinary experts. Besides, beyond the scope of human and animal diseases, the factors that affect the overall transmission, e.g., the market, economy, social, and behavioral factors, are difficult to understand. Interventions addressing diseases emerging from poultry farms or abattoirs have implications beyond the health of the population. These implications can impact livelihood, need infrastructural change, and will also have cultural influence. Therefore, zoonoses were considered a multidisciplinary problem, which resulted in the emergence of One Health's concept. According to the WHO, One Health is considered "an approach to designing and implementing programs, policies, legislation, and research in which multiple sectors communicate and work together to achieve better public health outcomes." In these cases, an interdisciplinary theoretical framework typically functions as a catalyst for sector-wise role stratification, sharing of resources, and multilayered integration is needed to design a roadmap to combat such problems.

3.3 Strategic Communication

In multisectoral approaches, the key to success is effective strategic communication. The gap in understanding common terms and contexts in interdisciplinary work, as described above, can only be overcome by targeted strategic communication models. Strategic communication is an emerging discipline that has proven increased research productivity over a decade [48]. The discipline of mass and media communication has a key role to play in this regard. Strategic imperatives for using mass and media communication in interdisciplinary studies have already been advanced for creating effective interdisciplinary alliances by removing the barriers of interdisciplinary communications. Five critical strategic imperatives of application of mass communication in promoting interdisciplinary studies and research include: "embracing an interdisciplinary approach" in mass communication studies, provisioning "internships" across the different fields of study, improving "learning in the classroom," "stressing global literacy" in college, and "addressing issues of discipline and assessment in the curriculum" [49]. Increasingly, the arts and sciences disciplines are working more closely, either using arts as an integral part of the design of the research tools to capture data such as the use of participatory theater or to elucidate well-being status using paintings to express emotions.

3.4 Community/end-User Engagement

Beneficiary participation and use are the keys to the success of applying interdisciplinary interventions. Thorough knowledge of context, culture, and market plays a crucial role in tailoring solutions to user communities. Research and analytical techniques help achieve end-user engagement at different stages of intervention or product development. For example, User Acceptability Testing, Feasibility Studies are research techniques typically applied before an intervention or product at scale to understand the target beneficiary's scope of uptake is used. Market analysis and behavioral surveys are typically conducted even earlier when the service or product has not been designed. These studies help understand end-user needs, the availability of alternative solutions, and the end-user's attitude to a proposed solution. Brand lift and campaign analytics studies are considered late-stage research studies, typically applied right before or during product or service marketing or branding. While research techniques mentioned above enable understanding the scope of community engagement, several techniques that engage the community in all phases of research and development also exist. Some important ones are participatory research techniques engaging community champions and frontline service delivery agents/providers, community-based open data collection and feedback as is conducted under the Citizen Science approach, and engagement of community agencies (e.g., schools, local governance) in research planning, designing, and scaling up [50–52]. Furthermore, with the rapid growth of digitalization, media is increasingly used as a channel for raising mass awareness and engaging debates. Innovative global health and media partnerships are using the power of social media platforms, audio-visual and digital tools like short films, blogs, and vlogs to reach out to a large section of the population. Internet access, mobile technology, and the overarching growth of information technology are critical enablers in this regard. One such campaign, "Bite Ko Mat Lo Lite" (Do Not Take Bite Lightly) that reached over 150 million Indians in 2020 to provide health education on the prevention of mosquito-borne diseases, has been highlighted as a best practice for providing health education at scale during the pandemic [53, 54]. Besides mass and social media, the role of mid-media is gaining currency in community-based research (CBR), engaging both marginalized and non-marginalized populations. For example, Nukkad Natak (street theater) puppetry shows have been used in India for health promotion. Now, these techniques are also being used in CBRs. One such example is the use of theater in participatory research [55, 56]. Art exhibitions are getting more attention in developed countries to express health and environmental research in a more accessible way.

3.5 Combined Structured Evaluation

IDR evaluation is complex. The degree of convergence, the "sensible balance" in weaving perspectives, and the "effectiveness in advancing understanding and inquiry" play an important role in setting the goals and criteria for evaluation [57]. There is no one size fits all technique to evaluate the IDRs. Several frameworks for the evaluation of IDR have been advanced in recent years. The commonality exists in recognizing a structured approach tailored to the variability in goals, metrics, levels of integration, interaction, and management, and direct and indirect impact assessment needs [57–59]. The evaluation should be combined, focusing on all the integration aspects of IDR [60]. Economic evaluation plays a crucial role in this. It helps in choosing the most cost-effective components in the case of multisectoral interventions and helps in overall advocacy for multisectoral action through information on return on investments and funding gaps.

3.6 Co-Development

A participatory approach in each step of the research process is quintessential for IDR. An IDR should have co-generated research questions, garnering the expertise and experience of all the disciplines. One of the critical steps in inclusivity is the engagement of implementing agencies and end-user representatives in the design, process, and dissemination of IDRs. Structured collaborative approaches like Community-based Participatory Research (CBPR) using focus group discussion (FGDs), key informant interview (KII), and healthcare provider interview for data collection and community or beneficiary engagement in intervention mapping have been successfully used in the past for intervention co-development [61]. Besides, Lakhanpaul et al. [13] applied the motivation, awareness, resources, knowledge, skills (MARKS) framework as a structured approach of community engagement and co-development in the PANChSHEEL [13]. Successful collaboration also requires appropriate opportunities for cross-learning. The leadership role of a dedicated knowledge partner in the overall cross-learning, acculturation, and capacity building is quintessential for the success of the participatory approach in IDR. Participation and knowledge translation also go hand-in-hand with policy influencing and advocacy in interdisciplinary sciences. However, unlike the monodisciplinary approach, the application of interdisciplinary science requires shared financing, strong intersectoral coordination, accountability, and the highest-level political will to ensure sustainability in funding and coordination. A classic example of the above is India's National Nutrition Mission or the POSHAN Abhiyaan. Launched in March 2018, the "multi-ministerial convergence" mission focuses on addressing the recognized problems, e.g., anemia, low birth weight, stunting, and undernutrition. Through this mission, the Indian Government has established a robust mechanism to bring synergy between various schemes previously implemented discretely [62].

4 Conclusion

Interdisciplinarity is the elephant that requires urgent attention to achieve the SDGs. The interdisciplinarity research, integration, and evaluation approaches require a practical application and evidence for use across sectors and problems. The 6 C's of interdisciplinary research [Convergence, Catalysis or Catalytic Framework, Communication, Community Engagement, Combined (Structured) Evaluation, and Co-development] provide a comprehensive framework to address the issues of interdisciplinarity in research.

Core Messages

- The interdisciplinary science approach is the key to achieving the 2030 Sustainable Development Goals (SDGs).
- Health, engineering, and environment play a concerted role in overcoming the barriers to attaining SDGs.
- Epistemological incompatibility and understanding gaps are the major barriers to the interdisciplinary approach.
- 6 Cs (Convergence, Catalysis, Communication, Community Engagement, Combined Evaluation, Co-development) are important.
- Motivation, awareness, resources, knowledge, skills (MARKS) framework helps community engagement and co-development.

References

1. Oluoko-Odingo AA (2009) Determinants of poverty: lessons from Kenya. Geo J 74:311–331. https://doi.org/10.1007/s10708-008-9238-5
2. Fuso Nerini F, Tomei J, To LS, Bisaga I, Parikh P, Black M, Borrion A, Spataru C, Castán Broto V, Anandarajah G, Milligan B, Mulugetta Y (2018) Mapping synergies and trade-offs between energy and the sustainable development goals. Nat Energy 3:10–15. https://doi.org/10.1038/s41560-017-0036-5
3. Parikh P, Diep L, Hofmann P, Tomei J, Campos L, Teh T-H, Mulugetta Y, Milligan B, Lakhanpaul M (2020) Mapping synergies and trade-offs between sanitation and the sustainable development goals. UCL Open Environ J. UCL Press. https://doi.org/10.14324/111.444/000054.v1
4. Ejemot-Nwadiaro RI, Ehiri JE, Arikpo D, Meremikwu MM, Critchley JA (2015) Hand washing promotion for preventing diarrhoea. Cochrane Database Syst Rev. https://doi.org/10.1002/14651858.CD004265.pub3
5. Luby SP, Rahman M, Arnold BF, Unicomb L, Ashraf S, Winch PJ, Stewart CP, Begum F, Hussain F, Benjamin-Chung J, Leontsini E, Naser AM, Parvez SM, Hubbard AE, Lin A, Nizame FA, Jannat K, Ercumen A, Ram PK, Das KK, Abedin J, Clasen TF, Dewey KG, Fernald LC, Null C, Ahmed T, Colford JM (2018) Effects of water quality, sanitation, handwashing, and nutritional interventions on diarrhoea and child growth in rural Bangladesh:

a cluster randomised controlled trial. Lancet Glob Health 6:e302–e315. https://doi.org/10. 1016/S2214-109X(17)30490-4

6. Stepien J (2018) Urban green space as a tool for cohesive and healthy urban community. In: Zielinski T, Sagan I, Surosz W (eds) Interdisciplinary approaches for sustainable development goals. Springer

7. (2019) PANChSHEEEL project conference. In: University College of London. https:// mediacentral.ucl.ac.uk/Play/18337

8. Gelting RJ, Chapra SC, Nevin PE, Harvey DE, Gute DM (2019) "Back to the Future": time for a renaissance of public health engineering. Int J Environ Res Public Health 16:387. https:// doi.org/10.3390/ijerph16030387

9. Mihelcic JR, Naughton CC, Verbyla ME, Zhang Q, Schweitzer RW, Oakley SM, Wells EC, Whiteford LM (2017) The grandest challenge of all: the role of environmental engineering to achieve sustainability in the world's developing regions. Environ Eng Sci 34:16–41. https:// doi.org/10.1089/ees.2015.0334

10. University College of London (2020) UCL Grand Challenges

11. Kapur K, Suri S (2020) Towards a malnutrition-free India: best practices and innovations from POSHAN Abhiyaan. Observer Research Foundation

12. PANChSHEEEL: participatory approach for nutrition in children, strengthening HEEE linkage | UCL Great Ormond Street Institute of Child Health—UCL—University College London. https://www.ucl.ac.uk/child-health/research/population-policy-and-practice-research-and-teaching-department/champp-child-and-2. Accessed 14 Nov 2021

13. Lakhanpaul M, Sharma S, Roy S, Santwani N, Prakash Pattanaik S, Dang P, Chaturvedi H, Kumar Pandya P, Singh Muniya T, Agrahari N, Karal Nair N, Jegannathan S, Mistry N, Saxena A, Karve A, Chariar VM, Sundararaman T, Warwick I, Mehta R, Singhal A, Pelton J, Maheshwari S, Strommer S, Kashyap A, Khandelwal R, Puri Kapur R, Khandelwal A, Narain R, Dhir S (2019) Participatory approach for nutrition in children strengthening health education engineering and environment linkages

14. Mills A (1983) Vertical vs horizontal health programmes in Africa: Idealism, pragmatism, resources and efficiency. Soc Sci Med 17:1971–1981. https://doi.org/10.1016/0277-9536(83) 90137-5

15. World Health Organization (2018) Multisectoral and intersectoral action for improved health and well-being for all: mapping of the WHO European Region—governance for a sustainable future: improving health and wellbeing for all

16. Ministry of Women & Child Development Government of India (2014) Multi-sectoral program to address maternal & child under-nutrition—letters to Secretaries of concerned states UTs on multi-sectoral programme

17. Zhou X, Moinuddin M (2017) Sustainable development goals interlinkages and network analysis: a practical tool for SDG integration and policy coherence

18. Manikam L, Bou Karim Y, Boo YY, Allaham S, Marwaha R, Parikh P, Lakhanpaul M (2020) Operationalising a one health approach to reduce the infection and antimicrobial resistance (AMR) burden in under-5 year old urban slum dwellers: the childhood infections and pollution (CHIP) consortium. One Health 10:100144. https://doi.org/10.1016/j.onehlt.2020. 100144

19. Forecasting Healthy Futures—Malaria No More. https://www.malarianomore.org/our-impact/international-programs/forecasting-healthy-futures/. Accessed 13 Nov 2021

20. Grygoryan R (2019) The relativity concept for human physiology and health assessment. Eur Sci J 34:51–58

21. Merril R (2010) Foundations of epidemiology. In: Introduction to epidemiology, 5th ed. pp 1–22

22. Batta Gori G (1989) Epidemiology and the concept of causation in multifactorial diseases. Regul Toxicol Pharmacol 9:263–272. https://doi.org/10.1016/0273-2300(89)90065-2

23. MacMahon B, Pugh TF (1970) Epidemiology: principles and methods. Little, Brown

24. Sekhon JS (2007) The neyman-rubin model of causal inference and estimation via matching methods. In: The Oxford handbook of political methodology
25. World Health Organzation (2017) International classification of health interventions
26. Kivits J, Ricci L, Minary L (2019) Interdisciplinary research in public health: the 'why' and the 'how.' J Epidemiol Community Health 73:1061–1062. https://doi.org/10.1136/jech-2019-212511
27. Langdon EJ, Wiik FB (2010) Anthropology, health and illness: an introduction to the concept of culture applied to the health sciences. Rev Lat Am Enfermagem 18:459–466. https://doi.org/10.1590/S0104-11692010000300023
28. Parikh P, Parikh H, McRobie A (2013) The role of infrastructure in improving human settlements. Proc Inst Civ Eng—Urban Des Plan 166:101–118. https://doi.org/10.1680/udap.10.00038
29. Singh A, Singh S (2008) Diseases of poverty and lifestyle, well-being and human development. Mens Sana Monographs 6:187. https://doi.org/10.4103/0973-1229.40567
30. Wang W, Chen J, Sheng H-F, Wang N-N, Yang P, Zhou X-N, Bergquist R (2017) Infectious diseases of poverty, the first five years. Infect Dis Poverty 6:96. https://doi.org/10.1186/s40249-017-0310-6
31. Organisation for Economic Co-Operation and Development (2003) DAC guidelines and reference series: poverty and health
32. World Health Organzation WHO Global Coordination Mechanism on the Prevention and Control of NCDs
33. Bloom DE, Cafiero ET, Jane-Llopis E, Abrahams-Gessel S, Bloom LR, Fathima S, Feigl AB, Gaziano T, Mowafi M, Pandya A, Prettner K, Rosenberg L, Seligman B, Stein AZ, Weinstein C (2011) The global economic burden of non-communicable diseases. Geneva
34. Fahy N, Hervey T, Greer S, Jarman H, Stuckler D, Galsworthy M, McKee M (2017) How will Brexit affect health and health services in the UK? Evaluating three possible scenarios. Lancet 390:2110–2118. https://doi.org/10.1016/S0140-6736(17)31926-8
35. Bekker MPM, Greer SL, Azzopardi-Muscat N, McKee M (2018) Public health and politics: how political science can help us move forward. Eur J Pub Health 28:1–2. https://doi.org/10.1093/eurpub/cky194
36. Abrams M, Nuzum R, Zezza M, Ryan J, Kiszla J, Guterman S (2015) The affordable care act's payment and delivery system reforms: a progress report at five years
37. Reference terms: engineering. In: ScienceDaily
38. Broy M, Cengarle M, Geisberger E (2012) Cyber-physical systems: imminent challenges, pp 1–28
39. Rahimifard S, Trollman H (2018) UN sustainable development goals: an engineering perspective. Int J Sustain Eng 11:1–3. https://doi.org/10.1080/19397038.2018.1434985
40. McCullers JA, Dunn JD (2008) Advances in vaccine technology and their impact on managed care. P & T: Peer-Reviewed J Formulary Manage 33:35–41
41. Venables R (2005) Civil engineering—Jubilee River. In: Engineering for sustainable development: guiding principles. The Royal Academy of Engineering
42. Yates T, Allen J, Leandre MJ, Lantagne D (2017) WASH interventions in disease outbreak response
43. Mara D, Lane J, Scott B, Trouba D (2010) Sanitation and health. PLoS Med 7:e1000363. https://doi.org/10.1371/journal.pmed.1000363
44. Marriott NG, Gravani RB (2006) principles of food sanitation, 5th edn. Springer
45. Brister E (2016) Disciplinary capture and epistemological obstacles to interdisciplinary research: lessons from central African conservation disputes. Stud Hist Philos Sci Part C: Stud Hist Philos Bio Biomed Sci 56:82–91. https://doi.org/10.1016/j.shpsc.2015.11.001
46. What is Interdisciplinary Research? | NSF—National Science Foundation. https://nsf.gov/od/oia/additional_resources/interdisclplinary_research/definition.jsp. Accessed 13 Nov 2021
47. Champions Project. https://www.championsproject.co.uk/. Accessed 13 Nov 2021

48. Werder KP, Nothhaft H, Verčič D, Zerfass A (2018) Strategic communication as an emerging interdisciplinary paradigm. Int J Strateg Commun 12:333–351. https://doi.org/10.1080/1553118X.2018.1494181
49. Petrausch RJ (2009) Five strategic imperatives for interdisciplinary study in mass communications/media studies in the US and UK. Interdiscip Learn Teach Higher Edu: Theo Pract 1:124–133. https://doi.org/10.4324/9780203928707
50. Lindsay J, Rogers BC, Church E, Gunn A, Hammer K, Dean AJ, Fielding K (2019) The role of community champions in long-term sustainable urban water planning. Water (Switzerland) 11:476. https://doi.org/10.3390/w11030476
51. Mahajan S, Kumar P, Pinto JA, Riccetti A, Schaaf K, Camprodon G, Smári V, Passani A, Forino G (2020) A citizen science approach for enhancing public understanding of air pollution. Sustain Cities Soc 52:101800. https://doi.org/10.1016/j.scs.2019.101800
52. Roitman S, Webster C, Landman K (2010) Methodological frameworks and interdisciplinary research on gated communities. Int Plan Stud 15:3–23. https://doi.org/10.1080/1356347 1003736886
53. Malaria No More India Concludes The Series of Webinars Under The Mega Campaign Bite Ko Mat Lo Lite—Malaria No More. https://www.malarianomore.org/news/press-releases/news-malaria-no-more-india-concludes-series-of-webinars-under-the-mega-campaign-bite-ko-mat-lo-lite/. Accessed 13 Nov 2021
54. RBM Partnership to End Malaria Social and Behaviour Change Working Group Case Studies (2021) Malaria social and behaviour change during the COVID-19 pandemic RBM partnership to end malaria social and behaviour change working group case studies
55. Meeks S, Shryock SK, Vandenbroucke RJ (2017) Theatre involvement and well-being, age differences, and lessons from long-time subscribers. Gerontologist. https://doi.org/10.1093/geront/gnx029
56. MeHeLP INDIA Participatory models of theatre practice
57. Institute of Medicine, National Academy of Engineering, National Academy of Sciences, Committee on Science Engineering and Public Policy, Committee on Facilitating Interdisciplinary Research (2005) Facilitating interdisciplinary Research. The National Academy Press
58. Klein JT (2008) Evaluation of interdisciplinary and transdisciplinary research. Am J Prev Med 35:S116–S123. https://doi.org/10.1016/j.amepre.2008.05.010
59. McLeish T, Strang V (2016) Evaluating interdisciplinary research: the elephant in the peer-reviewers' room. Palgrave Commun 2:16055. https://doi.org/10.1057/palcomms.2016.55
60. Repko AF, Szostak R (2016) Interdisciplinary research: process and theory. SAGE Publications
61. Lakhanpaul M, Culley L, Robertson N, Alexander EC, Bird D, Hudson N, Johal N, McFeeters M, Hamlyn-Williams C, Manikam L, Boo YY, Lakhanpaul M, Johnson MRD (2020) A structured collaborative approach to intervention design using a modified intervention mapping approach: a case study using the management and interventions for asthma (MIA) project for South Asian children. BMC Med Res Methodol 20:271. https://doi.org/10.1186/s12874-020-01148-y
62. Government of India POSHAN Abhiyaan—PM's overarching scheme for holistic nourishment. In: 2020

Kaushik Sarkar is a Global Health professional turned Artificial Intelligence (AI) practitioner who aims to benefit the billion's health using the untapped opportunities of applying AI through strategic innovations and transdisciplinary solutions. He is the founding Director incubating the Institute for Malaria & Climate Solutions (IMACS) of Malaria No More and heads Malaria No More India. Besides, Kaushik is one of the Founder-Directors of Aceso Global Health Consultants Limited, India. He has advanced several strategic and technology innovations, including precision programming and climate-based forecasting models, to strengthen health systems and access to services by the most marginalized. He also led the establishment of the Strategic Support Unit and convened multisectoral platforms—Malaria Action Coalition and India Interagency Expert Committee on Malaria and Climate to accelerate India's progress towards malaria elimination.

Professor Monica Lakhanpaul is an academic researcher and practicing pediatric consultant. She is a Professor of Integrated Community Child Health at UCL Great Ormond Street Institute of Child Health, UCL Pro-Vice-Provost for South Asia, and adjunct Professor at Public Health Foundation India. She is committed to improving the lives of those in the most vulnerable communities through holistic, cross-sectoral interdisciplinary interventions that encompass health, environmental and educational factors. She is the recipient of the Asian Women of Achievement Award, Royal Society of Public Health award, BSA media fellowship. Her research tackles some of the most pressing issues facing marginalized, minority, and vulnerable families globally, such as early years, nutrition, development, and mental health. She uses participatory research, citizen science, and arts-based approaches, ensuring that communities are involved in co-developing holistic, integrated solutions. She has over 180 publications to her name.

Priti Parikh is a fellow of the Institution of Civil Engineers and an associate professor at UCL. She has 15 years of engineering industry experience in Asia, Africa, and the UK on infrastructure delivery. She created an MSc program in Engineering for International Development (EFID) and the EFID student hub and heads the interdisciplinary EFID Research Centre addressing the Sustainable Development Goals through infrastructure provision. Dr. Parikh won the UCL Provost Education Award for her contribution to Engineering Education and the BBOXX/Royal Academy of Engineering Senior Research fellowship on solar energy access. Dr. Parikh is a key influencer and has been recognized as an Engineers Without Borders Changemaker and serves as chair of the editorial panel for the Institution of Civil Engineers Engineering Sustainability Journal. She is on the board of trustees for the Happold Foundation and Engineers Against Poverty.

Super-Spreading in Infectious Diseases: A Global Challenge for All Disciplines

16

Richard A. Stein

"We live in evolutionary competition with microbes—bacteria and viruses. There is no guarantee that we will be the survivors."

Joshua Lederberg

Summary

From the Plague of Athens, the earliest pandemic recorded in history, to COVID-19, infectious disease outbreaks have relentlessly impacted societies, cultures, and economies. For a long time, it has been assumed that the spread of pathogens in a population is homogeneous, with each infected host having approximately equal probabilities of encountering and infecting susceptible secondary contacts. More recently, it was shown that many outbreaks are shaped by heterogeneities, which may occur in the spatial or temporal dimensions and can arise through various mechanisms. These heterogeneities exist at the level of individuals, groups of individuals, and species, and were described in the interaction of human, animal, and cellular populations with bacteria, viruses, parasites, or vectors. Extreme cases of these transmission heterogeneities are known as super-spreading events. Transmission heterogeneities and super-spreading are governed by factors that depend on the host, the pathogen, and the environment. While super-spreading events were documented for many outbreaks studied to date, they are usually identified retrospectively. This makes it challenging to incorporate them into the management of ongoing epidemics or pandemics. Previous studies point toward co-infection, immune suppression, pathogen virulence, airflow dynamics, and high numbers of social contacts as

R. A. Stein (✉)
Department of Chemical and Biomolecular Engineering, NYU Tandon School
of Engineering, 6 MetroTech Center, Brooklyn, NY 11201, USA
e-mail: steinr01@nyu.edu

© The Author(s), under exclusive license to Springer Nature Switzerland AG 2022   347
N. Rezaei (ed.), *Multidisciplinarity and Interdisciplinarity in Health*,
Integrated Science 6, https://doi.org/10.1007/978-3-030-96814-4_16

some of the factors involved in super-spreading. A better understanding of super-spreading, from epidemiology to the cellular processes and the mechanistic details, promises to support the development of a framework to identify transmission heterogeneities early during outbreaks, incorporate them into epidemic and pandemic preparedness plans, and reshape the future of public health.

"We live in evolutionary competition with microbes—bacteria and viruses. There is no guarantee that we will be the survivors". Joshua Lederberg.

Super-spreading in infectious diseases

[Adapted from the Association of Science and Art (ASA), USERN; Made by Nastaran Hosseini].

The code of this chapter is *01101110 01101110 01101001 01101111 01100101 01110010 01110100 01000101 01110110 01101110 01101101*.

Keywords

COVID-19 · Epidemics · Heterogeneities · HIV · Infectious diseases · Influenza · Malaria · Pandemics · Public health · SARS · Super-shedding · Super-spreading

1 Introduction

Since the dawn of history, infectious disease outbreaks have emerged and reemerged in a manner that has been difficult to predict accurately [1], and they continuously shaped humanity [2]. Epidemics and pandemics have decimated societies, contributed to the decline of empires, and decided the fate of wars, sometimes creating more casualties than war-inflicted wounds [2–4]. At the same time, they catalyzed advances in science, medicine, economy, and politics [3]. At least 30 new human infectious diseases have emerged in the past three decades [5], and other infectious diseases, thought to be under control, re-emerge regularly [6, 7]. Finally, an increase in human mobility, by as much as 1000-fold since the 1800s in some high-income countries, led to an increase in the frequency and the spread of outbreaks [8].

The risk of an infectious disease is shaped by characteristics of the infectious host, the pathogen, the susceptible host, and the environment [9]. Historically, models that described the spread of infectious diseases have assumed that populations of hosts and pathogens mix homogenously and that every infected individual has approximately the same probability of infecting susceptible individuals [10–15]. However, more recent studies described marked population-level heterogeneities for many infectious diseases, which sometimes change very dynamically over time [10, 16–18]. Generally, about 20% of the hosts in a population contribute to about 80% of the transmission potential at the population level, a relationship known as the 20/80 rule [17, 19–21]. These heterogeneities are influenced by a multitude of biological and social factors [22, 23], such as age [24, 25], sex [26], genetic variation [27, 28], pathogen evolution [29], susceptibility to disease [30, 31], position in space [32], and individual behavior [33, 34]. It is critical to incorporate these heterogeneities into prophylactic and therapeutic initiatives when developing public health intervention measures.

Heterogeneities in transmission were described at the level of *individuals*, *groups of individuals*, and *species*. For example, infected individuals who have disproportionately more secondary contacts than most others in the population, simply due to their increased social mobility, may initiate super-spreading events [20]. Heterogeneities at the level of groups are illustrated by a study of >15,000 Scottish sheep farms over a four-year period, which revealed that <20% of the farms contributed with >80% of the transmission potential of infectious diseases [35]. Finally, a study that found that American robins (*Turdus migratorius*), even though they represented on average only 3.7% of the avian abundance at five testing sites, accounted for >43% of the feedings by mosquitoes that transmit West Nile virus, illustrates heterogeneities at the level of species. Mosquitoes fed on robins nearly 17-times more often than it would be expected if no feeding preferences existed [36]. Heterogeneities were documented not only for the pathogen burden in a population of hosts, whereby ∼20% of the hosts contribute to at least 80% of the transmission events [17], but also with respect to the infectiousness of a host for the vector [37]. For example, a study that quantitated the contribution of human hosts with different levels of *Leishmania donovani* parasitemia found that 3.2% of the most infected individuals

(>1000 parasites/ml) were responsible for infecting a mean of 62% of the *Phlebotomus orientalis* sand flies [37]. Extreme cases of these transmission heterogeneities are called super-spreading events [38, 39].

While transmission heterogeneities are critical for understanding the dynamics of outbreaks, and have been extensively studied in many epidemics and pandemics, individuals who initiate super-spreading events can easily become the object of stigmatization [40, 41]. Even though the term "super-spreader" has been and continues to be widely used in the scientific literature and beyond, to refer to individuals who create more secondary contacts than most others in the population, it is preferable, as much as possible, to refer, instead, to "super-spreading" and "super-spreading events" instead to avoid, singling out individuals thought to have initiated transmission chains [41].

An example of stigmatization revolves around the concept "patient zero" which, even though it has attracted fascination throughout history [42] and may convey a sense of excitement and perhaps closure when investigating an outbreak [43], can also lead to the vilification and stigmatization of individuals, only to be revealed, later on, that "patient zero" was actually infected by someone else, who had not yet been identified at the time as the source of the outbreak [44]. A widely publicized example of stigmatization related to the "patient zero" label occurred in the wake of the HIV/AIDS pandemic. A 1984 study, which intended to examine transmissibility, not the origin of the newly recognized condition [45], linked several AIDS patients with Kaposi Sarcoma and opportunistic infections through sexual contact. This study reported that transmission was consistent with an infectious agent [46]. In the manuscript, the term "patient O" was used as an abbreviation to indicate that one of the patients resided "Out[side]-of-California" [45, 47, 48]. The letter "O" was subsequently misinterpreted, and the patient became "patient 0" or "patient zero" [45, 47, 48], identified by journalist Randy Shilts by name in his 1987 book *And the Band Played on: Politics, People, and the AIDS Epidemic*, as Air Canada flight attendant Gaétan Dugas [45, 49, 50]. Shilts noted in his book that "there's no doubt that Gaétan Dugas played a key role in spreading the new virus from one end of the United States to the other" and referred to him as "the Quebeçois version of Typhoid Mary" [50]. A prominent newspaper's headline in October 1987, titled "THE MAN WHO GAVE US AIDS," and claiming that Dugas triggered the "'*gay cancer*' epidemic" in the US, captured the climate in the wake of Shilts' book [45, 51].

More recently, a 2016 study that examined historical blood samples, including that of "patient zero" from the 1984 study, revealed that Gaétan Dugas was neither the first person with AIDS to be studied in the USA nor the first one to display symptoms [47] and that HIV-1 arrived in New York City from the Caribbean around 1970 and reached San Francisco around 1975 [47, 52]. Nevertheless, it was relevantly noted that, based on the narrative in the book, Gaétan Dugas was presented as "a problem rather than a person" [53].

2   Super-Spreading in Infectious Diseases

Super-spreaders are individuals that create many more secondary contacts during an infectious disease outbreak than most other infected individuals in the same population. The number of secondary contacts that make a super-spreader has been somewhat variable in the studies across the literature. For example, super-spreading was defined as transmitting the virus to ≥ 3 [54], >8 [55], ≥ 8 [56], ≥ 10 [57, 58], or >10 secondary contacts [59] during the 2002–2003 SARS epidemic, or to 6 [60] or 10 [61] secondary contacts during the COVID-19 pandemic.

Two dimensions of super-spreading were described. Super-*spreaders* are hosts that have many more opportunities to create secondary contacts than most other hosts in the same population [62, 63]. For example, a patient hospitalized in the pre-symptomatic phase of a respiratory infectious disease for unrelated health conditions and subsequently transferred between hospital wards or between hospitals may create many secondary contacts among healthcare workers, other patients, and visitors at various locations [20]. Super-*shedders*, on the other hand, are infectious hosts that disperse a larger number of microorganisms than most other hosts in the same population [63]. For example, some cattle disperse more *Escherichia coli* than others and become super-shedders for the bacteria [62, 64]. Probably one of the most extensively described super-spreading events in history was associated with Typhoid Mary, on her real name Mary Mallon, an Irish immigrant who worked for several families in New York [65]. An asymptomatic carrier of *Salmonella typhi* and the first known example of a healthy carrier of a pathogen in the USA [66], Mary Mallon was responsible for several outbreaks of typhoid fever [65]. She is thought to have infected at least 57 people, three of which died [67–69].

Super-spreading events are usually identified retrospectively, and it is very challenging to predict them. However, collecting and integrating information from various outbreaks is powerfully positioned to build a framework that will help more accurately identify and possibly predict super-spreading events, in an attempt to make informed decisions during future outbreaks.

2.1   Super-Spreading During the 2002–2003 SARS Outbreak

The severe acute respiratory syndrome (SARS), caused by SARS-CoV, was first reported as an atypical pneumonia in November 2002 in the Foshan municipality from the Guangdong Province, China, and subsequently spread to 33 countries worldwide, leading to 8447 cases and 813 deaths [70–74]. During the 2002–2003 SARS outbreak, super-spreading events facilitated the worldwide spread of the virus [55, 75].

On February 21, 2003, a 64-year-old Chinese medical doctor from the Guangdong Province traveled to Hong Kong [70] and spent a night in room 911 [76] on the 9th floor of Hotel Metropole [77, 78]. He was admitted to the hospital on

February 22, 2003, with fever and respiratory symptoms [77]. During his stay in Hotel Metropole, he unknowingly infected at least 16 other guests on the same floor [79], and they carried the virus internationally to their home countries [76]. Among these were three visitors from Singapore, two from Canada, and one from Vietnam, Ireland, and the USA each [80, 81].

Hong Kong experienced two major outbreaks over the following weeks. One of these was a hospital-based outbreak in the Prince of Wales Hospital, and the other one was a community outbreak in the Amoy Gardens residential complex [82, 83]. The outbreak at the Prince of Wales Hospital started when a 26-year-old patient who had visited an acquaintance staying on the 9th floor of Hotel Metropole was admitted to the hospital on March 4, 2003, with fever and pneumonia [84, 85]. By March 25, 156 patients and healthcare workers developed SARS, and they were all traceable to this patient [86–88]. In addition to overcrowding in the hospital ward, the use of a nebulized bronchodilator is thought to have contributed to the extensive dissemination of the virus around the index patient [84, 87].

The second major outbreak in Hong Kong occurred in the Amoy Gardens, a residential complex with 19 buildings [89], and accounted for 329 cases [84, 90], or 1.7% of the total population of the complex [91], and 42 deaths [90, 92]. The outbreak started when a 33-year-old patient from the Princess of Wales Hospital, who was undergoing hemodialysis, visited his brother in unit 7 of block E, on the 16th floor of the apartment complex on two occasions, on March 14 and March 19 [84, 93]. It is thought that when he flushed the toiled, large quantities of aerosols entered by hydraulic action into the vertical drainage pipes or sanitary risers, which collect waste from toilets, sinks, bathtubs, and floor drains, and the virus spread to other apartments [84, 92]. An investigation of the outbreak revealed that in many of the apartments, the U-shaped water traps in the floor drains had not been filled with water for long periods. This caused the water seals in the traps to dry out and facilitated communication with the vertical drainage pipes [94]. The negative pressure caused by the exhaust fans in the bathrooms is thought to have helped aerosols or fine droplets containing the virus to enter the bathrooms [80, 94]. The virus was also transported by the wind to adjacent buildings and caused additional exposures [93]. On March 21, 2003, several residents started showing symptoms, and by April 15, there were 321 cases, ∼41% of them in block E [80, 93].

The distribution of the infection in the affected apartments was not random [90], and the infection rate was lower in blocks that were further away from Block E [95]. An analysis of the nasopharyngeal viral load among hospitalized patients from the Amoy Gardens outbreak revealed that the viral load was higher in patients living in adjacent units of the same block as the index patient and decreased in patients living further away [96]. An epidemiologic analysis revealed that the shape of the outbreak was consistent with a single source [94].

Another super-spreading event was involved in the transmission of SARS-CoV on board of Air China flight 112 on a three-hour flight from Hong Kong to Beijing on March 15, 2003 [97]. The flight carried 120 individuals, including 112 passengers, six flight attendants, and two pilots. A 72-year-old man on the flight was symptomatic and had been having fever since March 11. In early March, he visited on several occasions his brother, who was admitted to the Prince of Wales Hospital in Hong Kong and later died. Among the persons on that flight, 16 developed laboratory-confirmed SARS, two had a diagnosis of probable SARS, and four were reported to have SARS but could not be interviewed [98].

Several SARS super-spreading events occurred in Toronto. Early during the outbreak, a 78-year-old woman with a history of coronary heart disease and type 2 diabetes [99], who had stayed on the 9th floor of Hotel Metropole in Hong Kong, returned to Toronto on February 23, 2002 [100, 101]. She developed a fever, non-productive cough, myalgia, and anorexia [99]. After visiting her family doctor on February 28, she was prescribed an antibiotic [102], but her cough and dyspnea became more severe, and she died at home on March 5, 2003 [99, 103]. At that time, she was not yet recognized as having SARS [104]. Her 44-year-old son became infected [102, 105], was admitted with respiratory symptoms to the Scarborough Hospital, Grace Division in a Toronto suburb [106] on March 7, diagnosed with community-acquired pneumonia [107], and died on March 13, 2003 [99, 100, 104]. In the hospital, he infected two other patients in the emergency department and a healthcare worker [100, 108] who transmitted the virus to others. Ultimately 128 individuals became infected in association with this community hospital outbreak, including hospital staff, patients, and visitors [100, 108, 109]. Four additional family members developed SARS and were admitted to four different hospitals [104].

A 77-year-old patient was exposed to the above-mentioned individual on March 7, in the emergency department of the Scarborough Hospital, where he was being treated for congestive heart failure. He was admitted there for three days. He returned to Scarborough Hospital on March 14 with fever, dyspnea, and pulmonary infiltrates. His exposure status was not yet known at the time. On March 16, after his renal and respiratory functions deteriorated, the patient was transferred to York Central Hospital for emergency renal dialysis [100, 110], and he died on March 29 of multi-organ failure [111]. On March 21, his 77-year-old wife was admitted to York Central Hospital with dyspnea, chest discomfort, and fever. She stayed there until March 26, when she was transferred to a nursing home and was then read-mitted to York Central Hospital on March 29 with respiratory symptoms. Subsequent transmission at York Central Hospital affected hospital staff, patients, and visitors who were identified with probable or suspected SARS [110].

2.2 Super-Spreading During the MERS Outbreak

The Middle East respiratory syndrome coronavirus (MERS-CoV) was first identified in September 2012, in Saudi Arabia, in the sputum of a 60-year-old man who developed severe pneumonia and acute kidney injury in June 2012 [112–115]. The virus subsequently spread to 27 countries, including South Korea [112]. As of November 2019, 2494 confirmed cases were reported to the World Health Organization, including 858 deaths [116, 117]. The infection can range from asymptomatic to mild upper respiratory illness, severe respiratory distress, or death [116] and has a case fatality rate of $\sim$ 35–43% [118–120].

One of the distinguishing features of the outbreak in South Korea was the hospital-to-hospital spread, in addition to the intra-hospital transmission [121]. The outbreak in South Korea started when a 68-year-old South Korean businessman, "patient 1," with hypertension and hyperlipidemia [122], returning to South Korea from Saudi Arabia, the United Arab Emirates, and Qatar [123], developed myalgia on May 11 [124]. He visited a local clinic on May 12, 14, and 15 [124], developed a non-productive cough on May 15 and was admitted with pneumonia to the 8th floor [125] of the Pyeongtaek St. Mary's Hospital, where he stayed until May 17 [122, 124]. He then visited another clinic, and on May 18, he was admitted to the Samsung Medical Center (SMC) in Seoul [122, 124], where MERS was confirmed on May 20 [123, 124]. The same day, he was transferred to the National Medical Center of Korea for treatment [122, 126].

Between May 18 and June 4, MERS developed in 36 people at Pyeongtaek St. Mary's Hospital [122, 127]. Of these, 26 are thought to have been first-generation cases [124]. One of them was "patient 14," who was infected while sharing a ward with "patient 1" at Pyeongtaek St. Mary's Hospital and was later admitted to the Samsung Medical Center without knowing that he had been exposed [128]. His exposure to "patient 1" occurred between May 15 and May 17, and he developed a fever on May 21 [124]. When he visited the emergency room at the Samsung Medical Center on May 27, he stayed until May 29 in three different zones of the emergency room and exposed several other individuals [124, 128]. After being notified on May 29 about his possible exposure to "patient 1," he was immediately transferred to a negative-pressure isolation room. MERS was confirmed on May 30, and he was transferred to the Seoul National University Hospital [124, 128]. Several clusters were established at additional hospitals, mostly as a result of nosocomial transmission [129, 130].

It is relevant that even though "patient 1" infected 36 people in St. Mary's Hospital [126, 127], he did not cause any confirmed secondary cases at Samsung Medical Center, while "patient 14" caused 82 secondary cases in the emergency room at Samsung Medical Center [112, 128]. Moreover, they were both treated at Samsung Medical Center around day 7 of their illness, when they were highly infectious: "patient 1" between May 17–20, and "patient 14" between May 27–29 [112]. It was proposed that in addition to the time from the onset of disease, other factors, including symptomatology, the duration of contact, the pattern of behavior inside the hospital, and the kinetics of viral shedding, could all explain these differences [128].

During the 2015 MERS outbreak from South Korea, 86 healthcare institutions accepted MERS patients, and 186 individuals had laboratory-confirmed MERS [124, 127, 131]. The outbreak involved three transmission phases, and nosocomial spread occurred in 12 hospitals, four clinics, and two ambulances [124, 127]. While 166 patients did not lead to any secondary infections, five super-spreaders (<3%) contributed to ~ 79–82% of the infections [132–134]. Of these five super-spreading events, two, at St. Mary's Hospital and Samsung Medical Center, were related [112]. One of the major differences between the outbreak in Saudi Arabia and the one in South Korea was that while transmission in Saudi Arabia was maintained by a zoonotic source and persisted for >6 years, in South Korea it was sustained by nosocomial transmission and lasted for only two months [135].

2.3 Super-Spreading in COVID-19

SARS-CoV-2, previously known as 2019-nCoV, was first reported in December 2019 in Wuhan, China [136]. It subsequently spread to many other countries and was declared a pandemic by the World Health Organization on March 11, 2020 [137]. As of March 2022, COVID-19 caused >452,000,000 infections and >6,000,000 deaths worldwide [138].

Several super-spreading events were described early on during the COVID-19 pandemic; these include a temple ceremony in Ningbo, China; a food market in Beijing; an apartment block in Melbourne, Australia; and a nightclub in Seoul [139–142]. On January 19, 2020, a 64-year-old woman from Ningbo, China, developed a fever. The same day she was picked up by a tour bus and joined a temple ceremony in which 348 pilgrims participated. On January 29, 2020, she was confirmed to have COVID-19. This person transmitted the virus to 28 individuals, of which 27 developed COVID-19, and one remained asymptomatic. Aggressive testing and contact tracing helped successfully stop the spread of the virus [61].

A study that included 9120 confirmed COVID-19 cases reported between January 15 and February 29, 2020, in mainland China, outside of Hubei Province, reconstructed 643 transmission clusters formed by 1407 transmission pairs and identified 34 individuals as super-spreaders. Five of the super-spreading events occurred within households. In this study, males aged 18–64 years old had a higher risk of being infected outside of their households, a finding that highlighted the importance of non-pharmaceutical interventions to mitigate the pandemic [143]. In a study that examined information about all 135 COVID-19 cases between January 21 and February 26, 2020, available from public sources in Tianjin, China, a 57-year-old man was identified as a super-spreader. He transmitted the infection to six secondary contacts at the workplace. It was suspected that transmission heterogeneity in Tianjin could be explained by the asymptomatic transmission and high contagiousness during the incubation period [60].

In another example of super-spreading, a barkeeper who worked at an après-ski bar in Ischgl, a ski town in the Austrian Alps, was confirmed with COVID-19 and is thought to have been the source for infections later reported in other countries, including Germany, Iceland, Norway, and Denmark. Several cases were reported between March 9–16 at University Hospital Münster, a tertiary care center in Germany. Of the 90 COVID-19 patients, 36 (~40%) visited Ischgl [144]. On March 8, 2020, a midwife working at the University of Regensburg maternal and perinatal center returned from a skiing vacation in Ischgl. She developed acute respiratory symptoms during the March 9 night shift and transmitted the infection to 36 staff members, including physicians, nurses, and midwives [145, 146].

Another super-spreading event occurred when a British man who attended a conference in Singapore in late January 2020, and is thought to have become infected there, subsequently traveled to the French skiing resort of Contamines-Monjoie, in Haute-Savoie. He spent four days at the resort before returning to the UK on January 28. At least 21 individuals were exposed to the SARS-CoV-2 at the resort, and for 13 of them, the infection was confirmed by a positive test. Of the people who stayed at the chalet, COVID-19 was confirmed in five of the ones who returned to the UK, six who remained in France, and one individual who traveled to Mallorca and became the second case in Spain [147]. A super-spreading event was documented in March 2020 at a choir in Skagit County, WA, where 87% of choir attendees were infected by a single person following a single 2.5 h joint choir practice [148].

From the beginning of the COVID-19 pandemic, several instances were documented when individuals suspected of initiating super-spreading events were blamed. This includes the British businessperson returning from Singapore, who was labeled as a "super-spreader" in the headline of several major newspapers [147]. A woman in her 60 s, who was a member of the Shincheonji church in Daegu, South Korea [149], experienced COVID-19 symptoms on February 7 and attended services on February 9 and February 16 [150, 151]. On February 18, it was confirmed that she had COVID-19 [152]. Afterward, several church members developed the infection. As of mid-March 2020, 5006 confirmed cases were linked to the church, and at the time, they represented >61% of the cases confirmed in the country [150]. This created an epicenter of the outbreak in South Korea [149], which led to anger directed at the church [41, 150, 151]. Finally, ~30 deaths were linked to a 70-year-old Sikh priest from Punjab, India, who died after returning from a trip to Italy and Germany without quarantining, and 40,000 people were subsequently quarantined [41, 153, 154].

2.4 Super-Spreading in Ebola

Ebola virus, the causative agent of the Ebola viral disease [155], formerly known as Ebola hemorrhagic fever, is a member of the *Filoviridae* family, together with the Marburg virus [156]. Ebola virus was first reported in humans in 1976 in two simultaneous outbreaks, in Nzara, Sudan, and in Yambuku, the Democratic Republic of the Congo (then Zaire) [157–162]. Several epidemiologic and ecological studies implicated bats as a key reservoir [157, 163–165].

Between September 5–7, 2014, the traditional funeral of a prominent pharmacist in Sierra Leone was linked to 28 laboratory-confirmed Ebola viral disease cases. Among those who attended the funeral, 57% had direct contact with the pharmacist before he died and 75% reported touching his body after he died [166]. An epidemiologic investigation found a link between the first case of the Ebola virus from Sierra Leone, reported on May 25, 2014 at the Kenema General Hospital, and 13 women who attended the funeral of a traditional healer who had been treating Ebola virus disease patients from Guinea [167]. Eventually, 365 Ebola deaths were linked to that funeral [168]. Phylogenetic analyses based on the sequences of 99 Ebola virus genomes from 78 patients in Sierra Leone and three patients from Guinea that sought to interrogate the transmission pattern of the virus during the first few weeks of the 2014 epidemic revealed that the virus probably spread from central Africa around 2004, and in 2014 it crossed from Guinea to Sierra Leone [167]. The transmission of the Ebola virus outbreak in Guinea in 2014 had a highly skewed transmission distribution, with $\sim 72\%$ of the infected individuals not generating any secondary contacts [169, 170]. The analysis of burial data from Sierra Leone illustrated that a small number of super-spreading events drove the epidemic and that the prompt identification and isolation of super-spreaders could prevent $\sim 61\%$ of the transmission events. The study also revealed that individuals from certain age groups, show higher transmissibility and may become super-spreaders. In the analysis, individuals <15 years and >45 years old showed higher transmissibility [171].

In early April 1995, a small nosocomial cluster of Ebola virus disease among nursing staff from the Kikwit II Maternity Hospital, one of the two major hospitals from Kikwit, the Democratic Republic of Congo, was initially misdiagnosed as epidemic dysentery. A similar nosocomial cluster was identified later that month in Kikwit General Hospital, the other major hospital in Kikwit. This second outbreak occurred among operating room staff who performed surgery on a Kikwit II Maternity Hospital technician. Typhoid-associated abdominal perforation was suspected but was not confirmed during the procedure, and an appendectomy was performed. After the pain worsened, the patient underwent another laparotomy, which revealed diffuse bleeding in the abdominal cavity. He died two days later, and a few days afterward, several hospital staff members developed a viral hemorrhagic fever. After specimens were sent to the Institute of Tropical Medicine and the CDC, the Ebola virus was confirmed as the etiologic agent in 14 patients. The nosocomial outbreak at Kikwit General Hospital resulted in four generations of cases in the hospital and the community. This outbreak led to 315 cases. Two index patients are thought to have been the source of the infection for >50 of the cases, and both of them presented gastrointestinal hemorrhage. The lack of samples from the two index cases made it challenging to study these super-spreading events [172].

Another super-spreading event occurred during the 2000 Ebola virus disease outbreak from the Masindi District in Uganda. The attack rate was 26% among 73 index family members, but it was 52% in the 15–49-year, economically active age group. A super-spreading event appears to have existed, based on the observation that an adult male in the 3rd generation created ten 4th generation cases, while two other males in the 3rd generation only transmitted the virus to one or two secondary contacts in the 4th generation [173].

During the 2014–15 Ebola virus outbreak in Guinea, epidemiologic analyses linked five infections to a 2-year-old toddler who became ill on December 2, 2013, died in Meliandou on December 6, and was suspected to be the first case. This toddler's sister, mother, grandmother, a nurse, and the village midwife became infected and died. A healthcare worker from the Guéckédou hospital became infected as a result and subsequently transmitted the virus to several family members in the Guéckédou Farako District, and 15 others at the Macenta hospital.

A doctor from the Macenta hospital transmitted the infection to several other individuals who traveled to Kissidougou and Nzérékoré and further expanded the outbreak [174].

2.5   Super-Spreading in Measles

Measles or rubeola [175] is a highly contagious, acute viral disease of childhood [176, 177] that before the advent of vaccination caused >2 million deaths worldwide annually [178]. Despite the availability of a safe and effective vaccine, it remains a cause of death among children globally [177]. A global milestone was achieved in 2016 when the number of measles-related deaths dropped under 100,000 worldwide for the first time [179].

In a measles outbreak that occurred in 1989 in a high school from Honkajoki, a small rural Finnish municipality, one super-spreader transmitted the virus to 22 secondary contacts within a single day. The index case in this outbreak was an 18-year-old high school student who was unvaccinated because he did not belong to the birth cohort that received the vaccine. After he became ill, 22 students developed measles in one generation. This included eight students who were vaccinated once and one student who was vaccinated twice [180].

In 2011, the largest measles outbreak in a decade started in North America, when a teacher who returned to Quebec, Canada, on March 24, 2011, after a one-week vacation in the Caribbean, developed a fever nine days later, on April 2, 2011. He started coughing on April 4 and developed a rash on April 7, when he stopped working. The teacher had been vaccinated in childhood with a single dose, was susceptible to infection, and triggered an outbreak that affected 110 students and four additional staff members within the school. The school outbreak was almost over by May 20, but transmission continued in the adjacent town, lasting until October 4 and causing 678 cases. The index high school accounted for 72% of the 678 cases. During this outbreak, the attack rates were 5.8% among 2-dose recipients who had their first immunization at 12 months and 2% among 2-dose recipients who had their first immunization at ≥ 15 months [181–183].

2.6   Super-Spreading in Influenza Outbreaks

Influenza viruses are among the most common human respiratory pathogens. They cause seasonal, endemic infections and, more rarely, pandemics with unpredictable periodicity [184]. Even though influenza viruses have been studied for nearly a century, and it is known that droplets, aerosols, and direct contact are involved in the transmission, relatively little is known about the relative contribution of these different transmission modalities [185, 186]. Influenza virus strains can be grouped into types A, B, C, and D, of which A and B are clinically the most relevant for humans, and type A is the only one with pandemic potential [187–189].

An influenza A super-spreading event was documented in March 1977, during an outbreak on board of a Boeing 737 jet flying from Anchorage to Kodiak, with a stop in Homer. In Anchorage, 24 passengers and five crew members boarded the aircraft. In Homer, six passengers got off, and 31 passengers boarded. However, one of the engines failed during takeoff, and this led to a 4.5 h delay. During the delay, 30 persons remained on the plane, including the index patient, and the other 23 passengers left and returned for various time periods. A smaller plane then flew most of the passengers from Homer to Kodiak and then returned and took 15 passengers and five crew members back to Anchorage, from where they were flown the same evening by a direct flight to Kodiak. Within 72 h, 72% of the passengers and crew who were on board when the engine malfunctioned became ill with cough, fever, myalgia, fatigue, and a sore throat. The attack rate was highest among those who were on the plane for >3 h (86%), followed by those who spent 1–3 h (56%) and < 1 h (53%). None of the six passengers who got off in Homer became ill. The index case, a 21-year-old woman who joined the flight in Homer, became ill 15 min after boarding. This outbreak was caused by an H3N2 virus, and the fact that the airplane ventilation system was not operative during the delay was thought to account for the high attack rate [190].

Another instance of super-spreading as part of an influenza A/H1N1 2009 outbreak was documented in August 2009, during a summer camp in France. Six adults and 26 children, 6 to 11 years old, participated in the camp. Of the 32 participants, 29 used two successive buses, one for 30 min and the other one for 90 min, and a 3.5 h train ride. The index case started to exhibit, from the time of departure, asthenia, fever, and cough, and he spent part of the travel time with his nose close to the wagon vent, which was located near the base of the window. Even though the air-conditioning was not operational during the trip, the ventilation was working, and the fact that 90% of the air was recycled but not renewed is thought to have contributed to the spread of the virus. Within two days, 21 children and three adults were affected by the outbreak, indicating a single-point exposure. Two children and one adult who took alternative transportation means to the camp site were not affected. This outbreak indicated that super-spreading could occur in confined spaces that lack sufficient air renewal [191].

2.7   Super-Spreading in Other Species

Super-spreading was described in the interaction between several animal species and viruses, bacteria, or parasites that colonize or infect them. One example is the interaction between water buffaloes and *Brucella abortus*, a Gram-negative non-motile bacterium of the genus *Brucella* that causes brucellosis, the most widespread zoonosis worldwide [192–194]. *Brucella abortus* affects mainly cattle [195] and water buffaloes [196], but also other species, including sheep, goats, pigs, bison, horses [197], and humans [197, 198]. A study that examined 500 water buffaloes from four herds in Caserta, southern Italy, revealed that of the 101 animals that shed the bacteria, 84% were low shedders ($\leq 10^3$ CFU/ml milk), and $\sim 16\%$

were high shedders or super-shedders ($\geq 10^4$ CFU/ml milk). Longitudinal analysis revealed that removing the high-shedding animals was sufficient to stop the transmission of the bacterium within the herd [199].

Another example of heterogeneities in transmission was described in the interaction between cattle and *Escherichia coli*, the most common commensal bacterium in the intestinal tract of warm-blooded vertebrates [200, 201]. *E. coli* is one of the first microbial species to colonize the intestinal tract of infants as part of the human intestinal microbiota [202]. However, several *E. coli* strains are pathogenic for humans [203, 204]. One of these, *E. coli* O157:H7, first recognized as a human pathogen in 1982 [205], belongs to the group of enterohaemorrhagic *E. coli* (EHEC) and is an important foodborne pathogen [203, 206]. *E. coli* O157:H7 has been involved in human disease worldwide, and healthy cattle are the primary reservoir [207, 208]. In cattle, the bacteria preferentially colonize the recto-anal junction [209], a region where the columnar epithelium that lines the rectum transitions to the stratified squamous epithelium that lines the anus [209, 210].

Human *E. coli* O157:H7 infection may result in asymptomatic carriage, mild non-bloody diarrhea, hemorrhagic colitis, or potentially life-threatening hemolytic-uremic syndrome or thrombotic thrombocytopenic purpura [211–214]. Cattle shedding >10^4 CFU of *E. coli* O157:H7 per gram of feces are defined as super-shedders [215, 216]. Studies examining bacterial shedding by cattle on 56 Scottish farms revealed that 9% of the animals contributed >96% of the bacteria shed by all the animals tested [217]. A cross-sectional survey of cattle farms in Scotland revealed that ~80% of the *E. coli* O157 transmission arises from the 20% of most infectious animals and that preventing infection in <5% of the animals with the highest infectiousness could decrease transmission by at least one-third [64].

A study that examined *E. coli* O157 shedding at a Scottish abattoir revealed that animals shed the bacteria in the cooler months (January-March) more than in the warmer months (May–July) (11.2 vs 7.5%). This is the reverse of the seasonality of human infections with the bacterium, which peak during the summer. Even though the proportion of high-shedding cattle was comparable between the warmer and the colder months (0.7 vs 0.6%), the high-shedding animals dispersed almost sixfold more bacteria during the warmer months compared to the winter months. It was this trend, the one that followed the seasonality of the human infection [218].

3   Transmission Heterogeneities in Vector-Borne Infectious Diseases

Vector-borne infectious diseases, such as schistosomiasis, malaria, yellow fever, dengue, and plague, are shaped by interactions between three players that need to be considered simultaneously—the pathogen, the vector, and the vertebrate host, as well as by interactions between each of these factors and their environments [219]. Heterogeneities were described at the level of the host–vector contact [17, 220] and the level of the vector–pathogen interaction [221–223]. This additional complexity makes it more challenging to interrogate the dynamic interactions that shape transmission heterogeneities for vector-borne infectious diseases.

3.1 Schistosomiasis

Transmission heterogeneities were extensively documented for schistosomiasis, a vector-borne disease caused by *Schistosoma* trematodes. Schistosomiasis affects >250 million people in tropical and subtropical regions of the developing world, and 90% of the worldwide cases are in sub-Saharan Africa [33, 224]. The parasitic disease affects multiple organs [225] and is a significant cause of morbidity and mortality [226]. It ranks second to malaria among parasitic diseases in terms of the number of people affected globally [224, 227, 228]. The first causative organism of human schistosomiasis, *Distomum haematobium*, was discovered in 1851 by Theodor Bilharz [228–230] and renamed *Schistosoma haematobium* in 1858 [226].

Most human schistosomiasis is caused by *S. haematobium, S. japonicum,* and *S. mansoni* [226], while *S. intercalatum* and *S. mekongi* are less common causes of systemic human disease [225]. *S. mansoni* and *S. japonicum* mostly cause intestinal schistosomiasis, and *S. haematobium* causes urogenital schistosomiasis [231, 232]. Schistosoma transmission requires an intermediate snail host, water contamination, and human contact with water that contains infected snails [228, 233–235].

In the 1980s, it was anticipated that school- and community-based programs distributing praziquantel, the drug of choice to treat schistosomiasis, would help eliminate the infection, but elimination was not successful in certain high-risk communities. Sustained transmission resulted mostly from super-spreading events, fueled by children who did not attend school, persistent contacts between humans and snails, sustained transmission due to human movement, and the existence of additional mammalian reservoirs [236–244].

At least two levels of transmission heterogeneities were described in schistosomiasis. The first one involves heterogeneities in the interaction between the parasite and its intermediate snail hosts. The second one involves heterogeneities at the interface between human hosts and the water containing infected snails [245]. This is exemplified by a study in late 1987 that found heterogeneities in the *S. haematobium* distribution in a population of the *Bulinus globosus* intermediate snail host in the Nyamakari River from eastern Zimbabwe. Only 11% of the snails that were examined were infected, and 89% had no patent infection. The proportion of single-sex and mixed-sex infections, $\sim 8\%$ and $\sim 3\%$, respectively, was higher than expected based on a random distribution of the infection among the snails [245]. The second level of heterogeneity is exemplified by an analysis of the *S. haematobium* epidemiology in a rural community from eastern Zimbabwe, which found that 90% of the water contacts over a two-week period were made by only 37% of the population, and >50% of the water contacts were made at two of five sites that were studied [246]. Another study that examined human schistosomiasis transmission in communities from Zimbabwe and Mali underscored that missing individuals or water contact sites that contribute markedly to transmission could render prophylactic measures ineffective [235]. A study in Kenya that used a geographic information system to examine the spatial distribution of urinary schistosomiasis reported high infection rates around a water contact site that had many snails shedding *S. haematobium* cercariae and showed that different age groups had significant clustering at different distances from the high-risk water source [247]. In another study that examined a farming community from northern Senegal, it was reported that clusters of infection in a region co-endemic for *S. mansoni* and *S. haematobium* could be related to heterogeneities in the self-reported use of different water contact sites [248].

3.2   Malaria

Malaria, a disease caused by protozoan parasites that belongs to the *Plasmodium* species, is transmitted to humans by the bite of infected female *Anopheles* mosquitoes. The *Plasmodium* life cycle occurs in a sexual phase in the mosquito vector and an asexual phase in the hepatocytes and erythrocytes of the human intermediate host [249]. The five parasite species that cause human disease are *P. falciparum, P. ovale, P. vivax, P. malariae*, and *P. knowlesi* [250]. Malaria is endemic in 76 countries [251]. In 2017, it affected about 291 million people and caused about 435,000 deaths worldwide [249].

Several factors shape the individual and household risk to develop malaria, including host genetic factors [252], occupation [253, 254], behavioral factors [255], travel [256], gender [257], bed net use [252], and type of housing [255]. Malaria distribution in endemic regions is not random [258]. Heterogeneities in malaria transmission occur at several levels, including mosquito biting rates [259], the spatial pattern of infection [260], parasite virulence [261], human infectiousness [260, 262], spatial and temporal variations in the mosquito and parasite transmission [263], and seasonality [264, 265]. For example, in a study of humans bitten by *Anopheles funestus* and *Anopheles gambiae* in Western Kenya, polymerase chain reaction analyses revealed that <20% of the human hosts contributed to >50% of the blood meals, and 42% of the participants did not receive any bites [266].

Areas with higher malaria transmission are known as hotspots [267, 268]. Malaria hotspots have temporal and spatial components [269, 270]. Some hotspots are stable over time and may predict the risk of future malaria reemergence, while others are unstable and may not emerge again in the same location [267]. Hotspots may change from one year to the next and from one season to another [271]. For example, a study in western Kenya that included 64,000 people examined foci of self-reported malaria and found that hotspots changed over short time periods. Over the 2.5 years of data collection, $\sim 35\%$ of the households were located in a hotspot during any six-month interval, and only $\sim 1.5\%$ of the households were located in a hotspot during three or more six-month intervals. Households located in a hotspot in the fall were more likely to be located in a hotspot during the following fall, but the opposite was true for households located in a hotspot during the spring [271].

In terms of spatial distribution, hotspots were identified at several scales, including within and between individual households [272], villages [273], larger areas [252], regions within a country [252, 274, 275], and countries [267, 276]. Some spatial heterogeneities and hotspots are associated with the macro-spatial scale, such as temperature and precipitation, and others are associated with the micro-spatial scale, such as wind direction, proximity to larval sites, elevation, and land use [267, 277, 278]. A longitudinal analysis of 256 homes in the Kilifi District on the Kenyan coast between January 1998 and June 2009 collected data on febrile malaria episodes, asymptomatic parasitemia, and antibody titers, and described two types of clustering. Hotspots of asymptomatic parasitemia, stable over time, had an average radius of 1 km. Hotspots of febrile malaria, unstable over time, had an average radius of 1.3 km. In this analysis, the 20% of the homes where a febrile malaria episode occurred during one month of the dry season experienced 65% of all the febrile malaria episodes during the subsequent year [279]. In a study conducted in a hyperendemic Malian village, the use of geographic information systems identified variations in *P. falciparum* infection and transmission at the individual household scale, with a 1–3 m resolution [272].

Heterogeneities in malaria transmission were reported in studies from several countries. A study in two villages from southern Zambia found that 25% of the households contributed with $\sim 78\%$ of the *Anopheles arabiensis* mosquitoes [280]. Studies of African children under age 15 from >90 communities revealed that 20% of them received $\sim 80\%$ of all the infections [281]. An analysis of malaria in Mâncio Lima, the main urban malaria hotspot of the Amazon Basin of Brazil, revealed that 20% of the urban residents contribute $\sim 86\%$ of the vivax malaria burden in the community [282]. Finally, a study that used remote sensing technologies in southern Zambia between April 2007 and December 2008 to generate large-scale spatial risk maps and target malaria control interventions showed that interventions directed at only $\sim 24\%$ of the households can target households in the top 80th percentile of the predicted malaria risk [283]. Understanding the causes and consequences of heterogeneities, and targeting hotspots, are important facets of developing prophylactic interventions [272] and of implementing malaria control [256, 267]. However, difficulties in predicting future hotspots make this task particularly challenging [271].

A longitudinal study of natural malaria infection known as the Dielmo Project, initiated in 1990 [284], performed a long-term analysis of the host–pathogen interaction among residents from a Senegalese village. The number of malaria episodes differed markedly among the individuals examined. While some children experienced only one malaria attack in their first two years of life, other children suffered up to 20 episodes, and one child suffered 40 episodes between ages three and seven [285]. Another study that followed children under ten from Senegal for 3–5 years found that $\sim 55\%$ of the malaria episodes documented during the study were experienced by only 23% of the cohort person-years. This increased susceptibility appeared to be malaria-specific, as it was not observed for non-malarial infectious fevers [286].

4   Heterogeneities and Super-Spreading at the Cellular Level

Heterogeneities were described not only at the level of human and animal populations, but also at the level of cellular populations. For example, in a study that examined single cells infected with the influenza virus, comparisons of the transcriptomes revealed large variations in the amount of viral mRNA expressed in individual cells. Eight hours after the initiation of the infection, $\sim 50\%$ of all viral mRNA was derived from $\sim 8\%$ of the infected cells [287].

Diverse outcomes may result after the encounter of a bacterium with a macrophage. Studies on the susceptibility of macrophages to *Salmonella typhimurium* revealed that some macrophages are more susceptible to infection than others. Despite a large number of contacts between macrophages and the bacteria, at the lowest multiplicity of infection, most macrophages remained uninfected, and only a few contacts led to successful infection events [288, 289]. While some macrophages engulfed and lysed the ingested bacteria, others remained permissive for intracellular bacterial replication, and yet others allowed the formation of non-replicating persisters [288, 290–293].

Another example of cellular heterogeneities is provided by the interaction between the human immunodeficiency virus (HIV) and the $CD4^+$ T cells. At any given time, most of the $CD4^+$ T cells are in a quiescent state, which is characterized by low metabolic rates, low transcription levels, and small size [294], a state that is poorly permissive for viral replication [295] but can serve as a latent viral reservoir [296]. In the endocervix, quiescent $CD4^+$ T cells outnumber active $CD4^+$ T cells by $\sim 70{:}1$ [297], and in most human lymphoid tissues, the quiescent cells represent 95% or more of the total $CD4^+$ T cells [298]. HIV replication is inefficient in quiescent cells [299], and the virus preferentially replicates in activated $CD4^+$ T cells [295, 300, 301], creating more viral particles, which facilitates the dissemination of the virus [299]. In rhesus macaques, activated $CD4^+$ T cells in the vaginal tissues were found to contain fourfold to sevenfold more simian immunodeficiency virus (SIV) RNA than resting $CD4^+$ T cells [302]. A study that used a non-human primate model of SIV infection reported that the more rare, activated $CD4^+$ T cells in tissues were surrounded by 12-times more viral particles as compared to the resting $CD4^+$ T cells [297].

Transmission heterogeneities were also reported in the interaction between bacteriophages and plasmids from the bacterial cells that they infect. Bacteriophages were discovered in 1915 [303] and were first shown to lyse bacteria in 1917 [304], but their ability to transfer bacterial traits was first described only in 1952 [305]. In bacteriophage biology, significant attention has focused on the ability of bacteriophages to transduce bacterial genes, but their contribution to transformation events has received relatively less attention [306, 307]. A study that examined the ability of a library of lytic bacteriophages infecting an *Escherichia coli* strain to promote the transformation of various plasmids identified a subset of natural lytic phages that released larger amounts of intact plasmid DNA upon lysis than most other lytic phages. Although the majority of phage lysates yielded <100 transformants per 10^8 cells lysed, two isolates were ~50-fold more efficient in promoting plasmid transformation and were called "super-spreaders" at the cellular level [306].

5 Mechanisms of Super-Spreading and Super-Shedding

Studies of various outbreaks provided insights into some of the mechanisms that underlie transmission heterogeneities. Overall, super-spreading and super-shedding are shaped by factors related to the pathogen, the host, and the environment. Some of these factors may belong to more than one of these categories (Table 1).

In 1960, it was reported that certain asymptomatic infants that are nasally colonized with *Staphylococcus aureus* and develop an infection with a respiratory virus, such as an adenovirus or an echovirus, may be surrounded by "clouds" of bacteria and, as a result, shed large numbers of the bacteria into the environment and become contagious. These infants were shown to cause explosive *S. aureus* outbreaks in nurseries, and the term "cloud baby" was coined to refer to them [308]. A similar phenomenon was later described in adults, and the contagious individuals were referred to as "cloud adults." Adult volunteers who were nasal carriers of *S. aureus*, after being experimentally infected with a rhinovirus, dispersed twofold more bacteria into the air, and peak dispersal increased up to 34-fold [309]. In another study, an adult experimentally infected with a rhinovirus dispersed 40-fold more *S. aureus* that the person was colonized with than in the absence of the rhinovirus infection [310]. It was speculated that the upper respiratory infection could have led to swelling of the nasal turbinates, creating aerosols due to the high-speed turbulent airflow [310]. Several upper respiratory viral infections were reported to facilitate the transmission of microbes that colonize the nose, including *Neisseria meningitidis*, *S. pneumoniae*, *S. pyogenes*, and *H. influenzae* [88, 311].

A study of mice coinfected with the gastrointestinal helminth *Heligmosomoides polygyrus* and a self-bioluminescent strain of the respiratory bacterial pathogen *Bordetella bronchiseptica* provided insights into the mechanisms of super-shedding. The coinfected mice shed significantly more helminth eggs for a longer time and had a higher lung bacterial load than animals infected with one of the pathogens alone [312]. One potential mechanism to explain this is the antagonism between the immune response to *Bordetella bronchiseptica*, which is primarily Th1-mediated [313], and the response to *Heligmosomoides polygyrus*, which primarily depends on the Th2-mediated signaling pathways [314, 315], which might impair the clearance of the two pathogens [312].

Table 1 Super-spreading and super-shedding events are shaped by factors related to the pathogen, the host, and the environment

Factors that shape transmission heterogeneities	Categories	Examples
Host	Behavioral	21 of 28 (75%) attendees to a funeral who developed Ebola virus disease reported touching the body of the deceased
	Physiological	American robins, even though they represent on average only ∼ 3.7% of the avian abundance, account for > 43% of the feedings by mosquitoes that transmit the West Nile virus
	Immunological	People with depressed immunity may have increased SARS-CoV loads, shed the virus for a longer time, and become super-spreaders for SARS
	Stress	Transportation stress increases the fecal shedding of *E. coli* by calves
	Large number of social contacts	Infected individuals who had several contacts with others during the pre-symptomatic phase of the illness sometimes contributed to the spread of SARS-CoV or SARS-CoV-2 to several other individuals
	High shedding	Some cattle shed many more *E. coli* than most other cattle on the same farm
Pathogen	Co-infection	Individuals nasally colonized with *Staphylococcus aureus* became super-spreaders for the bacteria when they were also infected with a respiratory virus
	Genetics	Certain *Mycobacterium tuberculosis* genotypes are more virulent than others
	Pathogen load	A high viral load is one of the several factors that may be responsible for some of the super-spreading events during the COVID-19 pandemic
Environment	Building airflow dynamics	During the 2002–2003 SARS outbreak from the Amoy Gardens residential complex, airflow dynamics contributed to the spread of the virus among many apartments
	Unrecognized disease	People with infectious diseases that are unrecognized, are recognized late, have unusual presentations, or are misdiagnosed, have contributed to super-spreading events during the SARS epidemic and the COVID-19 pandemic
	Crowding	A distance of ≤ 1 m between hospital beds during the 2002–2003 SARS outbreak was associated with super-spreading events

(Prepared with data from [20, 36, 38, 61, 84, 90, 93, 104, 144, 166, 180, 308, 322, 334–342)

The contribution of co-infection to super-shedding emerges from several human studies. For example, HIV-1 RNA concentration in the seminal fluid was eightfold higher in HIV-1-positive men with urethritis than in men without urethritis [316], and cervicovaginal HIV-1 shedding was higher in women with *Neisseria gonorrhoeae* and *Chlamydia trachomatis* infections [317]. In another study, cattle coinfected with *Fasciola hepatica*, known as the common liver fluke, had an increased shedding of *E. coli* O157 [318]. *F. hepatica* was shown to alter host immunity to other pathogens [319]. For example, in mice coinfected with *F. hepatica* and the respiratory pathogen *Bordetella pertussis*, the *B. pertussis*-specific protective Th1 response was markedly suppressed, and clearance of the bacteria from the lungs was delayed [320].

The contribution of immune suppression to super-spreading also emerged from studies of the SARS outbreak in Amoy Gardens. The index patient, who stayed with his brother on two occasions, on March 14 and March 19, 2003 [84, 96], had renal failure and was receiving hemodialysis [84], known to depress immunity [321–323]. He had diarrhea and used the bathroom in his brother's apartment [80, 84]. Patients with SARS were shown to shed the virus in their stool [324–326]. The spread of the virus in the apartment complex was amplified by the water drainage system [327], ultimately 329 people were infected, and 42 died [80, 92, 324, 328].

The contribution of immune suppression to super-spreading emerged from another study, which showed that mice super-shedding *Salmonella typhimurium* had increased inflammation at gastrointestinal and systemic sites, and a neutrophil-dependent blunting of the IL-2 mediated Th1 immune response in secondary lymphoid organs [329]. The link between the inability to develop an effective immune response and the persistent *S. typhi* carrier state was reported in a study that interrogated the host response to typhoid fever during acute disease, convalescence, and recovery among naturally infected individuals from the Mekong Delta region in Vietnam [330]. Ninemonths after treatment, 25% of the participants, even though they had recovered clinically, exhibited in several cell types a distinct gene expression signature that was closer to newly admitted patients than to most convalescent patients. The dysregulation of several genes with potential involvement in the immune response indicated the possibility of long-term immune response perturbations after infection [330].

The SARS epidemic [75, 84, 331] and the COVID-19 pandemic [61] revealed that some of the factors potentially associated with super-spreading include high viral loads [332], shedding the virus for longer times [333], amplification of the virus through environmental control systems, such as drainage system and aerosol flows [334], and the use of a nebulized bronchodilator [326].

6 Conclusion

While historically it was assumed that the dynamics of infectious diseases in populations is a homogeneous process, in more recent years, it has become increasingly apparent that transmission heterogeneities are a defining feature of many outbreaks. Heterogeneities in transmission shape the interaction of human, animal, and cellular populations with various vectors, bacteria, parasites, and viruses; they occur at the spatial and temporal levels; and they are influenced by factors related to the host, the pathogen, and the environment. Many infectious diseases studied to date, including influenza, Ebola viral disease, measles, malaria, SARS, MERS, and COVID-19, exhibit these heterogeneities, and they represent a critical facet of outbreak dynamics. Extreme cases of transmission heterogeneities are called "super-spreading events," a term that refers to instances when an infected host creates many more secondary contacts than most other hosts in the same population. Super-spreading events have far-reaching medical and public health implications. Therefore, incorporating them into epidemiologic analyses is critical for visualizing the dynamics of outbreaks and for implementing more effective disease control measures. A major challenge is that super-spreading events are usually only identified retrospectively. Characterizing super-spreading events from previous outbreaks, building a framework to understand the multiple factors that shape these heterogeneities, and interrogating their mechanistic bases, are emerging as critical facets of epidemic and pandemic preparedness.

Core messages

- Temporal and spatial heterogeneities in transmission were documented in many infectious disease outbreaks.
- Extreme cases of transmission heterogeneities are called super-spreading events and are caused by super-spreader and/or super-shedder hosts.
- Super-spreading was described at the level of individuals, groups, and species.
- Factors related to the host, the pathogen, and the environment contribute to super-spreading.

A challenge during infectious disease outbreaks is that super-spreading events are usually identified only retrospectively.

Acknowledgements   I thank Timothy J. Cardozo, M.D., Ph.D., NYU Langone Health, Department of Biochemistry and Molecular Pharmacology, New York City, NY, USA, and Oana Ometa, Ph.D., Babes-Bolyai University, Faculty of Political, Administrative and Communication Sciences, Cluj-Napoca, Romania, for valuable comments and conversations during the preparation of this manuscript.

References

1. Fauci AS, Touchette NA, Folkers GK (2005) Emerging infectious diseases: a 10-year perspective from the National Institute of Allergy and Infectious Diseases. Emerg Infect Dis 11(4):519–525. https://doi.org/10.3201/eid1104.041167
2. Cunha BA (2004) Historical aspects of infectious diseases, part I. Infect Dis Clin North Am 18 (1):XI–V. https://doi.org/10.1016/S0891-5520(03)00098-9
3. Huremović D (2019) Brief history of pandemics (pandemics throughout history). In: Psychiatry of pandemics, pp 7–35. https://doi.org/10.1007/978-3-030-15346-5_2
4. Balloux F, van Dorp L (2017) Q&A: what are pathogens, and what have they done to and for us? BMC Biol 15(1):91–91. https://doi.org/10.1186/s12915-017-0433-z
5. Mukherjee S (2017) Emerging infectious diseases: epidemiological perspective. Indian J Dermatol 62(5):459–467. https://doi.org/10.4103/ijd.IJD_379_17
6. van Doorn HR (2014) Emerging infectious diseases. Medicine (Abingdon) 42(1):60–63. https://doi.org/10.1016/j.mpmed.2013.10.014
7. Khabbaz R, Bell BP, Schuchat A, Ostroff SM, Moseley R, Levitt A et al (2015) Emerging and reemerging infectious disease threats. Mandell, Douglas, Bennett's Principles Pract Infect Dis 158–177:e156. https://doi.org/10.1016/B978-1-4557-4801-3.00014-X
8. Tatem AJ, Rogers DJ, Hay SI (2006) Global transport networks and infectious disease spread. Adv Parasitol 62:293–343. https://doi.org/10.1016/S0065-308X(05)62009-X
9. Trauer JM, Dodd PJ, Gomes MGM, Gomez GB, Houben RMGJ, McBryde ES et al (2019) The Importance of heterogeneity to the epidemiology of tuberculosis. Clin Infect Dis (an official publication of the Infectious Diseases Society of America) 69(1):159–166. https://doi.org/10.1093/cid/ciy938
10. Kong L, Wang J, Han W, Cao Z (2016) Modeling heterogeneity in direct infectious disease transmission in a compartmental model. Int J Environ Res Public Health 13(3). https://doi.org/10.3390/ijerph13030253
11. Rodriguez DJ, Torres-Sorando L (2001) Models of infectious diseases in spatially heterogeneous environments. Bull Math Biol 63(3):547–571. https://doi.org/10.1006/bulm.2001.0231
12. Bansal S, Grenfell BT, Meyers LA (2007) When individual behaviour matters: homogeneous and network models in epidemiology. J R Soc Interface 4(16):879–891. https://doi.org/10.1098/rsif.2007.1100
13. Del Valle SY, Hyman JM, Chitnis N (2013) Mathematical models of contact patterns between age groups for predicting the spread of infectious diseases. Math Biosci Eng 10(5–6):1475–1497. https://doi.org/10.3934/mbe.2013.10.1475
14. Bolzoni L, Real L, De Leo G (2007) Transmission heterogeneity and control strategies for infectious disease emergence. PLoS ONE 2(8):e747. https://doi.org/10.1371/journal.pone.0000747
15. Bolker B, Grenfell B (1995) Space, persistence and dynamics of measles epidemics. Philos Trans R Soc Lond B Biol Sci 348(1325):309–320. https://doi.org/10.1098/rstb.1995.0070
16. Yates A, Antia R, Regoes RR (2006) How do pathogen evolution and host heterogeneity interact in disease emergence? Proc Biol Sci 273(1605):3075–3083. https://doi.org/10.1098/rspb.2006.3681

17. Woolhouse ME, Dye C, Etard JF, Smith T, Charlwood JD, Garnett GP et al (1997) Heterogeneities in the transmission of infectious agents: implications for the design of control programs. Proc Natl Acad Sci U S A 94(1):338–342. https://doi.org/10.1073/pnas. 94.1.338

18. Bansal S, Read J, Pourbohloul B, Meyers LA (2010) The dynamic nature of contact networks in infectious disease epidemiology. J Biol Dyn 4(5):478–489. https://doi.org/10. 1080/17513758.2010.503376

19. Woolhouse ME, Shaw DJ, Matthews L, Liu WC, Mellor DJ, Thomas MR (2005) Epidemiological implications of the contact network structure for cattle farms and the 20–80 rule. Biol Lett 1(3):350–352. https://doi.org/10.1098/rsbl.2005.0331

20. Stein RA (2011) Super-spreaders in infectious diseases. Int J Infect Dis 15(8):e510-513. https://doi.org/10.1016/j.ijid.2010.06.020

21. Stein RA, Katz DE (2017) Escherichia coli, cattle and the propagation of disease. FEMS Microbiol Lett 364(6). https://doi.org/10.1093/femsle/fnx050

22. López L, Burguerner G, Giovanini L (2014) Addressing population heterogeneity and distribution in epidemics models using a cellular automata approach. BMC Res Notes 7:234–234. https://doi.org/10.1186/1756-0500-7-234

23. Buonomo B, Chitnis N, d'Onofrio A (2018) Preface to the special issue on "Demographic and temporal heterogeneities in infectious disease epidemiology." Ricerche mat 67(1):3–6. https://doi.org/10.1007/s11587-018-0369-9

24. Anderson RM, May RM (1985) Age-related changes in the rate of disease transmission: implications for the design of vaccination programmes. J Hyg (Lond) 94(3):365–436. https:// doi.org/10.1017/s002217240006160x

25. Gambhir M, Swerdlow DL, Finelli L, Van Kerkhove MD, Biggerstaff M, Cauchemez S et al (2013) Multiple contributory factors to the age distribution of disease cases: a modeling study in the context of influenza A(H3N2v). Clin Infect Dis 57(Suppl 1):S23-27. https://doi. org/10.1093/cid/cit298

26. Ubeda F, Jansen VA (2016) The evolution of sex-specific virulence in infectious diseases. Nat Commun 7:13849. https://doi.org/10.1038/ncomms13849

27. Doeschl-Wilson AB, Davidson R, Conington J, Roughsedge T, Hutchings MR, Villanueva B (2011) Implications of host genetic variation on the risk and prevalence of infectious diseases transmitted through the environment. Genetics 188(3):683–693. https://doi.org/10. 1534/genetics.110.125625

28. Manning SD, Woolhouse ME, Ndamba J (1995) Geographic compatibility of the freshwater snail Bulinus globosus and schistosomes from the Zimbabwe highveld. Int J Parasitol 25 (1):37–42. https://doi.org/10.1016/0020-7519(94)00097-8

29. Kao RR (2006) Evolution of pathogens towards low R0 in heterogeneous populations. J Theor Biol 242(3):634–642. https://doi.org/10.1016/j.jtbi.2006.04.003

30. Rodrigues P, Margheri A, Rebelo C, Gomes MG (2009) Heterogeneity in susceptibility to infection can explain high reinfection rates. J Theor Biol 259(2):280–290. https://doi.org/10. 1016/j.jtbi.2009.03.013

31. Dwyer G, Elkinton JS, Buonaccorsi JP (1997) Host heterogeneity in susceptibility and disease dynamics: tests of a mathematical model. Am Nat 150(6):685–707. https://doi.org/ 10.1086/286089

32. Kong L, Wang J, Han W, Cao Z (2016) Modeling heterogeneity in direct infectious disease transmission in a compartmental model. Int J Environ Res Public Health 13(3):253. https:// doi.org/10.3390/ijerph13030253

33. Mari L, Ciddio M, Casagrandi R, Perez-Saez J, Bertuzzo E, Rinaldo A et al (2017) Heterogeneity in schistosomiasis transmission dynamics. J Theor Biol 432:87–99. https:// doi.org/10.1016/j.jtbi.2017.08.015

34. Alemu K, Worku A, Berhane Y (2013) Malaria infection has spatial, temporal, and spatiotemporal heterogeneity in unstable malaria transmission areas in northwest Ethiopia. PLoS ONE 8(11):e79966. https://doi.org/10.1371/journal.pone.0079966

35. Volkova VV, Howey R, Savill NJ, Woolhouse ME (2010) Sheep movement networks and the transmission of infectious diseases. PLoS ONE 5(6):e11185. https://doi.org/10.1371/journal.pone.0011185
36. Kilpatrick AM, Daszak P, Jones MJ, Marra PP, Kramer LD (2006) Host heterogeneity dominates West Nile virus transmission. Proc Biol Sci 273(1599):2327–2333. https://doi.org/10.1098/rspb.2006.3575
37. Miller E, Warburg A, Novikov I, Hailu A, Volf P, Seblova V et al (2014) Quantifying the contribution of hosts with different parasite concentrations to the transmission of visceral leishmaniasis in Ethiopia. PLoS Negl Trop Dis 8(10):e3288. https://doi.org/10.1371/journal.pntd.0003288
38. Lloyd-Smith JO, Schreiber SJ, Kopp PE, Getz WM (2005) Superspreading and the effect of individual variation on disease emergence. Nature 438(7066):355–359. https://doi.org/10.1038/nature04153
39. Paull SH, Song S, McClure KM, Sackett LC, Kilpatrick AM, Johnson PT (2012) From superspreaders to disease hotspots: linking transmission across hosts and space. Front Ecol Environ 10(2):75–82. https://doi.org/10.1890/110111
40. Kumar S, Jha S, Rai SK (2020) Significance of super spreader events in COVID-19. Indian J Public Health 64(Supplement):S139-s141. https://doi.org/10.4103/ijph.IJPH_495_20
41. Cave E (2020) COVID-19 Super-spreaders: definitional quandaries and implications. Asian Bioeth Rev 1–8. https://doi.org/10.1007/s41649-020-00118-2
42. Carinci F (2020) Covid-19: preparedness, decentralisation, and the hunt for patient zero. BMJ 368:bmj.m799. https://doi.org/10.1136/bmj.m799
43. MacKay R (2020) Patient zero: why it's such a toxic term. Available at: https://theconversation.com/patient-zero-why-its-such-a-toxic-term-134721. Last Accessed 4 July 2020
44. Vaaben L (2020) Patient Zero er historien om den onde, dumme smittespreder Available at: https://www.information.dk/moti/2020/03/patient-zero-historien-onde-dumme-smittespreder. Last Accessed 4 July 2020
45. McKay RA (2014) "Patient Zero": the absence of a patient's view of the early North American AIDS epidemic. Bull Hist Med 88(1):161–194. https://doi.org/10.1353/bhm.2014.0005
46. Auerbach DM, Darrow WW, Jaffe HW, Curran JW (1984) Cluster of cases of the acquired immune deficiency syndrome. Patients linked by sexual contact. American J Med 76 (3):487–492
47. Worobey M, Watts TD, McKay RA, Suchard MA, Granade T, Teuwen DE et al (2016) 1970s and "Patient 0" HIV-1 genomes illuminate early HIV/AIDS history in North America. Nature 539(7627):98–101. https://doi.org/10.1038/nature19827
48. Cohen J (2016) Infectious disease. 'Patient Zero' no more. Science (New York, NY) 351 (6277):1013. https://doi.org/10.1126/science.351.6277.1013
49. Darrow WW (2017) And the band played on: before and after. AIDS Behav 21(10):2799–2806. https://doi.org/10.1007/s10461-017-1798-2
50. Shilts R (1987) And the band played on: politics, people, and the AIDS epidemic (hereafter Band New York: St. Martin's, p 412
51. Crimp D (1987) How to have promiscuity in an epidemic. 43:237–271. https://doi.org/10.2307/3397576
52. Anonymous (2016) How researchers cleared the name of HIV Patient Zero. Nature 538 (7626):428. https://doi.org/10.1038/538428a
53. Highberg NR (2004) "Because we were just too scared": rhetorical constructions of patient zero. Med Humanit Rev 18(1–2):9–26
54. Yu IT, Xie ZH, Tsoi KK, Chiu YL, Lok SW, Tang XP et al (2007) Why did outbreaks of severe acute respiratory syndrome occur in some hospital wards but not in others? Clin Infect Dis 44(8):1017–1025. https://doi.org/10.1086/512819

55. Shen Z, Ning F, Zhou W, He X, Lin C, Chin DP et al (2003) (2004) Superspreading SARS events, Beijing. Emerg Infect Dis 10(2):256–260. https://doi.org/10.3201/eid1002.030732

56. Al-Tawfiq JA, Rodriguez-Morales AJ (2020) Super-spreading events and contribution to transmission of MERS, SARS, and COVID-19. J Hosp Infect. https://doi.org/10.1016/j.jhin.2020.04.002

57. Gopalakrishna G, Choo P, Leo YS, Tay BK, Lim YT, Khan AS et al (2004) SARS transmission and hospital containment. Emerg Infect Dis 10(3):395–400. https://doi.org/10.3201/eid1003.030650

58. (CDC) CfDCaP (2003) Severe acute respiratory syndrome—Singapore, 2003. MMWR Morb Mortal Wkly Rep 52(18):405–411

59. Wallinga J, Teunis P (2004) Different epidemic curves for severe acute respiratory syndrome reveal similar impacts of control measures. Am J Epidemiol 160(6):509–516. https://doi.org/10.1093/aje/kwh255

60. Zhang Y, Li Y, Wang L, Li M, Zhou X (2020) Evaluating transmission heterogeneity and super-spreading event of COVID-19 in a Metropolis of China. Int J Environ Res Public Health 17(10). https://doi.org/10.3390/ijerph17103705

61. Lin J, Yan K, Zhang J, Cai T, Zheng J (2020) A super-spreader of COVID-19 in Ningbo city in China. J Infect Public Health 13(7):935–937. https://doi.org/10.1016/j.jiph.2020.05.023

62. Chase-Topping M, Gally D, Low C, Matthews L, Woolhouse M (2008) Super-shedding and the link between human infection and livestock carriage of Escherichia coli O157. Nat Rev Microbiol 6(12):904–912. https://doi.org/10.1038/nrmicro2029

63. Woolhouse M (2017) Quantifying Transmission. Microbiol Spectr 5(4). https://doi.org/10.1128/microbiolspec.MTBP-0005-2016

64. Matthews L, Low JC, Gally DL, Pearce MC, Mellor DJ, Heesterbeek JA et al (2006) Heterogeneous shedding of Escherichia coli O157 in cattle and its implications for control. Proc Natl Acad Sci U S A 103(3):547–552. https://doi.org/10.1073/pnas.0503776103

65. Adriani KS (2019) [The uncomfortable history of Mary Mallon and typhoid fever]. Ned Tijdschr Geneeskd 163

66. Marineli F, Tsoucalas G, Karamanou M, Androutsos G (2013) Mary Mallon (1869–1938) and the history of typhoid fever. Ann Gastroenterol 26(2):132–134

67. Brooks J (1996) The sad and tragic life of Typhoid Mary. CMAJ 154(6):915–916

68. Soper GA (1939) The curious career of Typhoid Mary. Bull N Y Acad Med 15(10):698–712

69. Mary Mallon (Typhoid Mary) (1939). Am J Public Health Nations Health 29 (1):66–68. https://doi.org/10.2105/ajph.29.1.66

70. Cherry JD (2004) The chronology of the 2002–2003 SARS mini pandemic. Paediatr Respir Rev 5(4):262–269. https://doi.org/10.1016/j.prrv.2004.07.009

71. Luk HKH, Li X, Fung J, Lau SKP, Woo PCY (2019) Molecular epidemiology, evolution and phylogeny of SARS coronavirus. Infect Genet Evol 71:21–30. https://doi.org/10.1016/j.meegid.2019.03.001

72. Cleri DJ, Ricketti AJ, Vernaleo JR (2010) Severe acute respiratory syndrome (SARS). Infect Dis Clin North Am 24(1):175–202. https://doi.org/10.1016/j.idc.2009.10.005

73. Xu R-H, He J-F, Evans MR, Peng G-W, Field HE, Yu D-W et al (2004) Epidemiologic clues to SARS origin in China. Emerg Infect Dis 10(6):1030–1037. https://doi.org/10.3201/eid1006.030852

74. Rosling L, Rosling M (2003) Pneumonia causes panic in Guangdong province. BMJ 326 (7386):416. https://doi.org/10.1136/bmj.326.7386.416

75. Li Y, Yu IT, Xu P, Lee JH, Wong TW, Ooi PL et al (2004) Predicting super spreading events during the 2003 severe acute respiratory syndrome epidemics in Hong Kong and Singapore. Am J Epidemiol 160(8):719–728. https://doi.org/10.1093/aje/kwh273

76. Yan W (2015) Going batty: Studying natural reservoirs to inform drug development. Nat Med 21(8):831–833. https://doi.org/10.1038/nm0815-831

77. Sampathkumar P, Temesgen Z, Smith TF, Thompson RL (2003) SARS: epidemiology, clinical presentation, management, and infection control measures. Mayo Clin Proc 78(7): 882–890. https://doi.org/10.4065/78.7.882
78. Cherry JD, Krogstad P (2004) SARS: the first pandemic of the 21st century. Pediatr Res 56(1):1–5. https://doi.org/10.1203/01.Pdr.0000129184.87042.Fc
79. Fleck F (2003) WHO says SARS outbreak is over, but fight should go on. BMJ 327 (7406):70. https://doi.org/10.1136/bmj.327.7406.70-c
80. Hung LS (2003) The SARS epidemic in Hong Kong: what lessons have we learned? J R Soc Med 96(8):374–378. https://doi.org/10.1258/jrsm.96.8.374
81. Update: Outbreak of severe acute respiratory syndrome–worldwide, 2003 (2003). MMWR Morbidity and mortality weekly report 52(12):241–246, 248
82. Leung GM, Hedley AJ, Ho LM, Chau P, Wong IO, Thach TQ et al (2004) The epidemiology of severe acute respiratory syndrome in the 2003 Hong Kong epidemic: an analysis of all 1755 patients. Ann Intern Med 141(9):662–673
83. Tam DK, Lee S, Lee SS (2007) Impact of SARS on avian influenza preparedness in healthcare workers. Infection 35(5):320–325. https://doi.org/10.1007/s15010-007-6353-z
84. Tomlinson B, Cockram C (2003) SARS: experience at Prince of Wales hospital, Hong Kong. Lancet 361(9368):1486–1487. https://doi.org/10.1016/s0140-6736(03)13218-7
85. Hui DSC, Zumla A (2019) Severe acute respiratory syndrome: historical, epidemiologic, and clinical features. Infect Dis Clin North Am 33(4):869–889. https://doi.org/10.1016/j.idc.2019.07.001
86. Wong RS, Hui DS (2004) Index patient and SARS outbreak in Hong Kong. Emerg Infect Dis 10(2):339–341. https://doi.org/10.3201/eid1002.030645
87. Lee N, Hui D, Wu A, Chan P, Cameron P, Joynt GM et al (2003) A major outbreak of severe acute respiratory syndrome in Hong Kong. N Engl J Med 348(20):1986–1994. https://doi.org/10.1056/NEJMoa85
88. Harrison LH, Armstrong CW, Jenkins SR, Harmon MW, Ajello GW, Miller GB Jr et al (1991) A cluster of meningococcal disease on a school bus following epidemic influenza. Arch Intern Med 151(5):1005–1009
89. Yu IT, Qiu H, Tse LA, Wong TW (2014) Severe acute respiratory syndrome beyond Amoy Gardens: completing the incomplete legacy. Clin Infect Dis 58(5):683–686. https://doi.org/10.1093/cid/cit797
90. Li Y, Duan S, Yu IT, Wong TW (2005) Multi-zone modeling of probable SARS virus transmission by airflow between flats in Block E Amoy Gardens. Indoor Air 15(2):96–111. https://doi.org/10.1111/j.1600-0668.2004.00318.x
91. Yip C, Chang WL, Yeung KH, Yu IT (2007) Possible meteorological influence on the severe acute respiratory syndrome (SARS) community outbreak at Amoy Gardens Hong Kong. J Environ Health 70(3):39–46
92. Meng X, Huang X, Zhou P, Li C, Wu A (2020) Alert for SARS-CoV-2 infection caused by fecal aerosols in rural areas in China. Infect Control Hospital Epidemiol 1–1. https://doi.org/10.1017/ice.2020.114
93. McKinney KR, Gong YY, Lewis TG (2006) Environmental transmission of SARS at Amoy Gardens. J Environ Health 68(9):26–30; quiz 51–22
94. Yu IT, Li Y, Wong TW, Tam W, Chan AT, Lee JH et al (2004) Evidence of airborne transmission of the severe acute respiratory syndrome virus. N Engl J Med 350(17):1731–1739. https://doi.org/10.1056/NEJMoa032867
95. Lau JT, Lau M, Kim JH, Tsui HY, Tsang T, Wong TW (2004) Probable secondary infections in households of SARS patients in Hong Kong. Emerg Infect Dis 10(2):235–243. https://doi.org/10.3201/eid1002.030626
96. Chu CM, Cheng VC, Hung IF, Chan KS, Tang BS, Tsang TH et al (2005) Viral load distribution in SARS outbreak. Emerg Infect Dis 11(12):1882–1886. https://doi.org/10.3201/eid1112.040949

97. Bowen JT, Laroe C (2006) Airline networks and the international diffusion of severe acute respiratory syndrome (SARS). Geogr J 172(2):130–144. https://doi.org/10.1111/j.1475-4959.2006.00196.x

98. Olsen SJ, Chang HL, Cheung TY, Tang AF, Fisk TL, Ooi SP et al (2003) Transmission of the severe acute respiratory syndrome on aircraft. N Engl J Med 349(25):2416–2422. https://doi.org/10.1056/NEJMoa031349

99. Poutanen SM, Low DE, Henry B, Finkelstein S, Rose D, Green K et al (2003) Identification of severe acute respiratory syndrome in Canada. N Engl J Med 348(20):1995–2005. https://doi.org/10.1056/NEJMoă34

100. Varia M, Wilson S, Sarwal S, McGeer A, Gournis E, Galanis E et al (2003) Investigation of a nosocomial outbreak of severe acute respiratory syndrome (SARS) in Toronto Canada. Cmaj 169(4):285–292

101. Maskalyk J, Hoey J (2003) SARS update. CMAJ: Canadian Medical Association journal = journal de l'Association medicale canadienne 168(10):1294–1295

102. MacDougall H (2007) Toronto's Health Department in action: influenza in 1918 and SARS in 2003. J Hist Med Allied Sci 62(1):56–89. https://doi.org/10.1093/jhmas/jrl042

103. Braden CR, Dowell SF, Jernigan DB, Hughes JM (2013) Progress in global surveillance and response capacity 10 years after severe acute respiratory syndrome. Emerg Infect Dis 19 (6):864–869. https://doi.org/10.3201/eid1906.130192

104. Svoboda T, Henry B, Shulman L, Kennedy E, Rea E, Ng W et al (2004) Public health measures to control the spread of the severe acute respiratory syndrome during the outbreak in Toronto. N Engl J Med 350(23):2352–2361. https://doi.org/10.1056/NEJMoa032111

105. Skowronski DM, Petric M, Daly P, Parker RA, Bryce E, Doyle PW et al (2006) Coordinated response to SARS, Vancouver. Canada. Emerg Infect Dis 12(1):155–158. https://doi.org/10.3201/eid1201.050327

106. Wong O (2004) Severe acute respiratory syndrome (SARS). Occup Environ Med 61(1): e1–e1

107. Raboud J, Shigayeva A, McGeer A, Bontovics E, Chapman M, Gravel D et al (2010) Risk factors for SARS transmission from patients requiring intubation: a multicentre investigation in Toronto. Canada. PloS one 5(5):e10717–e10717. https://doi.org/10.1371/journal.pone.0010717

108. McDonald LC, Simor AE, Su IJ, Maloney S, Ofner M, Chen KT et al (2004) SARS in healthcare facilities, Toronto and Taiwan. Emerg Infect Dis 10(5):777–781. https://doi.org/10.3201/eid1005.030791

109. Ofner-Agostini M, Wallington T, Henry B, Low D, McDonald LC, Berger L et al. (2008) Investigation of the second wave (phase 2) of severe acute respiratory syndrome (SARS) in Toronto, Canada. What happened? Canada communicable disease report = Releve des maladies transmissibles au Canada 34(2):1–11

110. Dwosh HA, Hong HH, Austgarden D, Herman S, Schabas R (2003) Identification and containment of an outbreak of SARS in a community hospital. CMAJ 168(11):1415–1420

111. Loeb M, McGeer A, Henry B, Ofner M, Rose D, Hlywka T et al (2004) SARS among critical care nurses, Toronto. Emerg Infect Dis 10(2):251–255. https://doi.org/10.3201/eid1002.030838

112. Hui DS (2016) Super-spreading events of MERS-CoV infection. Lancet 388(10048):942–943. https://doi.org/10.1016/s0140-6736(16)30828-5

113. Zaki AM, van Boheemen S, Bestebroer TM, Osterhaus AD, Fouchier RA (2012) Isolation of a novel coronavirus from a man with pneumonia in Saudi Arabia. N Engl J Med 367 (19):1814–1820. https://doi.org/10.1056/NEJMoa1211721

114. Ramadan N, Shaib H (2019) Middle East respiratory syndrome coronavirus (MERS-CoV): a review. Germs 9(1):35–42. https://doi.org/10.18683/germs.2019.1155

115. de Groot RJ, Baker SC, Baric RS, Brown CS, Drosten C, Enjuanes L et al (2013) Middle East respiratory syndrome coronavirus (MERS-CoV): announcement of the coronavirus study group. J Virol 87(14):7790–7792. https://doi.org/10.1128/JVI.01244-13

116. Killerby ME, Biggs HM, Midgley CM, Gerber SI, Watson JT (2020) Middle East respiratory syndrome coronavirus transmission. Emerg Infect Dis 26(2):191–198. https://doi.org/10.3201/eid2602.190697

117. Park M, Thwaites RS, Openshaw PJM (2020) COVID-19: Lessons from SARS and MERS. Eur J Immunol 50(3):308–311. https://doi.org/10.1002/eji.202070035

118. Alsolamy S, Arabi YM (2015) Infection with Middle East respiratory syndrome coronavirus. Can J Respir Ther 51(4):102

119. Majumder MS, Rivers C, Lofgren E, Fisman D (2014) Estimation of MERS-coronavirus reproductive number and case fatality rate for the spring 2014 Saudi Arabia outbreak: insights from publicly available data. PLoS Curr 6. https://doi.org/10.1371/currents.outbreaks.98d2f8f3382d84f390736cd5f5fe133c

120. Hajjar SA, Memish ZA, McIntosh K (2013) Middle East respiratory syndrome coronavirus (MERS-CoV): a perpetual challenge. Ann Saudi Med 33(5):427–436. https://doi.org/10.5144/0256-4947.2013.427

121. Ki M (2015) 2015 MERS outbreak in Korea: hospital-to-hospital transmission. Epidemiol Health 37:e2015033. https://doi.org/10.4178/epih/e2015033

122. Kim KM, Ki M, Cho SI, Sung M, Hong JK, Cheong HK et al (2015) Epidemiologic features of the first MERS outbreak in Korea: focus on Pyeongtaek St Mary's Hospital. Epidemiol Health 37:e2015041. https://doi.org/10.4178/epih/e2015041

123. Choi JY (2015) An outbreak of Middle East respiratory syndrome coronavirus infection in South Korea, 2015. Yonsei Med J 56(5):1174–1176. https://doi.org/10.3349/ymj.2015.56.5.1174

124. Oh MD, Park WB, Park SW, Choe PG, Bang JH, Song KH et al (2018) Middle East respiratory syndrome: what we learned from the 2015 outbreak in the Republic of Korea. Korean J Intern Med 33(2):233–246. https://doi.org/10.3904/kjim.2018.031

125. Jo S, Hong J, Lee SE, Ki M, Choi BY, Sung M (2019) Airflow analysis of Pyeongtaek St Mary's Hospital during hospitalization of the first Middle East respiratory syndrome patient in Korea. R Soc Open Sci 6(3):181164. https://doi.org/10.1098/rsos.181164

126. Lee JY, Kim YJ, Chung EH, Kim DW, Jeong I, Kim Y et al (2017) The clinical and virological features of the first imported case causing MERS-CoV outbreak in South Korea, 2015. BMC Infect Dis 17(1):498. https://doi.org/10.1186/s12879-017-2576-5

127. Middle East Respiratory Syndrome Coronavirus Outbreak in the Republic of Korea, 2015 (2015). Osong Public Health Res Perspect 6(4):269–278. https://doi.org/10.1016/j.phrp.2015.08.006

128. Cho SY, Kang JM, Ha YE, Park GE, Lee JY, Ko JH et al (2016) MERS-CoV outbreak following a single patient exposure in an emergency room in South Korea: an epidemiological outbreak study. Lancet 388(10048):994–1001. https://doi.org/10.1016/s0140-6736(16)30623-7

129. Park GE, Ko JH, Peck KR, Lee JY, Lee JY, Cho SY et al (2016) Control of an outbreak of Middle East respiratory syndrome in a Tertiary hospital in Korea. Ann Intern Med 165(2):87–93. https://doi.org/10.7326/m15-2495

130. Kim KH, Tandi TE, Choi JW, Moon JM, Kim MS (2017) Middle East respiratory syndrome coronavirus (MERS-CoV) outbreak in South Korea, 2015: epidemiology, characteristics and public health implications. J Hosp Infect 95(2):207–213. https://doi.org/10.1016/j.jhin.2016.10.008

131. Kang CK, Song KH, Choe PG, Park WB, Bang JH, Kim ES et al (2017) Clinical and epidemiologic characteristics of spreaders of Middle East respiratory syndrome coronavirus during the 2015 outbreak in Korea. J Korean Med Sci 32(5):744–749. https://doi.org/10.3346/jkms.2017.32.5.744

132. Chun BC (2016) Understanding and modeling the super-spreading events of the Middle East respiratory syndrome outbreak in Korea. Infect Chemother 48(2):147–149. https://doi.org/10.3947/ic.2016.48.2.147

133. Kim Y, Ryu H, Lee S (2018) Agent-based modeling for super-spreading events: a case study of MERS-CoV transmission dynamics in the Republic of Korea. Int J Environ Res Public Health 15(11):2369. https://doi.org/10.3390/ijerph15112369

134. Frieden TR, Lee CT (2020) Identifying and interrupting superspreading events-implications for control of severe acute respiratory syndrome coronavirus 2. Emerg Infect Dis 26(6): 1059–1066. https://doi.org/10.3201/eid2606.200495

135. Willman M, Kobasa D, Kindrachuk J (2019) A comparative analysis of factors influencing two outbreaks of Middle Eastern respiratory syndrome (MERS) in Saudi Arabia and South Korea. Viruses 11(12). https://doi.org/10.3390/v11121119

136. Zheng J (2020) SARS-CoV-2: an emerging coronavirus that causes a global threat. Int J Biol Sci 16(10):1678–1685. https://doi.org/10.7150/ijbs.45053

137. Yu J, Chai P, Ge S, Fan X (2020) Recent understandings toward coronavirus disease 2019 (COVID-19): from bench to bedside. Front Cell Dev Biol 8:476. https://doi.org/10.3389/fcell.2020.00476

138. JHCR. C (2020) Center JHCR. Available at: https://coronavirus.jhu.edu/map.html. Last Accessed March 10, 2022

139. Bouffanais R, Lim SS (2020) Cities—try to predict superspreading hotspots for COVID-19. Nature 583(7816):352–355. https://doi.org/10.1038/d41586-020-02072-3

140. Kang CR, Lee JY, Park Y, Huh IS, Ham HJ, Han JK et al. (2020) Coronavirus disease exposure and spread from nightclubs, South Korea. Emerg Infect Dis 26(10). https://doi.org/10.3201/eid2610.202573

141. News B (2020) Coronavirus Australia: Melbourne locks down tower blocks as cases rise. Available at: https://www.bbc.com/news/world-australia-53289616. Last Accessed on 28 July 2020

142. Feng E (2020) Lockdowns ordered as COVID-19 cluster found near Beijing food market. Available at: https://www.npr.org/2020/06/15/876962147/lockdowns-ordered-as-covid-19-cluster-found-near-beijing-food-market. Last Accessed on 27 July 2020

143. Xu XK, Liu XF, Wu Y, Ali ST, Du Z, Bosetti P et al (2020) Reconstruction of transmission pairs for novel coronavirus disease 2019 (COVID-19) in mainland China: estimation of super-spreading events, serial interval, and hazard of infection. Clin Infect Dis. https://doi.org/10.1093/cid/ciaa790

144. Correa-Martínez CL, Kampmeier S, Kümpers P, Schwierzeck V, Hennies M, Hafezi W et al (2020) A pandemic in times of global tourism: superspreading and exportation of COVID-19 cases from a ski area in Austria. J Clin Microbiol. https://doi.org/10.1128/jcm.00588-20

145. Kabesch M, Roth S, Brandstetter S, Häusler S, Juraschko E, Weigl M et al (2020) Successful containment of Covid-19 outbreak in a large maternity and perinatal center while continuing clinical service. Pediatr Allergy Immunol. https://doi.org/10.1111/pai.13265

146. Preßler J, Fill Malfertheiner S, Kabesch M, Buntrock-Döpke H, Häusler S, Ambrosch A et al (2020) Postnatal SARS-CoV-2 infection and immunological reaction: a prospective family cohort study. Pediatr Allergy Immunol. https://doi.org/10.1111/pai.13302

147. Hodcroft EB (2020) Preliminary case report on the SARS-CoV-2 cluster in the UK, France, and Spain. Swiss Med Wkly 150 (9–10). https://doi.org/10.4414/smw.2020.20212

148. Hamner L, Dubbel P, Capron I, Ross A, Jordan A, Lee J et al (2020) High SARS-CoV-2 attack rate following exposure at a Choir Practice—Skagit County, Washington. MMWR Morb Mortal Wkly Rep 69(19):606–610. https://doi.org/10.15585/mmwr.mm6919e6

149. Kang YJ (2020) Characteristics of the COVID-19 outbreak in Korea from the mass infection perspective. J Prev Med Public Health 53(3):168–170. https://doi.org/10.3961/jpmph.20.072

150. Kim S, Jeong YD, Byun JH, Cho G, Park A, Jung JH et al (2020) Evaluation of COVID-19 epidemic outbreak caused by temporal contact-increase in South Korea. Int J Infect Dis: IJID: Official Publ Int Soc Infect Dis 96:454–457. https://doi.org/10.1016/j.ijid.2020.05.036

151. Korean Society of Infectious D, Korean Society of Pediatric Infectious D, Korean Society of E, Korean Society for Antimicrobial T, Korean Society for Healthcare-associated Infection C, Prevention et al. (2020) Report on the Epidemiological Features of Coronavirus

Disease 2019 (COVID-19) Outbreak in the Republic of Korea from January 19 to March 2, 2020. J Korean Med Sci 35(10):e112–e112. https://doi.org/10.3346/jkms.2020.35.e112

152. Shim E, Tariq A, Choi W, Lee Y, Chowell G (2020) Transmission potential and severity of COVID-19 in South Korea. Int J Infect Dis: IJID: Official Publ Int Soc Infect Dis 93:339–344. https://doi.org/10.1016/j.ijid.2020.03.031

153. BBC (2020) Coronavirus: India 'super spreader' quarantines 40,000 people. Available at: https://www.bbc.com/news/world-asia-india-52061915 Last Accessed 8 July 2020

154. Quadri SA (2020) COVID-19 and religious congregations: Implications for spread of novel pathogens. Int J Infect Dis: IJID: Official Publ Int Soc Infect Dis 96:219–221. https://doi.org/10.1016/j.ijid.2020.05.007

155. Towner JS, Amman BR, Sealy TK, Carroll SA, Comer JA, Kemp A et al (2009) Isolation of genetically diverse Marburg viruses from Egyptian fruit bats. PLoS Pathog 5(7):e1000536. https://doi.org/10.1371/journal.ppat.1000536

156. Feldmann H, Sprecher A, Geisbert TW (2020) Ebola. N Engl J Med 382(19):1832–1842. https://doi.org/10.1056/NEJMra1901594

157. Olival KJ, Hayman DT (2014) Filoviruses in bats: current knowledge and future directions. Viruses 6(4):1759–1788. https://doi.org/10.3390/v6041759

158. Pourrut X, Kumulungui B, Wittmann T, Moussavou G, Delicat A, Yaba P et al (2005) The natural history of Ebola virus in Africa. Microbes Infect/Institut Pasteur 7(7–8):1005–1014. https://doi.org/10.1016/j.micinf.2005.04.006

159. Carroll SA, Towner JS, Sealy TK, McMullan LK, Khristova ML, Burt FJ et al (2013) Molecular evolution of viruses of the family Filoviridae based on 97 whole-genome sequences. J Virol 87(5):2608–2616. https://doi.org/10.1128/jvi.03118-12

160. Muhlberger E (2007) Filovirus replication and transcription. Future virology 2(2):205–215. https://doi.org/10.2217/17460794.2.2.205

161. Breman JG, Heymann DL, Lloyd G, McCormick JB, Miatudila M, Murphy FA et al (2016) Discovery and description of Ebola Zaire virus in 1976 and relevance to the West African epidemic during 2013–2016. J Infect Dis 214(suppl 3):S93-s101. https://doi.org/10.1093/infdis/jiw207

162. Kadanali A, Karagoz G (2015) An overview of Ebola virus disease. North Clin Istanb 2(1):81–86. https://doi.org/10.14744/nci.2015.97269

163. Leroy EM, Kumulungui B, Pourrut X, Rouquet P, Hassanin A, Yaba P et al (2005) Fruit bats as reservoirs of Ebola virus. Nature 438(7068):575–576. https://doi.org/10.1038/438575a

164. Leroy EM, Epelboin A, Mondonge V, Pourrut X, Gonzalez JP, Muyembe-Tamfum JJ et al (2009) Human Ebola outbreak resulting from direct exposure to fruit bats in Luebo, Democratic Republic of Congo, 2007. Vector borne and zoonotic diseases (Larchmont, NY) 9(6):723–728. https://doi.org/10.1089/vbz.2008.0167

165. Nakayama E, Saijo M (2013) Animal models for Ebola and Marburg virus infections. Front Microbiol 4:267. https://doi.org/10.3389/fmicb.2013.00267

166. Curran KG, Gibson JJ, Marke D, Caulker V, Bomeh J, Redd JT et al (2014) (2016) Cluster of Ebola virus disease linked to a single funeral—Moyamba District, Sierra Leone. MMWR Morb Mortal Wkly Rep 65(8):202–205. https://doi.org/10.15585/mmwr.mm6508a2

167. Gire SK, Goba A, Andersen KG, Sealfon RS, Park DJ, Kanneh L et al (2014) Genomic surveillance elucidates Ebola virus origin and transmission during the 2014 outbreak. Science 345(6202):1369–1372. https://doi.org/10.1126/science.1259657

168. World Health Organization (2015) Sierra Leone: a traditional healer and a funeral. https://www.who.int/news/item/01-09-2015-sierra-leone-a-traditional-healer-and-a-funeral. Last Acessed 11 Nov 2021

169. Faye O, Boëlle PY, Heleze E, Faye O, Loucoubar C, Magassouba N et al (2015) Chains of transmission and control of Ebola virus disease in Conakry, Guinea, in 2014: an observational study. Lancet Infect Dis 15(3):320–326. https://doi.org/10.1016/s1473-3099(14)71075-8

170. Althaus CL (2015) Ebola superspreading. Lancet Infect Dis 15(5):507–508. https://doi.org/
     10.1016/s1473-3099(15)70135-0
171. Lau MS, Dalziel BD, Funk S, McClelland A, Tiffany A, Riley S et al (2017) Spatial and
     temporal dynamics of superspreading events in the 2014–2015 West Africa Ebola epidemic.
     Proc Natl Acad Sci U S A 114(9):2337–2342. https://doi.org/10.1073/pnas.1614595114
172. Khan AS, Tshioko FK, Heymann DL, Le Guenno B, Nabeth P, Kerstiens B et al. (1999) The
     reemergence of Ebola hemorrhagic fever, Democratic Republic of the Congo, 1995.
     Commission de Lutte contre les Epidemies a Kikwit. J Infect Dis 179(Suppl 1):S76–86.
     https://doi.org/10.1086/514306
173. Borchert M, Mutyaba I, Van Kerkhove MD, Lutwama J, Luwaga H, Bisoborwa G et al
     (2011) Ebola haemorrhagic fever outbreak in Masindi District, Uganda: outbreak description
     and lessons learned. BMC Infect Dis 11:357. https://doi.org/10.1186/1471-2334-11-357
174. Baize S, Pannetier D, Oestereich L, Rieger T, Koivogui L, Magassouba N et al (2014)
     Emergence of Zaire Ebola virus disease in Guinea. N Engl J Med 371(15):1418–1425.
     https://doi.org/10.1056/NEJMoa1404505
175. Vassantachart JM, Yeo AH, Vassantachart AY, Jacob SE, Golkar L (2020) Art of
     prevention: the importance of measles recognition and vaccination. Int J Womens Dermatol
     6(2):89–93. https://doi.org/10.1016/j.ijwd.2019.06.031
176. Krawiec C, Hinson JW (2020) Rubeola (Measles). In: StatPearls. StatPearls Publishing
     Copyright © 2020, StatPearls Publishing LLC., Treasure Island (FL)
177. Abad CL, Safdar N (2015) The Reemergence of Measles. Curr Infect Dis Rep 17(12):51.
     https://doi.org/10.1007/s11908-015-0506-5
178. Wolfson LJ, Strebel PM, Gacic-Dobo M, Hoekstra EJ, McFarland JW, Hersh BS (2007) Has
     the 2005 measles mortality reduction goal been achieved? A natural history modelling study.
     Lancet 369(9557):191–200. https://doi.org/10.1016/s0140-6736(07)60107-x
179. Krishnamoorthy Y, Sakthivel M, Eliyas SK, Surendran G, Sarveswaran G (2019)
     Worldwide trend in measles incidence from 1980 to 2016: a pooled analysis of evidence
     from 194 WHO Member States. J Postgrad Med 65(3):160–163. https://doi.org/10.4103/
     jpgm.JPGM_508_18
180. Paunio M, Peltola H, Valle M, Davidkin I, Virtanen M, Heinonen OP (1998) Explosive
     school-based measles outbreak: intense exposure may have resulted in high risk, even
     among revaccinees. Am J Epidemiol 148(11):1103–1110. https://doi.org/10.1093/oxford
     journals.aje.a009588
181. De Serres G, Boulianne N, Defay F, Brousseau N, Benoît M, Lacoursière S et al (2012)
     Higher risk of measles when the first dose of a 2-dose schedule of measles vaccine is given at
     12–14 months versus 15 months of age. Clin Infect Dis 55(3):394–402. https://doi.org/10.
     1093/cid/cis439
182. Seward JF, Orenstein WA (2012) Editorial commentary: a rare event: a measles outbreak in
     a population with high 2-dose measles vaccine coverage. Clin Infect Dis 55(3):403–405.
     https://doi.org/10.1093/cid/cis445
183. De Serres G, Markowski F, Toth E, Landry M, Auger D, Mercier M et al (2013) Largest
     measles epidemic in North America in a decade–Quebec, Canada, 2011: contribution of
     susceptibility, serendipity, and superspreading events. J Infect Dis 207(6):990–998. https://
     doi.org/10.1093/infdis/jis923
184. Taubenberger JK, Morens DM (2008) The pathology of influenza virus infections. Annu
     Rev Pathol 3:499–522. https://doi.org/10.1146/annurev.pathmechdis.3.121806.154316
185. Killingley B, Nguyen-Van-Tam J (2013) Routes of influenza transmission. Influenza Other
     Respir Viruses 7 Suppl 2 (Suppl 2):42–51. https://doi.org/10.1111/irv.12080
186. Fabian P, McDevitt JJ, DeHaan WH, Fung ROP, Cowling BJ, Chan KH et al (2008)
     Influenza virus in human exhaled breath: an observational study. PLoS ONE 3(7):e2691–
     e2691. https://doi.org/10.1371/journal.pone.0002691

187. Saunders-Hastings PR, Krewski D (2016) Reviewing the history of pandemic influenza: understanding patterns of emergence and transmission. Pathogens 5(4):66. https://doi.org/10.3390/pathogens5040066
188. Ghebrehewet S, MacPherson P, Ho A (2016) Influenza. BMJ (Clinical research ed), vol 355, pp i6258–i6258. https://doi.org/10.1136/bmj.i6258
189. Suzuki A, Mizumoto K, Akhmetzhanov AR, Nishiura H (2019) Interaction among Influenza viruses A/H1N1, A/H3N2, and B in Japan. Int J Environ Res Public Health 16(21):4179. https://doi.org/10.3390/ijerph16214179
190. Moser MR, Bender TR, Margolis HS, Noble GR, Kendal AP, Ritter DG (1979) An outbreak of influenza aboard a commercial airliner. Am J Epidemiol 110(1):1–6. https://doi.org/10.1093/oxfordjournals.aje.a112781
191. Pestre V, Morel B, Encrenaz N, Brunon A, Lucht F, Pozzetto B et al (2012) Transmission by super-spreading event of pandemic A/H1N1 2009 influenza during road and train travel. Scand J Infect Dis 44(3):225–227. https://doi.org/10.3109/00365548.2011.631936
192. Ducrotoy MJ, Muñoz PM, Conde-Álvarez R, Blasco JM, Moriyón I (2018) A systematic review of current immunological tests for the diagnosis of cattle brucellosis. Prev Vet Med 151:57–72. https://doi.org/10.1016/j.prevetmed.2018.01.005
193. Mancilla M (2016) Smooth to rough dissociation in Brucella: the missing link to Virulence. Front Cell Infect Microbiol 5:98–98. https://doi.org/10.3389/fcimb.2015.00098
194. Hull NC, Schumaker BA (2018) Comparisons of brucellosis between human and veterinary medicine. Infect Ecol Epidemiol 8(1):1500846. https://doi.org/10.1080/20008686.2018.1500846
195. Kim H, Jeong W, Jeoung HY, Song JY, Kim JS, Beak JH et al (2012) Complete genome sequence of Brucella abortus A13334, a new strain isolated from the fetal gastric fluid of dairy cattle. J Bacteriol 194(19):5444. https://doi.org/10.1128/jb.01124-12
196. Borriello G, Capparelli R, Bianco M, Fenizia D, Alfano F, Capuano F et al (2006) Genetic resistance to Brucella abortus in the water buffalo (Bubalus bubalis). Infect Immun 74(4):2115–2120. https://doi.org/10.1128/iai.74.4.2115-2120.2006
197. Gheibi A, Khanahmad H, Kashfi K, Sarmadi M, Khorramizadeh MR (2018) Development of new generation of vaccines for Brucella abortus. Heliyon 4(12):e01079–e01079. https://doi.org/10.1016/j.heliyon.2018.e01079
198. Galińska EM, Zagórski J (2013) Brucellosis in humans–etiology, diagnostics, clinical forms. Ann Agric Environ Med 20(2):233–238
199. Capparelli R, Parlato M, Iannaccone M, Roperto S, Marabelli R, Roperto F et al (2009) Heterogeneous shedding of Brucella abortus in milk and its effect on the control of animal brucellosis. J Appl Microbiol 106(6):2041–2047. https://doi.org/10.1111/j.1365-2672.2009.04177.x
200. Conway T, Cohen PS (2015) Commensal and pathogenic Escherichia coli metabolism in the gut. Microbiol Spectr 3(3). https://doi.org/10.1128/microbiolspec.MBP-0006-2014
201. Lasaro M, Liu Z, Bishar R, Kelly K, Chattopadhyay S, Paul S et al (2014) Escherichia coli isolate for studying colonization of the mouse intestine and its application to two-component signaling knockouts. J Bacteriol 196(9):1723–1732. https://doi.org/10.1128/JB.01296-13
202. Eggesbø M, Moen B, Peddada S, Baird D, Rugtveit J, Midtvedt T et al (2011) Development of gut microbiota in infants not exposed to medical interventions. APMIS 119(1):17–35. https://doi.org/10.1111/j.1600-0463.2010.02688.x
203. Reilly A (1998) Prevention and control of enterohaemorrhagic Escherichia coli (EHEC) infections: memorandum from a WHO meeting. WHO Consultation on Prevention and Control of Enterohaemorrhagic Escherichia coli (EHEC) Infections. Bull World Health Organ 76(3):245–255
204. Stromberg ZR, Van Goor A, Redweik GAJ, Wymore Brand MJ, Wannemuehler MJ, Mellata M (2018) Pathogenic and non-pathogenic Escherichia coli colonization and host inflammatory response in a defined microbiota mouse model. Dis Model Mech 11(11):dmm035063. https://doi.org/10.1242/dmm.035063

205. Weir E (2000) Escherichia coli O157:H7. CMAJ 163(2):205
206. Poirier K, Faucher SP, Béland M, Brousseau R, Gannon V, Martin C et al (2008) Escherichia coli O157:H7 survives within human macrophages: global gene expression profile and involvement of the Shiga toxins. Infect Immun 76(11):4814–4822. https://doi.org/10.1128/IAI.00446-08
207. Munns KD, Selinger LB, Stanford K, Guan L, Callaway TR, McAllister TA (2015) Perspectives on super-shedding of Escherichia coli O157:H7 by cattle. Foodborne Pathog Dis 12(2):89–103. https://doi.org/10.1089/fpd.2014.1829
208. Lim JY, Yoon J, Hovde CJ (2010) A brief overview of Escherichia coli O157:H7 and its plasmid O157. J Microbiol Biotechnol 20(1):5–14
209. Mir RA, Schaut RG, Looft T, Allen HK, Sharma VK, Kudva IT (2020) Recto-Anal junction (RAJ) and fecal microbiomes of cattle experimentally challenged with Escherichia coli O157:H7. Front Microbiol 11:693. https://doi.org/10.3389/fmicb.2020.00693
210. Lim JY, Li J, Sheng H, Besser TE, Potter K, Hovde CJ (2007) Escherichia coli O157:H7 colonization at the rectoanal junction of long-duration culture-positive cattle. Appl Environ Microbiol 73(4):1380–1382. https://doi.org/10.1128/aem.02242-06
211. Su C, Brandt LJ (1995) Escherichia coli O157:H7 infection in humans. Ann Intern Med 123 (9):698–714. https://doi.org/10.7326/0003-4819-123-9-199511010-00009
212. Rahal EA, Kazzi N, Nassar FJ, Matar GM (2012) Escherichia coli O157:H7-Clinical aspects and novel treatment approaches. Front Cell Infect Microbiol 2:138. https://doi.org/10.3389/fcimb.2012.00138
213. Nangaku M, Nishi H, Fujita T (2007) Pathogenesis and prognosis of thrombotic microangiopathy. Clin Exp Nephrol 11(2):107–114. https://doi.org/10.1007/s10157-007-0466-7
214. Sabouri S, Sepehrizadeh Z, Amirpour-Rostami S, Skurnik M (2017) A minireview on the in vitro and in vivo experiments with anti-Escherichia coli O157:H7 phages as potential biocontrol and phage therapy agents. Int J Food Microbiol 243:52–57. https://doi.org/10.1016/j.ijfoodmicro.2016.12.004
215. Wang O, McAllister TA, Plastow G, Stanford K, Selinger B, Guan LL (2018) Interactions of the hindgut mucosa-associated microbiome with its host regulate shedding of Escherichia coli O157:H7 by Cattle. Appl Environ Microbiol 84(1). https://doi.org/10.1128/aem.01738-17
216. Mir RA, Brunelle BW, Alt DP, Arthur TM, Kudva IT (2020) Supershed Escherichia coli O157:H7 has potential for increased persistence on the rectoanal junction squamous epithelial cells and antibiotic resistance. Int J Microbiol 2020:2368154. https://doi.org/10.1155/2020/2368154
217. Omisakin F, MacRae M, Ogden ID, Strachan NJ (2003) Concentration and prevalence of Escherichia coli O157 in cattle feces at slaughter. Appl Environ Microbiol 69(5):2444–2447. https://doi.org/10.1128/aem.69.5.2444-2447.2003
218. Ogden ID, MacRae M, Strachan NJ (2004) Is the prevalence and shedding concentrations of E. coli O157 in beef cattle in Scotland seasonal? FEMS Microbiol Lett 233(2):297–300. https://doi.org/10.1016/j.femsle.2004.02.021
219. Powell JR (2019) An evolutionary perspective on vector-borne diseases. Front Genet 10:1266–1266. https://doi.org/10.3389/fgene.2019.01266
220. Bockarie MJ, Davies JB (1990) The transmission of onchocerciasis at a forest village in Sierra Leone. II. Man-fly contact, human activity and exposure to transmission. Ann Trop Med Parasitol 84(6):599–605. https://doi.org/10.1080/00034983.1990.11812515
221. Medley GF, Sinden RE, Fleck S, Billingsley PF, Tirawanchai N, Rodriguez MH (1993) Heterogeneity in patterns of malarial oocyst infections in the mosquito vector. Parasitology 106(Pt 5):441–449. https://doi.org/10.1017/s0031182000076721
222. Pichon G, Awono-Ambene HP, Robert V (2000) High heterogeneity in the number of Plasmodium falciparum gametocytes in the bloodmeal of mosquitoes fed on the same host. Parasitology 121(Pt 2):115–120. https://doi.org/10.1017/s0031182099006277

223. Mitri C, Vernick KD (2012) Anopheles gambiae pathogen susceptibility: the intersection of genetics, immunity and ecology. Curr Opin Microbiol 15(3):285–291. https://doi.org/10.1016/j.mib.2012.04.001

224. Ciddio M, Mari L, Sokolow SH, De Leo GA, Casagrandi R, Gatto M (2017) The spatial spread of schistosomiasis: a multidimensional network model applied to Saint-Louis region, Senegal. Adv Water Resour 108:406–415. https://doi.org/10.1016/j.advwatres.2016.10.012

225. Nelwan ML (2019) Schistosomiasis: life cycle, diagnosis, and control. Curr Ther Res Clin Exp 91:5–9. https://doi.org/10.1016/j.curtheres.2019.06.001

226. Verjee MA (2019) Schistosomiasis: still a cause of significant morbidity and mortality. Res Rep Trop Med 10:153–163. https://doi.org/10.2147/rrtm.S204345

227. Makaula P, Sadalaki JR, Muula AS, Kayuni S, Jemu S, Bloch P (2014) Schistosomiasis in Malawi: a systematic review. Parasit Vectors 7(1):570. https://doi.org/10.1186/s13071-014-0570-y

228. Colley DG, Bustinduy AL, Secor WE, King CH (2014) Human schistosomiasis. Lancet 383 (9936):2253–2264. https://doi.org/10.1016/s0140-6736(13)61949-2

229. Ross AG, Bartley PB, Sleigh AC, Olds GR, Li Y, Williams GM et al (2002) Schistosomiasis. N Engl J Med 346(16):1212–1220. https://doi.org/10.1056/NEJMra012396

230. Mitreva M (2012) The genome of a blood fluke associated with human cancer. Nat Genet 44 (2):116–118. https://doi.org/10.1038/ng.1082

231. Grimes JE, Croll D, Harrison WE, Utzinger J, Freeman MC, Templeton MR (2014) The relationship between water, sanitation and schistosomiasis: a systematic review and meta-analysis. PLoS Negl Trop Dis 8(12):e3296. https://doi.org/10.1371/journal.pntd.0003296

232. Brindley PJ, Hotez PJ (2013) Break out: urogenital schistosomiasis and Schistosoma haematobium infection in the post-genomic era. PLoS Negl Trop Dis 7(3):e1961. https://doi.org/10.1371/journal.pntd.0001961

233. Gryseels B, Polman K, Clerinx J, Kestens L (2006) Human schistosomiasis. Lancet 368 (9541):1106–1118. https://doi.org/10.1016/s0140-6736(06)69440-3

234. Collins JJ 3rd, King RS, Cogswell A, Williams DL, Newmark PA (2011) An atlas for Schistosoma mansoni organs and life-cycle stages using cell type-specific markers and confocal microscopy. PLoS Negl Trop Dis 5(3):e1009. https://doi.org/10.1371/journal.pntd.0001009

235. Woolhouse ME, Etard JF, Dietz K, Ndhlovu PD, Chandiwana SK (1998) Heterogeneities in schistosome transmission dynamics and control. Parasitology 117(Pt 5):475–482

236. Ross AG, Sleigh AC, Li Y, Davis GM, Williams GM, Jiang Z et al (2001) Schistosomiasis in the People's Republic of China: prospects and challenges for the 21st century. Clin Microbiol Rev 14(2):270–295. https://doi.org/10.1128/cmr.14.2.270-295.2001

237. Lin DD, Hu GH, Zhang SJ (2005) Optimal combined approaches of field intervention for schistosomiasis control in China. Acta Trop 96(2–3):242–247. https://doi.org/10.1016/j.actatropica.2005.07.018

238. Yu Q, Zhao GM, Hong XL, Lutz EA, Guo JG (2013) Impact and cost-effectiveness of a comprehensive Schistosomiasis japonica control program in the Poyang Lake region of China. Int J Environ Res Public Health 10(12):6409–6421. https://doi.org/10.3390/ijerph10126409

239. French MD, Churcher TS, Gambhir M, Fenwick A, Webster JP, Kabatereine NB et al (2010) Observed reductions in Schistosoma mansoni transmission from large-scale administration of praziquantel in Uganda: a mathematical modelling study. PLoS Negl Trop Dis 4(11): e897. https://doi.org/10.1371/journal.pntd.0000897

240. Koukounari A, Gabrielli AF, Toure S, Bosque-Oliva E, Zhang Y, Sellin B et al (2007) Schistosoma haematobium infection and morbidity before and after large-scale administration of praziquantel in Burkina Faso. J Infect Dis 196(5):659–669. https://doi.org/10.1086/520515

241. Doenhoff MJ, Cioli D, Utzinger J (2008) Praziquantel: mechanisms of action, resistance and new derivatives for schistosomiasis. Curr Opin Infect Dis 21(6):659–667. https://doi.org/10.1097/QCO.0b013e328318978f
242. Carlton EJ, Hubbard A, Wang S, Spear RC (2013) Repeated Schistosoma japonicum infection following treatment in two cohorts: evidence for host susceptibility to helminthiasis? PLoS Negl Trop Dis 7(3):e2098. https://doi.org/10.1371/journal.pntd.0002098
243. Zhou YB, Liang S, Jiang QW (2012) Factors impacting on progress towards elimination of transmission of schistosomiasis japonica in China. Parasit Vectors 5:275. https://doi.org/10.1186/1756-3305-5-275
244. King CH (2009) Toward the elimination of schistosomiasis. N Engl J Med 360(2):106–109. https://doi.org/10.1056/NEJMp0808041
245. Woolhouse ME, Chandiwana SK, Bradley M (1990) On the distribution of schistosome infections among host snails. Int J Parasitol 20(3):325–327
246. Chandiwana SK, Woolhouse ME (1991) Heterogeneities in water contact patterns and the epidemiology of Schistosoma haematobium. Parasitology 103(Pt 3):363–370
247. Clennon JA, King CH, Muchiri EM, Kariuki HC, Ouma JH, Mungai P et al (2004) Spatial patterns of urinary schistosomiasis infection in a highly endemic area of coastal Kenya. Am J Trop Med Hyg 70(4):443–448
248. Meurs L, Mbow M, Boon N, van den Broeck F, Vereecken K, Dièye TN et al (2013) Micro-geographical heterogeneity in Schistosoma mansoni and S. haematobium infection and morbidity in a co-endemic community in northern Senegal. PLoS Negl Trop Dis 7(12): e2608. https://doi.org/10.1371/journal.pntd.0002608
249. Talapko J, Škrlec I, Alebić T, Jukić M, Včev A (2019) Malaria: the past and the present. Microorganisms 7(6). https://doi.org/10.3390/microorganisms7060179
250. Fletcher TE, Beeching NJ (2013) Malaria. J R Army Med Corps 159(3):158–166. https://doi.org/10.1136/jramc-2013-000112
251. Dhiman S (2019) Are malaria elimination efforts on right track? An analysis of gains achieved and challenges ahead. Infect Dis Poverty 8(1):14–14. https://doi.org/10.1186/s40249-019-0524-x
252. Bousema T, Drakeley C, Gesase S, Hashim R, Magesa S, Mosha F et al (2010) Identification of hot spots of malaria transmission for targeted malaria control. J Infect Dis 201(11):1764–1774. https://doi.org/10.1086/652456
253. Parker BS, Paredes Olortegui M, Peñataro Yori P, Escobedo K, Florin D, Rengifo Pinedo S et al (2013) Hyperendemic malaria transmission in areas of occupation-related travel in the Peruvian Amazon. Malar J 12:178. https://doi.org/10.1186/1475-2875-12-178
254. Naidoo S, London L, Burdorf A, Naidoo RN, Kromhout H (2011) Occupational activities associated with a reported history of malaria among women working in small-scale agriculture in South Africa. Am J Trop Med Hyg 85(5):805–810. https://doi.org/10.4269/ajtmh.2011.11-0092
255. Ghebreyesus TA, Haile M, Witten KH, Getachew A, Yohannes M, Lindsay SW et al (2000) Household risk factors for malaria among children in the Ethiopian highlands. Trans R Soc Trop Med Hyg 94(1):17–21. https://doi.org/10.1016/s0035-9203(00)90424-3
256. Baidjoe AY, Stevenson J, Knight P, Stone W, Stresman G, Osoti V et al (2016) Factors associated with high heterogeneity of malaria at fine spatial scale in the Western Kenyan highlands. Malar J 15:307. https://doi.org/10.1186/s12936-016-1362-y
257. Chirebvu E, Chimbari MJ, Ngwenya BN (2014) Assessment of risk factors associated with malaria transmission in Tubu village, northern Botswana. Malar Res Treat 2014:403069–403069. https://doi.org/10.1155/2014/403069
258. Esayas E, Woyessa A, Massebo F (2020) Malaria infection clustered into small residential areas in lowlands of southern Ethiopia. Parasite Epidemiol Control 10:e00149. https://doi.org/10.1016/j.parepi.2020.e00149

259. Smith DL, Dushoff J, McKenzie FE (2004) The risk of a mosquito-borne infection in a heterogeneous environment. PLoS Biol 2(11):e368. https://doi.org/10.1371/journal.pbio.0020368
260. Hansen E, Buckee CO (2013) Modeling the human infectious reservoir for malaria control: does heterogeneity matter? Trends Parasitol 29(6):270–275. https://doi.org/10.1016/j.pt.2013.03.009
261. Gupta S, Hill AV, Kwiatkowski D, Greenwood AM, Greenwood BM, Day KP (1994) Parasite virulence and disease patterns in Plasmodium falciparum malaria. Proc Natl Acad Sci U S A 91(9):3715–3719. https://doi.org/10.1073/pnas.91.9.3715
262. Bonnet S, Gouagna LC, Paul RE, Safeukui I, Meunier JY, Boudin C (2003) Estimation of malaria transmission from humans to mosquitoes in two neighbouring villages in south Cameroon: evaluation and comparison of several indices. Trans R Soc Trop Med Hyg 97(1): 53–59. https://doi.org/10.1016/s0035-9203(03)90022-8
263. Mbogo CM, Mwangangi JM, Nzovu J, Gu W, Yan G, Gunter JT et al (2003) Spatial and temporal heterogeneity of Anopheles mosquitoes and Plasmodium falciparum transmission along the Kenyan coast. Am J Trop Med Hyg 68(6):734–742
264. Roca-Feltrer A, Schellenberg JR, Smith L, Carneiro I (2009) A simple method for defining malaria seasonality. Malar J 8:276. https://doi.org/10.1186/1475-2875-8-276
265. Giha HA, Rosthoj S, Dodoo D, Hviid L, Satti GM, Scheike T et al (2000) The epidemiology of febrile malaria episodes in an area of unstable and seasonal transmission. Trans R Soc Trop Med Hyg 94(6):645–651. https://doi.org/10.1016/s0035-9203(00)90218-9
266. Scott TW, Githeko AK, Fleisher A, Harrington LC, Yan G (2006) DNA profiling of human blood in anophelines from lowland and highland sites in western Kenya. Am J Trop Med Hyg 75(2):231–237
267. Stresman G, Bousema T, Cook J (2019) Malaria hotspots: is there epidemiological evidence for fine-scale spatial targeting of interventions? Trends Parasitol 35(10):822–834. https://doi.org/10.1016/j.pt.2019.07.013
268. Kangoye DT, Noor A, Midega J, Mwongeli J, Mkabili D, Mogeni P et al (2016) Malaria hotspots defined by clinical malaria, asymptomatic carriage, PCR and vector numbers in a low transmission area on the Kenyan Coast. Malar J 15:213–213. https://doi.org/10.1186/s12936-016-1260-3
269. Kang SY, Battle KE, Gibson HS, Cooper LV, Maxwell K, Kamya M et al (2018) Heterogeneous exposure and hotspots for malaria vectors at three study sites in Uganda. Gates Open Res 2:32. https://doi.org/10.12688/gatesopenres.12838.2
270. Rouamba T, Nakanabo-Diallo S, Derra K, Rouamba E, Kazienga A, Inoue Y et al (2019) Socioeconomic and environmental factors associated with malaria hotspots in the Nanoro demographic surveillance area. Burkina Faso. BMC Public Health 19(1):249. https://doi.org/10.1186/s12889-019-6565-z
271. Platt A, Obala AA, MacIntyre C, Otsyula B, Meara WPO (2018) Dynamic malaria hotspots in an open cohort in western Kenya. Sci Rep 8(1):647. https://doi.org/10.1038/s41598-017-13801-6
272. Gaudart J, Poudiougou B, Dicko A, Ranque S, Toure O, Sagara I et al (2006) Space-time clustering of childhood malaria at the household level: a dynamic cohort in a Mali village. BMC Public Health 6:286. https://doi.org/10.1186/1471-2458-6-286
273. Oduro AR, Conway DJ, Schellenberg D, Satoguina J, Greenwood BM, Bojang KA (2013) Seroepidemiological and parasitological evaluation of the heterogeneity of malaria infection in the Gambia. Malar J 12:222. https://doi.org/10.1186/1475-2875-12-222
274. Ernst KC, Adoka SO, Kowuor DO, Wilson ML, John CC (2006) Malaria hotspot areas in a highland Kenya site are consistent in epidemic and non-epidemic years and are associated with ecological factors. Malar J 5:78. https://doi.org/10.1186/1475-2875-5-78
275. Tewara MA, Mbah-Fongkimeh PN, Dayimu A, Kang F, Xue F (2018) Small-area spatial statistical analysis of malaria clusters and hotspots in Cameroon; 2000–2015. BMC Infect Dis 18(1):636. https://doi.org/10.1186/s12879-018-3534-6

276. Toty C, Barré H, Le Goff G, Larget-Thiéry I, Rahola N, Couret D et al (2010) Malaria risk in Corsica, former hot spot of malaria in France. Malar J 9:231. https://doi.org/10.1186/1475-2875-9-231

277. Midega JT, Smith DL, Olotu A, Mwangangi JM, Nzovu JG, Wambua J et al (2012) Wind direction and proximity to larval sites determines malaria risk in Kilifi District in Kenya. Nat Commun 3:674. https://doi.org/10.1038/ncomms1672

278. Garske T, Ferguson NM, Ghani AC (2013) Estimating air temperature and its influence on malaria transmission across Africa. PLoS ONE 8(2):e56487. https://doi.org/10.1371/journal.pone.0056487

279. Bejon P, Williams TN, Liljander A, Noor AM, Wambua J, Ogada E et al (2010) Stable and unstable malaria hotspots in longitudinal cohort studies in Kenya. PLoS Med 7(7):e1000304. https://doi.org/10.1371/journal.pmed.1000304

280. Norris LC, Norris DE (2013) Heterogeneity and changes in inequality of malaria risk after introduction of insecticide-treated bed nets in Macha. Zambia. Am J Trop Med Hyg 88 (4):710–717. https://doi.org/10.4269/ajtmh.11-0595

281. Smith DL, Dushoff J, Snow RW, Hay SI (2005) The entomological inoculation rate and Plasmodium falciparum infection in African children. Nature 438(7067):492–495. https://doi.org/10.1038/nature04024

282. Corder RM, Ferreira MU, Gomes MGM (2020) Modelling the epidemiology of residual Plasmodium vivax malaria in a heterogeneous host population: a case study in the Amazon Basin. PLoS Comput Biol 16(3):e1007377. https://doi.org/10.1371/journal.pcbi.1007377

283. Moss WJ, Hamapumbu H, Kobayashi T, Shields T, Kamanga A, Clennon J et al (2011) Use of remote sensing to identify spatial risk factors for malaria in a region of declining transmission: a cross-sectional and longitudinal community survey. Malar J 10:163. https://doi.org/10.1186/1475-2875-10-163

284. Trape JF, Rogier C, Konate L, Diagne N, Bouganali H, Canque B et al (1994) The Dielmo project: a longitudinal study of natural malaria infection and the mechanisms of protective immunity in a community living in a holoendemic area of Senegal. Am J Trop Med Hyg 51(2):123–137

285. Trape JF, Pison G, Spiegel A, Enel C, Rogier C (2002) Combating malaria in Africa. Trends Parasitol 18(5):224–230. https://doi.org/10.1016/s1471-4922(02)02249-3

286. Mwangi TW, Fegan G, Williams TN, Kinyanjui SM, Snow RW, Marsh K (2008) Evidence for over-dispersion in the distribution of clinical malaria episodes in children. PLoS ONE 3(5):e2196. https://doi.org/10.1371/journal.pone.0002196

287. Russell AB, Trapnell C, Bloom JD (2018) Extreme heterogeneity of influenza virus infection in single cells. Elife 7:e32303. https://doi.org/10.7554/eLife.32303

288. Gog JR, Murcia A, Osterman N, Restif O, McKinley TJ, Sheppard M et al (2012) Dynamics of Salmonella infection of macrophages at the single cell level. J R Soc Interface 9 (75):2696–2707. https://doi.org/10.1098/rsif.2012.0163

289. Achouri S, Wright JA, Evans L, Macleod C, Fraser G, Cicuta P et al (2015) The frequency and duration of Salmonella-macrophage adhesion events determines infection efficiency. Philos Trans R Soc Lond B Biol Sci 370(1661):20140033–20140033. https://doi.org/10.1098/rstb.2014.0033

290. Avraham R, Hung DT (2016) A perspective on single cell behavior during infection. Gut Microbes 7(6):518–525. https://doi.org/10.1080/19490976.2016.1239001

291. Helaine S, Cheverton AM, Watson KG, Faure LM, Matthews SA, Holden DW (2014) Internalization of Salmonella by macrophages induces formation of nonreplicating persisters. Science 343(6167):204–208. https://doi.org/10.1126/science.1244705

292. McIntrye J, Rowley D, Jenkin CR (1967) The functional heterogeneity of macrophages at the single cell level. Aust J Exp Biol Med Sci 45(6):675–680. https://doi.org/10.1038/icb.1967.67

293. Helaine S, Cheverton AM, Watson KG, Faure LM, Matthews SA, Holden DW (2014) Internalization of Salmonella by macrophages induces formation of nonreplicating

persisters. Science (New York, NY) 343(6167):204–208. https://doi.org/10.1126/science.1244705

294. Zack JA, Kim SG, Vatakis DN (2013) HIV restriction in quiescent CD4⁺ T cells. Retrovirology 10:37–37. https://doi.org/10.1186/1742-4690-10-37

295. Han Y, Lassen K, Monie D, Sedaghat AR, Shimoji S, Liu X et al (2004) Resting CD4+ T cells from human immunodeficiency virus type 1 (HIV-1)-infected individuals carry integrated HIV-1 genomes within actively transcribed host genes. J Virol 78(12):6122–6133. https://doi.org/10.1128/JVI.78.12.6122-6133.2004

296. Pan X, Baldauf H-M, Keppler OT, Fackler OT (2013) Restrictions to HIV-1 replication in resting CD4+ T lymphocytes. Cell Res 23(7):876–885. https://doi.org/10.1038/cr.2013.74

297. Zhang Z-Q, Wietgrefe SW, Li Q, Shore MD, Duan L, Reilly C et al (2004) Roles of substrate availability and infection of resting and activated CD4+ T cells in transmission and acute simian immunodeficiency virus infection. Proc Natl Acad Sci USA 101(15):5640–5645. https://doi.org/10.1073/pnas.0308425101

298. Doitsh G, Galloway NLK, Geng X, Yang Z, Monroe KM, Zepeda O et al (2014) Cell death by pyroptosis drives CD4 T-cell depletion in HIV-1 infection. Nature 505(7484):509–514. https://doi.org/10.1038/nature12940

299. Gonzalez SM, Aguilar-Jimenez W, Su R-C, Rugeles MT (2019) Mucosa: key interactions determining sexual transmission of the HIV infection. Front Immunol 10:144–144. https://doi.org/10.3389/fimmu.2019.00144

300. Vatakis DN, Nixon CC, Zack JA (2010) Quiescent T cells and HIV: an unresolved relationship. Immunol Res 48(1–3):110–121. https://doi.org/10.1007/s12026-010-8171-0

301. Card CM, Ball TB, Fowke KR (2013) Immune quiescence: a model of protection against HIV infection. Retrovirology 10:141–141. https://doi.org/10.1186/1742-4690-10-141

302. Zhang Z, Schuler T, Zupancic M, Wietgrefe S, Staskus KA, Reimann KA et al (1999) Sexual transmission and propagation of SIV and HIV in resting and activated CD4+ T cells. Science 286(5443):1353–1357. https://doi.org/10.1126/science.286.5443.1353

303. Twort FW (1915) An investigation on the nature of ultra-microscopic viruses. The Lancet 186(4814):1241–1243. https://doi.org/10.1016/S0140-6736(01)20383-3

304. D'Herelle F (1917) Sur un microbe invisible antagoniste des bacilles dysenteriques. C R Acad Sci 165:373–375

305. Zinder ND, Lederberg J (1952) Genetic exchange in Salmonella. J Bacteriol 64(5):679–699

306. Keen EC, Bliskovsky VV, Malagon F, Baker JD, Prince JS, Klaus JS et al (2017) Novel "superspreader" bacteriophages promote horizontal gene transfer by transformation. MBio 8 (1). https://doi.org/10.1128/mBio.02115-16

307. Sugiura C, Miyaue S, Shibata Y, Matsumoto A, Maeda S (2017) Bacteriophage P1vir-induced cell-to-cell plasmid transformation in Escherichia coli. AIMS Microbiol 3 (4):784–797. https://doi.org/10.3934/microbiol.2017.4.784

308. Eichenwald HF, Kotsevalov O, Fasso LA (1960) The "cloud baby": an example of bacterial-viral interaction. Am J Dis Child 100:161–173. https://doi.org/10.1001/archpedi.1960.04020040163003

309. Bassetti S, Bischoff WE, Walter M, Bassetti-Wyss BA, Mason L, Reboussin BA et al (2005) Dispersal of Staphylococcus aureus into the air associated with a rhinovirus infection. Infect Control Hosp Epidemiol 26(2):196–203. https://doi.org/10.1086/502526

310. Sherertz RJ, Reagan DR, Hampton KD, Robertson KL, Streed SA, Hoen HM et al (1996) A cloud adult: the Staphylococcus aureus-virus interaction revisited. Ann Intern Med 124(6):539–547. https://doi.org/10.7326/0003-4819-124-6-199603150-00001

311. Sherertz RJ, Bassetti S, Bassetti-Wyss B (2001) "Cloud" health-care workers. Emerg Infect Dis 7(2):241–244. https://doi.org/10.3201/eid0702.010218

312. Lass S, Hudson PJ, Thakar J, Saric J, Harvill E, Albert R et al (2013) Generating super-shedders: co-infection increases bacterial load and egg production of a gastrointestinal helminth. J R Soc Interface 10(80):20120588. https://doi.org/10.1098/rsif.2012.0588

313. Gueirard P, Minoprio P, Guiso N (1996) Intranasal inoculation of Bordetella bronchiseptica in mice induces long-lasting antibody and T-cell mediated immune responses. Scand J Immunol 43(2):181–192. https://doi.org/10.1046/j.1365-3083.1996.d01-30.x

314. Maruszewska-Cheruiyot M, Donskow-Lysoniewska K, Piechna K, Krawczak K, Doligalska M (2019) L4 stage Heligmosomoides polygyrus prevents the maturation of dendritic JAWS II cells. Exp Parasitol 196:12–21. https://doi.org/10.1016/j.exppara.2018.10.010

315. Strandmark J, Steinfelder S, Berek C, Kuhl AA, Rausch S, Hartmann S (2017) Eosinophils are required to suppress Th2 responses in Peyer's patches during intestinal infection by nematodes. Mucosal Immunol 10(3):661–672. https://doi.org/10.1038/mi.2016.93

316. Cohen MS, Hoffman IF, Royce RA, Kazembe P, Dyer JR, Daly CC et al (1997) Reduction of concentration of HIV-1 in semen after treatment of urethritis: implications for prevention of sexual transmission of HIV-1. AIDSCAP Malawi Res Group. Lancet 349(9069):1868–1873. https://doi.org/10.1016/s0140-6736(97)02190-9

317. Ghys PD, Fransen K, Diallo MO, Ettiègne-Traoré V, Coulibaly IM, Yeboué KM et al (1997) The associations between cervicovaginal HIV shedding, sexually transmitted diseases and immunosuppression in female sex workers in Abidjan. Côte d'Ivoire. Aids 11(12):F85-93. https://doi.org/10.1097/00002030-199712000-00001

318. Howell AK, Tongue SC, Currie C, Evans J, Williams DJL, McNeilly TN (2018) Co-infection with Fasciola hepatica may increase the risk of Escherichia coli O157 shedding in British cattle destined for the food chain. Prev Vet Med 150:70–76. https://doi.org/10.1016/j.prevetmed.2017.12.007

319. Moreau E, Chauvin A (2010) Immunity against helminths: interactions with the host and the intercurrent infections. J Biomed Biotechnol 2010:428593–428593. https://doi.org/10.1155/2010/428593

320. Brady MT, O'Neill SM, Dalton JP, Mills KH (1999) Fasciola hepatica suppresses a protective Th1 response against Bordetella pertussis. Infect Immun 67(10):5372–5378

321. Lisowska KA, Pindel M, Pietruczuk K, Kuźmiuk-Glembin I, Storoniak H, Dębska-Ślizień A et al (2019) The influence of a single hemodialysis procedure on human T lymphocytes. Sci Rep 9(1):5041. https://doi.org/10.1038/s41598-019-41619-x

322. Lim WH, Kireta S, Russ GR, Coates PT (2007) Uremia impairs blood dendritic cell function in hemodialysis patients. Kidney Int 71(11):1122–1131. https://doi.org/10.1038/sj.ki.5002196

323. Kuroki Y, Tsuchida K, Go I, Aoyama M, Naganuma T, Takemoto Y et al (2007) A study of innate immunity in patients with end-stage renal disease: special reference to toll-like receptor-2 and -4 expression in peripheral blood monocytes of hemodialysis patients. Int J Mol Med 19(5):783–790

324. Peiris JSM, Chu CM, Cheng VCC, Chan KS, Hung IFN, Poon LLM et al (2003) Clinical progression and viral load in a community outbreak of coronavirus-associated SARS pneumonia: a prospective study. Lancet (London, England) 361(9371):1767–1772. https://doi.org/10.1016/s0140-6736(03)13412-5

325. Isakbaeva ET, Khetsuriani N, Beard RS, Peck A, Erdman D, Monroe SS et al (2004) SARS-associated coronavirus transmission, United States. Emerg Infect Dis 10(2):225–231. https://doi.org/10.3201/eid1002.030734

326. Abdullah ASM, Tomlinson B, Cockram CS, Thomas GN (2003) Lessons from the severe acute respiratory syndrome outbreak in Hong Kong. Emerg Infect Dis 9(9):1042–1045. https://doi.org/10.3201/eid0909.030366

327. Tsui PT, Kwok ML, Yuen H, Lai ST (2003) Severe acute respiratory syndrome: clinical outcome and prognostic correlates. Emerg Infect Dis 9(9):1064–1069. https://doi.org/10.3201/eid0909.030362

328. Leung TF, Wong GWK, Hon KLE, Fok TF (2003) Severe acute respiratory syndrome (SARS) in children: epidemiology, presentation and management. Paediatr Respir Rev 4(4):334–339. https://doi.org/10.1016/s1526-0542(03)00088-5

329. Gopinath S, Hotson A, Johns J, Nolan G, Monack D (2013) The systemic immune state of super-shedder mice is characterized by a unique neutrophil-dependent blunting of TH1 responses. PLoS Pathog 9(6):e1003408. https://doi.org/10.1371/journal.ppat.1003408

330. Thompson LJ, Dunstan SJ, Dolecek C, Perkins T, House D, Dougan G et al (2009) Transcriptional response in the peripheral blood of patients infected with Salmonella enterica serovar Typhi. Proc Natl Acad Sci U S A 106(52):22433–22438. https://doi.org/10.1073/pnas.0912386106

331. Xie SY, Zeng G, Lei J, Li Q, Li HB, Jia QB (2003) Analyses on one case of severe acute respiratory syndrome 'super transmitter' and chain of transmission. Zhonghua Liu Xing Bing Xue Za Zhi 24(6):449–453

332. Beldomenico PM (2020) Do superspreaders generate new superspreaders? A hypothesis to explain the propagation pattern of COVID-19. Int J Infect Dis: IJID: Official Publ Int Soc Infect Dis 96:461–463. https://doi.org/10.1016/j.ijid.2020.05.025

333. Wong G, Liu W, Liu Y, Zhou B, Bi Y, Gao GF (2015) MERS, SARS, and Ebola: the role of super-spreaders in infectious disease. Cell Host Microbe 18(4):398–401. https://doi.org/10.1016/j.chom.2015.09.013

334. Mohammed AA (2007) Risk factors associated with superspreading events in SARS. Thorax 62(7):649–649

335. Luo T, Comas I, Luo D, Lu B, Wu J, Wei L et al (2015) Southern East Asian origin and coexpansion of Mycobacterium tuberculosis Beijing family with Han Chinese. Proc Natl Acad Sci USA 112(26):8136–8141. https://doi.org/10.1073/pnas.1424063112

336. Bach SJ, McAllister TA, Mears GJ, Schwartzkopf-Genswein KS (2004) Long-haul transport and lack of preconditioning increases fecal shedding of Escherichia coli and Escherichia coli O157:H7 by calves. J Food Prot 67(4):672–678. https://doi.org/10.4315/0362-028x-67.4.672

337. Zhong N-S, Wong GWK (2004) Epidemiology of severe acute respiratory syndrome (SARS): adults and children. Paediatr Respir Rev 5(4):270–274. https://doi.org/10.1016/j.prrv.2004.07.011

338. Amer H, Alqahtani AS, Alzoman H, Aljerian N, Memish ZA (2018) Unusual presentation of Middle East respiratory syndrome coronavirus leading to a large outbreak in Riyadh during 2017. Am J Infect Control 46(9):1022–1025. https://doi.org/10.1016/j.ajic.2018.02.023

339. Hui DS, Azhar EI, Memish ZA, Zumla A (2020) Human coronavirus infections—severe acute respiratory syndrome (SARS), Middle East Respiratory Syndrome (MERS), and SARS-CoV-2. Ref Module Biomed Sci: B978-970-912-801238-801233.811634-801234. https://doi.org/10.1016/B978-0-12-801238-3.11634-4

340. Kim S-H, Ko J-H, Park GE, Cho SY, Ha YE, Kang J-M et al (2017) Atypical presentations of MERS-CoV infection in immunocompromised hosts. J Infect Chemother 23(11):769–773. https://doi.org/10.1016/j.jiac.2017.04.004

341. Venegas-Vargas C, Henderson S, Khare A, Mosci RE, Lehnert JD, Singh P et al (2016) Factors associated with Shiga toxin-producing Escherichia coli shedding by dairy and beef cattle. Appl Environ Microbiol 82(16):5049–5056. https://doi.org/10.1128/AEM.00829-16

342. Wang O, McAllister TA, Plastow G, Stanford K, Selinger B, Guan LL (2017) Host mechanisms involved in cattle Escherichia coli O157 shedding: a fundamental understanding for reducing foodborne pathogen in food animal production. Sci Rep 7(1):7630–7630. https://doi.org/10.1038/s41598-017-06737-4

Richard A. Stein, M.D., Ph.D., received his M.D. from the "Iuliu Haţieganu" University of Medicine and Pharmacy, Cluj-Napoca, Romania, and his Ph.D. in Biochemistry from the University of Alabama at Birmingham. He is currently an Industry Associate Professor at the NYU Tandon School of Engineering. His articles were published in several medical and biomedical journals. In 2015, he published his first book, Super-spreading in Infectious Diseases. He is a co-editor of the 3rd edition of Foodborne Diseases, published in 2017. He was interviewed by CNN, The Wall Street Journal, EMaxHealth, Orlando Sun-Sentinel, and Dagbladet Information for articles on epidemics and pandemics. He served on the editorial board of The American Journal of Infection Control, Biologicals, and The European Journal of Internal Medicine and is currently an Editorial Board member for GERMS and a Senior Editor for the International Journal of Clinical Practice. He is a regular contributor to Genetic Engineering and Biotechnology News.

Possibility of Changes in Travel Behavior as a Consequence of the Pandemic and Teleworking

Ireneusz Celiński and Grzegorz Sierpiński

"Some things are in our control and others not. Things in our control are opinion, aspiration, desire, aversion, and, in a word, whatever are our own actions. Things not in our control are body, property, glory, privilege, and, in one word, whatever are not our own actions."

Epictetus

Summary

The article addresses the problem of changes in transport behavior during the current period of increased epidemic risk. Historically, the epidemics reported in Europe lasted for one and two to even three years and more. This phenomenon might recur; for example, in the 14th and the seventeenthcentury, local epidemic recurrences were observed every few or dozen or so years in various locations of Europe (e.g., London, Marseille). Given the repeatability of the phenomenon, the problem discussed in this article still is and will continue to be topical. New waves of the epidemic will probably emerge in the years to come. The organizational solutions used contemporarily to weaken the epidemic spread dynamics include teleworking, telelearning, and teleshopping. These are all means to isolate the potential sources of infection. The article's authors have linked the various forms of teleactivity with the overall body of problems related to the consequential change in transport behavior patterns. They claim that the

I. Celiński · G. Sierpiński (✉)
Department of Transport Systems, Traffic Engineering and Logistics, Faculty of Transport and Aviation Engineering, Silesian University of Technology, Krasińskiego Str. 8, 40-019 Katowice, Poland
e-mail: grzegorz.sierpinski@polsl.pl

I. Celiński
e-mail: ireneusz.celinski@polsl.pl

N. Rezaei (ed.), *Multidisciplinarity and Interdisciplinarity in Health*,
Integrated Science 6, https://doi.org/10.1007/978-3-030-96814-4_17

spatial distribution of places where teleactivity occurs should contact the transport system's parameters and the traveling behavior patterns observed to date. Concerning the preceding, a theoretical model for modal split and selection of the spatial distribution of places considered convenient for teleworking and telelearning has been proposed in this paper. The model is linked with adequate transport behavior changes in terms of the problem of transport sustainability. What the authors have taken into account in the model in question is the parameters of the transport network, traffic, traveling population's transport behavior patterns observed.

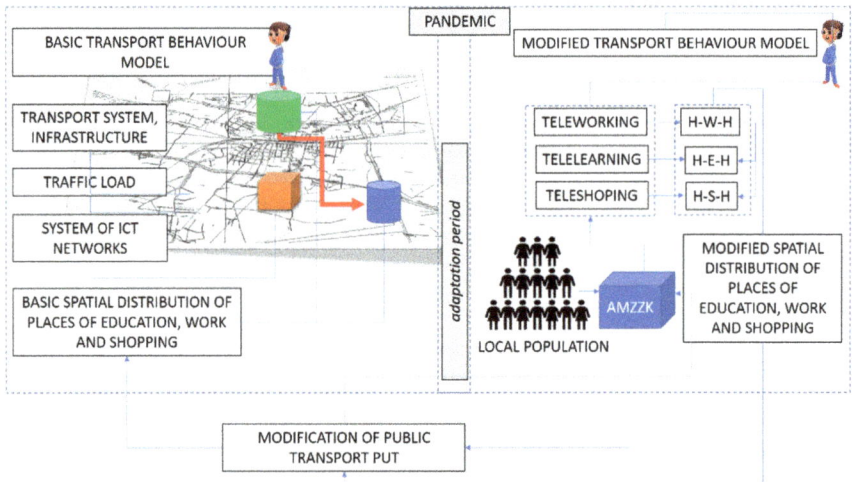

Simplified model diagram: H-Home, E-Learning (education), and S-shopping

The code of this chapter is *01101111 01,100,011 01,101,000 01,100,101 01,101,110 01,111,001 01,010,100 01,101,100 01,101,111 01,100,111.*

Keywords

COVID-19 · Pandemic · Travel behavior · Teleactivity · Telework · Upper Silesian Voivodeship

1 Introduction

In the spirit of Chinese philosophy, a black and white distinction between dependent and non-dependent things is debatable. Scientific development is an ongoing effort to change this distinction whether an epidemic threat is such a case in which a human can do something? In the old days, there was little he could do beyond insulation and hygiene. Today, there are opportunities on the horizon apart from insulation. It is up to us whether further activities can be carried out in a stoic way or global hysteria conditions.

Further, increase of the epidemic risk is predicted to occur throughout the next two decades [1–16]. The coronavirus disease (COVID-19), which currently poses the main issue, is one of many threats facing contemporary societies in this sphere [3, 17, 18]. One can also expect secondary epidemic outbreaks [19–22]. The current crisis in Europe has revealed that the social and economic structures are only moderately prepared for the hazards of this kind. This is particularly true of transport, which virtually froze completely at the peak of the epidemic in areas such as air transport or passenger rail transport in international connections [23–26]. Collective public transport of agglomeration and regional range has proved to be only slightly better suited for the situation [23–26]. Besides being forced to adhere to the organizational and administrative restrictions imposed relatively quickly on the use of public transport (number of occupied places, number of runs, closed schools, and some workplaces), people began to perceive this mode of transport as a threat to their health and life. Only, the perspective of job loss has forced them to use these means of transport in cases where they cannot use individual transport [27–29]. Recent reports from China indicate that the second wave of infections can strike quickly [30, 31]. Secondary outbreaks in China emerged at a considerable distance from the primary ones [32].

In Europe, the recorded epidemics usually lasted from one to two years to three or even eight years in the last millennium. The most widespread and most prolonged of all the recorded epidemics took eight years. The Black Death epidemic struck almost the entire Europe in the years 1346–1353. For the seventeenth century epidemics, local recurrences were reported with the frequency of several up to a dozen or so years [33–38]. Such events were observed equally often in antiquity, e.g., in 431 BC, the epidemic referred to as the Plague of Athens broke out. This implies that such events have taken place systematically throughout the entire known history of humanity.

What appears to be particularly important in the context addressed in this article is diverse epidemics involved a significantly different level of development of the transport system. In times dominated by navigation, the Plague of Athens was brought from Ethiopia, and its sources on the European continent were reported in the port of Athens, Piraeus. Consequently, as in most cases, the disease transmission media were the intercontinental (land and sea) transport routes. Over short distances, the disease is also spread by "channels" linked with the transport system.

Insofar as hundreds of years ago, this was a slow process that took weeks. Nowadays, given the current level of development of high-speed railways and international aviation (or national in the USA), the epidemic (its secondary sources) may spread to any point of the globe over one day. In the past, when journeys took many weeks, the death of the infected people was possible to spontaneously extinguish a potential new outbreak during a long and tedious travel. Nowadays, the infected can make intercontinental journeys whose duration is shorter than the onset of the disease. The epidemic outbreaks thus propagated between continents are further replicated locally via regional and agglomeration-wide transport systems, being transferred to persons involved in socioeconomic activities. Contemporary transport behavior patterns do not take such risks into account. So far, people have freely used the available means of public transport while resorting to telework under completely different conditions, originating inside their organizations instead of being imposed under administrative regulations related to infectious diseases. The main threat connected with public transport was traveling safety (road incidents and accidents, assaults, theft). However, the reality has changed drastically and irreversibly over the recent months. Public transport poses a threat that is extremely difficult to tackle [39–41].

Some of the contemporary solutions applied to eliminate or slow down the dynamics of the spread of infectious disease include teleworking, telelearning, and teleshopping [42–44]. Such teleactivities limit physical contact between people, thus eliminating or reducing the dynamics of disease dissemination. This makes it easier to manage hospitalization, boiling down to a smaller number of patients. On the one hand, teleactivity minimizes the number of contacts at workplaces (especially in large companies), and on the other hand, the number of contacts in means of public transport. While the contacts in the former case are typically limited to those between persons familiar with one another, and so are relatively easy to track (potential quarantine), in the latter case, one must deal with considerable randomization of this process, especially in non-work and non-education-related transfers (other motivations). Moreover, not all public transport users and persons of different groups use personalized smart cards in transport, so they are generally difficult to identify [45–47]. In this way, infectious diseases can spread randomly via means of public transport. Consequently, isolation is the only effective form of combat against the spread of a virus in the absence of efficient vaccines (at the time when this paper is being written, such vaccines are being introduced into the market; however, their application properties and efficacy over the population are not yet known) [48–51].

Contemporary means of transport are in no way prepared for such epidemic threats. They are not analyzed in terms of the organization of public transport lines and other service parameters. Open-rolling stock interiors with a single air-conditioning circuit are the predominant solution. They also lack efficient means of disinfection for contact surfaces. All this makes problems addressed in this paper legitimate and topical, both for the present and probably for the next few decades. Passenger counting and filling ratio control in means of transport leaves much to be desired (according to the authors' own observations) despite their being equipped with appropriate counters. The obligation to wear protective devices in public places is not respected by everyone (apart from whether or not it is legitimate in specific cases).

Therefore, one should examine how transfers can be organized under the changed external conditions arising in the transport system's environment (epidemic threat). Taking the seriousness of the current situation into account, such actions should consider the sustainability of transport at the same time, while the applicable administrative restrictions, if justified, should be respected.

Concerning the problems of the epidemic threat mentioned above, the authors of this article have developed a model for the division of transport-related tasks and transport behavior changes. These changes are a byproduct of the epidemic threat and implementing the teleactivity during the pandemic. What is being created in relation to the epidemic, but not with the transport system specifically in mind (even in detachment from the changes in this system during this period), is a certain spatial distribution of places where teleworking and telelearning are performed. This distribution affects the traffic parameters in the transport network.

It constitutes a specific preselection modal split: a split between people who are immobilized and those who have retained the privilege of mobility. Assuming such a perspective, the division into means of transport in the mobile people group is subsequent.

As approached in this article, the resulting change in transport behavior has been linked with transport sustainability, which can be inhibited or even thwarted by the pandemic. Over the last two decades, motivated by sustainable transport development, the outflow of passengers from public transport was successfully contained in many agglomeration areas [52–55]. The effect achieved on certain routes was even that of passengers abandoning individual transport in favor of public transport [56]. In this way, the trend that appeared in the 1990s and continued ever since has been reversed in most cases. It is particularly the case at locations where the time of traveling by means of public transport has been reduced compared to that of traveling by individual transport means [56]. However, the pandemic threat creates a potential for the results of several decades of work performed by promoters of the transport sustainability idea and by the institutions which organize and finance public transport to be thwarted in the coming years. It also entails a risk of squandering considerable financial resources accumulated and used for sustainable transport development. Due to the possibility of contamination in public transport, many people perceive individual transport as safer. One can also expect that the sales of new and used cars will rise in the coming months, following a temporary crisis, and people will start abandoning public transport in massive numbers. Similarly, an increased vehicle sales rate has been observed in recent years after a temporary collapse [57]. This trend may become particularly evident, bearing in mind that an outbreak of viruses more lethal than the present one can be regarded as probable in the future, even considering the improvement in hygiene over the past few decades. This time, it is evident that the temporary improvement in terms of hygiene is not permanent and has not been implemented for good in all public facilities.

The scenario assuming the future unsustainable transport behavior changes may be limited by appropriately designing the teleactivity in combination with a legitimate public transport offer. This means that the measures aimed at triggering the behavioral change from the activities performed in a stationary manner in favor of the teleactivity and the measures related to the public transport offer modeling should be linked, starting from the stage of the preselection split of travel to the basic modal split stage. In this context, not without relevance is also the road network parameters. With the current situation in mind, certain cross-sections may be overloaded despite the general drop in global traffic volumes. The same applies to information and communications technology (ICT) networks that are not adapted to the observed changes, especially when considering e-learning accounted for 100% of the population in the age group subject to the relevant regulations (all schools and universities being closed). It may explain specific problems in certain locations: Many e-learning courses were either being limited or suffered functionality problems, and there was network congestion related to individual online activities [58–61].

This chapter discusses some solutions to the problems connected with the emergence of this kind of scenario: problems involving abrupt changes in the transport behavior patterns among the inhabitants of agglomerations and metropolitan areas during the pandemic. They trigger changes to the modal split resulting from the teleactivity and the outflow of traffic streams from the public transport. In this context, the problem of international transport, whose nature is different, has not been addressed.

The remedy proposed by the authors is a specific model enabling changes in the observed transport behavior patterns to be shaped based on the distribution of places intended for teleworking and telelearning (preselection split). The said model considers the parameters related to the transport network, the traffic in this network, and relevant ICT networks, as well as the currently observed transport behavior patterns. The model has been tested in a simplified form using sample input data from the Upper Silesian conurbation.

Graphical Abstract is a simplified flowchart of such a model. The model algorithm, referred to as AMZZK or legitimate transport behavior change model algorithm, allocates the places intended for teleactivity, which is effective from the transport system perspective (preselection split, also taking the ICT system parameters into account). Otherwise, such changes in the distribution of activity in the socioeconomic system will result from decisions that disregard the transport system. What is required in both cases of the transport behavior changes (either legitimate and not relevant to the transport system) are the data representing the rolling stock parameters (filling ratio of means of transport) and the offer of transport services in terms of the number of runs. The crucial difference is that the AMZZK algorithm takes the transport system and other networks (ICT) in the infrastructure into consideration. In contrast, the transport behavior-related decisions currently being observed (direct effect of the teleactivity location) are made in detachment from the altered transport system characteristics. This problematic state of matters will not surface until students massively return to schools and people to workplaces in the second half of 2020, facing no significant collective transport organization changes.

The authors' observations imply that even now (June 2020), it would be difficult to adhere to the applicable restrictions in the use of public transport. Distances of 1.2 m are not and will not be maintained. This situation should also be compared with the five persons per square meter standard applied so far. Assuming the social distance of 2 m (which is nothing drastic at all), the maximum occupancy ratio drops 125 times! It decreases 45 times for the distance of 1 m! Moreover, even these distances cannot be maintained during passenger exchange. It makes public transport dysfunctional.

By analogy with the European history of epidemics over the last millennium, it can be claimed that the problem remains valid regardless of the current situation. It may also happen that further epidemics (or other catastrophes, e.g., climatic, war, economic collapse) are much more dangerous to the existence of contemporary societies (e.g., the Asian flu viruses of the turn of the previous century). Considering this aspect, one should be prepared to shape the future transport behavior patterns by taking all this into account, as well as the specificity of the threat and its impact on how the transport system functions, putting special emphasis on collective transport. The gist of the problem is that the shift of circumstances between the normal and the pandemic situation may either be abrupt or too slow in public transport adaptation.

2 Historical Outline (Context)

Archaeological and written sources document the history of pandemics. They have been observed on Earth for at least several thousand years. The proofs of their emergence include mass graves uncovered in archaeological works. Unlike the graves associated with armed conflicts, the gender and age distribution of those buried at such sites is different. What also strikes is the intentionality of cremation of the corpses. One of the first recorded epidemics is the one that struck the territory of contemporary China ca. 5000 years ago [1]. In addition to the demographics of the deceased, the cremation of the corpses, as well as the type of their physical damage, also the discontinuation of the settlement process in such a place in consecutive years (periods) speaks for the non-military nature of the deaths, apart from the fact that armed conflicts also accompanied the development of various epidemics in the world's history. An example is the Plague of Athens of 471 BC mentioned above. In this case, written sources imply the disease-induced nature of the deaths (related to the plague and not directly to the armed conflict in the Peloponnese). This plague is confirmed to have been brought to Athens by sea (with individual outbreaks initially appearing in Piraeus's port), probably from the source of present-day Ethiopia [62]. Another pandemic known from the written sources is the Antonine Plague of 165–180 AD, which spread over ancient Rome [63]. It was brought from Asia (the Middle East) following the military campaign of the Roman legions. Unlike in the two previous cases, the number of victims was counted not in thousands and hundreds of thousands, but probably in millions (up to 2,000 inhabitants of Rome alone died every day). Characteristically, the plague recurred with an average frequency of 4 to 8 years in this period. Subsequent major plagues appeared in cycles of several decades. In the years 1346–1353, Europe was plagued by a disease referred to as the Black Death, which, like the previous epidemics, emerged in Europe due to transport to and from Asia. There are diverse estimates concerning the plague's victims, but some imply its mortality to the range at 60% in some regions of Europe. The fact is that it took another 200 years for Europe to recover its economic potential. Another known pandemic emerged in Central America in the years 1545–1548, and its toll was counted in many millions as well. Yet, another plague struck London between 1665 and 1666 and Marseille in 1720–1723 (these are only some of the places where it occurred). As the epidemic often emerges in port cities (Piraeus, London, Marseille, Philadelphia), this highlights transport as a rather obvious context of its cause. The epidemics whose victims count in thousands and millions repeat themselves in cycles of several decades (influenza, polio, etc.). In more contemporary times, these were the following successive epidemics:

- the Spanish flu (1918–1919);
- the Asian flu (1957);
- the Hong Kong flu (1968);
- the AH1N1 influenza pandemic (2009–2010); and
- COVID-19 (2020–?)

Among the above epidemics, the Spanish flu, also associated with the transport context, is believed to have been brought to Europe with the Allied troops that arrived in Europe connected with the ongoing armed conflict [64]. Similarly, the current COVID-19 pandemic came from China, mainly by air transport to Italy, from where it has spread across entire Europe. The characteristics of the epidemics recorded so far are as follows:

- extensive area coverage resulting from international trade contacts or armed conflicts (typically of a large scale) and a large number of deaths recorded in total and per day;
- high infectiousness reflected in rapid disease spread, proportional to the speed of movement of the currently available means of transport;
- transport context of the disease with the spread reported in the documents in almost every case, e.g., port cities or towns situated on trade routes, movements of large groups of people associated with military conflicts (with troops moving between continents);
- duration of many years, ranging from 1 to 8 years, with secondary waves/outbreaks;
- the incubation time of pathogens transmitted through transport makes it difficult (maybe impossible? in a big perspective) to control.

3 Teleworking and Telelearning: Transport Connotations

An efficient way to combat the spread of contagious diseases is to isolate the infected persons (hospitalization, quarantine) and those exposed to the disease (quarantine) from the rest of the population or perform large-scale vaccination. Isolation (preventive, in this case) also applies to healthy people living in large communities on a daily basis (workplaces, schools, shops, kindergartens, crèches, etc.).

Vaccinations are problematic for several reasons. There may be a temporary shortage of the vaccine profiled for a specific new (unknown) infection source, lasting until it is created. It may take years to develop an effective vaccine within a reasonable time frame, or it may never be created at all [48–51]. Vaccination campaigns will additionally encounter some social resistance [65, 66].

Moreover, people in contact with the pathogen can use public transport in the asymptomatic stage of the disease. Additionally, given the current safety measures and different approaches of transport administration bodies to this problem (the matter of compliance with administrative regulations, e.g., the 1–2 m social distance mentioned above), sick persons may be present in transport. For all these reasons, the isolation using teleactivity is and is likely to remain the most effective means of addressing such risks. Therefore, these activities must be synchronized with the transport system's parameters, especially if it changes considerably, for example, in terms of the public transport offering.

In the contemporary means of transport, such isolation is particularly difficult to achieve in the absence of compartments in the rolling stock (mainly open interiors) and the closed air circulation/conditioning systems. Therefore, a potential non-transport solution introduces various types of teleactivity, such as teleworking, telelearning, teleshopping, tele-entertainment, and tele-x (unknown (nowadays) activity related to transport).

No carrier is basically in possession of a rolling stock capable of providing transport services in time of the epidemic on a level comparable to that before the epidemic. Any restrictions related to passengers' distance reduce the filling ratio geometrically (if they are to be interpreted literally). The space in the vehicles such as buses and trams is not suited for such actions. This is directly due to the nature of the propagation of pathogens and the design of the contemporary means of transport.

The problem with implementing various forms of teleactivity during a pandemic is that such solutions disrupt the transport behavior patterns observed formerly and do not take the transport network parameters into account, including the existing and legitimate division of transport tasks. A significant part of the current transfers is canceled, but also:

- their frequency changes, while the transfer process itself (in a reduced number) is often stochastic (e.g., management on duty, work schedule changes);
- the times of the day when travels are performed change (changed transport peak hours, work schedules, equally often the place of work changed to a different company office);

- atypical, mainly local, new transfer chains develop (e.g., with limited access to a shop, when looking for a smaller queue, when looking for a place to park);
- the priority of interaction between different means of transport changes (empty trains block individual vehicles at level crossings, redundant on a working day personal car block other personal cars, moving every day, etc.);
- buses use separate lanes, while their occupancy ratio is only slightly different than that of individual vehicles, a complete contradiction to the idea of sustainable transport development; and

In a nutshell, it all boils down to a "controlled chaos" emerging, where its control is only of formal nature. For instance, finding a parking space is related to the global traffic reduction in the network. Most vehicles are permanently parked (Stay Home campaign) [67]. Similar are the problems of air transport and non-flying vehicles [68].

The spatial distribution of teleworking and telelearning takes the socioeconomic network parameters into account (traffic aggregators and generators in the network) but not those of the transport network or the ICT network. What also matters in this context is the problem of the shopping place changing. During an epidemic, entire attractors in the socioeconomic system are closed down, while the generators are left operational. In this sense, the traffic structure changes dramatically as it ceases to be symmetrical in time and space. At the peak of the restrictions, this poses no significant problems (they may be unusual and local), while everyone stays home in isolation, but real problems begin with the economy being revived, and the education resumed. From the public transport perspective, conventional education's resumption will be particularly critical (September 2020).

The restrictions imposed on transport means during the pandemic are arbitrary [69–71]. Pathogen propagation models indicate that pathogens travel beyond the safe distance range recommended by public transport administration bodies [72–76]. Consequently, the social distance ranging from 1–2 m is insufficient. Some models imply pathogen propagation over distances as large as 10 m or even greater, and masks reduce these distances. Time for the aerosol to float is another problem, especially in public places.

Apart from the fact that not even these implications are reflected in the public transport organization regulations introduced after the pandemic outbreak, there are simply no efficient solutions for public transport nowadays. When passengers exchange, it is completely impossible to maintain any distance-related standards with the current process implemented. Therefore, in this aspect, the transport network can be adapted to the distribution of the teleactivity-related places and the locations associated with the base state activity (predating the epidemic outbreak, regular). As proposed in this article, an alternative approach is to adapt the teleactivity to the capabilities of the transport and ICT networks. The ICT network's existing infrastructure is not 100% prepared for the e-learning of the school-age population. For example, a person formerly driving an individual vehicle who has switched to teleworking during the pandemic will no longer perform home-work-home transfers as well as the related one. Alternatively, they will occasionally perform them on days and in times of the day that are difficult to predict (on-duty work, maintenance of technical equipment, etc.), regardless of whether they previously used to travel by individual or public means of transport. Such a person is probably also going to start shopping at another place if they used to do this on the occasion of commuting to work. This change will also occur irrespective of whether or not this person was traveling by private or public means of transport so far. In the former case, the variation of transfers is more flexible. Besides the home-work-home transfers, the home-other-home type of transfer will change. Transfer chains will be simplified, shortened, and eliminated, depending on the nature of administrative restrictions. A person commuting to a place of education by public transport is likely to completely "disappear" from the transport network for the time of remote learning. The network's transport congestion peaks may also change, mainly due to schools and large production facilities being closed. In this aspect, a separate problem is the development of pedestrian transfers over the pandemic period in densely populated urban areas [77]. The only solution in this respect is personal protection devices, such as masks and scarves. The problem which remains, however, is the necessity of touching impulse motion sensors, etc. Therefore, during a period marked with restrictions, it should be expected that significant changes in transport behavior patterns will primarily apply to the following motivations: home-work-home, home-education-home, and home-other-home. The daily and weekly traffic volume characteristics will change significantly. The latter will concern local road networks of poorer parameters. However, these changes will primarily consist of reducing transfers or shortening the chain in terms of the obligatory routes. Similarly, a certain group of routes related to professions responsible for maintaining production continuity, ensuring safety, or health care must be maintained and cannot be moved outside the existing transfer chains. Therefore, the primary division will differentiate between socially required and non-required transfers (these not being equivalent to the obligatory transfers).

Several different research problems arise in this context:

- What is the relationship between these behavior patterns and the traffic load on the transport network (individual transport), especially in the context of a shift between means of transport, from public to individual, as envisaged for many users (already a considerable share of road intersections function at the verge of or even beyond the threshold of their capacity)?;
- What is the relationship between these changes and the reduced filling ratio of the means of public transport, including those comprising multimodal services? In this respect, canceling individual routes requires individual carriers' coordination due to the necessity of en-route line changing. These changes consider the imposed organizational restrictions that may become more stringent in light of the recent studies [78, 79]. The social distance of 1–2 m may be insufficient, not to mention that it is typically not maintained in buses, trams, and trains. Moreover, there are no changes currently envisaged in the sphere of structural solutions of public means of transport assumed to address the epidemic threat. The filling ratio is only being restricted geometrically (if strictly and literally adhered to);

- How does traffic load change, and how will it change in the successive months to come in main and local roads (probably shift of part of the traveling population streams from public transport to individual transport, the latter being perceived as more secure during the pandemic)? Given the overall traffic volume reduction during the pandemic, the changes in transport behavior patterns may cause overloading of specific road network cross-sections (in the vicinity of large-format stores, car parks, pharmacies, hospitals, etc.);
- What is the difference between the transport behavior patterns during a pandemic and those observed outside such a period, and not only concerning volume but also in terms of distribution in time?

Figures 1, 2, 3 and 4 illustrate selected problems related to the process of changing stationary activity into teleactivity. First and foremost, numerous transfer chains, including indirect ones, will be reduced or altered. What will also change with the commuters abandoning public transport is the load on specific road network cross-sections. Most important or even critical for the road network condition will be the change in the modal split. The local road network load characteristics will change as well. Everything will depend primarily on the main three factors: i) the spatial distribution of teleactivity; ii) changes in the offer of public transport services; and iii) development of the pandemic.

There are further factors in the background, such as the current changes in the road network and ICT networks, which also affect the problem addressed in this article. As of now, neither road nor ICT networks are adapted to the present situation. Let us also consider the seemingly trivial problem of car parks, for example, where all residents are forced to park nearly 24 h a day in the time of the pandemic. Outside this period (pandemic), a significant part of the vehicles is usually in motion instead of occupying the parking space.

Fig. 1 Schematic illustration of the problem occupancy limit and doors bottleneck

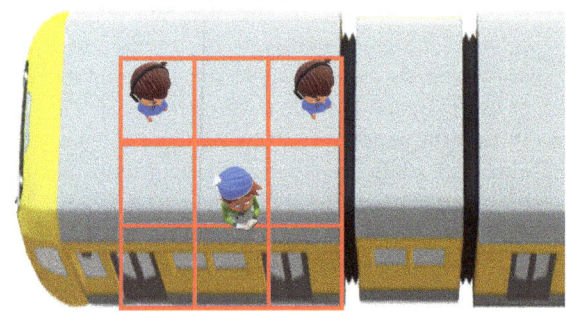

GEOMETRIC OCCUPANCY RATIO REDUCTION

DOOR AS THE SYSTEM'S BOTTLENECK

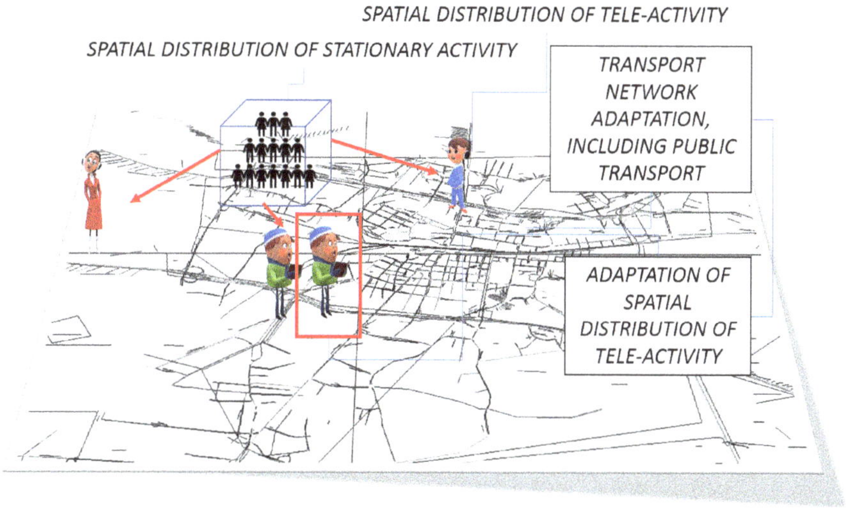

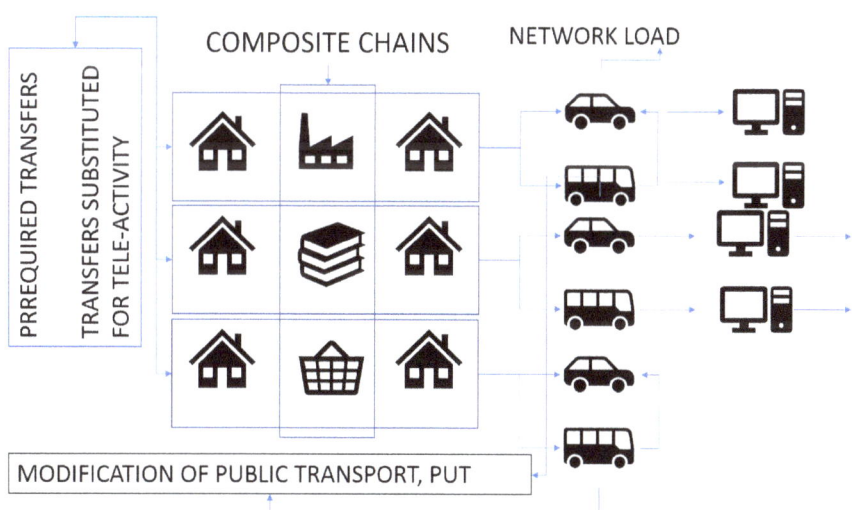

Fig. 2 Different approaches to the problem (two stages approach)

Fig. 3 Problems of the pandemic time—alternative concept

As the experience of the last few months (March–June 2020) has shown, during the epidemic, the range of services offered by public transport entities operating in the area of both rail and bus transport will also change (reduced number of runs and permissible filling ratio of the rolling stock). This alone means that the former transport behavior patterns cannot remain to be valid in the existing framework.

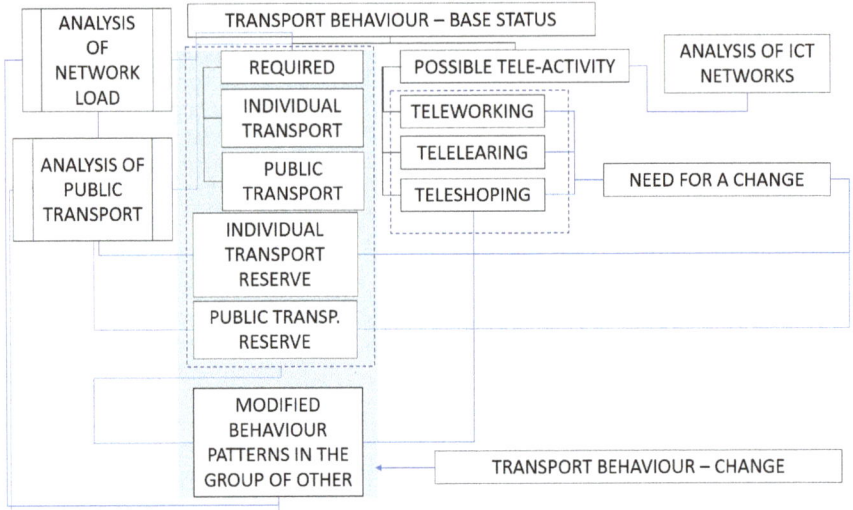

Fig. 4 Diagram of analysis of the legitimate transport behavior change model algorithm (AMZZK)

The change in the offer concerns the number of seats in the rolling stock considered safe by the public transport administration body. This is yet another problematic aspect that may change significantly in the next few years.

The problems that may arise in the sphere of individual transport are connected with the road network load on selected routes, additional parking-related problems, etc. There will be a local change in the traffic distribution over the network (the traffic structure will be altered at some junctions). In connection with teleworking and telelearning, some atypical road network loads may emerge in places where large stores are situated, near hospitals, and in other characteristic places, even if the total traffic volume across the transport network becomes significantly reduced.

4 Input Data

The model proposed for analyzing and shaping the changes in the transport behavior patterns during a pandemic should take various data into account. The model's input data are transport network infrastructure data, traffic data, ICT network data, data on the offer of public transport services, and the traffic behavior patterns observed so far. Some additional data that may also come into play concern the spatial distribution of disease outbreaks, etc. This type of analysis is difficult to perform due to the dynamics of these processes. It requires complex models to be developed.

The transport network-related data concern the network of road, rail, and other connections which can be used to perform transfers during a pandemic. It should be noted that, in critical situations, part of the network may even become closed for traffic or be subject to other significant restrictions affecting traffic parameters (e.g., in the vicinity of hospitals, stores, etc.). What also proves to be important in this context is the following data on the traffic in the transport network: traffic volumes in individual cross-sections of the network, number of launched runs of public transport means, permissible number of places (allowed filling ratio), actual filling ratio (in practice, the counting of passengers is not performed after the filling has changed). Another important problem, which has only been touched upon in this paper up to this point, is the data on the ICT networks. These data influence the possibility of effective implementation of teleactivity, and so, they should also be taken into account in the analysis of the changes in the transport behavior patterns during a pandemic.

Public transport data have not been included in the network traffic data. Nevertheless, their importance in the future should be stressed. In case of a considerable threat, it will not be possible to adhere to the existing standards (every second seat occupied by a passenger). Traffic organization will also have to be adapted to the actual demand during a pandemic. Therefore, the right way to proceed is not to reduce the number of runs but rather adapting the timetable to the current needs in a completely new manner. Many people have changed their behavior (by changing their work to another company seat, working on an on-duty basis, changing work times during a day, etc.).

The data concerning transport behavior patterns include the actual parameters of people's transfers during the pandemic, which differ from those made when they perform a work-related activity in a stationary mode outside the pandemic period. Knowing the transport behavior data from before the pandemic period, they can be compared with each other. Accordingly, one should determine which transport behavior patterns observed so far should be removed from the network based on the imposed administrative (by government, local authorities) and organizational restrictions (by carriers). This is done by introducing teleworking or telelearning solutions corresponding to these transport behavior patterns and ICT networks and local transport networks' capabilities. Among other transport behavior patterns discussed, one should consider how many of all the transfers can be performed by the modified means of transport adapted to the pandemic period and with a limited number of runs, and how many will be substituted for individual means of transport. In both cases, there may be drastic limitations. In the latter case, one should consider the available road network capacities (given that they have already been exceeded in many cross-sections). A schematic representation of the analysis and the modification of changes in the traffic behavior patterns during a pandemic has been provided in Fig. 4.

4.1 Shortened Model Proposal

A shortened model for a change of the transport behavior patterns in terms of the description of variables has been proposed below. The transport network data are most conveniently described using a graph containing peaks (junctions in the transport network) and arcs (road connections):

$$G = (V, E).$$

V is a set of road network junctions, and E, a set of road connections in the transport network.

Every arc in the graph describing the network has predefined flow capacities and traffic data:

$$C = \{c1, c2 \ldots, c_i. \ldots\}$$

where parameter c_i describes the flow capacities at individual arcs.
The following are the parameters that describe traffic in sections:

$$Q = \{q1, q2, \ldots, q_i. \ldots\}$$

where parameter q_i describes the traffic volumes at individual arcs.
The ICT networks-related data should be spatially aggregated in such a manner that area of the analysis is spatially delimited, where each of the spatial territories is to be described by the number of connections for which the capacity for effective teleworking or telelearning is to be provided:

$$N = \{n1, n2, \ldots, n_o, \ldots\}$$

where parameter n_o describes the number of places available for teleworking and telelearning in individual territories (o-number of the spatial regime).
The data concerning public transport should be described in a form appropriate for the timetable organization. To attain the right level of detail of the description, one should consider all lines, the runs launched for individual lines, and the maximum permissible occupancy ratios.

$$L = \{l1, l2, \ldots, l_p\},$$

l_p is the number of public transport lines subject to analysis, p, the number of lines.

$$K = \{k1, k2, \ldots, k_r\}.$$

k_r is the number of runs analyzed for corresponding public transport p lines;

$$N = \{n1, n2, \ldots, n_s\}.$$

where, n_s is the number determining the permissible occupancy ratio for the public transport lines analyzed and for the given run.

Noted that this is one of the most critical parameters of the transport system and method in the pandemic period.

Another element is to acquire data concerning the transport behavior patterns: set of transfers for each network user:

$$Z = \{z1, z2, \ldots, z_x\}.$$

where x—consecutive transport network user,
description of individual transsfers for consecutive transport network users:

$$DP = \{x, y, \text{ts,tk,m,ma,mi1,mi2,t1,t2,mot}).$$

where:
x—transfer's start point (WGS84),
y—transfer's end point (WGS84),
ts—transfer start time [Coordinated Universal Time (UTC)],
tk—transfer end time [Coordinated Universal Time (UTC)],
m—means of transport used (PC, Bus, Tramway, Bike, etc.),
ma—alternative means of transport (@PC: Taxi, Bus, Tramway, Bike, etc.),
mi1, mi2—consecutive means of transport in a multimodal transfer chain (e.g., PC $\rightarrow$ Train $\rightarrow$ Plane),
$t1$, $t2$—acceptable change waiting times, from-to [Coordinated Universal Time (UTC)],
mot—motivation (h-w-h, h-e-h, h-o-h, etc.).

It is crucial to divide the activities formerly performed at the absorbent locations into the necessary and unnecessary ones during a pandemic. The data on the necessary (essential) work are linked with the definition of specific persons and the transfers typically performed by these persons, and they are necessary to maintain the efficiency of the socioeconomic system. These include the occupations which are responsible for the supply and production of basic food products (bakeries, meat factories, beverage bottlers), but also those which represent work related to safety and health care (hospitals, clinics, police), maintenance of the power grid and water supply systems, etc. This can be noted as follows:

ACT = NECACT + TELE-ACT
where:
NEC-ACT—necessary activity (hospital staff, policeman, fireman, etc.)
TELE-ACT—teleactivity, all not requiring direct presence at the workplace.

Such a breakdown is relatively simple to perform using central databases (personal identification numbers—PESEL) and other data sources (statistical business numbers—REGON).

What one should also establish is what other, besides the aforementioned, transfers are necessary (permanent health care, rehabilitation, social assistance, deliveries, etc.).

Then, the necessary transfers should be distributed over the transport network and the means of collective transport (what matters in this respect is the proportions of the modal split, i.e., how many transfers are to be completed by individual means and how many by collective transport). This distribution is implemented depending on the permissible filling ratios of the public transport means as well as the potential road network load. The remaining transfers can be distributed by keeping a reserve in the road network and the filling ratio of the means of public transport.

The above preselection and modal split solve the problems with the division of the necessary transfers and their distribution over individual and collective transport networks. Assuming that these transfers' adequate parameters are maintained, the remaining transfers, which used to be made before the pandemic, may be qualified as intended for teleactivity (work, education), provided that the parameters of the ICT networks allow it. The gist of the problem is that such a distribution can cause local disturbances in the road network. Persons engaged in teleactivity will perform their transfers by following the home-other-home motivation near their residence place, thus generating a load differently from before, affecting the road network's local fragments (perhaps also the railway network). The foregoing particularly applies to some non-standard periods in which these transfers will be made.

It is also critical to use the reserve in terms of the permissible filling ratio of the means of public transport and the road network's capacity. In the former case, it is known that the restrictions imposed so far are inconsistent with the models of pathogen propagation by an infected person [72–76]. What appears to be the threat in the case of the road network is the depletion of the flow capacity in selected cross-sections on a large number of travels shifted from public transport to individual transport. In this context, it should be noted that such transport may also be preferred by the administration in times of a pandemic. Moreover, such preference makes sense when the filling ratio of each personal vehicle (PC) is one, or when vehicles are only used by persons from families living together in shared households. The carpooling and carsharing schemes become groundless without introducing drastic hygiene standards to be followed between consecutive passenger exchanges. Therefore, it is clear that the pandemic has been changing almost everything in transport networks.

Such transfer behavior patterns have not been studied during a pandemic yet. It is not known what percentage of routes has changed for specific transfers. This is because the role of individual transport has increased, perhaps because of relatives and family members being driven around more frequently. Hence, there is the need for a survey of the modified behavior patterns in order to draw reasonable conclusions.

In summary, the actions comprising the method proposed by the authors can be schematically brought down to the following:

- taking into account the changes and limitations of public transport during a pandemic, changes in the number of runs, routes, restrictions on the permissible filling ratio, changes at the time of passenger exchange, etc.;
- breakdown of travels into necessary and permissible;
- limiting the set of permissible travels by analyzing the capabilities of ICT networks;
- network loading due to the necessary and permissible travels which cannot be replaced by teleworking and telelearning against individual travels and travels made using public transport;
- calculating possible reserves in the network of individual and public transport after covering the necessary travels and those which cannot be covered using the ICT network resources, and utilizing them for purposes of other motivations; and

In this respect, other motivations are assigned a lower priority than the necessary transfers and are not covered by the ICT network resources.

5 Conclusion

The chapter outlines the context of transport issues in the pandemic period. The problem discussed in this article is interdisciplinary in nature (biology, physics, medicine, sociology, psychology, vehicle construction, etc.) and so is transport itself. Besides the strictly transport-related problems addressed in this article (transport behavior, modal and pre-selection split, transport network, etc.), it also touches upon some medical issues (model of pathogen propagation in means of transport, protection of the traveling population in means of transport, movement of sick people whom medical services cannot transport due to the shortage of the available means), the capacity of ICT networks, as well as sociological and psychological problems related to the sense of fear in times of danger. The latter is particularly important because of people's accumulation in small and closed areas in means of transport and mass service points [80–83].

In the sphere of transport, the problem concerns the design of transport means, traffic scheduling problems, optimization of public transport lines, shaping a substantiated modal split, shaping and studying altered transport behavior patterns, building new traffic models under changed conditions external to the transport network.

In conclusion, an obligatory set of tools and solutions should be proposed for the modified (intended) breakdown of transport behavior patterns during a pandemic, particularly since secondary and next waves of the disease are expected, or a new threat may emerge. They are as follows:

- change in the organization of the public transport network, modification of lines, stops, and the number of runs (causing a forced change in transport behavior patterns);

- structural change of means of transport including in-vehicle isolation of passengers (per compartment);
- changing the passenger exchange process;
- autonomization of vehicles (elimination of drivers); and
- introducing an event control and recording system.

History implies that this kind of threat tends to come back, so now is the time to introduce changes in the organization of public transport and the design of the means of transport. Both these efforts will trigger a change in transport behavior patterns. Only, the next epidemic will verify society's preparation for this type of threat. In particular, in the context of the preparation and adaptation of transport systems on this issue, it is not difficult to imagine a much worse scenario than the present one (Historia Magistra vitae).

Core Messages

- The history of the epidemics shows the repetition of the phenomena.
- The lack of preparation concerns very different elements of the transport process, e.g., the passenger exchange.
- The important thing is that the current situation is destroying decades of efforts to develop sustainable transport.
- A partial solution to the problems is the optimization of the spatial distribution of teleactivity.
- This distribution of teleactivity makes it possible to adjust the needs in terms of the current movement.

References

1. Jarus O (2020) 20 of the worst epidemics and pandemics in history. Live Science. Available via https://www.livescience.com/worst-epidemics-and-pandemics-in-history.html. Accessed 26 July 2020
2. WHO Coronavirus Disease (COVID-19) Dashboard (2020) World Health Organization. Available via https://covid19.who.int/. Accessed 26 July 2020
3. Susskind D, Manyika J, Saldanha J, Burrow S, Rebelo S, Bremmer I (2020) How will the world be different after COV-19? Finance Dev 57(2):26–29
4. Ross AGP, Crowe SM, Tyndall W (2005) Planning for the next global pandemic. Int J Infect Dis 38:89–94
5. Gostin LO, Friedman EA (2015) A retrospective and prospective analysis of the West African Ebola virus disease epidemic: robust national health systems at the foundation and an empowerd WHO at the apex. Lancet 385(9980):1902–1909
6. Jones KE, Patel NG, Levy MA, Storeygard A, Balk D, Gittleman JL, Daszak P (2008) Global trends in emerging infectious diseases. Nature 451:990–993

7. Alirol E, Getaz L, Stoll B, Chappuis F, Loutan L (2011) Urbanisation and infectious diseases in a global world. Lancet Infectious Dis 11(2):131–141
8. Ross AG, Olds GR, Farrar J, Cripps AW, McManus DP (2013) Enteropathogens and chronic illness in returning travellers. N Engl J Med 368:1817–1825
9. Watts S (1999) Epidemics and history: disease, power and imperialism epidemics and history: disease, power and imperialism. Yale University Press, New Haven
10. Morse SS, Mazet JA, Woolhouse M, Parrish CR, Carroll D, Karesh WB, Zambrana-Torrelio C, Dr LWI, Daszak P (2013) Prediction and prevention of the next pandemic zoonosis. Lancet 380(9857):1956–1965
11. Global Pandemic Emergency Facility (2020) The World Bank. Available via https://www.worldbank.org/en/topic/pandemics/brief/fact-sheet-pandemic-emergency-financing-facility. Accessed 26 July 2020
12. Report of National Intelligence Council (2008) Global trends 2025: a transformed world, Washington DC
13. Crosby AW (2003) America's forgotten pandemic: the influenza of 1918. Cambridge University Press, Cambridge
14. Almond D (2006) Is the 1918 influenza pandemic over? Long-term effect of in utero influenza exposure in the post-1940 U.S. population. J Political Econ 114 (4):672–712
15. Potter C (2001) A history of influenza. J Appl Microbiol 91(4):572–579
16. Workshop Summary (2007) Ethical and legal considerations in mitigating Pandemic Disease Institute of Medicine. Forum on Microbial Threats. US National Academies Press, Washington DC
17. Wu Z, McGoogan JM (2020) Characteristics of and important lessons from the coronavirus disease (COV-19) outbreak in China: summary of a report of 72314 cases from the Chinese center for disease control and prevention. JAMA 323(13):1239–1242
18. Wang D, Hu B, Hu C, Zhu F, Liu X, Zhang J, Wang B, Xiang H, Cheng Z, Xiong Y, Zhao Y, Lib Y, Wang X, Peng Z (2020) Clinical characteristics of 138 hospitalized patients with 2019 novel coronavirus-infected pneumonia in Wuhan, China. JAMA 323(11):1061–1069
19. Buheji M (2020) Future foresight of post COVID-19 generations. Int J Youth Econ 4(1):1–3
20. Dziugys A, Bieliunas M, Skarbalius G, Misiulis E, Navakas R (2020) Simplified model of Covid-19 epidemic prognosis under quarantine and estimation of quarantine effectiveness. Available via https://www.medrxiv.org/content/10.1101/2020.04.28.20083428v4. Accessed 26 July 2020
21. Roosa K, Lee Y, Luo R, Kirpich A, Rothenberg R, Hyman JM, Yan P, Chowell G (2020) Short-term forecasts of the COVID-19 epidemic in Guangdong and Zhejiang, China. J Clin Med 9(2):596
22. Li L, Yang Z, Dang Z, Meng C, Huang J, Meng H, Wang D, Chen G, Zhang J, Peng H, Shao Y (2020) Propagation analysis and prediction of the COVID-19. Infectious Dis Modelling 5:282–292
23. Impact of COVID-19 on air passenger transport (2020) Eurostat. Available via https://ec.europa.eu/eurostat/web/products-eurostat-news/-/DDN-20200616-2. Accessed 26 July 2020
24. Public transport and covid-19 (2020) UITP. Available via https://www.uitp.org/public-transport-and-covid-19. Accessed 26 July 2020
25. Impact on freight and passenger transport of the global Coronavirus (COVID-19) (2020) IRU. Available via https://www.iru.org/resources/tools-apps/flash-info/impact-on-freight-and-passenger-transport-of-the-global-coronavirus-covid-19-outbreak-889. Accessed 26 July 2020
26. Covid-19 pandemic: The continuity of passenger transport services is crucial (2020) UIC. Available via https://uic.org/com/IMG/pdf/cp_covid19-joint_statement_en2.pdf. Accessed 26 July 2020
27. Employment and unemployment statistics during the COVID-19 crisis (2020) OECD. Available via http://www.oecd.org/sdd/labour-stats/OECD-employment-and-unemployment-statistics-during-the-COVID-19-crisis.pdf. Accessed 26 July 2020

28. The impact of the coronavirus (COVID-19) pandemic on The Employment Situation for March 2020 (2020) U.S. Bureau of Labor Statistics, March 2020, Washington DC. Available via https://www.bls.gov/cps/employment-situation-covid19-faq-march-2020.pdf. Accessed 26 July 2020
29. EU labour force survey explanatory notes (2020) Eurostat. Available via https://ec.europa.eu/eurostat/documents/1978984/6037342/EU-LFS-explanatory-notes-from-2016-onwards.pdf. Accessed 26 July 2020
30. A second wave of COVID-19 in China? (2020) CGTN Insight. Available via https://news.cgtn.com/news/2020-06-19/A-second-wave-of-COVID-19-in-China–RrSVGLFLeU/index.html. Accessed 26 July 2020
31. Coronavirus: Fear of second wave in Beijing after market outbreak (2020) BBC News Available via https://www.bbc.com/news/world-asia-china-53034924. Accessed 26 July 2020
32. How Bad Is China's 'Second Wave' Coronavirus (2020) Forbes. Available via https://www.forbes.com/sites/kenrapoza/2020/06/16/how-bad-is-chinas-second-wave-coronavirus/. Accessed 26 July 2020
33. Benedictow O (2005) The black death—The greatest catastrophe ever. Hist Today 55(3): 42–49
34. Herlihy D (1997) The black death and the transformation of the west. Harvard University Press
35. Byrne J (2012) Encyclopedia of the black death. ABC-CLIO
36. Pamuk S (2007) The black death and the origins of the 'Great Divergence' across Europe, 1300–1600. Eur Rev Econ Hist 11(3): 289–317
37. Scott S, Duncan CJ (2001) Biology of plagues evidence from historical populations. Cambridge University Press, Cambridge
38. Robbins H (1928) A comparison of the effects of the black death on the economic organization of France and England. J Political Econ 36 (4):447–479
39. COVID-19 Coronavirus & travelers (2020) IATA. Available via https://www.iata.org/en/youandiata/travelers/health/. Accessed 26 July 2020
40. The risk of getting sick on a plane is lower than you might think—if you know what to watch out for (2020) Insider. Available via https://www.insider.com/what-are-health-risks-plane-air-travel-2020-6. Accessed 26 July 2020
41. Coronavirus FAQs: How risky is it to fly? Is there any way to reduce the risks? (2020) NPR. Available via https://www.npr.org/sections/goatsandsoda/2020/05/15/848706362/coronavirus-faqs-how-risky-is-it-to-fly-is-there-any-way-to-reduce-the-risks?t=1595079749250. Accessed 26 July 2020
42. Nicholas AJ (2016) Management and telework faculty and staff—Articles & Papers 60. Available via https://digitalcommons.salve.edu/fac_staff_pub/60. Accessed 26 July 2020
43. Gray M, Hodson N, Gordon G (1993) Teleworking explained. John Wiley and Sons, Chichester
44. Olsen M, Primps S (1984) Working at home with computers: work and non-work issues. J Soc Issues 40(3):97–112
45. Share of individuals who purchased tickets for events online in Great Britain in 2019, by age and gender (2020) Statista. Available via https://www.statista.com/statistics/286113/event-tickets-online-purchasing-in-great-britain-by-demographic/. Accessed 26 July 2020
46. E-ticketing (2020) IATA. Available via https://www.iata.org/en/programs/stb/e-ticketing/. Accessed 26 July 2020
47. Ambrose PJ, Johnson GJ (1998) A trust based model of buying behaviour in electronic retailing. AMCIS 1998 Proc 91:263–265
48. Veljkovic V, Perovic V, Paessler S (2020) Prediction of the effectiveness of COVID-19 vaccine candidates. F1000Research 2020 9:365
49. EU Strategy for COVID-19 vaccines (2020) Available via https://ec.europa.eu/info/sites/info/files/communication-eu-strategy-vaccines-covid19_en.pdf. Accessed 26 July 2020

50. Coronavirus vaccine trials have delivered their first results—but their promise is still unclear (2020) Nature. Available via https://www.nature.com/articles/d41586-020-01092-3. Accessed 26 July 2020

51. Treatments and a vaccine for COVID-19: the need for coordinating policies on R&D, manufacturing and access (2020) OECD. Available via https://www.oecd.org/coronavirus/policy-responses/treatments-and-a-vaccine-for-covid-19-the-need-for-coordinating-policies-on-r-d-manufacturing-and-access-6e7669a9/. Accessed 26 July 2020

52. Silitonga S, Sulistio H, Djakfar L, Wicaksono A (2011) Modal split model for public transport development in Indonesia. J Appl Sci Res 7(12):2036–2041

53. Santos G, Maoh H, Potoglou D, Brunn T (2013) Factors influencing modal split of commuting journeys in medium-size European cities. J Transp Geogr 30:127–137

54. Kottenhoff K, Freij BK (2009) The role of public transport for feasibility and acceptability of congestion charging—the case of Stockholm. Transp Res Part A 43(3):297–305

55. Modal split of passenger transport (2020) UN meth. Available via https://www.un.org/esa/sustdev/natlinfo/indicators/methodology_sheets/consumption_production/modal_split_passenger_transport.pdf. Accessed 26 July 2020

56. One more million passenger (2020) Silesian Railway. Available via https://www.kolejeslaskie.com/milion-pasazerow-wiecej-w-2018-roku/. Accessed 26 July 2020

57. Passenger cars in the EU (2020) Eurostat. Available via https://ec.europa.eu/eurostat/statistics-explained/index.php/Passenger_cars_in_the_EU. Accessed 26 July 2020

58. New update from BEREC on internet capacity during the COVID-19 crisis (2020) BEREC. Available via https://berec.europa.eu/eng/news_and_publications/whats_new/7203-new-update-from-berec-on-internet-capacity-during-the-covid-19-crisis. Accessed 26 July 2020

59. Keeping the Internet up and running in times of crisis (2020) OECD. Available via http://www.oecd.org/coronavirus/policy-responses/keeping-the-internet-up-and-running-in-times-of-crisis-4017c4c9/. Accessed 26 July 2020

60. Wiederhold BK (2020) Social media use during social distancing. Cyberpsychol Behav Soc Netw 23(5):275–276

61. King DL, Delfabbro PH, Billieux J, Potenza MN (2020) Problematic online gaming and the COVID-19 pandemic. J Behav Addictions 9(2):184–186

62. Littman RJ (2009) The plague of Athens: epidemiology and paleopathology. Mt Sinai J Med 76(5):456–467

63. Smith CA (1996) Plague in the ancient world: a study from thucydides to Justinian. Student Hist J. Available via http://people.loyno.edu/~history/journal/1996-7/Smith.html. Accessed 26 July 2020

64. Patterson KD, Pyle GF (1991) The geography and mortality of the 1918 influenza pandemic. Bull Hist Med 65(1):4–21

65. Durbach N (2005) The anti-vaccination movement in England, 1853–1907. J Royal Soc Med 98(8):384–385

66. The Anti-Vaccine Movement in 2020 (2020) Available via https://www.mcgill.ca/oss/article/covid-19-pseudoscience/anti-vaccine-movement-2020. Accessed 26 July 2020

67. Will COVID-19 change the parking business? (2020) The Hustle. Available via https://thehustle.co/covid-19-business-of-parking-lots/. Accessed 26 July 2020

68. Problems in parking for grounded fleets (2020) IATA. Available via https://www.airlines.iata.org/news/problems-in-parking-for-grounded-fleets. Accessed 26 July 2020

69. Principles for covid-19 public transport operations (2020) Available via https://www.infrastructure.gov.au/transport/files/covid19_public_transport_principles_29052020.pdf Accessed 26 July 2020

70. Aloi A, Alonso B, Benavente J, Cordera R, Echániz E, González F, Ladisa C, Lezama-Romanelli R, López-Parra A, Mazzei V (2020) Effects of the COVID-19 lockdown on urban mobility: empirical evidence from the City of Santander (Spain)

71. Anzai A, Kobayashi T, Linton NM, Kinoshita R, Hayashi K, Suzuki A, Yang Y, Jung S-M, Miyama T, Akhmetzhanov AR, Nishiura H (2020) Assessing the impact of reduced travel on exportation dynamics of novel coronavirus infection (COVID-19). J Clin Med 9:601

72. Guo ZD, Wang ZY, Zhang SF, Li X, Li L, Li C, Cui Y, Fu R-B, Dong Y-Z, Chi X-Y, Zhang M-Y, Liu K, Cao C, Liu B, Zhang K, Gao Y-W, Lu B, Chen W (2020) Aerosol and surface distribution of severe acute respiratory syndrome coronavirus in hospital wards, Wuhan, China, 2020. Emerging Infectious Dis 26(7):1583–1591

73. Faridi S, Niazi S, Sadeghi K, Naddafi K, Yavarian J, Shamsipour M, Jandaghi NZS, Sadeghniiat K, Nabizadeh R, Yunesian M, Momeniha F, Mokamel A, Hassanvand MS, MokhtariAzad T (2020) A field indoor air measurement of SARS-CoV-2 in the patient rooms of the largest hospital in Iran. Sci Total Environ 725:138401

74. Ong SWX, Tan YK, Chia PY, Lee TH, Ng OT, MBBS, Wong MSY, Marimuthu K (2020) Air, surface environmental, and personal protective equipment contamination by severe acute respiratory syndrome coronavirus 2 (SARS-CoV-2) from a symptomatic patient. JAMA 323 (16):1610–1612

75. Liu ZQ, Ye Y, Zhang H, Guohong X, Yang J, Wang JL (2020) Analysis of the spatio-temporal characteristics and transmission path of COVID-19 cluster cases in Zhuhai. Tropical Geogr 1–13

76. Re-spacing Our Cities For Resilience (2020) International transport forum. Available via https://www.itf-oecd.org/sites/default/files/respacing-cities-resilience-covid-19.pdf. Accessed 26 July 2020

77. Transport Policy Responses to the Coronavirus Crisis (2020) International transport forum. Available via https://www.itf-oecd.org/covid-19/policy-responses. Accessed 26 July 2020

78. Public transport closures during the COVID-19 pandemic (2020) Available via https://ourworldindata.org/grapher/public-transport-covid. Accessed 26 July 2020

79. Ahorsu D, Chung-Ying L, Vida I, Mohsen S, Griffiths MD, Pakpour AH (2020) The fear of COVID-19 scale: development and initial validation. Int J Ment Heal Addict 1–9. https://doi.org/10.1007/s11469-020-00270-8

80. LeDux J (2015) Anxious: using the brain to understand and treat fear and anxiety. J Undergrad Neurosci Educ 14(2):R22–R23

81. Gaëtan M, Lotte G, Duijndam S, Salemink E, Engelhard IM (2020) Fear of the coronavirus (COVID-19): predictors in an online study conducted in March 2020. J Anxiety Disorders 74:102258

82. Qiu J, Shen B, Zhao M, Wang Z, Xie B, Xu Y (2020) A nationwide survey of psychological distress among Chinese people in the COVID-19 epidemic: implications and policy recommendations. General Psychiatry 33:e100213

83. Wang C, Pan R Wan X, Tan Y, Xu L, McIntyre RS, Choo FN, Tran B, Ho R, Sharma VK, Hoe C (2020) A longitudinal study on the mental health of general population during the COVID-19 epidemic in China. Brain Behav Immunity 87:40–48

Ireneusz Celiński serves as an assistant professor at the Faculty of Transport and Aviation Engineering of the Silesian University of Technology. He has been the author and co-author of over 230 publications, including articles in scientific journals and chapters in monographs at home and abroad. He has conducted many scientific and research work with transport public and private companies in the country and abroad. His scientific interests combine traffic engineering problems, visual techniques, contemporary technologies, microprocessors, eye tracking, and behavioral analysis.

Grzegorz Sierpiński is an associate professor at the Faculty of Transport and Aviation Engineering of the Silesian University of Technology. He is also the Head of the Department of Transport Systems and Traffic Engineering and Rector's Proxy for the Priority Research Area: Smart Cities and Future Mobility. He has been the author and co-author of over 230 publications, including articles in scientific journals and chapters in monographs at home and abroad and the scientific editor of several monographs. He has conducted many scientific and research works and managed two international projects under the ERANET program. He combines road traffic engineering problems (including traffic analysis and forecasting, modeling of transport systems, and optimization of transport networks) with shaping travel behaviors in cities.

Bringing the Two Cultures of the Arts and Sciences Together in Complex Health Interventions

18

Brian Brown and Monica Lakhanpaul

"A good many times I have been present at gatherings of people who, by the standards of the traditional culture, are thought highly educated and who have with considerable gusto been expressing their incredulity at the illiteracy of scientists. Once or twice I have been provoked and have asked the company how many of them could describe the Second Law of Thermodynamics. The response was cold: it was also negative. Yet I was asking something which is the scientific equivalent of: Have you read a work of Shakespeare's? I now believe that if I had asked an even simpler question - such as, What do you mean by mass, or acceleration, which is the scientific equivalent of saying, Can you read? - not more than one in ten of the highly educated would have felt that I was speaking the same language"

Charles Percy Snow [1, p. 14–15]

B. Brown (✉)
Faculty of Health and Life Sciences, De Montfort University, Leicester L1 9BH, UK
e-mail: brown@dmu.ac.uk

Integrated Science Association (ISA), Universal Scientific Education and Research Network (USERN), Leicester, UK

M. Lakhanpaul
University College London, London, UK
e-mail: m.lakhanpaul@ucl.ac.uk

Whittington NHS Trust, London, UK

Integrated Science Association (ISA), Universal Scientific Education and Research Network (USERN), London, UK

© The Author(s), under exclusive license to Springer Nature Switzerland AG 2022
N. Rezaei (ed.), *Multidisciplinarity and Interdisciplinarity in Health*,
Integrated Science 6, https://doi.org/10.1007/978-3-030-96814-4_18

Summary

In the mid-twentieth century, a divide between the 'two cultures' of the arts and sciences was often seen as an impediment to effective science and policy. However, a number of philosophers of science have emphasized the common sources of imagination and creativity as well as the versatility of logic and method that cut across the two cultures. This chapter draws on our experience of being involved in a number of initiatives to try and blend different disciplines, from the point of view of promoting the 'health humanities' as a way of bringing health research and creative activities together and from our involvement in practical projects in our native UK and abroad. Drawing on the authors' experience of bringing different subject areas together in interdisciplinary projects in the healthcare field, we will explore how this dialog can be promoted most productively. We suggest that arts scholars and practitioners have a great deal to contribute to the feasibility, exploratory, and developmental stages of a project. This can help shift the narrative of complex interventions away from a focus on the evaluation of a particular intervention and toward a richer understanding of how it is made up in the first place, from the collective work of active, interpreting, meaning-making individuals in a specific cultural context. Thus, a dialog between the arts and sciences can help to achieve a more explicit and rigorous framework for integrating different disciplines and developing a more comprehensive approach to the conceptualisation and design of complex interventions.

Birthing a Better Future Projects Ali Ferguson 'Not just Blue'
The code of this chapter is *01101000 01100100 01101001 01101100 01101110 01000011 01110010 01100101.*

Keywords

Complex interventions · Co-production · Health humanities ·
Interdisciplinarity · Two cultures

1   Introduction: The Two Cultures of the Arts and Sciences

In framing this chapter as a dialog between the 'two cultures' of the arts and
sciences, we are, as some readers will have spotted, alluding to the work of Charles
Percy Snow (1905–1980), a famous son of our native Leicester in the UK. In many
of his essays and novels, he dealt with what he saw to be the divide between
specialists in the arts and the sciences and felt this was a major obstacle to
addressing humanity's problems. C.P. Snow's words on the subject have been
widely quoted but bear repeating here because they have structured a great many
subsequent discussions about the different cultures of the arts and humanities versus
those of the sciences.

Snow bemoaned the way that, in his view, the arts and humanities had been
privileged in British education and public life at the expense of science and engineering. The humanities, especially the language and literature of classical antiquity, he felt, had dominated the education of elite groups – the sort of people who
became politicians, civil servants, government advisers, and university lecturers –
to the extent that our progress, our ability to address future problems, and our
security were compromised. Assessments nearer the present day have suggested
that Snow perhaps exaggerated the divide [2], but it has nevertheless entered our
language and common sense like a persistent cultural trope.

2   The Common Wellsprings of Creativity in the Arts
  and Sciences

By contrast to Snow, some influential twentieth century thinkers such as Jacob
Bronowski [3], Frances Yates [4], and Gerald Holton [5] have stressed the common
imaginative wellspring of art and science, often focusing on the Renaissance period
in Europe. In their various ways, these thinkers have promoted the notion that by
doing things, making artifacts, objects, architecture, dramatic productions, and the
like, our species has driven forward the human story. For example, Bronowski was
concerned to stress that it was the process of making things and the concomitant
relationship between hand, eye, and brain, which was significant in human evolution. Yates, taking a more cerebral approach, in 'The art of Memory' contended
that it was the systematic cultivation of memory by philosophers, intellectuals,

politicians, and performers that had shaped human history. This process of doing things, *practice*, if you will, is a feature that unites many different aspects of human endeavor and one to which we shall return later in our consideration of work in which we have involved ourselves.

The divisions between the arts and sciences bemoaned by C.P. Snow are themselves of comparatively recent date. The notion of a 'scientist' him or herself is a relatively modern invention, being proposed by William Whewell in the 1830s [6]. Such activity had been going on long before Whewell, of course, but had often been termed natural philosophy. The idea of a scientist, someone who studies this sort of thing by occupation or vocation, was a relatively new invention at the time. The focus was on observation, experimentation, and description of phenomena rather than the more discursive interest in the philosophical background from which interest in the phenomena emerged.

The following century saw increasing consolidation of the scientist's role, for example, in universities, larger teaching hospitals, and industry, such that science could be said to have developed its own culture, community, and world view. By the mid-twentieth century, some influential theorists of the history and philosophy of science, such as Thomas Kuhn [7], highlighted differences between the arts and sciences that they tied into how the arts and the sciences were institutionalized in radically different ways.

In Kuhn's view, and other philosophers and historians of science such as Imre Lakatos [8], typically within most branches of science, there is some degree of consensus over the problems on which scientists should be working, the theoretical and methodological tools which should be deployed to solve them and how to judge proposed solutions or adjudicate between competing explanations. This tendency toward standardization can be seen particularly clearly in the end product of much science, the published peer-reviewed journal articles. The way these are written evinces a kind of anonymous universalism, as if theory and previous work suggested the research questions, or as if the method was applied impersonally, and the conclusions followed ineluctably from the data. This then is 'normal science' within what Kuhn termed a paradigm, or Lakatos called a 'research program'. This is rather different from art. Here, there is often a much greater emphasis on practitioner self-differentiation, and a greater value is placed on innovation. Rather than the written word, artists' labor often yields instead the exhibition, performance, installation, or other experience. While art students and critics often seek to contextualize their own and others' work, the aspects of novelty, rupture, questioning, and shock are often significant. Particular prestige attaches to those who have coined new ways of looking at the human or natural world or new movements.

As Lakatos [8] and Feyerabend have pointed out, however, if science adhered to a single rigorous working way, there would be little scope for genuine novelty or discovery. As Feyerabend [9, p. 57] put it:

Lakatos realized and admitted that the existing standards of rationality, standards of logic included, were too restrictive and would have hindered science had they been applied with determination. He therefore permitted the scientist to violate them (he admits that science is not "rational" in the sense of *these* standards). However, he demanded that research programmes show certain features *in the long run* — they must be progressive.... I have argued that this demand no longer restricts scientific practice. Any development agrees with it.

Feyerabend pointed out that science itself has been remarkably versatile in terms of the methods it includes, the kinds of evidence it accepts, and the arguments it makes. Indeed, he argued that careful study of what scientists do leads to the position that there is no single set of principles, methodologies, or working assumptions that characterize scientific activity [10]. For example, in his view, Galileo would never have ended up supporting a heliocentric theory of the cosmos if he had adhered to the standards of reason and logic dominating intellectual inquiry in his time. Feyerabend's position has sometimes been characterized as methodological anarchism. For Feyerabend, the choice of a particular theory or method is as much esthetic as it is based on rational criteria.

In some ways, Feyerabend took the opposite line to that of C.P. Snow. He felt it was tragic that some scientists, especially physicists, had not steeped themselves in philosophy.

The withdrawal of philosophy into a "professional" shell of its own has had disastrous consequences. The younger generation of physicists, the Feynmans, the Schwingers, etc., may be very bright; they may be more intelligent than their predecessors, than Bohr, Einstein, Schrödinger, Boltzmann, Mach and so on. But they are uncivilized savages, they lack in philosophical depth [11, p. 385].

The foregoing lays out some intellectual and scientific backgrounds to the divisions, differences, and similarities between the arts and the sciences in the present day. There have been attempts to characterize the work of science by philosophers, historians, and scientists themselves in a way that shows how it is distinctive, and efforts to characterize and remedy the perceived divide between the apparently different intellectual communities in the sciences and the arts and humanities. At the same time, a strong strand in this scholarship has been the common feature of creativity that has animated all approaches to understanding nature, and the human condition within it, that are common to all disciplines. In the present, a renewed enthusiasm for interdisciplinary work has characterized science policy and the agendas of funding bodies, so this has provided some novel opportunities to get science and the arts and humanities working together in productive ways. The following section describes some of our attempts to do so in the U.K. and other nations. These illustrate points of confluence and divergence between differing worldviews and ways of working as well as some new opportunities for rapprochement between disciplines in tackling human challenges facing us in the twenty-first century.

3 Synergies Between the Sciences and Arts: The Health Humanities

In an effort to bring together interests in the arts, humanities, and health care over the last few years, the author B.B. has been involved in a new movement, which we have characterized as the 'health humanities'. Prior to our intervention, the medical humanities have been around for many years. Historically, these have foregrounded the value of arts and humanities to medical education and explored how medicine and its work can be enhanced by studying philosophy and history of medicine, anatomical illustration, ethics, literature, and drama. This has demonstrable value, and the medical humanities have been a central node for this kind of work. However, in our broader public culture, the arts and humanities are among the best known and most popular routes toward health, well-being, resilience, and social connectedness. Indeed, as we have argued elsewhere [12], it is as if they are a 'shadow', informal, and not necessarily medically-driven 'health and social service' in many parts of the world.

There are new features to our promotion of the health humanities that are noteworthy. As Crawford and Brown write [12, p. 41], the health humanities offer a 'superordinate evolution' that advances innovation, mutuality, and dialog between congruent traditions. In this way, it seeks to inspire, not to control or govern innovation. Klugman [13, p. 419] suggests that health humanities may even 'stave off the decline of the broader humanities' in higher education where STEM subjects (typically defined as science, technology, engineering, and mathematics/medicine) are deemed to more clearly tie in with economic needs. Skylar [14] reports how health humanities inform medical training in new and compelling ways, focusing on the use of arts and humanities to foreground the perspectives of patients, their families, and social conditions, or environments.

Central to our approach is the idea that some stakeholder groups have so far had little voice in the medical humanities. More healthcare disciplines other than medicine are seeking to apply arts and humanities insights. These include nurses, occupational therapists, physiotherapists, and even biomedical scientists. Moreover, there is a whole range of non-professional carers, perhaps working in charities and NGOs, support organizations, or as care assistants. We are detecting a growing interest in arts and humanities activity in these groups. Informal carers, who are undertaking the care of their family members or their loved ones, represent a large group with little voice in the medical humanities curriculum. We are also seeking to include the large number of people involved in the creative therapies, for example, art therapy, dance and movement therapy, music therapy, poetry therapy, bibliotherapy, and many more, who have a wealth of experience in making their interventions work in practical settings. Taken together, these bodies of experience open up the possibility, as we have argued [12, 15], for bottom-up, 'practice-based evidence' to be brought to bear on the situation. Moreover, we hoped that research in the health humanities would foreground the health and well-being experience and cultural capital of broader populations of ordinary people rather than the vested interests of arts and humanities organizations or bodies.

So far, this developing field has spurred a broadening in the scope, inclusion, and application of previously established and more narrowly configured disciplines. As Tess Jones and her colleagues argue [16, p. 932], the move to the health humanities is not simply a matter of semantics and 'splitting hairs'. Given the range of subjects, health professions, stakeholders, and practice environments it involves, the health humanities represent a 'more encompassing, contemporary, and accurate label' [17, p. 6]. To date, multiple networks, research units, projects, and taught courses in the health humanities have emerged worldwide. Importantly, this 'burgeoning' [18] field has brought diverse academics, creative practitioners, and professionals in health and social care and education to work more closely with the public to find new applications and social innovations through the arts and humanities in an interdisciplinary and non-hierarchical way. As such, the field has shifted beyond medical conceptions of health and well-being, rejecting a pecking order for who controls or mandates applying the arts and humanities to improve the human condition.

Some of these ideas came into practice with our 'Creative Practice as Mutual Recovery' project, which ran from 2013 to 2018 (http://cpmr.mentalhealth.org.uk). It involved 14 separate projects that examined how a selection of different creative practices in the arts and humanities can promote good mental health and well-being. It explored how this is possible within and between groups of health service users, family carers, arts practitioners, and professionals in health and social care and education in a shared practice we termed 'mutual recovery'. One of the projects at the Royal College of Music, for example, found that a ten-week program of group drumming reduced depression by as much as 38% and anxiety by 20% while improving social resilience by 23% and mental well-being by 16% [19]. It is creative public health in action, adding to a raft of evidence for how the arts and humanities can promote mutual recovery alongside and without necessarily a prescription as such. Notably, mental health service users co-designed the program from start to finish. In a separate investigation of how medieval medical remedies can tackle antibiotic resistance, ancientbiotics forges new possibilities for applying historical texts to contemporary public health problems such as dealing with infections [20].

Through the creative practice under the mutual recovery project, we could foreground some health humanities features as we see them and as we have tried to promote them over the past decade. First, we attempted to bring service users, carers, and practitioners together in creative activity because getting people together to work with a common goal has valuable health benefits. It has been brought to public attention recently in the U.K. with increased attention to the adverse effects of loneliness [21, 22]. Second, we have sought to integrate health humanities concern with the rich body of practical knowledge originating in creative therapies and practice. It has helped break through the barriers that have often existed between these approaches. Third, as mentioned above, we have tried to incorporate informal carers, a massive number of people amounting to between 6.5 and 8.5

million in the U.K. [23], a group who have not figured prominently in the medical humanities hitherto. Fourthly, we have sought to privilege the creative activity rather than the medical aspects of the problem. Thus, rather than selecting people based on their putative diagnosis, we have included people with a wide variety of problems and symptoms, and recognized that carers and professionals might have vulnerabilities or sources of distress in their lives. Therefore, we have sought independence from the predominantly medical frame that governs much medical humanities and social science work and instead foreground experience, creativity, and the esthetics of the work. In this way, we hope to open up new spaces for collaboration, creation, and inquiry.

To take another example, the author M.L. has contributed to two pieces of work related to minority populations in the U.K. These projects have made extensive use of film as a way of enabling community members and creating dissemination opportunities. The U.K. Neon project (Nurture Early for Optimal Nutrition, https://www.acesoghc.com/neon) [24] is a project which works with the Bangladeshi community residing in the U.K. It concerns infant feeding, using a participatory learning and action approach with community members. From the formative qualitative data and through work with the community and community facilitators, short films were created to help share the research findings and raise awareness of the topic addressed, from the perspective and with the voices from the community themselves. It, therefore, provides something engaging for community members and avoids the usual format of expert talking heads so often found in media of this kind. Similarly, the management of interventions for asthma study (MIA) [25, 26] was grounded in extensive collaborative work, interviews, and focus groups exploring the meaning of asthma and its role in children's lives in South Asian communities. Likewise, films were made to reflect the experiences described and showcase participants' contributions (e.g., https://vimeo.com/105418355.) This formed part of the study's outcome that was an intervention planning framework to tackle asthma pathways. It culminated in an exemplary integrated, multifaceted approach to asthma, named ACT that stands for awareness, context (cultural and organizational), and training.

These projects have highlighted the value of film to engage target audiences and foreground the experiences of people who have contributed to the studies providing a lasting tribute to their involvement. The human aspects of illness experience, feeding children, or caring for someone with asthma can often be lost in conventional scientific reports on the subject, even if they are illustrated with quotes from participants. Moreover, creating videos about the issue can become a further focus of engagement and commitment for the people involved, further building social relationships and social capital. The arts and humanities then have various functions that add value and enhance the process of inquiry and intervention where health is concerned.

More recently, M.L. and her Ph.D. student Diana Margot Rosenthal (D.M.R.) have been working on a project, 'Walk In My Shoes', with mothers with young children (<5 years) living in temporary accommodation (T.A.) in London. Participants were asked to draw their feelings or thoughts in a multimedia book.

They were told this should have no restrictions and would be an open book; many took the opportunity to visually document their experiences while living in T.A. and the multifaceted barriers they face. The aim was to convert these drawings into illustrations to be displayed in a future exhibition. Also, the project used 'citizen science' approaches, where the participants collected and interpreted data themselves. The mothers took photographs of their housing and neighborhood environments, which will be showcased as part of the exhibition along with the illustrations to promote public awareness of child homelessness and the difficulties in addressing the Healthy Child program recommendations when living in T.A. Photographs were captured utilizing a mobile app created by D.M.R. and tailored based on collaborative dialog with the participants. Furthermore, through UCL's Train and Engage Fund in Public Engagement, some mothers were enabled to attend sewing workshops to learn transferrable skills from quilting to embroidery. D.M.R. and M.L. worked with mothers to translate the data collected via 'citizen science' approaches into a patchwork map quilt, creating a representation of the challenges faced in attempting to engage with healthcare services and supporting their children to meet developmental milestones under the 'no recourse to public funds' status as many are migrants. This illustrates the potential of creative approaches to engage and offer something back to the communities with whom research is conducted. All too often, the participant fills in a questionnaire, answers questions in an interview, or provides a specimen, and that is the extent of their involvement. These studies, by contrast, offer something back to participants over a more sustained period. The pleasure and sense of mastery involved in creative activity, the community, and social relationships fosters the sense that one contributes to a greater whole are all benefits that the use of the arts and creative activity enable.

So far, we have been discussing projects largely based in the U.K. As we have mentioned, there is interest beyond the U.K.'s borders in these fusions of the arts and sciences, so in the next section, we will consider some of the efforts we have made to try out these approaches in other countries.

4   Making the Health Humanities Global

The medical humanities have gained a foothold in universities and teaching hospitals in many different nations. Recently, however, through the generous support of the U.K.'s Global Challenges Research Fund, we have been able to put some of our principles into practice in an international context. Several empirical projects in India have been funded through this channel, and it is worth examining them, so the reader can see how we attempted to implement the fusion between the arts and sciences in practice.

The first of these was titled 'An Exploration of Mental Health and Resilience Narratives of Migrants in India Using Community Theater Methodology' and ran in a low-income neighborhood in Pune, Maharashtra, India, between 2017 and 2019.

As the title suggests, our focus was on community drama as a way of both eliciting experiences of migration and enabling an appreciation of these stories to contribute to resilience and community cohesion. Drama and performance seemed to be a valuable and vital way of tackling the issue because, as our Indian partners proudly told us, India has an unbroken, living theater tradition stretching back millennia, originating in early Sanskrit theater. A Web site detailing the project sited at http://mhri-project.org/ India has many internal migrants: 450 million according to the World Bank [27], many of whom have moved from rural areas to cities in search of work following several years of challenging conditions in the farming industry. Working in a slum area ('the Basti'), our team first undertook interviews with community members seeking to elicit their migration stories. Early in the project, we decided to frame the issue in terms of resilience because much previous work had been undertaken, focusing on deprivation and poor health outcomes for migrants, and we wanted to do something to counter the implied deficit model in much of this research. While not ignoring the inequalities and hardships faced by residents, it became clear that a high degree of resourcefulness was being exercised as people solved problems, sought to gain a livelihood, and took care of themselves and one another. From the outset, the project team involved academics from both countries interested in health and social welfare and theater practitioners from the U.K. and India. Theater workshops and activities were conducted in the Basti neighborhood, and the stories from the narrative interviews were used to inform the construction of a play that was developed and performed with contributions from local people.

Of particular interest to this chapter is the dialog between project members and stakeholders with differing backgrounds. For example, our Indian colleagues have a great deal of experience with epidemiological and survey work, but qualitative research in health and welfare has made only limited inroads into academic and policy circles in India. Accordingly, some of the early conversations concerned the nature and value of the qualitative approaches and what we hoped to achieve with them. It highlighted why there are various methodological and epistemological stances within the study of health and social care, with some enclaves cleaving firmly to one approach or another.

Further, differences emerged between those with a background in drama and those with backgrounds in health or social sciences. For dramatists, 'research' had a much more expansive and elastic definition. 'Research' was everywhere. It included everything from exploratory conversations about the feasibility of the project, field visits to the neighborhood, interactions with local people, and the work to formulate, co-construct, rehearse, and refine the play, as well as all the collateral activities, are done with the theater group and people in the neighborhood. These included making banners, model birds, and improvising music from objects found in the area. All of these were 'research' in some sense or other. To social scientists, however, research was somewhat more circumscribed. It was something you did with interview schedules, recording devices, questionnaires, and protocols. It was an activity you had perforce to detail in advance and submit to committees for ethical scrutiny. Even with free-form approaches like participant observation, there is a good deal of planning and pre-specification. Where one will go, when one will go there, how access will be negotiated, whom one will speak to, what one will ask them, and so on, even if this is accompanied by a good deal of informal hanging around. Experiences like this highlighted differences between different practical and intellectual traditions we were trying to bring together. These were readily resolved in the dialog which took place throughout the project

A further source of productive dialog between project members of different backgrounds concerned differing understandings of community theater on theater practitioners from the U.K. and India. The U.K. team was steeped in the ethos of co-production and participatory work. It involved local people from the community in every stage, from the conception and construction of the drama to its final performance. By contrast, our Indian theater partners had in mind a somewhat different model of community involvement; 'Ah, like a talent show' was their response, reflecting the familiarity and popularity of that format in India, whereas the U.K. team had in mind the creation of an entire play based on community involvement. As the project progressed, some neighborhood activities did indeed include events that resembled a talent show, where local people took the stage to play music, sing, and otherwise perform. This was a format that everyone enjoyed and understood. Therefore, it was undertaken with pleasure by both the amateur performers and their audience. This provided a valuable way of gaining a foothold for the main feature, a play involving local people based on community members' narratives of resilience, detailing their ingenuity in tackling the challenges they faced in making a new life themselves in the city.

Accordingly, differences within communities of scholars and practitioners in the arts and sciences might be as prominent as those between them. There are competing schools of thought and different ways of thinking about and undertaking their craft tasks, which sometimes, but not always, cleave along national or cultural lines. Trends and intellectual fashions in one part of the world may bypass those of another, yet some ideas may become part of a sort of planetary vulgate. For example, common to many of these approaches is the notion that research is something that reduces uncertainty. 'You don't know what's going to happen until you try it', as a performer acquaintance puts it. The research activity may be more systematized in sciences and social sciences, but it performs some of the same functions for the artist or dramatist. The project also has highlighted how human hardship prompts different kinds of responses. Whenever B.B. mentions it at conferences, a question usually asked (often by a very earnest, serious young man) is that given the manifest poverty and deprivation of the people, surely the priority is to alleviate their material hardship, and that theater, music, and dance are a somewhat frivolous response. B.B.'s answer to this is generally framed in the following terms. Just because people are in straitened circumstances does not mean that cultural life somehow goes away. Indeed, perhaps it is even more important. Activities such as drama, music, and performance are especially vital when one's social position is marginal. The camaraderie, solidarity, and gains from engaging in a shared activity with a common purpose are durable and can lead to people's ability to campaign more effectively for other causes that may make a material difference to their lives. The arts then are not merely an add-on once material needs have been met but can be a way of addressing those needs in the first place. Moreover, the artifacts created and the show itself enhanced the neighborhood's amenity value; the show involving local people was performed elsewhere, including in a large auditorium in front of a conference audience. This allowed participants to 'tread the boards' on a big stage, an opportunity which they probably would not have had otherwise.

The second project started much more recently and was interrupted by the precautions imposed due to the COVID-19 pandemic in early 2020, so at the time of writing, the project's overall shape and form are still uncertain. However, at the proposal stage, we began with the notion of 'mental health literacy'. In Kerala, where the project is based, the population seems to be doing well on many indices. A predominantly literate, well-educated population enjoys higher life expectancies and incomes than in many parts of India, having followed a trajectory that other Indian states are keen to emulate. However, mental health problems have appeared as a prominent source of burden for the population. At the same time, among many people, the notion of mental illness carries a strong stigma and can adversely affect everything from a person's employment prospects to their perceived suitability as a marriage partner. Starting from the premise that the situation might be ameliorated if people were able to be kinder and more compassionate toward each other and have better conversations about mental health problems, we began planning the project. Like the resilience project in Pune, a team was assembled, including scholars and practitioners from the humanities, social sciences, health sciences, and theater practitioners from both the U.K. and India. The early stages of the project (http://mehelp-india.org/) were characterized by some debate about cross-cultural issues concerning 'mental health literacy' itself. By definition, mental health literacy refers to 'as knowledge and beliefs about mental disorders which aid their recognition, management, or prevention' [28, p. 186]. It will be bound, for example, in 'the ability to recognize specific disorders; knowing how to seek mental health information; knowledge of risk factors and causes, of self-treatments, and of professional help available; and attitudes that promote recognition and appropriate help-seeking'. This entrains some assumptions concerning the medical model of mental disorders and the desirability of help seeking from healthcare practitioners. Immediately, this elicited skepticism from some members of the team, both U.K. and Indian. To what extent does this concept map intelligibly onto the Indian experience? There is a large body of literature where researchers and authors argue for the importance of context, culture, and local frame of reference in making sense of what in the West is called 'mental disorder' [29, 30]. Fernando [31] argues that the over application of Western notions represents a barrier to better and more culturally sensitive mental health research and practice. In light of this, the participatory community theater approach is precious in eliciting and interpreting interest experiences.

Similar to the resilience project in Pune, the initial data are an extensive collection of interviews with mental health service users, informal carers, and health professionals. Like the Pune project, this body of data will be used to construct pieces of drama with a local theater company, and these, in turn, will yield a further layer of response and discussion from the audiences. It is hoped that this will yield fresh insights moving repose the concept of mental health literacy in a more culturally literate way and has implications as to how a kinder, and more compassionate approach can be facilitated. This need not involve adopting a particular kind of professionalized medical world view.

At the same time, some researchers and practitioners in India often express a good deal of enthusiasm for the concepts of mental disorder described in the American Psychiatric Association's Diagnostic and Statistical Manual (DSM-5; American Psychiatric Association, [32]) or the World Health Organization's International Classification of Diseases (ICD 11, [33]). It is as if these represent a modern, scientifically rigorous, diagnostically precise, and technologically hopeful way of making sense of the situation and offer an advance over what they see to be superstition and backward-looking folkways. This contrasts with the U.K. and U.S. A., where there is much lively debate about medicalization and the value and ontology of these systems of categories. It is more than a peripheral issue; some of the most trenchant critiques of DSM-5 have come from the very heart of the U.S. medical establishment, such as the then director of the National Institute of Mental Health Thomas Insel's comments on DSM-5's publication [34]. In a cross-cultural context, this kind of discussion is likely to continue for some time to come. The relevance of this point to the role of the arts and humanities in healthcare research is that the creative aspects of a project provide a focus that circumvents these contention and uncertainty points. The arts, in this case, community theater, and storytelling, and ultimately the creation of films, focus on the everyday experiences of distress and disorientation and family and community responses to these, rather than prescribing a particular theory of their etiology. In this way, they are valuable in providing common ground between practitioners, scientists of different orientations, and laypeople, whether they be sufferers, carers, or members of wider family and community networks.

A third example, from M.L.'s work, concerns the PANChSHEEEL project, funded by the UK's Global Challenges Research Fund (GCRF) and Medical Research Council (MRC); the study is a collaboration between University College London (UCL), Save the Children, Jawaharlal Nehru University (JNU), Delhi and the Indian Institute of Technology (IIT), Delhi. The project is an interdisciplinary, cross-sector study, designed to explore health, education, engineering, and environment) (HEEE) factors that influence Infant and Young Child Feeding (IYCF) practices and nutrition in India. The project focused on 'what' food was fed to children and 'how' hygienically it was administered to reduce infections and improve sanitation. The project's trajectory was documented in a photo book, using illustrations and own-words stories from the community, to characterize the factors that contribute to Infant and Young Child Feeding (IYCF) practices and nutrition (https://www.pahus.org/panchsheeel-project,   and   https://www.ucl.ac.uk/child-health/research/population-policy-and-practice-research-and-teaching-department/champp-child-and-2). Also, a film was created, giving more information on the problems themselves and providing more detail about our research approach. It has been shown to policymakers, funders, and students. The project team is now working with partners in India to develop an electronic exhibition and then, in due course, a physical exhibition called 'The Early Years-A window of Opportunity' to showcase more of the photos from this project but will also include artwork from an exhibition Zero2 Expo, Birthing a Better Future (https://www.zero2expo.com/). This latter project was initially developed in the U.K. and is concerned with the first 1001 days; a period which was chosen because it covers the point of conception through to the age of two. It presents short sections of literature extracted from publications and perspectives from academics on the subject, while at the same, time linking to pictures or paintings created by creative artists on the theme to make it more accessible and engaging to the audiences. An example of the kind of work involved is provided in graphical abstract.

The forthcoming exhibition will bring together the PANChSHEEEL project with an India version of Zero2 Expo and photos and drawings taken from the CHIP (Childhood Infection and Pollution https://www.acesoghc.com/chip) project, all with a common focus on the early years of children's lives and the sustainable development goals. So far, the indications are that these activities are widely believed to be worthwhile. An evaluation of the earlier Zero2 Expo exhibition [35] indicates that viewers felt it had an important role in raising awareness and found it thought-provoking.

The examples from beyond the U.K. we have discussed have largely come from India, but the fusion of the arts and sciences in these projects suggests a good deal of potential in other nations where research and interventions with marginalized communities are contemplated, the arts form a useful point of contact. Even where people live in difficult circumstances, their cultural traditions are often a source of pride and considerable enjoyment. It may involve visual art, music, performance, drama, dance, textiles and needlecraft, and much else besides. It also means a different kind of relationship can be cultivated between researchers and participants, such that the latter have opportunities to contribute, rather than being the abject recipients of largesse or expertise from the more privileged classes. The opportunities created in arts-based methods to treat participants as experts and co-creators in their own right also tie in with current calls in the U.S. and U.K. to 'decolonize' the university system. The legacies of such projects further democratize the research process. Increasingly, questions are being asked about the value for money that research represents. In many nations, it is funded from general taxation, often levied on people who earn far less than the relatively well-paid academics who implement research projects. The traditional output in the form of academic papers is often not terribly accessible, both in terms of style and because they are kept behind publishers' paywalls. A research project's creative and artistic legacies may have greater reach and longevity than the academic outputs.

5 Enriching the Story of Complex Interventions: The Complementarities of Art and Science

A further important point about the advantages of fusing the arts and sciences in the ways we have discussed here concerns what this means for so-called complex interventions. The term 'complex interventions' is generally taken to refer to interventions with several interacting components. While studies to determine the effectiveness of medications are still popular, where an intervention group is compared to a control group, increasing numbers of researchers and clinicians plan and implement complex interventions. The desire is to see how treatment packages, sometimes involving various interventions, may prompt desirable changes in diverse real-world settings. In the U.K., teams of experts in research methods commissioned by the research councils have developed guidance and recommendations about how complex interventions should be tackled (e.g., the U.K.'s Medical Research Council [36, 37]). The greater part of this documentation concerns the evaluation of complex interventions based around the underlying belief that the randomized controlled trial is the ideal evaluative technique and how this experimental logic can be applied to different kinds of intervention. By this logic, it is recommended that complex interventions are planned to facilitate evaluation, with clear study protocols, explicitly defined outcomes, and a range of possible outcome measures that could be linked statistically with the independent variables. This kind of approach reprises well-known concerns around experimental designs and the hypothetico-deductive logic of inquiry, familiar with the health sciences' research methods. The craft of the methodology expert here is enabling them to apply to the multiple variables and variegated real-world settings of complex interventions.

What is less well developed in guidance of this kind, however, is the imagination involved in designing an intervention in the first place. It is a task of eliciting from prospective participants and other stakeholders what the key issues are and how they might be most effectively addressed, examining the factors that might sustain an effective activity beyond the lifetime of the study, facilitating a conversation about problems which may be difficult or hard to talk about, and so on. It is here that the arts-based activities we have discussed in this chapter come into their own. Dealing with migration challenges, mental health problems, childhood nutrition, asthma, or insecure housing can be expressed, represented, researched, and tackled via creating artifacts, artworks, and performances.

The arts and humanities-based approaches remind us that people are more complex than the usual issues addressed in healthcare effectiveness reports. Measures of pain, disability and symptomatology, indices of well-being, morbidity and mortality, days off work, or hospital time are only a partial capture of the human condition. More often than not, people inhabit rich cultural environments where they and their families enjoy art, music, performance, and literature. Sometimes, this is the culture they make themselves; at others, it is enjoyed and shared through popular media; sometimes, it is the high culture of the museum or stage performance. But, all these involve activities, experiences, and indeed 'variables' that healthcare evaluation has so far barely touched upon.

This is not to undermine the value of experimental methods in determining the efficacy of interventions. After all, if we were to undertake some treatment, it would be good to know that we were likely to do better than a control group. As the MRC [37] documents, there are examples of widely-used invasive surgical procedures which, on more sober evaluation, perform no better than a placebo treatment in terms of reducing pain and impairment. Given the risks attendant on surgery and the time and cost involved, this raises legitimate questions about why anyone would bother. Despite the value in adjudicating between effective and less effective treatments, a focus on study design at this level does not necessarily lead to the development of insightful or innovative interventions in the first place. This is why we would argue for incorporating arts and humanities approaches and activities at the outset of a project when it is still being conceptualized and discussed informally. Our recent success in obtaining funding for work of the kind we have detailed above suggests that these issues are being taken seriously by funding bodies, at least in the U.K., and the reviewers and committees who evaluate proposals are increasingly persuaded of the value of incorporating arts, humanities, and conventionally scientific approaches. Especially, where the work involves people, the conceptualization, implementation, and uptake of innovation deriving from research can be enhanced through these approaches.

To illustrate what we mean, let us consider the project we did on community theater migration and resilience in Pune. There are a variety of factors that are believed to be involved in migration. As we might well imagine, there are socioeconomic and demographic elements. In our study, issues, such as challenging conditions in farming, ranging from drought, poor wages and employment opportunities, unemployment, and so on, were mentioned as drivers. In India and other parts of the world, various structural factors drive the movement from the countryside to cities. However, the life narrative interviews and community play elicited a variety of other issues. First was the ingenuity, resilience, and sheer hard work that went into moving, finding somewhere to live, finding a way to make a living, and bringing up one's family together with the hope that one's children might get a better education and thus avoid the hardship one had undergone oneself. A further factor mentioned by several participants was difficulties within the family unit. Domestic violence was mentioned, but this was not just between partners. Intergenerational violence emerged as a theme, and the mother-in-law was identified as a particularly culpable figure. It does not mean that intrafamilial violence is the only or even the major cause of migration. However, it is significant in terms of how it figures in people's moving accounts and why they did so. The participants' stories also underscored how they were unlike many popular media images of migrants or refugees as victims [38] or fragile, suffering, helpless, or needy [39]. Rather than being helpless or abject, people described considerable practical and intellectual agility in their own lives and migration journeys. This leads to the implication that migration is a kind of moral and political act; it is a way of saying that hardship and violence are unacceptable that one is off to seek better fortune elsewhere.

Consequently, the approach we undertook enabled us to effectively reframe the key questions of migration away from a focus on deficits and needs toward the celebration and enhancement of resilience and ingenuity. This was a direction not necessarily anticipated at the outset when the proposal was being constructed and highlighted the fruitfulness of the approach in opening up new vistas on the subject. Therefore, incorporating arts and humanities into the question of migrant welfare has yielded some useful reframing of the issues and the opportunity to construct further interventions that address aspects of the situation that have hitherto been overlooked. In our case, ingenuity and intrafamilial violence was relatively novel concerns. Moreover, these can prompt researchers to consider new outcome measures to gage a more rounded picture of what interventions might achieve.

Thus, the evaluative focus of much-existing discussion of complex interventions can usefully be extended and placed in context using the interdisciplinary synthesis; we are advocating here as part of the preliminary work to scope out the situation or as part of the evaluation itself. Further, it provides new perspectives on human beings' nature in their social context and the research process itself.

6 Conclusion

This chapter helps us to appreciate the value that can be gained using interdisciplinary syntheses. We have summarized key differences between research in the arts and research in the sciences in Table 1. It is a very rough characterization, of course, but reflects our impression that the sciences involve the definition, manipulation, comparison, and evaluation, whereas the arts are often concerned with the processes which take place rather earlier in the engagement, with how we see the world, with how people will react, with what our capabilities might be when we have not quite tried them out yet. As we have suggested several times in this chapter, perhaps all the approaches to knowledge and experience spring from common creative processes.

Table 1 Differing conceptions of research in the arts and the sciences

Research in the sciences	Research in the arts
Research tends to follow the format of prior work in the field, in what might be called 'normal science' or a 'research program'	There is an emphasis on novelty, breaks with the past, and practitioner self-differentiation
Research is often more planned, with well-formulated research questions and accepted methods in a particular field	Research is more spontaneous and involves finding questions or hypotheses in the first place
Research is undertaken by people who have undergone a long 'apprenticeship' of degrees, doctorates, and research teams' membership before undertaking their own projects	New research can be done by people, very early in their careers, or by laypeople

(continued)

Table 1 (continued)

Research in the sciences	Research in the arts
There is often an agreed set of criteria for judging solutions	The question of criteria for judging outcomes may itself be up for debate and discussion
A focus on comparison, evaluation, adjudication between competing explanations	A focus on how the field may be conceptualized, what the problems are, to begin with, how the activities may be designed
A focus on definitions (often operational) and measurement	A focus on meaning and aesthetic experience
Of interest to a narrow, technically skilled audience	Tries to appeal to a broader public outside the narrow, technical discipline-specific community

Core Messages

- Differences between the arts and the sciences originate in intellectual and institutional systems in the nineteenth century.
- There is common ground of creativity in arts and sciences; differences are institutional rather than epistemological.
- Integrating the arts and sciences can yield novel sources of ideas and research methods.
- Including the arts in research offers something back to the participants and lay audiences as works of art.
- Involving the arts in health projects enhances their power to enrich 'complex interventions'.

References

1. Snow CP (1959) Two cultures Cambridge: Cambridge University Press
2. Physics N (2009) Across the great divide (editorial). Nat Phys 5:309. https://doi.org/10.1038/nphys1258
3. Bronowski J (1975) The ascent of man London: Little, Brown
4. Yates F (1966) The art of memory. Routledge and Kegan Paul, London
5. Holton G (1978) The scientific imagination: case studies. Harvard University Press, Cambridge
6. Whewell W (1840) The philosophy of the inductive sciences founded upon their history. John W. Parker, London
7. Kuhn TS (1962) The Structure of Scientific Revolutions. University of Chicago Press, Chicago

8. Lakatos I (1978) The methodology of scientific research programmes: philosophical papers, vol 1. Cambridge University Press, Cambridge
9. Feyerabend P (1978) Science in a free society. New Left Books, London
10. Feyerabend P (1975) Against method. New Left Books, London
11. Lakatos I, Feyerabend P (1999) For and against method. University of Chicago Press, Chicago
12. Crawford P Brown B (2020) Health humanities: democratising the arts and humanities applied to healthcare, health and well-being. In: Bleakley A (ed) Handbook of medical humanities, Routledge, pp 401–409
13. Klugman CM (2017) How health humanities will save the life of the humanities. J Med Humanit 38:419–430
14. Skylar D (2017) Health humanities and medical education: joined by a common purpose. Acad Med 92:1647–1649
15. Crawford P, Brown B, Baker C, Tischler V, Abrams B (2015) Health Humanities. Palgrave, London
16. Jones T, Blackie M, Garden R, Wear D (2017) The almost right word: the move from medical to health humanities. Acad Med 92(7):932–935
17. Jones T, Wear D, Friedman LD (2014) Health humanities reader. Rutgers University Press, New Brunswick, NJ.
18. Purser A (2017) Dancing intercorporeality: a health humanities perspective on dance as a healing art. J Med Human. Open Access. J Med Humanit https://doi.org/10.1007/s10912-017-9502-0
19. Fancourt D, Perkins R, Ascenso S, Carvalho LA, Steptoe A, Williamon A (2016) Effects of group drumming interventions on anxiety, depression, social resilience and inflammatory immune response among mental health service users. PLoS ONE 11(e0151136):1–16
20. Harrison F Roberts AEL Gabrilska R Rumbaugh KP Lee C Diggle SP (2015) A 1000 year old antimicrobial remedy with anti-staphylococcal activity. mBio, 6:3. https://doi.org/10.1128/mBio.01129-15
21. Richard A, Rohrmann S, Vandeleur CL, Schmid M, Barth J, Eichholzer M (2017) Loneliness is adversely associated with physical and mental health and lifestyle factors: Results from a Swiss national survey. PLoS ONE, 7(12):e0181442. https://doi.org/10.1371/journal.pone.0181442.
22. BBC (2018) How should we tackle the loneliness epidemic? Available at: http://www.bbc.co.uk/news/uk-42887932. Last accessed: 17 Mar 2018
23. Carers UK (2019) Facts and figures https://www.carersuk.org/news-and-campaigns/press-releases/facts-and-figures. Accessed 21/6/2020
24. Lakhanpaul M, Benton L, Lloyd-Houdley O, Manikam L (2020) Nurture Early for Optimal Nutrition (NEON) programme: qualitative study of drivers of infant feeding and care practices in a British-Bangladeshi population. BMJ Open 10(6):e035347. https://doi.org/10.1136/bmjopen-2019-035347
25. Lakhanpaul M, Bird D, Culley L, Hudson N, Robertson N, Johal N, McFeeters M, Hamlyn-Williams C, Johnson M (2014) The use of a collaborative structured methodology for the development of a multifaceted intervention programme for the management of asthma (the MIA project), tailored to the needs of children and families of South Asian origin: a community-based, participatory study, Health Services and Delivery Research 2(28) https://europepmc.org/books/n/ukhsdr0228/pdf/. Accessed 27 7July 2020
26. Hudson N, Culley L, Johnson M, McFeeters M, Robertson N, Angell E, Lakhanpaul M (2016) Asthma management in British South Asian children: an application of the candidacy framework to a qualitative understanding of barriers to effective and accessible asthma care. BMC Public Health 16:510. https://doi.org/10.1186/s12889-016-3181-z
27. World Bank (2019) Internal migration in India grows, but inter-state movements remain low, https://blogs.worldbank.org/peoplemove/internal-migration-india-grows-inter-state-movements-remain-low. Accessed 22 June 2020

28. Jorm AF, Korten AE, Jacomb PA, Christensen H, Rodgers B, Pollitt P (1997) "Mental health literacy": a survey of the public's ability to recognise mental disorders and their beliefs about the effectiveness of treatment. Med J Aust 166(4):182–186

29. Gopalkrishnan N (2018) Cultural diversity and mental health: considerations for policy and practice, frontiers in public health, 6(179). https://doi.org/10.3389/fpubh.2018.00179

30. Fernando S (2015) Race and culture in psychiatry. Routledge, Hove

31. Fernando S (2014) Globalization of psychiatry—a barrier to mental health development. Int Rev Psychiatry 26:551–557. https://doi.org/10.3109/09540261.2014.920305

32. American Psychiatric Association (2013) Diagnostic and statistical manual, 5th Edition (DSM 5) American Psychiatric Association, Washington

33. World Health Organisation (2018) International classification of diseases 11th revision. World Health Organisation, Geneva

34. Pickersgill M (2014) Debating DSM-5: diagnosis and the sociology of critique. J Med Ethics 40:521–525. https://doi.org/10.1136/medethics-2013-101762

35. Cupp MA, Florschutz A, Beckingham A, Kisan V, Manikam L, Lakhanpaul M (2018) Birthing a better future: a mixed-methods evaluation of multimedia exposition conveying the importance of the first 1001 days of life. The Lancet 392(supplement 2):s27. https://doi.org/10.1016/S0140-6736(18)32191-3

36. Medical Research Council (2000) MRC framework for the development and evaluation of RCTs for complex interventions to Improve Health, Medical Research Council, London

37. Medical Research Council (2019) Developing and evaluating complex evaluations London: Medical research Council

38. Hameleers M (2019) Putting our own people first the content and effects of online right-wing populist discourse surrounding the european refugee crisis. Mass Commun Soc 22(6): 804–826

39. Amores JJ Arcilia C (2019) Deconstructing the symbolic visual frames of refugees and migrants in the main Western European media. In: 7th international conference on technological ecosystems for enhancing multiculturality, TEEM 2019; Leon; Spain; 16 October 2019 through 18 October 2019; Code 154355

Brian Brown holds the Health Communication chair at De Montfort University's Faculty of Health and Life Sciences. While focusing on language and communication in health, his work has also embraced theoretical and epistemological perspectives in social and health sciences, mental health issues, interpersonal relationships, qualitative methodologies, and sociology and health studies. He has completed fourteen books and around 90 articles and book chapters. Most notably, his books have included the groundbreaking health humanities (with Paul Crawford et al., Palgrave, 2014), evidence-based health communication (With P. Crawford and R. Carter, Open University Press, 2006), and evidence-based research: dilemmas and debates in health care (with P. Crawford and C. Hicks, Open University Press, 2003). Recently, he has been working with colleagues in the U.K. and internationally to find new ways for the arts and humanities to contribute to the vitality of healthcare research.

Monica Lakhanpaul is an academic researcher and practicsing paediatric consultant. She is a Professor of Integrated Community Child Health at UCL Great Ormond Street Institute of Child Health, UCL Pro-Vice-Provost for South Asia and adjunct Professor at Public Health Foundation India. She is committed to improving the lives of those in the most vulnerable communities through holistic, cross-sectoral interdisciplinary interventions that encompass health, environmental, and educational factors. She is recipient of the Asian Women of Achievement Award, Royal Society of Public Health award, BSA media fellowship. Her research tackles some of the most pressing issues facing marginaliszed, minority, and vulnerable families globally such as early years, nutrition, development, and mental health. She uses participatory research, citizen science, and arts based ensuring that communities are involved in co- developing holistic- integrated solutions. She has over 180 publications to her name.

Thinking Deeper, Wider, Further: Visual tools for the Pandemic 3.0 and the Game-Changing Pathways Ahead

19

Joe Ravetz

"To change something, build a new model that makes the existing model obsolete."

Buckminster Fuller

Summary

This chapter demonstrates the principles of a synergistic *Science-3.0*, with a live example of visual thinking in action. The case study is the COVID-19 pandemic, which raises urgent questions on new modes of scientific knowledge and inquiry in controversial situations. In this case, it seems there are options: "the science" may be part of a reactive crisis management approach—or possibly, part of a transformative pathway for an emerging agenda, the *"Pandemic 3.0."* This raises the agenda and the potential for a more diverse and pluralistic *Science-3.0*, better able to engage with *wider* communities and *deeper* societal challenges.

J. Ravetz (✉)
Manchester Urban Institute, Manchester University, Manchester, UK
e-mail: joe.ravetz@manchester.ac.uk

School of Environment Education and Development, Manchester University, Oxford Rd, Manchester M13 9PL, UK

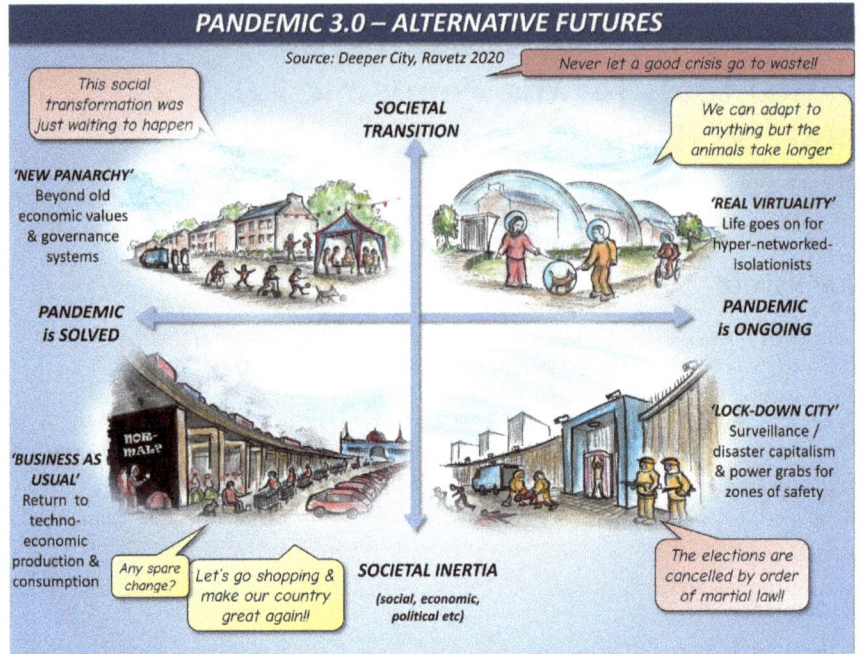

Alternative futures

[Graphics © Joe Ravetz under Creative Commons License: 'Attribution-Non-Commercial-Share-Alike' 4.0 International (CC BY-NC-SA 4.0). Specific restrictions imposed by this licence are on https:// creativecommons.org/licenses/by-nc-sa/4.0/].

The code of this chapter is *01101111 01101100 01101111 01101110 01000101 01110110 01110100 01110101 01101001.*

Keywords

COVID-19 · Crisis management · Participatory gaming · Transition pathways · Visual thinking

---

1   Introduction

The COVID-19 pandemic raises huge questions on the role of science in controversial and urgent situations—an ideal moment to rethink common assumptions about science, knowledge, and society. Whichever way (if and when) the situation is resolved, one thing seems clear that the former accepted boundaries of scientific disciplines and policy sectors are all up for debate. In this case, it seems "the science" has a choice: it can continue to support a reactive crisis management approach, or just possibly, seek to enable the transformative pathways forward—what we frame here as a *"Pandemic 3.0."* This raises the stakes and the potential for a more diverse and pluralistic *Science-3.0,* one which is able to engage with *wider* communities and *deeper* societal challenges.

This short-illustrated chapter demonstrates the principles and practice of a synergistic *Science*-3.0, with a live example of visual thinking in action. As a chapter in a volume on "integrated science," this is quite deliberate in putting up a set of cartoons as the primary content. Many other sources on this topic have epidemiological graphs, public health, or system diagrams: here, we are concerned with the cross-cutting nature of the problem and the synergistic scope of any forward pathways. So, the visual examples shown here are not so much "results," more like a work in progress, with the aim that others can take it forwards in the future.

In this chapter, the next section outlines some of the agenda of visual thinking and its uses. The third section is a demonstration of visual thinking for Pandemic 3.0 and the 'Corona-Games' series. There follows an outline agenda for a synergistic *Science-3.0,* in which the visual dimension is inter-connected with others. A short conclusion then points to implications for further action.

2 Visual Thinking and Its Uses

The COVID-19 pandemic is an opportunity to try new forms of creative thinking for an existential challenge. As a "grand societal challenge," it calls for a parallel *"grand knowledge challenge"*: a synergistic cognitive space, where problems and responses, conflicts and controversies, uncertainties and ignorance, opportunities and risks can each inter-connect.

Such an extended frame of multiple knowledges calls for multiple channels for learning, thinking, creating, and producing. Visual arts, for example, are a long way from the reductive rationality of most modern science but have the potential to connect in diverse ways to other levels of human and experience: and in practical terms, they also connect more easily with the written text [1]. Visual thinking can mobilize and empower alternative modes of thinking: it can demonstrate tacit and experiential knowledge and the creative personal dimension. It can bridge some of the gaps between problems and responses and with a design thinking approach in forwarding pathways. Visual thinking offers one kind of "trading zone" between different forms of analysis and experience [2]. A cartoon in the right time and place can represent a powerful cross-cutting truth, with different modes of knowledge working in parallel, which is hardly possible in mainstream scientific terms.

Visual thinking methods can be applied to deliberation processes, which again is about the experience and inter-subjective exchange, as much as technical information. For instance, "graphic facilitation" is now established as a valuable technique in process-focused meetings and workshops. In parallel, the methods of "visual synergistics" have been developed (by the author) from standard foresight methods, where visual material (from on or off-site) can be a powerful catalyst for new insights, for small interviews or larger group meetings. Overall, we can define three strands of visual/rational thinking combinations:

- visualization IN process—real-time production of images to support dialogue or deliberation for visioning, consensus building, conflict mediation, road-mapping, and many other forms of negotiation.
- visualization OF process—representing and capturing dialogue, debate, argument: for example, a cartoon at the right place and time can communicate a complex set of nuances and ironies, hardly possible in any other way.
- visualization AS process—catalyzing creativity, mobilizing public agendas and dilemmas, empowering marginalized communities, co-producing shared experience, and heritage.

Overall it seems clear that visual thinking can go beyond technical information, to include lived experience on many levels of the human psyche. Figure 1 explores some of that agenda.

3 Demonstration: Pandemic 3.0 and the Corona-Games

Here we demonstrate some of the above ideas, with a live example of work in progress (for more, see www.urban3.net/mind-games). This builds on the ideas of the *Deeper City*, an in-depth exploration of collective intelligence, where visual thinking is integral to its methods and tools [3].

This focuses on the concept of a *Pandemic 3.0—defined as "the study of whole systems in pandemic-induced disruption and transformation towards a collective health intelligence"* [4]. This extended frame centers on public health and epidemiology in relation to social, technical, economic, ecological, political, and cultural layers (the "STEEP" combination used by the foresight community). And it seems the vital qualities of *collective intelligence* can be mapped on three main *Modes* of system organization, ranging from the technical to the *co-evolutionary*. In summary,

- *Mode-I* systems are framed as technical problems to be fixed by functional solutions: so, the *Pandemic-1.0* is basically about epidemiological modeling, public health, and medical care systems;
- *Mode-II* systems are framed with *evolutionary* "winner takes all" competition: for the *Pandemic-2.0,* we look to markets and smart innovations, along with the typical side-effects of waste and inequality.
- *Mode-III* systems are framed as *co-evolutionary* "winners are all," with synergies between many layers of logic and value—social, technical, economic, ecological, political, cultural, and others. A *Pandemic 3.0* system mobilizes

Fig. 1 Wider visions *(Graphics © Joe Ravetz under Creative Commons License: 'Attribution-Non-Commercial-Share-Alike' 4.0 International (CC BY-NC-SA 4.0). Specific restrictions imposed by this licence are on* https://creativecommons.org/licenses/by-nc-sa/4.0/)

deeper forms of *collective intelligence* across *wider* communities of interest to bring all these together. The capacity for collaborative learning, thinking, innovation, and co-production seems to be the key to steering a pandemic crisis towards opportunities for transformation.

In practical terms, all three *Modes* are needed to work in parallel. While *Mode-I* does the basic "problem fixing," *Mode II* works with incentives and social psychology, and *Mode-III* brings all layers together for a *Pandemic 3.0* level of collective intelligence.

4   Visualizing the Pandemic 3.0

In the COVID-19 crisis, other challenges such as climate change or growing inequality are unlikely to disappear; rather, they could escalate as new structures of power and exploitation emerge. If this pandemic can be contained or resolved, then we can get back to work on these challenges and others: but if it continues (as seems likely) to be messy and divisive, or indeed as the next pandemic arrives, then we could face new challenges alongside the old.

The sketches below show three angles on this global crisis of critical danger and opportunity. They start with the saying "never let a good crisis go to waste—and then ask, if new systems of *Mode-III* collective intelligence for social-political-economic cooperation can emerge from this crisis, how to let these grow and flourish? And how to counter or bypass the forces of "winner takes all" populism, of exclusion and intolerance, hijack of truth, and expropriation of livelihoods? This is a brief sketch for a planet-sized challenge, which draws on current thinking on *"collective intelligence and the pathways from smart to wise"* [3]. For the Pandemonic 3.0, such pathways aim to turn crisis into opportunity—and if this crisis can be resolved, then to better prepare for the next.

4.1   Scenarios—Unknowns or Unknowable?

At this moment, it is an epidemiological unknown whether the COVID-19 virus can be contained or continues to diversify and re-emerge: but it is a deeper kind of unknown as to how social, economic, and political systems interact with this epidemiology. It is also a deeper unknown (perhaps "unknowable"), whether or not social-economic-political systems could return to the old normal or transform towards some kind of "new normal." We can map out the combinations as possible "what-if" scenarios, visualized in the 'Graphical Abstract'.

- new panarchy: we ask, what-if progress is resumed, and the pandemic solved while staying vigilant for the next one? Meanwhile, there is deeper and wider learning from the 2020 episode and a serious agenda to look beyond old-style hierarchies and extractive systems.
- business as usual: as the general direction of most official perspectives, this simply looks to the other side of the pandemic and aims to reconstruct the familiar game of techno-economic production and consumption.
- real virtuality: here everything has changed, with technology as the enabler for hyper-networked-isolationists, a new normal of video holograms, decontamination suits, and sterile pods. While humans are endlessly adaptable, this future brings huge challenges for individuals and communities and maybe opportunities.
- lock-down and out: a familiar techno-dystopia of "Blade-runner" type surveillance/disaster capitalism. Here the ongoing pandemic and its effects of disruption

and trauma is an open door for power-mongers and warlords who merge with the tech corporates. The graphic shows how "safe zones" can easily turn into zones of exclusion and oppression.

It gets more interesting, as it emerges these scenarios are not only neutral visions of a possible distant future—they are more like active and contested grabbing of the present and near future (about a week at the time of writing). It also gets more interesting to explore the scenarios not as distinct and separate, more like different angles on an inchoate combination of overlapping realities.

4.2 Alternative Games—From evolution to Coevolution

Here the players are not letting their crisis go to waste; rather, they are pushing their interests by whatever means, in Fig. 2. Here we sketch a typical process of learning, thinking, co-creation, and co-production—asking the question, how would different kinds of actors adapt and evolve with these challenges and opportunities? Again, we can map different levels of cognitive systems, from linear (Mode-I) to evolutionary (Mode-II) to co-evolutionary (Mode-III).

With a linear Mode-I response, seen on the left of Fig. 2, we plan ahead with the best available evidence, with enforcement on transmission paths, with full backups of medical equipment, and with fully functioning communications (seen in just a few countries so far). This is the framing of epidemiological analysis, such as the modeling study, which informed the UK response [5].

When the shortcomings of the linear emerge, then Mode-II evolutionary thinking then comes into play, with advanced risk management, socio-psycho "nudges" or incentives, and smart urban micro-engineering. But for those with the aim of political or economic advantage, the crisis is also an opportunity to accelerate "control." The sketch on the left shows a likely direction of travel towards a dystopian logic of digitally enhanced social engineering solutions.

In contrast, the co-evolutionary Mode-III shows a deeper level of aspirations—where the problem "frame" is about how to use such a crisis for social-economic-political transformation. Here we are talking not only problem fixing "solutions" but extended pathways, which combine all three Modes. We look for advanced systems of integrated tracking of cases and transmissions (Mode-I): and for the best social psychology communications, with incentives and resonance for hearts and minds (Mode-II). And most of all, we look for a co-evolutionary mesh-work structure (Mode-III), with a 'collective social intelligence' in the learning and thinking capacity of communities/organizations/networks. All these point towards a transformation in systems of mutual aid and collective empowerment. It also highlights some fundamental political choices, between a "bounce back" to structures of inequality and alienation - or a "bounce-forward" towards a collective intelligence transformation.

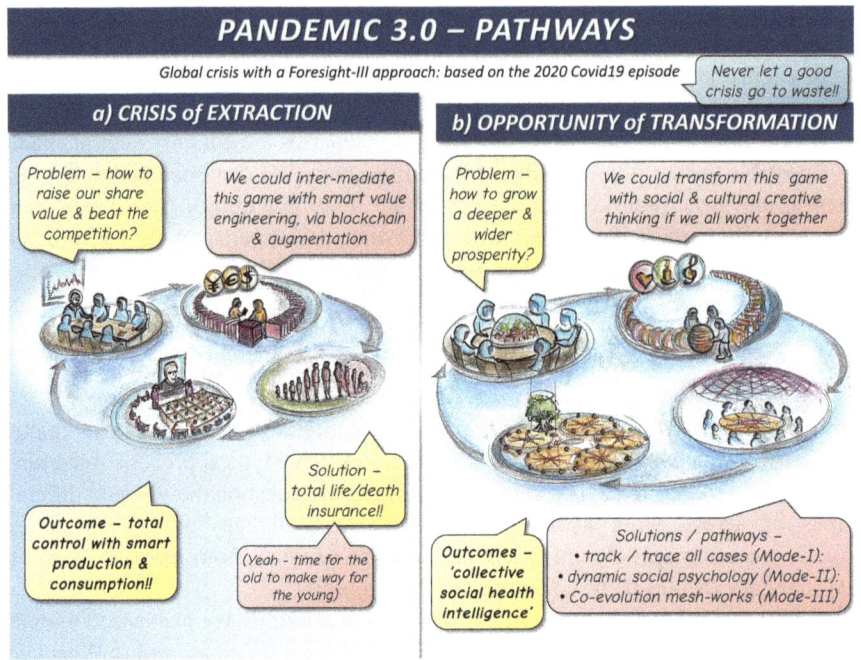

5 Corona-Games

To explore such ideas, which are deeper and wider than normal limits, we demonstrate work in progress on the *Corona-Games*, an application of the *Mind-Games* [3]. This is an ongoing experiment with visual foresight, with the aim of visualizing the "unthinkables and unknowables" to explore challenges of *deeper complexity* and the pathways from crisis to opportunity. The basic idea of the *Mind-Game* is simple and comes from many experiences running futures/foresight workshops. We track the different players and their typical roles in a situation (from beggars to billionaires), if possible, around an actual table (otherwise a virtual equivalent), with results drawn on flipcharts (or the online equivalent). The steps include:

- We explore the challenge in the form of a "game," based on the roles and interactions of the stakeholders. Where possible, we look for models in existing games, from chess to chequers, or from poker to pokemon.
- explore what happens through the game-play, either in concept or for real;

- track the results, the winners/losers, and the overall effects: in particular, the crucial question is whether there is a zero-sum result or rewards for cooperation?);
- experiment with game-changer ideas with new rules or new pieces; the general goal is to shift from "zero-sum" (*winner takes all*) towards a "positive sum" for transformation (*winners are all*); explore how the new games might work, and what are the overall results, for goals such as equality, sustainability, cooperation, etc.;
- finally, we explore the likely enablers or pathways towards these transformative game-changers.

From experience, this cartoon format can help to highlight deeper kinds of knowledge, in a safe space of "what-if" scenario thinking. Game theorists and sports scientists study highly complex formal games in a detailed technical sense, but many real-life games have a *deeper complexity*—where social, technical, economic, ecological, political, cultural, urban, and other layers are all mixed and entangled. To explore such *deeper* problems, and envision *deeper* pathways with *wider* communities of interest, we have to think out of the box, via experimentation with the games, to unlock the art of the possible.

The following section is a brief tour of the Corona-Games series in progress.

5.1 Introducing the Corona-Games

The Corona-Games provide an alternative view on pathways to turn the COVID-19 crisis into opportunity—with some notion on a Pandemic 3.0 system based on collective learning, thinking, and collaboration between all concerned. But first, we visualize the forces of power, wealth, knowledge, and ideology and how each plays its part in the game for different objectives (Fig. 3).

5.2 Deeper Threat Multiplier Game #1

We can explore the notion of *Deeper Threat Multipliers* (as used in the USA security/defense industry), visualized as a game-play in Fig. 4. This illustrates the scale of the challenge, where the COVID-19 may disrupt many systems and push them to thresholds and tipping points. The trigger could be in the Middle East, Southeast Asia, or somewhere not yet on the radar—the immediate causes could be climate chaos, political corruption, financial collapse, or just the "multiplier" effect of the pandemic on the whole combination. The game image here is the venerable strategy game of Go, where the game-play shows apparently random weak signals over a wide area, which suddenly emerge as an existential threat.

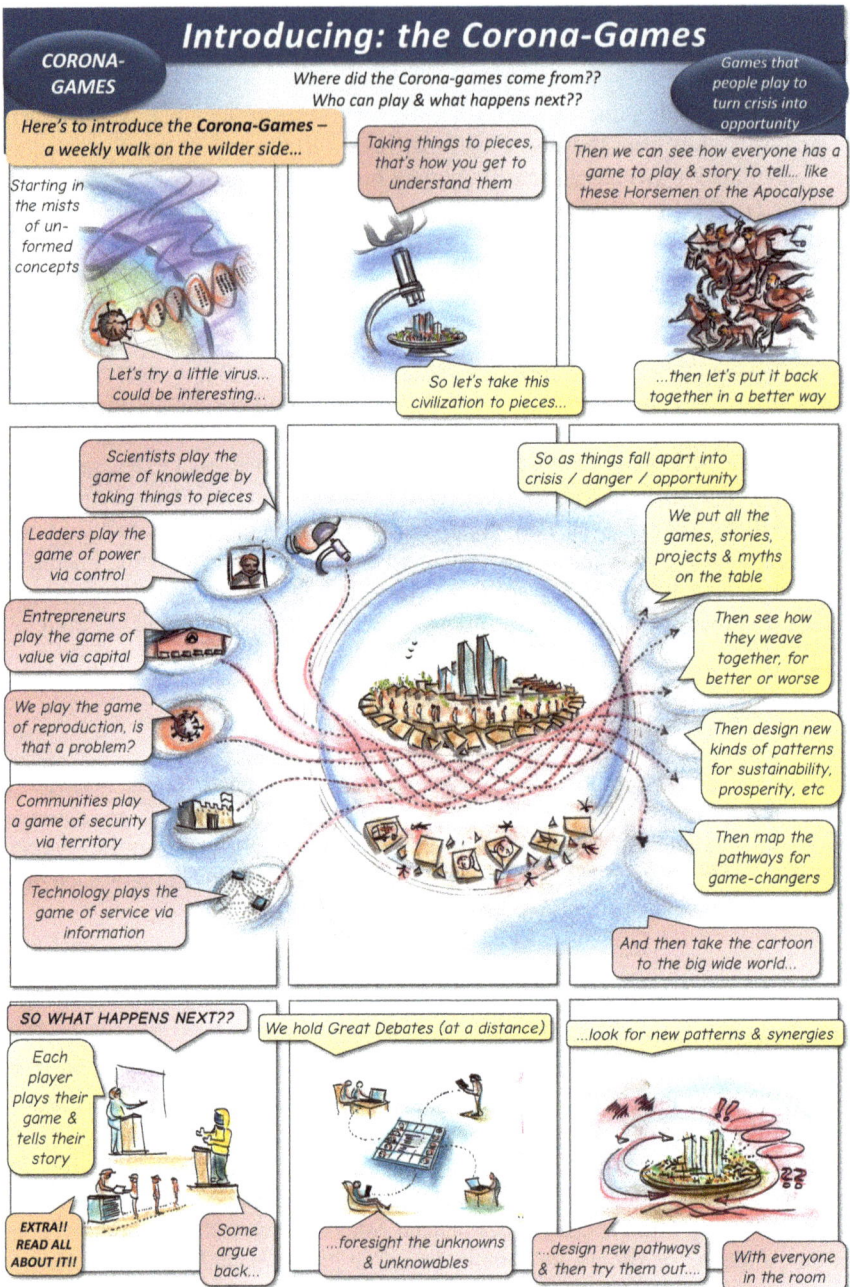

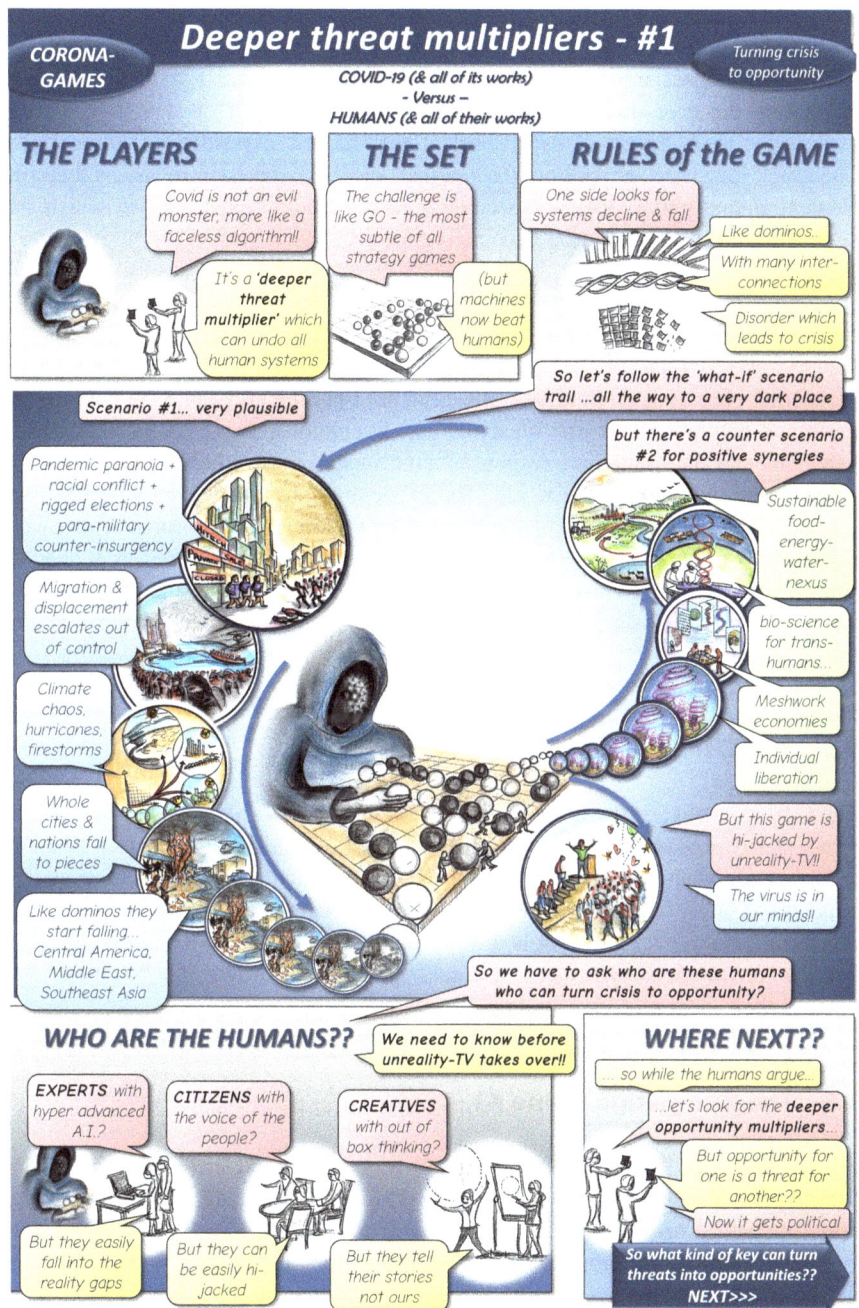

Fig. 4 Deeper threat multipliers *(Graphics © Joe Ravetz under Creative Commons License: 'Attribution-Non-Commercial-Share-Alike' 4.0 International (CC BY-NC-SA 4.0). Specific restrictions imposed by this licence are on* https://creativecommons.org/licenses/by-nc-sa/4.0/)

5.3   Risk Game #1

This game focuses more on the personal and political levels. With the game of navigating the everyday decisions (e.g. "whether to go out or stay in"), we can explore the many layers of psycho-social risk, hazard, exposure, or vulnerability. The quick tour here shows how the rational scientific approach to risk is then one element among many other perspectives—social, cultural, political—in which risk is woven into the experiential and inter-subjective fabric (Fig. 5).

5.4   The Cities Game #1

Meanwhile, life goes on in cities—meaning not only the grey areas on the map but the many layered matrix for lifestyles and livelihoods. It seems the COVID-19 pandemic and the immediate responses have sucked much vitality from cities and urban experience around the world. The existential question is then whether cities bounce back to the old systems or bounce-forward to a new normal, with a very different set of winners and losers? Such a game can be visualized as a head-on contest of COVID-19 versus the humans/citizens. In this game, the city may descend into fear and paranoia—or maybe it can adapt and find new ways to live and thrive (Fig. 6).

5.5   The Cities Game #2

The Cities Game #2 is the beginning of an inquiry on new challenges—such as how to adapt the physical city, from a place of paranoia and contagion to a place of security and liveability. Is the solution a smart segmented city, a carefully organized structure of safety and value, where citizens are allocated into high value/low risk or other combinations? The result is a new kind of axis of safety/contagion, which is overlaid on previous structures of wealth and power, where the "winners" can achieve their ideal city [4]; see Fig. 7.

5.6   The Knowledge Game #1

Finally, how can science-technology-innovation (STI), and the knowledge industries in general, adapt and shift the COVID-19 crisis towards the Pandemic 3.0 type opportunities? This last Fig. 8 shows a university library in chaos, where conventional divisions of subjects and fields and sectors are no longer relevant. Each player in this Knowledge Game #1 is looking for signs and threads which add up to some kind of coherent reality in a bigger picture. While the *"Experts"* look for robust scientific evidence in their field, other kinds of *"Knowledge But Not As We*

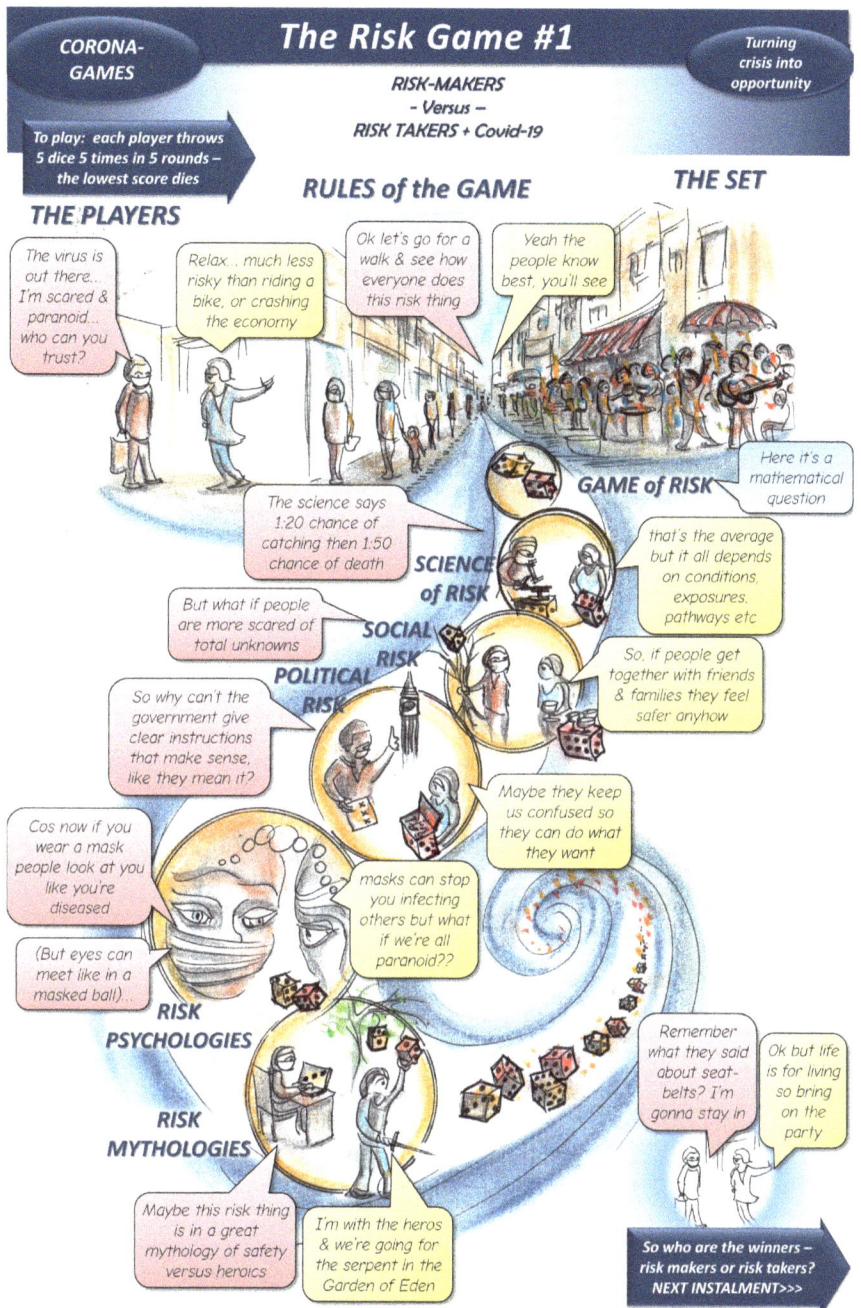

Fig. 5 The Risk Game *(Graphics © Joe Ravetz under Creative Commons License: 'Attribution-Non-Commercial-Share-Alike' 4.0 International (CC BY-NC-SA 4.0). Specific restrictions imposed by this licence are on* https://creativecommons.org/licenses/by-nc-sa/4.0/)

Fig. 6 Cities game #1 *(Graphics © Joe Ravetz under Creative Commons License: 'Attribution-Non-Commercial-Share-Alike' 4.0 International (CC BY-NC-SA 4.0). Specific restrictions imposed by this licence are on* https://creativecommons.org/licenses/by-nc-sa/4.0/)

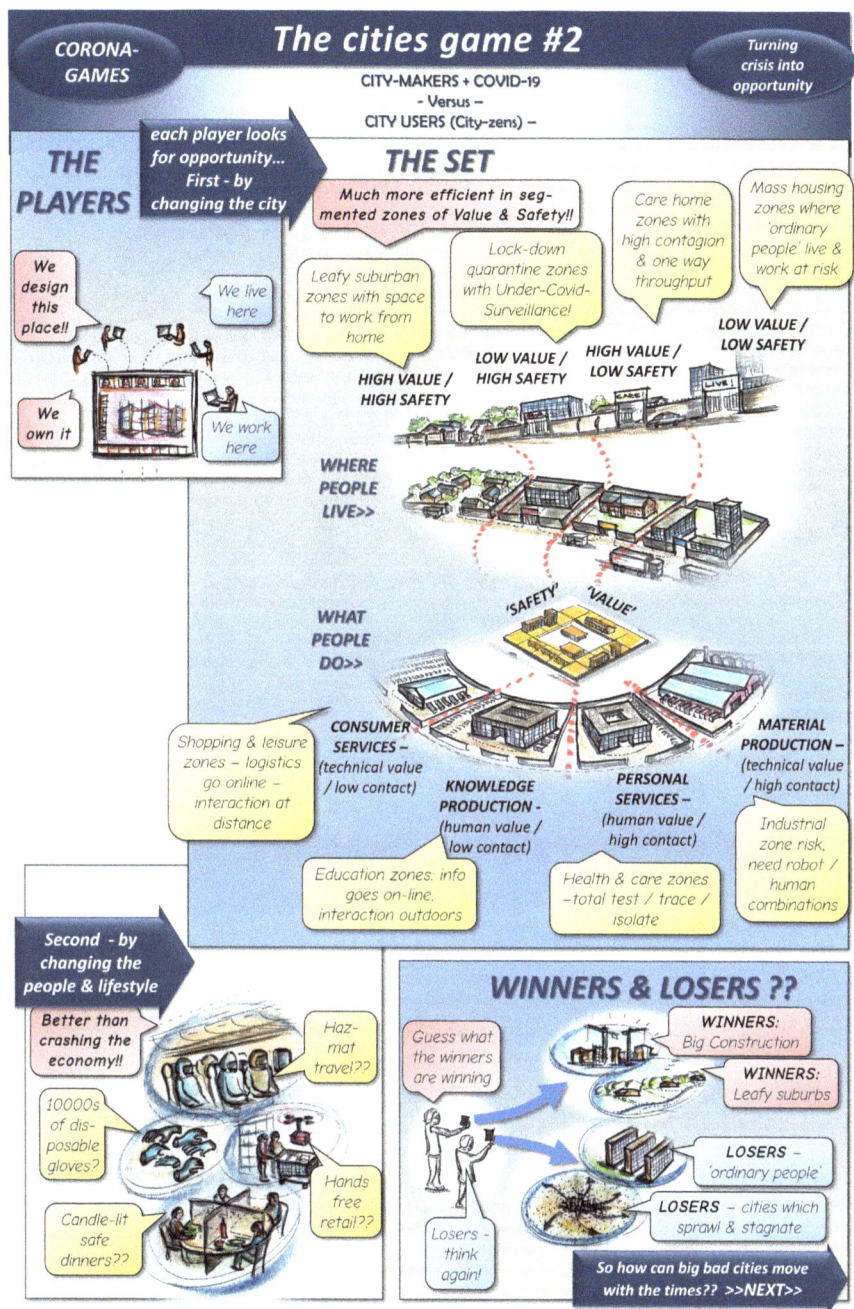

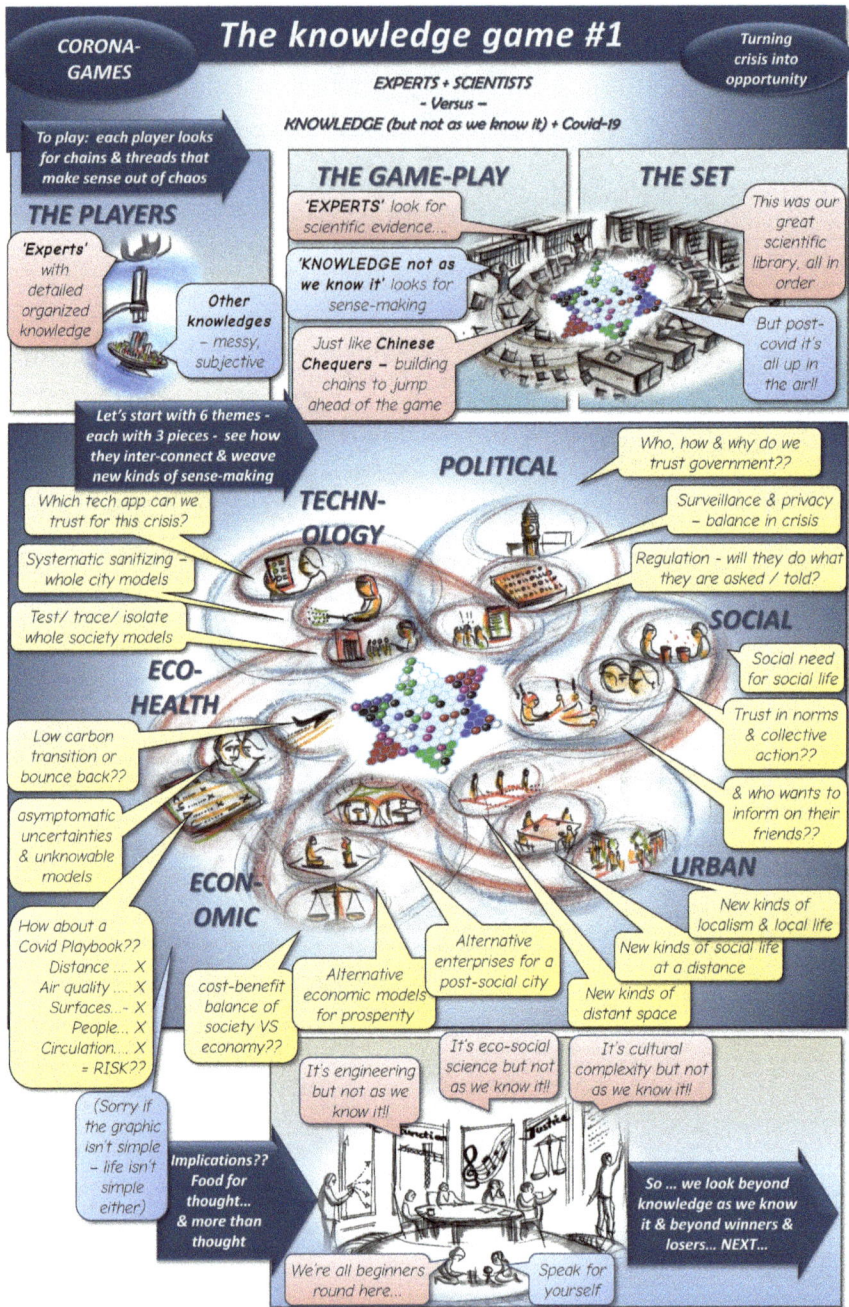

Know It" seek alternative experiences and rationalities. This visualization sees a traditional game of Chinese Chequers, where players compete to build moving chains that enable their pieces to leap across the board—an interesting metaphor (Fig. 8).

6 Towards a Synergistic Science-3.0

These adventures in visual thinking and foresight-gaming point towards a transformative agenda, framed as a *Science Mode-III,* or *Science-3.0* [1, 3, 6].

Generally, the direction of modern (Western) science was towards fundamental laws which could be framed with "elegance and parsimony." But for catastrophes of *deeper complexity,* such as COVID-19 among others, the debate is conflicted and controversial, and elegance or parsimony are very scarce. There are many insights pointing in similar directions, such as "transformative sustainability science," "Science-3.0," or framed by some as *"Mode-II"* science [7, 8]. Post-normal science focuses on Grand Societal Challenges, "wicked problems," "problematic knowledge," "critical heuristics," and the Rumsfeld "known-unknowns," where uncertainties multiply and controversies cannot be resolved [9–11]. There is a new awareness of the interactions between different worldviews and domains, and the "trading zones" and "boundary objects" between them. This might link, for instance, earth science to political processes, or "lay" knowledge of communities to theory-based technical analysis, or creative culture to financial innovation [12, 13].

The question is then whether to pull back from unknowns/unknowables towards safer "knowns" where the rational-reductive model applies—or, is it possible to swim in *deeper* waters, in a more dynamic interplay of knowns and unknowns, between analysis and experience? If there is more certainty on engineering, then maybe less on ethics; if more agreement on social trend data, then more controversy on causes. In this synergistic landscape and worldview, there is always the next stage of exploration in the cycle of discovery.

Visual thinking is a powerful catalyst for "out-of-box" thinking, as one kind of "trading zone" between analysis and experience. It helped, for instance, to explore the interconnections between global climate science and a creative-therapeutic-cathartic recovery process [6, 14–16]. From experience (as a graphic facilitator), a cartoon can represent a powerful cross-cutting insight, with *Mode-I, II, and III* types of knowledge working in parallel, almost impossible in normal scientific terms. Overall, for each *"Grand Societal Challenge"* as defined by the OECD or World Bank, it seems there's also a *"Grand Knowledge Challenge"*: a *nexus-connexus* of intractable controversies, existential risks, multiple uncertainties, pressing risks, and opportunities. And a question is left hanging: can such knowledge challenges continue to grow ever grander?

7 Conclusion

The rationalist reductive model of scientific knowledge is very effective for problems that are defined and tractable: but clearly, it may struggle with problems of deeper complexity, intractable uncertainty and controversy, and so-called "grand societal challenges." The COVID-19 pandemic has raised such a prospect: where the unfolding crisis showed that medical science had to interact somehow with politics, business, psychology, culture and society. So, with this example in mind., we can debate and envision the scope of a more synergistic and integrated *Science-3.0*. The key features of such an agenda could include:

- *Deeper* scope of knowledge to include multiple values/worldviews, with synergistic interconnections between policy, business, civil society, and others;
- *Wider* communities of practice and research, suited to systems of deeper complexity, beyond anyone form of analysis;
- *Further* insights on the cognitive dynamics and moving frontiers between rational, experiential, instrumental, and transcendent forms of knowledge.

Some parts of this picture may be realized through big data, the Internet of things, and artificial intelligence. Other parts may be realized through more advanced forms of deliberation, social learning, or action research. And a third dimension comes by linking rational-reductive research to other creative-experiential channels: these can include visual, film, performing, narrative, or public arts. With the example of the Pandemonics 3.0 agenda, and some creative methods to explore it, this short chapter shows a first demonstration and quick tour of the potential ahead.

Core Messages

- Global "grand challenges" call for a transformative "Science-3.0" model with extended boundaries and multiple channels.
- The "Pandemonics-3.0" example shows the potential of visual thinking and story-lining for such extended thinking.
- This calls for "deeper" layers of knowledge—both technical-economic, and social, psychological, political, cultural.
- It includes "wider" communities of interest: policy, business, civil society, academia, and citizens.
- It provides "further" horizons of change where "analysis" is inter-connected with "synthesis," between and beyond disciplines.

References

1. Ravetz J (2011) Exploring creative cities for sustainability with deliberative visualization. Sustainable city and creativity Ashgate Publishing, London, UK, p 339
2. Galison P (1997) Image and logic: a material culture of microphysics. University of Chicago Press
3. Ravetz J (2020) Deeper city: collective intelligence and the pathways from smart to wise. Routledge
4. Ravetz J (2020) Pandemic-3.0–from crisis to transformation–exploring the COVID-19 challenge. J Future Stud. https://jfsdigital.org/2020/08/18/pandemic-3-0-from-crisis
5. Ferguson NM, Laydon D, Nedjati-Gilani G, Imai N, Ainslie K, Baguelin M et al (2020) Impact of non-pharmaceutical interventions (NPIs) to reduce COVID-19 mortality and healthcare demand. Imperial College COVID-19 Response Team, p 20
6. Ravetz J, Ravetz A (2017) Seeing the wood for the trees: Social Science 3.0 and the role of visual thinking. Innovation: Euro J Soc Sci Res 30(1):104–120
7. Miedema F (2012) Science 3.0: real science, real knowledge. Amsterdam University Press
8. Nowotny H, Scott PB, Gibbons MT (2013) Rethinking science: knowledge and the public in an age of uncertainty. Wiley
9. Leach M, Stirling AC, Scoones I (2010) Dynamic sustainabilities: technology, environment, social justice. Routledge
10. Ravetz J (2004) The post-normal science of precaution. Futures 36(3):347–357
11. Stacey RD (2007) Strategic management and organisational dynamics: the challenge of complexity to ways of thinking about organisations. Pearson Education
12. Balducci A, Mäntysalo R (2013) Urban planning as a trading zone. Springer
13. Zlatev J, Racine TP, Sinha C, Itkonen E (2008) The shared mind: perspectives on intersubjectivity, vol 12. John Benjamins Publishing
14. Ravetz J (2013) Beyond the linear: the role of visual thinking and visualization. In: O'Riordan T, Lenton T (eds) Addressing tipping points for a precarious future. Oxford University Press for the British Academy, pp 289–299
15. Sousanis N (2015) Unflattening. Harvard University Press
16. Horn RE (2018) The little book of social messes: connecting the smudges in an age of wicked problems. Macro-VU

Joe Ravetz is the Leader of the Manchester Urban Institute on Future-Proof Cities. He has worked on sustainable cities, environmental policy, economic development, and foresight/systems analysis as Co-Director of the Collaboratory for Urban Resilience at the University of Manchester. From this, he developed the *Synergistics* methods for working with "*collective intelligence*" in urban, economic, and policy systems. His main publications include "*City-Region 2020,*" "*Environment and City,*" and the new *Deeper-City-Collective-Intelligence-and-the-Pathways-from-Smart-to-Wise*. He has advised policy (UNIDO, UN-Habitat, DG Regio, EU Parliament, BEIS, Defra, etc.) and is also a Principal of SAMI Consulting. Joe"s current projects include global urban/climate risk, urban policy innovation, citizen science, and pandemic research. As a former architect and planner, he also works as a visual thinker, systems designer, and foresight facilitator in many countries.

The Clinical Benefits of Art Therapy: Definition, History, and Outcomes with a Focus on Music Therapy

Niloufar Yazdanpanah, Helia Mojtabavi, Heliya Ziaei,
Zahra Rahimi Pirkoohi, Elham Rayzan, and Nima Rezaei

"I would teach children music, physics, and philosophy; but most importantly music, for the patterns in music and all the arts are the keys to learning."

Plato

Summary

Art therapy is a nonverbal means of communication and self-expression with possible benefits in individuals with various mental or physical illnesses. Notably, music is among the most documented art modalities and has been shown to affect the brains' reward circuitry. It offers complementary therapy for neuropsychiatric disorders, e.g., brain injury, stroke, movement disorders, autism, dementia, depression, anxiety, schizophrenia, personality disorder, and seizure. Also, music therapy contributes to the autonomic regulation of

N. Yazdanpanah · H. Mojtabavi · H. Ziaei · Z. R. Pirkoohi · E. Rayzan
Health and Art (HEART), Universal Scientific Education and Research Network (USERN),
Tehran, Iran

N. Yazdanpanah · H. Mojtabavi · H. Ziaei · E. Rayzan · N. Rezaei (✉)
Research Center for Immunodeficiencies, Children's Medical Center, Tehran University
of Medical Sciences, Dr. Qarib St, Keshavarz Blvd, Tehran 14194, Iran
e-mail: Rezaei_nima@tums.ac.ir

Z. R. Pirkoohi
Shahid Beheshti University, Tehran, Iran

N. Rezaei
Integrated Science Association (ISA), Universal Scientific Education and Research Network
(USERN), Tehran, Iran

Department of Immunology, School of Medicine, Tehran University of Medical Sciences,
Tehran, Iran

N. Rezaei (ed.), *Multidisciplinarity and Interdisciplinarity in Health*,
Integrated Science 6, https://doi.org/10.1007/978-3-030-96814-4_20

cardiovascular outcomes, such as blood pressure, heart rate, and heart rate variability. Although the use of music dates back to very early primitive societies, the question of whether musical interventions should be specifically tailored per patient disease or "does one size fit all" remains unanswered.

Art therapy

Photos are taken from different angles of clay artwork. Shamsa on a ring dish combined with musical instruments' design represents the sun along with the paths of the planets.

The poem written on the dish is from Masnavi-Book 4 and says:

(For) the shrill noise of the clarion and the menace of the drum somewhat resemble that universal trumpet

Hence philosophers have said that we received these harmonies from the revolution of the (celestial) sphere,

(And that) this (melody) which people sing with pandore and throat is the sound of the revolutions of the sphere

Some believe that Mulana's poem took inspiration from Pythagoras and Plato, who believed that earth sounds and songs result from the soul's motion. The motion of the planets reflects they possess musical expressions. These musical expressions are the source of joy; for this, when our body plays/listens to artwork, our soul makes her dance and forgets sadness.

[Adapted with permission from the Association of Science and Art (ASA), Universal Scientific Education and Research Network (USERN); Made by Heliya Ziaei].

The code of this chapter is *01101101 01110100 01101001 01110101 01000001 01110011*.

Keywords

Anxiety · Art therapy · Autism · Brain injury · Cancer · Cardiovascular · Dementia · Depression · Movement disorders · Music therapy · Neuropsychiatric disorders · Nonverbal · Pain · Personality disorder · Schizophrenia · Seizure · Stroke

1   Introduction

The creativity integrated with art-making processes accelerated by restitution or the convalescent period is the core idea of art therapy. It is a nonverbal means of communication and self-expression with possible benefits in individuals with various mental or physical illnesses [1]. Psychological and physiological aspects of art therapy include stabilizing heart rate (HR) and blood pressure (BP) and controlling the cortisol level [2, 3]. Besides, it positively affects decreasing pain, anxiety, and grief, improving individuals' self-expression and social communication [4, 5]. Considering the accepted pleasurable nature of art, it might be possible that neural response to art triggers the reward centers and circuits in the brain [6]. Several neuroimaging surveys have been conducted in an attempt to discover the brain regions which react to visual art perception. These studies indicated the bilateral medial orbitofrontal cortex, left ventral striatum, and right amygdala as brain regions involved in art perception [7–9]. Interestingly, the mentioned regions are also parts of the brains' reward circuitry [6]. To further investigate the direct effect of art perception on the brain, the observations about the deterioration of the artistic abilities in patients with dementia or acquired brain injuries proposed critical attention [1, 10]. There are records of cases that developed a sudden enhanced ability to perform art after a brain injury or even during the dementia course. As per currently available data, frontotemporal dementia (FTD) can induce de novo artistry tendency specially in cases that degeneration is broader in temporal lobes compared to the frontal lobes [11, 12] while in Alzheimer's disease, the artistic skills are expected to deteriorate [10, 13, 14]. Indeed, there are also reports of preserved artistic talents in Alzheimer's patients [15, 16]. Furthermore, the emergence of new-onset artistic propensity or alteration in art styles is reported in different conditions, e.g., amyotrophic lateral sclerosis [17], corticobasal degeneration [18], epilepsy [19], stroke [20–22], and Parkinson's disease (PD) [23–25]. The onset of artistic passion in PD has been attributed to their medications [23, 24].

2 Art Therapy

Literature imposes several modalities for art therapy, such as *visual arts therapies (VAT), drama therapy (DT), dance/movement therapy (DMT), poetry therapy (PT), activity-based interventions, and music therapy (MT)* [26].

2.1 Visual Arts Therapies (VAT)

It is an effective therapeutic method for children diagnosed with post-traumatic stress disorder (PTSD). VAT is defined as a "therapeutic process based on spontaneous or prompted creative expression using various art materials and art techniques such as painting, drawing, sculpture, clay modeling and collage" [27]. Drawing can help children with PTSD express their emotions verbally by diminishing anxiety levels, facilitating the recovery of memories, encouraging the kids to talk more precisely, and providing a relaxing and friendly therapeutic session for them [28, 29].

The application of art therapy for anxiety has been investigated through coloring mandala patterns. *Mandala paintings* are geometric configurations of different symbols used to increase attention and establish a sacred space as meditation guidance. It is demonstrated that coloring a mandala for 20 min accelerates the anxiety resolution process more effectively than free-form coloring [30, 31]. It is possible that coloring a fixed preset design might help alleviate the anxiety if we consider anxiety as inner chaos [30, 32].

Clay working is proposed to be helpful for individuals with depression and anxiety [33]. It is hypothesized that working with clay is a potential means for emotional discharge, providing a kind of manual control of one's emotions and thoughts. Hence, the communication and expression of conscious and unconscious thoughts might be facilitated [34]. Clay art therapy engages different brain regions and pathways, such as haptic perception regions, visual cortex, aesthetic judgment regions and related circuits, working memory, concentration, and decision-making centers [35–37]. The studies that investigated the effectiveness of clay art therapy for individuals with depression [33, 37] and anxiety [33] represented promising results.

In conclusion, VAT has been shown to significantly improve global cognitive functions in older adults with mild cognitive impairment (MCI) and dementia in addition to the preventive effect on older adults with normal cognitive abilities. It also improves psychological complaints such as anxiety and depression in people suffering cognitive decline [38].

2.2 Drama Therapy (DT)

It is "the intentional and systematic use of drama/theater process to achieve psychological growth and change" [39]. DT is conceptualized based on the "role theory" developed by Landy et al. In this theory, role is defined as "a basic unit of personality containing specific qualities that provide uniqueness and coherence to that unit… [it is] the container of all the thoughts and feelings we have about ourselves and others in our social and imaginary worlds" [40]. This theory helps individuals to both see themselves internally and from a social perspective. In addition, forming the role demands individuals to use the executive function, working memory, and attention. Drama therapy also increases one's capacity in theory of mind (ToM). Individuals seeking drama therapy are engaged in the set of performances by which understanding the other performer allows further entraining into the role section. It is hypothesized that DT can be applied to schizophrenia, autism spectrum disorders (ASD), and dementia [41].

2.3 Poetry Therapy (PT)

It has recently gained attention as a new field of study in neuroscience. The schizophrenic population can benefit from this treatment due to metaphoric processing, including figurative language and abstraction, which outlines language processing in the brain region [41].

2.4 Dance Movement Therapy (DMT)

It induces the mirror neuron system through body posture, positioning, and movement [41]. This is principally performed as a complementary treatment for demented individuals, especially those with cognitive impairments, who have problems in verbal self-expression and communication. Karkou and Meekums have carried out a systematic review study on the efficacy of dance movement therapy for demented patients. However, they failed to conclude whether this intervention benefits patients with dementia or not [42]. In addition, DMT has shown promising results in treating mild depression [41]. In comparison, another systematic review reached a statically significant increase in global cognition and memory in the elderly with no significant benefit on working executive function [43].

Within the available literature on art therapy, MT is among the most developed modality. Hence, we focus the rest of this chapter on the definition and application of musical intervention specified by disease categories.

2.5 Music Therapy

2.5.1 Definition

The World Federation of Music Therapy defined MT as ``the professional use of music and its elements as an intervention in medical, educational, and everyday environments with individuals, groups, families, or communities who seek to optimize their quality of life and improve their physical, social, communicative, emotional, intellectual, and spiritual health and well-being' [44]. MT is a skill with evidence-based methods of using music for gaining therapeutic benefits for patients' mental and physical health. The first step is to comprehensively evaluate the patient's needs and complaints by a MT specialist. Next, the therapist suggests the proper treatment, including writing, singing, dancing, and/or listening to the music based on short and long-term treatment goals and, finally, assesses the effectiveness of the treatment [45].

MT is believed to be the best alternative way of self-expression for individuals who are not capable of verbal communication [46]. Furthermore, it can help individuals develop social interactions, improve concentration and self-confidence, work as a means of expressing emotions, put up with bereavement and grief, induce relaxation, and enhance motor skills and general physical rehabilitation [47].

Thinking of how MT might work, the first question is how musical intervention could be classified as a therapeutic intervention in MT? Table 1 represents the criteria suggested by Bruscia [45] that define whether an experience is interpreted as musical or not. Considering these criteria, the music experiences in the clinical setting might classify into five levels: pre-musical, musical, extramusical, paramusical, and nonmusical [45].

There are four levels of music experience and, therefore, four methods of client's involvement: improvisational methods, recreative methods, compositional methods, and receptive methods. Each technique requires a different degree of participation and a spectrum of skills and stimulates different emotions, and finally, each method is used to reach different purposes [45]. In the improvisational method, the patient sings and makes up the melody and rhythm spontaneously. This process can be done alone, with the therapist's contribution, or within a group of peers; this is how a new avenue for nonverbal communication and self-expression is built. Conversely, in the recreative model, the patient either performs, sings, or reproduces a pre-composed existing piece of music used as a model. The compositional method

Table 1 Bruscia criteria to define output and input experience as musical

Whether the input or output is human or nonhuman, random or orderly, and controlled intentionally or unintentionally
Whether the input or output is auditory, vibrational, or visual
Whether the input or output is organized according to sound parameters or other parameters (e.g., motor patterns, speech patterns)
Whether the sounds make meaningful forms
Whether any aspect of the experience is aesthetic in nature

is based on the demonstration of patients' emotions, concerns, and fantasies in music by the composition of a song. Finally, in the receptive or passive method, the music is played to the patient, and the patient reacts to it in different manners such as verbal expression of feelings and remains silent and processes inner thoughts [45].

2.5.2 History

The initial spark of using music as a physical and mental health promoter appeared after World War I and II when different musicians started to play music around the veterans' hospitals. After that, medical staff noticed a total improvement in both the physical and mental state of hospitalized patients. The oldest and first recorded reference on MT was published in 1789 in Columbian Magazine, entitled ``Music Physically Considered' [48]. About fifteen years later, the first two medical dissertations on the therapeutic effects of music were written by Benjamin Rush students, the first one by Edwin Atlee (in 1804) and the second by Samuel Mathews (in 1806). Benjamin Rush is the pioneer in promoting MT. There are two more important records in the history of MT in the 1800s: the earliest formal MT intervention in an organizational setting and the first documentation of systematic experiments on the therapeutic effects of music that was performing music to modify clients' dreams in the psychotherapy process. Following the growing interest in MT, in the 1900s, the first associations in MT were established (National Society of Musical Therapeutics in the USA-1903, National Association for Music in Hospitals-1926, and the National Foundation of Music Therapy-1941). Regardless of their efforts in MT and funding the first educational courses, books, and journals, they failed to design a specific profession. In the 1940s, three pioneers in MT appeared; Ira Altshuler, Willem van de Wall; and the father of music therapy, Everett Thayer Gaston. Furthermore, in 1944, the first MT official curriculum was developed by Michigan State University. Today, the American Music Therapy Association is the greatest organization in this field, which is developed by the integration of the American Association for Music Therapy (AAMT, established in 1971) and the National Association for Music Therapy (NAMT, established in 1950) in 1998 [48].

2.5.3 Neurophysiology of Music

When a piece of music is played, we feel the urge to tap along, not with its rhythm but with the steady, regularly paced pulse known as music beat. Perceiving the beat and rhythm in the motor area of the brain makes this spontaneous synchronized movement to the music. Music perception is also associated with detectable activity in the supplementary motor area and basal ganglia of the brain, even in the absence of any motor movement. This auditory-motor interaction fired by musical stimulations can potentially ease many motor complaints such as tremors and tics. It can also act as a cue to induce gait and motor rehabilitation, especially after traumatic brain injuries and strokes [49]. To support the claim on the efficacy of music on the movement and gait problems [50], it has been suggested that the processing of music in subcortical structures of the brain induces muscular and autonomic responses in addition to music perception. Therefore, this provocation of motor and autonomic neurons potentially induces the "move to the beat' impulsion in individuals [51, 52].

To elucidate the action mechanism of MT, different aspects of music effects on the brain have been investigated. For instance, brain anatomical pathways of music perception and related emotions or the biochemical correlates to music hearing in the brain. The normal pathway for auditory perception starts from the cochlear nerve and then continues through the brain stem, the medial geniculate body of the thalamus, and finally reaches the auditory cortex. Besides, some neuronal pathways originating from the auditory part of the thalamus might reach the regions involved in the processing of emotional behavior, such as the amygdala and the medial orbitofrontal cortex [53, 54]. In addition to the evidence implying that music potentially triggers the amygdala and cingulate gyri in the limbic system that are subcortical structures [55, 56], there is evidence that cortical regions are involved in the emotional process of music as well [57].

To further support this notion, a 1999 survey by Blood et al. [58] revealed that cerebral blood flow as an indicator of brain activity increases in the frontal lobes by listening to harmonic music and in the temporal lobes by exposure to disharmonic music [58]. Furthermore, Koelsch et al. [55] have reached the same results through functional magnetic resonance imaging. Dopamine is the first neurotransmitter to be discussed in the matter of biochemical correlates of music. Dopamine has a pivotal contribution to the pathways triggered by naturally rewarding and pleasurable stimuli [59, 60]. Moreover, it has been discovered that dopamine might secrete in some areas of the brain which are activated in subjects exposed to harmonic music such as ventral striatum [55, 56, 61].

Music can potentially recruit the same reward and activate the same emotional neuronal pathways as their natural stimulants, such as food and sex, and the pathways that can be stimulated by drug abuse [62]. Hence, the provocation of these sensations might probably enhance the chance of showing better compliance for the intervention due to the distraction and ignorance of senses like fatigue [63]. Activating the pathways specialized in emotional behaviors such as the cortex of insula and cingulate gyri, hypothalamus, hippocampus, ventral striatum, amygdala, and prefrontal cortex happens with the musical intervention [53, 61, 62, 64]. Hypothalamus is involved in maintaining the physiological body homeostasis and the pathophysiologic mechanism of anxiety and depression by regulating the hypothalamus–pituitary–adrenal (HPA) axis [64, 65]. HPA axis modulates the secretion of endocrine and neuronal messengers and cytokines and, consequently, controls the neuroendocrine and immune responses [64, 65]. Nowadays, MT is applied in various medical conditions ranging from psychological complaints to neurologic disorders. Music and art categorized into the mind–body medicine can improve the holistic health of the mind and the body. Music influences the "neurovisceral integration" and is an effective intervention for a variety of medical conditions which can potentially coordinate the whole body. The main target of music is the autonomic branch of the nervous system. This impact ranges from the evoked and emotional responses to the psychoneuroimmunological regulations [66,

67]. Below, we illustrate mainly investigated disorders for the possible effects of a musical intervention.

3 The Plastic Brain

Neuroplasticity is the ability of the brain to change biologically and to reorganize itself. This process can occur in the molecular, neuronal, synaptic, and structural levels [68]. It is well documented that music stimulates different regions and pathways in the brain [61, 69, 70]. Furthermore, several studies have demonstrated variations in the brain's functional and structural features between musician and non-musician individuals [71–76]. Therefore, musicians' brains are a promising model for neuroplasticity studies [77]. The exact corresponding mechanism of music-induced neuroplasticity is not determined yet. However, some factors are suggested to be involved in this process. For instance, as mentioned earlier, music potentially stimulates the reward circuit and, consequently, induces a dopamine release [78] that is proposed to have a pivotal role in music-induced neuroplasticity [79]. In music-based therapeutic interventions, music is paired with a task or behavior that needs to be learned or recovered [68]. The music triggers the dopamine release by activating motivation, reward, and learning pathways—the dopamine release response transfers to the paired task or behavior. Consequently, the new stimulus starts to trigger the dopamine release independently as the primary stimulus is gradually deleted. After all, the paired stimulus mediates the neuroplasticity process, and the task or behavior will be learned [68]. The potential effect of music in inducing synchronized neuronal firing has been stated as another mechanism in mediating the neuroplasticity process. Neurogenesis and neural reorganization are also attributed to the effect of music on the secretion of steroid hormones [80]. Moreover, Angelucci et al. [64] observed that brain-derived neurotrophic factor (BDNF) levels significantly increased by musical exposure. Adding this finding to the fact that BDNF has a core role in synaptic plasticity, it can be hypothesized that BDNF might be involved in music-induced neuroplasticity.

Furthermore, hypothalamic functions are affected through the secretion of BDNF and nerve growth factor (NGF) that have an important participation in the growth, survival, and regulation of the function of neurons [81]. Hence, music has a stimulatory effect on the hypothalamus, whereas BDNF and NGF influence the hypothalamus function. Interestingly, it has been indicated that the BDNF level increases while the NGF decreases in the hypothalamus of mice exposed to music [64, 82]. Thus, music can affect the hypothalamus functions by altering the level of BDNF and NGF. Besides, it has been suggested that both BDNF and NGF might potentially influence the HPA axis functions [83, 84]. Assuming that NGF can enhance the function of the HPA axis [84], music can attenuate the HPA axis activities by reducing the release of NGF. This might be a potential mechanism for the stress-reducing and relaxing effect of music.

4 Music as Medicine

4.1 Neuropsychiatric Disorders

4.1.1 Brain Injury

Acquired brain injury (ABI) is defined as an impairment in the brain functions due to miscellaneous causes such as head trauma, post-surgical complications, stroke, infection, inflammation, and different types of cranial hemorrhages [85, 86]. ABI can induce physical complications such as movement and walking problems, difficulty in sensations, speech, and language impairments in addition to emotional disruptions and cognitive impairments [85]. The patients with ABI tend to demonstrate depression symptoms, while the ABI recovery is often partial and restricted [87]. Consequently, depression is a major obstacle for the patients to return to their former routine social communications preventing them from proper participation in rehabilitation therapies [88]. On the other hand, the increasing rate of ABI places a significant burden on individuals' lives, healthcare systems, and economies [89]. A systematic review of the effect of music-based therapeutic interventions in patients with ABI reported a positive impact of music on individuals' gait, improvement of upper limb function, social communication, and the overall quality of life [85]. Considering the high expenses of ABI care and the promising results of music-based interventions in related clinical trials and observational studies, music-based therapeutic measures might be a good choice for ABI individuals.

4.1.2 Stroke

Music can significantly improve gross motor activity, fine motor function, and speech in addition to increasing mood and lowering anxiety levels during rehabilitation sessions among the ischemic stroke population [90]. However, stroke can impair rhythm perception, preventing further therapeutic interventions such as rhythmic auditory stimulation (RAS) [49]. RAS is the most commonly used MT intervention for the stroke population. Promising results appeared by combining gait exercises with RAS in comparison to exercises without rhythm. Additionally, musical interventions had improved all motor tests, including Fugl-Meyer, time up and go, action research arm, and box and blocks in post-stroke individuals [91].

4.1.3 Movement Disorders

PD is a prevalent neurodegenerative disorder affecting approximately 2% of over 65-year-old adults. PD manifests with significant motor impairments such as gait imbalance, stooped posture, pin rolling tremor, and rigidity [92]. There is a recently emerged tendency for using music-based movement therapy to help individuals suffering from PD. A significant positive effect on the walking velocity of individuals exists for gait-related music-based movement therapy. However, this effect was not observed in dance-related interventions [93]. Therefore, music-based movement therapies might be promising for the gait improvement but did not show any significant positive effects on the quality of life and the issue of the freezing gait. This therapy can probably decrease the falling issue by augmentation of the patients' balance in the partnered dance model [93]. However, a positive effect of dancing on the freezing gait is also reported by some other observations [94]. There is a debate on whether other methods and aspects of music-based therapies such as singing or playing musical instruments might benefit PD patients or not. It is of critical importance to address the effectiveness of these methods in future investigations.

4.1.4 Autism Spectrum Disorders

ASDs are neurodevelopmental disorders in which the social perception is severely impaired. The background of MT for individuals with ASDs and other communication disorders is rooted in the investigations for the musical interpretation of vocal interactions between mothers and newborn infants [95, 96]. MT might enhance communication skills and provide a chance for social interaction for ASD patients [97]. Improvisational methods in MT facilitate the communication, relationship, and emotional connection between verbal and nonverbal people. Receptive or listening method is another method in MT for ASD patients that is performed with a piece of music relevant to the issue the patient is coping with, and then expression and reflection of patients' feelings. To further support the effectiveness of music-based therapies for the ASD population, it is claimed that musical activities in groups are, especially, beneficial for ASD patients who cannot easily make social interactions. Hence, it is a good opportunity to make initial social interactions with their peers [98]. Besides, musical activities provide alternative solutions for various ASD problems, such as divergent sensory perceptions, impaired or atypical communication, and emotional or behavioral problems [99]. Considering the higher ability of the ASD population in music perception [100], the application of music-based activities in the educational curriculum of ASD kids seems to be an effective method to better engage the students with school programs [101]. Finally, music, in the form of auditory integration training (AIT), could have a possible role in inducing synaptic plasticity among children suffering from ASD by increasing plasma glial cell line-derived neurotrophic factor (GDNF) levels [102].

AIT is an alternative or complementary therapeutic method established by Dr. Guy Berard as a new therapeutic measure for clinical depression, self-harm and suicidal attempts, ASD, and dyslexia [103, 104]. AIT usually consists of 20 half-hour sessions, twice a day, for ten days. The beneficial effect of this method is reported in attenuation of auditory processing deficits, concentration enhancement, and decline in the susceptibility to auditory stimuli [105, 106]. Using filters in the AIT device, the sound frequency is dampened and modulated to an appropriate level for each patient [103]. Sensory processing impairment is reported in ASD kids with a calculated prevalence of 42–88%, in which excessive hypersensitivity to auditory triggers is the most prevalent [107]. This might be due to the hypothesis that autistics perceive some specific frequencies in an alternative manner, and the behavioral and learning complications might be rooted in this abnormality [103, 108]. There are reports of the effectiveness of AIT in the amelioration of autistic symptoms [105], while there are studies that failed to draw a conclusion on the efficacy of AIT among ASD children [106, 109–111]. Regarding studies investigating the effect of music on the neurotrophic factors [112–114] and the pivotal role of these factors in neuroplasticity processes [115], Al-Ayadhi et al. [105] conducted a survey to address the effect of AIT in plasma levels of human GDNF in ASD population. They observed an increase in GDNF level after the AIT [105]. Synaptic plasticity is considered a crucial mechanism in sensory processing and perception and cognitive function. Due to abnormal synaptic plasticity in ASD [116, 117], AIT might be useful in reducing ASD symptoms due to its effect on the expression of GDNF and consequently regulating the neuroplasticity process.

4.1.5 Dementia

Music, singing, and playing musical instruments have shown positive results in improving different aspects of the life of individuals with dementia. Singing within a group might enhance patients' social skills and facilitate the recall and remembrance of memories. This can also decrease anxiety by producing a sense of familiarity and safety [118] and inducing relaxation and emotional wellness [119]. Concerning music's potentiality in provoking the language centers in the brain, it might be possible that music can benefit individuals with speaking impairments [118]. Despite the exacerbation of cognitive functions with the progression of dementia, studies demonstrate that individuals remained responsive to music, and their perception of music remained preserved [120–122]. To assess the efficacy of music-based measures in dementia, a meta-analysis was conducted that represented moderate-quality evidence on the attenuation of depressive symptoms and behavioral problems. It also served low-quality evidence on the ineffectuality of these interventions on cognition but the improvement of quality of life and anxiety (together considered as emotional well-being) [118]. Moreover, this study suggested that music-based therapies probably have little or no effect on aggressive behaviors and agitation [118]. Besides, another meta-analysis demonstrated that music-based therapeutic interventions might positively impact disruptive behavior and anxiety of demented patients and a positive trend for depression, cognition, and quality of life [123], which were consistent with the earlier meta-analyses [124–126]. Taken together, MT as a non-pharmacological, noninvasive, easily delivered, well-tolerated, and affordable therapeutic measurement with no side effects and good compliance is a reliable method in improving the overall condition in demented patients.

4.1.6 Depression

People benefit from music in daily life to enhance their mood, balance their diverse emotions, and augment coping mechanisms in facing everyone's routine life [127, 128]. Music can decrease anxiety and increase mood and patients' daily function, and it is as effective as a psychological modality [129]. Although no statistically significant difference exists in the active versus receptive MT method in improving depressive symptoms, Atiwannapat et al. [130] concluded that the receptive group might get to the peak of the effect more rapidly, while the active group could potentially experience a higher peak effect.

4.1.7 Anxiety

Anxiety is a familiar feeling for every person. Many situations in life initiate an anxiety response in different parts of the body; for instance, increase in cortisol expression into the systemic circulation, elevation in BP, and increase in HR [131]. MT is a promising method in anxiety reduction in different conditions or during medical procedures. Music is effective in reducing anxiety during cystoscopy [132] and angiography [133], but there is no significant report about the anxiety reduction by music interventions in biopsy [134] and colposcopy [135]. Moreover, there are promising reports of the effectiveness of music-based interventions in decreasing anxiety in dementia [136, 137], pre and post-operative recovery period [138–140], and in critically ill patients with mechanical ventilation [141–143] alongside the same result is not observed in cancer patients [144] and pregnant women [145]. It is claimed that music distracts the patient from a stressful situation to something that is pacifying and pleasurable [146, 147]. Furthermore, music influences the central and autonomic nervous system and, thus, initiates a systemic relaxation in the body [148, 149]. The potential effect of music in inducing a sense of well-being due to the endorphins released by provoking the limbic system is also of critical importance [150].

4.1.8 Schizophrenia

Schizophrenia is a severe mental disease impairing the person's interpretation of reality remarkably. It severely affects thinking abilities leading to an abnormal and potentially harmful combination of hallucinations and delusions. Antipsychotic therapy is inevitable in schizophrenic patients. However, due to the prolonged course of the disease and various side effects of consumed medications, looking for additional therapeutic modalities seems necessary. When a patient is mentally too ill, verbal communication may not make the patient capable of continuing the treatment process. Hence, MT might be hope in the setting of schizophrenia treatment due to the claim that the initial parent-neonate communication has a strong musical nature [95, 151]. Schizophrenic patients may benefit from MT sessions as additional complementary care to improve their global and mental state, facilitate adaption to the social environment, and enhance their quality of life [152]. However, the reported effect was incompatible between included studies and generally varied through the quality and the number of MT sessions (it is calculated that minimally 20 ordered sessions are required to obtain the expected results) [152, 153].

4.1.9 Personality Disorder

People diagnosed with personality disorder have deep struggles in expressing their feelings, thoughts, and emotions. Art therapy provides a safe ground to explore inner conflicts and emotions with no need to use words to describe them [154]. The overall evidence in this regard is weak and calls for further investigations. However, a well-designed quantitative study has explored the effects of music on seven women with borderline personality disorder. This study supposed that improvisational music induces the sense of being "part of" rather than being "apart," which can potentially help people to strengthen their communication [155].

4.1.10 Seizure

Seizure is defined as a sudden uncontrolled discharge in a specific brain region. Pharmacological treatment is within the first line of patients' management. About one-third of epileptic patients suffer from recurrent episodes despite anti-epileptic drug administration. These are regarded as drug-resistant epilepsy. Randomized trials report statistically significant reduction in seizure recurrence and seizure activity after exposure to passive music, mainly Mozart. Moreover, when treating patients with epilepsy, clinicians have to consider musicality since some anti-epileptic drugs such as carbamazepine and oxcarbazepine can potentially alter pitch appreciation [156–159].

4.2 Pediatric and Neonatal Disorders

4.2.1 Prematurity

Preterm birth and prematurity are among the most prevalent cause of long-term morbidity and mortality, affecting one out of ten births. Children born prior to 37 weeks of gestational age are more vulnerable to disparate neurodevelopmental diseases in addition to various socio-economic problems. Prematurity exposes the developing brain to numerous noxious stimuli leading to disruptions in specific neural structures. Imaging studies have revealed reduced volume, especially in the thalamus, hippocampus, orbitofrontal lobe, and altered functional connectivity in frontostriatal pathways. Besides, other significant short-term effects of music include lowering infants' respiratory rate and mothers' anxiety levels [160, 161].

   Musical interventions include singing live lullabies by the infants' mothers to pre-recorded lullabies and harmonical pieces of music. Studies suggest that repeated listening to familiar music activates brain regions related to emotional processing in adults and newborns. It also enhances the functional connectivity between the auditory cortices and the subcortical brain regions in preterm infants at term equivalent age, as its short-term effect. Moreover, Lejeune et al. concluded that preterm infants with or without music intervention were significantly different in fear reactivity and anger reactivity at 12 and 24 months, respectively, compared to full-term infants. Interestingly, the difference between the preterm with music and the full-term controls was less than the routine preterm group and the controls [162].

4.3 Cardiovascular Health

Music enhances cardiovascular health by activating cardiac parasympathetic nervous system, which lowers systolic and diastolic blood pressure (SBP and DBP) and HR. Recent literature imposes heart rate variability (HRV) as an indicator of long-term cardiac health. This index simply analyzes the oscillation of successive heartbeats, which is also regulated through vagal activation, a branch of the parasympathetic system. Consequently, parasympathetic arousal due to music of any kind soothes the heart measurable through several indices as HR, SBP, DBP, and HRV [163, 164].

4.4 Pain

Pain is defined as ``an unpleasant sensory and emotional experience associated with actual or potential tissue damage or described in terms of such damage' [165]. Music stimulates both physiological and psychological responses in the human body [166]. A meta-analysis conducted in 2016 concluded that both active and receptive interventions of MT have favorable effects on pain. While no statistically significant difference was reported between active and receptive interventions in reducing pain levels, the active interventions tended to be more effective in decreasing the severity of pain. On the other hand, receptive interventions demonstrated a better reduction in palliative medicine use [167]. The alleviative effects of music on children were more noticeable [167]. In chronic pain conditions, music-induced statistically significant decline in self-reported pain, anxiety, and depressive symptoms. Besides, a better analgesic effect was observed for patient-chosen music compared to therapist-chosen music [168]. These results are in line with other studies in this field [147, 166, 169]. Gao et al. [170] performed a meta-analysis to investigate the effect of music in palliative care for terminally ill patients. They proposed that music might noticeably reduce pain and psychological burden of these patients, but the effect of music on their physical state requires further investigations. Since the fear of pain is the second-highest ranked reason of fear after fear of death [166], utilizing music-based intervention in palliative care needs critical attention since it can ease the overall mood significantly.

4.5 Cancer

Cancer places an excessive burden on individuals' mental health along with physical complications and social burdens. Based on a National Health Survey in 2010, cancer survivors have significantly poorer mental health compared to non-cancer adults; this issue deteriorates in patients who have other simultaneous chronic illnesses [171]. Besides the psychosocial interventions in routine cancer care, music-based interventions can help ameliorate the side effects of pharmacological and surgical treatments, improve the physical complaints, and provide mental support for patients. Active music-based interventions might benefit the quality of life enhancement, but the same effect was not reported for receptive or passive methods among the cancer population. Furthermore, very low quality of evidence for an effect on depression, mood enhancement, and boosting physical function in cancer patients exists [172]. Based on a systematic review and meta-analysis in evaluating anxiety, depression, and pain of cancer patients, MT might be beneficial in improving the symptoms. They suggest an optimum duration of one to two months of MT in order to enhance the patients' quality of life. Moreover, music can relieve stress during hospitalization and make it easier to be familiar with the hospital atmosphere. It can also be used as an adjuvant treatment in patients during the treatment phase in various stages of the disease and in surviving patients as well [173].

Lastly, the matter of anxiety and depression is of critical value in cancer patients. It is estimated that about 25% of cancer patients potentially demonstrate depression and anxiety symptoms [174]. The reduction in the depression score and symptoms following art therapy interventions in depressed patients is reported from various surveys [175–177].

5 Conclusion

The role of music in everyday life is increasing day by day. Several medical conditions are affected by musical interventions ranging from decreasing stress and anxiety to improving motor functions in impaired patients with stroke or PD. The use of music dates back to very early primitive societies; however, the question of whether musical interventions should be specifically tailored per patient disease or "does one size fit all" remains unanswered [178]. Although a remarkable number of researchers have lately focused on this field, art therapy is still taking its first steps in science. Further research is crucial to address the issue.

Core Messages

- Music stimulates different regions and pathways in the brain, especially the reward circuit.
- Musical exposure induces the release of BDNF, a neurotrophic factor implicated in neuroplasticity.
- Music has been a subject of therapy in various neuropsychiatric disorders, namely autism and schizophrenia.
- Music enhances cardiovascular outcomes through autonomic regulation of cardiovascular function.
- High-quality research is needed to fit music into the category of complementary therapy.

References

1. Malchiodi CA (2011) Handbook of art therapy. Guilford Press
2. Clow A (2006) Normalisation of salivary cortisol levels and self-report stress by a brief lunchtime visit to an art gallery by London City workers. J Holistic Healthcare 3
3. Stuckey HL, Nobel J (2010) The connection between art, healing, and public health: a review of current literature. Am J Public Health 100(2):254–263. https://doi.org/10.2105/AJPH.2008.156497
4. Geue K, Goetze H, Buttstaedt M, Kleinert E, Richter D, Singer S (2010) An overview of art therapy interventions for cancer patients and the results of research. Complement Ther Med 18(3–4):160–170. https://doi.org/10.1016/j.ctim.2010.04.001
5. Leckey J (2011) The therapeutic effectiveness of creative activities on mental well-being: a systematic review of the literature. J Psychiatr Ment Health Nurs 18(6):501–509. https://doi.org/10.1111/j.1365-2850.2011.01693.x
6. Lacey S, Hagtvedt H, Patrick VM, Anderson A, Stilla R, Deshpande G et al (2011) Art for reward's sake: visual art recruits the ventral striatum. Neuroimage 55(1):420–433. https://doi.org/10.1016/j.neuroimage.2010.11.027
7. Di Dio C, Macaluso E, Rizzolatti G (2007) The golden beauty: brain response to classical and renaissance sculptures. PLoS One 2(11):e1201. https://doi.org/10.1371/journal.pone.0001201
8. Kawabata H, Zeki S (2004) Neural correlates of beauty. J Neurophysiol 91(4):1699–1705. https://doi.org/10.1152/jn.00696.2003
9. Kirk U, Skov M, Christensen MS, Nygaard N (2009) Brain correlates of aesthetic expertise: a parametric fMRI study. Brain Cogn 69(2):306–315. https://doi.org/10.1016/j.bandc.2008.08.004
10. Flaherty AW (2011) Brain illness and creativity: mechanisms and treatment risks. Can J Psychiatry Revue canadienne de psychiatrie 56(3):132–143. https://doi.org/10.1177/070674371105600303
11. Miller BL, Cummings J, Mishkin F, Boone K, Prince F, Ponton M et al (1998) Emergence of artistic talent in frontotemporal dementia. Neurology 51(4):978–982. https://doi.org/10.1212/wnl.51.4.978
12. Drago V, Foster PS, Trifiletti D, FitzGerald DB, Kluger BM, Crucian GP et al (2006) What's inside the art? The influence of frontotemporal dementia in art production. Neurology 67(7):1285–1287. https://doi.org/10.1212/01.wnl.0000238439.77764.da
13. Crutch SJ, Isaacs R, Rossor MN (2001) Some workmen can blame their tools: artistic change in an individual with Alzheimer's disease. Lancet (London, England) 357(9274):2129–2133. https://doi.org/10.1016/s0140-6736(00)05187-4
14. Crutch SJ, Rossor MN (2006) Artistic changes in Alzheimer's disease. Int Rev Neurobiol 74:147–161. https://doi.org/10.1016/s0074-7742(06)74012-0
15. Fornazzari LR (2005) Preserved painting creativity in an artist with Alzheimer's disease. Eur J Neurol 12(6):419–424. https://doi.org/10.1111/j.1468-1331.2005.01128.x
16. Maurer K, Prvulovic D (2005) Carolus horn—when the images in the brain decay. https://doi.org/10.1159/000085608
17. Liu A, Werner K, Roy S, Trojanowski JQ, Morgan-Kane U, Miller BL et al (2009) A case study of an emerging visual artist with frontotemporal lobar degeneration and amyotrophic lateral sclerosis. Neurocase 15(3):235–247. https://doi.org/10.1080/13554790802633213
18. Kleiner-Fisman G, Black SE, Lang AE (2003) Neurodegenerative disease and the evolution of art: the effects of presumed corticobasal degeneration in a professional artist. Mov Disord Official J Mov Disord Soc 18(3):294–302. https://doi.org/10.1002/mds.10360
19. Finkelstein Y, Vardi J, Hod I (1991) Impulsive artistic creativity as a presentation of transient cognitive alterations. Behav Med (Washington, DC) 17(2):91–94. https://doi.org/10.1080/08964289.1991.9935164

20. Annoni JM, Devuyst G, Carota A, Bruggimann L, Bogousslavsky J (2005) Changes in artistic style after minor posterior stroke. J Neurol Neurosurg Psychiatry 76(6):797–803. https://doi.org/10.1136/jnnp.2004.045492
21. Lythgoe MF, Pollak TA, Kalmus M, de Haan M, Chong WK (2005) Obsessive, prolific artistic output following subarachnoid hemorrhage. Neurology 64(2):397–398. https://doi.org/10.1212/01.Wnl.0000150526.09499.3e
22. Thomas-Anterion C, Creac'h C, Dionet E, Borg C, Extier C, Faillenot I et al (2010) De novo artistic activity following insular-SII ischemia. Pain 150(1):121–127. https://doi.org/10.1016/j.pain.2010.04.010
23. Kulisevsky J, Pagonabarraga J, Martinez-Corral M (2009) Changes in artistic style and behaviour in Parkinson's disease: dopamine and creativity. J Neurol 256(5):816–819. https://doi.org/10.1007/s00415-009-5001-1
24. Schrag A, Trimble M (2001) Poetic talent unmasked by treatment of Parkinson's disease. Mov Disord Official J Mov Disord Soc 16(6):1175–1176. https://doi.org/10.1002/mds.1239
25. Walker RH, Warwick R, Cercy SP (2006) Augmentation of artistic productivity in Parkinson's disease. Mov Disord Official J Mov Disord Soc 21(2):285–286. https://doi.org/10.1002/mds.20758
26. Beard RLJD (2012) Art therapies and dementia care: a systematic review. 11(5):633–656
27. Avrahami D (2006) Visual art therapy's unique contribution in the treatment of post-traumatic stress disorders. J Trauma Dissociation 6(4):5–38. https://doi.org/10.1300/J229v06n04_02
28. Gross J, Hayne H (1998) Drawing facilitates children's verbal reports of emotionally laden events. J Exp Psychol Appl 4(2):163–179. https://doi.org/10.1037/1076-898X.4.2.163
29. Lev-Wiesel R, Liraz R (2007) Drawings vs. narratives: drawing as a tool to encourage verbalization in children whose fathers are drug abusers. Clin Child Psychol Psychiatry 12(1):65–75. https://doi.org/10.1177/1359104507071056
30. Curry NA, Kasser T (2005) Can coloring mandalas reduce anxiety? Art Ther 22(2):81–85. https://doi.org/10.1080/07421656.2005.10129441
31. van der Vennet R, Serice S (2012) Can coloring mandalas reduce anxiety? A replication study. Art Ther 29(2):87–92. https://doi.org/10.1080/07421656.2012.680047
32. Grossman FG (1981) Creativity as a means of coping with anxiety. Arts Psychother 8 (3):185–192. https://doi.org/10.1016/0197-4556(81)90030-7
33. de Morais AH, Dalécio MAN, Vizmann S, Bueno VLRdC, Roecker S, Salvagioni DAJ et al (2014) Effect on scores of depression and anxiety in psychiatric patients after clay work in a day hospital. Arts Psychother 41(2):205–210. https://doi.org/10.1016/j.aip.2014.02.002
34. Jang H, Choi S (2012) Increasing ego-resilience using clay with low SES (Social Economic Status) adolescents in group art therapy. Arts Psychother 39(4):245–250. https://doi.org/10.1016/j.aip.2012.04.001
35. Elbrecht C, Antcliff LR (2014) Being touched through touch. Trauma treatment through haptic perception at the Clay Field: a sensorimotor art therapy. Int J Art Ther 19(1):19–30. https://doi.org/10.1080/17454832.2014.880932
36. Bastos AG, Guimarães LS, Trentini CM (2013) Neurocognitive changes in depressed patients in psychodynamic psychotherapy, therapy with fluoxetine and combination therapy. J Affect Disord 151(3):1066–1075. https://doi.org/10.1016/j.jad.2013.08.036
37. Nan JKM, Ho RTH (2017) Effects of clay art therapy on adults outpatients with major depressive disorder: a randomized controlled trial. J Affect Disord 217:237–245. https://doi.org/10.1016/j.jad.2017.04.013
38. Masika GM, Yu DSF, Li PWC (2020) Visual art therapy as a treatment option for cognitive decline among older adults. A Syst Rev Meta-analysis 76(8):1892–1910. https://doi.org/10.1111/jan.14362
39. Jones P (2007) Drama as therapy volume 1: theory, practice and research. Routledge
40. Landy RJ (1994) Drama therapy: concepts, theories and practices. Charles C Thomas Publisher

41. Frydman JS (2016) Role theory and executive functioning: constructing cooperative paradigms of drama therapy and cognitive neuropsychology. Arts Psychother 47:41–47. https://doi.org/10.1016/j.aip.2015.11.003

42. Karkou V, Meekums B (2017) Dance movement therapy for dementia. Cochrane Database Syst Rev (2). https://doi.org/10.1002/14651858.CD011022.pub2

43. Meng X, Li G, Jia Y, Liu Y, Shang B, Liu P et al (2020) Effects of dance intervention on global cognition, executive function and memory of older adults: a meta-analysis and systematic review. 32(1):7–19

44. Therapy WFoM (2011) About WFMT. http://www.wfmt.info/wfmt-new-home/about-wfmt/. Accessed 21 July 2020

45. Bruscia KE (2014) Defining music therapy. Barcelona Pub

46. van der Steen JT, van Soest-Poortvliet MC, van der Wouden JC, Bruinsma MS, Scholten RJ, Vink AC (2017) Music-based therapeutic interventions for people with dementia. Cochrane Database Syst Rev 5(5):Cd003477. https://doi.org/10.1002/14651858.CD003477.pub3

47. Therapy WFoM (2007) What are the benefits of music therapy? https://www.wfmt.info//wp-content/uploads/2014/05/What-are-the-benefits-of-music-therapy.pdf. Accessed 21 July 2020

48. Association AMT History of Music Therapy. https://www.musictherapy.org/about/history/. Accessed 21 July 2020

49. Patterson KK, Wong JS, Knorr S, Grahn JAJAopm, rehabilitation (2018) Rhythm perception and production abilities and their relationship to gait after stroke. 99(5):945–951

50. Hurt CP, Rice RR, McIntosh GC, Thaut MH (1998) Rhythmic auditory stimulation in gait training for patients with traumatic brain injury. J Music Ther 35(4):228–241. https://doi.org/10.1093/jmt/35.4.228

51. Phillips-Silver J, Trainor LJ (2005) Feeling the beat: movement influences infant rhythm perception. Science 308(5727):1430. https://doi.org/10.1126/science.1110922

52. Todd NP, Cody FW (2000) Vestibular responses to loud dance music: a physiological basis of the "rock and roll threshold"? J Acoust Soc Am 107(1):496–500. https://doi.org/10.1121/1.428317

53. Boso M, Politi P, Barale F, Enzo E (2006) Neurophysiology and neurobiology of the musical experience. Funct Neurol 21(4):187–191

54. Phelps EA, LeDoux JE (2005) Contributions of the amygdala to emotion processing: from animal models to human behavior. Neuron 48(2):175–187. https://doi.org/10.1016/j.neuron.2005.09.025

55. Koelsch S, Fritz T, DY VC, Müller K, Friederici AD (2006) Investigating emotion with music: an fMRI study. Human Brain Mapp 27(3):239–250. https://doi.org/10.1002/hbm.20180

56. Menon V, Levitin DJ (2005) The rewards of music listening: response and physiological connectivity of the mesolimbic system. Neuroimage 28(1):175–184. https://doi.org/10.1016/j.neuroimage.2005.05.053

57. Limb CJ (2006) Structural and functional neural correlates of music perception. Anat Rec A Discov Mol Cell Evol Biol 288(4):435–446. https://doi.org/10.1002/ar.a.20316

58. Blood AJ, Zatorre RJ, Bermudez P, Evans AC (1999) Emotional responses to pleasant and unpleasant music correlate with activity in paralimbic brain regions. Nat Neurosci 2(4):382–387. https://doi.org/10.1038/7299

59. Bressan RA, Crippa JA (2005) The role of dopamine in reward and pleasure behaviour–review of data from preclinical research. Acta Psychiatr Scand Suppl 427:14–21. https://doi.org/10.1111/j.1600-0447.2005.00540.x

60. Burgdorf J, Panksepp J (2006) The neurobiology of positive emotions. Neurosci Biobehav Rev 30(2):173–187. https://doi.org/10.1016/j.neubiorev.2005.06.001

61. Koelsch S (2014) Brain correlates of music-evoked emotions. Nat Rev Neurosci 15(3):170–180. https://doi.org/10.1038/nrn3666

62. Blood AJ, Zatorre RJ (2001) Intensely pleasurable responses to music correlate with activity in brain regions implicated in reward and emotion. Proc Natl Acad Sci USA 98(20):11818–11823. https://doi.org/10.1073/pnas.191355898

63. Lim HA, Miller K, Fabian C (2011) The effects of therapeutic instrumental music performance on endurance level, self-perceived fatigue level, and self-perceived exertion of inpatients in physical rehabilitation. J Music Ther 48(2):124–148. https://doi.org/10.1093/jmt/48.2.124

64. Angelucci F, Ricci E, Padua L, Sabino A, Tonali PA (2007) Music exposure differentially alters the levels of brain-derived neurotrophic factor and nerve growth factor in the mouse hypothalamus. Neurosci Lett 429(2):152–155. https://doi.org/10.1016/j.neulet.2007.10.005

65. Arborelius L, Owens MJ, Plotsky PM, Nemeroff CB (1999) The role of corticotropin-releasing factor in depression and anxiety disorders. J Endocrinol 160(1):1–12. https://doi.org/10.1677/joe.0.1600001

66. Ellis RJ, Thayer JF (2010) Music and autonomic nervous system (Dys) function. Music Percept 27(4):317–326. https://doi.org/10.1525/mp.2010.27.4.317%JMusicPerception

67. Vuust P, Ostergaard L, Pallesen KJ, Bailey C, Roepstorff A (2009) Predictive coding of music—brain responses to rhythmic incongruity. Cortex 45(1):80–92. https://doi.org/10.1016/j.cortex.2008.05.014

68. Stegemöller EL (2014) Exploring a neuroplasticity model of music therapy. J Music Ther 51(3):211–227. https://doi.org/10.1093/jmt/thu023

69. Zatorre RJ, Chen JL, Penhune VB (2007) When the brain plays music: auditory–motor interactions in music perception and production. Nat Rev Neurosci 8(7):547–558. https://doi.org/10.1038/nrn2152

70. Alluri V, Toiviainen P, Jääskeläinen IP, Glerean E, Sams M, Brattico E (2012) Large-scale brain networks emerge from dynamic processing of musical timbre, key and rhythm. Neuroimage 59(4):3677–3689. https://doi.org/10.1016/j.neuroimage.2011.11.019

71. Elbert T, Pantev C, Wienbruch C, Rockstroh B, Taub E (1995) Increased cortical representation of the fingers of the left hand in string players. Science 270(5234):305–307. https://doi.org/10.1126/science.270.5234.305

72. Amunts K, Schlaug G, Jäncke L, Steinmetz H, Schleicher A, Dabringhaus A et al (1997) Motor cortex and hand motor skills: structural compliance in the human brain. Hum Brain Mapp 5(3):206–215. https://doi.org/10.1002/(sici)1097-0193(1997)5:3%3c206::Aid-hbm5%3e3.0.Co;2-7

73. Pantev C, Oostenveld R, Engelien A, Ross B, Roberts LE, Hoke M (1998) Increased auditory cortical representation in musicians. Nature 392(6678):811–814. https://doi.org/10.1038/33918

74. Koelsch S, Schröger E, Tervaniemi M (1999) Superior pre-attentive auditory processing in musicians. NeuroReport 10(6):1309–1313. https://doi.org/10.1097/00001756-199904260-00029

75. Gaser C, Schlaug G (2003) Brain structures differ between musicians and non-musicians. J Neurosci Official J Soc Neurosci 23(27):9240–9245. https://doi.org/10.1523/JNEUROSCI.23-27-09240.2003

76. Bengtsson SL, Nagy Z, Skare S, Forsman L, Forssberg H, Ullén F (2005) Extensive piano practicing has regionally specific effects on white matter development. Nat Neurosci 8(9):1148–1150. https://doi.org/10.1038/nn1516

77. Münte TF, Altenmüller E, Jäncke L (2002) The musician's brain as a model of neuroplasticity. Nat Rev Neurosci 3(6):473–478. https://doi.org/10.1038/nrn843

78. Owesson-White C, Belle AM, Herr NR, Peele JL, Gowrishankar P, Carelli RM et al (2016) Cue-evoked dopamine release rapidly modulates D2 neurons in the nucleus accumbens during motivated behavior. J Neurosci Official J Soc Neurosci 36(22):6011–6021. https://doi.org/10.1523/jneurosci.0393-16.2016

79. Tramo MJ (2001) Music of the hemispheres. Science 291(5501):54. https://doi.org/10.1126/science.10.1126/SCIENCE.1056899

80. Fukui H, Toyoshima K (2008) Music facilitate the neurogenesis, regeneration and repair of neurons. Med Hypotheses 71(5):765–769. https://doi.org/10.1016/j.mehy.2008.06.019

81. Cunha C, Brambilla R, Thomas K (2010) A simple role for BDNF in learning and memory? 3(1). https://doi.org/10.3389/neuro.02.001.2010

82. Kim H, Lee MH, Chang HK, Lee TH, Lee HH, Shin MC et al (2006) Influence of prenatal noise and music on the spatial memory and neurogenesis in the hippocampus of developing rats. Brain Dev 28(2):109–114. https://doi.org/10.1016/j.braindev.2005.05.008

83. Naert G, Ixart G, Tapia-Arancibia L, Givalois L (2006) Continuous i.c.v. infusion of brain-derived neurotrophic factor modifies hypothalamic-pituitary-adrenal axis activity, locomotor activity and body temperature rhythms in adult male rats. Neuroscience 139(2): 779–789. https://doi.org/10.1016/j.neuroscience.2005.12.028

84. Scaccianoce S, Lombardo K, Nicolai R, Affricano D, Angelucci L (2000) Studies on the involvement of histamine in the hypothalamic-pituitary-adrenal axis activation induced by nerve growth factor. Life Sci 67(26):3143–3152. https://doi.org/10.1016/S0024-3205(00) 00899-7

85. Magee WL, Clark I, Tamplin J, Bradt J (2017) Music interventions for acquired brain injury. Cochrane Database Syst Rev (1). https://doi.org/10.1002/14651858.CD006787.pub3

86. Rudd AG, Bowen A, Young GR, James MA (2017) The latest national clinical guideline for stroke. Clin Med (Lond) 17(2):154–155. https://doi.org/10.7861/clinmedicine.17-2-154

87. Matsuzaki S, Hashimoto M, Yuki S, Koyama A, Hirata Y, Ikeda M (2015) The relationship between post-stroke depression and physical recovery. J Affect Disord 176:56–60. https://doi.org/10.1016/j.jad.2015.01.020

88. Giles GM, Manchester D (2006) Two approaches to behavior disorder after traumatic brain injury. J Head Trauma Rehabil 21(2):168–178. https://doi.org/10.1097/00001199-20060 3000-00009

89. James SL, Theadom A, Ellenbogen RG, Bannick MS, Montjoy-Venning W, Lucchesi LR et al (2019) Global, regional, and national burden of traumatic brain injury and spinal cord injury, 1990–2016: a systematic analysis for the Global Burden of Disease Study 2016. Lancet Neurol 18(1):56–87. https://doi.org/10.1016/S1474-4422(18)30415-0

90. Yakupov E, Nalbat A, Semenova M, Tlegenova KJN, Physiology B (2019) Efficacy of music therapy in the rehabilitation of stroke patients. 49(1):121–128

91. Šuriņa S, Duhovska J, Mārtinsone K (2019) Music therapy for stroke patients: a systematic review with meta-analysis. In: Proceedings of the international scientific conference, vol IV, p 300

92. Opara J, Małecki A, Małecka E, Socha T (2017) Motor assessment in Parkinson`s disease. Annals Agric Environ Med AAEM 24(3):411–415. https://doi.org/10.5604/12321966. 1232774

93. de Dreu MJ, van der Wilk AS, Poppe E, Kwakkel G, van Wegen EE (2012) Rehabilitation, exercise therapy and music in patients with Parkinson's disease: a meta-analysis of the effects of music-based movement therapy on walking ability, balance and quality of life. Parkinsonism Relat Disord 18(Suppl 1):S114-119. https://doi.org/10.1016/s1353-8020(11) 70036-0

94. Kalyani HHN, Sullivan K, Moyle G, Brauer S, Jeffrey ER, Roeder L et al (2019) Effects of dance on gait, cognition, and dual-tasking in Parkinson's disease: a systematic review and meta-analysis. J Parkinsons Dis 9(2):335–349. https://doi.org/10.3233/jpd-181516

95. Trevarthen C (1999) Musicality and the intrinsic motive pulse: evidence from human psychobiology and infant communication. Musicae Scientiae 3(1_suppl):155–215. https://doi.org/10.1177/10298649000030S109

96. Stern DN (1998) The interpersonal world of the infant: a view from psychoanalysis and developmental psychology. Karnac Books

97. Geretsegger M, Elefant C, Mössler KA, Gold C (2014) Music therapy for people with autism spectrum disorder. Cochrane Database Syst Rev 2014(6):Cd004381. https://doi.org/10.1002/ 14651858.CD004381.pub3

98. Darrow A-A, Armstrong T (1999) Research on music and autism implications for music educators. Update Appl Res Music Educ 18(1):15–20. https://doi.org/10.1177/87551 2339901800103

99. Bhat A, Srinivasan S (2013) A review of "music and movement" therapies for children with autism: embodied interventions for multisystem development. Front Integr Neurosci 7:22

100. Heaton P (2003) Pitch memory, labelling and disembedding in autism. J Child Psychol Psychiatry 44(4):543–551. https://doi.org/10.1111/1469-7610.00143

101. Hess KL, Morrier MJ, Heflin LJ, Ivey ML (2008) Autism treatment survey: services received by children with autism spectrum disorders in public school classrooms. J Autism Dev Disord 38(5):961–971. https://doi.org/10.1007/s10803-007-0470-5

102. Al-Ayadhi L, El-Ansary A, Bjørklund G, Chirumbolo S, Mostafa GA (2019) Impact of auditory integration therapy (AIT) on the plasma levels of human glial cell line-derived neurotrophic factor (GDNF) in autism spectrum disorder. J Mol Neurosci MN 68(4):688–695. https://doi.org/10.1007/s12031-019-01332-w

103. Bérard G (1993) Hearing equals behavior. Keats Pub

104. Sokhadze EM, Casanova MF, Tasman A, Brockett S (2016) Electrophysiological and behavioral outcomes of berard auditory integration training (AIT) in children with autism spectrum disorder. Appl Psychophysiol Biofeedback 41(4):405–420. https://doi.org/10. 1007/s10484-016-9343-z

105. Al-Ayadhi L, El-Ansary A, Bjørklund G, Chirumbolo S, Mostafa GA (2019) Impact of auditory integration therapy (AIT) on the plasma levels of human glial cell line-derived neurotrophic factor (GDNF) in autism spectrum disorder. J Mol Neurosci 68(4):688–695. https://doi.org/10.1007/s12031-019-01332-w

106. Sinha Y, Silove N, Wheeler D, Williams K (2006) Auditory integration training and other sound therapies for autism spectrum disorders: a systematic review. Arch Dis Child 91 (12):1018. https://doi.org/10.1136/adc.2006.094649

107. Baranek GT (2002) Efficacy of sensory and motor interventions for children with autism. J Autism Dev Disord 32(5):397–422. https://doi.org/10.1023/A:1020541906063

108. Markram H, Rinaldi T, Markram K (2007) The intense world syndrome—an alternative hypothesis for autism. 1(6). https://doi.org/10.3389/neuro.01.1.1.006.2007

109. Dawson G, Watling R (2000) Interventions to facilitate auditory, visual, and motor integration in autism: a review of the evidence. J Autism Dev Disord 30(5):415–421. https:// doi.org/10.1023/A:1005547422749

110. Weitlauf AS, Sathe N, McPheeters ML, Warren ZE (2017) Interventions targeting sensory challenges in autism spectrum disorder: a systematic review. Pediatrics 139(6). https://doi. org/10.1542/peds.2017-0347

111. Sinha Y, Silove N, Hayen A, Williams K (2011) Auditory integration training and other sound therapies for autism spectrum disorders (ASD). Cochrane Database Syst Rev (12). https://doi.org/10.1002/14651858.CD003681.pub3

112. Yeh S-H, Lin L-W, Chuang YK, Liu C-L, Tsai L-J, Tsuei F-S et al (2015) Effects of music aerobic exercise on depression and brain-derived neurotrophic factor levels in community dwelling women. Biomed Res Int 2015:135893. https://doi.org/10.1155/2015/135893

113. Kabra A, Sharma R, Kabra R, Baghel US (2018) Emerging and alternative therapies for parkinson disease: an updated review. Curr Pharm Des 24(22):2573–2582. https://doi.org/ 10.2174/1381612824666180820150150

114. Lee S-M, Kim B-K, Kim T-W, Ji E-S, Choi H-H (2016) Music application alleviates short-term memory impairments through increasing cell proliferation in the hippocampus of valproic acid-induced autistic rat pups. J Exerc Rehabil 12(3):148–155. https://doi.org/10. 12965/jer.1632638.319

115. Calabrese F, Rossetti AC, Racagni G, Gass P, Riva MA, Molteni R (2014) Brain-derived neurotrophic factor: a bridge between inflammation and neuroplasticity. 8(430). https://doi. org/10.3389/fncel.2014.00430

116. Chung L, Bey AL, Jiang Y-H (2012) Synaptic plasticity in mouse models of autism spectrum disorders. Korean J Physiol Pharmacol 16(6):369–378. https://doi.org/10.4196/kjpp.2012.16.6.369
117. Bourgeron T (2015) From the genetic architecture to synaptic plasticity in autism spectrum disorder. Nat Rev Neurosci 16(9):551–563. https://doi.org/10.1038/nrn3992
118. van der Steen JT, Smaling HJA, van der Wouden JC, Bruinsma MS, Scholten R, Vink AC (2018) Music-based therapeutic interventions for people with dementia. Cochrane Database Syst Rev (7). https://doi.org/10.1002/14651858.CD003477.pub4
119. Brotons M, Koger SM (2000) The impact of music therapy on language functioning in dementia. J Music Ther 37(3):183–195. https://doi.org/10.1093/jmt/37.3.183
120. Jacobsen J-H, Stelzer J, Fritz T, Chételat G, La Joie R, Turner R (2015) Why musical memory can be preserved in advanced Alzheimer's disease. Brain J Neurol 138. https://doi.org/10.1093/brain/awv135
121. Norberg A, Melin E, Asplund K (1986) Reactions to music, touch and object presentation in the final stage of dementia. An exploratory study. Int J Nurs Stud 23(4):315–323. https://doi.org/10.1016/0020-7489(86)90054-4
122. Baird A, Samson S (2009) Memory for music in Alzheimer's disease: unforgettable? Neuropsychol Rev 19(1):85–101. https://doi.org/10.1007/s11065-009-9085-2
123. Zhang Y, Cai J, An L, Hui F, Ren T, Ma H et al (2017) Does music therapy enhance behavioral and cognitive function in elderly dementia patients? A systematic review and meta-analysis. Ageing Res Rev 35:1–11. https://doi.org/10.1016/j.arr.2016.12.003
124. Chang YS, Chu H, Yang CY, Tsai JC, Chung MH, Liao YM et al (2015) The efficacy of music therapy for people with dementia: a meta-analysis of randomised controlled trials. J Clin Nurs 24(23–24):3425–3440. https://doi.org/10.1111/jocn.12976
125. Li H-C, Wang H-H, Chou F-H, Chen K-M (2015) The effect of music therapy on cognitive functioning among older adults: a systematic review and meta-analysis. J Am Med Dir Assoc 16(1):71–77. https://doi.org/10.1016/j.jamda.2014.10.004
126. Vasionytė I, Madison G (2013) Musical intervention for patients with dementia: a meta-analysis. J Clin Nurs 22(9–10):1203–1216. https://doi.org/10.1111/jocn.12166
127. Juslin PN, Liljeström S, Västfjäll D, Lundqvist L-O (2010) How does music evoke emotions? Exploring the underlying mechanisms. In: Handbook of music and emotion: theory, research, applications. Series in affective science. Oxford University Press, New York, NY, US, pp 605–642
128. Koelsch S, Jäncke L (2015) Music and the heart. Eur Heart J 36(44):3043–3049. https://doi.org/10.1093/eurheartj/ehv430
129. Aalbers S, Fusar-Poli L, Freeman RE, Spreen M, Ket JC, Vink AC et al (2017) Music therapy for depression. Cochrane Database Syst Rev 11(11):CD004517–CD004517. https://doi.org/10.1002/14651858.CD004517.pub3
130. Atiwannapat P, Thaipisuttikul P, Poopityastaporn P, Katekaew W (2016) Active versus receptive group music therapy for major depressive disorder—a pilot study. Complement Ther Med 26:141–145. https://doi.org/10.1016/j.ctim.2016.03.015
131. Scott A (2004) Managing anxiety in ICU patients: the role of pre-operative information provision. Nurs Crit Care 9(2):72–79. https://doi.org/10.1111/j.1478-5153.2004.00053.x
132. García-Perdomo HA, Montealegre Cardona LM, Cordoba-Wagner MJ, Zapata-Copete JA (2018) Music to reduce pain and anxiety in cystoscopy: a systematic review and meta-analysis. J Complement Integr Med 16(3). https://doi.org/10.1515/jcim-2018-0095
133. Lieber AC, Bose J, Zhang X, Seltzberg H, Loewy J, Rossetti A et al (2019) Effects of music therapy on anxiety and physiologic parameters in angiography: a systematic review and meta-analysis. J Neurointerventional Surg 11(4):416–423. https://doi.org/10.1136/neurintsurg-2018-014313
134. Song M, Li N, Zhang X, Shang Y, Yan L, Chu J et al (2018) Music for reducing the anxiety and pain of patients undergoing a biopsy: a meta-analysis. J Adv Nurs 74(5):1016–1029. https://doi.org/10.1111/jan.13509

135. Abdelhakim AM, Samy A, Abbas AM (2019) Effect of music in reducing patient anxiety during colposcopy: a systematic review and meta-analysis of randomized controlled trials. J Gynecol Obstet Human Reprod 48(10):855–861. https://doi.org/10.1016/j.jogoh.2019.07. 007

136. Ing-Randolph AR, Phillips LR, Williams AB (2015) Group music interventions for dementia-associated anxiety: a systematic review. Int J Nurs Stud 52(11):1775–1784. https:// doi.org/10.1016/j.ijnurstu.2015.06.014

137. Brown Wilson C, Arendt L, Nguyen M, Scott TL, Neville CC, Pachana NA (2019) Nonpharmacological Interventions for anxiety and dementia in nursing homes: a systematic review. Gerontologist 59(6):e731–e742. https://doi.org/10.1093/geront/gnz020

138. Hole J, Hirsch M, Ball E, Meads C (2015) Music as an aid for postoperative recovery in adults: a systematic review and meta-analysis. Lancet (London, England) 386(10004):1659–1671. https://doi.org/10.1016/s0140-6736(15)60169-6

139. Bradt J, Dileo C, Shim M (2013) Music interventions for preoperative anxiety. Cochrane Database Syst Rev (6). https://doi.org/10.1002/14651858.CD006908.pub2

140. van der Heijden MJ, Oliai Araghi S, van Dijk M, Jeekel J, Hunink MG (2015) The effects of perioperative music interventions in pediatric surgery: a systematic review and meta-analysis of randomized controlled trials. PLoS One 10(8):e0133608. https://doi.org/10.1371/journal. pone.0133608

141. Umbrello M, Sorrenti T, Mistraletti G, Formenti P, Chiumello D, Terzoni S (2019) Music therapy reduces stress and anxiety in critically ill patients: a systematic review of randomized clinical trials. Minerva Anestesiol 85(8):886–898. https://doi.org/10.23736/s0375-9393.19. 13526-2

142. Horne-Thompson A, Grocke D (2008) The effect of music therapy on anxiety in patients who are terminally ill. J Palliat Med 11(4):582–590. https://doi.org/10.1089/jpm.2007.0193

143. Bradt J, Dileo C (2014) Music interventions for mechanically ventilated patients. Cochrane Database Syst Rev (12). https://doi.org/10.1002/14651858.CD006902.pub3

144. Nightingale CL, Rodriguez C, Carnaby G (2013) The impact of music interventions on anxiety for adult cancer patients: a meta-analysis and systematic review. Integr Cancer Ther 12(5):393–403. https://doi.org/10.1177/1534735413485817

145. Corbijn van Willenswaard K, Lynn F, McNeill J, McQueen K, Dennis CL, Lobel M et al (2017) Music interventions to reduce stress and anxiety in pregnancy: a systematic review and meta-analysis. BMC Psychiatry 17(1):271. https://doi.org/10.1186/s12888-017-1432-x

146. Mitchell M (2003) Patient anxiety and modern elective surgery: a literature review. J Clin Nurs 12(6):806–815. https://doi.org/10.1046/j.1365-2702.2003.00812.x

147. Nilsson U (2008) The anxiety- and pain-reducing effects of music interventions: a systematic review. AORN J 87(4):780–807. https://doi.org/10.1016/j.aorn.2007.09.013

148. Gillen E, Biley F, Allen D (2008) Effects of music listening on adult patients' pre-procedural state anxiety in hospital. Int J Evid Based Healthc 6(1):24–49. https://doi.org/10.1111/j. 1744-1609.2007.00097.x

149. Lai H-L, Chen C-J, Chang F-M, Hsieh M-L, Huang H-Y, Chang S-C (2006) Randomized controlled trial of music during kangaroo care on perinatal anxiety and preterm infants' responses. Int J Nurs Stud 43:139–146. https://doi.org/10.1016/j.ijnurstu.2005.04.008

150. Lee OK, Chung YF, Chan MF, Chan WM (2005) Music and its effect on the physiological responses and anxiety levels of patients receiving mechanical ventilation: a pilot study. J Clin Nurs 14(5):609–620. https://doi.org/10.1111/j.1365-2702.2004.01103.x

151. Stern D (2010) The issue of vitality. Nordic J Music Ther 19(2). https://doi.org/10.1080/ 08098131.2010.497634

152. Geretsegger M, Mössler KA, Bieleninik Ł, Chen XJ, Heldal TO, Gold C (2017) Music therapy for people with schizophrenia and schizophrenia-like disorders. Cochrane Database Syst Rev 5(5):Cd004025. https://doi.org/10.1002/14651858.CD004025.pub4

153. Gold C, Solli HP, Krüger V, Lie SA (2009) Dose-response relationship in music therapy for people with serious mental disorders: systematic review and meta-analysis. Clin Psychol Rev 29(3):193–207. https://doi.org/10.1016/j.cpr.2009.01.001

154. Haeyen S (2018) People diagnosed with personality disorders in art therapy: what is the scientific evidence? Effectiveness of art therapy in personality disorders. In: Art therapy and emotion regulation problems. Springer, pp 95–119

155. Kenner J, Baker FA, Treloyn SJNJoMT (2020) Perspectives on musical competence for people with borderline personality disorder in group music therapy. 29(3):271–287

156. D'Alessandro P, Giuglietti M, Baglioni A, Verdolini N, Murgia N, Piccirilli M et al. (2017) Effects of music on seizure frequency in institutionalized subjects with severe/profound intellectual disability and drug-resistant epilepsy. 29:399–404

157. Bedetti C, Principi M, Di Renzo A, Muti M, Frondizi D, Piccirilli M et al. (2019) The effect of Mozart's music in severe epilepsy: functional and morphological features. 31(3):467–474

158. Babaie A, Ju S, Bajwa S (2020) Music therapy as an adjunctive treatment for refractory epilepsy (744). AAN Enterprises

159. Maguire MJPN (2017) Epilepsy and music: practical notes. 17(2):86–95

160. Anderson DE, Patel AD (2018) Infants born preterm, stress, and neurodevelopment in the neonatal intensive care unit: might music have an impact? 60(3):256–266. https://doi.org/10.1111/dmcn.13663

161. Bieleninik Ł, Ghetti C, Gold CJP (2016) Music therapy for preterm infants and their parents: a meta-analysis. 138(3)

162. Lejeune F, Lordier L, Pittet MP, Schoenhals L, Grandjean D, Hüppi PS et al (2019) Effects of an early postnatal music intervention on cognitive and emotional development in preterm children at 12 and 24 months: preliminary findings. 10:494

163. Gäbel C, Garrido N, Koenig J, Hillecke TK, Warth MJCmr (2017) Effects of monochord music on heart rate variability and self-reports of relaxation in healthy adults. 24(2):97–103

164. Mojtabavi H, Saghazadeh A, Valenti VE, Rezaei NJCTiCP (2020) Can music influence cardiac autonomic system? A systematic review and narrative synthesis to evaluate its impact on heart rate variability. 101162

165. Pain IAftSo (1994) IASP terminology. https://www.iasp-pain.org/terminology?navItemNumber=576#Pain. Accessed 21 July 2020

166. Engwall M, Duppils GS (2009) Music as a nursing intervention for postoperative pain: a systematic review. J Perianesthesia Nurs Official J Am Soc PeriAnesthesia Nurses 24(6):370–383. https://doi.org/10.1016/j.jopan.2009.10.013

167. Lee JH (2016) The effects of music on pain: a meta-analysis. J Music Ther 53(4):430–477. https://doi.org/10.1093/jmt/thw012

168. Garza-Villarreal EA, Pando V, Vuust P, Parsons C (2017) Music-induced analgesia in chronic pain conditions: a systematic review and meta-analysis. Pain Physician 20(7):597–610

169. Good M, Albert JM, Anderson GC, Wotman S, Cong X, Lane D et al (2010) Supplementing relaxation and music for pain after surgery. Nurs Res 59(4):259–269. https://doi.org/10.1097/NNR.0b013e3181dbb2b3

170. Gao Y, Wei Y, Yang W, Jiang L, Li X, Ding J et al (2019) The effectiveness of music therapy for terminally Ill patients: a meta-analysis and systematic review. J Pain Symptom Manag 57(2):319–329. https://doi.org/10.1016/j.jpainsymman.2018.10.504

171. Naughton MJ, Weaver KE (2014) Physical and mental health among cancer survivors: considerations for long-term care and quality of life. N C Med J 75(4):283–286. https://doi.org/10.18043/ncm.75.4.283

172. Bradt J, Dileo C, Magill L, Teague A (2016) Music interventions for improving psychological and physical outcomes in cancer patients. Cochrane Database Syst Rev (8). https://doi.org/10.1002/14651858.CD006911.pub3

173. Stanczyk MM (2011) Music therapy in supportive cancer care. Rep Pract Oncol Radiother 16(5):170–172. https://doi.org/10.1016/j.rpor.2011.04.005

174. Sellick SM, Crooks DL (1999) Depression and cancer: an appraisal of the literature for prevalence, detection, and practice guideline development for psychological interventions. Psychooncology 8(4):315–333. https://doi.org/10.1002/(sici)1099-1611(199907/08)8:4%3c315::Aid-pon391%3e3.0.Co;2-g

175. Goodwin PJ, Leszcz M, Ennis M, Koopmans J, Vincent L, Guther H et al (2001) The effect of group psychosocial support on survival in metastatic breast cancer. N Engl J Med 345 (24):1719–1726. https://doi.org/10.1056/NEJMoa011871

176. Sephton SE, Sapolsky RM, Kraemer HC, Spiegel D (2000) Diurnal cortisol rhythm as a predictor of breast cancer survival. J Natl Cancer Inst 92(12):994–1000. https://doi.org/10.1093/jnci/92.12.994

177. Bar-Sela G, Atid L, Danos S, Gabay N, Epelbaum R (2007) Art therapy improved depression and influenced fatigue levels in cancer patients on chemotherapy. Psychooncology 16(11):980–984. https://doi.org/10.1002/pon.1175

178. Sleight PJNHJ (2013) Cardiovascular effects of music by entraining cardiovascular autonomic rhythms music therapy update: tailored to each person, or does one size fit all? 21(2):99–100

Niloufar Yazdanpanah is currently a medical student at Tehran University of Medical Sciences (TUMS). She is a young researcher and the managing director of the Network of Immunity in Infection, Malignancy, and Autoimmunity (NIIMA) in the Universal Scientific Education and Research Network (USERN). She is the editorial assistant of Prof. Nima Rezaei in the Translational Immunology book series. She is one of the young managers of USERN, in charge of USERN Advisory Board Affairs.

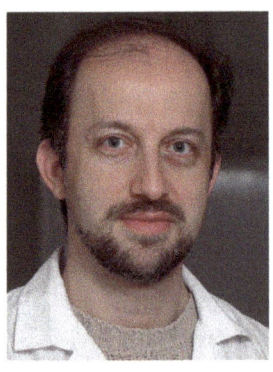

Nima Rezaei gained his medical degree (M.D.) from TUMS in 2002 and subsequently obtained an M.Sc. in Molecular and Genetic Medicine and a Ph.D. in Clinical Immunology and Human Genetics from the University of Sheffield, UK. He also spent a short-term fellowship in Pediatric Clinical Immunology and Bone Marrow Transplantation in the Newcastle General Hospital. Since 2010, Prof. Rezaei has worked at the Department of Immunology and Biology, School of Medicine, TUMS; he is now the full professor and vice-dean of International Affairs, School of Medicine, TUMS, and the co-founder and Head of the Research Center for Immunodeficiencies. He is also the founding president of USERN. He has edited more than 30 international books, has presented more than 500 lectures/posters in congresses/meetings, and has published more than 800 articles in international scientific journals.

When Combining Arts and Sciences Assists Medical Devices Uses: DeafSpace and Cochlear Implants

21

Andrée-Anne Blacutt and Stéphane Roche

"Only art and science make us suspect the existence of life to a higher level, and maybe also instill hope thereof."

Ludwig van Beethoven

Summary

Over the past two decades, patient-centered approaches have gradually been adopted to engage patients as a partner in the decision-making process and take into account their specificities, values, and experiences. The Montreal Model combined with the Disability Creation Process (DCP) Model is an innovative way of building a real professional-patient learning interaction and partnership for addressing disability as a component of the environment. As regards deafness, in particular, cochlear implants have become the principal way of supporting people who are severely deaf. With the aid of cochlear implants, a kind of electronic processor, it would be possible for these people to detect sound better. This chapter aims to demonstrate, through a few concrete examples (e.g., the Atomic Box project), to what extent the combination of art and science makes it possible to develop socially innovative responses (based on the two models mentioned above), and thus to assist the adoption of implants by hearing-impaired users.

A.-A. Blacutt
Faculty of Graduate and Postdoctoral Studies, Laval University, Quebec, Canada
e-mail: andree-anne.blacutt-grenier.1@ulaval.ca

S. Roche (✉)
Research Center for Geospatial Data and Intelligence, Laval University, Quebec, Canada
e-mail: stephane.roche@scg.ulaval.ca

Pavillon Louis-Jacques-Casault, Université Laval Québec, 1055, avenue du Séminaire, Québec G1V 06, Canada

© The Author(s), under exclusive license to Springer Nature Switzerland AG 2022
N. Rezaei (ed.), *Multidisciplinarity and Interdisciplinarity in Health*,
Integrated Science 6, https://doi.org/10.1007/978-3-030-96814-4_21

Atomic box: Artistic and design creative process

The code of this chapter is *01101110 01100100 01101110 01100101 01010101 01110010 01100001 01100100 01100111 01110011 01101110 01110100 01101001 00100000.*

Keywords

Cochlear implant · Cognition · Deaf gain · Deafness · Design · DeafSpace · Empowerment · Social innovation

1 Introduction

Paternalistic approaches were used for a very long time in the medical domain, where patients were only seen as sick people easily classified in simple categories, not as capable of sharing their knowledge and understanding of their condition [1]. Over the past two decades, patient-centered approaches have gradually been adopted. Patients' specificities, values, and experiences thus began to be taken into account in the diagnostics and treatments, as well as the understanding of their disabilities [2]. Even when their specific status is taken into account, in most cases, patients are still not considered stakeholders or partners in the health ecosystem, and therefore still not actively involved in the care process. However, numerous researches, particularly with respect to chronic diseases, have demonstrated the higher values of integrating the patient in the form of partnership. Indeed, in these particular cases where patients have to compose their whole life with a pathology, the evolution of the disease itself is intimately linked to their way of life, then their knowledge and experiences are essential in the decision-making [3].

In response, more holistic approaches of the patients in/with their health environment have been developed, aiming to consider patients as a real partner (equal to) of health professionals. The idea is then not only to take into account the patient and its own experiences but to really build a partnership between patient and professional where learning interactions and knowledge exchanges are used to feed the dynamic of care [4]. The Montreal Model is one of those partnership approaches [5]. It is focused not only on the patient but also on the patient's vision, the ones who have the most global vision on their condition [1]. Disabilities are not, strictly speaking, chronic diseases. However, like patients suffering from chronic illnesses, the whole life of these people is marked by their disability and, by doing so, their way of life, their practices, their perceptions of their environment influence the evolution of their situation of disability. Also, an approach like that of the Montreal Model seems entirely relevant to us to be implemented in the field of disabilities. This is particularly the case when combined with the Disability Creation Process—DCP [6]. In the DCP model, disability is, indeed, represented from the environment viewpoint rather than from the limited capacities of people (their body or their senses) [7]. Consequently, engaging the disabled person as a partner in the strategy and the measures to be implemented to modify the environment so as to adapt it to their specific skills appears to be an interesting approach [8].

In this sense, interdisciplinary and intersectoral methods have emerged to support those innovative approaches, particularly by combining design practices, social sciences, engineering, and health sciences [9]. A very few researches have explored the specific combination of arts and sciences, particularly for designing more relevant and useful medical devices and improving their users' adoption by engaging users as design collaborators [10].

This is precisely what this chapter deals with. It focuses on the ontological, epistemological, and practical issues (approaches, methods, technics, and data) of combining arts and sciences to assist in the design and development of medical devices through a holistic approach of the users in/with their environment. More specifically, the chapter is dedicated to hearing-impaired people and the case of cochlear implant: "an electronic device that allows people with severe to profound deafness to have better access to sound. It consists of an internal part inserted under the skin, behind the ear during surgery performed under general anesthesia; and an external part, a voice processor connected to the antenna. Although the helping nature of these implants is widely demonstrated, a certain number of technological, cognitive and social limits ... remain" [11].

Research on disabilities is now extended to see the role of assistive technologies, the environment, and their interactions [12]. As a matter of fact, among the works and research done on deafness, particularly those dealing with DeafSpace, are probably most consistent with the Montreal Model and the DCP one [13].

The science of deafness is structured in different dimensions, importantly cognition, culture, and innovation. DeafSpace offers an effective approach by not only thinking of but also by taking all these dimensions into account [14]. With this approach, the deaf culture is integrated with a universal design. This integration allows "to rethink the environment (rooms and housing, urban places and spaces...) from 360° learning approach that aims at improving deaf people skills (peripheral, transparency, reflection, vibration and shared sensory reach)" [11, 15].

Therefore, this chapter is specifically based on an experiment (prototype of DeafSpace) conducted in the context of the main author's Ph. D. thesis in design and social innovation. This thesis addresses the following issue. A cochlear implanted person often faces the illusion of being normal hearing. This is one of the reasons why a large part of the deaf community is extremely cautious about implantation. So, to what extent does this alter their (implanted deaf persons) perception of their condition, skills, and needs? This project consists of setting up a process of social innovation that allows people with disabilities, with hearing impairment from mild to profound, to understand the "trans-deafness" issue better. This research particularly implies the improvement of the communication of designers and users with a hearing impairment and implanted. The focus is specifically on the voice of people with hearing loss, in order to encourage their engagement, according to an approach inspired by the Montreal Model. In terms of social innovation, the thesis proposes to break with the general idea prevalent in the medical ecosystem that cochlear implants are only medical solutions ("auditory bandages"). As a contextual element, it is important to emphasize that one of the roots of this thesis project is a personal one. The first author, an artist by training, is the mother of a little six-year-old deaf girl (congenital deafness), implanted cochlear in both ears since the age of two.

The first section of this chapter is dedicated to analyzing the pros and cons of the cochlear implant solution. The main technological, cognitive, social, and economic limitations and issues are then explained and, a "kind of" SWOT synthesis is finally provided (Table 1). Section 2 is dedicated to presenting the conceptual basis of the Montreal Model of partnership and the DCP. In Sect. 3, we demonstrate, through a concrete example "The Atomic Box Project" based on the "360° learning approach" of DeafSpace, to what extent the combination of art and science makes it possible to develop socially innovative responses to cochlear implants limitations and issues, and thus to assist the adoption of implants by hearing-impaired users. The conclusion provides an occasion to open up the discussion and, to move toward more innovative projects.

Table 1 SWOT synthesis of cochlear implants

(Main) **Strengths**	(Main) **Weaknesses**
– Improving speech perception – Encouraging learning to speak for children under two years old – Supporting autonomy (mobility, for instance) – Improving security	– Integrating cochlear implant into the family life is still a challenge – The cost of the cochlear implant can be a brake depending on the country social health care system – The difficulty of practicing sports (cochlear implant hanging system does not allow the body to move freely)
(Main) **Opportunities**	(Main) **Threats**
– Adding communication accessories such as the communication system allowing students to hear their teachers directly via their cochlear implant – Contributing to social inclusion and integration	– Explantation – Appropriation of data by the health system

2   Pros and Cons of Cochlear Implants

The history of cochlear implants development dates back to the end of the fifties [16]. The cochlear implant story is roughly divided into the first clinical trials (1951–1976), the industrial development (1077–1997), and the contemporary period. According to Prof. Chouard [17], one of the most eminent researchers in cochlear implants, the first implant was performed in 1957 in Paris by Charles Eyriès, a Parisian otologist and anatomist. The device was designed and manufactured by André Djourno, a professor of medical physics, also in Paris [18]. It was not until 1961 that William House, a Los Angeles-based otologist took up from where Djourno and Eyriès had left off; "He (House) standardized the operation by placing the electrode in a stable position, threading it through the round window in the cochlea. He developed a reliable device and started to implant it in a growing number of patients" [17]. Worldwide criticisms have immediately emerged. House's critics accused him of endangering healthy cochlear structures of the few remaining auditory nerve fibers [19]. But the main weakness of House's single-channel implant was that it did not allow patients to understand the spoken word without lipreading [17]. According to Chouard and Mac Leod [20], the cochlear implant should have multiple electrodes.

From then, due to the technological requirements, a new collaboration between Chouard's research team and Jean Bertin's private Research and Development Company has started. Many improvements have been made, especially to endow the implant with two electrodes and improve the connection with the internal part of the implant. The first implantation of a multichannel cochlear implant took place in Paris in September 1976. It was carried out by Prof. Chouard's team. Before the hearing returned the next day, despite the inconvenient volume of the transmitter, the other five patients were quickly operated on. The Bertin company deposited on

March 16, 1977, patent No. 77/07824. This patent, issued from Mac Leod's physiological requirements, included two concurrent claims: (a) the sequential transmission to the cochlea of an undetermined number of frequency bands; and (b) the transmission of all audible audio information. This was the real beginning of the industrial development of cochlear implants. Progress has been achieved competitively in four main countries Australia, Austria, France, and the USA until a consensus was reached at the end of the nineties.

2.1 Concepts and Operating Principle of Cochlear Implant

Modern cochlear implants are surgically placed under the skin, behind the ear. There are several manufacturers and different models, but they all consist of an external part and an internal part (Fig. 1).

The external part includes:

- one or more microphones pick up the sound environment and transform it into an electrical signal;
- a processor that filters the sound information received, particularly to process human voice as a priority and distribute it over different channels. The electrical pulses are then directed to the transmitter through a thin conducting wire;
- an electromagnetic induction transmitter held by a magnet placed behind the outer ear. The transmitter supplies the internal part of the device with the energy necessary for the operation of the device as well as the electrical signals processed by the processor.

The internal part consists of:

- a receiver and stimulator placed under the skin are connected to the bone to be held in place. It converts signals into electrical pulses and sends them through a cable to the electrodes;
- a group of electrodes up to 22 electrodes for the most recent models cross the cochlea and transmit electrical impulses to the nerves of the tympanic ramp, which relay them to the brain.

Operation principles are quite simple (Fig. 2):

(i)   the sound information is delivered to a device installed behind the roof (external ear);

(ii)   the sound information is processed by a microprocessor embedded in the device, and the electrical signal is obtained by the cochlea via a wire connecting an antenna (placed under the subject's skin);

(iii) the cochlea carries the signal to the implanted electrodes (up to 22 electrodes); and

(iv) electrodes stimulate the auditory nerve, which sends information to the brain.

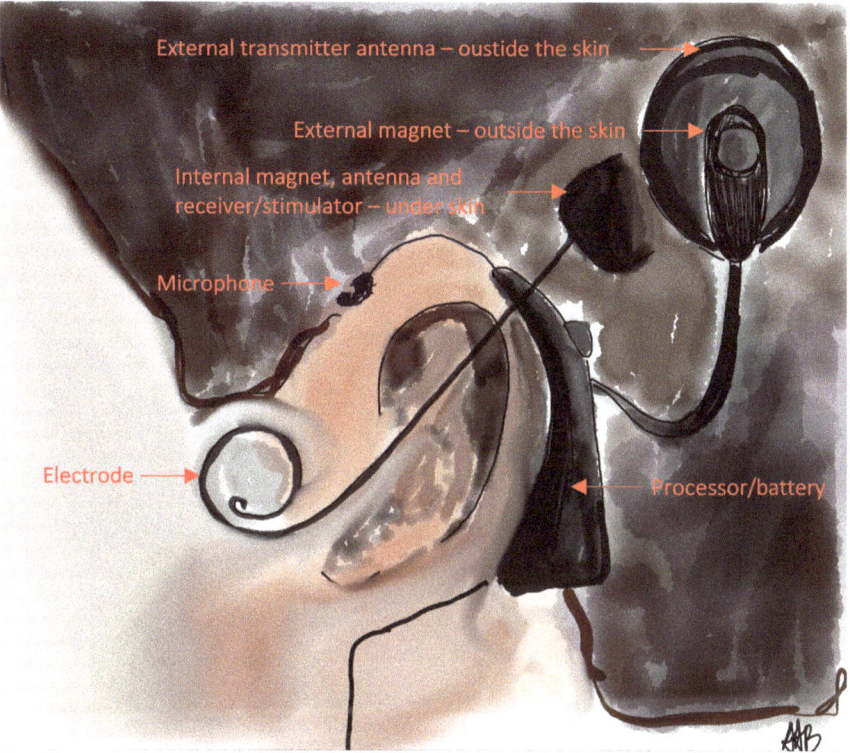

Fig. 1 Internal and external components of cochlear implant *(Made by Andrée-Anne Blacutt, inspired by NIHR Health Technology Assessment Program)*

Four main companies share the current market: Cochlear, Advanced Bionics, MED-EL, and Optical Medical. Their conceptual principles and technologies are still very close. Differences between those cochlear implants lie essentially in the services and accessories accompanying the technologies (connectivity with a smartphone, for example) and in the design of the implant itself.

2.2 Strengths, Weaknesses, Opportunities, and Threats

Initially, the sensation of hearing sounds in people who have received cochlear implants differs from that in people with normal hearing and those who are paired with external hearing devices.

Indeed, the cochlear implant has not been fully integrated yet. A hearing education team of experts in the treatment and rehabilitation of hearing loss-associated communication disorders can help patients practice audiology and/or speech. After a period of adaptation, the results are often excellent: many adults who have

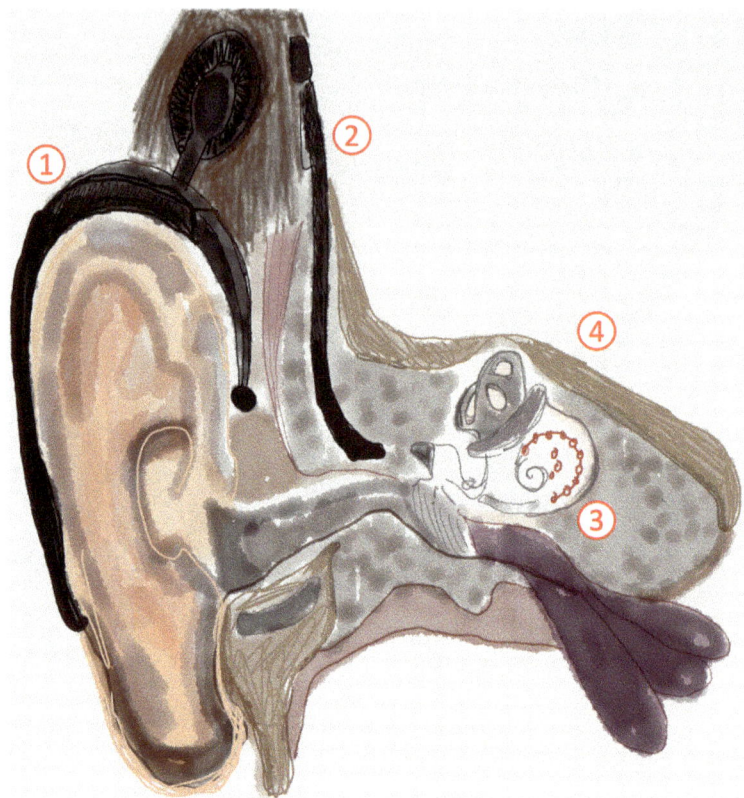

Fig. 2 Operating of cochlear implant *(Made by Andrée-Anne Blacutt, inspired by CHU de Québec, Université Laval)*

become deaf or children implanted very early are able to use their phones. Researches indicate that approximately one-third of implanted children achieve excellent results with an understanding equivalent to normal-hearing children. Another third acquires an understanding of correct speech, and the last third encounters difficulties, very often correlated with late implantation or existing disorders other than deafness [21].

Cochlear implants are mostly, in nature, helpful; however, they might not be ethically, cognitively, socially, and economically feasible, and some weaknesses and threats remain to be addressed. Deafness can naturally decrease or level down, but the hearing cells do not regenerate. The brain compensates for hearing loss mainly by mental support. With sustained efforts and adapted removable amplifying devices, deaf children can speak and sing nursery rhymes, except for profound and congenitally deafness. Therefore, the implantation of young children is a difficult decision for parents [22].

The current policy of implanting pre-lingual deaf children is highly contested by part of the deaf community [23]. Indeed, the deaf community sees cochlear implantation as a devaluation of sign language in favor of the oral language, even as a negation of the deaf culture. Part of the deaf community even interprets the choice of parents of deaf children who choose to implant their baby as an attempt to "repair the deafness" or to "repair their child" by making them disabled, hearing-impaired both in the deaf world and in the hearing world. This community fears that the choice to have a deaf child operated on risks hindering their integration into the deaf community without ensuring their perfect integration into the hearing community [24].

Besides, the number of explanations is increasing; more and more implanted people who have reached adulthood wish to have their cochlear implant removed because they consider themselves as deaf people and they reclaim their deaf identity. They consider cochlear implantation as a violation of the physical and psychological integrity of a deaf person [25]. Our understanding of biotechnology and biomedical engineering has advanced greatly that offer people with deafness, especially people who are severely or congenitally deaf, cochlear implants to effectively joy hearing. On the other hand, we see these implants that negatively affect patients, and psychological symptoms are likely to be caused in some cases. Therefore, cochlear implant's effectiveness depends on sustainable and integrated support that indicates the whole healthcare ecosystem and its parts to manifest their active presence and make their contribution [26–28].

3 Montreal Model and Disability Creation Process

According to the above discussion, it is clear that the cochlear implantation process has to be encompassed in the deaf gain trend: "Deaf gain" is a fairly new term that Deaf people use when discussing all of the benefits of being deaf and being involved in the Deaf community. Rather than seeing deafness as hearing "loss," many deaf people prefer to see it as Deaf "gain" [14]. Therefore, DeafSpace is where support and commitment to cochlear implanted users must be expressed and deployed [15]. There is then a need for a conceptual framework and technics and methods to support this complex process. The Montreal Model of a partnership relationship between patients and healthcare professionals, on the one hand, and the way of defining and considering disabilities by the DCP, on the other hand, brings innovative answers and solutions to the front.

3.1 The Montreal Model

The Montreal Model of a partnership relationship between patients and healthcare professionals has been developed for a decade by a University of Montreal-based team to improve the population health and healthcare service quality delivered by the health system. This patient partnership model focuses on patient engagement at different levels of healthcare as well as the training of health professionals [1]. It is based on recognizing patients' experiential knowledge, or simply knowledge of disease understood from living with a disease, and complementary to the scientific and medical/health knowledge. The model approach is typically centered on the same kind of holistic "360° learning vision" that the one proposed within the DeafSpace concept: a vision developed by and for (with) patients about all the health ecosystem components (specialists, administration, nurses, all stakeholders, etc.). Therefore, this model makes central the use of what patients see, perceive, understand, and know. It differs from classic patient-centered approaches because it engages a learning, co-management, or co-decision relationship to develop the skills of the patient and, in return, the ones of the healthcare professionals [29].

By considering patients as essential partners in all decisions that concern them and as experts in the organization of care, the Montreal Model offers relevant perspectives for managing chronic diseases [30]. In this sense, it is fully in line with the theoretical foundations of deaf gain that also claim the central role played by deaf people's specific skills and knowledge. Thus, we could hypothesize that the Montreal Model, perfectly suited to the management of chronic diseases, constitutes an extremely relevant avenue for managing deafness and the necessary monitoring of cochlear implantation in particular.

3.2 DCP

The authors of the chapter have previously described the model and their goals as follows: "universal design is being more and more mobilized in order to support the conception, development, and implementation of more inclusive assistive technologies [31] and, to make urban spaces more accessible for people with different forms of disabilities. More precisely, together with the Disability Creation Process (DCP) model, human-centric and UX design approaches have emerged as interesting and useful ways of thinking about the management of disability. Indeed, the use of the DCP model aims at defining disability in terms of the characteristics of the environment rather than as the limits of the person themselves (their bodies or their senses). This model aims at explaining the causes and consequences of diseases, trauma, and other effects on people's integrity and development. The DCP is a model that does not put the responsibility of disability on the person. From this model, the understanding and explanation of the disability phenomenon are based on the interaction between three conceptual domains: personal factors (organic system, capability of the person and identity factor), environmental factors (facilitators vs. obstacles/social vs. physical), and life habits (social participation situation vs. disabling situation/current activities vs. social roles)" [7, 11]. Altogether, consistent with the deaf gain and DeafSpace concepts, there is a big room in the DCP model, on the one hand, making it necessary to actively and wholly involve deaf people and their relatives in cochlear implantation management and, on the other hand, identifying *deafness not as a disability but rather as an opportunity* as important.

3.3 Assisting Cochlear Implanted Users Through a Deaf Gain-Based Combination of Art and Science

The usefulness of intersectoral and interdisciplinary approaches and methods in the search for solutions to assist in the design and development of medical devices through a holistic approach of users in/with their environment has already been extensively been demonstrated [32–35]. As described in the previous sections, if sciences and technologies are required to make cochlear implantation a success, sciences and technologies are certainly not sufficient. Deaf gain concepts and Deaf space approaches bring together sciences, design, and arts. People with deafness have a community of their own, with an affinity to their culture, using Deaf with a capital "D". Aaron Williamson, for instance, a deaf performance artist, says, "Why had all the doctors told me I was losing my hearing, but not one of them told me I was gaining deafness?" [36]. However, the relevance of combining arts and sciences, in particular, is not trivial and then often still underestimated [37], even regarding the design of things and objects [38, 39].

A very few concrete works, and most of the time unpublished, have attempted to reconcile arts and sciences to assist in the process of cochlear implantation in particular. Some works took place in the Department of American Sign Language and Deaf Studies, College of Arts and Sciences at Gallaudet University, specifically those who conceptualized DeafGain and DeafSpace. Basic elements are available in the SAGE Deaf Studies Encyclopedia [40]. Robinson [41] explains how to use Deaf history as a framework for teaching sign language to their students in a liberal art program. As a matter of fact, most of the papers dealing with Deaf studies in connection with the arts and sciences take signed languages as their main objects [42–44]. Geography concepts mixed with literature have also been explored to feed some of the basic foundations of DeafSpace. Rosen's work [45] about the Sensescape model is a very well-known example. Rosen explores the geographies in the DeafWorld, defined as "… the sites where different institutions create and imprint their ideologies, practices, and properties pertaining to their sensory notions of the deaf body onto brick-and-mortar spaces in the DeafWorld" [45]. The theoretical approach of Sensescape provided a basis for the description of DeafWorld institutional geographies. The fact remains that in the DeafGain/DeafSpace approach, the consideration to be given to deaf art (produced by and for the deaf community) is a major component of deaf culture. And this component must therefore be taken into account in assisting the cochlear implantation process. The following example is an illustration of this.

Kim Auclair is a Quebec City-based IT deaf entrepreneur, author, and artist. She has tried, from birth, to hide her deafness and to pretend that this issue did not exist. One year ago, after a "long journey," she was implanted in one ear. Since then, she has documented her hearing experience with her cochlear implant. She shares her hearing learning curve almost every day: her difficulties in dealing with sounds she has never heard before. She expresses how she misses the world of silence sometimes too. Most interestingly, she explains to what extent her implantation has helped her accept and now better live with her own deaf identity. Most of its personal information sharing is expressed in graphic form through social and classical media (Fig. 3). This artistic way of expressing and sharing her own speeches has contributed to improving the development of a better understanding of her own sensoriality.

4 Art and Science/Research-Creation

Human creativity is both a science and an art: "each of which in turn contributes to our understanding of nature and complexity" [46]. According to Richard Gregory, science is explanatory and art evocative [47]. Therefore, what seems to differentiate art from science is that art is usually done voluntarily with no expectation of return, while science is, in a certain sense, a contract indicating that some kind of return is expected [46]. Many concrete examples are published in the special issue of Inter, "a Québec-based journal of international renown that encourages artists to speak about art," dedicated to "Technocorps and cybermilieux" [48]. Research cannot be reduced to its scientific understanding, even if the recent evolution of higher education institutions has consolidated this vision in its methods as well in its forms of presentation. Research is the engine and vehicle of disciplinary (and more and more interdisciplinary) practices from which knowledge is produced and transmitted. So, research requires the researcher's personal, intellectual, and emotional commitment, the mobilization of his/her imagination, and his/her creative energy [49].

Among the profound debates that have fed the interactions between arts and sciences for decades, one way has recently emerged that tried to reconcile art and sciences, at least from a methodological point of view. Research-creation, which is not research about arts, has emerged as a new field [50]. Research-creation is characterized by the fact that the practice of art becomes an essential component of both the research process and the results it produces [51].

Moreover, Höök [52] defends the necessity of a qualitative shift as a requirement for design methods. She considers that the current "predominantly symbolic language-oriented design" stance should be replaced by a more "experiential, felt, aesthetic stance permeating the whole design, use and support cycle." Höök proposed an alternative form of design embedded in a process that "… allows designers to "examine" and improve on connections between sensation, feeling, emotion, subjective understanding, and values" [52]. This position fits very well with the needs mentioned above for supporting cochlear implantation.

The combination of design, research-creation, and cognitive science is precisely how we choose to address the project presented in this chapter. As illustrated in Fig. 4, research-creation combines the arts and sciences in a double movement. In the specific case of addressing cochlear implantation issues based on a DeafSpace conceptualization and operation, research-creation mainly contributes to the construction of speeches through developing a better understanding of deaf people's sensoriality [51]. From the right part of Fig. 4, art brings "tools" and perspectives to science in order to build self-understanding and better attention to what deaf people

Kim Auclair
J'augmente la visibilité des entreprises en démarrage dans les médias. Sourde,
j'entends grâce à un implant cochléaire.
masurdite.com/produit/bande-dessinee-ma-surdite
Abonnés : riomarti, webaquebec, sylviebougie et 7 autres

⊞ PUBLICATIONS   ☑ IDENTIFIÉ(E)

Fig. 3 Kim Auclair's personal hearing experience

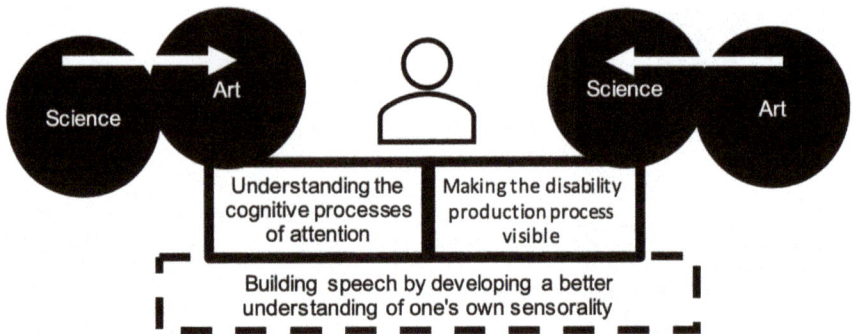

Fig. 4 Combining art and science

live and feel. Mobilization of an artistic creation process helps deaf people find a way of improving their self-confidence and the ability to assert themselves and express their deafness condition [52]. On the other side of Fig. 4, science (and in our specific case, cognitive science and design) provides art with a broader and richer understanding of deaf people's attention processes. Attention refers to a cognitive process of selecting and focusing on the major stimuli to respond properly [53, 54]. This cognitive capacity is of great importance given that we use it on a daily basis and that it allows us, in particular, to react to situations of risks or dangers, for example. Understanding those attention processes is a unique way for deaf people to acquire tools for better knowing themselves. This growth in their own knowledge about their deafness condition helps to limit brain fatigue [15]. Indeed, the effort of attention is extremely demanding.

The combination of art on the one hand, and cognitive science and science of design, on the other hand, make it possible to develop a means of "visiblizing" and of representing the situation of handicap (spectacle of digital and interactive "visiblization," public performance art/science, fiction design workshops, etc.). All while respecting the integrity of the deaf person by explicitly and actively mobilizing them in the DCP. Last but not least, in the specific case of cochlear implanted deaf people, the advantages of this combination are especially related to monitoring the improvement of their DeafSpace representation to understand their own deaf experiences and DeafGain better; and to their empowerment.

5 The Atomic Box Project

The Atomic Box is a prototype of a learning and gaming place developed by the first author. It aims at making cochlear implanted deaf persons conscious of their one body and senses into a stimulating sensorial environment. The design of the Atomic Box is based on the mesh of the characteristics and specific needs of people

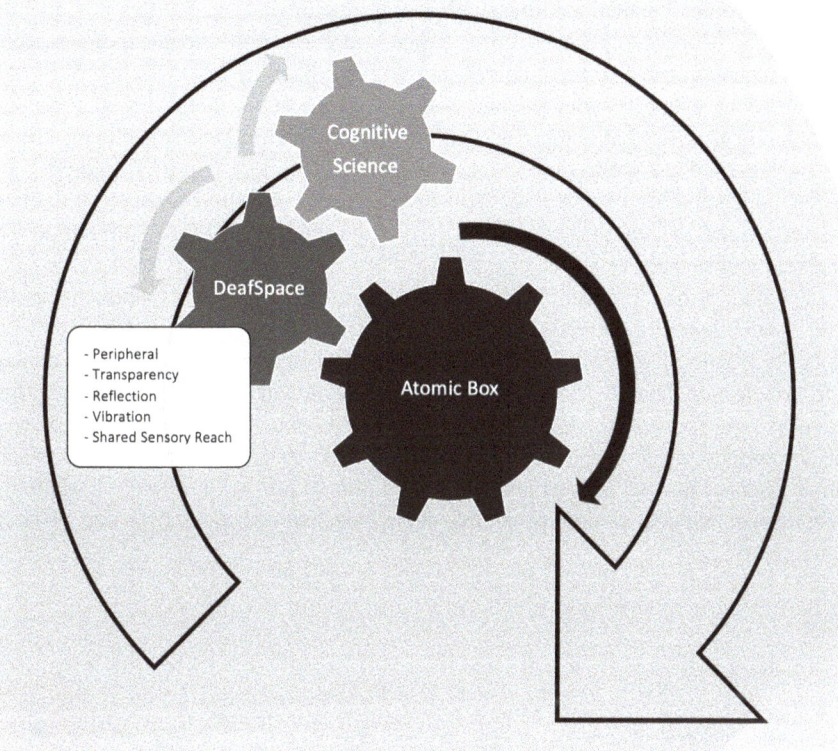

with hearing loss. It is based on an artistic creation process that combines music, instrument, and vocal production produced by the participant. So, gamification and sound production (musical-vocal) are at the forefront of this project.

From a theoretical basis, the Atomic Box takes up the concepts and principles of the 360° learning approach for DeafSpace (Fig. 5). For deaf people, the 360° learning approach is an aid to improve:

(i)   the capacity of feeling and sensation;
(ii) the ways of expressing experiences;
(iii) communication with confidence using touch; and
(iv) the observation skills, especially regarding the environment and relationships between human beings and objects [55].

Conceptually, DeafSpace is the architecture "tailored to deaf vision and culture in space. Buildings, hallways, stairs, and other spatial arrangements are designed to deaf people's ways of seeing and being in their environments. The first experiment of planning and building a DeafSpace was initiated by Hansel Bauman (architect) in 2005 with the ASL Deaf Studies Department at Gallaudet University, in collaboration with his brother Dirksen Bauman who is a professor of Deaf Studies at Gallaudet University. Mobility and Proximity, Space and Proximity, Acoustics (vibration), Lights (shadow, transparency, color), and sensory reach (refers to the needs of deaf people to be spatially oriented and visually aware of the activities in their surroundings) are essential components of the environment in DeafSpace" [11, 15].

As regards DeafSpace, 360° makes accessible "the five required markers for a better understanding of space for deaf people: peripheral, transparency, reflection, vibration, and shared sensory reach" [15]. Figure 6 and Table 2 summarize five markers and respective skills.

Together, those five DeafSpace markers form, in a way, the components of the 360° spatial learning equation. As written by Hansel Bauman [15]: "A sense of self —or "personhood"—for deaf individuals is reinforced through eyes-to-eyes contact and a sense of shared sensory experiences and creative endeavors." Providing adapted DeafSpace to the deaf community should contribute to developing a sense of community by providing multiple opportunities to view peers engaged in creative endeavors and to display and share this work with others. Essentially, "DeafSpace aims to create spaces that have maximum visual access and promote interaction" [15].

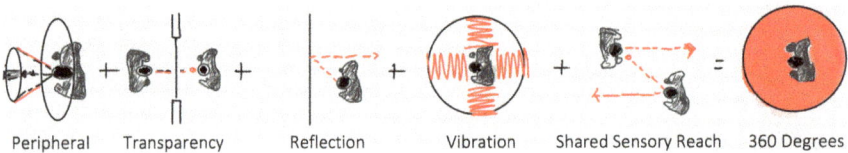

| Peripheral | Transparency | Reflection | Vibration | Shared Sensory Reach | 360 Degrees |

Fig. 6 Five markers of DeafSpace *(Made by Andrée-Anne Blacutt, inspired by Dangermond Keane Architecture)*

Table 2 Five required markers for a better understanding of space for deaf people [15]

DeafSapce marker	Properties
Peripheral	Rhythmic, repetitive, and intuitive visual cues support the deaf person's peripheral vision
Transparency	The degree of visual connections and the degree of openness and of the enclosure of spaces make the deaf person's space transparent
Reflection	The extension of vision allows deaf people to see behind them and to access depth and perspectives
Vibration	The characteristics of the floor surfaces allow deaf people to feel the presence of others and initiate contacts
Shared sensory reach	The interdependency of deaf individuals is understood when they navigate their environment, they indeed extremely depend on one another to extend their spatial and orientation skills and spatial reasoning capabilities

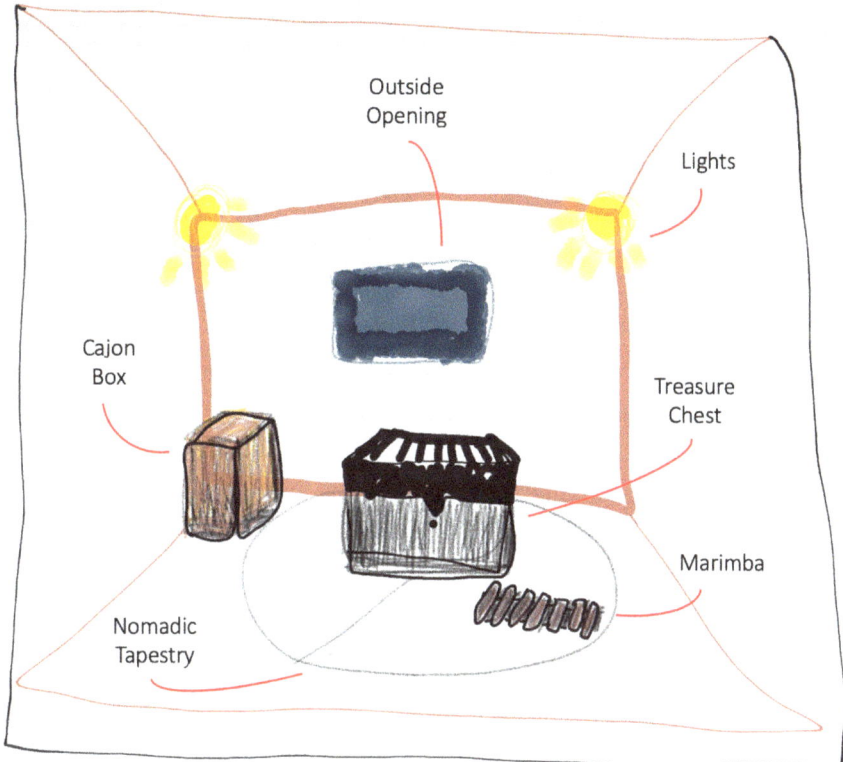

Fig. 7 Atomic box's treasure chest sketch

This is basically the main rationale of the Atomic Box project, offering to deaf individuals with a cochlear implant a DeafSpace prototype for supporting and improving their self-connection to their deafhood and their connections to their pairs, as well as to all the healthcare professionals and relatives involved in the management of the implantation process. Therefore, in the Atomic Box, a process of gamification invites participants to play around the treasure chest (Fig. 7), through which participants will come into contact with the components of the Atomic Box room (markers). At the very beginning of the process, the person hands over five personally meaningful objects either through the challenges those objects represent the curiosity they provoke or the desire to integrate them into their lives (e.g., a musical instrument, a telephone, a doorbell, a fire alarm, etc.). An illustration is provided in the box below.

The Atomic Box is the result of an artistic, creative process fed by design methods and cognitive sciences knowledge (the full design process is available here: https://vimeo.com/644592549). Four main phases have structured this process (Fig. 8).

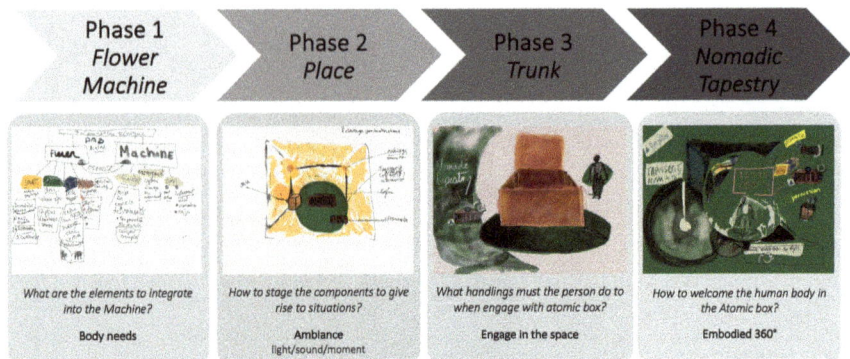

Fig. 8 Phases of the atomic box artistic creation process

(i) Phase 1 Flower Machine, The first phase was centered around classifying the elements included in the Flower Machine. The name Flower Machine refers to the symbolism linked to the pentagram flower, often associated with humans; one head, two hands, and two legs; but also five senses, and in the case of DeafSpace, five Makers. This symbol has been widely used by John Dewey, who adopted naturalism, rejecting any dualism opposing the body and the thought. Moreover, he has delivered a series of lectures at Harvard regarding the issue of arts, not reduced to the objects and commodities they produce, but rather considered as a way of transfiguring lived human experiences [56]. This also refers to Michel Foucault's "Machine à guérir" (Healing Machine) theory [57]. This phase was dedicated to doing an inventory of expected components of the environment (DeafSpace) as well as to characterizing those components;

(ii) Phase 2 place, The second phase was dedicated to naming the place and defining its characteristics. This phase aimed at relating the components of the Atomic Box to create a fertile place for sound production;

(iii) Phase 3 Trunk, Engaging the participant with the definition of senses was the aim of the third phase. The game learning strategy was developed at this stage (the one of the Treasure Chest in particular). Suggestions of meaningful objects (entering the Atomic Box) for people took place at this phase, too; and

(iv) Phase 4 Nomadic Tapestry, is the final phase, the one where all the components of the DeafSpace are coordinated to build the 360° learning environment. Figure 9 is one of the 50 and more sketches that have been produced to design the creative artistic process. The Atomic Box provides to users Nomadic Tapestries that allow to move and change different categories of markers stuck on its surface (contrast markers, movable markers, markers-actuators that trigger sound reactions) as well as projection plans, sources of sound captures (in/out), sound objects, textiles and visuals of Augmented-Reality (see also Fig. 10).

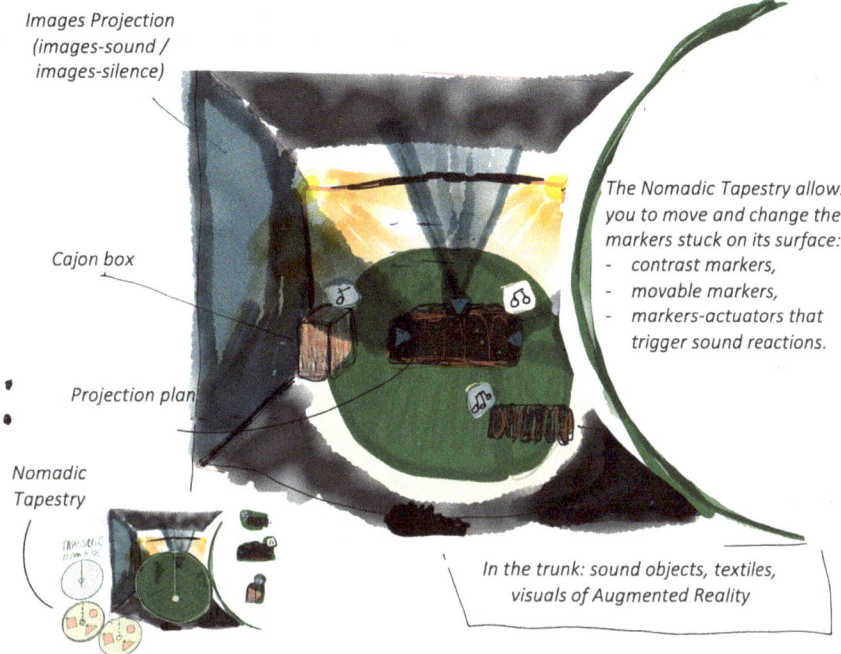

Images Projection
(images-sound /
images-silence)

Cajon box

Projection plan

Nomadic
Tapestry

The Nomadic Tapestry allows
you to move and change the
markers stuck on its surface:
- contrast markers,
- movable markers,
- markers-actuators that
  trigger sound reactions.

In the trunk: sound objects, textiles,
visuals of Augmented Reality

Fig. 9 The atomic box DeafSpace sketch

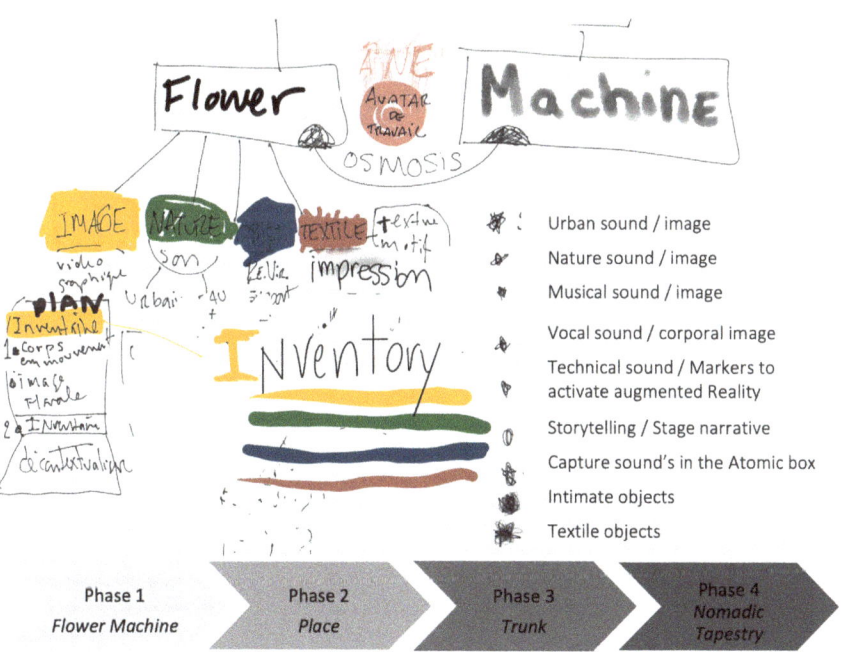

Urban sound / image

Nature sound / image

Musical sound / image

Vocal sound / corporal image

Technical sound / Markers to
activate augmented Reality

Storytelling / Stage narrative

Capture sound's in the Atomic box

Intimate objects

Textile objects

Phase 1	Phase 2	Phase 3	Phase 4
Flower Machine	Place	Trunk	Nomadic Tapestry

Fig. 10 Objects and markers inventory sketch

6 Conclusion

In the context of the Integrated Science Series that aims to publish research involving at least two academic fields to offer innovative views, approaches, methods, and knowledge for addressing twenty first-century complex issues, this chapter focuses on deafness rather as an opportunity to develop a new form of sensoriality, than as an individual disability. To do this, we have basically developed a prototype of 360° learning DeafSpace, based on an artistic creation process. The process that has supported the design of the Atomic Box (Fig. 11) is indeed grounded in an artistic vision and a creative way of formalizing ideas. It is,

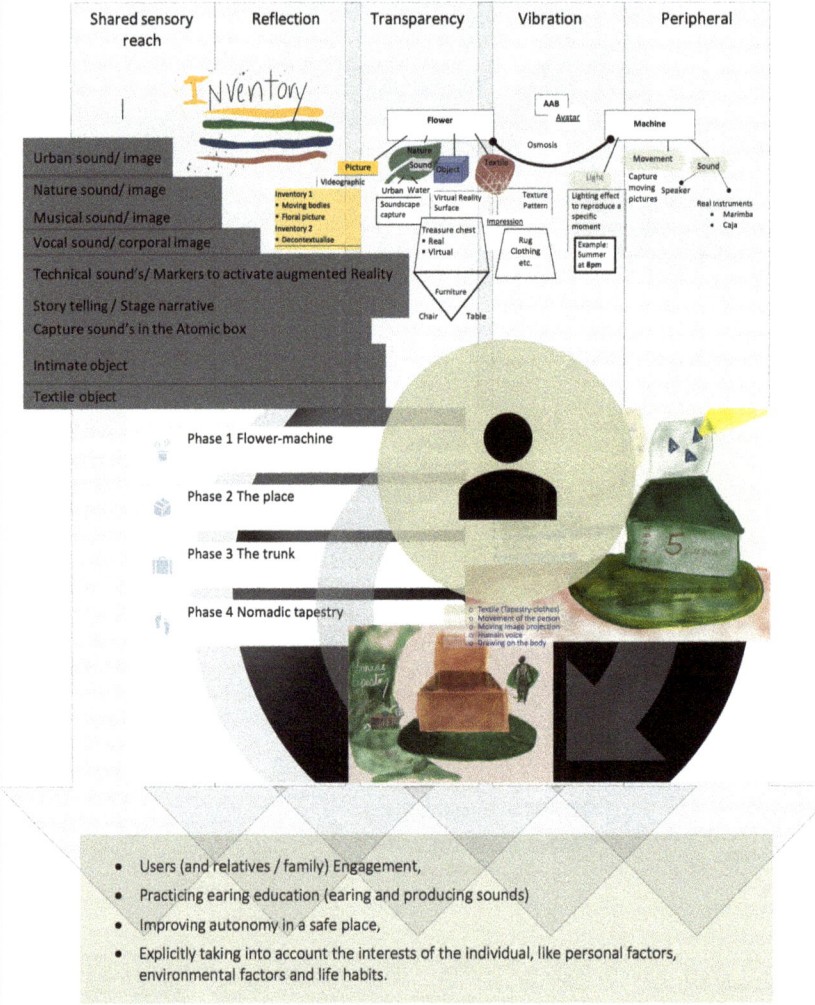

Fig. 11 Atomic box synthesis

moreover, fed by cognitive science knowledge—attention process and DeafGain—and science of design.

In this chapter, we have discussed, thanks to the Atomic Box project concrete example, how the combination of art and science makes it possible to develop socially innovative responses (based on the Montreal Model of a partnership relationship between patients and healthcare professionals, on the one hand, and on the Disability Creation Process—DCP on the other hand), and thus to assist the adoption process of implants by hearing-impaired users. In the domain of deafness, we saw how science, engineering, and medicine, even when they produce together highly sophisticated technologies, and the cochlear implant is certainly the best example, are not capable of addressing complex issues that Deaf studies bring to the front. Deaf studies definitely need academic fields cross-feeding.

This project shows how art is soluble in science and vice versa as long as the approach and the experimental devices are used to remain open throughout the process and that the focus is on the process itself rather than only on objects and artifacts that are produced. The error would be to reduce the role of the arts to the task of shaping and final illustration or valuing results. The creative process, as well as the artistic mindset, constitute extraordinary ways of enriching the scientific approach. As an extension of this research, projects in the fields of participatory sciences and contributory mapping are already envisaged in the context of a smart and inclusive city. In this area, limits of processes that are too focused on the map or the objects themselves have been demonstrated. The level of commitment of the stakeholders gains a lot in approaches favoring creation, iteration, and recursiveness, especially when facing a large variety of stakeholders and inclusion issues.

Besides, other modalities of articulation of cognitive sciences, design, and art are planned within the framework of the first author's thesis. The Atomic Box leads to another project to refocus the target and specify the breaking point in the human/cochlear implant relationship. So, the next research project will be based on design-fiction workshops. Design fiction is an approach that aims to bring out innovative ideas by imagining a future situation as if it was already real. We will engage with a reflection on the possibilities based on the scripting processes of the future. To get there, it will be necessary to project individuals into a sensory future. What if worldwide human beings suddenly woke up in a situation of hearing impairment, for instance? What would you feel? What would you do first? Who would you talk to first? What would your first desire be in relation to sound?, etc. The goal will be to simulate a situation in favor of creating new ways to facilitate the integration of the cochlear implant with more fluidity in the body and in the person's life. After having worked on a learning space, we want to explore the intimate relationship between the person and his/her cochlear implant in more depth.

The richness of this work goes through the Tango that the disciplines (cognitive science, design, and art) dance together. Because indeed, if the participants in the project agree to play the game without losing their uniqueness, then the entities at work keep their flavor. On the one hand, art preserves and maintains the mystery of the effectiveness of language, and science, on the other hand, infuses the precision of analysis. Each discipline reflects on the other, and each project toward the other.

For the process to be profitable, each discipline must not only stay on its own track but also look in the same direction as the other.

Core Messages

- Deafness has to be defined from the environment viewpoint rather than from the capacities of the person herself.
- Rather than seeing deafness as hearing "loss," many deaf people prefer to see it as deaf "gain."
- Cochlear implantation has to be encompassed with more fluidity in the body, senses, and the person's life.
- Combining design, research-creation, and cognitive science is designing DeafSpace: spatial Deaf vision and culture.
- Combining art, cognitive science, and design provides means of "visibilizing" the situation of deafness.

References

1. Pomey M-P, Flora L, Karazivan P, Dumez V, Lebel P, Vanier M-C, Débarges B, Nathalie Clavel and Emmanuelle Jouet (2015) Le « Montréal Model »: Enjeux du partenariat relationnel entre patients et professionnels de la santé. Santé Publique 2015/HS(S1):41–50. https://doi.org/10.3917/spub.150.0041
2. Philippe K, Dumez V, Flora L, Pomey M-P, Del Grande C, Ghadiri DP, Fernandez N, Jouet E, Vergnas OL, Lebel P (2015) The patient-as-partner approach in health care: a conceptual framework for a necessary transition. Acad Med 90(4):437–441. https://doi.org/10.1097/ACM.0000000000000603
3. Catherine T-T (2015) L'Éducation thérapeutique du patient: la maladie comme occasion d'apprentissage. De Boeck, Paris
4. Angela C (2002) The autonomous patient: ending paternalism in medical care. Stationery Office (for the Nuffield Trust), London
5. Lucie M, Leclère B, Le Glatin C, Moret L (2020) Patient involvement in healthcare workers' practices: how does it operate? A mixed-methods study in a French university hospital. BMC Health Serv Res 20(1):391. https://doi.org/10.1186/s12913-020-05271-w
6. Mélanie L, Desrosiers J, Tribble D-C (2007) Comparing the disability creation process and international classification of functioning, disability and health models. Can J Occup Ther 74:233–242
7. Fougeyrollas P, Boucher N, Edwards G, Yan G, Luc N (2019) The disability creation process model: a comprehensive explanation of disabling situations as a guide to developing policy and service programs. Scand J Disab Res 21(1):25–37. https://doi.org/10.16993/sjdr.62
8. Fougeyrollas P, Beauregard L Disability (2009) An interactive person-environment social creation. In: Albrecht GL, Seelman K, Bury M (eds) Handbook of disability studies. SAGE Publication Inc. https://doi.org/10.4135/9781412976251
9. Ellis K, Garland-Thomson R, Kent M, Robertson R (eds) (2018a) Interdisciplinary approaches to disability: looking towards the future, vol 2. Routledge

10. Ellis K, Garland-Thomson R, Kent M, Robertson R (eds) (2018b) Manifestos for the future of critical disability studies, vol 1. Routledge
11. Blacutt AA, Roche S (2020) When design fiction meets geospatial sciences to create a more inclusive smart city. Smart Cities 3(4):1334–1352. https://doi.org/10.3390/smartcities3040064
12. Gitlow L, Flecky K (eds) (2019) Assistive technologies and environmental interventions in healthcare: an integrated approach, Wiley Blackwell
13. Edwards C, Harold G (2014) DeafSpace and the principles of universal design. Disabil Rehabil 36(16):1350–1359. https://doi.org/10.3109/09638288.2014.913710
14. Bauman H (2014) Deafspace: an architecture toward a more livable and sustainable world. In: Bauman H-DL (ed) Deaf gain: raising the stakes for human diversity. University of Minnesota Press, p 375–401
15. Bauman H (2018) Deafspace. In: Ellen L, Lipps A (eds) The senses: design beyond vision. Princeton Architectural Press, New-York
16. Eshraghi AA, Nazarian R, Telischi FF, Rajguru SM, Truy E, Gupta C (2012) The cochlear implant: historical aspects and future prospects. Anat Rec (Hoboken) 295(11):1967–1980. 10.1002/ar.22580
17. Chouard CH (1978) Entendre sans oreille. Robert Laffont, Paris, 210p
18. Djourno A, Eyriès C (1957) Auditory prosthesis by means of a distant electrical stimulation of the sensory nerve with the use of an indwelt coiling. Presse Med 65(63):1417
19. Clark GM, Hallworth RJ, Zdanius K (1975) A cochlear implant electrode. J Laryngol Otol 89(8):787–792. https://doi.org/10.1017/s0022215100081020
20. Chouard CH, MacLeod P (1976) Implantation of multiple intracochlear electrodes for rehabilitation of total deafness: preliminary report. Laryngoscope 86:1743–1751
21. Contrera KJ, Choi JS, Blake CR, Betz JF, Niparko JK, Lin FR (2014) Rates of long-term cochlear implant use in children. Otol Neurotol 2035(3):426–430. https://doi.org/10.1097/MAO.0000000000000243
22. Maia TG (2020) Cochlear implants in congenitally deaf children: a discussion built on rights-based arguments. Am Ann Deaf 164(5):546–559. https://doi.org/10.1353/aad.2020.0002
23. Thomas B, Hodges AV, Goodman KW (1996) Ethics of cochlear implantation in young children. Otolaryngol Head Neck Surg 114(6):748–755. https://doi.org/10.1016/S0194-5998(96)70097-9
24. Bergheimer S, Lindenmeyer C (2018) Les effets subjectifs de l'implant cochléaire dans les liens intra et intergénérationnels. Dialogue 222(4):53–65. https://doi.org/10.3917/dia.222.0053
25. Vennetier S (2020) Implantation cochléaire et régulation juridique des relations entre les sourds et la médecine de l'oreille dans les années 1990 et 2000. L'exemple de deux associations françaises de défense des sourds. Développement Humain, Handicap et Changement Soc 26(1):23. https://doi.org/10.7202/1068188ar
26. Bennette RJ, Jayakody DMP, Eikelboom RH, Taljaard DS, Atlas MD (2016) A prospective study evaluating cochlear implant management skills: development and validation of the cochlear implant management skills. Clin Otolaryngol 41(1):51–58. https://doi.org/10.1111/coa.12472
27. House WF (1976) Cochlear implants. Ann Otol Rhinol Laryngol 85(3):3–3. https://doi.org/10.1177/00034894760850S303
28. Jan M, Marangos N, Ziegler E (2005) Reliability of cochlear implants. Otolaryngol-Head Neck Surg 132(5):746–750. https://doi.org/10.1016/j.otohns.2005.01.026
29. Pomey M-P, Dumez V, Boivin A, Rouly G, Lebel P, Berkesse A, Descoteaux A, Jackson M, Karazivan P, Clavel N (2019) The participation of patients and relatives in Quebec's health system: the montréal model. In: Pomey MP, Denis JL, Dumez V (eds) Patient engagement. Organizational behavior in healthcare. Palgrave Macmillan, Cham
30. Antoine B, Gauvin A-P, Dumez V, Macaulay AC, Lehoux P, Abelson J (2018) Patient and public engagement in research and health system decision making: a systematic review of evaluation tools. Health Expect 21(6):1075–1084. https://doi.org/10.1111/hex.12804

31. Desmond D, Layton N, Bentley J, Boot FH, Borg J, Dhungana BM, Gallagher P, Gitlow L, Gowran RJ, Groce N, Mavrou K, Mackeogh T, McDonald R, Pettersson C, Scherer MJ (2018) Assistive technology and people: a position paper from the first global research, innovation and education on assistive technology (GREAT) summit. Disabil Rehabil Assist Technol 13(5):437–444. https://doi.org/10.1080/17483107.2018.1471169
32. Lid IM (2013) Universal design and disability: an interdisciplinary perspective. Disabil Rehabil 36(16):1344–1349. https://doi.org/10.3109/09638288.2014.931472
33. Patrick F (2006) Quebec model of disability creation process. In: Albrecht GL (ed) Encyclopedia of disability, vol III. SAGE Publications, Chicago, pp 1325–1327
34. Johan B, Larsson S, Östergren P-O (2009) The right to assistive technology: for whom, for what, and by whom? Disab Soc 26(2):151–167. https://doi.org/10.1080/09687599.2011.543862
35. Fougeyrollas P, Dumont C (2009) Construction identitaire et résilience en réadaptation. Frontières. 22(1–2):22. https://doi.org/10.7202/045023ar
36. Bauman HD, Murray JJ (2014) Deaf gain: an introduction. In: Bauman H-DL (ed) Deaf gain: raising the stakes for human diversity. University of Minnesota Press, p xv–xlii
37. Yuan G-q, Ben D-n (2008) Combining art and science of the integration of designs. In: 9th International conference on computer-aided industrial design and conceptual design, Kunming, pp 910–912. https://doi.org/10.1109/CAIDCD.2008.4730709
38. Ashby MF, Johnson K (2013) Materials and design: the art and science of material selection in product design. Butterworth-Heinemann, 416p
39. Frechin J-L (2019) Le design des choses à l'heure du numérique. FYP éditions, 250p
40. Gertz G, Boudreault P (2016) The SAGE deaf studies encyclopedia. SAGE, 1128p
41. Octavian R (2016) Seeking that which might constitute our common humanity: deaf studies, social justice, and the liberal arts. Sign Lang Stud 17(1):89–95. https://doi.org/10.1353/sls.2016.0027
42. Morgan R, Kaneko M (2017) Being and belonging as Deaf South Africans: multiple identities in SASL poetry. Afr Stud 76(3):320–336. https://doi.org/10.1080/00020184.2017.1346342
43. Morgan R, Meletse J (2017) Rainbow: constructing a gay Deaf black South African identity in a SASL poem. Afr Stud 76(3):337–359. https://doi.org/10.1080/00020184.2017.1346344
44. Michael R (2018) The sign language interpreted performance: a failure of access provision for Deaf spectators. Theatr Top 28(1):63–74. https://doi.org/10.1353/tt.2018.0009
45. Rosen RS (2018) Geographies in the American DeafWorld as institutional constructions of the deaf body in space: the sensescape model. Disab Soc 33(1):59–77. https://doi.org/10.1080/09687599.2017.1381072
46. Masini EB (1996) The relationship between art and science. Leonardo 29(1):19–22
47. Gregory R (1979) Similarities between perception and science. In: Nodine EE, Fisher DL (eds) Perception and pictorial representation. Praeger
48. Donguy J, La Chance M (eds) (2018) Technocorps and cybermilieux. Inter. Special issue Number 128, Winter 2018, pp 3–67
49. Hirt LL (2015) Recherche-création en design à plein régime: un constat, un manifeste, un programme. Sci Des 1(1):37–44. https://doi.org/10.1177/1046496406297042
50. Stévance S, Lacasse S (2019) Research-creation in music and the arts: towards a collaborative interdiscipline. Routledge, London, p 188
51. Laurier D, Lavoie N (2013) Le point de vue du chercheur-créateur sur la question méthodologique: une démarche allant de l'énonciation. Rech Qual 32(2):294–319
52. Kristina H (2019) Designing with the body: somaesthetic interaction design. CHIRA. https://doi.org/10.7551/mitpress/11481.001.0001
53. Samar VJ, Parasnis I, Berent GP (1998) Learning disabilities, attention deficit disorders, and deafness. In: Marschark M, Clark MD (eds) Psychological perspectives on deafness, vol 2. Lawrence Erlbaum Associates Publishers, pp 199–242
54. Petersen SE, Posner MI (2012) The attention system of the human brain: 20 years after. Ann Rev Neurosci 21(35):73–89. https://doi.org/10.1146/annurev-neuro-062111-150525

55. McClary S, Walser R (1994) Theorizing the body in African American music. Black Music Res J 14(1):75–84. https://doi.org/10.2307/779459
56. John D (2005) Art as experience. Perigee Books, The Berkley Publishing Group, New York, p 384
57. Foucault M, Barret Kriegel B, Thalamy A, Béguin F, Fortier B (1979) Les machines à guérir: aux origines de l'hôpital moderne. Pierre Mardaga éditeur, Bruxelles, 184p

Andrée-Anne Blacutt is an artist by training and a doctoral candidate in design and social innovation at Laval University. The processes of sensory perception are at the center of her research and creation process. She has completed a master's degree in Visual Arts based on a narrative visual and sound system of narrative motifs. Since 2001, Andrée-Anne Blacutt has presented several projects that multiply the way human beings can be contextualized differently according to space and time. Theater, painting, design, television, and music are among her creative domains. She is currently working on a project to improve the quality of mobility of people as a value to foster a model of interaction between art and science.

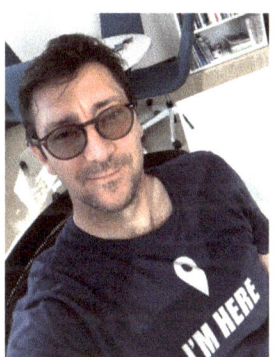

Stéphane Roche, an engineer and geographer, is a full professor of geospatial sciences at Laval University. He explores the complexity of human societies' spatial organization and digital transition issues. He has developed strong expertise on the social dimensions of geospatial technologies, Public Participation GIS —PPGIS, Volunteered Geographic Information—VGI, and Geodesign. His current work mainly focuses on the human, cultural, and inclusive dimensions of smart cities. He has been the Research and Academic Affairs Vice-Rector at the INRS (National Institute for Scientific Research); vice-dean for studies and research at the Faculty of Forestry, Geography and Geomatics at Laval University; Director of the Department of geomatics sciences, Scientific director of the Network of Centers of Excellence (RCE) GEOIDE, as well as director of the Geomatics Research Center—CRG. He previously served as the board of directors of the INRS, CIRANO, OURANOS, FUTUR EARTH, Quebec INNOV, Venice International University.

Art, Medicine, and Public Health: Synergizing Humanistic and Medical Strategies in Managing a Pandemic

22

Stephen E. Kekeghe

> *"Wherever the art of medicine is loved, there is also a love of humanity."*
>
> Hippocrates [1]

Summary

The intersection between the humanities and the medical profession has been acknowledged since classical times. Scholars have continued to establish the applicability of literary and humanistic approaches to advance medical practice. This is because a purely biomedical or bioscientific method to healthcare has consistently elicited poor therapeutic relationships. Disciplines in the humanities like literature, psychology, philosophy, history, and theology have facilitated better healthcare delivery. This study examines the interplay of humanistic and biomedical strategies in managing a pandemic disease like the ravaging coronavirus (COVID-19). Selected pathographies (narratives) by COVID-19 patients and caregivers were subjected to qualitative and critical analysis. Unethical experiences in the patient-caregiver relationship like dehumanization, verbal assault, and discriminatory utterances were evaluated. The essay reveals that indispensable humanistic approaches are required by physicians and other healthcare professionals during patients' consultation, diagnosis, and treatment. It also highlights the import of linguistic and communicative tools in creating curtailment awareness on a pandemic condition, and in this case, COVID-19. The discussion reveals that healthcare providers that rely completely on biomedical knowledge are mechanical and less empathic. The study, therefore, concludes as follows: First, the diagnostician, nurses, and laboratory technicians,

S. E. Kekeghe (✉)
Department of English, Ajayi Crowther University, Oyo, Nigeria
e-mail: se.kekeghe@acu.edu.ng

investigating the patient's condition and administering of therapeutics, should incorporate humanistic approaches like empathy, healthy discourse, and humorous rhetoric to humanize the patient. Second, resources of language and communication should be effectively deployed, both in creating awareness on the containment of a pandemic and in the medicalization process.

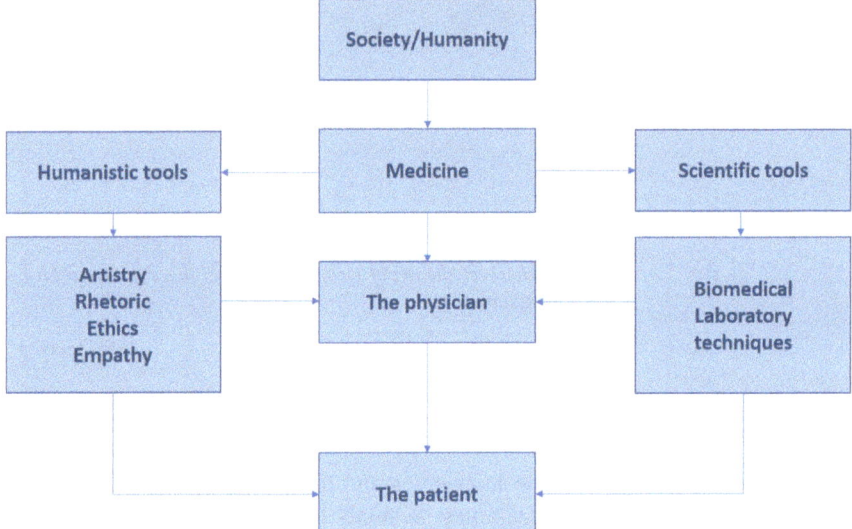

A model on the interplay of humanistic and bioscientific tools in medical practice.

The code of this chapter is *01100101 01,110,011 01,100,001 01,101,111 01,101,110 01,010,010 01,101,110 01,100,101 01,100,011.*

Keywords

Art of medicine · Healing words · Medical discourse · Medical humanities · Public health communication

1 Introduction

Contemporary academic studies have established the need to incorporate the humanities, especially literature, psychology, history, ethics, and philosophy, in modern medical education in order to facilitate effective healthcare delivery. The argument of such scholars and professionals is premised on the thesis that, since medicine is concerned with human lives, it is impossible to separate medical practice from humanistic cultures. Significantly, the humanities have maintained a symbiotic relationship with the medical field right from the evolution of human societies. In classical Greek society, for instance, Apollo, one of the Olympian deities, is venerated as the god of medicine, poetry, and music [2–4]. Hippocrates, the Greek philosopher, writer, and physician, is also reputed as the father of medicine and the inventor of the "case note tradition," used by medical doctors to record and document their observations of patients' health conditions [5]. Aristotle, another famous philosopher, also contributed appreciably to the advancement of the medical field through his literary, philosophical narratives. Besides his popular postulations in *Poetics* which underscores his cathartic method to mental health, Aristotle also has some outstanding, idealistic accounts that foreground the biological composition of man, which create a considerable connection between the humanities and medical sciences. The lyre by David to heal Saul of some psycho-spiritual difficulties attests to the medicinal import of poetry and music [6], which has come to be known as poetry therapy.

Given the rapid decline in therapeutic relationships, scientists have acknowledged the imperativeness of collaboration between medical and humanistic cultures. C. P. Snow's 1959 lecture, "Two Cultures and the Scientific Revolution," reveals that it is crucial for the sciences and the arts to synergize for social advancement, and this became a major springboard in the evolution of the medical humanities. Snow emphasizes that the disparity that is created between the arts and sciences is a major limitation in the total knowledge pool of humanity [7]. His lecture, therefore, bridged the gap between the scientific and artistic culture and revolutionized the general outlook of medical practice. G. S. Rousseau admits that the humanities and medicine once shared a productive, symbiotic relationship that was temporarily terminated in the twentieth century due to the devastating effect of the First World War that put an end to different forms of liberal arts education [8]. McManus, however, reveals that the quests to include the humanities in medical education began in the 1960s [9]. By the late 1980s and early 1990s, many medical doctors in Europe and North America who perceived the absence of empathy and compassion in general medical practice believed that the interface between literature and medicine would help humanize medical practice. On this note, Beveridge reveals:

"In fact, recent years have seen a resurgence of interest in the relationship between medicine and the arts. There have been the publications of the insistence that reading literature can help doctors better understand the "narrative" of their patients and the creation of the centre for Arts and Humanities in Health and Medicine at Durham University. Medical schools such as Glasgow and Birmingham now offer modules in the humanities. Such developments spring from the beliefs that it is beneficial for doctors to be exposed to the arts; that somehow it makes them better clinicians."[10].

The interdisciplinary study of medicine and literature is the most developed area of the medical humanities. The Russian physician-writer, Anton Chekhov (1860–1904), once declared: "medicine is my lawful, wedded wife, and literature is my mistress" [11]. The appointment of Joanne Trautmann-Banks in 1972 to the Faculty of the Pennsylvania State University College of Medicine, at Hershey, as the first person with a Ph.D. in literature to hold such a position in a medical school, became a major catalyst to the evolvement of literature and medicine as an academic subspecialty [12]. As a professor of literature, Trautmann-Banks employed literary approaches in medical education, and it marked a turning point in the medical curricula of universities in Europe and North America. The establishment of relevant academic journals like *Lancet, Medical Humanities, Arts Medica, Body Electric, Healing Muse, Journal of Poetry Therapy, Journal of Literature and Medicine* and *Atrium* helps to expand the academic compass of the discipline of the Medical Humanities, right from the 20th Century till date.

Faith McLellan and Anne Hudson Jones observe that medical students introduced to literary texts are better equipped with the experiences of life and its complexities, which positively impacts therapeutic relationships [13]. Evans also highlights the importance of the humanities in medical education [14]. According to Beveridge, literature helps "deepen the understanding of suffering and confer wisdom on clinical practice" [15]. Omobowale establishes that literary texts can facilitate the effective teaching of biomedical ethics [16]. Beveridge further illustrates that "several literary devices have clinical resonances" and that "the techniques involved in understanding and analyzing a novel can be applied to the understanding of a patient discourse" [17]. There are acclaimed physician-writers and scholars who believe that a purely biomedical or bioscientific model allows human beings access to a limited perspective. For instance, physician-writers like John Keats, Tobias Smollett, Georg Buchner, Anton Chekhov, Arthur Conan Doyle, Mikhail Bulgakov, Arthur Schnitzler, and William Carlos William maintained a dual vocation of literary and medical practice. In other words, humanistic tools are invaluable instruments in medical practice.

It is, however, discovered that interdisciplinary studies that intersect the humanities and medicine are mostly anchored on literary representations of medical themes: illnesses, diseases, treatment, recuperation, and ethics. For instance, the depictions of medical episodes and ethics in literature have been investigated by scholars across the globe. However, the significance of the humanities in public health seems to have received inadequate attention. This essay, therefore, examines the import of humanistic tools in the management of a pandemic disease. A pandemic is the outbreak of a disease, which spreads across regions, countries, and continents. This implies that the disease, usually contagious, manifests in people in high numbers. Coronavirus (COVID-19) is a good example of a pandemic, given its status of wide intercontinental spread. Some narratives by COVID-19 patients and healthcare professionals are analyzed, establishing the chaotic effects of the gap created between the humanities and medical practice by physicians and caregivers.

2 The Humanity in the Management of a Pandemic: The Patient-Doctor Relationship

In this section, selected narratives by COVID-19 patients and caregivers are critically discussed, highlighting inadequate operations of humanistic strategies in the medical procedure. Humanitarianism is core to the practice of medicine. Apart from the creation of public health awareness which entails language and communicative resources, the therapeutic relationship between the physician and the patient requires the effective deployment of humanistic tools. A good diagnostician or physician recognizes that the patient is the most important person in the medical procedure and needs to enjoy the invaluable humanitarian conduct of the care-givers. Any individual that swears to the Hippocratic Oath of medicine to be a healer of human beings has sworn to maintain the responsibility of loving humankind. Such a person, therefore, should harbor the emotional ability to respond compassionately to human's physical and mental difficulties. This idea is foregrounded in Hippocrates' assertion: "wherever the art of medicine is loved, there is also a love of humanity," which is the epigraph of this chapter. Like conventional art, medicine requires a period of apprenticeship, which involves the ability to synthesize complex human experiences. The physician needs to master the art of imagination and creativity in addition to his traditional biomedical knowledge so as to understand and manage different human conditions effectively. So, he or she should be able to blend multifaceted data through cognition, under-standing, perceiving emotions, and instinct to professional judgment. In order to illustrate the intersection between humanistic and biomedical knowledge in medical practice effectively, a model has been designed in this chapter to offer helpful explanations, as shown in graphical abstract.

The model highlights that medicine is a discipline that bestrides domains of the humanities and sciences. That makes it crucial for the medical practitioner to draw on humanistic and bioscientific resources for effective therapeutic relationships. This claim has been supported by a good number of scholars. Saunders admits that the practice of clinical medicine is both "an art and a science." [18]. Donabedian illustrates "the art of medicine" and establishes that medical practice requires technical and interpersonal skills to foster an effective therapeutic process [19]. Newton believes that the physician must manifest empathic feelings in managing the patient such that their therapeutic relationships should be for the overall benefit of the patient [20]. Analysis of the selected COVID-19 patients' narratives underscores that it is imperative to incorporate artistic and humanistic tools in managing such a disease.

Text A is the narrative of a COVID-19 patient in Warri, Nigeria, who laments her dehumanization by physicians and nurses. The patient laments the unprofes-sionalism and lack of empathy exhibited by healthcare officials in Central Hospital, Warri, when they saw her manifesting disease symptoms, like coughing. She narrates:

Text A

When I got to Warri Central Hospital, I coughed and when the doctors heard me cough, they all ran away. I was confused. There was one nurse among them, who started shouting that I couldn't stay there. In the midst of the shouting, my friend— who the news report said was my boyfriend—went to one of the doctors and asked what was actually happening, but the doctor could not say anything. In fact, they didn't know what drugs to administer to me. At the end of the day, I called one of the doctors and asked: 'what was the medication for a strong cough?' He couldn't answer; he was just looking at me. They didn't even put me on bed. After everything, the doctor then said I should be kept in one of their waiting rooms. When I didn't get any attention, I went home.

When I got to my house, the doctor called me that I should please come back and I told them that I was not going to come back after I had been treated like a refugee the previous day. But the doctor continued with his plea, urging me to come back so that I can be attended to.

When I got there, I was tossed up and down. I waited till 7:00 pm on Saturday evening before a doctor came to me and was even talking to me from a distance. He didn't want to come close. He kept saying that I had coronavirus. After everything, the man carried out some test on me and asked me to go. I asked him when I was going to get my result and he said they would put a call across to me. I said okay and went home.

On Tuesday, I was in my house when someone said some medical personnel were downstairs looking for me. I told them to come upstairs because I was feeling too weak to walk. But my mum persuaded me to go downstairs and meet them. I asked them what the problem was and they said my result showed I was Coronavirus positive…

Yesterday, they brought me a cup of tea that had a lot of sugar. I told them that I couldn't take the tea. I waited till afternoon for my lunch. When it was 1:00 pm, I went to them and requested for my lunch and was asked what I wanted to eat. I told them I would prefer rice and plantain. I waited till 6:00 pm yet there was no food for me. When I was not getting any food, I put a call across to my mum. When my mum came and they saw that she was creating scene, that was when they dropped my food at the bathroom entrance and asked me to get the food from there."

Till this very moment, I have not seen my result. I want Governor Okowa himself to come to my aid. I don't have Coronavirus. I want to be out of this place, I'm tired of staying in this isolation centre…".

From the excerpt above, it is evident that the healthcare professionals featured in the narrative—doctors and nurses—contravened the Hippocratic Oath of medicine, emphasizing beneficence, non-maleficence, and justice for the patient. The physician who handles the human condition is by his vocation, a humanist, and his care must be for the overall benefit of the patient. With appropriate kitting with personal protection equipment (PPE), the healthcare providers featured in the narrative above should have displayed empathizing rhetoric and compassionate conduct to humanize the patient who seeks medical help. Table 1 summarizes the patient's feelings of distress as manifested in her utterance in the excerpt above.

As highlighted in Table 1, the patient wishes to be treated as a human being. Sadly, the same people who should help her regain her humanity, facilitate her losing it. Doctors and other healthcare professionals are expected to overcome negative thoughts and emotions such as intolerance and prejudice that potentially interfere with the therapeutic relationship. This is because, besides experiences of physical suffering, the COVID-19 patient manifests emotional distress, which the caregiver should help ameliorate, not aggravate. So, relying strictly on biomedical

Table 1 Emotional distress of the patient who narrated Text A

Feeling	Why/When
Dehumanization (lack of empathy)	When the doctors heard me cough, they all ran away
Feeling of distress	I was confused
Dehumanization	There was one nurse among them, who started shouting that I couldn't stay there
Feeling of reproach triggered by dehumanization	I told them that I was not going to come back after I had been treated like a refugee the previous day
Dehumanization	they dropped my food at the bathroom entrance and asked me to get the food from there
Deep feeling of emotional distress	I want to be out of this place, I'm tired of staying in this isolation center

knowledge will offer too little for the patient's recovery. According to Kekeghe, "The art and humanities have been constantly deployed in creating awareness on the ethics of medicine and other domains of human health experiences" [21]. As recounted in the excerpt above, the narrator was discriminated against even before her test came positive. In her words, "when the doctors heard me cough, they all ran away. I was confused. There was one nurse among them, who started shouting that I couldn't stay there." The patient lamented that "[she] had been treated like a refugee the previous day." So, in addition to the physical pains of the patients, her distressing mood must have been aggravated by the same people that were supposed to offer her care. This is against the Hippocratic Oath of medicine, which emphasizes that the therapeutic process should be patient-centered. The utterance and conduct of the physician in handling the patient's condition should manifest compassion and hope. Optimistic and empathizing rhetoric is meant to light up the inner life of the patient, which will create room for quick recoverability. It is crucial to note that the physician, with appropriate kitting, should show serious empathy in managing the health condition of the patient, broken down by a pandemic disease such as COVID-19.

Text B is a narrative of a COVID-19 survivor in Lagos, Nigeria. Here, the patient reveals some cases of dehumanization, which she suffers at the hand of impassionate nurses and physicians before and during her experiences in the isolation center. Though recovered, she recalls her suffocating encounters in the diagnostic and therapeutic processes. Her account is shown below:

Text B

"Life finds ways of throwing LEMON at me. I've struggled with coming forward, but I want to inspire hope. I returned to Nigeria from the UK post-Commonwealth event (I totally enjoyed) and fell ill. As a responsible person, I self-isolated. Days after, I TESTED POSITIVE FOR COVID-19..."

The nurses eventually came out and treated me like a plague. I sat in the ambulance feeling rejected. No questions about how I felt. So many questions about my travel history. Same information I had provided to NCDC and Lagos State Government during profiling. Lack of data sharing!

After two hours, I was taken to my space. I felt lonely, bored and disconnected from the outside world. Few days after, another patient came in.

This is another case that shows the over-concentration on the illness instead of the patient. The nurses and the doctor in this account focus more on the illness instead of the physical and emotional distress of the patient. The caregivers, as shown in the narrative above, are mainly concerned about how many persons the patient might have infested instead of her health difficulties. In other words, the patient is constructed as a mere robotic object for analysis of COVID-19 transmission, and this leads to her dehumanization by the healthcare professionals.

Table 2 conveys the strained therapeutic relationship between the patient and the healthcare givers. Besides the patient's physical suffering, she was further devastated by the poor treatment she received at the hands of the caregivers. Contrary to her expectation of good medical attention, which made her invite the Nigerian Center for Disease and Control (NCDC), the patient was subjected to a series of dehumanization by the caregivers (verbal and conduct). Rather than expecting empathic utterance from the healers, "[she] sat in the ambulance feeling rejected. No questions about how [she] felt. So many questions about [her] travel history." This over-concentration on the sickness, rather than the sufferer's feeling, is a common medical error committed by physicians who consciously or unconsciously discard the indispensable humanistic strategies in medical and clinical practice. A healthy, humanizing dialog should be established between the physician and the patient in the consultative and diagnostic process. Since the treatment is patient-centered, the patient should be made to express herself freely. A. H. Hawkins observes that pathographies (narratives by patients) reveal that patients should be given the autonomy to convey their health condition in the clerking-in or diagnostic process [22]. He decries the too scientific and mechanical procedure, whereby physicians focus primarily on the patients' health condition instead of the emotional

Table 2 Dehumanization of the patient who narrated Text B

Feeling	Why/When
Dehumanizing language and conduct	The nurses … treated me like a plague
The negative effect of the discrimination on her mindscape	I sat in the ambulance feeling rejected
The patient, manifesting feeling of worthlessness	No questions about how I felt
The patient manifests feeling of deep distress	I felt lonely, bored and disconnected from the outside world

needs of the patient who suffers from the condition. Such over-concentration on the sickness instead of the sick, Hawkins observes, leads to the dehumanization of the patient or sufferer.

The art and humanities play a very significant role in medical practice. For instance, music and poetry have been identified as potent instruments deployed for terminally ill patients who are undergoing medication. On this note, the Welsh poet-physician, Dannie Abse, declares: "one knows that patient can be consoled in reading poetry and many doctors have no doubt seen, as I have, a volume of poems at the bedside of a terminally ill patient." [23]. Abse regards poetry as having a placebo effect on the patient. The implication is that the compassionate and optimistic remarks the physician provides for the patient signal the process for recuperation. In other words, like poetry, a humanizing utterance of medical doctors and nurses can offer a COVID-19 patient, who leaves with the pain and fear of the virus that ravages his or her psycho-physical being, a mental rejuvenating effect. For instance, a Nigerian medic in the United Kingdom recommends music for COVID-19 patients. She particularly expresses her experiences regarding the condition of a patient: "…If he [the patient] does well tomorrow, I'm going to ask his nurse to play KWAM 1 for him. They do music therapy to help reduce ICU delirium and my guy needs some faaji tunes abeg." So, the contemporary physician should step up beyond the traditional biomedical principles and deploy humanistic strategies like humorous language, euphemistic (not exaggerative) description of their health condition, and empathic disposition in managing the patient. It is particularly very important in the management of the ravaging COVID-19 pandemic.

Short films or kits are valuable artistic tools deployed in creating awareness of the containment of a pandemic condition. There are a number of such kits that are constantly utilized in the campaign against COVID-19 across the globe. It is a very significant humanistic strategy that is incorporated in the management of a pandemic disease. The implication is that in managing and treating a pandemic condition, the humanistic and artistic culture cannot be divorced from biomedical practice. This idea is foregrounded in a short film entitled "Covid-19 Self-Isolate Promo" gotten from *Barbados* online. In this skit, two aged women, Idaliha and Mavis, engage in a telephone conversation on the imperative of self-isolation. Their dialog reveals that the elderly are more vulnerable to coronavirus disease, which underscores the theme of the skit that highlights the import of self-isolation to contain the virus. To self-isolate suggests that physicians and healthcare practitioners have been overwhelmed by the virus outbreak, and it is crucial to deploy humanistic and artistic tools to curtail its spread. This short film is represented pictorially in Fig. 1. Art plays a significant role in the management of pathological and epidemiological conditions, especially a pandemic one that involves a communal and intercontinental spread. Given the resources of dialogs, humor, and spectacles that characterize a dramatic or an artistic product, if used to create public health awareness, it leaves a positive, lasting impact on the viewers.

Fig. 1 Short skit (film) on COVID-19

3 Containing a Pandemic: Language, Communication, and Public Health

In the curtailment of the coronavirus pandemic, language and communication should be used effectively, both in clinical practice (patient-caregiver relationship) and in creating awareness that borders on the spread and control of the virus. By implication, the humanities play an essential role in managing a pandemic, either clinically or socially. Kirsten Ostherr, in her article, "Humanities as Essential Services," argues that humanistic strategies "can be a vital part of the pandemic response through immediate, translational, frontline work" [24]. For instance, the creation of public health awareness, contact tracing, consultative, and diagnostic encounters like clerking-in with patients and medicalization are experiences that involve effective communication. Ostherr emphasizes further:

   ...research in the medical humanities has long shown that health cannot be attained and illness cannot be vanquished through biomedical and technical interventions alone. This pandemic has made the human fragility of our response infrastructure abundantly clear, and we need to understand how our decisions

about whose life matters will shape the future to come. Vaccines won't help if huge sections of the population believe that they are part of a government or corporate conspiracy. Ventilators won't save the lives of patients who are unable to access healthcare due to systemic racism. We need translational humanities now to complete our technological and biomedical response.

What role can the humanities play in addressing these issues right now? … Media scholars can draw on their knowledge of contagion films to alert health organizations to harmful visual iconographies and suggest alternatives. Literary scholars can identify how narratives are being used to spread misinformation, and they can advise health communicators how to create compelling counternarratives to challenge the fictions of conspiracy theorists. Creative writers can draw on their narrative expertise to craft compelling stories that help us imagine a path forward and the steps we could take to get there—a "science fiction prototyping" for pandemic response [25].

The creation of awareness at the frontline and the physician's diagnostic encounters are common ways through which the import of the humanities is felt in a time of the pandemic. For example, the World Health Organization (WHO) and different national centers established to control diseases require the potency of humanistic tools like communicative strategies, short films, and musical and comic narratives for the sensitization of the public. The greatest success in managing a pandemic is not the cure of the infested patients; it is its curtailment so that the people will not contract it. This is where effective utilization of the tools of the humanities comes to play. Since language is very instrumental in conveying knowledge that borders on the containment of the coronavirus pandemic, it becomes imperative to employ a communal method of sensitization so that those who cannot speak or read any of the modern European languages could be reached through their indigenous languages. This is particularly important in African communities where there may be language barriers.

A good sensitization will help to correct misconceptions by the people who, due to their distrust of the politicians, feel that the coronavirus pandemic is another ploy by the politicians to loot the wealth of their countries. This is a common thought in Africa, especially in Nigeria. If the community and religious leaders are put on the frontline to create public health awareness, the spread of the virus will be significantly controlled. Audiovisual broadcasts and flyers deployed to campaign against the spread of COVID-19 involve the use of communicative strategies. As shown in Fig. 2, the creation of public health awareness, whether oral or written, requires language resources. The implication is that medical practice greatly intersects with the humanities for successful deliverability.

Text C is a brief account of a United Kingdom-based Nigerian nurse invited to extubate a Nigerian COVID-19 patient in the United Kingdom. The health management team perceived that the patient could not speak English fluently. In order to beat the communication barrier, a Nigerian nurse based in the United Kingdom was made to address the patient in the Yoruba language. Language is very important in relating to patients. The narrative is shown below:

Fig. 2 Flyer highlighting precautionary measures for COVID-19

Text C.

"They just called me for a Nigerian COVID patient they want to extubate but weren't sure if he didn't understand English or was just agitated/delirious. I walked in the room and greeted him in Yoruba. He stopped fidgeting and looked at me, his eyes lit up and he started crying.

"Man I hope his extubation goes well! I'm going to be six feet away yelling "E lati wuko! E MA WUKO!"

"It is So hard not being able to have family members here with people, especially when there is a communication or language barrier."

"…I asked him 'se gbo oyibo?' and he nodded vigorously! Propofol na bastard, it can make you forget your own name sef"

"…If he does well tomorrow, I'm going to ask his nurse to play KWAM 1 for him. They do music therapy to help reduce ICU delirium and my guy needs some faaji tunes abeg."

The patient's inability to speak English, according to the Medic, could be as a result of a drug, Propofol, which causes patients to forget even their own name. However, when the Nigerian female healthcare giver, who is of both Igbo and Yoruba identity, spoke to the patient in Yoruba, his eyes lit up, and he broke down in tears. Thus, language can be deployed by healthcare givers to facilitate effective communication with the patient. Language humanizes and dehumanizes, depending on the way it is deployed. As a show of empathy, the nurse even suggests using Yoruba music (which is in the domain of arts) to help light up the distressing mood of the patient.

4   Conclusion

The foregoing discussion highlights the significance of humanistic tools like empathy and communicative strategies in controlling and treating pandemics. Two narratives by COVID-19 patients and one by a caregiver were analyzed to show some evidence of strained therapeutic relationships in the process of consultation, diagnosis, and treatment. The study argues that the Hippocratic Oath of medicine, which is patient-centered, embodies ethical issues that highlight the humanness and compassion which the physician should manifest in handling a patient.

Core Messages

- Healthcare providers should show some empathy to humanize the patient, especially during a pandemic condition.
- Physicians should demystify the dreadful myth of a pandemic by using humorous rhetoric to light up the patient's mood.
- Caregivers should ensure that the humanistic and relational aspects of the medical practice are well suited for the patient's good.
- Communicative and other humanistic tools like films and music help create a public health awareness to contain a pandemic.

References

1. Stone L, Gordon J (2013) A is for Aphorism—'Whenever the art of medicine is loved, there is also a love of humanity.' Aust Fam Physician 42(11):824–825
2. McClellan F (1982) Literature and medicine. Lancet 348:1014–1016
3. Hunter KM (1986) Doctors' stories: Physician-writers: serving Apollo two ways at once. Connecticut Scholar 8:27–37
4. Jones AH (1990) Literature and medicine: tradition and innovations. In: Clarke B, Aycock W (eds) The body and the text: comparative essays in literature and medicine. Lubbock, Texas Teck University Press
5. Zahir II (2016) Hippocrates: philosophy and medicine. Eur Sci J 12(26):199–210
6. Samuel 16: 14–23. *The Bible*
7. Snow CP (1998) [1959] The two cultures. Cambridge University Press, Cambridge
8. Rousseau GS (1986) Literature and medicine: towards a simultaneity of theory and practice. Lit Med 5:152–181
9. McManus IE (1995) Humanity and the medical humanities. Lancet 348:1143–1145
10. Beveridge A (2009) The benefits of reading literature. In: Oyebode F (ed) Mindreadings: literature and psychiatry. London, RCPsych Publications, pp 1–14. (p. 2)
11. Jones AH (1997) Literature and medicine: physician-poets. Lancet 349:275–278: (p. 349)
12. Jones AH (1990) Literature and medicine: tradition and innovations. In: The body and the text: comparative essays in literature and medicine. In: Clarke B, Aycock W (eds) Lubbock, Texas Teck University Press, pp 22
13. McLellan F, Jones AH (1996) Why literature and medicine? Lancet 348:109–111
14. Evans M (2009) Roles of literature in medical education. In: Oyebode F (ed) Mindreadings: literature and psychiatry. RCPsych Publications, London, pp 15–24
15. Beveridge A (2009) The benefits of reading literature. In: Oyebode F (ed) Mindreadings: literature and psychiatry. London, RCPsych Publications, pp 1–14 (p.1)
16. Omobowale EB (2006) Literature and the teaching of biomedical ethics in Nigeria: a creative writer's perspective. Romanian J Bioethics 4(2):20–30
17. Beveridge, A. Ibid.
18. Saunders J (2000) The paradise of clinical medicine as an art and a science. Med Humanities 26:18–22
19. Donabedian A (1979) The quality of medical care: a concept in search of a definition. J Fam Pract 9:277–284
20. Newton BW (2013) Walking a fine line: Is it possible to remain an empathic physician and have a hardened heart? Font Hum Neurosci 7:233
21. Kekeghe SE (2021) Dramatic art, medical ethics and rehabilitation: patient-centered therapeutic relationship in Omobowale's The President's physician. Int J Literature Arts Special Issue: Illnesses, Diseases and Medicalisation in African Literature. 9(4):177–182
22. Hawkins AH (1993) In: Reconstructing illness: studies in pathography. Purdue University Press
23. Abse D (1998) More than a green placebo. The Lancet 351:362–363
24. Ostherr K (2020) Humanities as essential services. (https://www.insidehighered.com/views/2020/05/21/how-humanisties-can-be-part-front-line-response-pandemic-opinion. Retrieved July 18, 2020
25. Ostherr, Kirsten. Ibid.

Stephen Ese Kekeghe (Ph.D.) is a scholar of Literature and the Medical Humanities at the Department of English, Ajayi Crowther University, Nigeria. As one of the best students, Kekeghe obtained his B.A. in English from Delta State University, Abraka, Nigeria; his M.A. in Literature and his Ph.D. in Literature and Medicine from the Department of English, University of Ibadan, Nigeria. He topped his 2012 set of the M.A. class of English. Stephen is a writer and literary critic. His poetry collection, Rumbling Sky, is a joint winner of the ANA Prize for Poetry, 2021. His articles have been published in Matatu (Brill), Routledge Handbook of Minority Discourses in African Literature edited by Tanure Ojaide and Joyce Ashuntantang (Taylor and Francis, 2020) and other prestigious platforms. His Ph.D. thesis examines the literary portrayal of the various sociopolitical factors that induce madness in people. Dr. Kekeghe is a member of the Literary Society of Nigeria (LSN), Universal Scientific Education and Research Network (USERN), Madness and Literature Network (MLN), International Society for the Oral Literatures of Africa (ISOLA), Association of Nigerian Authors (ANA), and Nigerian Oral Literature Association (NOLA).

Big Data and Artificial Intelligence for E-Health

Houneida Sakly, Mourad Said, Jayne Seekins, and Moncef Tagina

"The day healthcare can fully embrace AI is the day we have a revolution in terms of cutting costs and improving care."

Pr.Fei-Fei Li, The stanford institute for human-centered artificial intelligence, May 6, 2020

Summary

Artificial intelligence (AI) and big data are active research topics in e-health. Big data in medicine comprises massive data that includes image data, metadata, and rich clinical information from electronic health records (EHRs). Inherent big data challenges include lack of labeled data, obstacles to data share among institutions, need for information technology framework for data management and procurement, and data security. In this chapter, we explore AI concepts and

H. Sakly (✉) · M. Tagina
COSMOS Laboratory, National School of Computer Sciences (ENSI),
University of Manouba, Monouba, Tunisia
e-mail: houneida.sakly@esiee.fr

M. Tagina
e-mail: moncef.tagina@ensi-uma.tn

H. Sakly
Integrated Science Association (ISA), Universal Scientific Education
and Research Network (USERN), Street Island of Djerba, Monastir, Tunisia

M. Said
Radiology and Medical Imaging Unit, International Center Carthage Medical,
Monastir, Tunisia

J. Seekins
Department of Radiology, Stanford University School of Medicine, Stanford, USA
e-mail: jseekins@stanford.edu

© The Author(s), under exclusive license to Springer Nature Switzerland AG 2022
N. Rezaei (ed.), *Multidisciplinarity and Interdisciplinarity in Health*,
Integrated Science 6, https://doi.org/10.1007/978-3-030-96814-4_23

big data in medicine and their impact on e-health. We discuss the promise of AI and new opportunities for cancer detection and prevention, precision in diagnostic imaging, drug discovery, clinical decision-making, and its potential role for COVID-19 and other future pandemics. We also examine potential barriers and challenges to clinical translatability and fairness in AI and ethical implications.

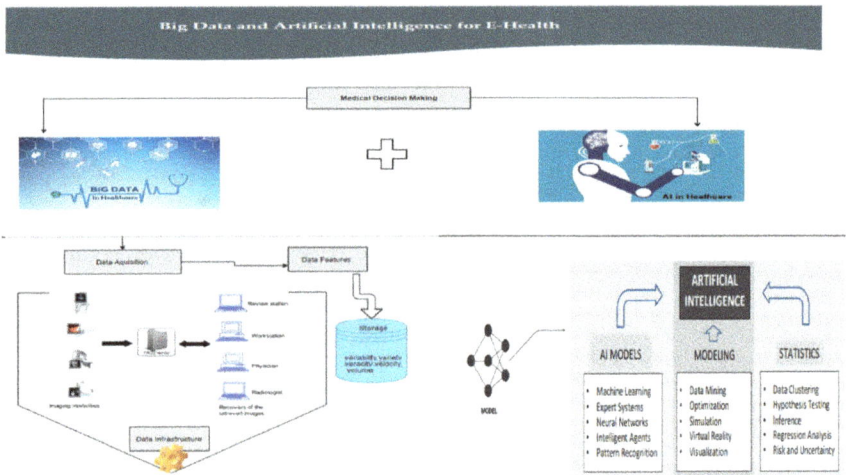

Big data and artificial intelligence for medical decision-making

The code of this chapter is *01101111 01101110 01100101 01100110 01101001 01110100 01100100 01111001 01101100 01110100 01101001 01101110 01100001 01101001 01000011*.

Keywords

Artificial intelligence · Big data · Data analysis · Deep learning · Massive computational medical imaging

1 Introduction

The healthcare industry generates and houses large data facilitated by advances in information technology, electronic health records (EHRs), digital imaging, and technological developments in medicine [1]. Big data also requires large storage and transfer capacity [2]. It works based on mandatory and manageable requirements for improving the quality of healthcare delivery by reducing costs and applying maintenance to address functional needs, including support for scientific choice, disorder monitoring, and population health management [3–5]. Big data is associated with computing standards which should adapt to the growing volume and unpredictability of information from many sources. Giant data incorporating organized, semi-organized, or unstructured data tends to have unpredictable synthetic, semantic, and hierarchical relationships. Epistemic uncertainties, including noisy data with artifacts, the input data errors, recurrence during the storage process, missing data concerning patient coordinates, or other variables such as their medial antecedent, could intercept in the pipeline of the data processing and influence experts' decision-making [6].

With recent advancements in computer and information technology, big data and artificial intelligence (AI) have been subjects of global focus for diagnostics and decision-making [7] using process automation which requires massive data processing systems [8]. Machine studying capability presents ailment models for discovering and developing novel therapeutics and prevention strategies for different medical prognoses. The computing device that is getting to know and AI can also acquire such capabilities, promoting its application to research and clinical settings.

In this chapter, the main concepts of AI, big data, and machine learning (ML) are presented in the context of e-health. The examples provided are exhaustive and have been chosen to cover a few important research areas, such as detecting and preventing brain, breast, and lung cancers, radiology and medical imaging, COVID-19 diagnosis, drug discovery, and healthcare decision-making. The implications for experts' roles and tasks for the way ahead and the hazards of data mining in healthcare are discussed [9], along with the impact of data mining on fraud detection, chronic disease management, and general decision-making [10]. Finally, the possible solutions to evaluate the engagement with the health industry are listed.

2   Big Data for E-Health Industry and Technological Trends

Due to the constraints imposed by the speed and quantity of data processing, big data can run on macrosystems with large-scale, distributed information processing, which frequently exceeds the capacity of portable computer systems and common software. Big data is featured by 5 V's [11, 12]:

  i. variability, lack of structure, consistency, and context;
 ii. variety, various types of data;
iii. velocity, data processing speed;
 iv. veracity, accuracy, noise, and uncertainty in the data; and
  v. volume, large datasets, as shown in Fig. 1.

Healthcare will attain the zettabyte (1021 gigabytes) scale and, quickly after, the yottabyte (1024 gigabytes) scale. Kaiser Permanente, the California-based healthcare network community with greater than 9 million members, keeps between 26.5 and 44 petabytes of doubtlessly EHR-rich data of images and annotations [1, 13].

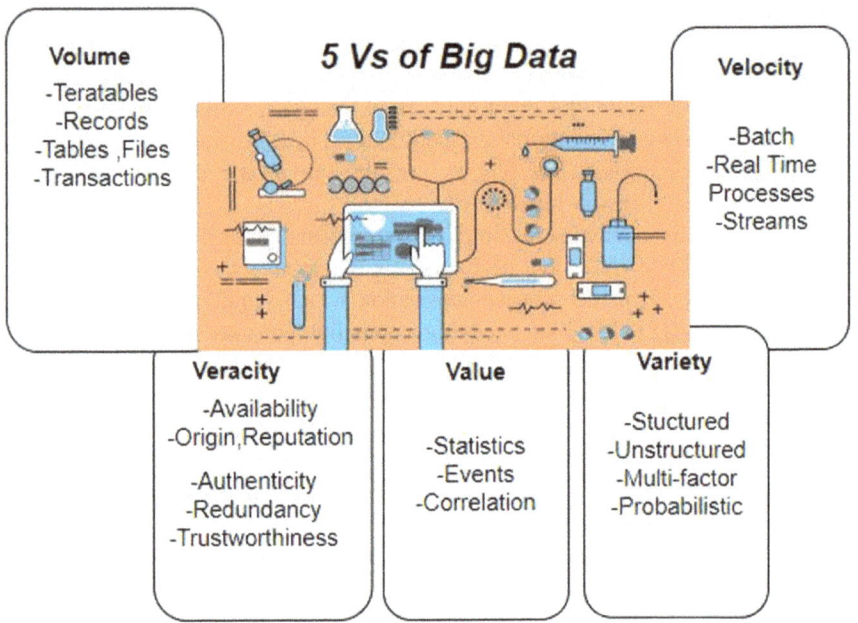

5 Vs of Big Data

Volume
-Teratables
-Records
-Tables ,Files
-Transactions

Velocity
-Batch
-Real Time
Processes
-Streams

Veracity
-Availability
-Origin,Reputation
-Authenticity
-Redundancy
-Trustworthiness

Value
-Statistics
-Events
-Correlation

Variety
-Stuctured
-Unstructured
-Multi-factor
-Probabilistic

Fig. 1 5 V's features of big data

Big data, thus, deals with digital health data that is massive and complicated to handle with conventional processing methods, requiring an effort to manage frequent data. Large volumes and heterogeneity in the types of medical data also limit data processing speed. Existing analytical methods can be used for the huge amount of current, but oddly unanimous, scientific skills and facts related to data exploration to get a deeper integration of the results, which can then be used at the level of daily patient care [14]. By uncovering associations and patterns within the medical records and other diagnostic data elements, an AI-enabled platform could potentially inform physicians and patients at each stage of the decision-making process and guide choosing the best available therapeutic intervention for the individual patient efficiently and at a reduced cost [15, 16].

2.1 Big data Analytics in Predictive Medicine: Promises and Challenges

By digitizing the process of medical prognosis, the healthcare sectors attempt to invest in the success of big data, including single physician workplaces and multi-provider healthcare businesses to networks that have many advantages to develop. The potential benefits revolve around the medical decision-making to prevent disease at advanced stages as well as the health managing of specific

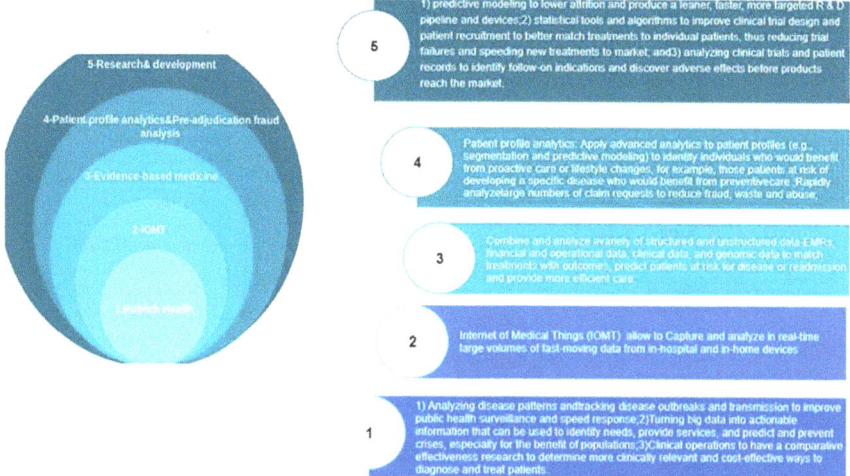

Fig. 2 Promoter axes in big data and healthcare

populations and effectively identifying healthcare fraud [1]. According to McKinsey & Company,[1] big data analytics can provide over \$300 billion a year in e-health's business economy in the United States, two-thirds of which cuts in spending of about 8% national health. McKinsey believes that big data crucially influences this market and could decrease waste and inefficiency in the areas, as seen in Fig. 2.

The complete big data framework encompasses large-scale research fields in the medical field, which are considered crucial for discovering data that can lead to uncertainties. The important challenges raised in developing these axes of big data are mainly the ethics of medicine and the precision of the medical decision, which contain the rights of ownership of personal information, confidentiality, and the control of their distribution and use or potential abuse by domain experts. Staff selection processes or credit ratings could be influenced by experts' perception of risk in terms of unauthorized knowledge of genomic information [17, 18]. The big data security debate revolves around the interoperability issue between distributed servers to exchange patient data, which can pose the problem of exchange security protocols and their active engagement with connected devices. The low funding of these servers and the lack of qualified IT specialists for handling the advanced tools are discussed in the following section. A summary that presents the weak side of big data in healthcare is presented in Fig. 3.

---

[1] https://www.mckinsey.com.

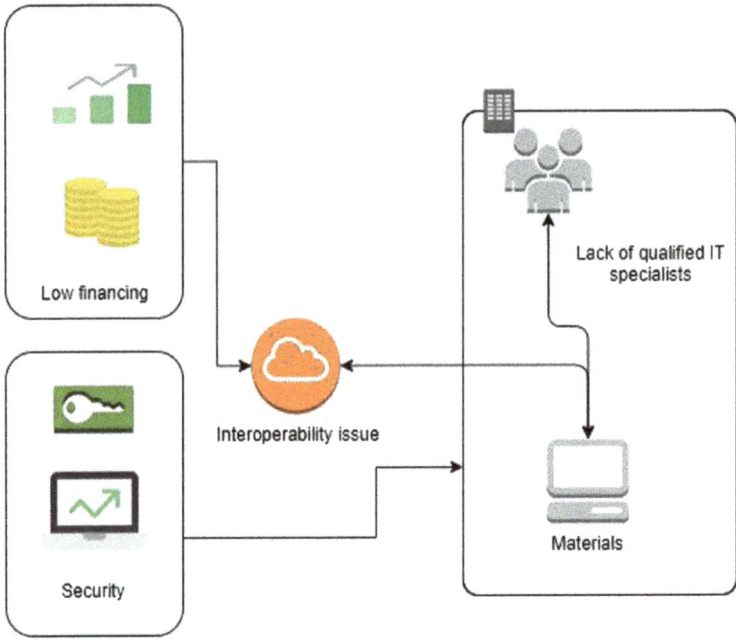

Fig. 3 Weak side of big data in healthcare

2.2 Framework and Tools

With the crucial need for big data exploitation in the medical field, healthcare companies need to select and put together the right tools, infrastructure, and practical strategies to harness huge data efficiently. The complicated big data processing relies on advanced techniques and technologies for information acquisition, storage, distribution, management, and analysis. Currently available analytical techniques could be applied to massive data. However, there are unresolved medical and medical services connected to patients' health and medical data in order to deliver correct analyses of outcomes for medical decision support. The process of big portions in healthcare information using regular statistics processing strategies and devices is considered a tough step in real time [19, 20]. Some statistical analytical tools for data evaluation are introduced in Table 1. This technical evaluation assists with inspection, cleaning, transformation, and storage to extract significant information, advises toward the quality feasible solutions, and guides proper choice-making in the indispensable conditions [21].

Table 1 Description of various big data analytical tools

Big data analytical tools	Features
Hadoop distributed file system (HDFS)	The system architecture of HDFS and the data read and write process. It also introduces key features, such as HA high reliability, metadata persistence, federated mechanisms, and common shell commands in HDFS [22, 23]
Map reduce—distributed off-line batch processing and yarn	Map Reduce offers a high-level parallel programming model to divide each task into sub-tasks to provide the final results in order to guarantee the monitoring of the process of each server or node at the same time. YARN provides the processing platform for data stored in HDFS based on Map-Reduce, focusing on batch processing methodology [24–26]
Spark2x—in-memory distributed computing engine	Spark2x mainly introduces three engines of spark: (1) Spark SQL: to deal with structured data; (2) Spark Streaming: A stream processing engine of micro-batch; and (3) Structured Streaming: A streaming data processing engine built on Spark SQL engine [27, 28]
HBase—Distributed NoSQL Database	HBase offers a non-relational database (i.e., NoSQL) model integrated on the top of HDFS. Big Healthcare data sets under various Hadoop frameworks provide random-based healthcare data access or querying capabilities to the end-users for the distribution process [29]
Hive—distributed data warehouse	The hive tool is used to structure, manage, process, and organize the huge healthcare data sets in HDFS. Hive query language (QL) allows the Hadoop to review, query, and analyze the large sets of immutable data located on top of the Hadoop [30]
Streaming—distributed stream computing engine	The Streaming is a Topology is a real-time application running, and a Worker runs processing component logic. This tool requires each message to be processed and turn off the reliable processing mechanism of the message to have a good performance [31]
Flink—stream processing and batch processing platform	A barrier is a special tuple that is periodically injected into a flow graph and flows through the flow graph with the data stream. Each barrier is the dividing line between the current and next capture. The main component of DataStream is Data Source, Transformations, and Data Sink [32]
Loader—data transformation	The main steps of the Loader job configuration include necessary information, the input configuration, conversion, and the output configuration [33]
Flume—massive logs aggregation	Flume consists of collecting logs generated by applications in the cluster to HDFS and collects clickstream logs in real-time to Kafka. After the sink taking the data and writing it to the destination, it will remove the event from the channel [34]

(continued)

Table 1 (continued)

Big data analytical tools	Features
Kafka—distributed message subscription system	Kafka's basic concept consists of Each Consumer belonging to multiple Consumer Groups. Kafka is a data streaming platform run as a cluster on one or more medical servers, and it stores streams of medical records. Each medical record consists of a key, a value, and a timestamp [35]
ZooKeeper—cluster distributed coordination service	The key feature of ZooKeeper is the final consistency Reliability, waiting for an unrelated nature. Zookeeper could coordinate between the synchronized process of multiple nodes in the cluster, send configuration attributes to a particular or all nodes in the cluster, then elects a leader node among the multiple nodes, and provides a reliable communication between and among the nodes in the cluster [36]
The FusionInsight solution	The FusionInsight solution consists of four sub-products, FusionInsight HD, FusionInsight MPPDB, FusionInsight Miner, FusionInsight Farmer, and one operating and maintenance system FusionInsight HD: an enterprise-level big data processing. environment, which is a distributed data processing system that provides large-capacity data storage, analysis and query, and real-time streaming data processing and analysis capabilities. This tool is a unified platform for Huawei's enterprise-class big data storage, query and analysis. Through real-time and non-real-time analysis and mining of massive information data, new value points and business opportunities are discovered [37]

3 AI and ML for Healthcare

3.1 AI

In generic terms, AI means a machine (or a process) that meets needs with advanced simulators and tools (or new data) and changes its behavior by its operation to improve the performance of index precision. The AI learning process consists of implementing datasets based on advanced concepts of mathematics and statistics [38]. A learning process is an iterative approach that allows you to construct an AI model based on data by adjusting parameters through trial and error mixed with reinforcement rules. The model accuracy is defined based on reducing the error rate between model predictions and experimental data. AI is considered a process that could potentially revolutionize patient care and usher in precision medicine. A trained AI model could perform disease detection, discrimination, and classification, potentially on par or beyond the performance of [39, 40]. Its clinical translation incorporates resources necessary for predictive analytics and decision-making, as shown in graphical abstract.

3.2 ML: Overview and Tools

AI in medicine applies predictive algorithms that can translate into clinical decision-making. ML, classified as a subfield of AI, proposes various algorithms that rely on learning strategies based on logistic regression, decision trees, or deep learning, which fall into supervised, unsupervised, and reinforcement learning techniques [41, 42]. The methods for clustering data are provided by unsupervised learning, while classification is done by supervised learning (Fig. 4).

Increasingly, patient medical data is stored in the form of EHRs in healthcare. In high-income countries, hospitals use the EHRs, containing confidential patient data and their medical antecedent, radiology exams, and clinical reports [43]. However, EHRs-based information might be inconsistent and noisy and have missing values and fields of unstructured textual content. Nevertheless, the very truth that this information is electronically at hand in massive volumes gives the possibility of using ML with the goal of contamination management. The prerequisites potentially require instant diagnostic and therapeutic movements. Early identification via

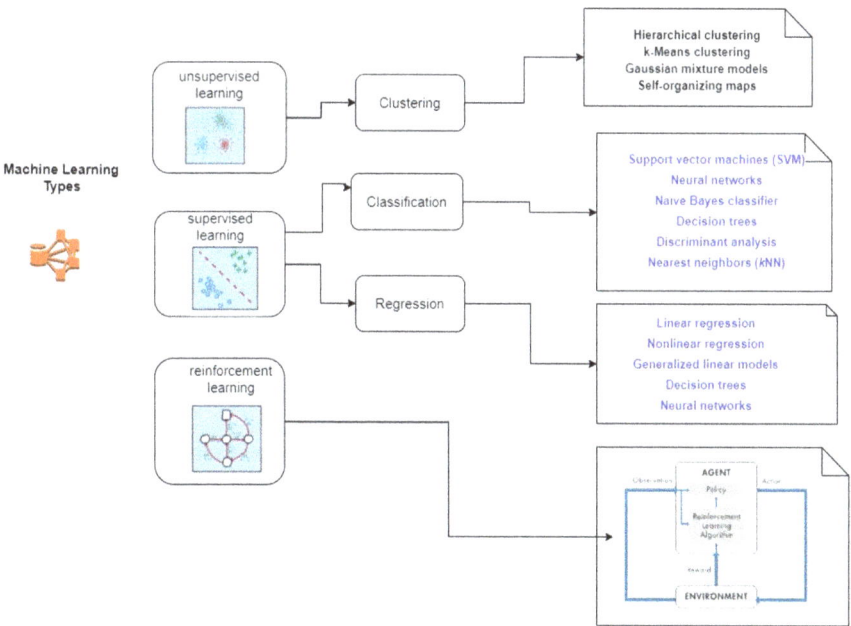

Fig. 4 Machine learning types of overview

Table 2 Framework for deep learning processing

Advanced tools and framework	Features
H2O.ai for AI in Healthcare[1]	• It presents new data-driven initiatives in unique ways to continuously improve patient outcomes, deliver care without increasing costs, and streamline clinical workflows; •There are new machine learning (ML) and Artificial Intelligence (AI) technologies available; and •H2O.ai is considered a leader tool in AI, entitled the main healthcare groups to supply AI options to enhance affected patient outcomes, manipulate claims, notice fraud, and predict obtained infections
Healthcare.ai[2]	The healthcare.ai software presents a streamlined model for healthcare based on machine learning. It encompasses computerized data preparation, function engineering, algorithm selection, and hyperparameter tuning, and model deployment
Anylogic[3]	AnyLogic is a leader in purpose simulation, which could be applied in the Healthcare Industry. AnyLogic models help AI practitioners in many different practical applications, including synthetic data generation, neural network training, and testing AI models [47, 48]
Chainer[4]	Chainer is considered the bridge the gap between algorithms and implementations of deep learning based on Cuda computation. This framework includes feed-forward nets, convnets, recurrent nets, and recursive nets and helps per-batch architectures. Flow statements control of Python without backpropagation sate, including forward computation, makes code intuitive and easy to debug [49]
RapidMiner[5]	• RapidMiner tool is considered among the advanced analytics and powerful platform services in artificial intelligence besides the Big Data storage. This solution has emerged in several Healthcare Industries • This tool helps the data scientist by depicting a depth insight into more than 1500 native algorithms, data preparation, and data science functions. This feature allows professionals to support any machine learning libraries and integrate Python and R codes. RapidMiner offers three different modalities to access their products: the main Platform, the automated data science, and the AI cloud [50, 51]
MIPAR[6]	MIPAR is a revolutionary image processing software. It is considered very powerful for extracting advanced measurements from medical images using the deep learning algorithms available for segmentation and classification [52]
Orange[7]	Orange is a machine learning and data visualization open tool. This tool could build data analysis workflows visually, with a large, diverse toolbox. The design works based on the scoring of various classification techniques on a dataset from medicine. It allows to generate testing and scoring given the data and a set of learners, does cross-validation and scores predictive accuracy, and outputs the scores for further examination [53]

(continued)

Table 2 (continued)

Advanced tools and framework	Features
IBM Watson[8]	Watson is a supercomputer created by IBM, named after company founder Thomas J. Watson. It is a "question answering machine," which combines artificial intelligence with the concept of big data. Watson is based on several key components: The Apache Unstructured Information Management Architecture frameworks, the infrastructure, and the various elements necessary to analyze unstructured data. Another major component is the popular Apache Hadoop framework for processing large data sets in a distributed computing environment [54, 55]

[1]https://www.h2o.ai
[2]https://healthcare.ai
[3]https://www.anylogic.com/features/artificial-intelligence/
[4]https://chainer.org
[5]https://rapidminer.com
[6]https://www.mipar.us/deep-learning.html
[7]https://orange.biolab.si
[8]https://www.ibm.com/cloud/watson-studio

ML-derived prediction modes is expected to improve patient care under "time is life" conditions [44]. Several works have focused on ML for clinical decision support by aiding the diagnostic process, severity prediction, and choosing an appropriate treatment [45]. The popularity of deep learning software tools is increased in response to the functional needs of experts in the field to implement these tools in the clinical routine effectively. These different algorithms attempt to improve the optimization aspect to obtain accurate results for the predictive model [46]. Some advanced frameworks for deep learning processing are presented in Table 2.

3.3 Inconvenience and Limits

3.3.1 Inductively network of AI
Machine-enabled learning is primarily data-driven, using large amounts of information to make assumptions instead of using intuition. This is one reason why ML might excel more in specific data-driven tasks, such as image-based diagnosis, compared to more nuanced activities that involve direct physician-patient interaction at an internist office, for example [55–57].

3.3.2 AI and Explored hypothesis
In emergency medicine, it has been shown that physicians generate an average of five hypotheses during the initial management of a patient [58]. The number of hypotheses simultaneously explored by the machine is theoretically unlimited. This means that faced with a painful organ, the machine could explore the dozens of

possible causes of such a symptom, while the clinician will have to focus on the intuitively generated hypothesis [59].

3.3.3 Deep training for Deep prediction

For an AI, the stage of learning is long and expensive. However, if it is well trained using high-quality data (that is, free of errors or approximations), not only will it not be wrong (or at least, with an acceptable error rate), but, moreover, it will always produce the same response to the same problem. As previously pointed out, the doctor makes mistakes, even when they reach the expert stage. His reasoning is also sensitive to many contextual factors such as fatigue, cognitive overload, interruptions, noise, or even emotions. AI is not mistaken if it has been well trained and shows consistency in its response [60].

3.3.4 Explainability in AI

Within the framework of implementing certain interview techniques, it is possible to understand very precisely how a doctor arrives, on the cognitive level, at a particular diagnosis, including in the event of error [60]. Most AI techniques (in particular neuronal networks) are characterized by a "black box" effect, making it impossible to explain how they arrive at the result. This situation poses problems of both an ethical (can we agree to treat a patient without being able to justify what we rely on to treat him?) and legal (who is responsible in the event of an error and how to establish "fault"?) nature. These topics are fertile grounds for research and future discussions [61].

3.3.5 Is AI Intelligent?

AI aims to reproduce certain cognitive activities specific to human beings, such as solving clinical problems. While it can generate high-performance results in certain situations, it cannot be considered "intelligent." The machine does not understand what it does and the result it produces. It performs a sequence of mathematical operations related to how it was designed and how it learned to operate. Therefore, it does not have a global reasoning mechanism; it is dedicated to solving a single problem that is well defined.

3.3.6 Vulnerability of AI in the Medical Ecosystem in Cybersecurity

Recent innovations relied upon AI-enabled experts to access the big data servers for the storage process. AI has become very effective within the cybersecurity realm. Among the raised cybersecurity challenges, it is the swiftness of the response time for the detection of valid issues. The production of a global security strategy and understanding the health ecosystem require tracking in real time from the healthcare system, such as healthcare information networks and systems, EHR, PHR, online prescription system. Traffic workflow and recommended smart medical devices as computerized systems based on ML could sustain and filter an excessive number of attacks. Analyzing behavior and preventing fraud with AI in cybersecurity make it incredibly valuable to recognize digital dangers within health information systems.

Cybersecurity aims to reduce the number of attacks and protect the entire data flow with security mechanisms. Unfortunately, this level of performance is unachievable in healthcare, considering the huge development of technical attacks in the cybersecurity field. In addition, many health systems are not yet sufficiently resilient against these predatory entities. Healthcare systems have emerged from the "health data capture" to the "data analysis" and "data sharing" stages, intending to improve medical decision-making. Loss of data availability could be triggered by factors such as network components misconnected within the EHR system, error in access control or authorization, and database storage components failure. Cybersecurity attacks could lead to a breach of confidentiality and data integrity violation. When a data breach occurs, recovery efforts by healthcare organizations can affect normal operations and potentially delay patient care [60–63].

4   Effects of ML and Big Data for Disease Implications and Therapeutic Strategy

AI applications hold promise in both precision health and medicine, including patient-specific therapies and disease prevention.

4.1   Cancer Prevention

Big data and macrolearning offer opportunities for improving prognosis and refining the theoretical models for dealing with the problem of cancer prevention and control. Several research works have focused on advancing research on cancer prevention and control [62]. Resteghini et al. [63] describe a methodology for predicting cancers of the head and neck using big data and ML. It is aimed at proposing new prognostic and predictive markers and therapeutics for such a complex condition. These approaches are based on integrating radiomics, genomics, clinical, and epidemiological data into extracting significant features from head and neck cancer. Low et al. [64] describe in their survey the impact of using deep learning and big data to prevent breast cancer using *the translation of big genomic data to cancer precision medicine*. Workman et al. [65] depict the transforming cancer drug discovery using big data, and AI. Lei et al. [66] describe the process of big data radiomics in the radiological assessment of advanced gastric cancer. These approaches have proven promising results for detecting lung cancer [67, 68].

4.2   Radiology and Medical Imaging

The rapid increase of huge data and statistics has proved the importance of using big data methodologies in the clinical setting. These strategies help analyze and derive insights from *high-volume, high-variety, and high-growth datasets*. With the technological know-how infrastructure of data archived digitally, radiology institutions are in particular placed to take benefit of big data approaches [69, 70]. On the technical side, there are already very interesting advances that allow the dose of irradiation to be reduced. Algorithms allow images to be reconstructed more quickly, which is fundamental for patient comfort. Some techniques help the radiologist organize his work, for example, by automatically highlighting the examinations to be treated as a priority. It is also very important for emergency management. Therefore, AI systems can detect lesions, which make the radiologist's job easier.

4.3 COVID-19 Pandemic

The SARS-CoV-2 pandemic has been classified as one of the most elusory diseases declared by the World Health Organization (WHO). Researchers are geared toward AI and big data to manage the enormous quantity of medical images without a history of data issued in real time. The potential applications of AI and big data include epidemic outbreak monitoring, trend forecasting, regular update of the situation by government institutions and agencies, and assistance with health facilities [71, 72]. While healthcare organizations have been placed in extremely urgent situations to adopt and define decision-making technologies to manage this virus, AI and big data help them obtain appropriate suggestions in real time to prevent its spread. Big data [73] describes a crucial role in preparing an adequate database to [74] suggest an accurate medical decision for COVID-19 cases based on deep benchmarking to distinguish similar symptoms of other pathologies such as lung cancer or pneumonia with that of COVID-19.

4.4 Drugs Discovery

The huge development of big data and AI allows the integration of various databases on diseases and data mining to discover therapeutic drug targets. AI-based medical decision-making and big data-based medical decision-making make a great difference in discovering the novel targeted drug. The deadlines of projects that reduce clinical attrition through better early decision-making need to be improved following the interpretation of big data in the drug discovery community. Massive volume of data and the way we ingest it are considered a barrier before the infrastructure integration to host them for the use of the data efficiently and productively [75, 76].

4.5 Psychiatry

Psychiatric disorders are complex phenomena rooted in social, cultural, and experiential factors. To address this complexity, computational psychiatry offers two ways for prognosis:

- Theory-based computational approaches, they use *mechanistic models* to formulate *explicit hypotheses at multiple levels of analysis*.
- ML approaches, they can make predictions from large-scale data.

Recently, several studies [77, 78] have focused on developing big data and learning approaches to alleviate the psychiatric disorders-related burden.

5 Conclusion

Data management makes healthcare outcomes more accurate for patients and reduces costs for providers who leverage a large amount of data and lead the industry into the future. AI approaches can help address the individual patient's needs in precision medicine while reducing costs to improve patient satisfaction. In big data, complex data issues require regulatory compliance for the healthcare ecosystem to innovate and keep opportunities to fight fraud, dirty data, waste, and abuse of medical data. At this stage, healthcare companies must consider these promises and ensure optimal healthcare experiences and skills to perform effective operations behind the scenes for medical decision-making. However, analyzing big data remains a time-consuming task in the healthcare industry. The outcome of data lakes must adapt to the real-time responses to the server's system. Finally, the vast volume of huge health data has the potential to expose data and make it extremely susceptible. Analysts and data scientists must take security considerations into account and manage datasets in a way that does not compromise privacy.

Core Messages

- Big data, machine learning, and AI approach help translate health data into meaningful and actionable insights.
- Big data and AI approaches apply to medical research and offer up new avenues for discovering therapeutics.
- ML and AI techniques can process all relevant, available data on medical conditions to help disease diagnosis.

References

1. Raghupathi W, Raghupathi V (2014) Big data analytics in healthcare: promise and potential. Health Inf Sci Syst 2:3. https://doi.org/10.1186/2047-2501-2-3
2. Sarkies MN, Bowles K-A, Skinner EH, Mitchell D, Haas R, Ho M, Salter K, May K, Markham D, O'Brien L, Plumb S, Haines TP (2015) Data collection methods in health services research: hospital length of stay and discharge destination. Appl Clin Inform 6:96–109. https://doi.org/10.4338/ACI-2014-10-RA-0097
3. Tan SS-L, Gao G, Koch S (2015) Big data and analytics in healthcare. Methods Inf Med 54:546–547. https://doi.org/10.3414/ME15-06-1001
4. Baghapour MA, Shooshtarian MR, Javaheri MR, Dehghanifard S, Sefidkar R, Nobandegani AF (2018) A computer-based approach for data analyzing in hospital's healthcare waste management sector by developing an index using consensus-based fuzzy multi-criteria group decision-making models. Int J Med Inf 118:5–15. https://doi.org/10.1016/j.ijmedinf.2018.07.001
5. Bates DW, Saria S, Ohno-Machado L, Shah A, Escobar G (2014) Big data in health care: using analytics to identify and manage high-risk and high-cost patients. Health Aff Proj Hope 33:1123–1131. https://doi.org/10.1377/hlthaff.2014.0041
6. Mayo C (2018) Community science and reaching the promise of big data in health care. Med Phys 45:e790–e792. https://doi.org/10.1002/mp.13140
7. Triantafyllidis AK, Tsanas A (2019) Applications of machine learning in real-life digital health interventions: review of the literature. J Med Internet Res 21:e12286. https://doi.org/10.2196/12286
8. Benke K, Benke G (2018) Artificial intelligence and big data in public health. Int J Environ Res Public Health 15:2796. https://doi.org/10.3390/ijerph15122796
9. Househ M, Aldosari B (2017) The hazards of data mining in healthcare. Stud Health Technol Inform 238:80–83
10. Sundermann AJ, Miller JK, Marsh JW, Saul MI, Shutt KA, Pacey M, Mustapha MM, Ayres A, Pasculle AW, Chen J, Snyder GM, Dubrawski AW, Harrison LH (2019) Automated data mining of the electronic health record for investigation of healthcare-associated outbreaks. Infect Control Hosp Epidemiol 40:314–319. https://doi.org/10.1017/ice.2018.343
11. Lee CH, Yoon H-J (2017) Medical big data: promise and challenges. Kidney Res Clin Pract 36:3–11. https://doi.org/10.23876/j.krcp.2017.36.1.3
12. Ramadan RA (2017) Big data tools-an overview. Int J Comput Softw Eng 2:1–15. https://doi.org/10.15344/2456-4451/2017/125
13. Viceconti M, Hunter P, Hose R (2015) Big data, big knowledge: big data for personalized healthcare. IEEE J Biomed Health Inform 19:1209–1215. https://doi.org/10.1109/JBHI.2015.2406883
14. Ambigavathi M, Sridharan D (2018) Big data analytics in healthcare. In: 2018 tenth international conference on advanced computing (ICoAC), pp 269–276
15. Scheen AJ (2015) Omics and big data, major advances towards personalized medicine of the future? Rev Med Liege 70:262–268
16. He KY, Ge D, He MM (2017) Big data analytics for genomic medicine. Int J Mol Sci 18:412. https://doi.org/10.3390/ijms18020412
17. Lunshof JE, Chadwick R, Vorhaus DB, Church GM (2008) From genetic privacy to open consent. Nat Rev Genet 9:406–411. https://doi.org/10.1038/nrg2360
18. Chadwick R, Levitt M, Shickle D (2014) The right to know and the right not to know. Genetic privacy and responsibility
19. Johri P, Singh T, Das S, Anand S (2017) Vitality of big data analytics in healthcare department, pp 669–673
20. Hansen MM, Miron-Shatz T, Lau AYS, Paton C (2014) Big data in science and healthcare: a review of recent literature and perspectives. Yearb Med Inform 9:21–26. https://doi.org/10.15265/IY-2014-0004

21. Alkhatib M, Talaei-Khoei A, Ghapanchi A (2016) Analysis of research in healthcare data analytics
22. Bhathal GS, Singh A (2019) Big data: hadoop framework vulnerabilities, security issues and attacks. Array 1–2:100002. https://doi.org/10.1016/j.array.2019.100002
23. Wu W, Lin W, Hsu C-H, He L (2018) Energy-efficient hadoop for big data analytics and computing: a systematic review and research insights. Future Gener Comput Syst 86:1351–1367. https://doi.org/10.1016/j.future.2017.11.010
24. Subramaniyaswamy V, Vijayakumar V, Logesh R, Indragandhi V (2015) Unstructured data analysis on big data using map reduce. Procedia Comput Sci 50:456–465. https://doi.org/10.1016/j.procs.2015.04.015
25. Ramsingh J, Bhuvaneswari V (2018) An efficient map reduce-based hybrid NBC-TFIDF algorithm to mine the public sentiment on diabetes mellitus—a big data approach. J King Saud Univ—Comput Inf Sci. https://doi.org/10.1016/j.jksuci.2018.06.011
26. Li R, Dong X, Gu X, Xue Z, Li K (2016) System optimization for big data processing. In: Buyya R, Calheiros RN, Dastjerdi AV (eds) Big data. Morgan Kaufmann, pp 215–238
27. Xu Y, Liu H, Long Z (2020) A distributed computing framework for wind speed big data forecasting on Apache Spark. Sustain Energy Technol Assess 37:100582. https://doi.org/10.1016/j.seta.2019.100582
28. Gupta GP, Khedwal J (2020) Framework for error detection & its localization in sensor data stream for reliable big sensor data analytics using Apache Spark streaming. Procedia Comput Sci 167:2337–2342. https://doi.org/10.1016/j.procs.2020.03.286
29. Sharma M, Bundele M (2019) Analysis of NoSQL schema design approaches using HBase for GIS data. Procedia Comput Sci 152:59–65. https://doi.org/10.1016/j.procs.2019.05.027
30. Rodger JA (2015) Discovery of medical big data analytics: Improving the prediction of traumatic brain injury survival rates by data mining patient informatics processing software hybrid hadoop hive. Inform Med Unlocked 1:17–26. https://doi.org/10.1016/j.imu.2016.01.002
31. AlNuaimi N, Masud MM, Serhani MA, Zaki N (2019) Streaming feature selection algorithms for big data: a survey. Appl Comput Inform. https://doi.org/10.1016/j.aci.2019.01.001
32. Toliopoulos T, Gounaris A, Tsichlas K, Papadopoulos A, Sampaio S (2020) Continuous outlier mining of streaming data in flink. Inf Syst 93:101569. https://doi.org/10.1016/j.is.2020.101569
33. Perçuku A, Minkovska D, Stoyanova L (2018) Big data and time series use in short term load forecasting in power transmission system. Procedia Comput Sci 141:167–174. https://doi.org/10.1016/j.procs.2018.10.163
34. Birjali M, Beni-Hssane A, Erritali M (2017) Analyzing social media through big data using infosphere biginsights and apache flume. Procedia Comput Sci 113:280–285. https://doi.org/10.1016/j.procs.2017.08.299
35. Wiatr R, Słota R, Kitowski J (2018) Optimising Kafka for stream processing in latency sensitive systems. Procedia Comput Sci 136:99–108. https://doi.org/10.1016/j.procs.2018.08.242
36. Erraissi A, Belangour A (2018) Meta-modeling of zookeeper and map reduce processing. In: 2018 international conference on electronics, control, optimization and computer science (ICECOCS), pp 1–5
37. Xiaozhu G (2015) FusionInsight: big results from big data—Huawei Publications. Operation Transformation Marketing Department
38. Vivancos D (2019) From big data to artificial intelligence 2019 Edition. Independently published
39. Yu K-H, Beam AL, Kohane IS (2018) Artificial intelligence in healthcare. Nat Biomed Eng 2:719–731. https://doi.org/10.1038/s41551-018-0305-z
40. Mintz Y, Brodie R (2019) Introduction to artificial intelligence in medicine. Minim Invasive Ther Allied Technol MITAT Off J Soc Minim Invasive Ther 28:73–81. https://doi.org/10.1080/13645706.2019.1575882

41. Roth JA, Battegay M, Juchler F, Vogt JE, Widmer AF (2018) Introduction to machine learning in digital healthcare epidemiology. Infect Control Hosp Epidemiol 39:1457–1462. https://doi.org/10.1017/ice.2018.265
42. Wiens J, Shenoy ES (2018) Machine learning for healthcare: on the verge of a major shift in healthcare epidemiology. Clin Infect Dis Off Publ Infect Dis Soc Am 66:149–153. https://doi.org/10.1093/cid/cix731
43. Obukhov A, Krasnyanskiy M, Nikolyukin M (2019) Implementation of decision support subsystem in electronic document systems using machine learning techniques. In: 2019 international multi-conference on industrial engineering and modern technologies (FarEastCon), pp 1–6
44. Luz CF, Vollmer M, Decruyenaere J, Nijsten MW, Glasner C, Sinha B (2020) Machine learning in infection management using routine electronic health records: tools, techniques, and reporting of future technologies. Clin Microbiol Infect Off Publ Eur Soc Clin Microbiol Infect Dis. https://doi.org/10.1016/j.cmi.2020.02.003
45. Sabarmathi G, Chinnaiyan R (2019) Reliable machine learning approach to predict patient satisfaction for optimal decision making and quality health care. In: 2019 international conference on communication and electronics systems (ICCES), pp 1489–1493
46. Sherkhane P, Vora D (2017) Survey of deep learning software tools. In: 2017 international conference on data management, analytics and innovation (ICDMAI), pp 236–238
47. Wallis L, Paich M (2017) Integrating artifical intelligence with anylogic simulation. In: 2017 winter simulation conference (WSC), pp 4449–4449
48. Yuyang J, Hongyan M (2018) Study on evacuation simulation of medical pension building based on anylogic. In: 2018 Chinese control and decision conference (CCDC), pp 2307–2312
49. Paton C, Kobayashi S (2019) An open science approach to artificial intelligence in healthcare. Yearb Med Inform 28:47–51. https://doi.org/10.1055/s-0039-1677898
50. Kitcharoen N, Kamolsantisuk S, Angsomboon R, Achalakul T (2013) RapidMiner framework for manufacturing data analysis on the cloud
51. Utmal M, Pandey RK (2015) Taxonomy on the integration of hadoop and rapid miner for big data analytics. In: 2015 international conference on computational intelligence and communication networks (CICN), pp 890–893
52. Sosa J (2020) Deep learning for micrograph analysis
53. Vaishnav D, Rao BR (2018) Comparison of machine learning algorithms and fruit classification using orange data mining tool. In: 2018 3rd international conference on inventive computation technologies (ICICT). pp 603–607
54. Chen Y, Elenee Argentinis J, Weber G (2016) IBM watson: how cognitive computing can be applied to big data challenges in life sciences research. Clin Ther 38:688–701. https://doi.org/10.1016/j.clinthera.2015.12.001
55. Winters-Miner LA, Bolding P, Hill T, Nisbet B, Goldstein M, Hilbe JM, Walton N, Miner G, Brown EW, Kohn MS (2015) Chapter 25—IBM watson for clinical decision support. In: Winters-Miner LA, Bolding PS, Hilbe JM, Goldstein M, Hill T, Nisbet R, Walton N, Miner GD (eds) Practical predictive analytics and decisioning systems for medicine. Academic Press, pp 1038–1040
56. Stoeklé H-C, Charlier P, Hervé C, Deleuze J-F, Vogt G (2018) Artificial intelligence in internal medicine: between science and pseudoscience. Eur J Intern Med 51:e33–e34. https://doi.org/10.1016/j.ejim.2018.01.027
57. Semigran HL, Levine DM, Nundy S, Mehrotra A (2016) Comparison of physician and computer diagnostic accuracy. JAMA Intern Med 176:1860–1861. https://doi.org/10.1001/jamainternmed.2016.6001
58. Pelaccia T, Tardif J, Triby E, Ammirati C, Bertrand C, Dory V, Charlin B (2014) How and when do expert emergency physicians generate and evaluate diagnostic hypotheses? a qualitative study using head-mounted video cued-recall interviews. Ann Emerg Med 64:575–585. https://doi.org/10.1016/j.annemergmed.2014.05.003

59. Pelaccia T, Forestier G, Wemmert C (2019) Deconstructing the diagnostic reasoning of human versus artificial intelligence. CMAJ 191:E1332–E1335. https://doi.org/10.1503/cmaj. 190506
60. Pelaccia T, Tardif J, Triby E, Charlin B (2011) An analysis of clinical reasoning through a recent and comprehensive approach: the dual-process theory. Med Educ Online 16:5890. https://doi.org/10.3402/meo.v16i0.5890
61. Tizhoosh HR, Pantanowitz L (2018) Artificial intelligence and digital pathology: challenges and opportunities. J Pathol Inform 9. https://doi.org/10.4103/jpi.jpi_53_18
62. Kantarjian H, Yu PP (2015) Artificial intelligence, big data, and cancer. JAMA Oncol 1:573–574. https://doi.org/10.1001/jamaoncol.2015.1203
63. Resteghini C, Trama A, Borgonovi E, Hosni H, Corrao G, Orlandi E, Calareso G, De Cecco L, Piazza C, Mainardi L, Licitra L (2018) Big data in head and neck cancer. Curr Treat Options Oncol 19:62. https://doi.org/10.1007/s11864-018-0585-2
64. Low S-K, Zembutsu H, Nakamura Y (2018) Breast cancer: The translation of big genomic data to cancer precision medicine. Cancer Sci 109:497–506. https://doi.org/10.1111/cas. 13463
65. Workman P, Antolin AA, Al-Lazikani B (2019) Transforming cancer drug discovery with big data and AI. Expert Opin Drug Discov 14:1089–1095. https://doi.org/10.1080/17460441. 2019.1637414
66. Tang L (2018) Radiological evaluation of advanced gastric cancer: from image to big data radiomics. Zhonghua Wei Chang Wai Ke Za Zhi Chin J Gastrointest Surg 21:1106–1112
67. Li L, Lu J, Xue W, Wang L, Zhai Y, Fan Z, Wu G, Fan F, Li J, Zhang C, Zhang Y, Zhao J (2017) Target of obstructive sleep apnea syndrome merge lung cancer: based on big data platform. Oncotarget 8:21567–21578. https://doi.org/10.18632/oncotarget.15372
68. Rabbani M, Kanevsky J, Kafi K, Chandelier F, Giles FJ (2018) Role of artificial intelligence in the care of patients with nonsmall cell lung cancer. Eur J Clin Invest 48:e12901. https://doi. org/10.1111/eci.12901
69. Morris MA, Saboury B, Burkett B, Gao J, Siegel EL (2018) Reinventing radiology: big data and the future of medical imaging. J Thorac Imaging 33:4–16. https://doi.org/10.1097/RTI. 0000000000000311
70. Kansagra AP, Yu J-PJ, Chatterjee AR, Lenchik L, Chow DS, Prater AB, Yeh J, Doshi AM, Hawkins CM, Heilbrun ME, Smith SE, Oselkin M, Gupta P, Ali S (2016) Big data and the future of radiology informatics. Acad Radiol 23:30–42. https://doi.org/10.1016/j.acra.2015. 10.004
71. Bragazzi NL, Dai H, Damiani G, Behzadifar M, Martini M, Wu J (2020) How big data and artificial intelligence can help better manage the COVID-19 pandemic. Int J Environ Res Public Health 17:3176. https://doi.org/10.3390/ijerph17093176
72. Grant-Kels JM, Sloan B, Kantor J, Elston DM (2020) Big data and cutaneous manifestations of COVID-19. J Am Acad Dermatol 83:365–366. https://doi.org/10.1016/j.jaad.2020.04.050
73. Chen C-M, Jyan H-W, Chien S-C, Jen H-H, Hsu C-Y, Lee P-C, Lee C-F, Yang Y-T, Chen M-Y, Chen L-S, Chen H-H, Chan C-C (2020) Containing COVID-19 among 627,386 persons in contact with the diamond princess cruise ship passengers who Disembarked in Taiwan: big data analytics. J Med Internet Res 22:e19540. https://doi.org/10.2196/19540
74. Vaishya R, Javaid M, Khan IH, Haleem A (2020) Artificial intelligence (AI) applications for COVID-19 pandemic. Diabetes Metab Syndr 14:337–339. https://doi.org/10.1016/j.dsx.2020. 04.012
75. Brown N, Cambruzzi J, Cox PJ, Davies M, Dunbar J, Plumbley D, Sellwood MA, Sim A, Williams-Jones BI, Zwierzyna M, Sheppard DW (2018) Big data in drug discovery. Prog Med Chem 57:277–356. https://doi.org/10.1016/bs.pmch.2017.12.003
76. Jing Y, Bian Y, Hu Z, Wang L, Xie X-Q (2018) Deep learning for drug design: an artificial intelligence paradigm for drug discovery in the big data era. AAPS J 20:58. https://doi.org/10. 1208/s12248-018-0210-0

77. Rutledge RB, Chekroud AM, Huys QJ (2019) Machine learning and big data in psychiatry: toward clinical applications. Curr Opin Neurobiol 55:152–159. https://doi.org/10.1016/j.conb. 2019.02.006

78. Tai AMY, Albuquerque A, Carmona NE, Subramanieapillai M, Cha DS, Sheko M, Lee Y, Mansur R, McIntyre RS (2019) Machine learning and big data: implications for disease modeling and therapeutic discovery in psychiatry. Artif Intell Med 99:101704. https://doi.org/10.1016/j.artmed.2019.101704

Houneida Sakly is a Ph.D. and Engineer in Medical Informatics. She is a Member of the research program deep learning analysis of Radiologic Imaging within Stanford University and a Member of MIT-Harvard Medical school Program. Her main field of research interest is the Data science (artificial intelligence, big data, blockchain, and Internet of Things) applied in healthcare. She is a Member of the Integrated Science Association (ISA) in Tunisia's Universal Scientific Education and Research Network (USERN). Currently, she is serving as a lead editor for the edited book "Trends of Artificial Intelligence and Big Data for E-Health," published by Springer, and a special issue with SAGE journals intituled "Intelligent Healthcare for Medical Decision Making: AI and Big Data for Cancer Prevention." Recently, she has won the best researcher award in the International Conference on Cardiology and Cardiovascular Medicine.

Moncef Tagina serves as the Director of the Doctoral School and President of the National School of Computer Sciences thesis committee in Tunisia (ENSI). He is also the co-founder of the COSMOS Laboratory in the ENSI (http://www.ensi-uma.tn) and Professor of High education with the responsibility of Research Master degree in artificial intelligence and decision support. He contributed to more than 100 publications in this field. He supervised more than 40 Ph.D. research projects in the field of AI and Robotics. He is a member of scientific committees and the organization of several conferences.

Artificial Intelligence in the Medical Context: Who is the Agent in Charge?

24

Emilio Maria Palmerini and Claudio Lucchiari

> *"Three-quarters of the sicknesses of intelligent people come from their intelligence. They need at least a doctor who can understand this sickness."*
>
> Marcel Proust

Summary

The application of predictive algorithms and deep learning artificial intelligence (AI) will transform the medical field. In recent years, medical AI has focused on the diagnosis and decision-making support. Both are complex tasks, and physicians often simplify them by an automatic mental process risking suffering from biases. AI could be a solution to this problem but still has issues to face. This chapter proposes an ontological and ethical framework to tackle one of the difficulties presented by AI: the responsibility problem. We will focus on artificial neural networks (ANNs) because of their diffusion. ANNs are opaque to external scrutiny. This opacity leads to uncertainties in attributing responsibilities in case of failure: nobody seems sufficiently in control to be held accountable for AI. We argue that considering AI as a tool is not an option because of the lack of local control. After accepting a definition of an agent, which can include AI, we attribute responsibility to the human decision-maker and not to the manufacturing process. Simple solutions can be devised to distribute responsibility between ANN and human decision-makers, and we list a few.

E. M. Palmerini · C. Lucchiari (✉)
Department of Philosophy, Università Degli Studi Di Milano, Milan, Italy
e-mail: claudio.lucchiari@unimi.it

Integrated Science Association (ISA), Universal Scientific Education and Research Network (USERN), Milan, Italy

© The Author(s), under exclusive license to Springer Nature Switzerland AG 2022
N. Rezaei (ed.), *Multidisciplinarity and Interdisciplinarity in Health*,
Integrated Science 6, https://doi.org/10.1007/978-3-030-96814-4_24

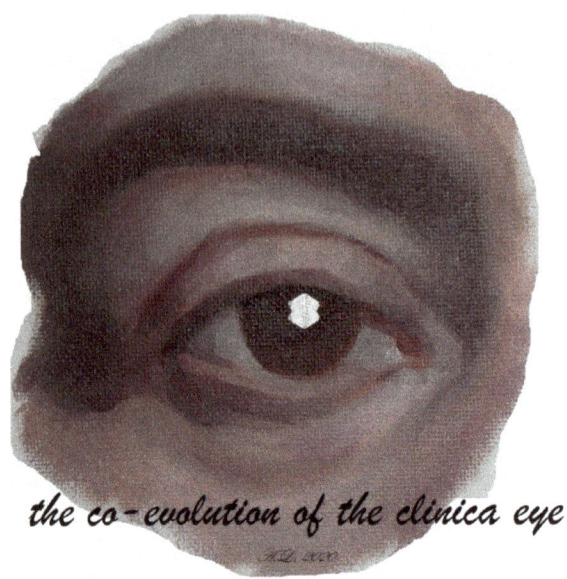

The co-evolution of the clinical eye
[Painted by Alice Lucchiari].
The code of this chapter is *01101111 01100001 01110010 01001101 01101100*.

Keywords

Artificial intelligence · Artificial neural network · Decision-making · Medicine · Responsibility

1 Introduction

The COVID-19 outbreak and the subsequent global crisis have once again shown the importance of using rational models to help professionals and decision-makers to make effective and efficient decisions. Despite the considerable development of medicine and technology during the last 50 years, the guidelines and procedures activated during the outbreak showed many weak points, which need attention from different perspectives. We will point here some of them. First, doctors need to face uncertainty and risk every day, and so they need tools to support them in reducing the related hazards. Since the human mind (also an expert one) may achieve a high but bounded rationality, especially when facing uncertainty [1], we argue that it is necessary to set the problem starting with the interaction between human (natural) and artificial intelligence (AI). However, uncertainty and risk cannot be eliminated even through the use of AI. This fact is due to many factors, the most important of which is that complexity makes it impossible to know all the factors implied so as to provide doctors with reliable estimations in every context. However, in most situations, a complex situation can be strongly simplified by discarding those variables whose contribution is marginal to the variance of the expected outcome. AI, in this sense, can be of great help. Though this does not mean that the uncertainty can be eliminated entirely, and then the final decision-maker (i.e., the agent in charge) will be called to take a risky decision anyway, with all the related practical, legal, and ethical concerns.

Moreover, every time we simplify a situation, eliminating or downgrading some aspects, we can add additional sources of risk. For instance, we could miss some aspects of the problem considered as marginal from a mathematical point of view, but that could have a significant impact from a psycho-social perspective. Furthermore, we could maximize short-term benefits without adequately considering middle- and long-term hazards.

A second important aspect is related to responsibility. Charging AI systems with high responsibilities generally leads to a diminished ability of the human decision-maker to plan and implement its own strategy, and this is basically due to two factors: first, systematic but inappropriate use of smart decision tools decreases the ability of humans to understand the whole scenario; second, in uncertain situation a human has psychological difficulties in taking responsibilities, being pushed to charge the AI whenever is possible. All this can generate a sort of need for the artificial system, which can be implicitly or explicitly recognized as infallible. In many sectors in which AI has been applied (e.g., engineering and logistics), it has been noticed the emergence of several biases linked to de-responsibility by the human manager, on the one hand, and a loss of competence on the other. Therefore, a zero-sum dynamic seems to occur, within which, by increasing the intelligence of one system (the artificial one), the intelligence of the other (the human) decreases, and vice versa.

As a matter of fact, AI is closely linked to the future of healthcare, public health, and safety in general. The development of AI in health care raises issues about accountability as AI tools evolve and increasingly affect patients' health and rights, as well as health professionals' responsibility (see Artwork). Recognizing the importance of the issue and taking into account the rapid development of technology, the European Parliament's Policy Department has drafted and presented on May 31 2016 a draft report on recommendations on civil law in the field of robotics [2], including the health sector. The purpose of the recommendations is to develop general principles governing the development of robotics and AI for civilian purposes, including healthcare. A resolution adopted by the European Parliament called for creating a new technology insurance system that could facilitate civil litigation, confirming the need for an extensive analysis of the possible impact of the inclusion of decision technologies in healthcare. Technological tools can have relevant effects on the health care system. For example, they can reduce health care costs by allowing medical professionals to shift their focus from treatment to prevention and make more budget resources available for a personalized approach, training, and research. However, a comprehensive and shared regulatory framework has not been recognized yet.

This chapter provides a synthetic overview of medical decision-making, and then we provide a multidisciplinary approach to the use of AI in the medical context. In particular, our main hypothesis is that before systematically including AI in medical activity, it is a strictly necessary theoretical analysis of the responsibility issues. Who will be the agent in charge? The doctor, the patient, the artificial tool, or the software programmer?

The paper is divided into three sections. Section 1 will introduce the basic concepts, the size of the medical error, and a cognitive perspective of medical decision-making. In part 2, we will describe a philosophical approach to AI, focusing on the ontological nature of AI tools and the related problem of controlling them. In Sect. 3, we will provide some possible solutions and introduce the final considerations in the frame of actual medical praxis.

2 The Medical Decision-Making

The work of doctors is pervaded by risk and uncertainty. Inefficient decision-making processes, due to cognitive or contextual factors, can lead to diagnostic delays, inappropriate therapeutic choices, and adverse events, which result in harm to the patient and psychological sequelae on health professionals.

Medical decision-making is the expression of preference between different treatment and care alternatives. Although doctors are involved in complex cognitive procedures daily, they generally spend little time reflecting on the processes underpinning their decisions. This means that they rarely develop decisional metacognition. In recent decades, many studies on this topic have been conducted, and converging evidence reveals that in addition to theoretical notions and technical expertise, health professionals would benefit considerably from a more in-depth knowledge of the decision science principles [3].

The study of medical decision-making has four main objectives [4]:

i. to identify the modulatory mechanisms that govern optimal decision-making processes;
ii. describe how practitioners make decisions and what are the intervening factors that negatively influence cognitive processes;
iii. develop smart tools and strategies to reduce the gap between typical and optimal decisions;
iv. improve communication to facilitate a productive exchange of information within the professionals).

Generally, the fallibility of decision-making processes should not be interpreted as an index of malfunctioning of the human mind but rather as the consequence of the way the mind habitually works. In the medical setting, an error can have severe or catastrophic consequences, which is why doctors generally experience a decision-making burden. However, the complexity of the decision-making environment poses a challenge, even for the most experienced doctors. In fact, the doctors' minds work precisely as any other mind and suffer from the same cognitive and emotional biases [5].

Medical decision-making, like any other decision-making process, includes a logical-analytical dimension (System II) and an intuitive-analogical dimension (System I), consisting of rapid, automatic, and implicit mechanisms [6]. However, these two systems should not be understood as mutually exclusive. On the contrary, these two components are variably imbricated within a cognitive continuum, and their balance depends both on external and internal factors. In other words, the cognitive processes that underlie the processing of decision-making dilemmas are the result of dynamic interaction between different elementary sub-processes, each of which can make use of different thinking systems, i.e., of System I and System II. Recent medical decision-making models have clarified how the efficiency of the decision-making process critically depends on a proper balance between these two components [7, 8].

A naïve attitude has led for many years to consider the intuitive-analogical modes of reasoning, represented by System I, as opposites of cognitive efficiency and the leading cause of the error. Several studies reveal how the relationship between the two cognitive dimensions is more complicated and should be situated in context [9]. Not always, anyway, procedures and guidelines lead to the best intervention, especially when epidemiological data are not reliable enough or change quickly. This fact leaves open the issue of individual medical decisions. Furthermore, this approach contrasts with the way the human mind works. In fact, so-called mindlines generally take the place of external guidelines determining specific cognitive pathways [10]. Mindlines are ready-to-use procedures that allow fast and effective decisions but also increase the likelihood of error in some cases.

Lucchiari and Pravettoni have proposed a balanced model in which System II operates as a monitoring model of System I and encourages critical review of decision-making processes [8]. Indeed, the evidence shows that decisions in medicine are frequently the result of an intuitive process rather than the systematic application of formal rules of thought. Although a doctor during her/his training is pushed to adopt probabilistic reasoning and to apply inferences according to the proper rules of clinical practice, often in daily professional activity, decisions are the product of intuitive and automatic processes that have been consolidated with experience. In clinical experience, in general, it is observed that novice or inexperienced doctors make use of System II. In contrast, more experienced doctors use it to deal with complex or unprecedented situations. Physicians often face complex problems, and they need to automatize most mental they use in practice. This way, several cognitive biases may implicitly affect a diagnosis [11], but doctors can consider them the result of a rational process.

Altogether, the study of medical decision-making pointed out that education and training are not enough to prevent medical errors efficiently [12]. Though different approaches might be followed, the use of cognitive-led smart technology is probably the most promising. Technological solutions might be the only reliable way to balance System I and System II, thus optimizing clinical outcomes.

2.1 Medical Decision-Making and AI

Medical decision-making is a task that is experiencing a constant increase in complexity. The more options a physician has, the more complicated the decision becomes. This is also due to the rise of awareness about the need for a personalized approach. Physicians should collect a lot of data about a patient. Physiological, genetic, psychological, and ethical data need to be collected, integrated and discussed to allow patients to make the best decision for themselves. Now doctors work in a shared context [13], where many actors are active, and much information needs to be collected and processed, an aspect linked to the so-called problem of "big data" [14]. This way, the role of technology becomes increasingly essential [15]. In particular, AI methods are vital for doctors to deliver effective early interventions, identify at-risk patients, make a diagnostic decision [16] as well as optimize health organizations [17].

Smart tools such as artificial neural networks (ANN) [18] and machine learning algorithms [19] are especially useful in the medical context due to their ability to learn through training. In this way, they can adapt to specific cases and needs, including psychological and cognitive factors.

An ANN is a mathematical tool that can simulate human information processes. There are many types of problems that ANNs can address, for example, prediction, adaptation, classification and pattern recognition, clustering, and dynamic time series [20]. Each ANN comprises several nodes, also called neurons, organized into layers and connected. Each connection's value determines the link's strength between two nodes. This value is calculated using a specific mathematical function, and they are updated over time during the training and learning phases. The output is contained in the final node, and it depends on the different connections within the network. Thanks to effective training, an ANN may learn how to connect several inputs to a specific output, for example, providing a diagnosis once that symptoms, signs, and other information are given as input.

The multi-layered architecture of a typical ANN includes a single input layer with numerous hidden and output layers (Fig. 1). Training is needed to create reliable connections between the layers [21]. To achieve this aim, hidden layers initially have random weights and are changed step-by-step during the training. Feed-forward and back-propagation algorithms are used to this end so that the weights of the hidden layers are repeatedly updated to tune the network and reduce the gap between the actual output and the desired one. The training phase allows estimating the reliability of the ANN, providing an estimate of the possible output error.

Three basic categories of ANNs can be described:

i. associative memories are designed to find patterns among complex input stimuli such as an image in an array of pixels. For example, they help work on a partial pattern provided as input to complete it to give rise to a meaningful pattern in the output layer;

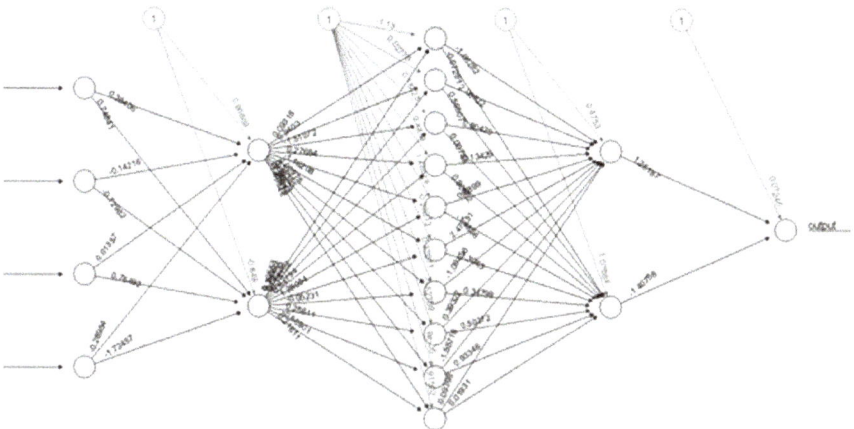

Fig. 1 Artificial neural network (ANN) example. This ANN have three hidden layers of neurons and one single output. Weights and biases of the network are visible. Image created with R

ii. mathematical function simulators: these ANNs are trained to determine which mathematical functions connect the input to output. Thanks to the training phase, by using back-propagation algorithms, the tool learns how to apply a specific mathematical formula to derive an output from an input even when the data provided is new. This way, these tools can interpolate and extrapolate rules from a dataset. They are particularly useful for predicting outputs when it is unclear what mathematical function should be applied. For this reason, these ANNs are considered black-box since the relationship between input and output cannot be described;

iii. classifiers: these ANNs are designed to categorize input based on similarities. In particular, the learning process that allows the tools to discover regularities in a dataset is self-organizing and then unsupervised. Consequently, there are no theorems or models for determining the optimal output.

2.2 AI and Natural Intelligence: A Couple of Dances

The use of AI systems must be accompanied by an increase of competence by the human decision-maker, who must be able to take full advantage of technology without falling into the trap of the technological bias, which attributes to the "machine" unlimited capacity. Instead, a virtuous interaction between intelligences can lead to a multiplication of cognitive abilities, but this requires the development of specific competencies, as well as a broader culture about man–machine interaction and related risks. Indeed, by combining the two spheres uncritically, the natural and the technological, new biases are generated, which should be known and managed exactly as is in the case with classic cognitive and emotional ones. On the one hand, it seems inevitable that risk management, particularly in situations of considerable uncertainty, takes advantage of AI; on the other hand, it is fundamental that the human decision-maker develops virtuous use of the available tools to increase the whole rationality. We need, therefore, to give course to a co-evolutionary process in which man and machine influence each other to reach the best solution possible, considering uncertainty and risks. In other words, we should reach a local maximum of a specific fitness landscape [22], which at the moment still seems a long way off.

To give a practical example of what we mean, we can consider the case of a football simulator, where a player needs to co-play with the AI engine to manage the complicated situation of the match. In fact, the player's attention needs to focus on one aspect at a time, while many other parameters are handled automatically by the computer. Being in a dynamic context, where many things happen at the same time, most of them totally out of human control, accidents can easily happen, in the sense that humans' intelligence and AI may conflict, leading to a wrong pass or an opponent's goal. However, the human player learns implicitly with a long practice of how to collaborate appropriately with the AI, and little by little will include in the game its own strategy and creativity, leaving to the computer the more complicated and annoying background computations. Also, the AI learns and adapts itself to the player's style. Thanks to practice and training, the player implicitly learns the AI "way of reasoning" and vice versa. It becomes more and more natural to pursue the same goal without any direct communication and any programming. After all, the human player needs to make the right decisions coherently with what the AI is already doing. They need to co-play in a situation where it is clear that the outcome (a win or a defeat) is in charge of the human agent (the player). We argue that the use of technologies in healthcare should be conceptualized similarly.

3   A Theoretical Approach to AI as a Tool: The Wrong Starting Point

AI is a tool. It is quite a simple statement and one accepted at face value. Indeed the ethical discussion around AI often starts with a statement like this [23]. AI is more exotic than, for example, a thermometer, but they share the same fundamental properties. They are both created to solve a specific problem improving the human capacity to modify his environment. Thermometers solve the problem of perceiving more nuances variations in temperature than our limited senses. AI solves more complex issues and can take over a vast spectrum of tasks. All tools use some proprieties of the physical world to work automatically. No human pushes the mercury (or its digital equivalent) in the thermometer as much as nobody directly operates a robot equipped with AI, but an automatic tool is not necessarily autonomous. What counts as genuine autonomy is problematic, but a simple way to differentiate between autonomy and automation is to consider the complexity of the task. If it is simple as monitoring temperature, it will hardly be regarded as an autonomous tool. Vehicles equipped with semi-autonomous driving software can surpass other vehicles, park, and maintain the lines. Parking and driving are complex tasks, which usually require human control and oversight, so we are more prone to consider a car as autonomous if it can perform such stunts. However, in a strict philosophical sense, neither a thermometer nor a semi-autonomous car is autonomous. If they are not autonomous, they must be tools.

However, as far as tools go, medical AI is quite peculiar. While humans control entirely more conventional tools, AI has achieved a level of automation close to autonomy [24–26]. In the medical context, AI is gaining momentum as a diagnostic tool [27, 28]. 75% of diagnostic errors are caused by cognitive factors, including human biases [29]. So, AI is put forward as decision-making support to reduce this error [27, 29, 30]. As such, it needs to be as autonomous as possible. Unfortunately, AI autonomy leads to several ethical problems [24, 26, 31]. Some voices have raised concerns about the allocation of responsibility or the lack of there of when uncontrollable AI is implemented in a high-risk moral context [31–35]. *Machine learning* seems to amplify the problem [31].

3.1 ANN and Structural Opacity

To analyze a tool from the ethical standpoint is mandatory to have a general comprehension of the tool itself, of its creation and application [36]. In the last few years, the number and model of AI created for medical applications have been quite vast and growing exponentially [37]. The most prominent model of AI seems to be an ANN, followed closely by a support vector machine (SVM) [27, 38]. For our purpose, ANNs are a good exemplification of the most advanced medical AI, so we will focus primarily on them. In the remainder of this article, we will consider ANN as if it will be implemented alone because the most ethically prominent problems are connected primarily with the machine learning module. However, it is useful to bear in mind that AI is often a multi-component system.

As we have already said, ANNs are computational models of the human brain [39]. As such, they are based on layers of interconnected nodes (called "neurons") capable of identifying and adapting to patterns in significant clusters of data (Fig. 2). Through this training process, ANN can categorize new data and make predictions about future events [39].

The AI of Arabasadi and colleagues [40] is an excellent example of how these ANNs will be used. To diagnose coronary artery disease (CAD), the most accurate tool is angiography. Angiography is, however, a relatively expensive and dangerous process. The ANN developed by the research team can correctly diagnose CAD without invasive interventions and with an accuracy of 93.85% [40]. It is easily understood why such ANN will be applied extensively after passing the necessary clinical tests. It will likely replace angiography in most cases. At that point, the diagnosis of AI will have the same value as the current diagnostic tool.

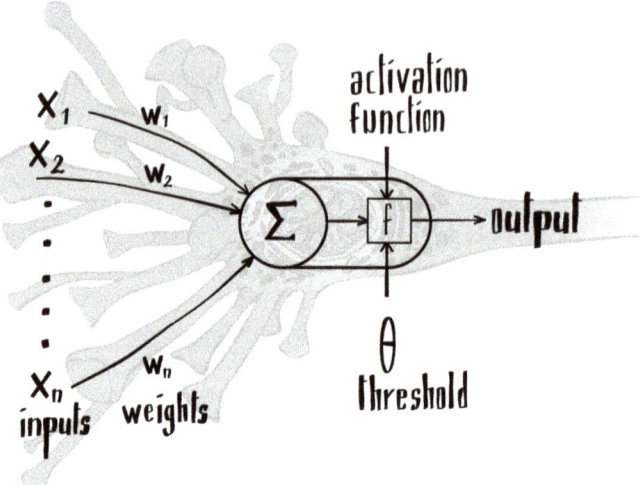

Fig. 2 An artistic depiction of a ANN "neural" node superimposed on a human neuron. Single node in the artificial network works as a neuron: it "shoots" when inputs reach a certain threshold. Image courtesy of Francesco Mazziotta

All software can be simplified in "if–then" statements. Examples are easy to find. We are writing on a word processor a complex set of lines like this one: if the user strikes "A" on the keyboard, the program writes "A" on the screen. An industrial AI operates commands as "if a cube appears on the conveyor belt, then grasp it." ANN (and other AI trained with machine learning technics) are not programmed in a traditional sense [39]. The difference between conventional programming and machine learning is both in scale, complexity, and opacity of the "if–then" set [26]. Opacity is our main concern. In traditional AI, a human programmer writes "manually" all the necessary if-thens: the full program can be a huge and complex structure or a few lines of code, but it is comprehensible in every part. It does not have to be opaque. It could be if, for example, some part of the code is proprietary or considered an industrial secret, but it can still be scrutinized when necessary [26, 41]. ANN learns the "if–then" set autonomously. A simple ANN for image classification of flowers can have up to one hundred and fifty thousand "if–then" statements, more technically called parameters. More complex ANNs can have up to millions of different settings. The parameters are expressed by small numbers, called weights, between 0 and 1. The variation between weights is usually too small to be meaningful in a semantic or narrative sense [26, 39]. Also, weights are intertwined non-logically: the importance of a single weight is situational, not absolute. Higher weights do not automatically translate into more important inputs. Take the flower classification from above: let us say that length of petals (a classical input) have a higher value than the numbers of petals. That does not mean the ANN will classify flowers primarily by lengths of petals. As a result, this complex internal structure is completely opaque for a human being [26]. This opacity is not an unintentional side effect: it is a consequence of ANN's high degree of autonomy [42].

Who speaks about problems in controlling AI often refers to this kind of opacity [26, 31, 34, 43, 44]. In the next paragraph, we will see why.

3.2 A Problem of Control

A tool or, at least, a good tool always behaves in the same way: it has well-defined parameters and offers reliable results. Controlling a well-known device is a matter of deciding when and how to use it. When malfunctions or mistakes happen, the attribution of responsibility is quite straightforward [25]. The person or group in control of the tool is responsible for the outcome. If the surgeon misuses a surgical robot, she may take responsibility for her errors. If the joystick of the robot malfunction (and the surgeon has no clue about it), the responsibility may be assigned to the manufacture [25]. With ANN, this attribution is more difficult. Nobody seems in control enough to be responsible for the outcome of the ANN [31, 32]. The final user does not know how the ANN thinks due to its opacity. In other words, the final user of an ANN has only general control over it [45]. *General control* can be defined as the control exerted upon a tool's purpose and objectives but not the sequence of actions, judgments, or events that the tool will use to reach these purposes. It is called *local control* if control is exerted over the complete sequence of actions, judgments, or activities. In the psychological and cultural understanding of moral action and responsibility, local and general control is not equally necessary for the attribution of responsibility.

Daniel Dennett explains the concept quite well, asking who deserves to be praised for the victory of Deep Blue, the famous computer which beat Kasparov [45]. Dennett correctly stresses that even if the IBM team (the creators of Deep Blue) can be praised for their work, Deep Blue chose the correct actions against Kasparov. The IBM team (and Kasparov himself) had only general control over Deep Blue: they decided when to use it and the parameters within which it operated. However, they did not have *local control* [45]. The computer was too complicated, and it takes decisions too fast to be comprehensible in real-time. Its moves were comprehensible only in hindsight. Without local control, praising the IBM team for the victory is similar to praising Kasparov's teachers for mastering the game: it does not imply they are directly responsible. Local control is often considered necessary for the attribution of moral responsibility. As a side note, Deep Blue highlights another problem with AI. Even if a solution for ANN opacity is found, the final user will be difficult to control simply because they will be superhumanly fast and extremely complex.

When blame or praise cannot be placed upon the human operator, the manufacture could be responsible for a failure in a tool (often because the production does not thoroughly test it or deliberately hides the same known problems) [25]. With ANN, this straightforward consideration of the manufacture is impossible: as we have already seen, "thoroughly testing" an ANN could be challenging if not straight impossible. Moreover, ANN capable of learning "on the field" can detect errors and problems entirely independent of the manufacturing process [35]. However, the need for continuous monitoring and updating is a challenge for medical practice [32]. Furthermore, the use of such systems in the clinical context it is not universally accepted as ethical [32, 35, 46]. The fact that the tool designer may choose which ANN output or behavior is permitted or not in a medical context could be detrimental to the autonomy of both patients and healthcare professionals [46]. Moreover, it would shorten the distance between clinic and research [46]: each upgraded version of the ANN will be adequately tested upon new patients.

4 AI as an Agent

AI, especially ANN, is a tool that lacks two interconnected characteristics typical of the tool category: it is not transparent, and it is not controllable in a local sense. How to consider AI is an ethical challenge because of these two missing "features." The standard ethical view sees tools as ethically neutral: the user is the proper ethical subject (agent), responsible for harm or help due to the tool [47]. The device itself is nothing but an extension of its user's moral interests and intentions. If ANN and AI are not neutral, and their actions cannot be blamed on a human operator, they do not fit this canonical framework for tools [47].

Even so, there are not many other options. The set of a possible moral agent is not very large [24]. In addition to individual human adults, corporations and groups could be considered as an agent if every member is an adult of our species [24]. Children and non-human animals are rarely considered: they do not have the *quid,* cognitive or otherwise, to be a full-blown moral agent [24, 31, 48]. It does not matter which ability or feature they are lacking. For some, this missing feature is *intentionality,* while for others, some complex clusters of cognitive skills like language or storytelling [46, 48]. The point is that only human adults are moral agents because only they can be fully responsible. So, if AI is not a tool (as far as the ethical theory is concerned) and it is not an agent, how could it be considered, and who is to blame for its failure?

Talbot, Jankins, and Purves proposed that AI could be considered more similar to a natural disaster than an agent [48]: we can find AI output as good or bad, but they are not agents. So AI cannot be evaluated from a moral and ethical standpoint [48]. When failures happen, and there is not any human to blame, there is no responsibility. Only prevention could be ascribed to someone.

However, even this option is not available in the medical context. The medical practice is high-risk with well-known and standardized ethical guidelines (information consent is one of them). If AI failures are implemented as an "artificial earthquake," they will harm the autonomy of patients and the accountability for their care.

4.1 Moral Agent

Floridi and Sanders present another possible solution in their 2004 article [24]. Instead of considering an agent in a theoretical vacuum, the authors employ an abstraction method to analyze a system [24]. This method utilizes a set of observable proprieties (called level of abstraction) to consider an entity [24] and the entity's transitions between states. A state is the set of values of the observable proprieties at a given time. Floridi and Sanders specify that a level of abstraction often is chosen in its rendering of the transitions between states at non-deterministic. In a level of abstraction granular enough, for example, one which considers how neurons function, even humans could not be regarded as agents in a traditional, non-deterministic sense.

In the medical context, a good level of abstraction to consider ANN is one based on the definition of three critical features: interactivity, adaptability, and autonomy. An entity to be interactive needs to act upon its environment and the environment to work upon the object [24]. ANN can be considered interactive in two ways. The first occurs during training, when patterns in the training database act upon the neuronal structure, and the ANN acts upon the database. The second way in which ANN could be considered interactive is more critical for our discussion. When "deployed on the field," ANN could continue to learn from new data and adapt to an environment and patient demographic. This kind of ANN acts upon the environment through the healthcare professional but are also acts upon the environment.

The same process of continuous learning grants adaptability. An entity is adaptable if it changes the rules of transitions between one state and another on account of the entity's interaction. This kind of adaptability is one of ANN's central features.

Autonomous entities are entities that change states without a direct response to an interaction. It is the most difficult criterion to meet. ANNs could be considered autonomous because the rules determining the output are completely hidden to a human observer. ANNs have at least two different internal states, each one connected with an output. When an ANN chooses between outputs, it changes its internal state in a not observable way. Implementing an ANN is often to improve medical diagnostic accuracy, which means that a human observer is not capable of distinguishing between the different outputs. So, if ANNs are autonomous, interactive, and adaptable, they are agents in this level of abstraction.

Floridi and Sanders use a quite simple argument to demonstrate the moral agenthood of an agent. If the agent's actions cause good or harm, and the agent itself is indistinguishable from a human in a level of abstraction, then the agent is a moral agent [24]. However, Floridi and Sanders don't consider enough moral agents to be held responsible for errors. To be a moral agent only means to be accountable for his own actions or events caused by his own actions. For the two philosophers, the problem is no longer so much control but the impossibility of participating in human social and cultural life: participation is inevitably the base of any ethical consideration. Floridi and Sanders say that AI cannot be blamed, and responsibility cannot be attributed, as much as full responsibility cannot be attributed to a child [24, 47]. The analogy, however, is misleading: our reactions are indeed different if the agent is an adult or a child, but this does not mean that we do not attribute responsibility in both cases. Nor does it mean that the responsibility assigned to the child is less relevant: the attribution of responsibility is linked to the skills that the individual possessed [46]. The higher the skills, the greater the responsibility that the owner can bear. Parents consider their children to be responsible for actions or events caused by them. They decide not to apply the same consequences to children preferring education to punishment. In conclusion, Floridi and Sanders consider only *complete* moral responsibility, similar to the legal or political one, but the ethical responsibility is far more complex.

4.2 Responsibility for AI

The problem of controlling AI can be reformulated as a responsibility attribution problem. The control becomes relevant with an autonomous machine like an ANN only if someone is responsible for the mistakes he makes. Although AI can be considered moral agents, they cannot take legal responsibility. Therefore, the problem of control becomes the problem of understanding how responsibility is distributed when AI is included in diagnostic decision-making.

There are multiple types of responsibilities (e.g., moral, political, social), but they all share three essential components. The first is the explanatory and causal component: you are only responsible for the actions of which you are the cause. In this sense, responsibility is a tool for understanding and explaining the social and ethical world. For example, responsibility is a criterion for distinguishing the victim from the executioner: the victim has no responsibility for what happened. An accomplice has only partial or accessory liability for the crime committed. Secondly, it is a normative concept [46]. When we say that someone is responsible for a fact, we want them to take their guilt (or merit). We want the thief to give us motivation and to show repentance (or at least justice).

Finally, in the attribution of responsibility, there is also a psychological and motivational component. By attributing responsibility to a subject, we take it for granted that he has a series of skills, including autonomy and judgment [46]. An adult human being is held responsible for his own actions until it is shown that he lacks autonomy (because manipulated by someone, for example) or the judgmental skills necessary to discern the effects of his actions. The relationship between these three components establishes the level of responsibility that can be attributed to a subject: a cat is responsible for having broken the vase because it is the cause of the event and is given enough autonomy to have decided to have broken the vase. Specifically, the cat is autonomous. However, the regulatory component is not present: we do not pretend that the cat provides us with an explanation of why he broke the vase, or we do not pretend that he assumes his faults. An earthquake, on the other hand, is responsible only in a metaphorical sense for having dropped the vase: it is the cause, but it does not possess any of the characteristics that would make it a responsible agent.

Responsibility is, therefore, a spectrum rather than a binary concept: at the lower end, responsibility is attributed only in a metaphorical sense. At the upper end, it is responsible in the full sense.

4.3 Functional Responsibility and Control

The ANNs cannot be the object of a full attribution of responsibility, but they are still in the spectrum of responsibility. ANNs, especially in the medical field, have the autonomy necessary to be considered agents: no one is in control of the diagnostic process. The ANNs can, therefore, be considered responsible in a practical sense [46, 49]. The concept of functional equivalence was created to explain how to recognize the capabilities of AI. In relating to AI, some aspects of it are considered as if they were the human equivalent.

A classic example is emotions: an ANN capable of simulating human emotions can be considered as if having those emotions [50]. An ANN capable of making a diagnosis can be regarded "as if" it has responsibility for the decisions it makes. In this way, it is no longer necessary to have control over the internal operations of the machine to be responsible for the errors it makes. The health professional is responsible not because she has complete control of AI but because she can monitor its functioning. In other words, responsibility must be spread within a network made up of AI and human supervisors.

Simple solutions can be devised to put this concept of network responsibility into practice [32]. One way to retain responsibility is by letting health care professionals decide to use ANN after a diagnose is already made. This *choice architecture* considers AI as a second diagnostic opinion instead of the primary diagnostic tool. Another way to achieve oversight is to design an opt-out choice for the healthcare professional. In this kind of choice architecture, even if the ANN (or AI) monitors and diagnoses autonomously, the healthcare professional could override the result, deciding when and how to use ANN suggestion. Moreover, stakeholder opinions could be considered to ensure control and attribution of responsibility in a more extended period with community engagement boards and representative oversight by community minorities [32]. Finally, the meaningful education of both patients and healthcare professionals about AI is critical to let patients autonomously decide if they consider it ethically acceptable to choose performance over responsibility and if they want to make an AI in charge of their health.

5 Conclusion

To conclude this pathway into the mixed world where artificial and natural agents live exchanging information and where the concept of responsibility is placed in a fluctuating space, we want to provide a brief excursus considering a broader context and taking into consideration the 2020 global crisis.

Behind the initial choice of the British government to avoid lockdown in favor of alleged generalized immunity, there seems to be an algorithm. Initially, the algorithm had discarded the total lockdown trusting with the British health care system resilience. When the same algorithm provides new data conflicting with its first prediction, the government quickly changes direction, trying to remedy it. The use of mathematical models is a novelty of the most recent epidemics, but they will have increasing importance in the next future. We feel that two relevant points need attention among others in this discussion:

- with a few exceptions, the predictions of such mathematical models have been translated into a rather ancient tool., i.e., the quarantine; and
- the systematic use of algorithms shad a dark light on responsibilities. In fact, it might be difficult for citizens to understand who is in charge of important decisions, which could profoundly impact their future.

The first aspect points out how the human decision-maker, being a physician or a politician, finds difficulty interacting with AI and, more generally, to a mathematical model. It seems that the systematic use of such systems adds data to a scenario that humans cannot readily appreciate. One of the possible consequences is that humans lose a portion of their creativity, restricting the number of possible options and being pushed toward conservative choices to adhere to the models' prediction. On the one hand, this is not surprising since, in uncertain contexts, human decision-makers show bounded rationality, while machines are expected to be "wiser." In this way, algorithms seem to absorb some of such uncertainty, and humans tend to trust them like new oracles.

On the other hand, it would be clear to decision-makers that AI tools only process data. This process gives rise to a possible scenario based on a series of parameters that some humans set in advance. So, the prediction of smart tools is associated with specific settings (i.e., *a-priori* human decisions). This way, by trying not to fall into so-called cognitive traps (biases), the combination of natural and AI seems to give rise to new hazards linked to new biases. At the same time, it does not allow decision-makers to think of new solutions, even though computational intelligence increased so much and gave rise to much larger decisional space (Fig. 3).

In extremely pragmatic terms, to address the above points, we feel that the human decision-maker should first of all learn to make decisions based on AI predictions without following the temptation to charge the AI of the decision. However, this means developing a comprehensive decision-making competence, including both normative and descriptive models. Secondly, it is necessary to train specific skills to allow decision-makers to collaborate with AI: they need to learn how to co-decide, though maintaining accountability and responsibility.

Unfortunately, the problem is complicated. The decision-maker often does not know how to use forecasting models directly but relies on an expert or even several experts who mediate the relationship between the artificial and the natural agents. These experts, e.g., doctors, statisticians, big data experts, and physicists, offer their

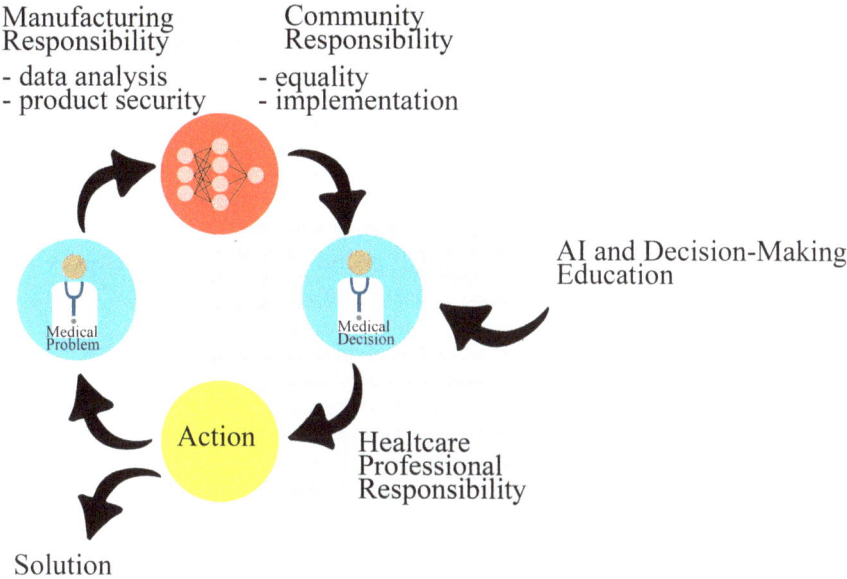

Manufacturing
Responsibility
- data analysis
- product security

Community
Responsibility
- equality
- implementation

AI and Decision-Making
Education

Healtcare
Professional
Responsibility

Action

Medical
Problem

Medical
Decision

Solution

Fig. 3 A schematic representation of a hybrid medical system, where different actors and formats of intelligence share responsibility for a decision flow

perspective, which is never completely neutral and can deserve more or less attention in a given situation. This way, the use of algorithms to simplify a complicated situation gives rise to new complications that the decision-maker cannot manage. This fact requires the decision-maker to a second level simplification by filtering out all the perspectives but one, which may be optimal, sub-optimal, or even frankly unproductive. There is, once again, a loss of rationality and the rise of simplification, within which trust becomes more important than utility maximization. Therefore, it seems essential to develop AI capable of providing data that can be used more efficiently by a human decision-maker to increase mutual intelligence and not lead to oversimplification or blind fiduciary delegation. This poses technical problems, which require greater integration between computer science, mathematics, psychology, and cognitive sciences.

The second point we raised, linked to responsibility, is strictly related to the main topic covered in this chapter. However, we focused our attention on physicians. At the same time, the problem addressed in this conclusive discussion refers to a decision-maker whose decision may impact a large population and not just one patient and her family. Furthermore, managing an epidemic crisis requires the decision-maker to consider several different aspects and utilities, ranging from health to wealth, also considering short-term and long-term consequences. The problem is much bigger, and it requires a broader analysis that is not coherent with the aims of this work. However, we feel that the main arguments discussed in this work might be useful to set a more extensive interdisciplinary discussion about general issues in the relationship between natural intelligence and AI. In particular, we argue that a human decision-maker should always be fully accountable, particularly for policymakers whose actions need to be evaluated by citizens. So, the use of algorithms and AI should be included in a decision frame where it is clear who is the person in charge and which perspective she takes in using predictions and estimating the cost/benefits ratio.

However, it is also necessary to analyze the concept of the co-evolution of artificial and natural from a psychological, philosophical, conceptual, and ethical point of view to understand how this process will affect our tomorrow's world.

Core Messages

- Medical decision often faces uncertainty and requires methods to reduce the impact of biases and errors.
- Artificial intelligence (AI) may improve medical decision-making and decrease error rates.
- AI, especially deep learning, has raised concerns about the allocation of responsibility and control.
- Considering AI as a moral agent instead of a tool could be part of the solution.
- Future research will focus on human–machine integrated systems in healthcare, and ethical issues are new challenges.

Acknowledgements This research was partially funded by the Department of Philosophy "Piero Martinetti" of the University of Milan under the Project "Departments of Excellence 2018–2022" awarded by the Ministry of Education, University and Research (MIUR).

References

1. Simon HA (1972) Theories of bounded rationality, pp 161–176
2. European Parliament resolution of 16 February 2017 with recommendations to the Commission on Civil Law Rules on Robotics
3. Beach MC, Sugarman J (2019) Realizing shared decision-making in practice. JAMA 322 (9):811–811. https://doi.org/10.1001/jama.2019.9797
4. Weinstein MC, Fineberg HV, Elstein AS, Heuhauser HS, Neutra RR (1980) Clinical decisions and limited resources, pp 261–263
5. Saposnik G, Redelmeier D, Ruff CC, Tobler PN (2016) Cognitive biases associated with medical decisions: a systematic review. BMC Med Inform Decis Mak 16(1):138–138. https://doi.org/10.1186/s12911-016-0377-1
6. Kahneman D (2012) Two systems in the mind. Bull Am Acad Arts Sci 55

7. Lucchiari C, Pravettoni G (2013) The role of patient involvement in the diagnostic process in internal medicine: a cognitive approach. Eur J Intern Med 24(5):411–415. https://doi.org/10.1016/j.ejim.2013.01.022

8. Lucchiari C, Pravettoni G (2012) Cognitive balanced model: a conceptual scheme of diagnostic decision making. J Eval Clin Pract 18(1):82–88. https://doi.org/10.1111/j.1365-2753.2011.01771.x

9. Evans KK, Georgian-Smith D, Tambouret R, Birdwell RL, Wolfe JM (2013) The gist of the abnormal: above-chance medical decision making in the blink of an eye. Psychon Bull Rev 20 (6):1170–1175. https://doi.org/10.3758/s13423-013-0459-3

10. Gabbay J, le May A (2008) Practice made perfect: discovering the roles of a community of general practice, pp 49–65

11. Holyoak KJ (2012) Analogy and relational reasoning. Oxford University Press. https://doi.org/10.1093/oxfordhb/9780199734689.013.0013

12. Sherbino J (2015) Education scholarship and its impact on emergency medicine education. West J Emerg Med 16(6):804–809. https://doi.org/10.5811/westjem.2015.9.27355

13. Gulbrandsen P (2020) Shared decision making: improving doctor-patient communication. BMJ:368-368. https://doi.org/10.1136/bmj.m97

14. Lohr S (2012) The age of big data. New York Times Feb.11. 20

15. Begoli E, Bhattacharya T, Kusnezov D (2019) The need for uncertainty quantification in machine-assisted medical decision making. Nat Mach Intell 1(1):20–23. https://doi.org/10.1038/s42256-018-0004-1

16. Lucchiari C, Folgieri R, Pravettoni G (2014) Fuzzy cognitive maps: a tool to improve diagnostic decisions. Diagnosis 1(4):289–293. https://doi.org/10.1515/dx-2014-0026

17. Berwick DM, Nolan TW, Whittington J (2008) The triple aim: care, health and cost. Health Affairs 27(3):759–769. https://doi.org/10.1377/hlthaff.27.3.759

18. Baxt WG (1992) Analysis of the clinical variables driving decision in an artificial neural network trained to identify the presence of myocardial infarction. Ann Emerg Med 21 (12):1439–1444. https://doi.org/10.1016/S0196-0644(05)80056-3

19. Kononenko I (2001) Machine learning for medical diagnosis: history, state of the art and perspective. Artif Intell Med 23(1):89–109. https://doi.org/10.1016/S0933-3657(01)00077-X

20. Demuth H, Beale M (1993) Neural network toolbox for use with matlab–user'S guide version 3.0

21. Hagan MT, Demuth HB, Beale MH, De Jesús O (1996) Neural network design

22. Kauffman S (1996) At home in the universe: the search for laws of self-organization and complexity

23. Patrick L, Jenkins R, Abney K (2017) Robot Ethics 2.0, vol 1. Oxford University Press, New York. https://doi.org/10.1093/oso/9780190652951.001.0001

24. Floridi L, Sanders JW (2004) On the morality of artificial agents. Mind Mach 14(3):349–379. https://doi.org/10.1023/B:MIND.0000035461.63578.9d

25. White T, Baum SD (2017) Liability for present and future technology. In: Patrick L, Jenkins R, Abney K (eds). Oxford University Press, New York

26. Burrell J (2016) How the machine 'thinks': understanding opacity in machine learning algorithms. Big Data Soc 3(1):1–12. https://doi.org/10.1177/2053951715622512

27. Jiang F, Jiang Y, Zhi H, Dong Y, Li H, Ma S, Wang Y, Dong Q, Shen H, Wang Y (2017) Artificial intelligence in healthcare: past, present and future. Stroke Vasc Neurol 2(4):230–243. https://doi.org/10.1136/svn-2017-000101

28. Krittanawong C, Zhang HJ, Wang Z, Aydar M, Kitai T (2017) Artificial Intelligence in precision cardiovascular medicine. J Am Coll Cardiol 69(21):2657–2664. https://doi.org/10.1016/j.jacc.2017.03.571

29. Dilsizian SE, Siegel EL (2014) Artificial intelligence in medicine and cardiac imaging: Harnessing big data and advanced computing to provide personalized medical diagnosis and treatment. Current Cardiol Rep 16(1). https://doi.org/10.1007/s11886-013-0441-8

30. Hosny A, Aerts HJWL (2019) Artificial intelligence for global health. Science 366(6468):955 LP-956. https://doi.org/10.1126/science.aay5189
31. Mittelstadt BD, Allo P, Taddeo M, Wachter S, Floridi L (2016) The ethics of algorithms: mapping the debate. Big Data Soc 3(2):1–21. https://doi.org/10.1177/2053951716679679
32. Cohen IG, Amarasingham R, Shah A, Xie B, Lo B (2014) The legal and ethical concerns that arise from using complex predictive analytics in health care. Health Aff 33(7):1139–1147. https://doi.org/10.1377/hlthaff.2014.0048
33. Datta A, Tschantz MC, Datta A (2015) Automated experiments on Ad privacy settings. Proc Privacy Enhancing Technol 2015(1):92–112. https://doi.org/10.1515/popets-2015-0007
34. Zarsky T (2016) The trouble with algorithmic decisions: an analytic road map to examine efficiency and fairness in automated and opaque decision making. Sci Technol Human Values 41(1):118–132. https://doi.org/10.1177/0162243915605575
35. Babic B, Gerke S, Evgeniou T, Glenn Cohen I (2019) Algorithms on regulatory lockdown in medicine. Science 366(6470):1202–1204. https://doi.org/10.1126/science.aay9547
36. Marckmann G (2014) Ethical assessment of medical technologies : a coherentist methodology. In: Battaglia F, Mukerji N, Nida-Rümelin J (eds). Pisa University Press, Pisa, pp 51–64. https://doi.org/10.1400/225024
37. Topol E (2019) Deep medicine. Basic Books, New York
38. Shahid N, Rappon T, Berta W (2019) Applications of artificial neural networks in health care organizational decision-making: a scoping review. PLoS ONE 14(2):1–22. https://doi.org/10.1371/journal.pone.0212356
39. Russel S, Norvig P (2010) Intelligenza Artificiale: un Approccio Moderno (I-II). Pearson Italia, Milano-Torino
40. Arabasadi Z, Alizadehsani R, Roshanzamir M, Moosaei H, Yarifard AA (2017) Computer aided decision making for heart disease detection using hybrid neural network-Genetic algorithm. Comput Methods Programs Biomed 141:19–26. https://doi.org/10.1016/j.cmpb.2017.01.004
41. Goodman B, Flaxman S (2017) European union regulations on algorithmic decision making and a "right to explanation." AI Mag 38(3):50–57. https://doi.org/10.1609/aimag.v38i3.2741
42. Borgesius F (2018) Artificial intelligence, and algorithmic
43. Lin P (2016) Is Tesla responsible for the deadly crash on auto-pilot? Maybe
44. Sütfeld LR, König P, Pipa G (2019) Towards a framework for ethical decision making in automated vehicles. PsyArXiv: 1–27. https://doi.org/10.31234/osf.io/4duca
45. Dennett D (2014) When HAL kills, who is to blame? computer ethics. In: Mukerij N, Julian N-R (eds) Battaglia F. Pisa University Press, Pisa, pp 203–214
46. Loh W, Loh J (2017) Autonomy and responsability in hybrid systems. In: Jenkins R, Abney K (eds) Patrick L. Oxford University Press, New York, pp 35–50
47. Sullins JP (2018) When is a robot a moral agent? In: Anderson SL (ed) Michael A. Cambridge University Press, Cambridge, pp 151–167
48. Talbot B, Jenkins R, Purves D (2017) when robots should do the wrong thing. In: Oxford University Press, New York
49. Wiegel V (2010) Wendell Wallach and Colin Allen: moral machines: teaching robots right from wrong. Ethics Inf Technol 12(4):359–361. https://doi.org/10.1007/s10676-010-9239-1
50. Anderson M, Anderson SL (2018) Machine Ethics. Cambridge University Press, Cambridge

Emilio Maria Palmerini graduated with an MPhil at the University of Milan. His research interests focus on the ethical challenges of a real-world application of artificial neural networks and machine learning in the medical context. More generally, he works on technology-related ethical problems in high-risk contexts and the socio-political aspect of the technology itself. He collaborates with the Department of Philosophy and the Philab (interdisciplinary research laboratory) at the University of Milan. He cooperates in different research lines within the Cognitive and Affective Research Studies (CARS) group led by Prof. Claudio Lucchiari.

Claudio Lucchiari Ph.D., serves as an Associate Professor in Cognitive Psychology, Department of Philosophy, Università Degli Studi di Milano, Milan, Italy, where he teaches "General Psychology" and "Mind and Brain" at the degree course of Philosophy. He is a member of the European Society for Cognitive and Affective Neuroscience and of the Integrated Science Association. Since 2019, Claudio has been an Associate Editor for Frontiers in Behavioral Neuroscience, Frontiers in Psychology, and Topic Editor for Brain Sciences. His research activities focus on neuro-cognitive and psycho-physiological aspects of learning, creativity, communication, conflict management, decision making in different domains, including health psychology, psycho-oncology, risk domains, medical decision-making, consumer choice, negotiation, and economy. He has authored more than 50 scientific articles at national and international levels, five books, and several essays in various publications and conference proceedings.

Ethical Deliberation on AI-Based Medicine

Sadra Behrouzieh, Mahsa Keshavarz-Fathi, Alfredo Vellido,
Simin Seyedpour, Saina Adiban Afkham, Aida Vahed,
Tommaso Dorigo, and Nima Rezaei

> *"AI is coming about and replacing routine jobs is pushing us to
> do what we should be doing anyway: the creation of more
> humanistic service jobs."*
>
> Kai-Fu Lee

Summary

In today's world, artificial intelligence (AI) is considered an inevitable part of
our life. AI in medicine has significantly evolved into a major player in the
diagnosis and treatment of diseases. However, this new phenomenon has faced

S. Behrouzieh · M. Keshavarz-Fathi · S. Seyedpour
School of Medicine, Tehran University of Medical Sciences, Tehran, Iran
e-mail: sadrabehrouzieh@gmail.com

S. Behrouzieh · M. Keshavarz-Fathi
Interest Group Department, Universal Scientific Education and Research Network (USERN),
Tehran, Iran

M. Keshavarz-Fathi
Cancer Immunology Project (CIP), Universal Scientific Education and Research Network
(USERN), Tehran, Iran

A. Vellido
Computer Science Department, Intelligent Data Science and Artificial Intelligence (IDEAI)
Research Center, Universitat Politècnica de Catalunya, Barcelona, Spain
e-mail: avellido@cs.upc.edu

S. Seyedpour
Nanomedicine Research Association (NRA), Universal Scientific Education and Research
Network (USERN), Tehran, Iran

criticism due to many ethical and moral challenges. To address these challenges, an understanding of ethical principles and various aspects of human communication is necessary. Humans are proven to affect each other by different means of communication, and this should be deliberated meticulously to study a human-AI relationship. A multi-dimensional comparison between humans and AI has demonstrated the influence of AI on many aspects, including autonomy, transparency, privacy, and the doctor-patient relationship. Finally, many people are concerned with the probability of the replacement of physicians by AI. This chapter will review the differences between humans and AI to discuss current challenges and future directions on ethical considerations of AI-based medicine.

S. A. Afkham
Innovation and Creativity Research Association for Transforming Education (I-CREATE), Universal Scientific Education and Research Network (USERN), Tehran, Iran

Network of Immunity in Infection, Malignancy and Autoimmunity (NIIMA), Universal Scientific Education and Research Network (USERN), Tehran, Iran

Biotechnology Department, Islamic Azad University, Tehran, Iran

A. Vahed
School of Pharmacy, Tehran University of Medical Sciences, Tehran, Iran

Health and Art (HEART), Universal Scientific Education and Research Network (USERN), Tehran, Iran

T. Dorigo
Istituto Nazionale Di Fisica Nucleare, Sezione Di Padova, Via F. Marzolo 8, 35131 Padova, Italy

Integrated Science Association (ISA), Universal Scientific Education and Research Network (USERN), Padova, Italy

N. Rezaei (✉)
Integrated Science Association (ISA), Universal Scientific Education and Research Network (USERN), Tehran, Iran
e-mail: rezaei_nima@tums.ac.ir

Research Center for Immunodeficiencies, Children's Medical Center, Tehran University of Medical Sciences, Tehran, Iran

Department of Immunology, School of Medicine, Tehran University of Medical Sciences, Tehran, Iran

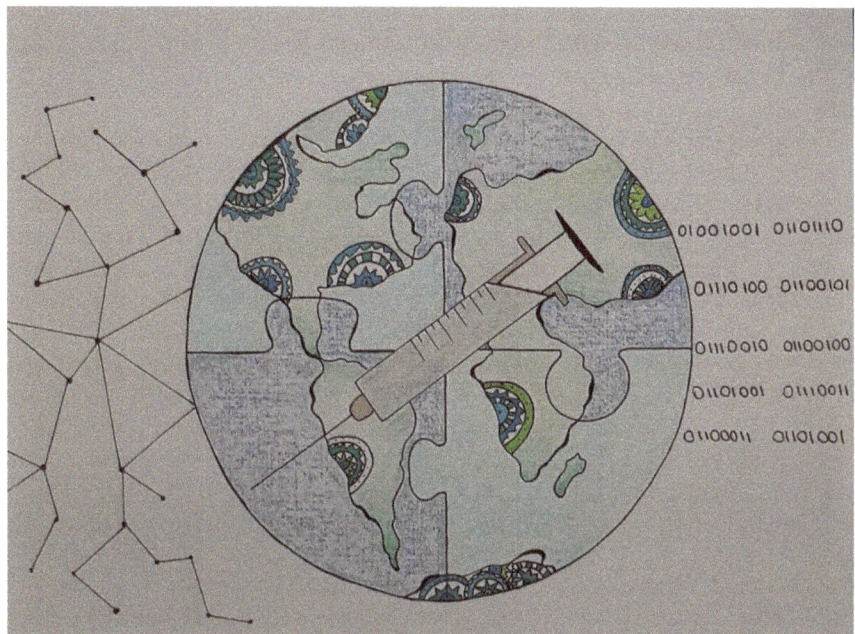

Ethical deliberation on AI-based medicine
[Adapted with permission from the Association of Science and Art (ASA),
Universal Scientific Education and Research Network (USERN); Made by Saina
Adiban Afkham].

The code of this chapter is *01101001 01,110,000 01,110,000 01,111,001*
01,110,010   01,110,100   01,101,111   01,001,111   01,110,101   01,101,110
01,110,100.

Keywords

Artificial intelligence · Deep learning · Doctor-patient relationship · Empathy ·
Healthcare · Human communication · Technology

1   Introduction

Humans have always been seeking a way to enhance the outcome of their efforts.
Since the emergence of artificial intelligence (AI) in the late 50s and with its strong
reemergence in the last two decades, various fields and professions have welcomed
its wide range of applications and started to increase their efficiency and productivity tremendously. Gradually, this technology has taken over certain tasks in each
profession. Therefore, several new questions, many of which are still debated, have
come to the fore. One of such domains, which is among the most crucial for human

life and has always followed innovations to contribute better, is medicine. Medicine and healthcare have benefited from AI in various ways [1, 2].

As AI has progressed and become commodified, it has increasingly stepped into new realms of communication and socialization, and this is exactly where the ethical questions have arisen. Medicine is part of the social contract and the way in which AI impacts it needs to be addressed from that viewpoint [3]. This discipline has four ethical principles at its core, namely autonomy, beneficence, non-maleficence, and justice, and some experts claim that AI violates them to some extent [4]. Human communication has always been an essential element of medicine, though some have admitted that since the emergence of AI, this communication has been overlooked qualitatively [5]. In addition, the defendants of human medicine have questioned the capability of AI to build an ethically proper bond between physicians and patients [6, 7]. Many similar claims have been made of late, but how relevant are they? Which side of the argument outweighs the other, if at all? Broadly speaking, does AI have enough potential to takeover humans' role in medicine anytime in future?

This chapter aimed to define the four basic principles of medical ethics, illuminate the concept of human communication, and assess the three models of doctor-patient communication within this concept. Furthermore, the differences between human and AI-based decision-making, and an outlook of situations where AI replaces humans, are among the subjects that will also be covered in this chapter.

2   Core Ethical Principles

Thomas L. Beauchamp and James F. Childress are two of the most recognized philosophers in medical ethics [4]. The four basic ethical principles introduced by these two philosophers are referred to as turning points in the history of medical ethics. These four principles include respect for autonomy, non-maleficence, beneficence, and justice, which account for a significant proportion of the moral challenges in today's medical practice [8]. It should be kept in mind that ethics are not a set of rules which take a piece of information and give the absolute answer. In fact, these principles are a common set of moral commitments and a common moral language between all, to be taken into consideration, and by means of the interrelations between principles generate a proper approach to the debated topic [4]. These ethical principles will be discussed in this section.

2.1   Respect for Autonomy

Autonomy means humans can make decisions based on deliberation, and it is a concept that can be divided into the autonomy of thought, of will/intention, and of action. The word "deliberation" means that autonomy is not defined for every person, as, for example, newborns clearly lack autonomy and need another person to decide on their behalf [4]. Autonomy requires us to obtain informed consent from patients before doing anything to help them. A way of showing respect to autonomy lies within the term "medical confidentiality," which, in short, means promising to keep the secrets of patients. Without such promises, patients are also far less likely to share their private and sensitive information that helps the physician to provide optimal care [8]. Good communication skills are necessary for respecting autonomy, and this mainly requires a good listening attitude. The doctor should carefully listen to patients, give them adequate information on the possible procedures, and finally, find out which intervention they prefer. However, some patients may prefer not to know about possible bad prognoses and simply let their physician decide on their behalf [9].

2.2 Non-maleficence "First Do not Harm" and Beneficence

The two principles, namely non-maleficence and beneficence, are discussed together as they are intertwined in practice. When doctors tend to help other people, they inevitably endanger them; therefore, they should avoid doing more harm than good. In other words, doctors expect themselves to contribute a net benefit when helping others. The term beneficence with non-maleficence discusses this expectation [8]. To produce a net benefit in the medical context, the first measure that comes to mind is the effective education and training of physicians. A way of choosing between multiple interventions and procedures is to analyze the empirical data related to past interventions and assess the overall benefit of each procedure [10]. Here, a combined definition called "empowerment," which means doing things to help patients be more in control of their healthcare should be introduced. This definition integrates autonomy and beneficence ending in the enhancement of the patients' autonomy [11]. Critics of these principles have suggested that autonomy may possibly conflict with the two latter ones. They argue that a patient may eventually decide to choose the procedure with a low net benefit, and the physician, according to autonomy, would be forbidden from changing their mind. Here, the only choice a physician has to explain all the possible consequences for the patient [9].

2.3 Justice: Distribute Health Resources Fairly

Aristotle pointed out many years ago that it is important to treat equals equally (what health economists call horizontal equity) and to treat the unequal unequally in proportion to the morally relevant inequalities (vertical equity) [12]. Justice can be summarized as the moral obligation to choose between competing claims on the basis of fair adjudication, and it is divided into three categories: distributive justice, rights-based justice, and legal justice [12]. A decent moral strategy for justice is first to distinguish whether it is an individual or an organization, profession, or society itself that has to make a decision. It is not a doctor's role to punish patients. For instance, withholding antibiotics from smokers who do not give up smoking is not a morally acceptable basis for rationing medical resources. Secondly, the cost is a moral issue at the center of distributive justice; therefore, if a cheaper drug is likely to produce as much benefit as a more expensive one, the physician should prescribe the cheaper one. Thirdly, the patients' rights are to be respected. For example, it is not morally acceptable to refuse to provide a sickness certificate for a patient whose lifestyle is disapproved by the physician [9].

3 Privacy and Confidentiality

Medical history is an essential piece of information received from patients for the physicians to support them better. Medical history includes sensitive information describing a patient's psychologic state, personal beliefs, high-risk behaviors, socioeconomic status, and other important but private information. The privacy of these data is certainly of great importance for the patient; therefore, the physician must be aware of potential invasion-of-privacy and breach-of-confidentiality lawsuits under state law for inappropriate use or disclosure of medical information [13]. Uncertainty about privacy and confidentiality damages the patient-physician relationship by weakening patients' willingness to communicate with the physician openly, and this, per se, disables physicians from helping patients [14]. As every medical student is familiar with, the Hippocratic oath obliges the physician to be cognizant of the confidentiality expected in the patient-physician communication: *"I will abstain [when treating patients] ...from whatever is deleterious and mischievous... Whatever, in connection with my professional practice, or not in connection with it, I may see or hear in the lives of men which ought not be spoken abroad I will not divulge, as reckoning that all such should be kept secret"* [13].

4 Human Communication

Human health has always been associated with their surroundings and environment. It is undeniable that the most important element within an individual's environment is the human society this individual lives in. Humans need communication and sympathy, and this need intensifies when they happen to suffer an illness or any kind of sorrow [15]. Therefore, in the medical and health-related context, physicians and nurses have always been expected to be the beacons of such sympathy [15]. There is evidence that patients who have experienced misconduct from health workers show less progress in defeating the illness when compared to patients with the same condition but benefit from appropriate attitudes of sympathy and healthy communication [16]. Several studies have proved that establishing a good relationship between patient and doctor helps the patient to communicate better and to provide more information about their history of illness willingly. As a result, the doctor can better assess the social and environmental factors associated with the patient's disease and deliver the optimal solution [17]. Overall, there are emotional aspects of human communication whose application in the medical context will be elaborated on in this section.

4.1 Emotions in Human Communication

Suffering from a disease puts great psychological pressure on patients and their families. The main psychological effect of a disease on patients and their families is emotional distress, which can further deteriorate the health status of patients if left undefeated [18]. Emotional distress is an unpleasant experience of emotional, social, or spiritual nature that hampers the ability to deal with the difficulties of the treatment process; it extends along a spectrum from common feelings like fear and sadness to deep depression and panic [18]. If left uncontrolled, it can generate psychological morbidity and also weaker biopsychosocial (e.g., fatigue and poorer response to treatment) and economic (e.g., due to longer stays at the hospital) outcomes [18]. Feelings of loneliness and isolation leave behind persistent sad memories, disturbing patients even long after recovery. Therefore, one of the most vital duties of the medical staff is to help patients and their families in overcoming this hardship [19]. Healthcare providers should keep in mind that this duty is challenging as well. When physicians attempt to inform about a disease or try to convey the unpleasant future consequences of it, they might unintentionally trigger severe effects on patients and their families if the necessary communication skills are absent. These effects range from depression to noncompliance with the treatment process [20]. Moreover, in the case of children, parents usually face a greater problem of acceptance, while children themselves face a greater problem of compliance, which may even be worsened by negative parental emotions [20].

Discussing physicians' obligations in establishing adequate communication with their patients, it is not always possible for them to alter patients' environment or their reaction to it, but at least, they can try their best to understand it [20]. An important point to be kept in mind is that the patients' stories about their diseases are not necessarily simple representations of underlying facts, yet they are psychologically crucial structures from the patients' inner worlds [21].

Several studies have proved that opening up emotions and providing empathic responses both have a direct relation with positive outcomes in terms of distress reduction [22], patient coherence [23], and symptom resolution [24]. Speaking of the doctor-patient relationship and moments of shared affection are called "connections" where the doctor feels an intense emotion and sense of sharing with the patient [25]. To have patients better adapted to their condition after diagnosis of a disease, especially a chronic one, physicians should help get them through the following stages of Maslow's hierarchy of needs:

i. The first stage, it includes the basic physiological needs vital for every human being;

ii. The second stage, it is the need for safety and feeling safe. In this stage, the medical staff should know how to make the patient feel safe, cared for, and not threatened. Scare tactics would probably exacerbate the patient's condition in this stage;

iii. The third stage, it is the need for belonging and love, which helps the patients regain confidence after they have rejected themselves as defective and lost hope in the treatment process;

iv. The fourth stage, it is known as "self-esteem and esteem from others;" and

v. The fifth stage to pass through is called "self-actualization," or simply put, the urge to grow. In this stage, the physician should support the patient in gaining the power to make life decisions [20].

To help patients more efficiently, health workers should be capable of understanding them better, and this requires them to be educated adequately and properly in terms of communication skills and capacities. Besides, the more they interact with patients, the more experience they would gain, and this will result in better performance over time [15].

The essential elements in human understanding, which are also of remarkable importance in the doctor-patient relationship, are illuminated below [26].

4.1.1 Pity

The most ancient word used to describe the feeling of pain elicited from seeing others suffer, and prompting a desire for its relief, is "pity." This concept ranges from contempt to compassion and emotional understanding. When people pity another person, they simply observe the person's suffering and feel sorry for him or her, but also tend to keep themselves and the other as totally separate individuals. Thus, they never see or feel the other person from their own perspective [26]. In other words, pity does not encompass the concept of "identification" [27]. Identification will be explained more in the next element of understanding.

4.1.2 Sympathy

The next level of understanding which is considered to be deeper than pity is sympathy. The root of the word "sympathy," consists of "sym-" meaning same or equal, and "-pathy" meaning pain and suffering, and as a whole, it means to share feelings [26]. While pity may seem a defense against identification, sympathy is identification [27]. The distinction between self and other is much more vibrant in pity than in sympathy [26]. Identification, in particular, means placing one's own self in the life of another self to observe their life history from their own perspective. In sympathy, identification may even reach the point of losing personal identity in fusion with the other [27]. A key point in the concept of identification in sympathy is that the sympathizing self, namely the physician, should hold enough self-awareness to remain sufficiently dissociated and strong in order to be able to help the other self (patient) to overcome the pain and suffering [27].

4.1.3 Empathy

The most complete concept for understanding is called "empathy." Empathy encompasses sharing and understanding one's emotions and thoughts, showing respect and providing support, which helps us better explore their feelings, and finally showing our concern [18, 28]. Simply put, when we empathize with others, we feel *for* them and *with* them [29]. In the medical context, in particular, empathy is often taught on the foundation of behaviorally-based micro-skills of listening and responding [15], which subsequently gives rise to the term "therapeutic empathy," which divides into cognitive (recognition), affective (sharing feels), and behavioral (expressing) empathy [30].

Patients always expect physicians to notice their endeavor complying with the instructions and orders. This expectation, if ever met, increases patients' confidence and trust in the treatment process. On the other hand, severe consequences like shame and anger may follow if this expectation is left unfulfilled [16, 21, 31].

4.2 Models of Doctor-Patient Communication

The term "therapeutic alliance" refers to a personal bond between the patient and the doctor toward achieving desired treatment goals, in which there is trust and patient involvement at its best. Achieving this personal bond requires an adequate doctor-patient relationship [30]. In an ideal doctor-patient relationship both sides have a series of duties to help maintain its strength. Physicians spend time, skills, and affection, and in return, the patient responds with compliance or verbal praise to other patients [32]. Empathic physicians devote their emotional labor to surface and deep acting, empowering the doctor-patient relationship, and reducing job burnout. Surface and deep acting are respectively defined as emotional expression and regulation of emotions [33, 34]. To best achieve the goal of this relationship, the physicians should be taught communication skills and knowledge [18, 35]. Using these communication skills becomes more significant, when it comes to sensitive topics like end-of-life care or breaking bad news [36]. Moreover, in the context of teaching doctor-patient relationships, interactional and integrational principles are of great importance [17]. Psychosomatic medicine that focuses mainly on psychosocial factors to improve the treatment process, if ever approved as a clinical approach, should concentrate on the doctor-patient relationship [17].

In 1995, Emanuel and Dubler indicated the doctor-patient relationship to comprise 6 C's: choice, competence, communication, compassion, continuity, and (no) conflict of interests [37]. In addition, earlier in 1956, Szasz and Hollender had introduced three basic models for the doctor-patient relationship: activity-passivity, guidance-cooperation, and mutual participation. The ultimate goal in developing such models was to decrease physician dominancy and increase patient control, trust, and satisfaction subsequently [38].

A more extensive and complex model for the doctor-patient relationship were presented by Dean and Street in 2014. This model covers three distinct stages in this order: 1 recognition, 2 exploration, and 3 therapeutic action. It is worth mentioning that the two former stages also have treatment values indirectly [18]. Besides, cognitive and communicative strategies aid the physician to pass through each of these stages successfully [18].

4.2.1 Recognition

In the first stage, patients expect physicians to understand their emotional distress, and physicians should make sure that patients feel heard, and this expectation is achieved [18]. Most of the time, patients hide their emotions for several reasons, such as embarrassment, not wanting to bother the doctor, or thinking that these problems are not within the boundaries of the physician's duty [39–42]. Given the circumstances, physicians have an important responsibility during both medical training and professional life. They are required to get sufficient education and experience on how to deal with patients' concerns [18]. Physicians should listen attentively and actively to the concerns to detect distresses more efficiently [43]. Moreover, a proper environment helps provide patients with opportunities and privacy to increase disclosure and expression of emotional problems [18, 44].

4.2.2 Exploration

In the second stage, doctors should let patients feel free to talk about anything they are concerned about, and then, by asking open-ended questions and clarifying emotions, try to find where the difficulties derived from [18]. Validating emotions and providing empathy are within the borders of this stage. As mentioned before, this stage has therapeutic value since it helps the patient talk about feelings freely [18].

4.2.3 Therapeutic action

The third and final stage encompasses the main therapeutic engagements of physicians. Here, physicians should make patients and their families "feel heard and known." They should present a detailed explanation about the treatment process with a proper manner of talking [18].

What is essential in all of these three stages are high emotional intelligence (EI). EI is one of the six domains a physician is expected to acquire, consisting of cognitive abilities, technical skills, integrative abilities, relationship skills, and habits of mind [45]. It originally has its roots in social intelligence, understanding and managing others, and showing adaptive behavior [32]. In 1997, Mayer and Salovey [46] introduced a four-dimensional definition of EI, involving i, appraisal and expression of emotion in oneself; ii, appraisal and recognition of emotion in others; iii, regulation of emotion in one self; and iv, the use of emotion to alleviate performance.

However, in the doctor-patient relationship, there may be a series of barriers, making it complicated for each side to communicate. The term "communication dynamics" refers to the personal differences in establishing a relationship. Different patients may show different preferences in a doctor-patient relationship and vice versa. In general, various activation patterns have been observed in both patients' and physicians' emotional responses, though only when these activations have been simultaneous has the communication shown significant progress [44].

5 AI and Medicine

Since its emergence in the late 50s, AI has undergone periods of enthusiasm and relative obscurity but, throughout medicine and healthcare, has been one of the domains to which it has paid more consistent attention, from early expert systems to the strong reemergence in the last two decades in the form of deep learning, an advanced form of machine learning. AI has, of late, attracted remarkable attention in society at large because it has become commodified and internalized as one of the main drivers of the large ICT companies of today. AI has always contributed significantly to the medical and healthcare domains, but the current interest of the private sector to aggressively introduce AI in these domains becomes a social concern that involves ethical issues at its core. It has been reported that, as in 2016, the biggest portion of investment in AI research could be found in healthcare applications [47]. Note that AI is often considered as an *umbrella* concept that encompasses areas such as machine learning, computational intelligence, or soft computing, to name a few [48].

In the context of medicine, one way to characterize AI is according to the following two subtypes [47]:

i. Algorithmic subtype: They just concern the use of algorithms that often represent data-based models and could be seen as the basis for data science or data mining. This subtype covers almost any application in medicine that has data as its basis. An increasingly popular example of this is AI for knowledge extraction from electronic health record (EHR) systems. Another example, which gets us closer to the next subtype, is the area of mobile health (m-health) as the "data provider" for AI-related algorithms. This includes the use of "intelligent assistants" such as Siri, Alexa, Bixby, and Cortana, to name a few, which allow users to access some aspects of primary care earlier and easier by responding to mental and physical health questions and are analyzed by healthcare conversational projects; and

ii. Physical subtype: Typical examples of this subtype are, for instance, AI-assisted robots assisting in or allowing performing telesurgery or its use for assisted living and palliative care, including intelligent prostheses for the care of the elderly and handicapped.

For the particular area of medical diagnosis, which is, arguably, the one to which more efforts have been put in the application of AI, many characterizations have been proposed. As an example, one of these outlines two different approaches [47]: flowchart-based approach and database approach. This flowchart-based method is based on replicating or imitating the human physician's diagnostic decision-making process and involves collecting heterogeneous data to generate an informed diagnosis. However, this method requires the AI to be fed with huge data, including symptoms and treatments. Besides, the outcome would most likely be narrower than what is expected because AI is not able to observe and gather cues that may easily be detected by a physician during a patient interview. Nonetheless, this approach attempts to approximate the physicians' need to comply with set medical guidelines and protocols and therefore may have lower barriers to acceptance in medical practice. In the database approach, AI-based models are taught (literally, they *learn*) via algorithms to perform pattern recognition in different forms. This knowledge-extraction-from-patterns method is known to be much more efficient in comparison with the former in terms of accuracy of performance, yet it faces challenges of its own which will be illuminated later in this text.

Ultimately, the goal of AI-based methods is to develop learning algorithms that use the available to extract patterns and hopefully, useful and novel knowledge that can be used for decision-making, conveying context-based information to others. To achieve this goal, AI is considered to possess several characteristics, which are as follows [48]:

- Understand and interpret data and context;
- Evaluate hazards and opportunities in the situation;
- Design plans and courses of action;
- Learn and adapt; and
- Share knowledge with other humans or AIs.

At the moment, AI has more than a few applications in healthcare, such as medical diagnostics and prognostics, patient monitoring, and learning healthcare systems, just to name a few. Among the numerous reasons for the extensive use of AI in medicine, we can mention knowledge extraction automation, the ability to analyze big data, and the provision of an additional layer in clinical decision-making [49].

The various roles for humans and AI in different control tasks, which can be either human-led or AI-led [48], are guidance, control, collaboration, fixing/mechanic, and modulation/by-standing. Accordingly, surgeon guidance AI system can be seen as a good example for a human-led guidance role, while the smart artificial limb is a well-known example for AI-led control role, where AI controls and human senses [48].

Considering an integrated approach to utilizing AI in medicine, Braun et al. have described three models of AI interaction in the healthcare system: the conventional AI-DSS (C-AI) systems, integrative AI-DSS (I-AI), and fully automated AI-DSS (F-AI). In the conventional model, the human feeds the AI with patients' and health-related data. The AI then produces outcomes that are transferred to the physician as recommendations. While in the integrative model, the AI can automatically gather information, make decisions, and share the output with the physician or add it to the patients' EHR. AI is not used as a supplementary recommendation in the fully automated model but as an alternative replacement for human-based design making [50].

5.1 Human Versus AI

The "Da Vinci" robotic surgical system is a prominent example of AI application in today's medical practice. This novel technology is widely used for its high precision and 3D magnified view, especially in urological and gynecological surgeries. It mimics a surgeon's hand movements causing minimum pain and blood loss and is considered a minimally invasive method [2, 51]. As an earlier example, in 1986, a decision support system called DXplain was developed by the University of Massachusetts, which, according to the symptoms, generated a list of probable diagnoses and could also be used for training medical students [47]. Further, at the University of Washington, a system known as GermWatcher was developed to trace hospital-acquired infections [47]. Additionally, in the critically technology-reliant field of radiology, new AI systems have been brought into use, which with the help of deep learning technology and studying several normal examples, gain the ability to detect histologic anomalies and malignancies, classify them in pre-defined stage categories, and eventually monitor the process of treatment. These systems have enabled radiologists to save a significant amount of time by helping them to detect pathologic lesions faster than ever [52, 53].

Healthcare-related examples of progress in AI as applied to medicine such as these, or countless others, have generated debate among practitioners about the future of these technologies and even the possibility of humans being eventually replaced by AI-based machines in medical tasks. Comparison between diverse human and AI characteristics may provide some answers to this argument [5, 6] (Fig. 1).

5.1.1 Emotions
To assess and compare emotions in humans and AI, the origins and basics of emotions in each are overviewed below.

Human
Only humans have the ability to intentionally "feel for" and act on behalf of others, although every individual experiences differently from others. Many human actions are directed toward or in response to others [54]. Unlike other primates, humans can express and convey their feelings with words. By establishing communication, they tend to share emotional experiences and develop empathy [55]. From the very time of birth, neonates can comprehend another's affective state and respond to it (intersubjective sympathy). The reactive crying of a newborn in response to the crying sound from other neonates was demonstrated as an example for intersubjective sympathy by Simner, M. L. in 1971 [56]. Rather than empathy, it is the sense of self-other overlap between two individuals that elicits the urge to help [54]. According to the psychoanalyst Theodor Reik (1949), human empathy involves four separate processes in the following order [54]:

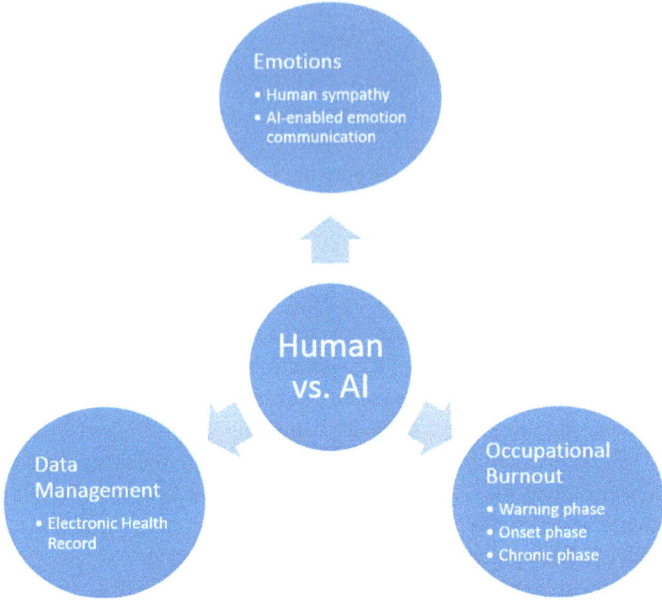

Major differences between humans and AI systems

   i. *Identification,* focusing on another person's contemplation and feel what they feel;

  ii. *Incorporation,* internalizing the other's experience and understanding what they feel;

 iii. *Reverberation,* experiencing the other's experience while giving cognitive response to it; and

 iv. *Detachment,* going back to the separate self to be able to express understanding and empathy

In addition, three major components of human empathy are sharing experiences between individuals, self-other awareness while going through identification, and mental flexibility to adopt the other's perception and emotion regulation. Human empathy is distinguished from that of the other mammals by the two latter components [54]. During an emotional encounter, areas of the brain associated with emotion processing, such as the ventromedial prefrontal cortex and amygdala, show higher activity. Besides, viewing facial expressions of pain engage these areas in the same way as the direct experience of pain. These are examples of the contagion effect of emotions between individuals [44, 57, 58].

Artificial intelligence

AI-enabled emotion communication had great improvement in the last decade. In the present-day model of this technology, several layers have been developed for receiving, analyzing, and storing emotional data to give proper responses. These layers are in the following order [59]:

i. *Data input:* In this layer, physiological data are gathered from the patient in various formats such as eye-tracking, facial expression, voice, heart rate, skin temperature, and so on;

ii. *Edge cloud:* In the next layer, the collected data are labeled and categorized to be sent to the following layer;

iii. *Remote cloud:* In this layer, the vital processes of computational analysis, cognition, and storage take place. In other terms, the AI *understands* the data in this layer; and

iv. *Feedback and decision control:* In this final layer, the system provides the user with personalized intelligence and humanized emotional feedback.

One of the most important achievements in this field is the system's ability to complement information about a particular patient by collecting new data during each interview and increasing the efficiency of the communication through time [59]. To elaborate more on the improvements made in this field, here are some of the recent works. Zhao et al. presented a method that inferred patients' emotions with radio frequency (RF) signals, extracted their heart rate data, and designed an algorithm to detect emotions [60]. Atkinson et al. proposed a method to recognize emotions based on EEG waves that worked on the basis of the brain-computer interface [61]. Hossain et al. proposed an emotion recognition system with enhanced auditory and visual data recognition accuracy using convolutional neural networks (CNN) and extreme learning machines (ELM) [62]. Trigeorgis et al. developed AI context-awareness by combining a CNN with a long short-term memory (LSTM) network [63].

5.1.2 Occupational burnout

Due to the nature of their jobs, healthcare workers are constantly exposed to others' suffering, fear, death, etc. [64]; therefore, they are vulnerable to professional burnout, a disturbing consequence to which AI systems are resistant. As stated by Schaufeli and Enzmann [65], this phenomenon is a defective cycle that keeps getting worse if physicians ignore its early symptoms. Occupational burnout encompasses different components and symptoms. Its components are listed below [66]:

- Emotional exhaustion due to depersonalization, which triggers the whole process

- Empathizing with traumatized patients leaves serious mental impacts on physicians, called "compassion fatigue" or "vicarious traumatization," which requires a noticeable amount of time to recover from [66].

Demanding attitudes of patients, involving anger in extreme cases [16]
Psychosocial fatigue

Generally, occupational burnout consists of three phases [66]:

i. Warning phase, it involves headaches and feelings of irritation, though there is no need for recovery;
ii. Onset phase, symptoms begin to last longer, quality of work declines, longer resting is needed; and
iii. Chronic phase, physical symptoms appear. Therapy is necessary to overcome this phase.

5.1.3 Data management

One of the main problems healthcare workers has always tried to cope with is the increasing amount of data related to patients, symptoms, and treatments available to them, now for instance in the form of EHR, which may sometimes exceed their capability and elicit further occupational burnout. Physicians, since they are humans, have limited capacity for handling large amounts of data and may, for instance, forget about the side effects of a specific medication or rare symptoms of a particular disease [2].

On the contrary, AI systems have the ability to search through large amounts of multi-modal (image, text, signal, and others) data, or through thousands of pages of medical articles in only a fraction of time, and come up with a decision, novel knowledge, or the latest information about any disease or treatment method. Moreover, these systems usually never forget (although catastrophic forgetting is one of the concerns of the newest deep learning-related developments); thus, they are assumed to be much better than humans in terms of data management [2].

Despite the aforementioned issues, there is a major problem with AI in this circumstance. The problem is that to be able to recognize a particular symptom; computers should be taught and exposed to many more instances than a human being usually needs. For example, a child may learn to recognize and categorize an object by seeing only two or three of a kind. However, present-day AI systems usually require studying thousands of images of such an object to finally recognize it with reasonable accuracy. This problem indicates that the learning processes of a human and an AI are different and that to train an AI to diagnose with the same effectiveness as a physician, a significant amount of time, data, and energy is required [1].

6 Challenges of AI

As a newly emerged technology, AI has always had proponents and opponents and has faced criticism in different stages and circumstances. According to Gartner hype cycle, the public's consciousness of and feedback to the emergence of any technology, e.g., AI, consists of five stages in the following order [7]:

   i. *The innovation trigger,* the moment a new concept appears;
  ii. *The peak of inflated expectations,* as the media coverage increases, its prominence escalates, and more people get to know about the new technology;
 iii. *The trough of disillusionment,* some of the boldest predictions soon turn out to be exaggerated, and the public starts to get disappointed; and
 iv. *The slope of enlightenment,* the new technology gradually gains real traction, and its trials begin to pay off.
  v. *The plateau of productivity:* The technology becomes an accepted part of current practice

With all the above in mind, there are numerous challenges that arise particularly from AI itself, some of which will be discussed in this section (Fig. 2).

6.1   Autonomy and Trust Issues

Mayer et al. defined "trust" as a condition in which a person, called trustor, puts him or herself *vulnerable* to the actions of another, namely trustee, regarding the expectation he or she has, that the trustee will act as requested by the trustor, even if no supervision is applied [67]. Vulnerability can be broken down into three dimensions [67]: *capability,* level of skills (automation); *opportunity,* level of

Fig. 2 Ethical concerns of AI in medicine

freedom (autonomy); *intent,* variable according to what AI learns from the environment. For the human-AI relationship, to align the intents of AI and humans, two distinct dimensions can be considered: automation and autonomy

If automation outweighs autonomy, a condition called "under reliance," the system is more capable than the opportunity given, leaving unwanted costs. On the other hand, if autonomy overshadows automation, a condition called "over reliance," the system is demanded to do more than it can; thus, disappointment and distrust occur. In terms of intent, mistrust, and disappointment occur if human and AI intents are not aligned [68]. Since, in most cases, the course of the disease is unpredictable, to reach a reasonable decision that meets the doctor's scientific and the patient's value-based expectations, it is necessary to engage in a dynamic dialog. In a medical setting where AI assists or replaces a doctor, it is crucial to ensure that the system will generate a decision to the best of its capacity. Overcoming this, challenge requires meticulous deliberation on machine learning and data analysis [69].

6.2   Ethics, Beneficence, and Non-maleficence

The concept of ethics is central to human life in society, and human intelligence has it as one of its systems of "checks and balances." This was never part of the founding ideas of AI, and ethics have only recently come at the forefront of the AI debate due precisely to the increasing societal impact of these technologies [3]. It could be argued that AI is yet quite unprepared to address ethical issues [70], but it is also true that some of these efforts have recently borne fruit in the form of ethical guidelines for the use of AI, such as the "Ethics Guidelines for Trustworthy AI" written by the European Union High-Level Expert Group on Artificial Intelligence [71], which state that "Trustworthy AI [...] should respect fundamental rights, applicable regulation and core principles and values, ensuring an ethical purpose." Unfortunately, these general guidelines do not cater for specific domains such as medicine and healthcare, for which a clear roadmap for the ethical use of AI and ML that involves players both from both medicine and AI is yet to be defined in practice. Another European project, *AI4People,* has provided guidelines in this topic [72] that, interestingly, even if not referring to medicine, still use the general healthcare-related concepts of beneficence, non-maleficence, autonomy, and justice.

The link between AI and the concepts of beneficence and non-maleficence in medicine is quite straightforward. Results derived from cohort studies are hardly ever decisive and often contain significant amounts of erroneous and uncertain data. Therefore, there is always a risk of jeopardizing patients' health, which conflicts with these two core ethical principles: beneficence and non-maleficence. However, this risk exists for both the humans and AI systems, and the main reason is that AI systems are modeling data that have been gathered and processed by humans and therefore include human bias that will unavoidably feed the models they generate and according to which patients' data are converted in potential medical decisions. In these circumstances, medical experts who are supposed to deal with AI systems should be aware of AI models' limitations and be willing to feed their expertise into the AI models so that these models become more similar to doctors' decision-making by taking into account the personal condition of a patient, which help develop models with more accuracy and effectiveness [69, 73].

6.3   Transparency and Accountability

One of the most commonly quoted limitations of AI-based data models is the one referring to their usual lack of interpretability and explainability, especially for complex nonlinear models. This is commonly referred to as the "black box syndrome." In an area such as medicine, the lack of interpretability and explainability (or transparency) of those models that aim to assist practitioners in their decision-making will limit the chances of adoption, in real practice, of any AI-relying systems [74]. Are patients in their right to know how their diagnoses, prognoses, or courses of treatment are modeled by an AI-based decision support system? Unfortunately, the answer is that it depends on the locally applied legislation. For instance, the European Union directive for General Data Protection Regulation (GDPR), from 2018, mandates a "right to explanation" of all decisions made by automated or artificially intelligent algorithmic systems [75], which involves providing requesting citizens with "meaningful information about the logic involved, as well as the significance and the envisaged consequences of such processing [automated decision-making] for the data subject." This is meant to have a direct impact on medicine, as medical institutions might not be willing to risk using an AI-based system that is not transparent enough, in order to avoid litigation costs. Here, several ethical questions come to mind [1, 47]:

- What if there is a discrepancy between the physician's and AI's decisions?

- This problem intensifies when a physician feels too experienced to obey the AI. On the contrary, medical students should be educated well enough to take responsibility and avoid overreliance.

  How much information does the physician require to know how an AI-based system works?

- This condition is sometimes compared with that in which a physician uses a given complex imaging technology (X-rays, MRI, PET, etc.). A physician is not necessarily aware of its mechanism but knows the instructions on how to use it and how to interpret its results. It is this latter part about which the physician should enquire in the case of an AI-based system.

  How much information is the patient allowed to have on how the decision is made?

- According to informed consent, a patient only requires disclosure of materially relevant information. Simply put, information impacts the patient's decisions.

This is not necessarily that clear-cut in case the physician has relied on an AI-based system to make an informed decision for the legal reason described in the previous paragraphs.

Who is in charge of the physician ignoring AI's decision?

– Regarding the law, the physician should be questioned if they mistakenly disregard what AI recommends, but, again, legislation is far from mature in this area, given the often unclearly defined figure of "data controller" vs "data analyst" in a medical context.

If eventually, an AI-based system is put in place to make decisions and autonomously take over certain medical procedures, a series of questions should be answered about who is in charge and who is legally liable. An ideal AI-based system, in these contexts, requires clarification of the following roles [76]: creator; distributor; owner; healthcare provider; supervisor; physician; regulating body; designated nationally; and governance infrastructure, to ensure accountability.

6.4   Data Privacy

At a time when the medicine is increasingly becoming data-centered and patients' data are managed through networked EHR, it comes as no surprise that concerns about privacy and data anonymization are on the rise. These concerns are further motivated by the ruthless incursions of big ICT companies in the health sector. This is exemplified by a recent scandal concerning the transference of a large body of patients' data from several hospitals in London, U.K., to a private corporation without due diligence in terms of patients' privacy preservation. Such data were merely meant to be used to test results for signs of acute kidney injuries through AI models. This scandal motivated a scathing letter from the Royal Statistical Society's executive director in Nature [77], in which three recommendations were distilled:

  i. due to society's increasing data trust deficit, data transference transparency and openness should be guaranteed;
 ii. data transference should be proportional to the medical task at hand; and
iii. governance mechanisms of data control should be strengthened or created when unavailable.

To make the best use out of AI, people should be informed about how their data are used to avoid challenges related to ethical concerns and data privacy. Privacy policies and relevant ethical standards should keep pace with the growing progress in medical AI. Besides, the government must promote legal frameworks such as the Health Insurance Portability and Accountability Act (HIPAA) [2].

6.5 Doctor-Patient Relationship

Although the AI industry has made plenty of progress recently, these technologies are still too immature to effectively communicate with humans in the way people do with each other. Therefore, a long-standing challenge for designers and programmers is to build an AI that communicates in ways that are indistinguishable from those of humans. However, the Chinese room argument discusses this challenge and provides a hypothesis that strong and weak AI systems are not different in understanding; an AI system is similar to a person who does not know the Chinese language but can use algorithms/instructions to translate a text from Chinese into English, and successfully pass the turning test, while does not have a mind to understand neither the question nor the response [78].

Another challenge has always been to define the extent to which a physician can depend on an AI-based system. The first level, diagnosis, mainly involves statistical data and calculations in which AI may often, but not necessarily, prevail over a human. After diagnosis, comes the treatment. The treatment process requires consultation and critical thinking, for which AI has not proved yet to be efficient. Humans are still superior to computers in terms of making value judgments and understanding social context. Thus, expert comments support the continuance of the doctor-patient relationship [49]. It has been argued that, in the end, an AI-based medical system deconstructs human beings to a set of medical concepts, thus failing to understand the true meaning of medical decisions [5]. In addition, Hubert Dreyfus's critique on AI claims that socialization and somatic experience are necessary for a clinical approach. As a result, just like a medical student who has only acquired context-free abstract concepts of medical practice and has almost no clinical experience, AI-based systems cannot fully replace humans in patient care and clinical context [5].

AI examining systems may distance the doctor from the patient. It can neither be responsible for a specific patient's preferences nor leave room for the physician to interpret the guidelines. Replacement of humans by AI systems is likely to transform many aspects of primary care into algorithmic models and consequently, personal differences between patients risk being overlooked, in a way that, for instance, all patients with similar symptoms will be classified in the same treatment plan [5]. Doctors at the bedside have the ability to interact with appropriate conversational expressions of empathy such as facial expressions or supportive touches on the hand to gain the patients' trust, yet, the AI system does not embody empathy. Besides, a human patient is unlikely to feel empathy for a computer system, even if humanized. After all, the main reason most patients tend to visit a physician is to feel empathized, understood, and reassured [5, 6]. Remarkably, to make decisions in a terminal condition, a human physician is preferred over an AI system because it involves not only clinical factors but also moral concerns for which the latter cannot be responsible [49]. Finally, physicians' duties include communication of findings, quality improvement, education, policymaking, etc., which AI systems are not yet capable of performing. Machines may excel at accuracy and performance but yet cannot care [7].

7   Conclusion

Like any other technology or radical technical advance, AI has encountered criticism and numerous challenges since its appearance. The use of AI in medicine, however, requires thoughtful deliberation on its ethical and moral aspects since it deals with human lives [7]; some of these ethical concerns overlap with yet unresolved issues involving governance, legislation, and privacy. Human communication makes up a core part of the treatment process since humans are proven to affect each other by means of interaction [26]. For an AI to contribute to this process alongside the physician, it requires sufficient social context understanding, empathizing, etc. Therefore, a roadmap to make AI compatible with such human communication in the context of medical practice is an imperative need. Researchers are still seeking ways for AI systems to become as interpretable, explainable, and transparent as possible in ways that would allow medical practitioners to use them at the point of care [48]. The presumption that physicians may be replaced with AI systems appears to be fragile, according to the existing evidence and statements made by experts in this field [5].

Over the years, AI has helped doctors remarkably in terms of accuracy, time effectiveness, etc.; therefore, in future, this technology is likely to grow to become an integral part of medicine, more likely in areas that comply with specific patterns and are amenable to partial or full automation. Yet, since AI is not expected to be able to take over humans' role in medicine completely, the new generation of medical students is to be trained to familiarize with the opportunities and limitations of AI and with ways to use this technology in the most efficient way possible to overcome the numerous challenges it involves. Regarding the four basic ethical principles and the relevant challenges, it is obvious that AI has to be deployed in a way to face the least possible conflict with them because these principles are fixed at the core of medicine. Undoubtedly, a major responsibility in ensuring compliance in this process lies in the hands of government and policymakers. The role of AI in future medicine is open to discussion, but this technology has reached this domain to stay; therefore, it is also researchers' responsibility to make sure that society as a whole keeps pace with it.

Core Messages

- The four basic ethical principles are at the core of today's medical practice.
- Ethical and human medicine are based on human communication, specifically the doctor-patient relationship.
- AI is growing to become part of medical practice, especially in areas that can seamlessly be integrated with it.
- Meticulous deliberation and law implementation are required to overcome challenges to integrating AI in real practice.
- AI's potential advantages appear to outweigh the disadvantages and support its promising future in medicine.

References

1. Thrall JH et al (2018) Artificial intelligence and machine learning in radiology: opportunities, challenges, pitfalls, and criteria for success. J Am Coll Radiol 15(3 Pt B): 504–508
2. Ahuja AS (2019) The impact of artificial intelligence in medicine on the future role of the physician. PeerJ 7:e7702
3. Vellido A (2019) Societal issues concerning the application of artificial intelligence in medicine. Kidney Dis (Basel) 5(1):11–17
4. Gillon R (1994) Medical ethics: four principles plus attention to scope. BMJ 309(6948):184–188
5. Karches KE (2018) Against the iDoctor: why artificial intelligence should not replace physician judgment. Theor Med Bioeth 39(2):91–110
6. Krittanawong C (2018) The rise of artificial intelligence and the uncertain future for physicians. Eur J Intern Med 48:e13–e14
7. Pesapane F et al (2020) Myths and facts about artificial intelligence: why machine- and deep-learning will not replace interventional radiologists. Med Oncol 37(5):40
8. Gillon R (2003) Ethics needs principles–four can encompass the rest–and respect for autonomy should be "first among equals." J Med Ethics 29(5):307–312
9. Gillon R (2015) Defending the four principles approach as a good basis for good medical practice and therefore for good medical ethics. J Med Ethics 41(1):111–116
10. Gillon R, Higgs R (2015) What is it to do good medical ethics? A kaleidoscope of views. J Med Ethics 41(1):1–4
11. Gillon R (1986) Do doctors owe a special duty of beneficence to their patients? J Med Ethics 12(4):171–173
12. Ruger JP (2015) Good medical ethics, justice and provincial globalism. J Med Ethics 41 (1):103–106
13. Liang BA (2002) Medical information, confidentiality, and privacy. Hematol Oncol Clin North Am 16(6):1433–1447
14. Gutierrez AM et al (2020) A right to privacy and confidentiality: ethical medical care for patients in United States immigration detention. J Law Med Ethics 48(1):161–168
15. Williams J, Stickley T (2010) Empathy and nurse education. Nurse Educ Today 30(8):752–755
16. Philip J et al (2007) Anger in palliative care: a clinical approach. Intern Med J 37(1):49–55
17. Schuffel W (1975) The doctor-patient relationship in the practice of medicine. Int J Psychiatry Med 6(1–2):183–193
18. Dean M, Street RL Jr (2014) A 3-stage model of patient-centered communication for addressing cancer patients' emotional distress. Patient Educ Couns 94(2):143–148
19. Howe EG (2016) Harmful emotional responses that patients and physicians may have when their values conflict. J Clin Ethics 27(3):187–200
20. Biermann J, Toohey B (1980) Emotional aspects: the patient's view. Diabetes Educ 6(4):16–19
21. Lucius-Hoene G et al (2012) Doctors' voices in patients' narratives: coping with emotions in storytelling. Chronic Illn 8(3):163–175
22. Duric V et al (2003) Reducing psychological distress in a genetic counseling consultation for breast cancer. J Genet Couns 12(3):243–264
23. Squier RW (1990) A model of empathic understanding and adherence to treatment regimens in practitioner-patient relationships. Soc Sci Med 30(3):325–339
24. Hojat M et al (2011) Physicians' empathy and clinical outcomes for diabetic patients. Acad Med 86(3):359–364
25. Matthews DA, Suchman AL, Branch WT Jr (1993) Making "connexions": enhancing the therapeutic potential of patient-clinician relationships. Ann Intern Med 118(12):973–977
26. Wilmer HA (1968) The doctor-patient relationship and the issues of pity, sympathy and empathy. Br J Med Psychol 41(3):243–248
27. Wilmer HA (1956) Rehabilitation: being is belonging. J Chronic Dis 4(2):212–215

28. Pollak KI et al (2010) Do patient attributes predict oncologist empathic responses and patient perceptions of empathy? Support Care Cancer 18(11):1405–1411
29. Wilmer HA (1957) Empathy and sensibility of heart. N Y State J Med 57(14):2410–2413
30. Schnur JB, Montgomery GH (2010) A systematic review of therapeutic alliance, group cohesion, empathy, and goal consensus/collaboration in psychotherapeutic interventions in cancer: uncommon factors? Clin Psychol Rev 30(2):238–247
31. Montello M (2003) Reading experience: Jodi Halpern's from detached concern to empathy. J Clin Ethics 14(4):286–289
32. Weng HC (2008) Does the physician's emotional intelligence matter? Impacts of the physician's emotional intelligence on the trust, patient-physician relationship, and satisfaction. Health Care Manage Rev 33(4):280–288
33. Larson EB, Yao X (2005) Clinical empathy as emotional labor in the patient-physician relationship. JAMA 293(9):1100–1106
34. Johnson HA, Spector PE (2007) Service with a smile: do emotional intelligence, gender, and autonomy moderate the emotional labor process? J Occup Health Psychol 12(4):319–333
35. Gallagher TH, Levinson W (2004) A prescription for protecting the doctor-patient relationship. Am J Manag Care 10(2 Pt 1):61–68
36. Hanson LC, Tulsky JA, Danis M (1997) Can clinical interventions change care at the end of life? Ann Intern Med 126(5):381–388
37. Emanuel EJ, Dubler NN (1995) Preserving the physician-patient relationship in the era of managed care. JAMA 273(4):323–329
38. Szasz TS, Hollender MH (1956) A contribution to the philosophy of medicine; the basic models of the doctor-patient relationship. AMA Arch Intern Med 97(5):585–592
39. Heaven CM, Maguire P (1997) Disclosure of concerns by hospice patients and their identification by nurses. Palliat Med 11(4):283–290
40. Cape J, McCulloch Y (1999) Patients' reasons for not presenting emotional problems in general practice consultations. Br J Gen Pract 49(448):875–879
41. Parle M, Jones B, Maguire P (1996) Maladaptive coping and affective disorders among cancer patients. Psychol Med 26(4):735–744
42. Street RL Jr et al (1995) Patients' predispositions to discuss health issues affecting quality of life. Fam Med 27(10):663–670
43. Robbins JM et al (1994) Physician characteristics and the recognition of depression and anxiety in primary care. Med Care 32(8):795–812
44. Finset A (2012) "I am worried, Doctor!" Emotions in the doctor-patient relationship. Patient Educ Couns 88(3):359–363
45. Weng HC et al (2008) Doctors' emotional intelligence and the patient-doctor relationship. Med Educ 42(7):703–711
46. Mayer JD, Salovey P, Caruso DR (2008) Emotional intelligence: new ability or eclectic traits? Am Psychol 63(6):503–517
47. Amisha et al (2019) Overview of artificial intelligence in medicine. J Family Med Prim Care 8 (7): 2328–2331
48. Abbass HA (2019) Social integration of artificial intelligence: functions, automation allocation logic and human-autonomy trust. Cogn Comput 11(2):159–171
49. Lysaght T et al (2019) AI-assisted decision-making in healthcare. Asian Bioethics Rev 11 (3):299–314
50. Braun M et al (2020) Primer on an ethics of AI-based decision support systems in the clinic. J Med Ethics
51. Hamet P, Tremblay J (2017) Artificial intelligence in medicine. Metabolism 69S:S36–S40
52. Liew C (2018) The future of radiology augmented with artificial intelligence: a strategy for success. Eur J Radiol 102:152–156
53. Hosny A et al (2018) Artificial intelligence in radiology. Nat Rev Cancer 18(8):500–510
54. Decety J, Jackson PL (2004) The functional architecture of human empathy. Behav Cogn Neurosci Rev 3(2):71–100

55. Harris PL (1999) Individual differences in understanding emotion: the role of attachment status and psychological discourse. Attach Hum Dev 1(3):307–324
56. Simner ML (1971) Newborn's response to the cry of another infant. Dev Psychol 5(1):136–150
57. Botvinick M et al (2005) Viewing facial expressions of pain engages cortical areas involved in the direct experience of pain. Neuroimage 25(1):312–319
58. Vollm BA et al (2006) Neuronal correlates of theory of mind and empathy: a functional magnetic resonance imaging study in a nonverbal task. Neuroimage 29(1):90–98
59. Li Y et al (2019) AI-enabled emotion communication. IEEE Network 33(6):15–21
60. Zhang J et al (2016) ReliefF-based EEG sensor selection methods for emotion recognition. Sensors (Basel) 16(10)
61. Melzer A, Shafir T, Tsachor RP (2019) How do we recognize emotion from movement? Specific motor components contribute to the recognition of each emotion. Front Psychol 10:1389
62. Hossain MS, Muhammad G (2018) Emotion recognition using deep learning approach from audio-visual emotional big data. Information Fusion 49
63. Trigeorgis G et al (2017) A deep matrix factorization method for learning attribute representations. IEEE Trans Pattern Anal Mach Intell 39(3):417–429
64. McCue JD (1982) The effects of stress on physicians and their medical practice. N Engl J Med 306(8):458–463
65. Bakker AB, Le Blanc PM, Schaufeli WB (2005) Burnout contagion among intensive care nurses. J Adv Nurs 51(3):276–287
66. Glebocka A, Lisowska E (2007) Professional burnout and stress among Polish physicians explained by the Hobfoll resources theory. J Physiol Pharmacol 58 Suppl 5(Pt 1):243–252
67. Blender DI, Maidlow S (1994) It starts with trust–building organizational effectiveness in health care organizations. Mich Hosp 30(3):28–33
68. Arrieta Valero I (2019) Autonomies in interaction: dimensions of patient autonomy and non-adherence to treatment. Front Psychol 10:1857
69. Beil M et al (2019) Ethical considerations about artificial intelligence for prognostication in intensive care. Intensive Care Med Exp 7(1):70
70. Moor JH (2006) The nature, importance, and difficulty of machine ethics. IEEE Intell Syst 21 (4):18–21
71. https://ec.europa.eu/futurium/en/ai-alliance-consultation/guidelines. Access Date: 25th July 2020
72. https://www.eismd.eu/ai4people. Access Date: 25th July 2020
73. Currie G, Hawk KE, Rohren EM (2020) Ethical principles for the application of artificial intelligence (AI) in nuclear medicine. Eur J Nucl Med Mol Imaging 47(4):748–752
74. Vellido A (2019) The importance of interpretability and visualization in machine learning for applications in medicine and health care. In: Neural computing and applications
75. Goodman B, Flaxman S (2016) EU regulations on algorithmic decision-making and a "right to explanation". AI Magazine 38
76. Pesapane F et al (2018) Artificial intelligence as a medical device in radiology: ethical and regulatory issues in Europe and the United States. Insights Imaging 9(5):745–753
77. Shah H (2017) The DeepMind debacle demands dialogue on data. Nature 547(7663):259
78. Anderson D, Copeland BJ (2002) Artificial life and the Chinese room argument. Artif Life 8 (4):371–378

Sadra Behrouzieh started studying medicine at the Tehran University of Medical Sciences (TUMS) in 2018. In 2020, he achieved a silver medal in the Philosophy of Medicine Exam at the 12th National Olympiad for Medical Sciences Students. He currently researches cancer immunology at the Cancer Immunology Project (CIP) interest group of Universal Scientific Education and Research Network (USERN), TUMS.

Nima Rezaei gained his medical degree (M.D.) from TUMS in 2002 and subsequently obtained an M.Sc. in Molecular and Genetic Medicine and a Ph.D. in Clinical Immunology and Human Genetics from the University of Sheffield, UK. He also spent a short-term fellowship in Pediatric Clinical Immunology and Bone Marrow Transplantation in the Newcastle General Hospital. Since 2010, Prof. Rezaei has worked at the Department of Immunology and Biology, School of Medicine, TUMS; he is now the full professor and Vice Dean of International Affairs, School of Medicine, TUMS, and the co-founder and Head of the Research Center for Immunodeficiencies. He is also the founding President of USERN. He has edited more than 30 international books, has presented more than 500 lectures/posters in congresses/meetings, and has published more than 800 articles in international scientific journals.

Toward an Integrative and Holistic Approach to the Discipline of Health Informatics

26

Andre Kushniruk, Elizabeth Borycki, and Helen Monkman

"Everything should be made as simple as possible but no simpler."

Albert Einstein

Summary

Health care is being transformed by the increased use of information technology at all healthcare system levels. It has included advances in supporting decision-making and processing information related to health for professionals, patients, and society in general. In parallel, these educational advances in health informatics have led to new conceptualizations of what health informatics entails as a field and what it encompasses in terms of content. Along these lines, the

A. Kushniruk (✉) · E. Borycki · H. Monkman
School of Health Information Science, University of Victoria, HSD A202,
3800 Finnerty Road, Victoria, Canada
e-mail: andrek@uvic.ca

E. Borycki
e-mail: emb@uvic.ca

H. Monkman
e-mail: monkman@uvic.ca

A. Kushniruk
Integrated Science Association (ISA), Universal Scientific Education and Research
Network (USERN), Victoria, Canada

School of Health Information Science, Faculty of Human and Social Development,
University of Victoria, Victoria, BC V8W 2Y2, Canada

authors describe an integrative and holistic approach to defining, conceptualizing, and operationalizing health informatics as a discipline that draws from many sub-fields yet forms a coherent body of discipline-specific knowledge. The chapter describes our nearly 40 years of experience with education and practice in health informatics at one of the world's well-established health information science schools at the University of Victoria in Canada. Although health informatics integrates a wide range of knowledge and methods from many sub-disciplines, the chapter argues for a coherent framework helpful in understanding health informatics advances and guiding education in the area. Challenges and future opportunities for health informatics as an academic discipline or field are also discussed.

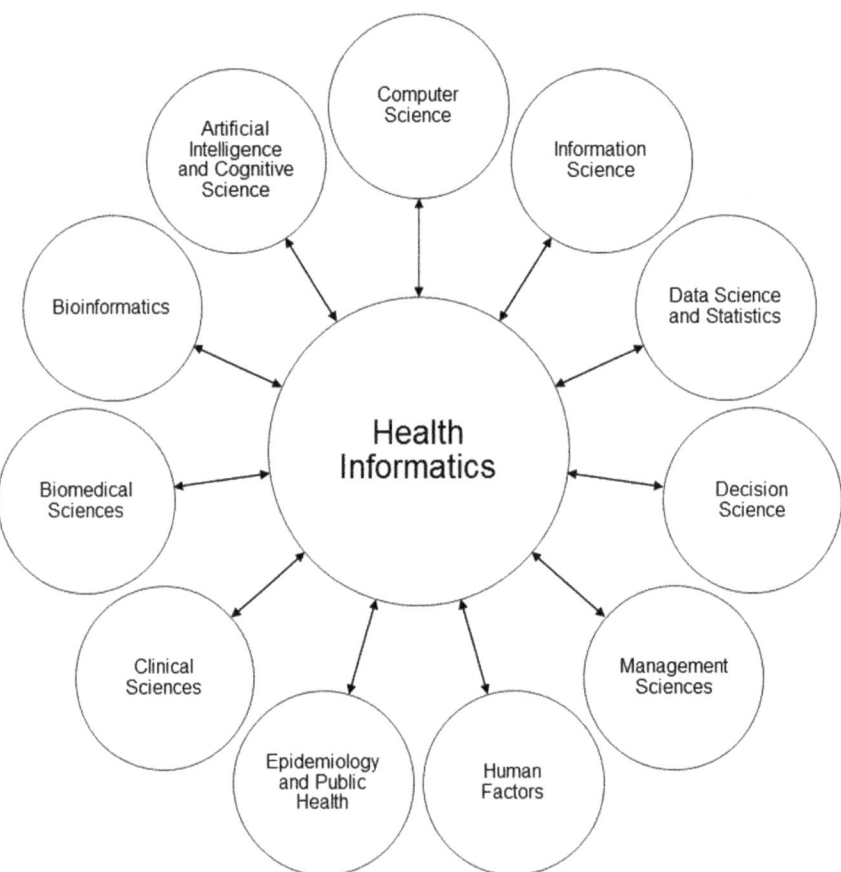

Fields of study that health informatics draws from and contributes to the code of this chapter is *01110000 01101001 01100001 01101100 01110100 01111010 01100001 01101001 01101111 01100101 01100011 01010011 01101001 01101110.*

Keywords

Biomedical informatics · Curricula · Education · Health informatics · Health information science

1 Introduction

The work of health professionals is being transformed through the use of a wide range of information technologies. It includes using information technology in the daily work of physicians, nurses, pharmacists, and all other allied health professionals. Besides, patients and society increasingly use a range of health-related information technologies, including health apps, Web sites, patient portals, medical devices, and online decision aids [1]. Indeed, the number of users now touched by the use of information technology in health care and the number of fields involved in this endeavor are rapidly growing. It has included electronic health record systems (EHRs), clinical decision support tools, and advanced technologies such as imaging systems and artificial intelligence (AI) in health care [2]. Keeping with this ever-increasing use and application of information technology, there would be a subsequent need for a structured and integrated disciplinary approach to health informatics. Such an approach should incorporate theory, methods, and content from a wide range of sub-disciplines as the field is complex, multi-faceted, and interdisciplinary. Also, over time advancements in health informatics continue to go beyond those of the sub-disciplines it draws from, creating its own unique body of knowledge, as well as feeding back to the disciplines it draws from.

By its very nature, the field of health informatics, which studies the use of technology in health care, involves a variety of scientific areas, specialties, practitioners, and researchers [3]. Health informatics, as the discipline of study focused on technologies in health care, requires inputs from a range of base disciplines that namely involve computer science, health information science, engineering science, social science, management science, and a range of other disciplines, such as the decision and data sciences [4]. It motivates the need for an integrative and holistic approach, as will be presented in this chapter. This approach to conceptualizing a field like health informatics should be extensible and modifiable enough to integrate new subfields as they become a part of the rapidly evolving healthcare informatics discipline.

2 What is Health Informatics?

Health informatics can be considered the study and practice of designing, implementing, deploying, and evaluating information technologies and medical devices for health professionals, patients, and laypeople. Therefore, health informatics's scope is broad and lies at the intersection and nexus of clinical science, information science, and management practice, research, and education. Since the early 1960s, health care has been transformed by the increasing use of technology [5, 6]. Also, there has been a massive explosion in the amount of medical and health-related data generated over the past several decades, requiring advances in information technology processing to make this data meaningful and transform it into actionable information and medical knowledge [7]. As a parallel development, advances from the computer science side in computing power have been impressive and have led to a broader range and increased complexity and capability for its potential application to individuals and populations' health. Furthermore, the advent of the Internet, the World Wide Web, and personal and mobile computing, as well as the miniaturization of devices, has opened up the potential for greater access to and use of healthcare information by health professionals as well as patients and the general population [8]. These rapid and profound changes have called for the need to develop an integrative and holistic health informatics discipline to effectively guide the education, practice, and research advancements in health information technology.

Health informatics is devoted to studying information processing (both by computers and by humans) and communication in health. This includes the use of health information and knowledge by healthcare workers, patients, and the population in general. A focus is on transforming health data to information and knowledge and includes ways of analyzing, designing, and displaying data, information, and knowledge to be of value for improving health and healthcare services. Health promotion has also become an important aspect as greater interest and laypeople's involvement in their own health continues to grow [8]. Also, the field aims to better understand phenomena around the concepts of communication, data, information, and technology design and implementation in health care.

One way of conceptualizing health informatics is the consideration of main knowledge domains that contribute to it. By definition, health informatics represents a large area of study and draws from (and reciprocally contributes back to) a wide range of relevant study fields. Graphical Abstract shows some of the fields involved, including, but not limited to computer science, information science, data science and statistics, decision science, management sciences, human factors, epidemiology and public health, clinical sciences, biomedical sciences, bioinformatics, AI, and cognitive science, and a host of allied health sciences (such as nutrition science, physiotherapy, and pharmacy). The many disciplines that contribute to health informatics may be considered the "base" or "sub" disciplines of the emergent health informatics discipline, with each contributing knowledge and practice to the discipline known explicitly as health informatics. Noteworthy is that health informatics borrows knowledge and methods from many of these disciplines; the discipline of health informatics feeds back knowledge and methods to each of these base disciplines (as indicated by the double-headed arrows in Graphical Abstract).

Another way to consider health informatics is to view it from an ontological perspective and develop an extensible framework that allows for accurate descriptions of the field and is modifiable in that it allows for the addition of new subfields and specialties as they arise from new and emerging areas of research that can contribute to health informatics as a field. Such a framework should allow for extension and coverage of the breadth and scope of health informatics. Figure 1 represents an ontological framework for considering health informatics as a discipline. At the top of Fig. 1 is the overarching concept of health informatics as the "parent class" for the discipline. At this level, we can consider health informatics to be the all-encompassing study dealing with information processing, information technology, and medical devices. As shown in Fig. 1, health informatics is a broad discipline that spans several subfields to integrate theory and practice from a wide range of relevant base disciplines.

Health informatics subclasses include, but are not limited to, ones shown in Fig. 1. For example, on the far left of the Figure is the subfield "Consumer Informatics." This rapidly evolving field deals with the use of health information and health information technologies by the consumers of health care, i.e., patients, their families, and the general population. "Bioinformatics" refers to the science involved in collecting, analyzing, and understanding complex human data and information at the level of the genetic code and thus could be considered another sub-field of health informatics. "Clinical Informatics" can be considered as another subclass or subfield of health informatics that studies and works with information technology in health care within the clinical practice setting (at the level of the patient and healthcare organization).

Some subclasses (or subfields) falling under the umbrella term "Clinical Informatics" include, but not limited to, medical informatics (the study of information processing and technology in medicine), nursing informatics (the study of information processing and technology in nursing), pharmacoinformatics (the study of

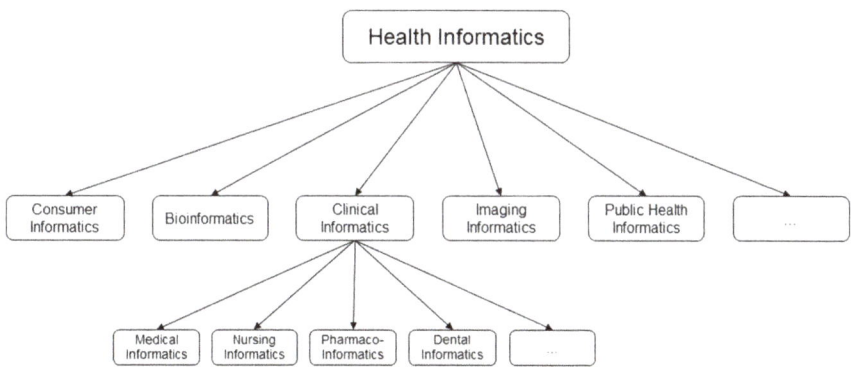

Fig. 1 Ontological framework for considering health informatics as a discipline

information processing and technology in the context of pharmacy), and dental informatics (the study of information and technology in dentistry). Other immediate subclasses of the superclass health informatics include "Imaging Informatics" (devoted to the emerging field studying the use of advanced techniques for computer-based imaging and interpretation of health data) and "Public Health Informatics" (focusing on the intersection between public health and information technology and dealing with population-level data).

The advantage of taking this type of approach to representing health informatics is that it allows for showing the relationship among sub-disciplines of health informatics and new subfields (and subclasses of those subfields) as they emerge over time. All the subfields of health informatics shown in the figure have (i.e., "inherit") in common a general characteristic that they all deal with information technology in health care in some way and context. However, each subfield represents a specialization of health informatics (e.g., in terms of theories and methods that advance each of them) and provides a context for a clearer discussion of each subfield's content and relationships. Note that the boxes in Fig. 1 with ellipses indicate that the new and emerging subfields and areas will become added to this ontology over time as technology and its uses in health informatics evolve, making the framework shown in the Figure highly extensible.

3   Toward a Framework for Development of the Academic Discipline of Health Informatics

In developing both a field of health informatics and curricula to educate and train both practitioners and researchers in this field, we have developed a framework that includes input from several key areas of study and practice. Health informatics involves input, knowledge, and skills from four foundational domains: health science, management science, information science and computer science, and decision and data science. Figure 2 presents a practical framework the authors applied and refined over several decades for considering health informatics to guide the design and implementation of health informatics curricula at the University of Victoria and Canada. The initial curriculum was developed at the School of Health Information Science at the University of Victoria, Canada, in 1981, as described consequently. It has evolved over the past decades through research and practical experiences in conducting health informatics education and experimenting with new programs [9]. The framework has been used to guide curriculum development and create courses related to the four domains in Fig. 2, tailored to health informatics.

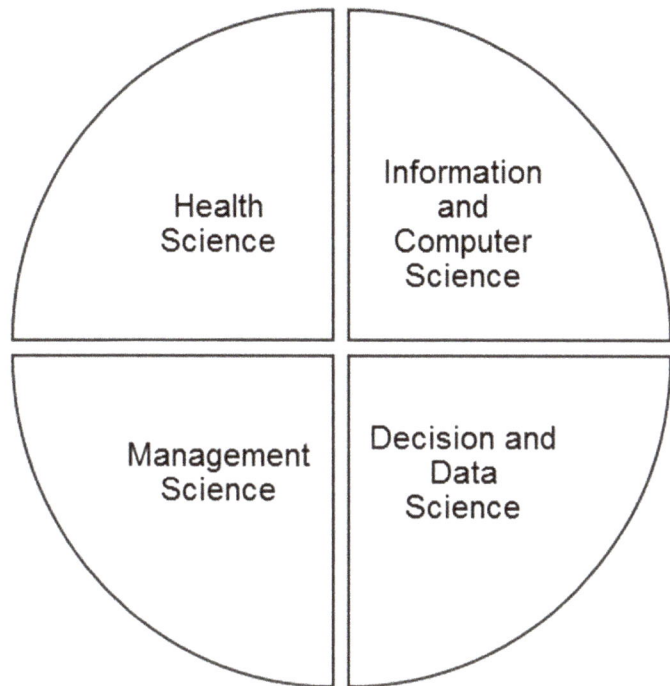

Fig. 2 Four main content domains of health informatics used for curriculum development at the School of Health Information Science (University of Victoria)

4 Application of the Framework: The Health Informatics Curriculum at the School of Health Information Science

The School of Health Information Science at the University of Victoria in Canada has pioneered an integrative and holistic view of health informatics, beginning with its formation and continuing to today [9]. The program began in 1981. Since then, it has provided a basis for the evolving overall framework, as outlined in Fig. 2. The School has graduated over 1000 alumni who hold undergraduate and graduate degrees in health informatics. Unlike other programs in this area, the School is dedicated to advancing health informatics and has treated the field as a separate and distinct area of study. It is currently the only School or Department in Canada devoted exclusively to health informatics at all degree levels. Students may obtain the following degrees:

- Bachelor of Science in Health Informatics;
- Combined Bachelor of Science in Computer Science and Health Informatics;
- Master of Science in Health Informatics;

- Combined Master of Nursing and Health Informatics Double Degree; and
- Ph.D. in Health Informatics

The graduates of the programs above have taken up a wide range of health informatics positions in Canada and internationally with high demand (close to 100% job placement upon graduation). There are entry-level positions as health informatics specialists, analysts, specialists in health information system training, health privacy consultants, health information system designers and evaluators, health program coordinators, testing analysts, and many more positions. At the intermediate level and senior levels, positions taken up by graduates have included health system managers, project managers, quality improvement specialists, standards directors, data and system architects, consultants, and clinical informatics managers. At the senior level, many of our graduates have obtained chief information officer positions for health authorities and provinces, chief privacy officer positions, chief medical and nursing informatics officers, chief health technology officer, and various vice president positions in healthcare organizations throughout the world with a focus on technology in health.

The programs' success mainly relies on a set of courses that fall into four sectors, as shown in Fig. 2. The courses focus on competencies (i.e., knowledge, skills, and abilities) related to the following four areas in the context of health informatics: health science, information and computer science, management science, and decision science and data science. Below is an overview of each of these four areas' main undergraduate courses and competencies, described in detail elsewhere [9].

4.1   Health Sciences

Courses at the University of Victoria undergraduate program in health informatics include a required biology course, a second course specifically designed to introduce students to biomedical fundamentals (for health informatics specialists), and an additional course that introduces students to the fundamentals of decision-making in the healthcare setting. At a high level, the competencies around health sciences include understanding the basics of biomedical fundamentals and acknowledging the complexity of health care that has its roots in cognitive processes and healthcare workflows.

4.2   Information and Computer Science

Courses include computer programming (which we are now teaching specifically in the context of computer programming for healthcare applications), human factors in health care, user interface and design thinking for health information systems and applications, requirements analysis in health care, and systems analysis and design (specifically tailored to the application in health care). Competencies around information and computer science include acquiring knowledge about the range of health information technologies, knowledge about computer programming in healthcare contexts, and understanding how technology can be effectively applied to improving health care.

4.3 Management Science

Courses include project management in health care, healthcare organizational behavior, change management, and healthcare strategy. Competencies around management science include understanding and applying theories and concepts from management and business to improve health care. Additionally, the courses focus on teaching students how management concepts can help optimize health work processes by applying technology.

4.4 Decision Science and data Science

Courses cover epidemiology, public health informatics, and AI in health care and health data science. Competencies around decision science and data science include applying methods from areas such as data science, decision science, and applied AI to improve healthcare processes and outcomes. Notably, the approach pioneered at the University of Victoria since its inception has been mostly incorporated and modified into Canada's national health informatics organization's framework for health informatics professional core competencies [10, 11]. Thus, the work at the University of Victoria has supported a strong and increasingly unified and unique Canadian perspective on health informatics as an integrative field focused not only on health professionals but also on improving patients' and citizens' health.

Another important aspect of programs and the approach pioneered at the School of Health Information Science focuses on experiential learning [12]. From the beginning of the program, it has included three, four-month mandatory cooperative (i.e., experiential learning) experiences that students must undertake within the four years of their Bachelor's degree in health informatics. Students must complete paid work experiences with a range of employers in these co-ops, including health authorities, hospitals, governmental organizations, and private healthcare IT companies. A hallmark of the co-op program is students' chance to try out and experience working in more than one of these different sectors, allowing them to chart and plan their own career path through the healthcare system. Also, the practical experience that students bring back to class upon return from their co-ops is invaluable. It raises the discussion level as students integrate relevant real-world experience with the more theoretical aspects of their health informatics education.

The graduate programs in health informatics at the School of Health Information Science are designed to allow students to customize their studies based on their prior educational and work background and experience [13]. Students entering the Masters of Science program in health informatics typically come from one of the following three backgrounds:

1. recent graduates of undergraduate bachelor degree programs in health informatics;
2. health professionals (e.g., nurses, physicians, pharmacists) who work in health informatics related positions; and
3. management and information technology specialists who work in the health informatics space but seeking credentialing and systematic education in the field

The School has pioneered virtual distance education and a blended learning approach at the graduate level since 2000, including a range of teaching methods such as synchronous online lectures, workshops, and virtual seminars [9]. These same teaching methods have proven helpful in current times as we move all of our courses (undergraduate and graduate) to the online platform. Graduates of the Ph.D. program in health informatics have become world-class experts in various aspects of health informatics and have taken up both high-level leadership positions and faculty positions in Canada and worldwide, setting up and supporting new and innovative health informatics educational programs [14].

5 Future Directions for Health Informatics as an Integrative Discipline

Emerging areas transform health care (including applied AI and data science) and promise to revolutionize health informatics and health care further. To address emerging areas, we have taken the approach that health informatics curricula (and the field of health informatics in general) must be flexible and agile enough to allow for influence from new subfields, extending the scope and coverage of the topics, competencies, and knowledge needed to work with health information and technology into the future. Toward this end, as health informatics is evolving, curricula and concepts around what is health informatics need to be reviewed and updated, making it fluid and moving target as a highly integrative field of study. Along these lines, many of the courses described above have accompanying laboratory components where new technologies and approaches are constantly updating the material covered. For example, as electronic health records (EHRs) have become more prevalent in Canada and worldwide, the School developed an innovative virtual platform enabling students to experiment with this technology remotely [15]. Also, as technologies move from the hospital to greater use by patients at home, mobile applications are becoming more prevalent [16]. Along these lines, a virtual smart home laboratory has been constructed to support student learning about the technologies used for virtual health care. The laboratory has been used to train students in designing and testing usable and safe health information technologies that can be used effectively by both health professionals and the lay public [1, 17, 18].

6 Conclusion

As health informatics evolves as a field, there is the need to develop descriptive and flexible frameworks for considering the field as a whole and guiding research, practice, consulting, and educational initiatives involving information processing and technology that will lead to effective education and healthcare outcomes. This chapter described a particular view of health informatics as an integrative and ever-evolving field of study and practice. It followed an approach to creating and modifying educational curricula based on that type of framework. The success of this approach to an integrative and holistic view of health informatics will be measured by the success of the integration of technology to improve the health and well-being of us all.

Core Messages

- Health informatics is a distinct discipline that integrates information technology (IT) into the health context.
- Health informatics involves an integrative approach to research, teaching, design, and evaluation of health IT.
- An integrative framework for considering health informatics as a field of study was presented in this chapter.

References

1. Borycki EM, Househ M, Kushniruk A, Nohr C, Takeda H (2012) Empowering patients: making health information and systems safer for patients and the public. Yearb Med Inf 21 (01):56–64
2. Cimino JJ, Shortliffe EH (2006) Biomedical informatics: computer applications in health care and biomedicine. Springer, Verlag
3. Kolokathi A, Hasman A, Chronaki C et al (2019) Education in biomedical and health informatics: a European perspective. Stud Health Technol Inf 264:1951–1952
4. Mantas J (2016) Biomedical and health informatics education—the IMIA Years. Yearb Med Inf Suppl 1(Suppl 1):S92–S102
5. Collen MF (1995) A history of medical informatics in the United States: 1950 to 1990. American Medical Informatics Association, Hartman Publishing, Bethesda MD
6. Wright A, Sittig DF, McGowan J, Ash JS, Weed LL (2014) Bringing science to medicine: an interview with Larry Weed, inventor of the problem-oriented medical record. J Am Med Inform Assoc 21(6):964–968
7. Shortliffe EH, Blois MS (2006) The computer meets medicine and biology: emergence of a discipline. In: Medical informatics: computer applications in health care and biomedicine. Springer, pp 3–40
8. Eysenbach G (2000) Consumer health informatics. BMJ 320(7251):1713–1716

9. Kushniruk A, Lau F, Borycki E, Protti D (2006) The school of health information science at the University of Victoria: towards an integrative model for health informatics education and research. Yearb Med Inf 159–165

10. COACH (2012) Health informatics professional core competencies November 2012. Digital Health Canada. Accessed from: https://digitalhealthcanada.com/wp-content/uploads/2019/07/Health-Informatics-Core-Competencies.pdf

11. El Morr C (2018) Introduction to health informatics: a Canadian perspective. Canadian Scholars, Toronto

12. Borycki EM, Frisch N, Kushniruk A, McIntyre M, Hutchinson D (2012) Integrating experiential learning into a double degree master's program in nursing and health informatics. In: 2012: 11th international congress on nursing informatics, June 23–27, 2012, Montreal, Canada. American Medical Informatics Association

13. Borycki EM, Frisch N, McIntyre M, Kushniruk A (2011) Design of an innovative double degree graduate program in health informatics and nursing: bridging nursing and health informatics competencies. Eur J Biomed Inf 7(2):31–39

14. Kushniruk A (2007) The school of health information science at the University of Victoria: celebrating 25 years of contributions to health informatics. Healthc Inf Manage Commun 20 (4):18–20

15. Borycki EM, Kushniruk AW, Joe R, Armstrong B, Otto T, Ho K, Silverman H, Moreau J, Frisch N (2009) The University of Victoria interdisciplinary electronic health record educational portal. Adv Inf Technol Commun Health, 49–54

16. Kushniruk A, Borycki EM, Armstrong B, Kuo MH (2012) Advances in health informatics education: educating students at the intersection of health care and information technology. Stud Health Technol Inf 172:91–99

17. Patel VL, Kushniruk A (1998) Understanding, navigating and communicating knowledge: issues and challenges. Methods Inf Med 37(04/05):460–470

18. Carvalho CJ, Borycki EM, Kushniruk A (2009) Ensuring the safety of health information systems: using heuristics for patient safety. Healthc Q 12:49–54

Andre Kushniruk is Professor and Director of the School of Health Information Science at the University of Victoria. He researches several areas, including evaluating the effects of technology, human–computer interaction in health care and other domains, and cognitive science. His work is known internationally, and he has published widely in health informatics and health informatics education. His work includes developing novel methods for conducting a video analysis of computer users. He is currently extending this research to the remote study of e-health applications and advanced information technologies. Kushniruk has held academic positions at many Canadian universities and has taught courses in human–computer interaction, database management, systems analysis, and design. Andre holds undergraduate degrees in Psychology and Biology and an M.Sc. in Computer Science, and a Ph.D. in Cognitive Psychology from McGill University.

Elizabeth Borycki is Professor in the School of Health Information Science at the University of Victoria, Victoria, British Columbia, Canada. Elizabeth joined the University of Victoria over 15 years ago. Before coming to the University of Victoria, she spent over 15 years working in healthcare in nursing, clinical, and health information technology roles. She has published over 200 articles, 40 book chapters, and ten edited books. Her research and publications have focused on health technology safety, virtual care (mobile, eHomecare, and telehealth), health information technology management and implementation, health technology competencies, and data science in health care. She received her Ph.D. from the Department of Health Policy, Management and Evaluation at the University of Toronto, a Master of Nursing from the University of Manitoba in geriatrics and community health nursing, and an Honours Bachelor of Science in Nursing from Lakehead University.

Helen Monkman is Assistant Professor in the School of Health Information Science at the University of Victoria. Her mission is to improve consumer health information systems by making them easier for people to use and the information therein easier to understand. Her work seeks to empower people and help them make better health decisions as well as have better conversations with their healthcare providers. Her research interests include human factors, user experience, usability, eHealth literacy, digital health literacy, information visualization, and how these factors impact consumer health information systems' use and understandability.

Noosha Samieefar, Sara Momtazmanesh, Hans D. Ochs,
Timo Ulrichs, Vasili Roudenok, Mohammad Rasoul Golabchi,
Mahnaz Jamee, Melika Lotfi, Roya Kelishadi,
Mohammad Amin Khazeei Tabari, Milad Baziar,
Sayedeh Azimeh Hosseini, Milad Rafiaei, Antonio Condino-Neto,
Elif Karakoc-Aydiner, Waleed Al-Herz, Morteza Shamsizadeh,
Niloofar Rambod Rad, Mohammadreza Fadavipour, Alireza Afshar,
Meisam Akhlaghdoust, Kiarash Saleki, Farbod Ghobadinezhad,
Zhila Izadi, Arash Khojasteh, Alireza Zali, and Nima Rezaei

"Medical education is not just a program for building knowledge and skills in its recipients... it is also an experience which creates attitudes and expectations."

Abraham Flexner

N. Samieefar
School of Medicine, Shahid Beheshti University of Medical Sciences, Tehran, Iran
e-mail: nooshasamieefar@sbmu.ac.ir

School of Advanced Technologies in Medicine, Shahid Beheshti University of Medical Sciences, Tehran, Iran

S. Momtazmanesh
Universal Scientific Education and Research Network (USERN), Tehran, Iran
e-mail: s-momtazmanesh@student.tums.ac.ir

School of Medicine, Tehran University of Medical Sciences, Tehran, Iran

© The Author(s), under exclusive license to Springer Nature Switzerland AG 2022
N. Rezaei (ed.), *Multidisciplinarity and Interdisciplinarity in Health*,
Integrated Science 6, https://doi.org/10.1007/978-3-030-96814-4_27

Summary

Innovative and more efficient learning methods have become one of medical education's main subjects of interest in recent years. It is particularly relevant due to the rapid implementation of novel information technologies (IT) and

H. D. Ochs
Department of Pediatrics, University of Washington, Seattle, WA, USA
e-mail: hans.ochs@seattlechildrens.org

Seattle Children's Research Institute, Seattle, WA, USA

Universal Scientific Education and Research Network (USERN), Washington, USA

T. Ulrichs
Institute for Research in International Assistance, Akkon University for Human Sciences, Berlin, Germany
e-mail: timo.ulrichs@akkon-hochschule.de

Universal Scientific Education and Research Network (USERN), Berlin, Germany

V. Roudenok
Belarusian State Medical University, Minsk, Belarus
e-mail: roudenok@bsmu.by

Universal Scientific Education and Research Network (USERN), Minsk, Belarus

M. R. Golabchi
USERN Office, Isfahan University of Medical Sciences, Isfahan, Iran

M. Jamee
USERN Office, Alborz University of Medical Sciences, Karaj, Iran

M. Lotfi
USERN Office, Zanjan University of Medical Sciences, Zanjan, Iran

R. Kelishadi
Child Growth and Development Research Center, Research Institute for Primordial Prevention of Non-Communicable Disease, Isfahan University of Medical Sciences, Isfahan, Iran
e-mail: Kelishadi@med.mui.ac.ir

USERN Office, Research Institute for Primordial Prevention of Non-Communicable Disease, Isfahan University of Medical Sciences, Isfahan, Iran

M. A. K. Tabari
Mazandaran University of Medical Sciences, Sari, Iran

USERN Office, Mazandaran University of Medical Sciences, Sari, Iran artificial intelligence in medicine (IT medicine), as well as the need to train highly qualified doctors who can work efficiently both under present conditions (COVID-19 pandemic) and possible natural or technogenic catastrophes (challenges) in the future. Integration, to make knowledge more long-lasting

M. Baziar
School of Medicine, Ardabil University of Medical Sciences, Ardabil, Iran

USERN Office, Ardabil University of Medical Sciences, Ardabil, Iran

S. A. Hosseini
Shahrekord University of Medical Sciences, Shahrekord, Iran

USERN Office, Shahrekord University of Medical Sciences, Shahrekord, Iran

Department of Medical Biotechnology, School of Advanced Technologies, Shahrekord University of Medical Sciences, Shahrekord, Iran

M. Rafiaei
USERN Office, Yazd University of Medical Sciences, Yazd, Iran

A. Condino-Neto
Department of Immunology, Institute of Biomedical Sciences, University of São Paulo, São Paulo, Brazil

Universal Scientific Education and Research Network (USERN), São Paulo, Brazil

E. Karakoc-Aydiner
Faculty of Medicine, Division of Pediatric Allergy and Immunology, Marmara University, Istanbul, Turkey

Universal Scientific Education and Research Network (USERN), Istanbul, Turkey

W. Al-Herz
Department of Pediatrics, Faculty of Medicine, Kuwait University, Kuwait City, Kuwait

Universal Scientific Education and Research Network (USERN), Kuwait City, Kuwait

M. Shamsizadeh
Department of Medical Surgical Nursing, School of Nursing and Midwifery, Hamadan University of Medical Sciences, Hamadan, Iran

USERN Office, Hamadan University of Medical Sciences, Hamadan, Iran

N. R. Rad
USERN Office, Islamic Azad University Medicine Faculty, Mashhad, Iran

M. Fadavipour
USERN Office, Abadan University of Medical Sciences, Abadan, Iran

Abadan University of Medical Sciences, Abadan, Iran

and education more enjoyable, has different aspects in medical education. An integrated medical curriculum that combines basic and clinical sciences pays attention to all dimensions of a patient, from physical to spiritual health, and provides a multidisciplinary learning program. In integrated medical education, learning is predominantly based on learners' independent searching and studying with case-based discussion classes.

A. Afshar
USERN Office, The Persian Gulf Marine Biotechnology Research Center, The Persian Gulf Biomedical Sciences Research Institute, Bushehr University of Medical Sciences, Bushehr, Iran

M. Akhlaghdoust · A. Zali
Functional Neurosurgery Research Center, Shohada Tajrish Comprehensive Neurosurgical Center of Excellence, Shahid Beheshti University of Medical Sciences, Tehran, Iran

USERN Office, Functional Neurosurgery Research Center, Shahid Beheshti University of Medical Sciences, Tehran, Iran

K. Saleki
Babol University of Medical Sciences, Babol, Iran

USERN Office, Babol University of Medical Sciences, Babol, Iran

F. Ghobadinezhad
School of Medicine, Kermanshah University of Medical Sciences, Kermanshah, Iran

F. Ghobadinezhad · Z. Izadi
USERN Office, Kermanshah University of Medical Sciences, Kermanshah, Iran

Z. Izadi
Pharmaceutical Sciences Research Center, Health Institute, Kermanshah University of Medical Sciences, Kermanshah, Iran
e-mail: izadi_zh@razi.tums.ac.ir

A. Khojasteh
Department of Oral and Maxillofacial Surgery, Shahid Beheshti University of Medical Sciences, Tehran, Iran

USERN Office, Shahid Beheshti University of Medical Sciences, Tehran, Iran

N. Rezaei (✉)
Research Center for Immunodeficiencies, Children's Medical Center, University of Medical Sciences, Dr. Qarib St, Keshavarz Blvd, 14194 Tehran, Iran
e-mail: Rezaei_nima@tums.ac.ir

N. Rezaei
Integrated Science Association (ISA), Universal Scientific Education and Research Network (USERN), Tehran, Iran

Department of Immunology, School of Medicine, Tehran University of Medical Sciences, Tehran, Iran

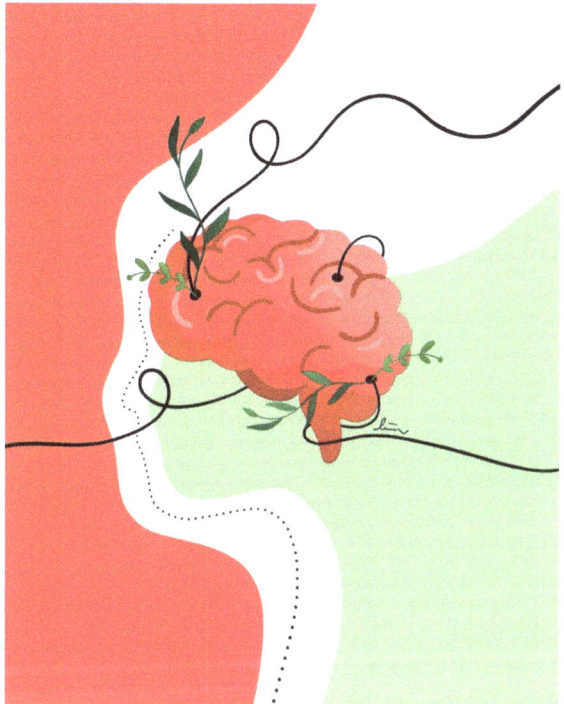

Integrated medical education

[Adapted with permission from the Association of Science and Art (ASA), Universal Scientific Education and Research Network (USERN); Made by Nastaran Hosseini].

The code of this chapter is *01101111 01100011 01110000 01110100 01101101 01100101 01111001 01100101 01000011 01101110*.

Keywords

Integrated medical education · Integrated medicine · Problem-based learning · Self-learning modules · Integrated sciences

1 Introduction

Over the past decades, medical education has undergone several changes. A greater emphasis is placed on deeper approaches to learning [1]. The complex and broad nature of the medical sciences indicates the importance of integration and interactive education [2].

Integrated curriculum is defined as "education that is organized in such a way that it cuts across subject-matter lines, bringing together various aspects of the curriculum into a meaningful association to focus upon broad areas of study." By the term "integrated medical education," we intend to combine different parts of a medical curriculum. Although it is controversial whether medical education should be integrated, the ultimate purpose of medical curricula, which is effective clinical practice, highlights the importance of integrating basic and clinical sciences from the first day of medical school. Furthermore, public health sciences should be considered another key factor in an efficient medical education model [3].

The concept of integrated medical education is broader than just a curriculum; it is the gathering of innovations [4]. Integrated medical education makes the learning environment more thrilling and increases students' adherence to education [3]. It has been shown that in this learning, the quality of training improves, and postgraduates have achieved better performance in clinical reasoning and can better link clinical practice with basics [5]. Furthermore, integration can decrease the amount of time spent on unnecessary repetition since similar topics of relative organs are taught simultaneously [6].

Moreover, the rapid development of information technologies (IT) and their use in medicine (telemedicine, digital medicine, artificial neural networks, augmented reality (AR) systems in surgery, development, and introduction of three-dimensional (3D) models of organs, etc.) requires integration of the IT medicine course in the curriculum.

However, there is a concern about neglecting some topics, especially basic sciences essentials. This is because, in this model, more attention is directed toward self-learning instead of lectures. On the other hand, more emphasis is on using visuals in learning, including pictures, diagrams, and animations, which is desirable for some but not all learners [3, 7].

Not all disciplines are easy to integrate, and integration concepts are not easy to understand [5]. This chapter reviews three aspects of integrated medical education (Fig. 1): vertical, horizontal, and spiral. Then, we investigate the advantages and disadvantages of some proposed educational models.

2 Considering Human Beings as an Integration of Different Dimensions in the Horizontal Axis (From Physical to Spiritual)

"That medicine is a social science sounds like a truism, yet it cannot be repeated often enough because in medical education we still act as if medicine were a natural science and nothing else. There can be no doubt that the target of medicine is to keep individuals adjusted to their environment as useful members of society, or readjust them when they have dropped out as a result of illness. It is a social goal..."

These sentences by Henry E. Sigerist emphasize that we should not consider the disease only due to pathological changes [8]. Now, health is defined as a concoction of physical, social, psychological, and spiritual well-being [9]. Accordingly, the

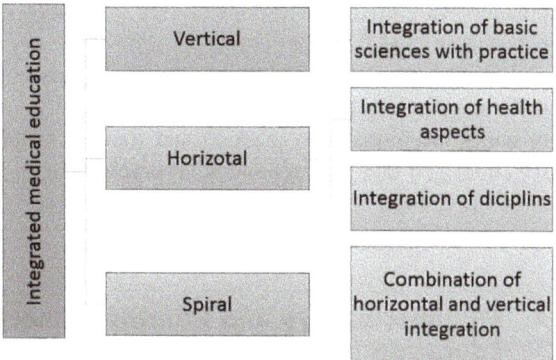

Fig. 1 Aspects of integrated medical education

concept of health as an integration of these dimensions must also be considered in medical education in the horizontal integration axis. For instance, in many infectious diseases, patients' social and behavioral history is a key component in diagnostic reasoning [3]. So, in the new curricula, public health courses are strongly recommended [10].

3 Vertical Integration of Medical Education (From Basic Sciences to Clinical Sciences)

In 1920, Flexner introduced the basic sciences into medical education and designed a medical curriculum with separate basic and clinical sciences. However, such a curriculum does not seem appropriate in the 21st century for our future health professionals [11]. The concept that medical students should eventually become professional practitioners underlines the importance of involving them with clinical judgment even before studying medicine [3].

A model of vertical integration is the Z model by Wijnen-Meijer. In this model, basic and clinical subjects are combined in all courses offered by medical school. However, the priority is in basic sciences during the first years, while clinical sciences are underscored in the last years of education [12].

The critical point is that just designing an integrated program would not lead to the desired outcomes. Students need to be guided by senior experts in the subjects (basic and clinical) or by a collaborative team of various specialists. Furthermore, seniors should be educated and monitored for their skills of education, assessment, and evaluation in accordance with the integrated model since such reforming may harbor a conceptual resistance in itself for both sides.

Vertical integration can be considered as a philosophy of education that affects the maturity and interaction of students with this profession and is also used for lifelong learning of professionals. In other words, the educational approach in vertical integration is the gradual increase of learner participation in the professional community through the gradual increase of knowledge-based participation in practice with graduate responsibilities in patient care and pursues a developed perspective [13].

Bringing practical aspects into the first years of medical education should not result in spending less time on basic sciences. The intention is to train practitioners to apply the basics into practice and find the connection between disease manifestations and underlying biological, psychological, and social factors, not omitting basics [11, 14].

4   Multidisciplinary and Interdisciplinary Medical Education (Horizontal Integration)

The other aspect of integrated medical education is bringing together distinct but relevant disciplines. The application is not only in basic sciences but also within clinical specialties. Integration of anatomy and physiology in basic sciences or separate internal medicine subspecialties in clinical sciences are examples of this approach [4].

The modules could be designed according to organ systems instead of medical departments with the collaborative teaching of professors from different specialties and subspecialties. However, pathophysiology and basic sciences should not be neglected [15]; otherwise, this multidisciplinary education is more suitable for residents in clinical disciplines, not for undergraduate medical students [4].

Another issue of organ-based education is that in some diseases, multiple organs are involved due to their anatomical proximity, a uniform pathogenetic mechanism of several diseases, a temporal cause-and-effect relationship between diseases, one disease as a complication of another, etc., as well as comorbid pathology, which increases considerably with age. Thus, in these subjects, case-based problem-solving models are preferred.

In conclusion, the challenges are bringing together all disciplines associated with the organ studied, linking clinical and basic sciences, convincing and gathering professors from different fields to participate in one integrated course, and finally providing the required sources of learning and teaching [15].

5   Integration of All: Spiral Integration

When all parts of a medical curriculum are joined together to create a unique educational model, the final goal of integration comes true. Spiral integration (Fig. 2) means that the medical curriculum integrates all disciplines and majors of clinical and basic sciences together [16]. Furthermore, it is noteworthy to pay attention to health and social medicine, palliative care education, and medical ethics through all the years of education [10].

It is also essential to take into consideration the issues related to the introduction and use of IT technologies (distance learning, electronic educational courseware, Learning Management System (MOODLE), etc.) in the educational process and the study of the basics of IT medicine—a rapidly developing branch, medicine of the future, specialists for which already need to be taught (prepared) now.

Of course, IT technologies in medical education will not wholly replace clinical practice for students—the main element (basis) in medical education. But it is necessary to use their potential to a full extent, especially during periods of (forced) complete transition of universities to distance learning, such as during the current quarantine of the COVID-19 pandemic or other possible emergencies in the world.

Harden compared the steps required to achieve the spiral integration of medical curriculum to the ladder stairs known as "Harden's integration ladder." The steps are defined as follows. The first step is "isolation/fragmentation;" the disciplines are taught discretely without any attention to other related subjects and disciplines. When the process turns to the second step "awareness," the teaching is still subject-based, but the teachers are now becoming aware of connections between disciplines. In the third step, "consultation/connection/harmonization," informal discussions between teachers of different disciplines are initiated. The fourth step, "infusion/nesting," is the master's beginning of aiming other contents to cover in their teaching. "Parallel teaching/temporal coordination" is the fifth step when related disciplines are taught concurrently but making connections between the subjects are assigned to the students themselves. "Joint teaching" is the sixth step, the beginning of integration in related subjects. The next step is called "concomitant program" that combines subject-based and integrated sessions, and then in the eighth step, "mixed program," the integrated part becomes prominent. In the "multidisciplinary/webbed" step, different disciplines are integrated with the focus of students learning, but subjects still have separate identities. "Interdisciplinary/monolithic" is the tenth phase in which the boundaries between subjects are removed. The 11th step of the ladder is when we reach an integrated spiral curriculum designated a "transdisciplinary" program. At this level, the integration is

Fig. 2 Spiral aspect of integrated medical education

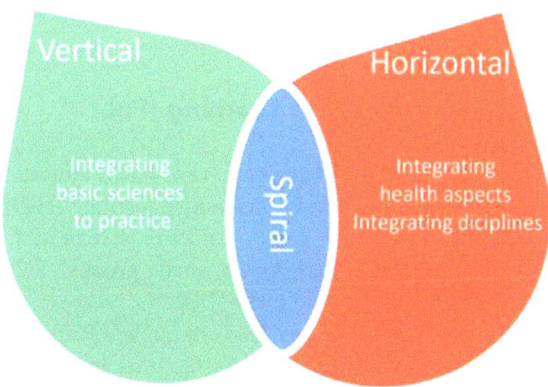

achieved in the mind of learners while the teachers provide the platform for creativity [17, 18].

6 Integrated Medical Education Models

6.1 Self-learning Modules (SLMs)

Even after graduation, educating students with self-directed training skills would guarantee ongoing independent learning. Not only after graduation but also in unforeseen (special) situations attributable to natural and man-made catastrophes (challenges), such as forced training of students in remote access (distance learning) in the case of quarantine in the country during the COVID-19 pandemic.

In this type of learning, the students are instructed to find the answers independently by using divergent strategies such as web-based SLMs and self-assessment quizzes. The student can harness this skill to prepare for collaborative group learnings and class discussions.

Four dimensions have been identified for SLMs. The first is learner control, making the model more inspiring and effective. Learners can adjust the pace of learning based on their own schedule and level of knowledge. The second aspect is personal autonomy, which will generate enthusiasm for learning by personal effort. Managing the skills individually and self-reliant approaches to learning are the other elements [7, 19].

The significance of this type of education is envisioned predominantly among basic science students who have been previously trained by various instructional methods and are now confronted with a multidisciplinary educational program. They have to integrate the different disciplines from anatomy to physiology [7]. On the other hand, some studies suggest that a previous knowledge base is required to use this model. Therefore, the application is superior among students of higher semesters [11].

However, the efficacy of this method in complex lessons such as physiology, which need discussion and masters' explanation, is disputable. A combination of seminars with student participation and masters' lectures seems more effective [20].

6.2 Problem-Based Learning (PBL)

In integrated medical education, case-based education and PBL, as an example, are given a lot of attention. In this model, learning is achieved through studying clinical synopses [4]. First, a case (problem) is reviewed, and then learning is pursued mostly through discussions in small groups. Subsequently, the discussion will continue with the whole class [21].

This education strategy leads to applied learning and builds a more long-lasting knowledge. This rises from the fact that in this method, students are exposed to scenarios that they would face in the future, and they could be able to match the real situation with the scenarios in their memory [4]. Additionally, in this model, team-working is reinforced, and students learn to assist each other in improving their abilities, resulting in better performance of their group [22].

Nevertheless, there are several challenges with this model. Students should be synched to the curriculum and study the topics before and after classes. Participation of students in debates is an essential factor for the efficacy of PBL. As a result, it is crucial that the setting created by the moderators during the class enables the students to make comments and not be scared of being wrong when commenting on the subjects.

However, we have to concede that case-based discussions are not sufficient for teaching some disciplines such as anatomy, and lectures are essential for teaching certain subjects [21].

6.3 Competency-Based Education (CBE)

"Competency-based education (CBE) is an approach to preparing physicians for practice that is fundamentally oriented to graduate outcome abilities and organized around competencies derived from an analysis of societal and patient needs. It deemphasizes time-based training and promises greater accountability, flexibility, and learner-centeredness" [23]. The healthcare professional should be educated with a medical curriculum that enables them to meet the community's health needs [24]. In this model, the skills are emphasized rather than knowledge.

The advantage is the adjustment of learning based on the learner's competence. The efficacy would improve, especially when PBL and CBE are combined simultaneously in medical education [25].

7 Experiences of Medical Universities in Running Integrated Programs

McMaster University was among the first medical universities to put forward an integrated curriculum in Canada. The model made a subsequently fair impression in medical education organizations around the world [10].

In 2014–2015's standards for accreditation of the United States medical education programs by the Liaison Committee on Medical Education (LCME), horizontal and vertical integration were considered in educational standards [26].

Arabian Gulf University (AGU) successfully applied the integration concepts in its educational program. Their curriculum is designed with consideration of integrating basic and clinical sciences, public health training and community exposure, and SLM and PBL [25].

A cross-sectional study at the University of British Columbia comparing the spiral integration with non-integrated medical education concluded that the integration leads to better retention of learnings. However, the proper implementation is time-demanding [27].

The Medical Faculty of the University of Hamburg started an integrated medical degree program in 2012. The main features of this program were the close integration of theoretical knowledge and practical skills, scientific orientation, and psychological and social skills training [28].

The experience of integration into the general dental curriculum in the world's universities in the 21st century was implemented in two models of horizontal and vertical integration. Among the universities offering horizontal model, we can mention Toronto, UK, Texas, Marquette, Ankara, Sheffield, Connecticut, Virginia, Harvard, and Boston, and for vertical model, Harvard, Japan, Baylor, Louisville, Germany, Netherlands, Pennsylvania, Toronto, Sweden, Switzerland, Indonesia, England, Baylor, Wales, and Germany [29].

Teaching staff at the University of Aberdeen considered integrating global health into the undergraduate medical curriculum through a student-selected component as a successful and practical method in 2016. This approach was the opportunity to expose students to the social determinants of health through interdisciplinary teaching [30].

The experience of curriculum integration from three faculties at a New Community College, Stella and Charles Guttman Community College, was published in 2015. The authors discussed an interdisciplinary course that concentrates on a critical issue that provides content and background for quantitative reasoning, reading, and writing to enhance students' developmental abilities. This integrated curriculum is provided in a learning community. Its aims comprise more information retention, enhanced transfer of information, and building developmental skills while learners obtain college credit. These tie in with the College's aims of enhancing retention and graduation rates. Initial outcomes from this curriculum were very positive [31].

There are several experiences of implementing integrated medical education in the curriculums of top Iranian medical universities. In Shiraz school of medicine, the horizontal basic science integration model and Early Clinical Exposure (ECE) for undergraduate medical students was established as an initial step towards a reformed medical education. According to a study, the course schedule and atmosphere of classes were considered suitable and encouraging by 87.81% and 83.86% of students, respectively, and 77.75% of the students were more eager to actively participate in lectures and relevant clinical activities held for educational purposes [32].

A similar experience of an integrated curriculum at Tehran University of Medical Sciences demonstrated that 53.7% of the students in the basic sciences phase were very satisfied or satisfied with the program (according to a five-point Likert scale from unsatisfied to very satisfied). However, student grade point averages (GPAs) and national comprehensive basic sciences exam results did not differ significantly between the traditional and renewed curriculum [33].

8 Conclusion

The most highlighted points of an effective and gratifying medical education program are active participation of medical students, integration of discrete disciplines along with the interconnection of basic and clinical sciences, provocation of self-learning skills, and the improvement of teamwork activities [34]. The power of integrating the basic and clinical sciences (vertical) along with distinct disciplines (horizontal) is an essential skill of a competent practitioner. However, the discussed models alone do not prepare the students to work in real life. Last but not least, we must perceive the patients as human beings with physical and social, with physiological and spiritual constituents. So, the educational curriculum must incorporate all these essentials. A certain qualification is also required of the moderator (teacher). He should be completely free to navigate the topic, know the clinical cases (problems), instantly assess and control the situation (process), quickly find the only solution to the problem, and provide alternative options. Experience shows that the most difficult stage is almost always finding and formulating the problem that will be discussed in the future. When the problem is found and formulated, the next steps are relatively easy. This is probably because most students in the classroom (especially in junior years) learn to solve the problems posed by the teacher or the authors of textbooks, but not to set and formulate their own ones. However, seeing and clearly identifying the problem is an important skill that every doctor should possess. As a result, it is crucial that the setting the moderators create in the class enables the students to express their comments and not be scared of being wrong when they are commenting on the subjects.

Shifting from the traditional curriculum to a new one is faced with difficulties and resistance. Medical instructors should hold an interdisciplinary method courageously and familiarize themselves with the lexicons of other fields, including the social sciences and humanities. Achieving a consensus in the curriculum design demands all stakeholders' collaboration and careful inter-organizational planning. The goal is to make education more engaging and efficient for posterity. Furthermore, there should be an agreement about objective ways to determine whether such change toward integrated medical education methods has led to better quality physicians and improved patient care. Moreover, lectures based on solving a particular problem will arouse much greater interest and motivation among a large student audience and will only improve the educational process.

Core Messages

- Innovative and more profound methods of learning have become the subject of interest in medical education.
- An integrated medical curriculum pays attention to all dimensions of a patient, from physical to spiritual.
- The power of integrating sciences (vertically) and disciplines (horizontally) is crucial to a competent practitioner.

- In integrated medical education, learning relies on learners' independent searching and case-based studying.
- Achieving a consensus in the curriculum design demands stakeholders' collaboration and inter-organizational planning.

References

1. Balasooriya CD, Hughes C, Toohey S (2009) Impact of a new integrated medicine program on students' approaches to learning. High Educ Res Dev 28(3):289–302
2. Mennin S (2010) Self-organisation, integration and curriculum in the complex world of medical education. Med Educ 44(1):20–30
3. Quintero GA, Vergel J, Arredondo M, Ariza M-C, Gómez P, Pinzon-Barrios A-M (2016) Integrated medical curriculum: advantages and disadvantages. J Med Educ Curricular Dev 3: JMECD.S18920
4. Hays R (2013) Integration in medical education: what do we mean? Educ Prim Care 24 (3):151–152
5. Muller JH, Jain S, Loeser H, Irby DM (2008) Lessons learned about integrating a medical school curriculum: perceptions of students, faculty and curriculum leaders. Med Educ 42 (8):778–785
6. Dent JHR, Hodges BD, Hunt D (2017) A practical guide for medical teachers
7. Khalil MK, Nelson LD, Kibble JD (2010) The use of self-learning modules to facilitate learning of basic science concepts in an integrated medical curriculum. Anat Sci Educ 3 (5):219–226
8. Breslow L (1972) A quantitative approach to the World Health Organization definition of health: physical, mental and social well-being. Int J Epidemiol 1(4):347–355
9. Samieefar N, Golmohammadi M, Rashedi R, Rezazadeh A, Heidarnia MA (2018) The relation of spiritual aspect of nutrition and health. Soc Determinants Health 4(4):208–216
10. Brauer DG, Ferguson KJ (2015) The integrated curriculum in medical education: AMEE Guide No. 96. Med Teach 37(4):312–322
11. Kadirvelu A, Gurtu S (2015) Integrated learning in medical education: are our students ready? Med Sci Educ 25(4):549–551
12. Wijnen-Meijer M, Ten Cate OTJ, Rademakers JJ, Van Der Schaaf M, Borleffs JC (2009) The influence of a vertically integrated curriculum on the transition to postgraduate training. Med Teach 31(11):e528–e532
13. Wijnen-Meijer M, van den Broek S, Koens F, Ten Cate O (2020) Vertical integration in medical education: the broader perspective. BMC Med Educ 20(1):1–5
14. Sánchez J, Andreu-Vázquez C, Lesmes M, García-Lecea M, Rodríguez-Martín I, Tutor AS et al (2020) Quantitative and qualitative evaluation of a learning model based on workstation activities. PloS One 15(8):e0236940
15. Shimura T, Aramaki T, Shimizu K, Miyashita T, Adachi K, Teramoto A (2004) Implementation of integrated medical curriculum in Japanese medical schools. J Nippon Med Sch 71(1):11–16
16. Yamani N, Rahimi M (2016) The core curriculum and integration in medical education. Res Dev Med Educ 5(2):50–54
17. Atwa HS, Gouda EM (2014) Curriculum integration in medical education: a theoretical review. Intellectual Property Rights: Open Access

18. Harden RM (2000) The integration ladder: a tool for curriculum planning and evaluation. Med Educ 34(7):551–557
19. Candy PC (1991) Self-direction for lifelong learning. a comprehensive guide to theory and practice. ERIC
20. Minhas PS, Ghosh A, Swanzy L (2012) The effects of passive and active learning on student preference and performance in an undergraduate basic science course. Anat Sci Educ 5 (4):200–207
21. Chang BJ (2016) Problem-based learning in medical school: a student's perspective. Ann Med Surg 12:88–89
22. Loftus S (2015) Understanding integration in medical education. Med Sci Educ 25(3):357–360
23. Frank JR, Mungroo R, Ahmad Y, Wang M, De Rossi S, Horsley T (2010) Toward a definition of competency-based education in medicine: a systematic review of published definitions. Med Teach 32(8):631–637
24. Gruppen LD, Burkhardt JC, Fitzgerald JT, Funnell M, Haftel HM, Lypson ML et al (2016) Competency-based education: programme design and challenges to implementation. Med Educ 50(5):532–539
25. Bindayna KM, Deifalla A (2020) The curriculum at the college of medicine and medical sciences at Arabian Gulf University: a way forward to meet the future medical education needs. J Med Educ Curric Dev 7:2382120520932904
26. Education LCoM (2007) Functions and structure of a medical school: standards for accreditation of medical education programs leading to the MD degree. Liaison Committee on Medical Education
27. Fraser S, Wright AD, van Donkelaar P, Smirl JD (2019) Cross-sectional comparison of spiral versus block integrated curriculums in preparing medical students to diagnose and manage concussions. BMC Med Educ 19(1):17
28. Rheingans A, Soulos A, Mohr S, Meyer J, Guse AH (2019) The Hamburg integrated medical degree program iMED. GMS J Med Educ 36(5)
29. Safarnavadeh M, Ebrahimpour Koumleh S, Mousapur N (1999) The experience of integration in the dental curriculum of the World's Accredited Universities and how to apply it in Iran. Future Med Educ J 8(1)
30. Aulakh A, Tweed S, Moore J, Graham W (2017) Integrating global health with medical education. Clin Teach 14(2):119–123
31. Saint-Louis N, Seth N, Fuller KS (2015) Curriculum integration: the experience of three founding faculty at a new community college. Int J Teac Learn Higher Educ 27(3):423–433
32. Rooholamini A, Amini M, Bazrafkan L, Dehghani MR, Esmaeilzadeh Z, Nabeiei P et al (2017) Program evaluation of an integrated basic science medical curriculum in Shiraz Medical School, using CIPP evaluation model. J Adv Med Educ Prof 5(3):148
33. Mirzazadeh A, Gandomkar R, Hejri SM, Hassanzadeh G, Koochak HE, Golestani A et al (2016) Undergraduate medical education programme renewal: a longitudinal context, input, process and product evaluation study. Perspect Med Educ 5(1):15–23
34. Arroyo-Jimenez MDM, Marcos P, Martinez-Marcos A, Artacho-Pérula E, Blaizot X, Muñoz M et al (2005) Gross anatomy dissections and self-directed learning in medicine. Clin Anat 18 (5):385–391

Noosha Samieefar is a seventh-year student in the doctor of medicine/master of public health (MD/MPH) dual degree program at Shahid Beheshti University of Medical Sciences. She has a commitment and engagement for volunteering and mentoring. As a medical student, she has volunteered at several scientific and executive positions like Managing Director and Research Manager of Network of Interdisciplinarity in Neonates and Infants (NINI), Universal Scientific Education and Research Network (USERN) Offices Manager, Executive Director and Research Deputy of Shahid Beheshti University of Medical Sciences USERN Office and Executive Director or Committee Member of about fifty scientific national and international events. Pediatrics and immunology and allergy is her field of interest in research and practice.

Nima Rezaei gained his medical degree (M.D.) from Tehran University of Medical Sciences (TUMS) in 2002 and subsequently obtained a M.Sc. in Molecular and Genetic Medicine and a Ph. D. in Clinical Immunology and Human Genetics from the University of Sheffield, UK. He also spent a short-term fellowship in Pediatric Clinical Immunology and Bone Marrow Transplantation in the Newcastle General Hospital. Since 2010, he has worked in the Department of Immunology and Biology, School of Medicine, TUMS; he is now Full Professor and Vice Dean of International Affairs, School of Medicine, TUMS, and Co-founder and Head of the Research Center for Immunodeficiencies. He is also founding President of USERN. He has edited more than 30 international books, has presented more than 500 lectures/posters in congresses/meetings, and has published more than 800 articles in international scientific journals.

Giving Voice to Social Values in Achieving Universal Health Coverage

28

Reza Majdzadeh, Haniye Sadat Sajadi, Remco van de Pas, and AbouAli Vedadhir

"Nearly all legislation involves a weighing of public needs as against private desires; and likewise a weighing of relative social values."

Louis D. Brandeis

R. Majdzadeh (✉)
Community-Based Participatory-Research Center, Knowledge Utilization Research Center, School of Public Health, Tehran University of Medical Sciences, Unit 7, 7th floor, No 1547, Research Institute Building of University of Medical Sciences, Northern Kargar, Tehran, Iran
e-mail: rezamajd@tums.ac.ir

R. Majdzadeh · H. S. Sajadi
Integrated Science Association (ISA), Universal Scientific Education and Research Network (USERN), Tehran, Iran
e-mail: hsajjadi@tums.ac.ir

H. S. Sajadi
Knowledge Utilization Research Center, University Research and Development Center, Tehran University of Medical Sciences, Tehran, Iran

R. van de Pas
Department of Public Health, Institute of Tropical Medicine Antwerp, Nationalestraat 155, 2000 Antwerp, Belgium
e-mail: rvandepas@itg.be

Department of Health Ethics and Society, Faculty of Health Medicine and Life Sciences, Maastricht University, 616, 6200 MD Maastricht, The Netherlands

A. Vedadhir
Department of Anthropology, Faculty of Social Sciences, University of Tehran, 14117-13118 Tehran, Iran
e-mail: vedadha@ut.ac.ir

Population Health Sciences, University of Bristol, Canynge Hall, Bristol BS8 2PS, UK

Summary

Access to relevant and affordable health services is a significant component of social protection, in particular, through universal health coverage (UHC). Despite the existing progress, achieving UHC is stalled or not on the right track. Worldwide, over 50% of people live without access to essential health services. Moreover, extended health emergencies such as the pandemic of coronavirus disease 2019 (COVID-19) put enormous pressure on the health systems around the world. There is a need to approach health and health care systems from a sociological view if we are seeking to address the current challenges of UHC. Social factors are important when determining health and strengthening the health care systems. Here, we explain why social values and considerations are central to achieving UHC. Critical social values and issues in the journey of attaining UHC are described. Then, some recommendations to align health and health care systems with a sociological lens are suggested. Strong collaboration, making, and implementing health policies with the aim of equity and regard to social inequalities, decent governance through more transparency, accountability, and responsiveness, investing more in common goods for health, and more public participation are social considerations that accelerate UHC. We deeply believe that achieving UHC is possible by giving voice to and applying social values and taking a multidisciplinary approach.

Priority areas of integration of socio-cultural perspectives for strengthening moving forward Universal health coverage (UHC)

The code of this chapter is *01101111 01101001 01000011 01101110 01101101 01110101 01101101 01111001 01110100*.

Keywords

Common goods for health · Health · Social values · Universal health coverage

1 Introduction

Universal health coverage (UHC) is a concept recently introduced in 2010. However, we listened to the earlier echoes in 2005 when the World Health Assembly resolution stated that financial issues should not deter access to health care [1]. UHC aims to ensure that all people and communities have access to their needed health services (including prevention, promotion, treatment, rehabilitation, and palliation) in good quality to be effective and without the risk of financial hardship [2]. UHC is currently the aspiration globally to frame health systems to improve both the subjective and objective well-being of a country's population. Undoubtedly, UHC would offer a high chance of achieving sustainable development goals (SDGs) of ending extreme poverty and increasing equity and shared prosperity [3]. It allows countries to invest in various forms of human capital, a foundational driver of sustainable development, and improved quality of life. Moreover, countries with advanced UHC have been better prepared to respond to global health risks, such as the current coronavirus disease 2019 (COVID-19) pandemic [4].

The global UHC monitoring report reveals that despite UHC progress, at least half the world's population still lacks access to essential health services [5]. Moreover, financial protection is going in the wrong direction. The population with out-of-pocket payments grew from 9.4% (2000) to 12.7% (2015), and 930 million people faced catastrophic health spending in 2015. In addition, the percentage of the impoverished population increased from 1.8% (2000) to 2.5% (2015) [6]. And we believe there is something wrong with further efforts, at both national and global levels, that are required to address UHC's current challenges in some countries. Beyond the health sector itself, the main important one is the high level of political leadership and commitment required to advance UHC. The UHC movement has gained global momentum, with the UN high-level meeting on UHC held in September 2019 [7, 8]. It affirms that the mainstay of UHC is an international commitment of high importance. Other efforts focus on issues like leaving no one behind, transparent and accountable regulating and legislating bodies, upholding the quality of care, inclusive economic growth, decent employment, moving together, and gender and income equality [9].

An in-depth look at the issues above implies that we need to approach health and health care systems from a sociological, anthropological, and political economic perspective. What makes this view or perspective important is the significant role of socio-cultural context and political economicy variables in determining or influencing individuals, groups, and society and strengthening the health care systems around the world. While no consistent taxonomy is likely to capture and model all the influence of society on health and illness, we can aggregate them into three broad groupings or paradigms which can be identified in the literature of sociology and anthropology: i, the social construction of health and illness; ii, the health lifestyles; and iii, the social determinants of health (SDH) and intersectionality [10].

1.1   The Social Construction of Health and Illness

In this view, the emphasis is on how meanings of phenomena are not necessarily rooted in the phenomena themselves but develop through social interactions in a sociocultural context. It dealt with many illnesses that are particularly embedded with cultural meaning that shapes how society responds to those afflicted and influences the experience of illness. Moreover, all illnesses are in some way socio-culturally constructed at the experiential level, based on how individuals come to understand and live with their health problems and find ways of overcoming them. Finally, medical knowledge and discourse about illness and disease are not necessarily given by nature but are constructed and developed by claims-makers and interested parties [11].

1.2   The Healthy Lifestyles

Healthy lifestyles are collective patterns of health-related practice based on choices (agency) from options available to people according to their life chances (structure). Such constellation of lifestyle practices can influence health as a leading determinant of people's lives, either positively or negatively, and even ambivalently. Chances are the probability that individuals will find a satisfactory account of life. These mainly depend on socioeconomic and/or class position, as well as age, gender, race and ethnicity, and other factors or forces that shape the lifestyle choices people make in their lives [12, 13]. As Micheal Mormot observed, "the result of unequal distribution of life chances is that health is unequally distributed. If you are born in the most fortunate circumstances, you can expect to have your healthy life extended by nineteen years or more, compared with being born into a disadvantage" [14].

1.3   The Social Determinants of Health (SDH)

Unlike the health lifestyles paradigm that asserts on sets of health practices that are determined by the dialectic between people's choice (agency) and chance (structure) in everyday life [15], SDH focuses on *the conditions in which people are born, grow, live, work, and age*. People's health lifestyles receive a direct and/or indirect effect of these conditions, which, in turn, are dependent on how much money, power, and resources are distributed at and penetrated at local, national, and global levels. Poor living and working conditions are critical examples of inherently social variables known to hurt health, threatening prospects of disease prevention and health maintenance. Viewed with the SDH, all diseases, including infections, genetic disorders, metabolic diseases, cancer, and degenerative conditions, appear complete only when the social context comes into play. More precisely, social factors determine the fate of exposure risk, host susceptibility, and disease prognosis. As William C. Cockerham accentuated, "social factors do more than influence health for large populations and the cultural meanings and lived experience of illness for individuals; rather, such factors have a *direct* causal effect on health and illness in all aspects. Society may indeed make you sick, or conversely, promote your health" [12].

Taking inspiration from sociological and anthropological accounts and perspectives, in this chapter, we shed light on why social values and considerations are important to achieve UHC, which is the defined goal of today's health systems. Then, we describe the major social values and issues that should be considered in our journey to achieve UHC. Finally, we bring up some key insights and recommendations to investigate health and health care systems from a sociological lens. We believe that achieving UHC is only possible by giving voice to and applying social values and taking an overtly multi- and inter-disciplinary approach.

2   Why Social Issues Matter to UHC

The UHC has focused on ensuring that health services are accessible and affordable for all. There are some countries where the implementation of the UHC could take them to a better place. This journey was, however, challenging for many countries. The challenges have become more complicated following the COVID-19 pandemic [4, 16], with substantial health, social, and economic impacts in all countries, regardless of their geographic position or whether rich or poor. We believe that achieving UHC requires some processes, policies, and solutions that, along with having technical features, should be socially sustainable, politically supported, and culturally acceptable. Hence, considering social values and technical issues is inevitable to address the existing and emerging challenges. We explain why social values and concerns matter to achieve UHC in the following.

2.1   Health and Well-Being as a Human Right

The UHC focuses on better health and well-being for all. *The enjoyment of the highest attainable standard of physical and mental health* is a human right; basic freedom needs to be met regardless of race, ethnicity, gender, religion, aspiration and belief, and socioeconomic condition. *The right to health* has some main aspects. First, it is an inclusive right, including a wide range of factors (i.e., safe drinking water and adequate sanitation, safe food, a healthy workplace without environmental issues, health-related education and information, access to health services, and financial contribution to the health expenditures) that can help us to take a healthy lifestyle and lead to a healthy life. Second, it includes a sense of agency and freedoms, both agency- and option-freedom (i.e., *free from non-consensual medical treatment, torture, cruel, inhuman, degrading treatment or punishment*; free from over-medicalization and social and structural iatrogenesis; and hegemony of the therapeutic state and pharmacracy, in Ivan Illich [17] and Thomas Szasz's words) [18, 19]. Third, it contains entitlements including, for instance, the right to a system of health protection for everyone, irrespective of gender, age, etc. Fourth is about the availability of health facilities, goods, and services that need to become accessible, culturally acceptable, high-caliber, and relevant and provided to all with no social discrimination and oppression [20]. Given the main aspects of the right to health, it is clear that the concept of UHC has an ethical basis that is rooted in social contract theory [21]. The world is, therefore, approaching to see and integrate many social values into the health care systems, as declared by scholars [22].

2.2   Equity as a Core Principle

Equal importance in quality health service, financial management, and assurance of health service with equity and access is another promise of UHC that comes to the concept of *equity* to our mind. This implies that UHC is per se a social value. Equity in health encompasses various dimensions, some related to means or processes, some related to ends or outcomes [23]: *equity in health care coverage (access, use of services) (often called horizontal equity: equal treatment for equal need); equity in health outcomes; equity in health financing (often called vertical equity, meaning that everyone contributes to health financing according to one's ability to pay); equity in financial protection.* It is now well recognized that biological and genetic endowment, although important in determining health, should not be the source of health disparities. Rather, race/ethnicity, culture, education, or other social advantages, which are mainly named the SDH, play a major role in shaping disparities in health outcomes. Hence, one requires further understanding of social concerns for identifying public policy solutions to support health and well-being [24]. Non-discrimination is a core principle. The policies that exclude certain individuals or groups are inconsistent with UHC. It is worth noting that although equity seems inherent to the pursuance of UHC, it is not a natural consequence of the implementation of UHC policies. Soors et al. provide the analytical insights that from an equity perspective, not only the three dimensions (known as the UHC cube) of population coverage, financial protection, and services must be considered in the implementation of UHC policies. Institutional design (structure), organizational practice (an agency of actors), particular context, politics, and path dependency have a great influence on whether UHC is being implemented equitably or not. Such a political economy approach is required to understand why, or not, UHC contributes to health equity at the national and local levels [25].

2.3 Community Participation

The dominant values of a health care system are also relevant for guiding UHC to be the main health system's goal. These values influence the dominant actors that should play a role in policymaking. As Andrew Twaddle described, there are three general modes of health care systems (democratic, professional, and market) that are based on and reproduce different social values. While the dominant value in democratic health care is equity, in the professional- and market-driven care systems, these values are effectiveness/quality and efficiency, respectively [26]. The dominant goals are respectively preservation and extension of democracy, enhancing knowledge and technical quality, and maximizing economic profit. In the first, the role of the people and political parties and their participation is more critical because they must be empowered to affect their health. In the second, professional and knowledge groups play a dominant role. Finally, in the third mode, the interaction of service recipients and service providers forms the health market. When a health care system follows UHC, people need to play a significant role and have dominant participation. This does not mean that people do not play a role without UHC, but the participation level is different from that of a system whose dominant role is not equity. Deliberation and representation of citizens are of core importance in implementing UHC policies and deciding on its benefit packages, including priority setting in health care. Active involvement of citizens enhances the legitimacy of UHC and the health system [27].

2.4 Progressive Universalism

The importance of social values is not limited to defining the goal for health or health care systems but also defining and applying strategies to achieve such a goal. One of the main strategies of UHC is "progressive universalism." It means countries from the beginning define a set of entitlements, even a limited set, and guarantee them for everyone. This is before extending the coverage with additional benefits. This service set should be recognized and appropriately financed to ensure these services adequately cover all and no one leaves behind. In this context, interventions against infectious diseases, conditions affecting reproductive, maternal, newborn, and child health, and non-communicable diseases (NCDs) have priority since they disproportionately affect vulnerable populations [28, 29]. Human rights scholars speak in this regard also about the core obligations, the essential set of entitlements, under the right to health to achieve UHC [30].

The progressive universalism shows how different UHC dimensions, including service coverage, population coverage, and financial protection, are proportionally benefiting the poorer part of the population and have an equity approach build-in its design. This is not just the socioeconomic status (SES) or socioeconomic position (SEP) of people [31], which is important in defining vulnerable populations. However, any other social source of discrimination or oppression such as ethnicity, gender, and residence place must be considered in the UHC. In practice, all indicators for tracking UHC are disaggregation for at least socioeconomic status, gender, and residents' place. However, these dimensions are defined and standardized for global monitoring. Each country should adopt stratified indicators according to inequity sources relevant to its own setting [5].

Additionally, based on the UHC definition, the UHC involves three coverages in health services, financial protection, and population. It is a dynamic and continuous process with a set of objectives to make changes in response to shifting demographic, epidemiological, and technological trends, as well as people's expectations and needs. Moving toward UHC and progressive universalism journey starts with a health needs assessment and defines a set of entitlements. It is a systematic method involving epidemiological, qualitative, and comparative methods to describe health problems of a population; identify inequalities in health and access to services; determine priorities for the most effective use of resources [32]. Setting the priorities cannot be made simply by appealing to mechanical sets of "what works" because measuring what works inevitably involves a set of value judgments. Even explicit and seemingly "scientific" criteria such as effectiveness and cost-effectiveness are embedded in views about, for example, the value of different health states. Hence, in addition to technical processes, broader social value judgments are inevitably brought to bear on decisions. Value judgments, which are an important element in any public justification of how priorities are set, are one of the major social issues [33, 34]. Setting priorities in health not only require considering technical considerations (such as effectiveness, cost, and etc.); it needs to consider how the public think and judge. Public values are important when we want to set out priorities. Sometimes some health interventions are cost-effective, but from the public's perspective, it is not acceptable because it is far from their values. One example was in the COVID-19 pandemic and the use of facilities in times of scarcity. What should be the social values and ethical principles for rationing the use of facilities such as mechanical ventilators? Medical staff in Italy was forced to cut off ventilators from those who did not have a good prognosis [35]. Alternatively, some countries did not prioritize the use of a ventilator for elderly patients that had to be hospitalized in an intensive care unit and rather chose to use the limited number of ventilators available for a younger patient group. These ethical criteria must be formed based on social values and deliberation with a community [36]. These revealed that the real participation or engagement of the people in all decision-making processes, including clinical decision-making at the triage level, is of much importance in the implementation of UHC.

2.5 Theory of Change

The health systems should be strengthened to progress toward UHC and boost the system performance. A health system serves to function as an organization around people, institutions, and financial, human, and scientific resources that are designed to improve, maintain, or restore the health and well-being of a given population. As health care systems are indeed social sub-systems embedded in macro-structural social contexts and shaped by human agency, social values are central to transforming and improving health systems. As such, health systems software—including values, goals, norms, ideas, and relationships—is considered a foundational focus of UHC efforts/policies/reforms. The international experience has also demonstrated that transformation in health care requires a systems approach [37]. Health care systems are dynamic and function as some complex adaptive systems, with its tipping points for change and unforeseen feedback loops and path dependency when interventions are introduced [38].

Moreover, to implement UHC policies in a better and efficient way, a complex process in itself, several factors need to be in place, work together, and lead to the desired outcomes. For this purpose, we need to use a theory of change (TOC), which is a tool to apply *critical thinking to the design, implementation, and evaluation of initiatives and programs intended to support change in their context.* The TOC helps to include not only *key variables, stakeholders, and pathways of change* but also *information about the context in which an intervention is implemented (including socio-cultural, political, and environmental conditions)* [31].

2.6 Investment in Society

The hazards and insecurities induced and introduced by modernization itself threaten the world. As Ulrich Beck [39–41], a German sociologist, argued in his work on "the global risk society." However, humans have always been subjected to a level of risk, which has typically been observed as made by non-human forces such as natural disasters. On the other hand, modern or late modern societies are exposed to various risks such as environmental pollution, climate change, newly discovered crime, and the outbreak of emerging diseases and illnesses that are mostly anthropogenic or manufactured. The human agency is highly involved in generating, regenerating, and mitigating these forms of risks. These risks are often externalized and not included in the design and preparedness of health care systems.

COVID-19 is a good example of the expressions of medical risks in our globalized world. It has strongly impacted countries with presumably developed health care systems as well. To provide an example, the United States of America (USA) was ranked as the best country to fight the outbreak of infectious disease as evaluated by the global health security index. However, nine months into the COVID-19 pandemic, the USA, with over 6 million cases and over 190.000 deaths, were the country affected worst by the viral pandemic. This indicates that in our globalized, modern societies, even wealthy nations can be unprepared in a time of emergency [42] and affected badly, and their health care systems overwhelmed by a global risk such as a viral pathogen [43].

These risks follow patterns of a lack of human capital and are distributed unequally in a population, even in the developed and modern reflexive societies, and influence all aspects of well-being and quality of life. Infectious diseases travel along veins of inequality [44]. These risks are multi-layered and complex, encompassing a spectrum from the micro to the global macro level and critical socio-cultural dimensions. Cosmo-political economic arrangements are central to framing and reframing these risks.

As the COVID-19 shows attention should be given to health resilience and health instead of the conventional health system approach. The latter, which is the ordinary course of a particular disease, was not successful and should be replaced with strengthening the whole system, including a whole governance approach, by considering the essential need to develop the public goods for health. Countries with a robust approach in using evidence for timely action and strong coordination as a kind of whole-of-governance approach were more effective and adaptive in recovering from the COVID-19 pandemic shock. These were countries, such as New Zealand [45] and Taiwan [46], with timely and substantial compliance of the whole society to preventive and control measures coherently. These were communities that have invested more in social cohesion, community awareness, and health literacy. In this view, the system's preparedness, such as knowledge communication, surveillance systems, contact tracing, and general laboratory capacities, should be strengthened.

In a similar line, paying attention to the community and community interventions is central to promoting health and building health resilience [47]. This delicate balance between health and its social aspects was evident in the lockdown effect on lower socioeconomic classes. These classes were disproportionally affected by the lockdown in several countries [44]. How to ease lockdown so that the economic, social, and health objectives can be considered balanced. If not done democratically, the community will not trust the government in its public health interventions and containment strategies. If there is no proper deliberation and participatory involvement of a broad range of citizens, neither health goals nor social development can be achieved [48].

All examples above reveal that the UHC has no place without considering socio-cultural aspects, values, and issues. This is needed in setting goals and determining strategies for achieving UHC, implementing it, and achieving it. The next will discuss solutions that should be considered for this link between the health and social sciences.

3 How Social Values Should Address for Achieving UHC

UHC is a great ambition or an "ideal type" in Max Weber's words [49] that the real circumstances and actions of many countries are far from achieving it. There are challenges in realizing UHC through building social, economic, and political capacities. It may not be possible to realize UHC without addressing these structural determinants and capacities, even though many health reforms have been developed and implemented over recent years. However, there is no doubt that moving toward such a goal is worthwhile. The willingness to achieve UHC is a goal itself. In this section, we illustrate how social science perspectives, values, and notions can facilitate the path to UHC at four international, national, health system, and service delivery levels (graphical abstract).

3.1 International Level

3.1.1 Strong Global Partnerships

The international level calls a need for an improved global health system. The UHC can only be realized with strong global partnerships and cooperation. Our new world is more interconnected than ever. Access to technology and knowledge has been improved, resulting in sharing ideas and fostering innovation. More coordinating policies are necessary for sustainable development and the reduction of health inequalities. The emergence of new economies in the east and the south (non-west) has made global exchanges no longer one-sided. Its speed has increased significantly, and we cannot consider limited geographical areas for the health care system's developments. A disease in one country is a risk that makes the disease's occurrence in other countries inevitable due to the intensity of international relations.

Another point is that with the entrance of emerging powers in the global economy and technology, we should expect more and more similar events in the world. These emerging powers must control these global risks/threats at the global level and be part of the solution. An example is the experience of the Far East countries to cope with avian influenza, severe acute respiratory syndrome (SARS), and Middle East respiratory syndrome-related coronavirus (MERS), which helped China and South Korea to manage and control the COVID-19 epidemic later. As a result, in the field of global health, we must move from the traditional model of unilateral knowledge production, dissemination, and consumption to a co-constructive cosmos-political approach. It means that the cooperation between the countries of the South and the North in a steady process will facilitate achieving UHC through being better prepared for the next dangers of the world. This may lead to a form or model of reciprocal and constructivist learning. We need to establish a global partnership between northern and southern countries for UHC as human well-being and rights (including economic and social rights) is fundamental social values for all people living on the globe.

3.1.2 Global Health Governance and Finance

However, global health governance architecture and progress toward UHC governance are a major challenge to ensure the targeting of global health objectives, including health security and UHC, in a coherent matter [50]. Global health financing mechanisms also influence UHC efforts at the national level [51]. Despite the international consensus on UHC, governance of its actual implementation at the global level is insufficient, inconsistent, and superficial. Albeit the technical and coordinating role of the World Health Organization (WHO), there is no international body, especially providing overall stewardship for implementing the UHC agenda. For instance, the function that UNAIDS and the Global Fund have in the international coordination of HIV/AIDS prevention, response, and mobilizing global funding is absent in the case of UHC.

Recent evidence indicates that in the absence of binding legal commitment, progress toward UHC will be constrained. For instance, the failure of universal coverage of essential health services to be front and center in the last decade might be due to global funding and financial challenges and lack of increase in international development assistance devoted to low-income countries. Accountability mechanisms for monitoring, reviewing, and action are not formally available to ensure compliance with these commitments [50, 52].

These observations are critical in light of the calculated price tag for SDG-3. More precisely, achieving the SDG health objectives in 67 low and middle-income countries (LMICs) containing 75% of the world's population would require an increase in investments from *US$ 134 billion annually to $371 billion, or $58 per person, by 2030*. Estimates also say that domestic resources can clean up 85% of the resulting costs. However, actions for SGDs would remain a dream in the 32 poorest countries that suffer from a trade gap of up to US$ 54 billion annually. The condition needs external aid for an average of US$1.7 billion [53]. External financing for advancing UHC persists in the years ahead. However, in 2015, only six high-income countries reached the target of 0.7% overseas development assistance (ODA), and in other countries, except a few, actions remained inefficient.

In 2014, the Working Group on Health Financing at the Chatham House Center on Global Health Security proposed *a set of policy responses encapsulated in 20 recommendations for how to make progress toward a coherent global framework for health financing*. These emphasize the notion of shared responsibility between developed and developing countries. Each government should a, commit at least 5% of its gross domestic product (GDP), progressively increase to a per capita spending of USD 86 on the heath; b, establish innovative financing mechanisms; and c, establish minimum entitlement packages through transparent priority setting processes. Globally, stakeholders should support institutions mandated to provide or finance global public goods and create a fruitful environment where countries can apply policies to investment in social sectors. Finally, it is necessary for

  i. developed countries to commit at least 0.15% of the GDP to Development
     Assistance for Health (DAH);
 ii. providers of DAH to provide clear criteria for the allocation of resources; and
iii. the government to assess and improve existing global pools of health finances
     [54, 55].

3.2 National Level

3.2.1 Alleviation of Any Form of Oppression

The UHC tries to see all people as universal regarding the concept of universalism,
but it does not mean that we ignore the socio-cultural and local context of societies.
The health needs of the population vary, and ironically, the specific health needs
can be identified and addressed through considering characteristics of different
socio-cultural subgroups. Thus, although the ultimate goal is equitable health for
all, irrespective of social stratifies, communities' diversity must be taken into
account to achieve UHC. The interventions required to achieve UHC must be
designed and implemented flexibly to address community diversity. *In other words,
to achieve the goal of leaving no one behind*, new ways to understand the complex
nature of health inequities, especially among the most vulnerable populations
around the world, are required. Intersectionality is seen as a promising approach to
produce and sustain unequal health outcomes. Not only can this approach acts on
individual factors (biology, socioeconomic status, sex, gender, and ethnicity, and
race) to define the subgroups of the community, but it focuses on the real roots of
health inequalities and the intersectional relationships, these factors are making.
Also, it can operate at different levels of society, thereby inferring groups and
geographical settings that might be at play when health is constructed (e.g., the
health status of poor migrant women in the highly masculine cultures, in G. Hofstede word [56]. By taking such approaches, health interventions to achieve UHC
can be designed, implemented, and evaluated with relevance to the entire population's real needs. In this case, there is hope for the sustainability and effectiveness of
health interventions. Moreover, it is of importance that citizens can generate a social
contract with their nation-state. This implies a fair and democratic generation and
allocation of resources for health services provision, best through a national pooled
fund. A proper accountable fiscal framework, funded by a diverse range of progressive taxation mechanisms and/or mandatory social insurance schemes, has
proven historically to be the best way to move forward with universal social protection for health mechanisms at the country level [21]. The Thai national assembly
model has indicated why it is important to have an inclusive, deliberative dialog
with citizens on developing such a social contract for health [48].

   In conclusion, while universalism should be pursued to achieve UHC, the
optimal use of socio-cultural means must be reflected to materialize this approach.
Social oppression and suppression should not be why a distinct subgroup of the
community is deprived of health [57]. Universalism can only be achieved by

recognizing the socio-cultural contexts and rights and diminishing discriminatory sources, including social oppressions and institutional discrimination [58]. If discriminatory distinctions can be overcome in society, we can expect to move toward UHC. In other words, the realization of equity requires creating a society in which there are no social benefits for certain groups. The difference between rich and poor, between the sexes, or between different races should not be the criteria to help merit groups. Rather, support should be directed to vulnerable groups who have difficulties accessing health services. Eventually, in the journey to achieve UHC, any discrimination against and between social subgroups should disappear.

3.2.2 The Dominance of Community Empowerment

For UHC to be achieved, more community participation will be required in democratic institutions such as citizen panels and citizens' jury and other platforms. In the absence of public participation, the desired result cannot be achieved. Community participation depends on socio-cultural contexts, civic institutions' capacity, and the political context in which the people's voice can be heard.

Community participation has different levels, from the information to the empowerment level. At the first level, i.e., inform, people are aware of decisions on their health. In the next level of consultation, they are asked about their social values and health services. At the involved level, people participate in the design of health programs. In collaboration, people are part of the process of doing work and have an active presence. Finally, at the empowerment level, people have the authority to decide on allocating health resources. Realizing community roles in each of these levels need its own specific approaches and techniques. This highlights the need for social sciences in involving people in the right place of acting in a health system. Fact sheets, Websites, and open houses are the primary modalities used to inform people, which is the lowest level of involvement. However, advisory councils must be formed with the presence of the people at the collaborative level. At the empowerment level, people should be able to decide on health interventions, which should be based on delegated decision mechanisms such as citizens' juries. As a result, UHC's goal as a goal is a structure that should be formed from collective wisdom, collective action, and community participation [59].

People and opinion leaders who have the advantages of power and resources should believe that UHC belongs to them and can fulfill their benefits. Active community participation in decision-making gives a collective identity. This needs a mutual understanding and responsibility of people and the public sector. The re-requisite of such capital in society is managing various social oppressions (such as gender, race, ethnicity, language, religion, and SES) and moving together [60, 61].

The communication strategies are also crucial for the persuasion of political leaders and community engagement toward UHC and their empowerment. These communicational and constructionist strategies and claim-making activities (e.g., naming, typifying, framing/reframing, and managing a problem) must be effective, relevant, contextualized, sustainable, and flexible to make communities and their leaders' advocate of the UHC and mobilize them for expressing demand for UHC and finally to make sound decisions [11, 62].

3.3 Health System Level

3.3.1 Decent Governance

Strengthening the health system governance of each country is essential to achieving UHC. Governance of the health system is the main key to improving performance in the health sector of a country and, along with improving health financing and delivery of services, is central to achieving UHC. Decent governance requires stronger mechanisms for transparency and accountability. Transparency is critical for accountability and for public trust in government [63]. It is an essential component of collective attention. It is crucial not to undermine dialog, build trust between stakeholders, and be accountable to people's concerns. For citizens to trust the health system, *they must know what governments are doing and have access to reliable information.* Websites are simple ways to deliver real-time data and add localized information on health topics. These Websites can be managed by the government, academia, or civil society; many result from collaboration among different actors, including the private sector [64].

Effective transparency requires proactive communication strategies that reach vulnerable and at-risk populations with the information they need in accessible formats. Transparency is also important at the international level to better coordinate global efforts, share experiences and lessons learned, and support countries to tailor their strategies to their own circumstances. Accountability has also become an essential aspect in improving the governance of the health system. General definitions of accountability *include the obligation of individuals or agencies to provide information about and/or justification for their actions to other actors*, along with *the imposition of sanctions for failure to comply and/or to appropriate engage in action.* Accountability is very critical since every human being must access health services and, in turn, pay taxes or their share from public money, which implies that health policymakers must demonstrate a strong sense of accountability [65].

3.3.2 Skilled Health Workforce

Over the last few decades, the global health workforce (HWF) gap has increased. By this gap, we mean the skilled HWF required to provide essential health care services across the world in an equitable manner. Due to demographic growth in different regions of the world, an aging workforce, and an epidemiological transition to chronic diseases worldwide, there is a need for and impetus required to invest in skilled health workers and their decent employment. The WHO estimates that 4.45 health workers per 1000 population are required to reach the SDGs' health-related targets. This amounts to a total global deficit of 17.6 million health workers relative to the current supply, with a projected deficit of 13.6 million health workers in LMICs alone [66]. It is of uttermost importance, even more in these pandemic times, that countries invest in the education and employment of a skilled, well-remunerated health workforce. Moreover, during their career, these health workers require proper career planning and postgraduate training. The COVID-19 pandemic has indicated the importance of a safe and enabling working environment for health workers, including the provision of personal protective equipment (PPE) [67].

3.4 Service Delivery Level: Considering Common Goods for Health

The COVID-19 has shown that the free health market, with all financial expenditures, was not successful in securing health for the community despite all technological illusions. Ultimately, the health free health system cannot provide resilience to deal with global health threats. Many of the essential interventions toward UHC are common goods for health (CGH). Therefore, CGH is the zero step in designing a strong health system.

The CGH is population-based functions or interventions that require collective financing, regardless of whether they are delivered by public or private sector providers. The CGH can contribute to health and economic progress. These public goods are non-rival and non-exclusionary that can affect all people, even those who did not choose them (social externality). CGH fall into the following categories:

- *policy and coordination;*
- *taxes and subsidies;*
- *regulations and legislation;*
- *information collection;*
- *population services;*
- *analysis; and*
- *communication* [68].

The examples for the CGH which are very related to the social aspects are managing community engagement, taxes on products with impact on health to create market signals leading to behavior change, subsidies affect the use of important public health interventions (e.g., TB, HIV, vaccinations), environmental regulations and guidelines (e.g., for biodiversity, water, and air quality), communication and dissemination, community behavior change communication, sewage treatment and control, and medical and solid waste management.

While for a product/service which benefits an individual, the method for choosing interventions in a free market is based on supply and demand with little or no government control, the financing of a CGH is complex; it does not purchase for any individual profit and preference. Since it has an externality, it needs a collective decision. The resource allocation criteria for CGH are not based on an individual's profit and preference. As Savedoff argued [69], an investment in CGHs entails people cooperating to:

- *See themselves as part of new forms of collective identity beyond individuals and families with reciprocal obligations;*
- Define *the problems that can be resolved by CGH become salient and identifiable, as part of the individual or collective interests;* and
- *Come to believe that CGH is in their own or their class's best interests.*

4 Conclusion

UHC is a crucial part of the global health agenda. However, it remains an ambitious program for many countries, as assessed by observing the very slow or even stagnation toward formulation and implementation of health policies and reforms required for sustainable transformation. A review of global UHC progress, the challenges involved, and the lessons learned from the COVID-19 pandemic suggests that UHC is a big aspiration that full public support is needed to achieve. If it is not considered a macro-social issue, the instability of change will fail all the successes. Achieving justice in any social sub-system, including equity in the health and health care system, requires an overhaul transformation in the dominant values and socio-political philosophy of governance and leadership of society and giving voice and role of the people. People's participation is instrumental and should not be considered a superficial phenomenon to target the roots of inequality.

We think that investment in the UHC and their implementation in the current uncertain and globalized world needs close collaboration at all levels (community members at the level of individuals and collectively, community with the government, organization with each other, governments, south-north and East and west of the globe). The trustworthiness synergism among all global societies is needed to respond to the existing and emerging global threats, including COVID-19. Of course, without collective action, no one can be immune from the coming shocks and crises. As a result, this is not the time for uni-directional and authoritative relations, reproduction of inequality, and undermining others by othering and stigmatization processes worldwide [70].

Such strong collaboration at different levels itself needs a fundamental change of thoughts, discourses, and practices. Instead of looking at the health system and its defined goals from a single technical or disciplinary perspective, an integrated multidisciplinary approach with key social science insights and imaginations should be applied to study and propose new health policies and measures. No one could expect the health sector to succeed in UHC by focusing on health policies and programs without a comprehensive approach to the whole society and governance.

Core messages

- UHC is a great ambition. Achieving UHC calls for integrating the social science perspectives, values, insights and notions and taking an overtly multidisciplinary/inter-disciplinary approach.
- Priority areas for strengthening moving forward UHC are a global partnership and change in the global governance and finance, alleviation of any form of social oppression and dominance of community empowerment at the national level, good health system governance, having capable health workforce in all aspects, and further investment in the common goods for health.

- The fundamental component of all social accelerators toward UHC is an investment in citizens' active involvement and paradigm shift in facilitating community-led decision-making at all levels within and out of the health care system.

References

1. Evans DB, Etienne C (2010) Health systems financing and the path to universal coverage. SciELO Public Health
2. https://www.who.int/healthsystems/universal_health_coverage/en/
3. https://www.worldbank.org/en/topic/universalhealthcoverage
4. Armocida B, Formenti B, Palestra F, Ussai S, Missoni E (2020) COVID-19: Universal health coverage now more than ever. J Glob Health 10 (1)
5. World Health Organization (2017) Tracking universal health coverage: 2017 global monitoring report
6. World Health Organization (2019) Primary health care on the road to universal health coverage: 2019 global monitoring report. WHO, Geneva
7. https://www.who.int/news-room/events/detail/2019/09/23/default-calendar/un-high-level-meeting-on-universal-health-coverage
8. https://www.un.org/pga/73/wp-content/uploads/sites/53/2019/05/UHC-Political-Declaration-zero-draft.pdf
9. https://www.uhc2030.org/fileadmin/uploads/uhc2030/Documents/UN_HLM/UHC_key_targets_actions_commitments_15_Nov_2019__1_.pdf.
10. Wainwright D (2008) The changing face of medical sociology. In: A sociology of health, pp 1–18
11. Conrad P, Barker KK (2010) The social construction of illness: key insights and policy implications. J Health Soc Behav 51(1_suppl):S67–S79
12. Cockerham WC (2013) Social causes of health and disease. Polity, Cambridge, UK
13. Cockerham WC (2013) Bourdieu and an update of health lifestyle theory. In: Medical sociology on the move. Springer, pp 127–154
14. Marmot M (2013) The health gap: the challenge of an unequal world. Bloomsbury, London, 2015. 44. Health Economics, Policy and Law
15. Cockerham WC (2020) Sociological theories of health and illness. Routledge, New York
16. Kickbusch I, Gitahi G (2020) COVID-19 (coronavirus): universal health coverage in times of crisis. Updated on 29:2020
17. Illich I (1975) Medical nemesis. Citeseer
18. Szasz T (2003) Pharmacracy: medicine and politics in America. Syracuse University Press
19. Szasz T (2007) The medicalization of everyday life: selected essays. Syracuse University Press
20. https://www.ohchr.org/Documents/Publications/Factsheet31.pdf
21. Swaan Ad (1988) In care of the state: health care, education and welfare in Europe and the USA in the modern era
22. Ooms G, Latif LA, Waris A, Brolan CE, Hammonds R, Friedman EA, Mulumba M, Forman L (2014) Is universal health coverage the practical expression of the right to health care? BMC Int Health Hum Rights 14(1):1–7

23. Cohen AB, Grogan CM, Horwitt JN (2017) The many roads toward achieving health equity. J Health Polit Policy Law 42(5):739–748
24. McGibbon E, McPherson C (2011) Applying intersectionality and complexity theory to address the social determinants of women's health. Women's Health and Urban Life 10(1) pp 59-86
25. Soors W, De Man J, Dkhimi F, Van de Pas R, Criel B, Ndiaye P (2016) Towards universal coverage in the majority world: the cases of Bangladesh, Cambodia, Kenya and Tanzania
26. Twaddle A (2004) How medical care systems become social problems. In: Handbook of social problems, pp 298–315
27. Norheim OF, Baltussen R, Johri M, Chisholm D, Nord E, Brock D, Carlsson P, Cookson R, Daniels N, Danis M (2014) Guidance on priority setting in health care (GPS-Health): the inclusion of equity criteria not captured by cost-effectiveness analysis. Cost Effectiveness Resou Allocation 12(1):18
28. Jamison DT, Summers LH, Alleyne G, Arrow KJ, Berkley S, Binagwaho A, Bustreo F, Evans D, Feachem RG, Frenk J (2013) Global health 2035: a world converging within a generation. Lancet 382(9908):1898–1955
29. World Health Organization (2014) Making fair choices on the path to universal health coverage. Final report of the WHO Consultative Group on Equity and Universal Health Coverage
30. Forman L, Beiersmann C, Brolan CE, McKee M, Hammonds R, Ooms G (2016) What do core obligations under the right to health bring to universal health coverage? Health Hum Rights 18(2):23
31. Paina L, Wilkinson A, Tetui M, Ekirapa-Kiracho E, Barman D, Ahmed T, Mahmood SS, Bloom G, Knezovich J, George A (2017) Using Theories of Change to inform implementation of health systems research and innovation: experiences of future health systems consortium partners in Bangladesh, India and Uganda. Health Res Policy Syst 15(2):109
32. Wright J, Williams R, Wilkinson JR (1998) Development and importance of health needs assessment. BMJ 316(7140):1310–1313
33. ClarkS W (2012) Social values in health priority setting: a conceptual framework. J Heal Organ Manag 26(3):293–316
34. Littlejohns P, Weale A, Chalkidou K, Teerwattananon Y, Faden R, Clark S (2012) Social values in health priority setting: a conceptual framework. J Health Organ Manag
35. Truog RD, Mitchell C, Daley GQ (2020) The toughest triage—allocating ventilators in a pandemic. N Engl J Med 382(21):1973–1975
36. Emanuel EJ, Persad G, Upshur R, Thome B, Parker M, Glickman A, Zhang C, Boyle C, Smith M, Phillips JP (2020) Fair allocation of scarce medical resources in the time of Covid-19. Mass Med Soc
37. De Savigny D, Adam T (2009) Systems thinking for health systems strengthening. World Health Organization
38. Van Olmen J, Marchal B, Van Damme W, Kegels G, Hill PS (2012) Health systems frameworks in their political context: framing divergent agendas. BMC Public Health 12 (1):774
39. Beck U (1998) Politics of risk society (Franklin J (ed)). Polity, Cambridge, UK
40. Beck U (2009) World at risk. Polity
41. Beck U, Lash S, Wynne B (1992) Risk society: towards a new modernity, vol 17. Sage
42. Lakoff A (2017) Unprepared: global health in a time of emergency. Univ of California Press
43. Dalglish SL (2020) COVID-19 gives the lie to global health expertise. The Lancet 395 (10231):1189
44. Pas van de R (2020) Globalization paradox and the coronavirus pandemic, clingendael report. Clingendael, Netherlands Institute of International Relations, Netherlands
45. Jamieson T (2020) "Go Hard, Go Early": preliminary lessons from New Zealand's response to COVID-19. Am Rev Public Adm 50(6–7):598–605

46. Huang IYF (2020) Fighting against COVID-19 through government initiatives and collaborative governance: Taiwan experience. Public Adm Rev
47. Wulff K, Donato D, Lurie N (2015) What is health resilience and how can we build it? Annu Rev Public Health 36:361–374
48. Rajan D, Mathurapote N, Putthasri W, Posayanonda T, Pinprateep P, de Courcelles S, Bichon R, Ros E, Delobre A, Schmets G (2019) Institutionalising participatory health governance: lessons from nine years of the National Health Assembly model in Thailand. BMJ Glob Health 4(Suppl 7):e001769
49. Weber M (1949) The methodology of the social sciences (trans: Shils EA, Finch HA (eds)). FreePress, Glencoe
50. Kickbusch I (2016) Global health governance challenges 2016–are we ready? Int J Health Policy Manag 5(6):349
51. Balabanova D, McKee M, Mills A, Walt G, Haines A (2010) What can global health institutions do to help strengthen health systems in low income countries? Health Res Policy Syst 8(1):22
52. World Health Organization (2017) Fourth global forum on human resources for health
53. Stenberg K, Hanssen O, Edejer TT-T, Bertram M, Brindley C, Meshreky A, Rosen JE, Stover J, Verboom P, Sanders R (2017) Financing transformative health systems towards achievement of the health sustainable development goals: a model for projected resource needs in 67 low-income and middle-income countries. Lancet Glob Health 5(9):e875–e887
54. Ottersen T, Elovainio R, Evans DB, McCoy D, Mcintyre D, Meheus F, Moon S, Ooms G, Røttingen J-A (2017) Towards a coherent global framework for health financing: recommendations and recent developments. Health Econ Policy Law 12(2):285–296
55. Røttingen J, Ottersen T, Ablo A, Arhin-Tenkorang D, Benn C, Elovainio R, Evans D, Fonseca L, Frenk J, McCoy D (2014) Shared responsibilities for health: a coherent global framework for health financing. Final Report of the Centre on Global Health Security Working Group on Health Financing
56. Hofstede G, Consequences CS (2001) Comparing values, behaviors, institutions and organizations across nations. Sage, Thousand Oaks, CA
57. McGibbon EA (2012) Oppression: a social determinant of health. Fernwood Pub
58. Pincus FL (2019) Race and ethnic conflict: contending views on prejudice, discrimination, and ethnoviolence. Routledge
59. Odugleh-Kolev A, Parrish-Sprowl J (2018) Universal health coverage and community engagement. Bull World Health Organ 96(9):660
60. Van Wormer K, Link R (2016) Minority groups and the impact of oppression. In: Social welfare policy for a sustainable future. SAGE Publications, Inc, Thousand Oaks, CA, pp 179–210
61. Vickers T (2016) Refugees, Capitalism and the British State: implications for social workers, volunteers and activists. Routledge
62. Loseke DR (2011) Thinking about social problems: an introduction to constructionist perspectives. Transaction Publishers
63. Mabillard V, Pasquier M (2015) Transparency and trust in government: a two-way relationship. In: Yearbook of Swiss administrative sciences, pp 23–34
64. Montero AG, Blanc DL (2020) Resilient institutions in times of crisis: transparency, accountability and participation at the national level key to effective response to COVID-19. United Nations Department of Economic and Social Affairs
65. Baez Camargo C (2011) Accountability for better healthcare provision: a framework and guidelines to define understand and assess accountability in health systems
66. Liu JX, Goryakin Y, Maeda A, Bruckner T, Scheffler R (2016) Global health workforce labor market projections for 2030. The World Bank
67. Van de Pas R (2020) A cosmopolitan outlook on health workforce development
68. Yazbeck AS, Soucat A (2019) When both markets and governments fail health. Health Syst Reform 5(4):268–279

69. Savedoff WD (2019) Why do societies ever produce common goods for health? Health Syst Reform 5(4):402–405
70. Gover AR, Harper SB, Langton L (2020) Anti-Asian hate crime during the CoViD-19 pandemic: exploring the reproduction of inequality. Am J Crim Justice 45(4):647–667

Reza Majdzadeh is an Epidemiology professor who runs two research centers of knowledge translation and community-based participatory research (CBPR) at Tehran University of Medical Sciences. His main area of interest is health inequalities. Most of his works are related to providing evidence for informed poli-cymaking. The result of the conjunction of these two is the production and utilization of the knowledge for closing health inequity gaps. He established several educational programs in Iran, including MD-MPH, MPH in various minors, and Healthy Technology Assessment (HTA). He has a robust research port-folio with the publication of more than 300 peer-reviewed papers. He was responsible for a health observatory, monitoring Universal Health Coverage (UHC), and evaluating the health sector reform at the national level, focusing on access to services and financial protection. He was in charge as the director of Iran's National Institute of Health Research and consulted the Eastern Mediterranean Regional Office of the World Health Organization on service package design and implementation for achieving UHC.

Haniye Sadat Sajadi is an associate professor at Tehran University of Medical Sciences. She received her Ph.D. in health services management on July 29, 2013. she has always been interested in conducting health policy and system research and working in multidisciplinary teams. Since she started working as a technical expert and now as an academic member, she has often been involved in producing relevant evidence to answer questions and issues posed by policymakers or managers, both at national and local levels. Her research interest includes health policy analysis, evidence-informed health policymaking, health system reforms, and quality improvement in healthcare. She has published widely in prestigious journals in the field of health policy and systems. She is now working on several projects to examine how sanctions and the recent global crisis, COVID-19, affect the health system and its primary functions.

Remco van de Pas is a public health doctor and a global health researcher. He has a position as senior research fellow global health policy at the Institute of Tropical Medicine, Antwerp, and a Global Health lecturer at Maastricht University. His teaching and research focus on global health governance, its political economy, and foreign policy with special attention on health workforce development and migration, health system strengthening, social protection and health financing, global health security, globalization, climate change, and its impact on health equity. Remco is a board member of MMI –Network Health for All!, a visiting research fellow at Clingendael, Netherlands Institute of International Relations, and editorial board member of the academic journals BMC Globalization and Health and BMC Human Resources for Health. He received a Ph.D. degree in 2020 with a thesis on global health policy and the international governance of the health workforce and labor migration. An overview of academic and other publications can be found via ResearchGate.

AbouAli Vedadhir (Ali) is currently Associate Professor of Anthropology and Health Studies at the University of Tehran and Honorary Senior Research Fellow of Population Health Sciences in the Bristol Medical School, University of Bristol, UK. His research interests include anthropology and global health; socio-cultural studies of reproductive and sexual health with a focus on assisted reproductive technoscience and selective reproduction; social construction of health and illness; medicalization studies; community-based nutrition and food policy; anthropological demography and the anthropological science and technology studies (STS) especially in the context of the Middle East, where he has researched since 1997. He has written and published various articles on health issues from an anthropological perspective. Ali has formerly held visiting and research fellow positions at Tehran University of Medical Sciences, McMaster University, University College London, and the University of Oxford. He is presently working on 'Social Science Research Contributions to Antimicrobial Resistance (AMR).'

Discrimination in Medical Research Sampling: Recommendations and Applications to Psychology

Gerald Young

"Injustice anywhere is a threat to justice everywhere."

Martin Luther King Jr

Summary

The American Medical Association (AMA) ethics code deals with the full scope of medical practice, research, and other practice aspects, through nine ethical principles and nine chapters of opinions, which are like standards. This chapter focuses on how research ethics is dealt with in the AMA code and how it can be improved, especially for research inclusiveness for minorities, gender, the elderly, the vulnerable, and different socioeconomic status (SES) strata, which the code has not sufficiently considered. Different groups have different physiologies, and the side effects of drugs might be different for underrepresented groups in medical research, such as in randomized clinical trials, or the drugs might be less effective for those groups. Informed consent procedures in recruiting research samples should refer to what the research sampling in medicine lacks and the research undertaken should address these concerns. Medical practitioners should be required ethically to inform patients of outgroups in medical research sampling that proposed medications and treatments might not be as effective as claimed for the typical sampled patient in medical research. The chapter includes related bioethical conundrums in medical research concerning discrimination, points out similar psychological research sampling deficiencies, and makes equivalent recommendations for this field as well.

G. Young (✉)
Glendon College, York University, Toronto, ON, Canada
e-mail: gyoung@glendon.yorku.ca

© The Author(s), under exclusive license to Springer Nature Switzerland AG 2022
N. Rezaei (ed.), *Multidisciplinarity and Interdisciplinarity in Health*,
Integrated Science 6, https://doi.org/10.1007/978-3-030-96814-4_29

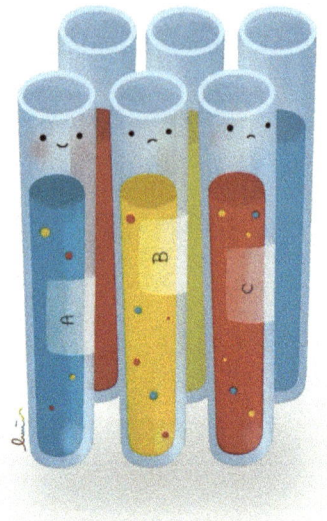

Discrimination in medical research sampling
[Adapted from the Association of Science and Art (ASA), USERN; Made by Nastaran Hosseini].
The code of this chapter is *01101001 01110010 01100101 01100110 01010000 01101111 01101110 01101111 01110011 01110011.*

Keywords

Elderly bias · Gender bias · Group bias · Informed consent · Medical ethics · Medical research · Medical research sampling · Psychological research sampling · Research ethics

1 Introduction

The chapter underscores a lack of representativeness in medical research sampling related to gender, the elderly, minority and related groups, and the vulnerable, leading to deleterious consequences for these groups. The chapter contributes to the literature by covering in one publication space the recent literature on bias in various aspects of medical research related to gender, minorities, races, elderly, vulnerable, and different socioeconomic status (SES) strata. This serves to skew medical research results to the majority studied, more often male, Western, Caucasian, not elderly, etc., which has health consequences for these outgroups.

The chapter makes recommendations relative to research ethics and informed consent procedures. Medical practitioners should be required ethically to inform patients that typical medical research sampling in research on medications and interventions might lead to results that do not deal with their particular group and they might not be as effective as claimed for the patient. The chapter concludes with further examples of what medical research ethics lacks in these regards and how the medical research process's current framework can accommodate them bioethically. In this regard, Guttierez [1] has highlighted that anti-racism policies should not be just "performed" but have real and reliable strategic institutional changes.

It is acknowledged that there are risks in over-reporting to patients all the material risks in proposed medicinal drugs and medical interventions, an increasing occurrence in the field taking place in order to avoid potential lawsuits [2]. However, the chapter maintained that this practice in the process of obtaining informed consent would not only protect the practitioner and patient but also will spur necessary systemic change in the medical community and the pharmaceutical companies involved in conducting medical research. The present chapter's proposals on informed consent are based on the principles, patient rights, and research chapter published in the American Medical Association's (AMA) ethics code [3]. Accordingly, revisions could be proposed to the AMA ethics code and also to the informed consent process with patients being recruited for medical research in light of the typical gaps in medical research sampling.

2 The AMA Ethics Code

2.1 Background

The AMA ethics code [3] describes nine fundamental principles from reasoned opinions held in nine chapters. The opinions include a list of patient rights; Chap. 7 covers research ethics. There are neither specific principles related to research nor specific patient rights related to research within the nine ethical principles and within the research chapter. However, the code does address issues related to making voluntary informed decisions and confidentially and the like (Table 1).

Without explicit advocacy for lack of discrimination in medical research sampling in principles of the AMA ethics code and in its research chapter, disproportionate sampling of specific population segments in such medical research has an increased probability of continuing, with the deleterious consequences for these populations continuing, as well. Given that these segments concern women, elderly, and different races, among others, to date, the potential harmful medical consequences of medical research have been on a vast scale.

Table 1 Principles of medical ethics of the American Medical Association [3]

Principle	Explanation
I	Give competent care, with compassion, and with respect for dignity/rights
II	Be professional, honest, and strive to report practitioners who are not (e.g., fraudulent)
III	Respect the law and take the responsibility to seek changes in laws that are contrary to the best interests of patients
IV	Respect the rights of patients, colleagues, and other health professionals and safeguard patient confidentiality and privacy (within the limits of the law)
V	Continue to study, apply, and advance scientific knowledge and duly inform patients, colleagues, and the public as required
VI	Feel free to choose to whom one should provide care (and in what context and with whom in association)
VII	Recognize one's responsibility to contribute to the improvement of public health
VIII	In caring for a patient, consider as paramount the responsibility to the person
IX	Support access to care for all

2.2 The AMA Ethics Code's Research Chapter

It is in Chap. 7 where research in the AMA medical ethics code is described [3] as an expository text/value statement about research and innovation, important terms that lie in the chapter title (Table 2). The chapter text states ethically-based research involves respect for *persons, beneficence,* and *justice.* These are standard ethical principles related to the ones proposed by Beauchamp and Childress [4] in their principlism approach but precede that approach in their construction. As used here, the term justice does not explicitly state that research should include full access in the sense of using fully representative samples in the research conducted, for example, concerning gender, age, and group.

The AMA [5] has instituted a program called All of Us (https://www.ama-assn.org/delivering-care/precision-medicine/about-genetics-personalized-medicine) toward obtaining representative samples for precision medicine. This effort should be propagated to all research endeavors sponsored by the AMA or conducted by any of its members.

2.3 Required Revisions Related to Medical Research Sampling

The present chapter does not simply advocate for an added article to the AMA ethics code to accomplish the goal of assuring full-scale representation of all relevant segments of the population in medical research to accommodate the lack to date in these regards. Rather, it takes the view that the whole AMA medical ethics

Table 2 Standards of research and innovation in the AMA 2017 medical ethics code [3]

#	Standard
1. Practitioner involvement	
1.1	Involvement
1.2	Informed consent
1.3	Study design/sampling
1.4	Conflicts of interest
1.5	Misconduct
2. Disseminating results	
2.1	Principles
2.2	Release of data from unethical studies
2.3	Patents/dissemination of research products
3. Special issues	
3.1	Ethical use of placebo controls
3.2	Research of emergency interventions
3.3	International research
3.4	Maternal–fetal research (NA)
3.5	Research using human fetal tissue (NA)
3.6	Research in gene therapy and genetic engineering (NA)
3.7	Safeguards in the use of DNA databanks
3.8	Research with stem cells
3.9	Commercial use of human biological materials

tome needs to be reworked to improve generally the code principles, and it should explicitly mention research when required, including in the patient list of rights. For the question of unbalanced or nonrepresentative sampling in medical research, a revised AMA medical ethic code based on the present proposal could mention clearly in standards relating to research that discrimination or lacks in establishing sample representativeness violate various medical ethical principles, such as those related to caring and beneficence/nonmaleficence, integrity, and respect/dignity/rights, and justice. For example, biased medical research sampling could be flagged in the section that contains opinions (standards) of a revised AMA ethics code and is related to the core principles in the following way; specifically, this type of discrimination could lead to more undocumented side effects for excluded segments or groups of the population concerning drugs not being researched representatively.

As for informed consent to patients when selecting drugs for their medical conditions, a revised AMA ethics code should insist that the practitioner ethically should explain to individuals of non-tested groups any risks. If they belong to groups excluded in the research sampling, then the actual side effects and even the possibility of death due to the use of the said medication would not have been fully explored by the extant research.

2.4 Recent Literature Review on Discrimination in Medical Research Sampling

The current literature review briefly deals with medical ethics before moving on to recent research on research sampling in medical research. The lacunae in the sampling procedures in medical research have been well-documented, and calls for changes are increasing to protect outgroups in the research sampling methodology in medicine.

2.5 Medical Ethics

Young and Wagner [6] examined medical ethics history and difficulties presented by the AMA medical ethics code. They noted that physicians are generally unable to elucidate the principles in the code. The authors emphasized ethics as a system and medical ethics in relation to science. Young and Wagner [6] referred to patient rights as possibly the next medical ethics horizon.

In this regard, patients being recruited for clinical trials of drugs should be explained the problems in research sampling representativeness, if any, in the research being conducted. If there are any, the medical researcher should explain how the study's future results might not apply to the person to the same extent as might be the case for other segments of the population that are being included in the study.

2.6 Biased Sampling in Medical Research

Before examining particular demographic sampling lacks in medicinal research and their consequences, the chapter examines the sampling question generally. Specifically, what should medical research do in the initial phases of research design and in conducting the research with respect to sampling?

Krauss [7] highlighted the following assumptions that need to be respected in randomized clinical trials (RCT) research related to initial sample selection: they need to be generated randomly. They need to be selected representatively with respect to characteristics that qualify the general population. It permits an equitable distribution of background traits of the general population in the selected sample. The eligibility and exclusion criteria need to be considered to avoid biased sample selection. Researchers need to keep track of participant refusers to ascertain whether they differ significantly on critical variables related to the hypotheses of the trial. Small sample size bias should be mitigated.

Krauss continued that multiple factors contribute to bias in medical research, including sampling bias. How he arrived at his conclusion was creative in that he analyzed well-received RCTs in medicine. Krauss selected the ten most-cited RCT studies worldwide, using the data engine Scopus, and considered studies such as these influence policy. The results appeared biased in participants' background traits unequally distributed between trial groups. Also, the trials were not necessarily fully blinded, among other factors.

This study illustrates that the literature has noted the lack of representativeness in medical research sampling.

The reasons that medical research sampling bias can negatively affect outgroups in medical research sampling relate to medical factors in these outgroups. Different groups have different physiologies and consequently different reactions to administered drugs, for example. The side effects of drugs might be different from one group to the next because of this. More tellingly, a drug's touted efficacy might not be equivalent from one group to the next. The side effects might be the only effect of the medicine, leaving the benefit/risk ratio tilted away from beneficence toward maleficence. This kind of argument about different physiologies in different population segments applies to women compared to men. Moreover, women constitute half of the population, so that gender bias in medical research sampling is considered before other biases in the sampling.

2.6.1 Gender Bias

Santos-Casado and Garcia-Avello [8] conducted a systemic review of clinical studies to investigate the impact of gender bias on the efficacy of long-acting antipsychotics, such as risperidone, in people diagnosed with schizophrenia. In the 40 trials analyzed, only 36.4% were females. Furthermore, only 6 of the 40 datasets had considered sex differences. Multiple international and national organizations are calling for equality in gender in medical clinical trials, for example, the National Institute of Health (NIH). Nevertheless, multiple factors have been postulated for the failure to reach this goal, including the extra monetary costs involved.

Brazil [9] called for sex equality in new medicine trials. She maintained that women respond differently to medications than men, a finding that stood out in Santos-Casado and Garcia-Avello [8]. Brazil [9] claimed that research on generic medications is almost exclusively conducted on men. Her call is timely.

In a bibliometric analysis, Cislak, Fromanowitz, and Saguy [10] showed that the research on gender bias in academia exhibits a bias against it! There are proportionately fewer female scientists; research samples have fewer women than they should, and articles that deal with gender bias do not get published in high-impact factor journals. The authors referred to these findings in terms of discrimination.

Medical research needs to be aware of this type of discrimination to change the gender profile of its researchers, samples, and publications on the topic. The AMA should be leading the way in this regard and call gender bias in medical research an unethical practice. However, other biases in medical research sampling should be equally considered in the proactive stance that the AMA should adopt in its research advisories, as per the following.

2.6.2 Elderly Bias

Wright [11] reviewed research on the bias against the elderly in clinical trials. The elderly constitute a fast-growing segment of the population. However, according to

Wright, half of the ongoing clinical trials are susceptible to a sampling bias by placing an upper age limit on the research. Also, in the remaining research undertaken, exclusion criteria often limit the upper ages in the elderly age distribution; these exclusionary criteria include comorbid conditions, cognitive impairment, and polypharmacy. According to Wright, the typical average age for a clinical trial is in the 40 s, so generalization of study results to a 75-year-old is problematic. The diseases reviewed for elderly bias in research sampling include heart disease, cancer, diabetes, and other conditions that afflict the elderly (e.g., Bourgeois et al. [12]).

According to Wright, the NIH has implemented policies to counter elderly bias in medical research as of 2019. Researchers need to justify any age-based exclusion criteria in their research applications, and the NIH will conduct appropriate follow-up in these regards concerning any research undertaken with their funds. The author argued that the FDA should encourage pharmaceutical companies to follow similar practices in the research that they sponsor, given that the amount of their grant allocations is 25 times as much as that of the NIH.

Jackson et al. [13] considered ageism rampant in society and not well-studied in the person's public health and well-being. They conducted a longitudinal observational study in England, with a sample taken of over 7000 from the nationally representative English Longitudinal Study of Aging. With confounding variables accounted for, perceived age discrimination and baseline health status were studied in relation to health outcomes six years later. The measures were self-report. Odds ratios were established. About one-quarter of the sample reported perceived age discrimination, and six years later, this group reported more heart disease, lung disease, stroke, diabetes, and depression. The authors called for effective interventions related to perceived age discrimination in public medical health and its deleterious effects.

Enzenbach et al. [14] sought to determine the reasons for self-selection bias in research on the elderly. They randomly selected over 9000 participants from the Life-Adult study in Leipzig and compared participants and non-participants in the study. Participants reported relatively better health, better education, better work status, and being younger age, non-smoker, married, and male, relative to non-participants and Leipzig citizens generally. The main reasons associated with study non-participation included health reasons and a lack of interest. The authors made suggestions toward accommodating selection bias in the research on the elderly.

This section of the chapter on medical research sampling bias related to the elderly emphasizes the main themes found on gender bias. It adds an important factor that accommodations could and should be made. Women represent half the population, so this topic was placed first in the review. However, in the current charged environment related to race, minorities, skin color, and the like, the next section is equally important. It too not only reviews the medical research sampling bias but also makes accommodatory recommendations. Racial bias often stems from implicit bias, which is considered in the next section, as well.

2.6.3 Racial Bias

Evans et al. [15] decried racism in contemporary American society and the medical system. Racism is a toxicant that has deleterious effects on Black Americans. Research indicates that the commercial artificial intelligence (AI) algorithms employed in predicting patient needs with uncontrolled illnesses are racially biased [16]. This hearkens to the racial profiling with similar AI programs on city areas of crime.

Sakran et al. [17] and DeAngelis [18] relate the lack of equality in care for persons of color to imply bias, which concerns the unconscious attribution of qualities to a group that does not fit the empirical facts about them. The authors call on mandated implicit bias training in the medical community to counter systemic racism and training programs that can alter the dynamics of prejudiced approaches to patients [19].

According to Evans et al. [15], Black Americans find it difficult to obtain a racially concordant physician. The grant application success rates are lower for Black biomedical researchers. The racial bias in medical research samples might be less striking than for gender bias, but it exists and has repercussions on Black American health. The AMA would do well to heed this editorial and address the need for equality throughout the medical system and its research.

Geneviève et al. [20] confirmed the racial bias in medicine, referring to structural racism in precision medicine. They described biases in the initial and later integration of health data and in delivering precision medicine. For example, data collection is biased racially on the first encounter with the healthcare system and researchers. The authors offer remedies for the structural racism evident in the medical system.

Psychological research suffers from the same demographic-blind biases in its research sampling. Roberts et al. [21] showed the racial inequalities endemic in psychological research and trends consistent with medicine's structural racism. They analyzed articles in developmental, social, and cognitive psychology in top-tier journals up to the year 2018. Publications that highlighted race were rare. The journal editors were Caucasian for the most part. In keeping with the present topic, the authors writing on racial topics were disproportionately Caucasian, and their research samples were biased racially. The authors made recommendations concerning discrimination for authors and journals. Similarly, Benuto et al. [22] found that psychological research sampling for RCTs on prolonged exposure therapy for post-traumatic stress disorder (PTSD) was not equivocally unbiased, with equitable group participation. Ruglass et al. [23] found that in a sample of women administered a standardized PTSD assessment instrument, the Clinician-Administered PTSD Scale (CAPS), African American and Latina groups scored differentially on certain items analyzed statistically one way not typically used (differential item functioning) but not as clearly another way typically used. In the same vein, but even more disconcerting, Qu et al. [24] found that in 80 developmental neuroscience publications referred to in five meta-analyses, 99% used samples from Western countries, 78% did not provide information on racial/ethnic background, and 82% ignored SES background.

2.6.4 Vulnerable Group Bias

The sampling gaps are endemic in medical research and have been decried for their negative effects in treating patients with drugs and interventions. Aside from the areas reviewed to this juncture in the present chapter (gender, elderly, and racial discrimination), other potential selection bias areas exist, such as for the vulnerable [25]. Targeting the vulnerable population gap in medical research sampling, Bracken-Roche et al. [26] described that the concept of vulnerability is vaguely defined, which complicates research on it and policy decision-making. On this question, they had examined major national and international research ethics policies and guidelines, and the sources differed in quality of the definitions, their application, and so on.

Concerning big data mining and use in relation to vulnerable populations, Jackson et al. [27] emphasized that these populations are understudied and under-consulted. For example, in biomedical data analysis, the research shows consistent inequalities in sampling related to vulnerability or those needing quality data. According to the authors, this lack in medical research sampling of vulnerable populations in data collecting, data analysis, and data use has significant implications for their wellness.

As for the effect of medical research sampling bias in COVID-19 times, the pandemic has significantly impacted public mental health, and the ongoing crisis in mental health should be considered in medical survey research sampling [28, 29]. Some argue that it is urgent to get high-quality data on the pandemic's effects on all vulnerable groups' mental health in the population. Without reliable data, decisions in public health could be harmful. Cogently, they also argue that acting from the basis of "misleading" data might be worse than acting without any data. Bias in mental health data will be a problem for people who are the neediest. The difficulty is that people in such groups are less likely to have the computer resources and motivation to participate in surveys on their health. Researchers must accommodate these difficulties so that policy deals with these needy groups.

2.7 SES Bias

To conclude the recent literature review on sampling bias in medical research demographics, one last area to consider is the lack of full representation of differential portions of socioeconomic strata (e.g., see Stringini et al. [30]). Granted, it is harder to recruit patients from all strata in the SES spectrum, but the same ethical concerns voiced about biased medical research sampling for other segments of the population apply to this aspect of the population, as well. The systemic and structural biases in the system are changing but need to be acknowledged in a revised AMA ethics code and in the informed consent process in recruiting patients for clinical trials.

3 Ethics

Young [31] has indicated how the AMA ethics code can be revised according to five basic principles, which speak to the ethical research process, including medical research sampling. The five principles concern: "life preservation; caring beneficence/nonmaleficence; relational integrity; respect for the dignity and rights of persons and people; and promoting and acting from justice in society." The principle "life preservation" is related to the prevention of suicide, violence, torture, etc. Superordinate principles concern personhood and responsibility. Supplementary principles include, for example, scientific literacy. The AMA ethics code could be revamped according to these principles. Moreover, each can be applied to medical research sampling in the following way:

- For life preservation, ethical medical research sampling dictates that extreme effort needs to be made to use representative samples in order to control differential mortality and morbidity rates over gender, age, racial/minority status, differential SES strata, etc.;
- For caring beneficence/nonmaleficence, ethical medical research sampling informs maximum concern for relative equality in representation for the outgroups hitherto not represented well in the research to date;
- For relational integrity, ethical medical research sampling calls for relating openly to solicited research participants about the lack of prior high-quality research about sampling, how that is being accommodated in the proposed research, and the limitations still evident in these regards. Further, the researcher should inform participants about the unknown effects of certain drugs and treatments because of the lack of proper representation in past research and that the present trial might not be sufficient to change that outcome evident in the research to date;
- For respect for human dignity and rights, clearly, the research conducted in medicine should be inclusive of all segments and groups in the population, such that the fair representation in the research sample studied can help generalize the results to all persons and peoples, which will facilitate offering them proper choices in their care. The language of the principle implies worldwide representativeness in research sampling across studies for each medical issue investigated; and
- For investments on justice in the community in general, proper medical research sampling ethics dictates that equal access takes place in research sampling for all persons and peoples such that the samples used are representative of different genders, ages, minorities and races, strata in SES, vulnerabilities of all types, and so on. This is how medical research can thrive, populations can develop better physical and mental health and general well-being, and quality of life can be optimized.

4   Conclusion

The present chapter has made recommendations for informed consent in research and practice based on the current lack of full representation of all population segments in medical research. In particular, this bias alters the beneficence/non-maleficence ratio in benefits to risks for women, Black Americans and other minorities, elderly, vulnerable, and lower strata in the SES spectrum.

The AMA ethics code refers to universal access in medicine but not to the representativeness of research samples. The present recent literature review on this topic refers to systemic and structural biases related to gender, elderly, and race. Medical research ethics principles should promote equality and prohibition of discrimination of any kind, including access to care and research sampling, leading to more reliable and valid results and medical research applications. The AMA should consider revising its medical ethics code in light of these issues. Crucially, the chapter advocates for a full upfront informed consent process for medical practitioners on the lack of medical research sampling related to patients being offered drugs or interventions that have not been studied in the research with representative samples of the population, especially for members of outgroups in the research being offered the medical services. It is acknowledged that this procedure might make practitioners uncomfortable, and the procedure could be viewed as an effort to mitigate the risk of legal consequences more than actual concern for the patients involved. That said, if practitioners become legally obligated to inform their patients of lacks and biases in medical research sampling related to their conditions and the recommendations for them, this might spur the medical community, its institutions, and related stakeholders, especially granting agencies and pharmaceutical underwriters of medical research, to follow through and vet medical research publications and research proposals for sample representativeness or efforts to mitigate lacks therein.

Psychology has the same concerns in its research base, and research on therapies is most often conducted with nonrepresentative samples in one way or another along the lines being analyzed demographically. Therapies in psychology need to be evidence-informed or -based, given the scientific integrity sought by the discipline. The field is facing a research replication crisis, and part of the problem might relate to demographic narrowing in the research. With broader research sampling that covers all aspects of the population with respect to gender, minority/race/ethnicity status, ages, and strata in SES, psychology will obtain a better picture of what is effective in its therapies and for which types of people in the population, and what types are not as effective, especially for different groups and segments in the population. In this regard, concerning informed consent, the argument raised that the procedure should include the limitations in the research sampling in the literature should apply to psychology patients, as much as medical ones. They need to be informed if they are outgroup members in the typical research conducted and what that means for their psychological services. This obligation will help instill the same kind of professional and community action predicted for the medical field and improve psychology research and practice.

The consequences of gender and race bias in psychology were underscored by Garb [32]. He defined bias in terms of differential validity over groups in diagnosis. Using this criterion and examining studies with good internal validity, he found that multiple disorders evidenced race and gender bias. Race bias was found for conduct disorder, antisocial personality disorder, comorbid substance abuse, mood disorders, eating disorders, post-traumatic stress disorder, and differential diagnosis of schizophrenia and psychotic affective disorders. Gender bias was evident for autism spectrum disorder, attention deficit hyperactivity disorder, conduct disorder, and antisocial and histrionic personality disorders. Different groups express symptoms differently, as well. There might be interactions between gender and race that were not considered in the review. The biases reported affect patient health care, and recommendations were made to improve diagnostic proficiency.

Moreover, the following questions are meant to initiate relevant policy decisions on discrimination in medical research sampling. What practical strategies should medical researchers be required to adopt in recruiting research subjects to accommodate discrimination issues? Researchers should start by becoming more aware of minority and discriminated populations in their catchment area. What kind of catchment area outreach could be initiated to counteract discrimination in participation in medical research? Researchers should start by encouraging engagement with marginalized and minority populations toward fostering greater trust in medical research. Medical researchers should document all the strategies used to recruit in minority neighborhoods and other discriminated outgroup recruiting venues, including by canvassing in the community and using electronic, social media, and internet sources. Mistrust of medical research might be mitigated by medical researchers who build community presence. Aside from reaching out to the discriminated, what practical steps should researchers take to alleviate the logistical difficulties that marginalized persons face, such as cell phone access, repeated visits to physicians/researchers, and the travel costs involved? Finally, should researchers have to demonstrate or document these and related steps toward alleviating the discrimination problem in medical research? This chapter maintains that the ethics approval of medical research investigations should be based on sufficient planning in the accommodation of lack of outgroup inclusion in the typical research in the area, or steps should be put in place to deny funding applications.

As for extensions of the present concerns into other bioethical research issues and conundrums, the AMA should address some of the following issues that characterize contemporary medical research. These suggestions extend beyond medical research sampling, but their implementation will improve that aspect of medical research, considered the starting point of unbiased, high-quality research.

For further work on representativeness in medical research, the field should query the whole range of institutional and researcher representativeness in the research process. Do the institutional review boards (IRBs) that vet research ethics query discriminatory sample representativeness in submitted proposals for research, and are the IRBs' members representative of the general population? Are the medical researchers conducting the research representative this way? Are the evaluation committees in national medical research grant agencies, such as the NIH, representative of the general population? Do pharmaceutical-based research evaluation committees consider sufficiently discrimination matters in medical research sampling? Should they be vetted for their lack in these regards, for example, on the lack of such vetting in the grants they approve? Are the editors-in-chief of medical research journals sufficiently representative, and the submission reviewers, as well? Indeed, are members of medical faculties through which researchers conduct their research similarly representative?

As for related issues on the ethics of representativeness in medical research, are the diseases researched in the medical community more prevalent in certain groups and less studied in groups that are not represented in the typical medical research? For research replications that are being conducted in medical science, are representative samples being used? Indeed, is there research that investigates less researched groups for diseases that afflict them disproportionately more often at the outset? Certainly, for all these questions being posed, one could find exemplars that set aside concerns, but it is the general trends in the data related to the questions that should be used in spurring more research along these lines.

The COVID-19 pandemic has added to the urgency of proper medical research sampling in ongoing medical research. Without appropriate research sampling, particular groups, including women, the elderly, certain minorities and racial groupings, the more vulnerable along multiple physical and psychological lines, and the disadvantaged socio-economically and otherwise, will not be properly included and vetted in the COVID-19 research and suffer accordingly.

Core Messages

- Medical research sampling is not representative demographically.
- Outgroups in medical research sampling include women, racial minorities, elderly, vulnerable, and strata in socioeconomic status.
- The danger is that members of these groups have different physiologies, and recommended pharmaceutical drugs and medical interventions do not account for their differences.
- The chapter recommends that informed consent procedures openly indicate these lacunae and that the medical profession, medical ethics code, and medical research should change in these regards.

References

1. Guttierez KJ (2020) The performance of "anti-racism" curricula. N Engl J Med 383:e75. https://doi.org/10.1056/NEJMpv2025046
2. Cyna AM, Simmons SW (2017) Guidelines on informed consent in anaesthesia: unrealistic, unethical, untenable…. Br J Anaesth 119(6):1086–1089. https://doi.org/10.1093/bja/aex347

3. American Medical Association (2017) Code of medical ethics. American Medical Association
4. Beauchamp TL, Childress JF (2012) Principle of biomedical ethics (7th ed). Oxford University Press
5. American Medical Association (2020) About genetics and personalized medicine. Published online https://www.ama-assn.org/delivering-care/precision-medicine/about-genetics-personalized-medicine
6. Young M, Wagner A (2020) Medical ethics. In: StatPearls [Internet], StatPearls Publ Co. Published on Jan: https://www.statpearls.com/kb/viewarticle/41747
7. Krauss A (2018) Why all randomised controlled trials produce biased results. Ann Med 50 (4):312–322. https://doi.org/10.1080/07853890.2018.1453233
8. Santos-Casado M, Garcia-Avello A (2019) Systematic review of gender bias in the clinical trials of new long-acting antipsychotic drugs. J Clin Psychopharmacol 39(3):264–272
9. Brazil R (2020) Why we need to talk about sex and clinical trials. Pharm J
10. Cislak A, Fromanowitz M, Saguy T (2018) Bias against research on gender bias. Scientometrics 115:189–200
11. Wright KC (2019) Research review: upper age limits in clinical trials—research lacking with older adults for evidence-based medicine. Today's Geriatr Med 12(2):30. Published online https://www.todaysgeriatricmedicine.com/archive/MA19p30.shtml
12. Bourgeois FT, Orenstein L, Ballakur S, Mandl KD, Ioannidis JPA (2017) Exclusion of elderly people from randomized clinical trials of drugs for ischemic heart disease. J Am Geriatr Soc 65(11):2354–2361
13. Jackson SE, Hackett RA, Steptoe A (2019) Associations between age discrimination and health and well-being: cross-sectional and prospective analysis of the English longitudinal study of ageing. Lancet Public Health 4(4):200–208
14. Enzenbach C, Wicklein B, Wirkner K, Loeffer M (2019) Evaluating selection bias in a population-based cohort study with low baseline participation: the LIFE-Adult-Study. BMC Med Res Methodol 19(13):5. https://doi.org/10.1186/s12874-019-0779-8
15. Evans MK, Rosenbaum L, Malina D, Morrissey S, Rubin EJ (2020) Diagnosing and treating systematic racism. N Engl J Med 383:274–276. https://doi.org/10.1056/NEJMe2021693
16. Obermeyer Z, Powers B, Vogeli C, Mullainathan S (2019) Dissecting racial bias in an algorithm used to manage the health of populations. Science 366(6464):447–453
17. Sakran JV, Hilton EJ, Sathya C (2020) Racism in health care isn't always obvious. Sci Am. Published online 9 July 2020. https://www.scientificamerican.com/article/racism-in-health-care-isnt-always-obvious/
18. DeAngelis T (2019) How does implicit bias by physicians affect patients' health care. APA Monit 50(3):33–37
19. Forscher PS, Mitamura C, Dix EL, Cox WTL, Devine PG (2017) Breaking the prejudice habit: mechanisms, time course, and longevity. J Exp Soc Psychol 72:133–146. https://doi.org/10.1016/j.jesp.2017.04.009
20. Geneviève LD, Martani A, Shaw D, Elger BS, Wangmo T (2020) Structural racism in precision medicine: leaving no one behind. BMC Med Ethics 21:17. https://doi.org/10.1186/s12910-020-0457-8
21. Roberts SO, Bareket-Shavit C, Dollins FA, Goldie PD, Mortenson E (2020) Racial inequality in psychological research: trends of the past and recommendations for the future. Perspect Psychol Sci 1–15. Published online 24 June 2020. https://doi.org/10.1177/1745691620927709
22. Benuto LT, Bennett NM, Casas JB (2020) Minority participation in randomized controlled trials for prolonged exposure therapy: a systematic review of the literature. J Trauma Stress 33 (4):420–431. https://doi.org/10.1002/jts.22539
23. Ruglass LM, Morgan-López AA, Saavedra LM, Hien DA, Fitzpatrick S, Killeen TK, Back SE, López-Castro T (2020) Measurement nonequivalence of the clinician-administered PTSD scale by race/ethnicity: implications for quantifying posttraumatic stress disorder severity. Psychol Assess. Published online 27 Aug 2020: https://doi.org/10.1037/pas0000943

24. Qu Y, Jorgensen NA, Telzer EH (2020) A call for greater attention to culture in the study of brain and development. Perspect Psychol Sci. Published online 10 Aug 2020. https://doi.org/10.1177/17456911620931461
25. Luków P (2020) Two concepts of vulnerability of research subjects. In: Medical research ethics: challenges in the 21st century. Springer Nature AG
26. Bracken-Roche D, Bell E, MacDonald ME, Racine R (2017) The concept of vulnerability in research ethics: an in-depth analysis of policies and guidelines. Health Res Policy Syst 15:8. https://doi.org/10.1186/s12961-016-0164-6
27. Jackson L, Kuhlman C, Jackson F, Fox PK (2019) Including vulnerable populations in the assessment of data from vulnerable populations. Front Big Data 2:19. https://doi.org/10.3389/fdata.2019.00019
28. Pierce M, McManus S, Jessop C, John A, Hotopf M, Ford T, Hatch S, Wessely S, Abel KM (2020) Says who? The significance of sampling in mental health surveys during Covid-19. Lancet Psychiatry 7(7):567–568. https://doi.org/10.1016/S2215-0366(20)30237-6
29. Holmes EA, O'Connor RC, Perry VH, Tracey I, Wessely S, Arseneault L, Ballard C, Christensen H, Silver RC, Everall I, Ford T, John A, Kabir T, King K, Madan I, Michie S, Przybylski AK, Shafran R, Sweeney A, Worthman CM, Yardley L, Cowan K, Cope C, Hotopf M, Bullmore E (2020) Multidisciplinary research priorities for the COVID-19 pandemic: a call for action for mental health science. Lancet Psychiatry 7(6):547–560. https://doi.org/10.1016/S2215-0366(20)30168-1
30. Stringini S, Carmeli C, Jokela M, Avendaño M, Muennig P, Guida F, Ricceri F, d'Errico A, Barros H, Bochud M, Chadeau-Hyam M, Clavel-Chapelon F, Costa G, Delpierre C, Fraga S, Goldberg M, Giles GG, Krogh V, Kelly-Irving M, Layte R, Lasserre AM, Marmot MG, Preisig M, Shipley MJ, Vollenweider P, Zins M, Kawachi I, Steptoe A, Mackenbach JP, Vineis P, Kivimäki M (2017) Socioeconomic status and the 25 × 25 risk factors as determinants of premature mortality: a multicohort study and meta-analysis of 1·7 million men and women. Lancet 389(10075):1229–1237. https://doi.org/10.1016/S0140-6736(16)32380-7
31. Young G (2020, in press) Toward a unified health work ethics code. Ethics Med Public Health
32. Garb HN (2021) Race bias and gender bias in the diagnosis of psychological disorders. Clin Psychol Rev 90:102087. https://doi.org/10.1016/j.cpr.2021.102087

Gerald Young is a full professor in Psychology at Glendon College, York University, Toronto, Canada. He is a Fellow of the Association for Psychological Science (APS) and the American Psychological Association (APA). He has received awards from the American Psychological Association and the Canadian Psychological Association (CPA), including lifetime achievement. Young is Editor-in-Chief of the journal Psychological Injury and Law, which he founded, and his work in that area has led to invited speaker addresses at scientific conferences. His most recent books are Revising the APA Ethics Code (Springer, 2017) and Causality and Development: Neo-Eriksonian Perspectives (Springer, 2019). Works in progress include the book: Causality and neo-stages in development: Toward unifying psychology; the articles: Toward a unified health work ethics code, promoting children and youth lives; and the book chapter: Discrimination in medical research. He has appeared as an expert witness for a case involving the Supreme Court of Canada. His practice covers rehabilitation and couples/families.

Integrated Science 2050: Multidisciplinarity and Interdisciplinarity in Health

Nima Rezaei, Amene Saghazadeh, Abdul Rahman Izaini Ghani, AbouAli Vedadhir, Aida Vahed, Alfredo Vellido, Alireza Afshar, Alireza Zali, Andre Kushniruk, Andrée-Anne Blacutt, Antonino Pennisi, Antonio Condino-Neto, Arash Khojasteh, Armando E. Soto-Rojas, Brian Brown, Bruna Velasques, Claudio Lucchiari, Daniel Atilano-Barbosa, Danielle Aprígio, Donald R. Kirsch, Donata Chiricò, Elham Rayzan, Elif Karakoc-Aydiner, Elizabeth Borycki, Emilio Maria Palmerini, Esther A. Balogh, Fabio Minutoli, Farbod Ghobadinezhad, Farid Farrokhi, Faruque Reza, Gerald Young, Grzegorz Sierpiński, Haniye Sadat Sajadi, Hans D. Ochs, Heikki Murtomaa, Helen Monkman, Helia Mojtabavi, Hélio A. Tonelli, Heliya Ziaei, Houneida Sakly, Hunkoog Jho, Ireneusz Celiński, Jafri Malin Abdullah, Jakub Šrol, Jayne Seekins, Joe Ravetz, Juan José Garrido Periñán, Juliana Bittencourt, Kaushik Sarkar, Kiarash Saleki, Luisa de Siqueira Rotenberg, Mahnaz Jamee, Mahsa Keshavarz-Fathi, Mariana Gongora, Mauricio Cagy, Meisam Akhlaghdoust, Melika Lotfi, Milad Baziar, Milad Rafiaei, Mohammad Amin Khazeei Tabari, Mohammad R. Khami, Mohammad Rasoul Golabchi, Mohammadreza Fadavipour, Moncef Tagina, Monica Lakhanpaul, Morenike Oluwatoyin Folayan, Morteza Shamsizadeh, Mourad Said, Niloofar Rambod Rad, Niloufar Yazdanpanah, Noosha Samieefar, Pedro Ribeiro, Prathip Phantumvanit, Priti Parikh, Remco van de Pas, Reza Majdzadeh, Riccardo Laudicella, Richard A. Stein, Roberto E. Mercadillo, Roya Kelishadi, Sadra Behrouzieh, Saina Adiban Afkham, Sara Momtazmanesh, Sayedeh Azimeh Hosseini, Sergio Baldari, Silmar Teixeira, Simin Seyedpour, Stéphane Roche, Stephen E. Kekeghe, Steven R. Feldman, Thayaná Fernandes, Timo Ulrichs, Tommaso Dorigo, Vasili Roudenok, Veeraraghavan J. Iyer,

Veronica K. Emmerich, Victor Marinho, Vladimíra Čavojová,
Waleed Al-Herz, Zahra Rahimi Pirkoohi, Zaitun Zakaria,
Zamzuri Idris, and Zhila Izadi

N. Rezaei · A. Saghazadeh
Integrated Science Association (ISA), Universal Scientific Education and Research Network
(USERN), Tehran, Iran

N. Rezaei · A. Saghazadeh · E. Rayzan · H. Mojtabavi · H. Ziaei · N. Yazdanpanah
Research Center for Immunodeficiencies, Children's Medical Center, Tehran University of
Medical Sciences, Tehran, Iran

N. Rezaei
Department of Immunology, School of Medicine, Tehran University of Medical Sciences,
Tehran, Iran

A. R. I. Ghani · F. Reza · J. M. Abdullah · Z. Zakaria · Z. Idris
Department of Neurosciences, School of Medical Sciences, Universiti Sains Malaysia,
Kubang Kerian, 16150 Kelantan, Malaysia
e-mail: faruque@usm.my

Z. Zakaria
e-mail: zakariaz@tcd.ie

Z. Idris
e-mail: neuroscienceszamzuri@yahoo.com

A. R. I. Ghani · J. M. Abdullah · Z. Zakaria · Z. Idris
Brain and Behaviour Cluster (BBC), School of Medical Sciences, Universiti Sains Malaysia,
Kubang Kerian, 16150 Kelantan, Malaysia

A. R. I. Ghani · F. Reza · J. M. Abdullah · Z. Zakaria · Z. Idris
Hospital Universiti Sains Malaysia (HUSM), Universiti Sains Malaysia, Kubang Kerian,
16150 Kelantan, Malaysia

A. Vedadhir
Department of Anthropology, Faculty of Social Sciences, University of Tehran, 14117-13118
Tehran, Iran
e-mail: vedadha@ut.ac.ir

Population Health Sciences, University of Bristol, Canynge Hall, Bristol BS8 2PS, UK

A. Vahed
School of Pharmacy, Tehran University of Medical Sciences, Tehran, Iran

A. Vahed · E. Rayzan · H. Mojtabavi · H. Ziaei · M. Fadavipour · M. Shamsizadeh ·
N. Rambod Rad · N. Yazdanpanah · N. Samieefar · R. Kelishadi · S. Momtazmanesh ·
Z. R. Pirkoohi
Health and Art (HEART), Universal Scientific Education and Research Network (USERN),
Tehran, Iran
e-mail: roya.kelishadi@gmail.com

S. Momtazmanesh
e-mail: s-momtazmanesh@student.tums.ac.ir

A. Vellido · S. A. Hosseini · T. Ulrichs · V. Roudenok · W. Al-Herz · Z. Izadi
Computer Science Department, Intelligent Data Science and Artificial Intelligence (IDEAI)
Research Center, Universitat Politècnica de Catalunya, Barcelona, Spain
e-mail: avellido@cs.upc.edu

S. A. Hosseini
e-mail: Sah.biotec@gmail.com

T. Ulrichs
e-mail: timo.ulrichs@akkon-hochschule.de

V. Roudenok
e-mail: roudenok@bsmu.by

W. Al-Herz
e-mail: wemh@hotmail.com

Z. Izadi
e-mail: izadi_zh@razi.tums.ac.ir

A. Kushniruk · E. Borycki · H. Monkman · Z. Izadi
School of Health Information Science, University of Victoria, HSD A202, 3800 Finnerty
Road, Victoria, Canada
e-mail: andrek@uvic.ca

E. Borycki
e-mail: emb@uvic.ca

H. Monkman
e-mail: monkman@uvic.ca

A.-A. Blacutt
School of Design, Laval University, Quebec, Canada
e-mail: andree-anne.blacutt-grenier.1@ulaval.ca

A. Pennisi
Department of Cognitive Sciences, University of Messina, Messina, Italy
e-mail: apennisi@unime.it

A. E. Soto-Rojas
Cariology, Operative Dentistry and Dental Public Health Department, Indiana University
School of Dentistry, Indianapolis, US
e-mail: arsoto@iu.edu

B. Brown
De Montfort University, Leicester, UK
e-mail: brown@dmu.ac.uk

B. Velasques · D. Aprígio · J. Bittencourt · M. Gongora · M. Cagy · P. Ribeiro
Brain Mapping and Sensory Motor Integration, Federal University of Rio de Janeiro, Rio de
Janeiro–RJ, Brazil

C. Lucchiari · E. M. Palmerini
Department of Philosophy, Università Degli Studi Di Milano, Milan, Italy
e-mail: claudio.lucchiari@unimi.it

D. Atilano-Barbosa
Institute of Neurobiology, Universidad Nacional Autónoma de México, Mexico City, Mexico

D. R. Kirsch
Department of Biology, Columbia University, 1150 Amsterdam Ave, New York, NY 10027, USA

D. Chiricò
Department of Humanities, University of Calabria, Rende, Italy
e-mail: donata.chirico@unical.it

D. Chiricò · E. A. Balogh · S. R. Feldman · V. K. Emmerich
Center for Dermatology Research, Department of Dermatology, Wake Forest School of Medicine, Winston-Salem, North Carolina, USA
e-mail: ebalogh@wakehealth.edu

S. R. Feldman
e-mail: sfeldman@wakehealth.edu

V. K. Emmerich
e-mail: vemmeric@wakehealth.edu

F. Minutoli · R. Laudicella · S. Baldari
Nuclear Medicine Unit, Department of Biomedical and Dental Sciences and Morpho-Functional Imaging, University of Messina, Messina, Italy
e-mail: fminutoli@unime.it

S. Baldari
e-mail: sbaldari@unime.it

F. Farrokhi
Community Oral Health Department, Tehran University of Medical Sciences, Tehran, Iran
e-mail: f-farrokhi@razi.tums.ac.ir

G. Young
Glendon College, York University, Toronto, ON, Canada
e-mail: gyoung@glendon.yorku.ca

G. Sierpiński · I. Celiński
Silesian University of Technology, Faculty of Transport and Aviation Engineering, Katowice, Poland
e-mail: grzegorz.sierpinski@polsl.pl

I. Celiński
e-mail: Ireneusz.Celinski@polsl.pl

H. S. Sajadi
Knowledge Utilization Research Center, University Research and Development Center, Tehran University of Medical Sciences, Tehran, Iran
e-mail: hsajjadi@tums.ac.ir

H. Murtomaa
Oral Public Health, University of Helsinki, Helsinki, Finland
e-mail: heikki.murtomaa@helsinki.fi

H. A. Tonelli
FAE Business School, Department of Neuropsychology and Caetano Marchesini Clinic, Curitiba, Brazil

H. Sakly · M. Tagina
COSMOS Laboratory, National Institute of Computer Sciences (ENSI), University of Manouba, Manouba, Tunisia
e-mail: houneida.sakly@esiee.fr

M. Tagina
e-mail: moncef.tagina@ensi-uma.tn

H. Jho
Graduate School of Education, Dankook University, Yongin, Korea
e-mail: hjho80@dankook.ac.kr

J. Šrol · V. Čavojová
Institute of Experimental Psychology, Centre of Social and Psychological Sciences, Slovak Academy of Sciences, Dúbravská cesta 9, 841 04 Bratislava, Slovakia
e-mail: vladimira.cavojova@savba.sk

J. Ravetz
Manchester Urban Institute, Manchester University, Manchester, UK
e-mail: joe.ravetz@manchester.ac.uk

School of Environment Education and Development, Manchester University, Oxford Rd M13 9PL, UK

J. J. G. Periñán
Department of Aesthetics and History of Philosophy, University of Seville, Sevilla, Spain
e-mail: jjgarper@us.es

J. Bittencourt
Veiga de Almeida University, Rio de Janeiro, Brazil

K. Sarkar · M. Lakhanpaul
UCL Great Ormond Street Institute of Child Health, UCL, London, England
e-mail: m.lakhanpaul@ucl.ac.uk

K. Sarkar
Aceso Global Health Consultants Limited, Delhi, India

K. Sarkar · M. Lakhanpaul · P. Parikh
Childhood Infection and Pollution Consortium, London, England
e-mail: priti.parikh@ucl.ac.uk

L. de Siqueira Rotenberg
Bipolar Disorders Research Program, Department of Psychiatry of São Paulo Medical School,
São Paulo, Brazil

M. Keshavarz-Fathi · S. Behrouzieh · S. Seyedpour
School of Medicine, Tehran University of Medical Sciences, Tehran, Iran
e-mail: seyedpoursimin@gmail.com

M. Keshavarz-Fathi · S. Behrouzieh
Interest Group Department, Universal Scientific Education and Research Network (USERN),
Tehran, Iran

M. Keshavarz-Fathi
Cancer Immunology Project (CIP), Universal Scientific Education and Research Network
(USERN), Tehran, Iran

M. R. Khami
Research Center for Caries Prevention, Dentistry Research Institute, Tehran University of
Medical Sciences, Tehran, Iran
e-mail: mkhami@tums.ac.ir

M. O. Folayan
Department of Child Dental Health Obafemi Awolowo University, Ife, Nigeria

M. Said
Radiology and Medical Imaging Unit, International Center Carthage Medical, Monastir,
Tunisia

P. Phantumvanit
Faculty of Dentistry, Thammasat University, Bangkok, Thailand

P. Parikh
Engineering for International Development (EFID) Research Centre, University College of
London, London, England

R. van de Pas
Department of Public Health, Institute of Tropical Medicine Antwerp, Nationalestraat 155,
2000 Antwerp, Belgium
e-mail: rvandepas@itg.be

Department of Health Ethics and Society, Faculty of Health Medicine and Life Sciences,
Maastricht University, 616, 6200 MD Maastricht, The Netherlands

R. Majdzadeh
Community-Based Participatory-Research Center, Knowledge Utilization Research Center,
School of Public Health, Tehran University of Medical Sciences, Tehran, Iran
e-mail: rezamajd@tums.ac.ir

R. A. Stein
NYU Tandon School of Engineering, Department of Chemical and Biomolecular
Engineering, 6 MetroTech Center, Brooklyn, NY 11201, USA
e-mail: steinr01@nyu.edu

LaGuardia Community College, Department of Natural Sciences, City University of New York,
New York, NY 11101, USA

R. E. Mercadillo
National Council for Science and Technology, Mexico City, Mexico
e-mail: remercadilloca@conacyt.mx

Area of Neurosciences, Department of Biology of Reproduction, Universidad Autónoma
Metropolitana, Iztapalapa Unit, Mexico City, Mexico

S. A. Afkham
Innovation and Creativity Research Association for Transforming Education (I-CREATE),
Universal Scientific Education and Research Network (USERN), Tehran, Iran

Network of Immunity in Infection, Malignancy and Autoimmunity (NIIMA), Universal Scientific
Education and Research Network (USERN), Tehran, Iran

Biotechnology Department, Islamic Azad University, Tehran, Iran

S. Teixeira · T. Fernandes · V. Marinho
Federal University of Parnaíba Delta, Parnaíba 64202-020, Brazil
e-mail: silmarteixeira@ufpi.edu.br

T. Fernandes
e-mail: thayana.fernandes@hotmail.com

V. Marinho
e-mail: victormarinhophb@hotmail.com

S. Teixeira · V. Marinho
The Northeast Biotechnology Network (RENORBIO), Federal University of Piauí, Teresina,
Brazil

S. Seyedpour
Nanomedicine Research Association (NRA), Universal Scientific Education and Research
Network (USERN), Tehran, Iran

S. Roche
Research Center for Geospatial Data and Intelligence, Laval University, Quebec, Canada
e-mail: Stephane.Roche@scg.ulaval.ca

S. E. Kekeghe
Department of English, Ajayi Crowther University, Oyo, Nigeria
e-mail: se.kekeghe@acu.edu.ng; kekeghestephen@gmail.com

S. R. Feldman
Department of Pathology, Wake Forest School of Medicine, Winston-Salem, North Carolina, USA

Department of Social Sciences and Health Policy, Wake Forest School of Medicine, Winston-Salem, North Carolina, USA

Department of Dermatology, University of Southern Denmark, Odense, Denmark

T. Dorigo
Istituto Nazionale Di Fisica Nucleare, Sezione Di Padova, Via F. Marzolo 8, 35131 Padova, Italy
e-mail: tommaso.dorigo@gmail.com

Integrated Science Association (ISA), Universal Scientific Education and Research Network (USERN), Padova, Italy

V. J. Iyer
Institute of Living/Hartford Hospital, Hartford, USA
e-mail: vjyster@gmail.com

Z. R. Pirkoohi
Shahid Beheshti University, Tehran, Iran

N. Rezaei (✉)
Children's Medical Center Hospital, Dr. Qarib St, Keshavarz Blvd, 14194 Tehran, Iran
e-mail: rezaei_nima@tums.ac.ir

M. Lakhanpaul
Whittington Hospital NHS Trust, London, England

A. Afshar
USERN Office, The Persian Gulf Marine Biotechnology Research Center, The Persian Gulf Biomedical Sciences Research Institute, Bushehr University of Medical Sciences, Bushehr, Iran
e-mail: Alireza.af2017@gmail.com

A. Zali · M. Akhlaghdoust
Functional Neurosurgery Research Center, Shohada Tajrish Comprehensive Neurosurgical Center of Excellence, Shahid Beheshti University of Medical Sciences, Tehran, Iran
e-mail: Dr_alirezazali@yahoo.com

USERN Office, Functional Neurosurgery Research Center, Shahid Beheshti University of Medical Sciences, Tehran, Iran

A. Condino-Neto
Department of Immunology, Institute of Biomedical Sciences, University of São Paulo, São Paulo, Brazil
e-mail: antoniocondino@gmail.com

Universal Scientific Education and Research Network (USERN), São Paulo, Brazil

A. Khojasteh
Department of Oral and Maxillofacial Surgery, Shahid Beheshti University of Medical
Sciences, Tehran, Iran
e-mail: arashkhojasteh@gmail.com

USERN Office, Shahid Beheshti University of Medical Sciences, Tehran, Iran

E. Karakoc-Aydiner
Marmara University, Faculty of Medicine, Division of Pediatric Allergy and Immunology,
Istanbul, Turkey
e-mail: elif_karakoc@hotmail.com

Universal Scientific Education and Research Network (USERN), Istanbul, Turkey

F. Ghobadinezhad
Student's Research Committee, School of Medicine, Kermanshah University of Medical
Sciences, Kermanshah, Iran
e-mail: zoroastrianfarbod@gmail.com

USERN Office, Kermanshah University of Medical Sciences, Kermanshah, Iran

H. D. Ochs
Department of Pediatrics, University of Washington, Seattle, Washington, USA
e-mail: hans.ochs@seattlechildrens.org

Seattle Children's Research Institute, Seattle, Washington, USA

Universal Scientific Education and Research Network (USERN), Washington, USA

J. Seekins
Department of Radiology, Stanford University School of Medicine, Stanford, CA, USA

K. Saleki
Student Research Committee, Babol University of Medical Sciences, Babol, Iran

USERN Office, Babol University of Medical Sciences, Babol, Iran

M. Jamee
USERN Office, Alborz University of Medical Sciences, Karaj, Iran

M. Lotfi
USERN Office, Zanjan University of Medical Sciences, Zanjan, Iran

M. Baziar
Student Research Committee, School of Medicine, Ardabil University of Medical Sciences,
Ardabil, Iran

USERN Office, Ardabil University of Medical Sciences, Ardabil, Iran

M. Rafiaei
USERN Office, Yazd University of Medical Sciences, Yazd, Iran
e-mail: rafiaeimilad@gmail.com

M. Amin Khazeei Tabari
Student Research Committee, School of Medicine, Mazandaran University of Medical
Sciences, Sari, Iran
e-mail: aminkhazeeitabari@gmail.com

USERN Office, Mazandaran University of Medical Sciences, Sari, Iran

M. R. Golabchi
USERN Office, Isfahan University of Medical Sciences, Isfahan, Iran
e-mail: m.golabchi1996@yahoo.com

M. Fadavipour
USERN Office, Abadan University of Medical Sciences, Abadan, Iran
e-mail: Mohammadfadawi@gmail.com

Abadan University of Medical Sciences, Abadan, Iran

M. Shamsizadeh
Department of Medical Surgical Nursing, School of Nursing and Midwifery, Hamadan
University of Medical Sciences, Hamadan, Iran

USERN Office, Hamadan University of Medical Sciences, Hamadan, Iran

N. Rambod Rad
USERN Office, Islamic Azad University Medicine Faculty, Mashhad, Iran

N. Samieefar
Student Research Committee, School of Medicine, Shahid Beheshti University of Medical
Sciences, Tehran, Iran

USERN Office, School of Advanced Technologies in Medicine, Shahid Beheshti University of
Medical Sciences, Tehran, Iran

R. Kelishadi
Child Growth and Development Research Center, Research Institute for Primordial
Prevention of Non-Communicable Disease, Isfahan University of Medical Sciences, Isfahan,
Iran
e-mail: roya.kelishadi@gmail.com

USERN Office, Research Institute for Primordial Prevention of Non-Communicable Disease,
Isfahan University of Medical Sciences, Isfahan, Iran

S. Momtazmanesh
Universal Scientific Education and Research Network (USERN), Tehran, Iran
e-mail: s-momtazmanesh@student.tums.ac.ir

School of Medicine, Tehran University of Medical Sciences, Tehran, Iran

S. A. Hosseini
Student Research Committee, Shahrekord University of Medical Sciences,, Shahrekord, Iran
e-mail: Sah.biotec@gmail.com

USERN Office, Shahrekord University of Medical Sciences, Shahrekord, Iran

Department of Medical Biotechnology, School of Advanced Technologies, Shahrekord University
of Medical Sciences, Shahrekord, Iran

T. Ulrichs
Institute for Research in International Assistance, Akkon University for Human Sciences,
Berlin, Germany
e-mail: timo.ulrichs@akkon-hochschule.de

Universal Scientific Education and Research Network (USERN), Berlin, Germany

V. Roudenok
Belarusian State Medical University, Minsk, Belarus
e-mail: roudenok@bsmu.by

Universal Scientific Education and Research Network (USERN), Minsk, Belarus

W. Al-Herz
Department of Pediatrics, Faculty of Medicine, Kuwait University, Kuwait City, Kuwait
e-mail: wemh@hotmail.com

Universal Scientific Education and Research Network (USERN), Kuwait City, Kuwait

Z. Izadi
Pharmaceutical Sciences Research Center, Health Institute, Kermanshah University of
Medical Sciences, Kermanshah, Iran
e-mail: izadi_zh@razi.tums.ac.ir

USERN Office, Kermanshah University of Medical Sciences, Kermanshah, Iran

"In the longer run and for wide-reaching issues, more creative solutions tend to come from imaginative interdisciplinary collaboration."

Robert J. Shiller

Summary

The authors of the *Integrated Science: Multidisciplinarity and Interdisciplinarity in Health* were asked *how you would see the future of your field 30 years later*. This chapter presents the authors' views on this subject in 2050.

The Al-Samah dance.

In the above poem, Hushang Ebtehaj says, a kind of patience is to the dear; no kind of patience is to the dearest. Below, we say, a sight of doubt is in the dear; no sight of doubt is in the dearest. Here, we mean ourselves discipline, field, or branch of the science of the dear and the integrated sciences of the dearest – which can be generalized to the meaning of the existence of the dear and the unified existence of all existences of the dearest. Indeed, the above poem indicates how much passion humankind has to think from above, and the below one is the reasoning behind this passionate performance.

[Adapted with permission from the Association of Science and Art (ASA), Universal Scientific Education and Research Network (USERN); Artwork made by Saba Sajaditabar; Nasta'līq prepared by http://nastaliqonline.ir/]

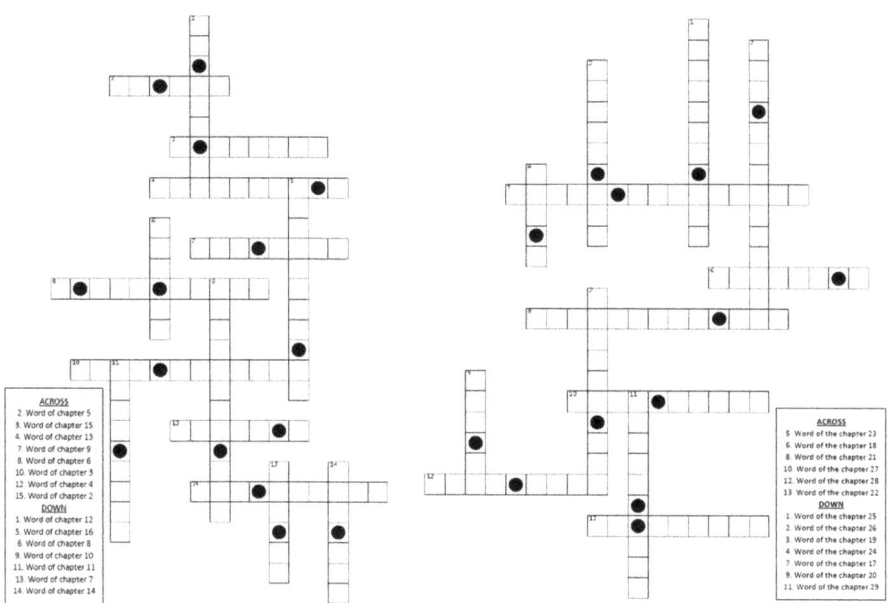

ACROSS
2. Word of chapter 5
3. Word of chapter 15
4. Word of chapter 13
7. Word of chapter 9
8. Word of chapter 6
10. Word of chapter 3
12. Word of chapter 4
15. Word of chapter 2

DOWN
1. Word of chapter 12
5. Word of chapter 16
6. Word of chapter 8
9. Word of chapter 10
11. Word of chapter 11
13. Word of chapter 7
14. Word of chapter 14

ACROSS
5. Word of the chapter 23
6. Word of the chapter 18
8. Word of the chapter 21
10. Word of the chapter 27
12. Word of the chapter 28
13. Word of the chapter 22

DOWN
1. Word of the chapter 25
2. Word of the chapter 26
3. Word of the chapter 19
4. Word of the chapter 24
7. Word of the chapter 17
9. Word of the chapter 20
11. Word of the chapter 29

✓ There is a binary code for each chapter;
✓ There is a meaningful word for each chapter;
✓ The meaningful word of the chapter is made out of unsorted letters of the binary code of the chapter;
✓ The puzzle is filled in with the words of the chapters according the clues provided;
✓ There is a mystery for the puzzle;
✓ The mystery of the puzzle contains two phrases A and B made out of letters in cells that contain a black circle;
✓ Phrase A is a three-word sentence and phrase B is a four-word sentence;
✓ The first word of phrase A and phrase B is the same;
✓ Solve it, put your name on the puzzle, and send the two-phrase mystery to nimarezaei.usern@gmail.com

The puzzle of Integrated Science: Multidisciplinarity and Interdisciplinarity in Health

Keywords

Aesthetic experience · AI · Art · Artificial intelligence · Behavior · Cognitive neuroscience · Consciousness · Education · Empathy · Epidemic · Ethics · Framework · Health · Integrated science · Interdisciplinarity · Multidisciplinarity · Pandemic · Social diversity

1   Introduction

This book is a volume included in the book series of Integrated Science.[1] The theme of this book has been multidisciplinarity (MD) and interdisciplinarity (ID) in health. The introductory Chap. 1 discussed how the healthcare systems apply an MD/ID approach, not only in practice and education but also in research. In the succeeding chapters of the book, the authors focused on different theoretical questions to critically assess practical concepts helpful in understanding how an MD/ID approach works.

Each of us is working in the discipline(s) different from each other, but all of us have concerns about health. What is the result when we all solve the same complex problem? Simply, we have a mixture of ideas suggested to solve the problem. We must choose those that can provide one full solution. Before this choosing, we first need to mix these ideas well. And it is not easy (if not impossible) to do unless an MD/ID approach acts on it. As the world has changed, the complexity of problems has grown, and interest in the ID approaches multiplied. Especially, the last two decades have been the flowering period of ID with various health teams, perspectives, curriculums, collaborations, fields, disciplines, digital tools, models, studies, educational and learning programs, and research associations began to emerge. This development denotes the intervening role of ID in the management of healthcare in different settings, for example, geriatric care [1], end-of-life care [2], nursing home residents' care [3], pediatric care [4, 5], general medical wards [6], rehabilitation and physical therapy [7], and perioperative care [8]. Also, ID interventions in a particular area focus on predicting and preventing lifestyle risks associated with obesity, cardiovascular diseases, and diabetes [9, 10]. That they are working with both the healthy and the happy is hopeful for sustainability.

ID without integration is by its own nature enough for success; however, with integrated ID, we can imagine a greater extent of cohesiveness to which concepts and practices, methods and data, etc., are achieved [11]. Hence, integrated ID knowledge has been added to the human person to offer a new or better scientific understanding of the science of complex problems, such as fighting *for life* [12, 13] when living is a *war*.

This last chapter is different from other chapters of the book. The question for this chapter is: *How would you see the future of integrated science: multidisciplinarity and interdisciplinarity in health 30 years later?*

We have collected the expert opinions and integrated them into a framework. It consists of several branches of ID knowledge. First, ID cognitive neuroscience is outlined under the major subsections: neurophysics, neuroaesthetics, neuropsychology, and neurobehavioral health, for their contribution to the understanding of consciousness, aesthetic experience, empathy, and social diversity and of the distinction of healthy from unhealthy behaviors. Second, ID is integrated into medical and health sciences, from global epidemics research, nuclear medicine, and dentistry to sociomedical sciences, medical ethics, and medical education to drug

---

[1] https://www.springer.com/series/16554.

discovery and health informatics. Third, artificial intelligence (AI) offers a means of integrating ID in healthcare, behavior engineering, and personalized medicine. This section ends with ethical and philosophical dilemma questions and future trends for AI-based integrated ID. Finally, the role of art in integrated ID healthcare is pointed out.

2 ID Cognitive Neuroscience

If I wanted to define it with a single expression, I would say that in 2050 we will move from cognitive science to the *spectrum of cognitive sciences*. In fact, I think that cognitive sciences will become the major field of integration between the sciences of nature and the humanistic social sciences in the next 30 years. On the one hand, the currently distributed articulations of neuroscience will flow into this field in a unitary perspective. I particularly believe in the ever-wider development of cognitive, affective, and social neurosciences, the most externalist perspective of this area of research. The current cognitive curvature of psychology, psychopathology, and psychiatry will increasingly approach the neo-naturalistic methods of renewed cognitive neuroscience. Language studies will all convert to neurolinguistics and neuropragmatics. In the same way, secondary sectors such as neuroaesthetics, neuroethics, studies on narratology and the performing arts, neuroeconomics and neuro-marketing, cognitive sociology, evolutionary psychology, and biopolitics will become essential sectors for the expansion of the *cognitive spectrum*. The epistemological and theoretical foundations of a global cognitive science will force the most advanced philosophies (philosophy of science, mind, and language) to configure application frames increasingly connected to the developments of the sciences of nature, on the one hand, and of AI from the other hand. In particular, the evolutionary pole will provide the tools to understand better the causal basis of any problem involving the nature of cognitive problems and the development of human and non-human animals. AI will develop the typical potential of technological and information technology (IT) innovation, definitively moving away from the simulation of human cognition and above all by providing application solutions based on increasingly efficient automatisms and a dizzying expansion of the use of big data and deep learning (Antonino Pennisi, Donata Chiricò 2021).

2.1 Neurophysics: The Understanding of Consciousness

The brain is regarded as the most complicated structure in our body, and currently, we have limited knowledge and little apprehension on how it works. Similarly, nobody has a solid grasp of quantum physics or the science of small objects, waves, and light. Nonetheless, the scientific institution should start to integrate the brain sciences with quantum physics (brain-physics). With that integration, we are optimistic, a new horizon and innovation will be made. Quantum wave detector is one of the possible future innovations which may enhance and expand our understanding in this field and indirectly contributes to the possibility of gaining better insight for the following subjects: a, brainwaves and how the brain works; b, brain energy field and its interaction with other energy fields; c, neurological and psychiatric diseases; d, consciousness, comatose state, near-death experience, dreaming, awake, and asleep; e, brain and cosmology; and last but not least; f, perhaps, the connection between physics and metaphysics (Zamzuri Idris, Zaitun Zakaria, Faruque Reza, Abdul Rahman Izaini Ghani, Jafri Malin Abdullah 2021).

2.2 Neuroaesthetics: The Understanding of Aesthetic Experience

For the last two decades, neuroimaging technologies have brought about a great understanding of how human brains work. With the development of the genome project, it is expected that human thoughts and actions can be explained by chemical and physiological changes in the brain. However, such a reductionist idea encounters a complex and ambiguous issue on brain activities. Despite the same active regions, different directional circuits and vectors bring about different results. The inseparable nature of cognition and emotion in the aesthetics of science also reflects similar features. As a deeper understanding of neural networks has a great progression in computer science, the network approach to the aesthetics derived from connectionism would be beneficial to understand how we feel science beautiful (Hunkoog Jho 2021).

2.3 Neurobehavioral Health: The Distinction of Healthy From Unhealthy Behaviors

2.3.1 Aging

The future is uncertain; however, behavioral neuroscience updates will elucidate neurobiological aspects for adapting species to the modern world. In focus, neuroscience applied to health in humans unveils possible behavioral phenotypes that can serve as role models in quality and duration of life. For example, for high cognitive and motor performance, recurrent use of tasks is applied to rehabilitation in a non-invasive way. The ID between technology and neuroscience maximizes these advances. Innovation as an adjunct to neurosciences may unravel the pathophysiological processes of some neurological diseases, help healthy aging, and perhaps extend our life in more years. About neurogenetics, this area is expected to be the next generation of unveiling the brain function, which can offer potentially revolutionary advances in disease prevention and treatment (Juliana Bittencourt, Victor Marinho, Silmar Teixeira, Danielle Aprígio, Mariana Gongora, Mauricio Cagy, Thayaná Fernandes, Pedro Ribeiro, Bruna Velasques 2021).

2.3.2 Psychiatric Symptoms

The research in neurobiology and theoretical studies in psychiatry is very promising. In the past 20 years, there has been tremendous progress in understanding the genetic underpinnings of psychiatric illnesses. With improved technologies such as positron emission tomography (PET) and other means of functional imaging, there is a larger scope for better understanding psychiatric diseases. Research and advocacy groups have helped further awareness of scientific data, research, and understanding psychiatric conditions. This has helped bridge some gaps between science and society. Empathic policies have helped with social rehabilitation and greater acceptance of psychiatric morbidities. This has also helped give a broader perspective and discourage a punitive approach to psychiatric problems and behaviors not under a person's volitional control. This also accepts that psychiatric symptoms are better managed by optimal treatment planning instead of confining people in jails/prisons. The work of such advocacy groups with legislative bodies has seen many changes in policymaking. This change continues to evolve with improved funding and clearer goals (Veeraraghavan J. Iyer 2021).

2.3.3 Obesity

Working with obesity can be frustrating due to numerous false beliefs about this condition, many of them prejudiced. The difficulty that patients themselves report controlling their eating impulses, associated with the widely disseminated idea among professionals attending patients with obesity, that weight gain results exclusively from higher caloric consumption versus lower energy expenditure, perpetuates frustration. Many of these professionals honestly seek to reflect on the reasons behind so many failures in dieting, weight regain, and eating behaviors that, in principle, seem irrational. These false beliefs, in part, may derive from a notion that decision-making consists of essentially rational processes and that emotions are nothing but undesirable noises obstructing our behavior. Yet, neuroscience has shown something else: instead of obstacles, healthy emotions are auxiliary to reason. On the other hand, aversive emotions, when excessively intense, may recruit dysfunctional regulatory strategies. Social cognition has provided a wide range of insights into how social regulation of emotion processes come into play in our daily lives. These processes are corrupted in several mental disorders, bringing distress to those who suffer from them. Recent evidence suggests that similar impairments affect the brain of individuals with obesity, which may explain much of their "irrational" behavior. This evidence comes at a good time to shift our beliefs about problematic eating behaviors in obesity and develop more efficient and empathetic treatment and prevention strategies, which, we imagine, will be the future of managing obesity (Hélio Tonelli and Luisa de Siqueira Rotenberg 2021).

2.4 Neuropsychology

2.4.1 The Understanding of Empathy

Wide and complex concepts intersecting affective, cognitive, and social components will be elaborated to comprehend empathy and similar abilities in future times. Biological and social scientists will assume that human responses to another's needs are the expression of biological properties embracing the natural history of our species and cultural scenarios built throughout human civilizations. We will realize that cognition cannot only be studied by transversal means since our behavior to a particular present moment necessarily involves our bodily and affective individual history and memory. More importantly, we may appreciate empathy as one of the most sublime ways humans have to know ourselves by recognizing ourselves in others. For these expectations, ID and transdisciplinary research will constitute our main objective to listen and engage with discussions focused on other elements besides the brain. Neuroimaging and experimental cognitive approaches will allow identifying brain anatomy and dynamics, but their interpretation will involve motivational evolved mechanisms and hormonal and neurochemical properties embracing the natural human history analyzed by evolutionary and physiological perspectives. Neuroimaging will design experiments according to real social circumstances derived from micro-social descriptions and community interventions elaborated by anthropological and psychosocial views. Insights resulting from transdisciplinary will be used to comprehend mental issues and propose new treatments considering physician-patient relations. Historical and sociological viewpoints will contribute to understanding transdisciplinary knowledge as part of broader social structures built during human history to be integrated into philosophical lines concerning new answers for brain-mind relations and nature-nurture properties (Roberto E. Mercadillo, Daniel Atilano-Barbosa 2021).

2.4.2 The Understanding of Social Diversity

In the upcoming times, open science and data sharing will impact the neuroscientific community by generating opportunities to consolidate worldwide transdisciplinary projects, especially by including theories and methodologies developed in non-western countries and institutions. Usually misunderstood in cognitive and social neuroscience, new approaches for research on human emotions and empathy will take advantage of the cultural and social diversity involving the human world. Moreover, current criticisms of neuroscientific reproducibility will provide possibilities to include novel methodologies and theories to consider individual differences in which subjectivity, social context, cognition, and brain anatomy and function may be integrated to comprehend the variable bio-psycho-social sphere constituting human beings. The development of multicultural and inclusive human neuroscience research will contribute to the discussion of social issues such as racism, immigration, and poverty, aiming over applied neuroscience focused on the implementation of social programs and public politics based on scientific knowledge (Daniel Atilano-Barbosa 2021).

3 Integrated ID in Health and Medical Sciences

3.1 Global Epidemics

Infectious disease outbreaks have shaped the fate of humanity from the earliest documented times and most certainly will continue to transform every facet of life on our planet. Many outbreaks studied to date unveiled marked heterogeneities in the population-level transmission of microbes. As a result of these heterogeneities, some contagious individuals create many more secondary contacts than most others in the same population and became known as super-spreaders or super-shedders. Even though isolating super-spreading or super-shedding hosts may limit an outbreak, transmission heterogeneities are usually identified only retrospectively. I envision and hope that, over the next 30 years, the field will introduce more sophisticated and ID toolsets to accurately and promptly identify and possibly predict super-spreading events during emerging outbreaks. Ideally, these toolsets will be built on powerful ID approaches situated at the convergence of fields as diverse as *-omics*, molecular genetics, single-cell analysis, immunology, epidemiology, airflow dynamics, and remote sensing technologies. As part of these advances, it is critical to rigorously characterize super-spreading events from different outbreaks, comprehensively understand and catalog the factors that caused or facilitated them, and systematically interrogate the mechanisms involved. Ultimately, these strides promise to transform the face of epidemic and pandemic preparedness (Richard A. Stein 2021).

In the field of traffic engineering and transport systems, the current situation poses a very serious threat. The last 30 years have been a constant struggle to reduce the outflow of passengers from public transport in favor of individual transport. During the epidemic, the concentration of people on public transport introduces an additional threat. Both the means of transport and their organization are not adapted to such threats. Without significant changes in this regard, people will return to individual transport, thus increasing traffic flows in this area and lowering the quality of traffic, which is already dreadful, considering the current congestion. The related economic issues will result in the impossibility of providing suitable capacity in public transport at certain times, especially if the epidemic entails a much higher fatality rate. For the same economic reasons, by 2050, the road infrastructure will not have taken over all the additional travel (from public transport and other origins). The solution is to introduce regular tele-activity on a larger scale than in the wake of the pandemic. Places intended for this type of activity should be correlated with the transport parameters and other technical infrastructure. Another way to proceed would be to develop effective solutions in the field of biological and medical sciences. This being the case, one could think about further development of transport in the spirit of its sustainability (Ireneusz Celiński, Grzegorz Sierpiński 2021).

At the moment, we talk about crossing boundaries when working across health and the environment. In the future, there will be no boundaries. When we think about health or the environment, they are considered in the same breath. Just as we have a right and left hand being of the same body, individual but working together health and environment will still be individual but always support each other. COVID-19 has highlighted the need to integrate health and environment domains. This ID approach will be vital in addressing the sustainable developmental goals (SDGs) in the 21st century (Kaushik Sarkar, Monica Lakhanpaul, Priti Parikh 2021).

The rationalist reductive model of scientific knowledge is very effective for problems that are defined and tractable: but it often struggles with problems of deeper complexity, intractable uncertainty, and controversy, or "grand societal challenges." The pandemic has raised such a prospect, as the unfolding crisis showed that medical science had to interact somehow with politics, business, and society. So, we can debate and envision the scope of a more synergistic and integrated mode-III science, or Science-3.0, with key features including:

- Deeper scope of knowledge to include multiple values/worldviews, with synergistic inter-connections between policy, business, civil society, and others;
- Wider communities of practice and research, suited to systems of deeper complexity, beyond anyone form of analysis; and
- Further insights on the cognitive dynamics and moving frontiers between rational, experiential, instrumental, and transcendent forms of knowledge

Some parts of this picture may be realized through big data, the internet of things, and AI. Other parts may be realized through more advanced forms of deliberation, social learning, or action research. And a third dimension comes by linking rational-reductive research to other creative-experiential channels, whether visual, performing, narrative, or public arts (Joe Ravetz 2021).

3.2 Nuclear Medicine

Nuclear medicine and molecular imaging reached clinical acceptance in assessing many oncological and non-oncological pathologies during the last decades. Steady improvements in detectors design and architecture and software and hardware technology implementation resulted in significant improvements in scintigraphy and PET image quality while reducing radiotracer dose and scanning time. Recently introduced digital PET detector technology yields higher intrinsic system sensitivity than the latest generation analog technology, enabling the detection of very small lesions with potential impact on disease outcome. Further advancements, including the widening use of PET/MR and the standardization of AI applications, open new and interesting molecular imaging scenery always addressed towards the personalized medicine aim. Also, the recent growing consensus about using the theragnostic approach is becoming more tangible, and advanced molecules, also using antibodies, are under survey. We expect and predict many more significant innovations are yet to come, making this a very exciting time to be in nuclear medicine and molecular imaging. We believe and encourage the ulterior dissemination of PET/CT with new clinical indications and the widening use of PET/MR that, in a futuristic way, will be directly used for several pathologies such as prostate cancer and head and neck cancer. Further, we can predict in the next decades a growing theragnostic use of radiolabeled "bullets" (such as antibodies, receptor ligands, antisense oligonucleotides, etc.). That will make nuclear medicine a "precision weapon" able to describe the illness in its different features, focus its extension, and treat it efficaciously, without any major adverse event. We rely on the radiopharmaceutical development in nuclear medicine will enable the (at least) phenotypic characterization of lesions and their treatment. Nowadays, evaluation of neoplastic lesions requires an ex-vivo histopathological analysis which is limited in its sight and repeatability; we believe that nuclear medicine will permit an in-vivo evaluation of each lesion of a single patient. In other words, nuclear medicine will non-invasively translate the data now obtainable mainly through histopathological sampling, with the advantages of repeatability and single-lesion characterization. The ability to assess several molecular features of lesions will guide therapeutic choices, monitor therapy efficacy, and increase therapeutic options, conveying radionuclide therapy to different targets, even changeable in the natural course of a specific pathology (Sergio Baldari, Fabio Minutoli, Riccardo Laudicella 2021).

3.3 Dentistry

Over the next 30 years, dentistry, as a profession, will continue advancing in the knowledge, treatment, and prevention of oral diseases. The current shift in practice from a curative approach to oral healthcare to the adoption of primary and secondary prevention strategies will be widespread. The use of technology for clinical practice will also be robust. Unfortunately, it is very unlike to see the same magnitude of change in the oral health disease profile of the vulnerable populations. Overcoming the current challenges of oral healthcare systems and oral health disparities and inequalities across the world requires a paradigm shift from treatment-oriented, individualistic dentistry to prevention-oriented, strong public health dentistry delivery systems. Achieving "health through oral health" means that our profession will take an important role in general health, including common risk factors with non-communicable diseases. In the future, dentists, in general, will become oral physicians. Our profession will have to improve the community's oral health through oral health promotion to solve the inequity and inequality issues. To achieve oral healthcare for all within the visions of the current SDGs and beyond, dentistry should go beyond its current borders and embrace an ID approach to take advantage of the resources of other disciplines to bridge oral healthcare access and improve the oral health quality of life of all persons by 2050 (Mohammad R. Khami, Morenike Oluwatoyin Folayan, Armando F. Soto-Rojas, Heikki Murtomaa, Prathip Phantumvanit, Farid Farrokhi 2021).

3.4 Medical Education

Medical education represents a complex world and its people, from medical students to medical professionals, pursue an undergraduate or graduate degree in this complexity. Looking over the literature, especially over the last three decades, medical educators declare that medical education suffers evident problems. The root of these problems is a lack of attention to moral values, basic needs, and social characters of young learners as humans terminated to older trainees neither able to deal with challenging clinical situations effectively nor appear to have self-esteem enough to cultivate the future thinking professionals. Disasters like the ongoing pandemic of COVID-19 are exacerbating the suffering of medical students from the ineffectiveness of time-education management. Therefore, the present time needs to apply modern integrated approaches in place of traditional approaches to medical education. Then, it can be expected to see medical professionals who do not need to force themselves to think on one side of the problem but are self-driven to form integrative strategies in practice. The next three decades would require such a powerful community of medical people (Noosha Samieefar, Sara Momtazmanesh, Hans D. Ochs, Timo Ulrichs, Vasili Roudenok, Mohammad Rasoul Golabchi, Mahnaz Jamee, Melika Lotfi, Roya Kelishadi, Mohammad Amin Khazeei Tabari, Milad Baziar, Sayedeh Azimeh Hosseini, Milad Rafiaei, Antonio Condino-Neto, Elif Karakoc-Aydiner, Waleed Al-Herz, Morteza Shamsizadeh, Niloofar Rambod Rad, Mohammadreza Fadavipour, Alireza Afshar, Meisam Akhlaghdoust, Kiarash Saleki, Farbod Ghobadinezhad, Zhila Izadi, Arash Khojasteh, Alireza Zali, Nima Rezaei 2021).

3.5 Sociomedical Sciences

The application of social sciences in health will change from an ID collaboration to a cluttered one. Social sciences will play a pivotal role in managing health systems. The presence of social sciences will not be a better understanding of health and diseases. Instead, the social perspectives, values, and notions will set health and economic goals, design health systems, define success indicators, and implement measures. The social approach to health and its ecological and economic drivers will dramatically affect health systems from global health governance at the macro to healthcare at the micro-level. The entrance of social sciences into health system management will shift the health systems from dependency on governmental institutions and economic growth to civil society and community actors' dominant role in circular economies that respect the planetary boundaries. Communities will contribute from policymaking to service delivery using digital technologies. Global attention to the sociological challenges of health is increasing (Reza Majdzadeh, Haniye Sadat Sajadi, Remco van de Pas, Abouali Vedadhir 2021).

3.6   Medical Ethics

Medical research sampling will become more representative of the populations studied, and this will happen worldwide. Women, minorities, the elderly, multiple types of vulnerable people, and different socioeconomic statuses will be better represented in medical research sampling. The medical profession will be more open in providing informed consent should the project of better medical research sampling not be completed by 2050, for example. It will make medical education more ethically focussed. The American Medical Association will have revised its medical ethics code to promote that representative medical research sampling takes place universally. Other medical ethics codes will also reflect this research practice. The granting agencies involved will have policies to ensure that researchers undertake appropriate medical research sampling demographically and otherwise. The pharmaceutical industry might not be fully compliant with these policies, but pressure from the medical and bioethical community might have it move somewhat in the right direction in these regards. The solutions are complex for this issue and require much vigilance and policy implementation, which must be ongoing. The field of psychology will have better engagement with the problem of biased re-search sampling compared to the full range of stakeholders in the medical pro-fession, but the solutions and policies will remain as difficult to implement for it as for the medical profession (Gerald Young 2021).

3.7   Health Informatics

Health informatics refers to the study and practice of the design, implementation, deployment, and evaluation of information technologies and medical devices used by health professionals, patients, and laypeople. Therefore, the scope of health informatics has been broad, and it uniquely lies at the intersection of clinical science, health sciences, information science, management practice, and computer science. The scope of the field will only continue to expand as new healthcare technologies appear and new approaches to streamlining healthcare emerge over the next several decades as healthcare becomes ever more digitized and virtual. Over the past several decades, health informatics has become recognized as being a new and distinct discipline (that has its own methods, theories, approaches, and research findings), having evolved much as computer science did from its roots in other fields (such as mathematics and engineering). This trend towards formalizing health informatics as science has characterized health informatics as a distinct discipline on its own. Health informatics has evolved as healthcare and technology have advanced and will continue to evolve as a field of study and profession into the future. In 30 years, major advances in the field will come from graduates of new educational and training programs designed to take a holistic and integrative approach to health informatics, viewing health informatics as an essential field of study for advancing and improving healthcare (Andre Kushniruk, Elizabeth Bor-ycki, Helen Monkman 2021).

3.8 Drug Discovery

It is highly risky, with success rates only in the 1% range, and fiendishly expensive, with each regulatory-approved drug requiring an investment of approximately $1.4 billion. Drug discovery also requires an extraordinarily high level of scientific skill. My chapter (Chap. "Drug Discovery in Big Pharma: Where "Birds" and "Fish" Collaborate to Find New Medicines") focused on the collaboration between business and science required to discover new drugs and the many difficulties associated with projects undertaken by these exceptionally different groups. Efforts to discover new drugs are the product of the system in which it operates. Three political systems were in competition during the 20th century: communism, fascism, and liberal democracy. By the end of the 20th century, only the liberal democratic system survived. Liberal democratic governments, to varying degrees, support the health of their citizens by funding basic biomedical research. The private sector uses government-provided scientific discoveries to find new drugs based on the profit motive. The 21st century has thus far has seen the expression of much dissatisfaction with the liberal democratic system with little or no discussion of what will replace it. If liberal democracy holds, drugs will be discovered much as before, except that scientific improvements over the next decades will yield better drugs as well as treatments for diseases previously thought to be incurable. But in any event, what gets done, and the way it gets done will depend upon the system in place. And whatever the system, drug discovery will likely be subject to the demands of political elites, who currently appear to be scientifically illiterate. In many current governments, we see significant fakery, manipulation of truth and fact, and the disaccreditation of competent science. The 20th-century German-American philosopher, Hannah, Arent, observed that "*If* everybody *always lies* to *you*, the consequence is not that *you believe* the *lies*, but rather that nobody believes anything any longer." Scientific progress becomes impossible in the absence of objective truth (Donald Kirsh 2021).

4   AI-Based Integrated ID

4.1   AI and Health

The future is bright. In a highly ethical world, adherence will be assured by empowering people. While human behavior is highly dependent on social interactions, electronic devices will easily mimic human contact touchpoints that enhance adherence behavior. People will trust their devices, and these devices, ever more intricately woven into our lives, will be programmed to gently and ethically nudge people to succeed in achieving their behavior goals, particularly ones that will better their health. Trusted devices will provide accountability; we see it happening already in a rudimentary—yet still effective form—with watches that encourage us to meet exercise goals. By 2050, these devices will act and feel more like our best friends, creating far greater perceived social pressure to engage in beneficial activities. Moreover, such devices will easily overcome one of our basic human limitations, our forgetfulness. By inserting our treatments into daily routine, holding us accountable, and being trustworthy—all the while also being our friendly, dependable companions—our electronic partners will assure that we adhere to behaviors designed to optimize our health and well-being (Veronica K. Emmerich, Esther A. Balogh, Steven R. Feldman 2021).

By 2050, AI is transforming our vision for manipulation in different axes in healthcare that will influence all aspects of diagnostics in the future. In 30 years, with the pressing and complex challenges in this field will be increased according to the excessive needs of experts, AI and big data can make a crucial difference to improve the quality of service and patient care. Urban intelligence with big data and AI to make predictions over the next 30 years is risky if the necessary security precautions are not considered in e-healthcare. Looking to the future 2050, the term "singularity" in AI is described as the point at which exponential technological advancements cross the threshold of "strong AI," and machines possess vast intelligence that exceeds human levels. A strong chance of this hypothesis may materialize by 2050, and the singularity is that human intelligence and AI will shape the whole medical field. In 2050, the human behavior of experts could be enhanced by AI approaches and massive data to be managed in real-time. AI and big data will sweep by 2050, although the design implications in healthcare will be an uphill task to implement (Houneida Sakly, Mourad Said, Jayne Seekins, Moncef Tagina 2021).

Over the years, AI has helped doctors remarkably in terms of accuracy, time effectiveness, etc.; therefore, in the future, this technology is likely to grow to become an integral part of medicine, more likely in areas that comply with specific patterns and are amenable to partial or full automation. Yet, since AI is not expected to be able to take over humans' role in medicine completely, the new generation of medical students are to be trained to familiarize with the opportunities and limitations of AI and with ways to use this technology in the most efficient way possible to overcome the numerous challenges it involves. Regarding the four basic ethical principles and the relevant challenges, it is obvious that AI has to be deployed to face the least possible conflict with them because these principles are fixed at the core of medicine. Undoubtedly, a major responsibility in ensuring compliance in this process lies in the hands of government and policymakers. The role of AI in future medicine is open to discussion, but this technology has reached this domain to stay; therefore, it is also researchers' responsibility to make sure that *society* as a whole keeps pace with it (Sadra Behrouzieh, Mahsa Keshavarz-Fathi, Alfredo Vellido, Simin Seyedpour, Saina Adiban Afkham, Aida Vahed, Tommaso Dorigo, Nima Rezaei 2021).

4.2   AI and Personalized Medicine

Future medicine will be personalized and extended. In 2050, the health problem will no longer be strictly speaking medical; it will be ethical. Scientific and technological development will make it possible to prevent and cure most diseases. This will lead to the stabilization of life expectancy around a value compatible with the survival of human society. However, this fact will make clinical decisions somewhat more complicated. In fact, more and more decision-making power will shift from the technical pole (the doctor) to the personal one, the individual citizen. AI will make decision-making parameters directly under the control of the individual, who will be called to make his own decisions, including values, preferences, and expectations, without the mediation of the doctor, who will become an effector or at most a consultant. However, this will not mean shifting the focus from doctors to technology, merely changing the nature of technical mediation from humans to machines. On the contrary, the development of AI would in itself be useless if a corresponding co-development of human intelligence did not accompany it. This process will eliminate the distinction between a natural and an artificial sphere of intelligence since one and the other will merge into an extended intelligence. In this way, the intelligence evolution will reach the apex, as far as we know, of the development of the homo genre since when the first instrument was built, thus expanding not only the panorama of possible behaviors and solvable problems but also the field of cognition and mind. This process engenders a virtuous circle in which *homo sapiens* transfers intelligence to new agents that, in turn, provide opportunities for the development of further intelligence. If, on the individual level, man–machine co-evolution will lead to new frontiers of personalized medicine, then the most crucial match will be played out on the social scale. If the core of the medical choice will be ethical, and if the focus will be personal, how will it be possible to reconcile the public with individual health-related interests? Will a real health democracy be possible? In 2050, integrated science will look for new ways of facing these questions (Emilio Maria Palmerini and Claudio Lucchiari 2021).

4.3   AI Trends

Electronic interventions become more able to promote better behavior and improve human well-being than even major advancements in biological sciences. Our common interest field is fed by research-creation, design and social innovation, and spatial reasoning, all challenged by the digital transition. Three main trends seem to be emerging. First, from a design point of view and, due to the technical and digital engineering progress, assistive technologies/robots should require more "anthropomorphization," not as simple human representation acts, but rather as operating and interacting modalities. The contamination of science by artistic processes will instill the need for symbolism and sensitivity that those technologies require to ensure that the mesh between non-living mechanics and the living is fertile for human-being. Moreover, design should probably offer methods and approaches to humanize and personalize even more technologies to make healthcare environments more inclusive. Second, in this same line, the design and planning of inclusive health spaces should further develop learning spaces using algorithmic personalization, physical/digital hybridization, and ambient technologies (smart sensors/actuators). Last but not least, the most repetitive part of scientific reasoning should be conducted by even better trained AI. This should leave more places for creativity, research-creation, and more sensitive thinking into the scientific methods and tools (Andrée-Anne Blacutt, Stéphane Roche 2021).

4.4 Dilemma Questions

4.4.1 Human Being Philosophy

In the future, philosophy will be vulnerable. In a future world increasingly governed by AI, philosophy will be necessary. The philosophical field of the search for a genuinely human meaning will be able to favor human lucidity. What makes us human? Can human beings be turned into machines? Is there an exclusively human existential meaning? Can machines be free? All these questions, and many others, will continue to inhabit human life, and philosophy will be responsible for searching for meaning to all human existence (Juan José Garrido Periñán 2021).

4.4.2 Ethics

When we look back at the state of psychological research 30 years ago and compare it to the current practice, we can see a great development that the researchers in this field were able to accomplish in terms of the increasing rigorousness of methodology, statistical analysis, open reporting practices, as well as expanding the range of topics covered by the psychological science. We expect that this trend will continue, and we will witness further expansion of open science with freely available knowledge and data from across all scientific disciplines. AI systems will automatically process and meta-analyze large quantities of data to obtain evidence for hypotheses or theories, while researchers will be freed to work on truly multidisciplinary challenges in large scientific collectives. Such developments will bring about new ethical dilemmas. With the advanced knowledge of cognitive processes, neuropsychology, and the human genome, we might be able to create artificial beings that could be used for experiments that would not be ethical to conduct on humans. It may be just a matter of time when we perfect these beings so that they will too gain consciousness, and so we will need to expand further the rights of non-human beings and animals for a dignified existence. When such efficient AI is available, what will become of scientists? Will they be reduced to mere generators of hypotheses and theories which AI will automatically test? Or will they remain the candle-bearers capable of shedding light on the unknown? (Jakub Šrol and Vladimíra Čavojová, 2021).

5   Art-Based Integrated ID

It is difficult to predict the future of work in the health humanities and research projects combining the arts and sciences. After all, 30 years ago, the field looked very different from the way it does today. Previously, the medical humanities were studied by a select group of medical students and medical academics, and disciplines such as art therapy or music therapy were often practiced in relative isolation. The last few years, as well have documented in our chapter (Chap. "Bringing the Two Cultures of the Arts and Sciences Together in Complex Health Interventions"), have seen much closer alliances between these disciplines in research projects and interventions, in a way which would have been hard to envisage 30 years ago. Consequently, we could see some substantial new developments in ID work and new individuals, groups, and clusters to challenge existing orthodoxies.

A promising strand of work concerns how experience, including experiences of creative activity and the arts, can get 'under the skin.' We have mentioned how immune system function could be shifted in a beneficial direction through music workshops. Similar work is being done on brain function about experiences such as music, poetry, or literature. This kind of work is ushering in a new understanding of the nervous system, mind, and body. The brain is no longer seen as a static entity, producing behavior and experience through immutable patterns of cellular wiring. Rather, the relationship between experience, action, and patterns of activation in the nervous system is a dynamic and dialectical one that offers genuine opportunities for change. In practical terms, there may be more examples of work that draws on local traditions of visual art performance and literature in different nations. This is beginning to be apparent in some of our projects, such as those we mentioned in India, which rely on musical and dramatic traditions from the local areas involved (Brian Brown, Monica Lakhanpaul 2021).

The exact effect of music in disparate areas is still quite a mystery wandering between myth and actual science. Further studies are required to replace art therapy, music in specific, with existing treatment modalities. Among investigated disorders, depression and anxiety have the greatest number of studies. Thus, these areas could be the first in replacing music therapy with pharmacological treatment. We believe that to improve the quality of research in the field of art therapy, more clinicians must accept these modalities as a science instead of seeing them as complementary therapy or pseudo-science (Niloufar Yazdanpanah, Helia Mojtabavi, Heliya Ziaei, Zahra Rahimi Pirkoohi, Elham Rayzan, Nima Rezaei 2021).

The Integrated Science series creates a healthy conversation for ID. I am optimistic that 30 years from now, the vision that informed this book project would have manifested tangibly for social engineering. It would remain an indispensable material in the humanities and sciences. Also, I am confident that the disparity between the artistic and humanistic cultures would have been thoroughly collapsed (Stephen Kekeghe 2021).

6 Conclusion

All authors of this chapter played a role in adding a chapter to our book *Integrated Science: Multidisciplinarity and ID in Health.* They also participated in an ID collaboration process; its resulting framework was constructed in this chapter. To summarize, this framework began with ID cognitive neuroscience. Then, we integrated ID in health and medical sciences. Finally, we found AI and arts as the means that facilitate such integrations.

Core Messages

- ID neuroscience includes neurophysics, neuroaesthetics, neuropsychology, and neurobehavioral health.
- ID neuroscience contributes to understanding consciousness, aesthetic experience, empathy, and social diversity.
- ID neuroscience contributes to the understanding of the distinction between healthy and unhealthy behaviors.
- The subject of integrated ID into medical, social, and health sciences is vast.
- AI offers a means of integration of ID in healthcare, behavior engineering, and personalized medicine.

References

1. Cohen D, Krajewski A (2014) Interdisciplinary geriatric resilience interventions: an urgent research priority. Topics in Geriatric Rehabilitation 30(3):199–206
2. Leclerc B-S, Blanchard L, Cantinotti M, Couturier Y, Gervais D, Lessard S et al (2014) The effectiveness of interdisciplinary teams in end-of-life palliative care: a systematic review of comparative studies. J Palliat Care 30(1):44–54
3. Nazir A, Unroe K, Tegeler M, Khan B, Azar J, Boustani M (2013) Systematic review of interdisciplinary interventions in nursing homes. J Am Med Dir Assoc 14(7):471–478
4. Liossi C, Johnstone L, Lilley S, Caes L, Williams G, Schoth DE (2019) Effectiveness of interdisciplinary interventions in paediatric chronic pain management: a systematic review and subset meta-analysis. Br J Anaesth 123(2):e359–e371
5. Rushton CH, Reder E, Hall B, Comello K, Sellers DE, Hutton N (2006) Interdisciplinary interventions to improve pediatric palliative care and reduce health care professional suffering. J Palliat Med 9(4):922–933
6. Pannick S, Davis R, Ashrafian H, Byrne BE, Beveridge I, Athanasiou T et al (2015) Effects of interdisciplinary team care interventions on general medical wards: a systematic review. JAMA Intern Med 175(8):1288–1298
7. Veerbeek JM, van Wegen E, van Peppen R, van der Wees PJ, Hendriks E, Rietberg M et al. (2014) What is the evidence for physical therapy poststroke? a systematic review and meta-analysis. PloS one 9 (2):e87987

8. Rosa VM, Fores JML, da Silva EPF, Guterres EO, Marcelino A, Nogueira PC et al. (2018) Interdisciplinary interventions in the perioperative rehabilitation of total laryngectomy: an integrative review. Clinics 73

9. Tapsell LC, Neale EP (2016) The effect of interdisciplinary interventions on risk factors for lifestyle disease: a literature review. Health Educ Behav 43(3):271–285

10. Campbell K, Waters E, O'Meara S, Summerbell C (2001) Interventions for preventing obesity in childhood. A Systematic Rev Obesity Rev 2(3):149–157

11. Grüne-Yanoff T (2016) Interdisciplinary success without integration. Eur J Philos Sci 6(3): 343–360. https://doi.org/10.1007/s13194-016-0139-z

12. Moradian N, Ochs HD, Sedikies C, Hamblin MR, Camargo CA Jr, Martinez JA et al (2020) The urgent need for integrated science to fight COVID-19 pandemic and beyond. J Transl Med 18(1):205. https://doi.org/10.1186/s12967-020-02364-2

13. Moradian N, Moallemian M, Delavari F, Sedikides C, Camargo CA Jr, Torres PJ et al (2021) Interdisciplinary approaches to COVID-19. Adv Exp Med Biol 1318:923–936

Index

A

Abilities, 12, 14, 21, 22, 42, 49, 55, 63, 114, 117, 119, 125, 131, 132, 146, 147, 153, 156, 177, 180, 181, 185–187, 189, 191, 192, 210, 220–227, 231, 274, 364, 418, 426, 427, 459, 461, 465, 467, 469, 496, 513, 547, 550, 551, 556, 573, 576, 578, 579, 581, 582, 587, 600, 617, 628, 655, 678

Abnormality, 87, 88, 116, 134, 227, 467

Abstraction, 64, 146, 159, 160, 461, 557

Acceptance, 118, 149, 180, 182, 190, 245, 573, 578, 677, 680

Accident, 55, 392, 552

Accommodation, 112, 118, 337, 423, 652, 657

Accountability, 27, 200, 206, 207, 214, 335, 336, 340, 547, 556, 560, 585, 586, 617, 624, 634, 637, 684

Accountable, 545, 557, 562, 625, 635, 637, 685

Accuracy, 125, 130, 135, 167, 181, 224, 225, 247, 249, 252, 258, 527, 532, 553, 578, 581, 582, 585, 587, 588, 685

Acetaminophen, 186

Acquired Immunodeficiency Syndrome (AIDS), 155, 180, 181, 185, 350, 634

Action, 11, 12, 16, 17, 22, 23, 44, 72–74, 116, 152, 154, 155, 157, 160, 167, 191, 221, 246, 279, 281, 284, 300, 301, 307, 314, 317, 326–330, 340, 352, 393, 397, 406, 422, 423, 437, 439, 454, 464, 466, 555–558, 562, 570, 575, 576, 578, 579, 583, 632–634, 636, 637, 639, 676, 680, 684, 688

Actionable, 539, 596

Action generation, 132

Action planning, 132

Action potentials, 91

Activation, 70, 75, 90, 114, 117, 118, 128, 129, 131, 134, 150–159, 226, 227, 256, 471, 576, 688

Activities, 7, 20, 22, 23, 25–27, 41, 43, 51, 52, 55, 75, 83, 85–90, 99, 112, 113, 117, 125, 129–131, 133, 147, 154, 159, 186, 225–230, 248, 251, 253, 255, 275, 281, 282, 296, 301, 319, 332, 333, 391, 394, 397, 400, 403, 405, 419–421, 425, 426, 429–431, 460, 463–467, 470, 492, 498, 535, 536, 548, 549, 555, 575, 580, 619, 636, 676, 685

Actor, 259, 443, 550, 561, 629, 637, 682

Actualization, 329

Acute, 24, 203, 230, 353, 355, 358, 367, 586

Acute radiation syndrome, 246

Adaptation, 55, 111, 123, 125, 130, 135, 258, 395, 408, 489, 550

Adaption, 148, 152, 469

Adaptive, 23, 119, 133, 205, 222, 223, 576, 631, 632

Addiction, 149, 150, 210, 213, 219, 221, 226, 227, 231

Adherence, 9, 27, 199–214, 612, 684

Adjudication, 571

Adjuvant, 29, 472

Administrative restriction, 391, 393, 399

Adolescent, 202, 224, 225, 272, 273, 276, 278

Adrenal, 114, 464

Adrenal gland, 113

Adrenaline, 113, 300

Adults, 27, 87, 109, 151, 193, 272, 273, 357, 359, 365, 460, 461, 466, 470, 471, 489, 556, 558, 652

Advancement, 241, 259, 284, 332, 335, 511, 527, 595, 596, 680, 685, 686

Advantage, 27, 87, 113, 210, 273, 304, 430, 443, 496, 528, 551, 552, 588, 598, 612, 617, 636, 678, 681

Milton Keynes UK
Ingram Content Group UK Ltd.
UKHW020128220823
427211UK00002B/6